AF324036

JOINT STATISTICAL PAPERS of AKAHIRA and TAKEUCHI

JOINT STATISTICAL PAPERS of AKAHIRA and TAKEUCHI

editors

Masafumi Akahira
University of Tsukuba, Japan

Kei Takeuchi
Meiji-Gakuin University, Japan

World Scientific
New Jersey • London • Singapore • Hong Kong

Published by

World Scientific Publishing Co. Pte. Ltd.

5 Toh Tuck Link, Singapore 596224

USA office: Suite 202, 1060 Main Street, River Edge, NJ 07661

UK office: 57 Shelton Street, Covent Garden, London WC2H 9HE

British Library Cataloguing-in-Publication Data
A catalogue record for this book is available from the British Library.

The editors and publisher would like to thank the following organizations and publishers of the various journals and books for their assistance and permission to reproduce the selected reprints found in this volume:

Institute of Mathematical Statistics
The University of Electro-Communications
Springer-Verlag
The Institute of Statistical Mathematics
Union of Japanese Scientists and Engineers
Blackwell Publishing Asia
Oldenbourg Wissenschaftsverlag GmbH
Physica-Verlag
Université Pierre et Marie Curie
Università degli Studi di Roma
VSP International Science Publishers
Blackwell Science Ltd
Marcel Dekker

While every effort has been made to contact the publishers of reprinted papers prior to publication, we have not been successful in some cases. Where we could not contact the publishers, we have acknowledged the source of the material. Proper credit will be accorded to these publications in future editions of this work after permission is granted.

JOINT STATISTICAL PAPERS OF AKAHIRA AND TAKEUCHI

ISBN   981-238-377-8

Printed in Singapore.

Foreword

Collaborative work, virtually inevitable in experimental and observational work, is increasingly common in the mathematical sciences too. There can, however, be few collaborations that have lasted so long and produced such an impressive and coherent body of work as that of Professors Akahira and Takeuchi celebrated in the present volume.

In the frequentist approach to statistical inference problems with an "exact" solution are important but essentially restricted to certain exponential family questions and to issues involving a transformational family, such as the scale and location model. This means that in more complex and realistic cases approximations cannot be avoided.

The approach almost universally adopted is to appeal to asymptotic arguments. It is supposed that the information available is large enough to justify, for example, local linearization of nonlinear dependencies and the invoking of the classical limit laws of probability theory. In more formal terms, the problem under study is imbedded in a sequence of problems in which the sample size, or more generally a measure of information, tends to infinity. Theorems based on the laws of large numbers and, typically, the Central Limit Theorem can then be established about the limiting behaviour of estimation, testing and prediction procedures. These results justify, in some sense, many widely used techniques such as maximum likelihood estimators and associated testing procedures, such as likelihood ratio tests.

There are broadly two difficulties with this approach. The passage to the limit is a fiction. Detailed results have to be applied for one or more specific sample sizes. Secondly, there are many different procedures that are equivalent in the limiting sense. While the different procedures will often give virtually the same answer in applications, this is not necessarily so and in any case some basis of choice between such alternative procedures is clearly desirable.

The first point is in principle best addressed by finding bounds on the approximations involved, broadly analogous to the Berry–Esseen bound associated with the Central Limit Theorem. While there is some fascinating work of this kind, stemming, for example, from the use of Stein's method, useful results are extremely limited in scope. Instead higher-order terms in asymptotic expansions can be exploited in some generality and this is the approach taken in the present work. Of course the issue still remains of the sample sizes for which the results give good approximations. The second issue of the choice between estimators and test statistics equivalent to the first order of asymptotic theory is subtler and needs very careful formulation. Professors Akahira and Takeuchi have made important contributions to both aspects; I find their results on the second aspect particularly striking.

Their broad approach stems from the Neyman–Pearson viewpoint emphasizing properties achieved in hypothetical repeated sampling. As such it is to be contrasted with studies proceeding from what might be called the Fisherian standpoint with its explicit emphasis on likelihood, sufficiency and conditioning. A detailed comparison of these approaches, and of both with Bayesian discussions, would probably be fruitful.

This is not the place to discuss or single out individual papers in the volume, the authors having in any case provided a valuable summary and introduction. The careful discussion of nonregular problems is, however, particularly to be noted.

Let me conclude by congratulating Professors Akahira and Takeuchi on their achievements; long may their joint work continue!

D. R. Cox
Nuffield College
Oxford
UK

January 2003

Introduction

This volume consists of 44 joint papers on statistical inference for about a quarter of a century from 1975 to 2001. In the latter half of 1970's, the higher order asymptotics was extensively developed by many people including Prof. J. Pfanzagl's group, Prof. J. K. Ghosh's one and ourselves. In particular, the second and third order asymptotic efficiencies of estimators are discussed using the Edgeworth expansion under suitable regularity conditions in the joint papers.

In cases when the regularity conditions do not hold, there are less well known "surprising" results such as the existence of zero variance unbiased estimators based on a sample of fixed size. The order of consistency and asymptotic efficiency of estimators are also strongly influenced from the conditions of non-regularity. Various non-regular cases are discussed in the joint papers.

The papers on both of the higher order asymptotics and non-regular estimation formed two monographs of Akahira and Takeuchi (1981, 1995). Other papers are related to the problem of prediction sufficiency, the Edgeworth type expansions and miscellaneous topics. More precisely following subjects and their implications are studied in the joint papers in this volume.

1. Prediction sufficiency

The concept of prediction sufficiency (adequacy) was defined and discussed by, among others, Skibinsky (1967), [1], and Torgersen (1977). It was shown by Bahadur (1955) that sufficiency as defined in terms of conditional expectations, under regularity conditions, implies "real sufficiency," i.e. sufficiency in terms of risk functions. In the paper [1], the converse is shown to hold provided the loss function depends on the unknown parameter, but no longer true if the loss function is independent of the unknown parameter. In the latter case, conditional independence still holds but ordinary sufficiency is not required. In the paper [15], we show the relation between prediction sufficiency and the sufficiency in terms of conditional expectations.

2. Edgeworth type expansions

In the papers [3] and [6], we give the Edgeworth type expansions of the distribution of the sum of independent and identically distributed (i.i.d.) random variables without higher order moments. For special cases of t-type density, we obtain the exact formulas for the asymptotic expansion of the density, and the results are generalized to the case when the density is approximated by a rational function. In [4], it is also shown that the asymptotic distribution for cases without finite variance are stable laws with fractional characteristic exponents, and the asymptotic expansion of the density for some case is given. In the paper [7], the above expansion is extended to the multidimensional case.

3. Second order asymptotic efficiency

In the papers [2] and [5], the second order asymptotic efficiency of asymptotically efficient estimator was discussed for one parameter and multiparameter cases. Under suitable regularity conditions, the maximum likelihood estimator (MLE) adjusted to be second order asymptotically median unbiased (AMU) is shown to be second order asymptotically efficient. And it is shown that the first order efficiency implies the second order efficiency, which phenomenon was also found by Pfanzagl (1979). The papers [9] and [13] demonstrate the second order efficiency of the generalized Bayes estimator (GBE). In [12], we define the discretized likelihood estimator (DLE) as a solution of the discretized likelihood equation, and show that the MLE as the limit of DLE is second order asymptotically efficient. Thus the structure of the higher order asymptotics is clarified by considering the DLE.

In the paper [28], from the viewpoint of the decision-theoretic approach, we get the second order asymptotic completeness of the MLE, and, in [10], discuss the problem of the second order asymptotic efficiency of the confidence intervals, that is, consider the asymptotic power of confidence interval up to the order $n^{-1/2}$ in the neighborhood of the true value of the parameter, where n is a size of sample.

In the papers [27] and [34], we obtain the Bhattacharyya type bound for the asymptotic variance of sequential estimation procedures under suitable regularity conditions, and show that the modified sequential maximum likelihood estimation procedure attains the bound if the stopping rule is properly determined, which can not be uniformly attained if the size of sample is fixed.

4. Third order asymptotic efficiency

Already three-quater century ago, R. A. Fisher (1925) suggested that the MLE was asymptotically best in the class of all asymptotically efficient estimators. He guessed that the MLE was asymptotically best in the sense that the MLE had the asymptotically minimum loss of information in the class, and calculated the asymptotic value of the loss of information of the MLE and minimum chi-square estimator in the multinomial distribution. Later, C. R. Rao (1961) corrected Fisher's calculation and defined the second order efficiency of the MLE, which corresponds to the third order asymptotic efficiency, in our terminology. However, it was not clear what the minimum of loss of information meant, and comparison of asymptotic variance or mean squared error up to the second order failed to establish uniform superiority of the MLE or any other estimators. This impasse was broken through by introducing the median-bias correction and considering the concentration probability of the estimator around the true value up to the order n^{-1} (see also [5], [8], Pfanzagl and Wefelmeyer (1985), Ghosh (1994)).

In the paper [8], it is shown that the MLE adjusted to be third order AMU is third order asymptotically efficient for a multiparameter exponential family of distributions. Similar results were obtained by Pfanzagl and Wefelmeyer (1978) and Ghosh, Sinha and Wieand (1980). In the papers [9] and [13] we show that, for any symmetric loss function, the GBE is third order asymptotically efficient in a restricted class $\mathbf{D}$ of third order AMU estimators for general one-parameter and multiparameter cases. In [16] and [28] it is seen that, for any class of estimators which admit Edgeworth expansions but are not necessarily AMU, we get the third order asymptotic completeness (or sufficiency) of the MLE together with the second order derivative of the log-likelihood function evaluated at the MLE. In [17], the concept of the asymptotic deficiency of Hodges and Lehmann (1970) is extended to the case when a common parameter is estimated from m sets of independent samples of each size n, and the asymptotic deficiencies of some asymptotically efficient estimators relative to the MLE based on the pooled sample are discussed in the presence of nuisance parameters (see also Akahira (1986)).

Further, under suitable regularity conditions, the third order asymptotic bound for the distribution of the sequential estimation procedures is obtained in [30]. And it is shown that the bias-adjusted maximum likelihood estimation procedure combined with appropriate stopping rule is uniformly third order asymptotically efficient in the sense that its asymptotic distribution attains the bound uniformly in stopping rules up to the third order, thus the asymptotic deficiency being zero.

For the curved exponential family of distributions, the second order efficiency from the Fisher-Rao approach is discussed by Ghosh and Subramanyam (1974), and also by Efron (1975) and Amari (1985) from the viewpoint of the differential geometry. Amari's approach attracted much attention in that he could successfully connect the curvature of differential manifold with the asymptotic deficiency of the estimator.

5. Non-regular unbiased estimation

For the lower bound for the variance of unbiased estimators, most famous is the so-called Cramér-Rao bound. But the Cramér-Rao bound and its Bhattacharyya extension assume a set of regularity conditions. Chapman and Robbins (1951), Kiefer (1952), and Fraser and Guttman (1951) obtained bounds with less stringent assumptions, but they still require the independence of the support of the parameter θ or almost equivalently that the distribution with $\theta \neq \theta_0$ is absolutely continuous with respect to that with $\theta = \theta_0$ when θ_0 is the specified parameter value at which the variance is evaluated.

In the paper [20], the Bhattacharyya bound is generalized to the non-regular case when the support of the density depends on the parameter θ while it is k-times differentiable with respect to θ within the support, and the bound is also shown to be sharp, which fails in the regular case (see also Takeuchi (1962)). In [26], we introduce the concept of one-directionality which includes both cases of location (and scale) parameter and selection parameter and other cases, and show that the bound for the variance of unbiased estimators is sharp in the sense that the actual infimum of the variance of unbiased estimators coincides with the bound for a specified θ_0, for this class of distributions, using the result to minimize the variance under the conditions of unbiasedness, linearly independence of functions, etc. in [21]. We also establish that for a wide class of non-regular distributions the infimum of the variance of unbiased estimators can be zero when the size of sample is not smaller than 2. In [25], we give the exact forms of locally minimum variance unbiased estimators and their variances in the case of a discontinuous density function. In the non-regular case when the amount of Fisher information is infinity, the paper [19] shows that the infimum of variance of unbiased estimators is equal to zero, and gives some examples.

It is obvious to observe that for usual "regular" case, any parametric function which has unbiased estimators must be continuous or differentiable in the original parameter (e.g. see Zacks (1971) and Lehmann (1983)). However, in a sequential case or a randomized sample size case when the size of sample is not bounded, the continuity of estimable function does not necessarily follow. In the case of sequential Bernoulli trials, the paper [37] gives a sufficient condition for a parametric function to be unbiasedly estimable and shows the existence of a discontinuous unbiasedly estimable function using non-randomized sample size procedures.

6. Non-regular asymptotic estimation and test

In the asymptotic theory of estimation, we try to compare the regular versus non-regular situation to clarify the significance and implication of each of the regularity conditions in [14]. Ibragimov and Has'minskii (1981) also proposed similar considerations for non-regularity. As an example, in the case of estimation of a location parameter in the symmetrically truncated normal distribution, it is shown in [11] that the maximum probability estimator of Weiss and Wolfowitz (1974) is asymptotically inadmissible and has smaller concentration probability than the midrange which is asymptotically efficient.

For a family of uniform distributions, it is shown in [44] that for any small $\varepsilon > 0$ the average mean squared error (MSE) of any estimator in the interval of θ values of length ε and centered at θ_0 can not be smaller than that of the midrange up to the order $o(n^{-2})$ as the size n of sample tends to infinity. And the asymptotic lower bound for the average MSE is also shown to be sharp. In the paper [18], we consider the problem to estimate a common parameter for the pooled sample from the uniform distributions with scales as nuisance parameters. And we compare the MLE with others and show that the MLE based on the pooled sample is not (asymptotically) efficient.

Fisher (1934) calculated the loss of information of the MLE of the location parameter in the double exponential distribution, and showed that the loss is of order $\sqrt{n}$, unlike of constant order in the regular case, hence the MLE is not second order asymptotically efficienct. The results are caused from the fact that the density admits the first order differentiability with respect to the parameter, but not the second order. Hence the first order asymptotic theory of regular estimation can be applied, but, in the second order the situation is non-regular. The paper [32] extends these results by obtaining the (asymptotic) losses of information of order statistics and related estimators, and by comparing them via their asymptotic distributions up to the second order. In [39], we obtain the Bhattacharyya type bound for the variance of unbiased estimators of a location parameter of the double exponential distribution and the loss of information of the MLE.

In [29], we consider the estimation problem of a location parameter on a sample of size n from the two-sided Weibull type density $f(x-\theta) = C(\alpha)\exp(-|x-\theta|^\alpha)$ for $x \in \mathbb{R}^1$, $\theta \in \mathbb{R}^1$ and $1 < \alpha < 3/2$, where $C(\alpha) = \alpha/\{2\Gamma(1/\alpha)\}$. Then we obtain the bound for the distribution of AMU estimators up to the 2α-th order, i.e. $n^{-(2\alpha-1)/2}$, and calculate the asymptotic distribution of the MLE up to the same order. And we show that the MLE is not 2α-th order asymptotically efficient, and give the amount of the loss of asymptotic information of the MLE. In [35], we consider the estimation of a location parameter θ of the density function with a support of a finite interval and contact of the power $\alpha - 1$ at both endpoints, where $1 < \alpha < 2$. Then we obtain the bound for the asymptotic distribution of AMU estimators of θ based on a sample from the density. It is also shown that the bias-adjusted MLE is not asymptotically efficient in the sense that its asymptotic distribution does not uniformly attain the bound.

In the regular case, it is known that the order of consistency, i.e. the order of convergence of consistent estimators is equal to $\sqrt{n}$, but in the non-regular case with i.i.d. sample it is not always so, and could be $n^{1/\alpha}$ $(0 < \alpha < 2)$, $(n \log n)^{1/2}$ etc., but are usually independent of the unknown parameter (see Akahira (1975a, 1975b)). However, in the non i.i.d. sample case, the order of consistency may depend on the parameter. In [23], we consider the first order autoregressive (AR) process with a parameter θ, and obtain the asymptotic means and variances of the log-likelihood ratio test statistic L_T under the null and alternative hypotheses in the case when $|\theta| \geq 1$. We also discuss the asymptotic distribution of L_T under both of the hypothesis. For the case of a two-sided Gamma type distribution, the paper [42] shows that the largest order of consistency is n^2 and there exists a test with under n^2 of consistency, and we obtain the asymptotic power function of the test.

In [36], we propose an amount of information between two distributions which is always well defined, symmetric and additive for independent sample and information contained in a statistic is always not greater than that in the whole sample and the equality holds if and only if the statistics is sufficient. And we discuss the relative (asymptotic) efficiency of a statistic (or an estimator) by the ratio of the amounts of information contained in the statistic and in the sample in a systematic and unified way both in regular and non-regular cases.

7. Others

For the process of continuous time observations are usually made on a finite number of discrete time points, information thus provided depends on the choice of observation points. In the paper [22], we assume that $X(\tau)$ is a continuous time simple Markov process with a parameter θ, and consider the problem choosing observation points $\tau_0 < \tau_1 < \cdots < \tau_T$ which provide with the maximum possible information on θ. If the observation points are equally spaced, that is, for $t = 1, \ldots, T$, $\tau_t - \tau_{t-1} = s$ is constant, we get the optimum value for s.

In the asymptotic theory of estimation, the concept of asymptotic expectation is widely used, and it is usually remarked that it can be different from the asymptotic value of expectation. But, the concept itself is not sufficiently accurately defined in the literature, especially when the asymptotic distribution does not exist. The paper [24] gives a rigorous definition of the asymptotic expectation, and shows its properties, e.g. its linearity and a Markov type inequality. And we obtain the necessary and sufficient conditions for the convergences in probability and distribution.

In the regular case when the dimension of the parameter is finite, asymptotic loss of information is constant order, but, for semiparametric models, it is shown in [31] that, under fairly regularity conditions, the asymptotic deficiency of the MLE or any regular best asymptotically normal estimator is infinity.

In [33], we consider the sampling properties of the bootstrap process, that is, the empirical process obtained from a sample of size n (with replacement) of a fixed sample of size n of a continuous distribution. And we give the cumulants of the bootstrap process up to the order n^{-1}, and discuss their unbiased estimation. We also demonstrate that the bootstrap process has an asymptotic minimax property for some class of distributions up to the order $n^{-1/2}$, and further we suggest the bootstrap method can be improved in the next order by taking a sample of size $n - 1$ instead of n.

Usually confidence interval is defined as an interval with preassigned confidence level $1 - \alpha$ for all the values of parameters. In [38], generalizing the concept, we consider interval estimation procedures with confidence coefficient varying according to the value of the unknown parameter, and associated procedure to estimate the actual level. Such a consideration also leads to more general procedures including conditional procedures when the ancillary statistic is involved.

For discrete distributions it is usually impossible to obtain a non-randomized test or confidence interval with exact given size, and an actual size is often quite different from the prescribed level. But a randomized procedure, which is quite nice in theory, is not easily acceptable to practitioners. Still, there is something to promote randomized procedures in practical applications. In the paper [41], for a family of one-parameter discrete exponential family of distributions, we derive the higher order approximation of randomized confidence intervals from the optimum test. Indeed, it is shown that they can be asymptotically constructed by means of the Edgeworth expansion. The usefulness is seen from the numerical results in the case of Poisson and binomial distributions.

The distribution of the sum of not identically but independently distributed random variables are difficult to calculate exactly, and the normal and Edgeworth type approximations can be applied when the number of independent random variables is not too small. But it is not always sufficiently accurate especially for the tail part. In such cases, large-deviation approximations were proposed in order to give better approximations especially for tails. In the paper [43], we obtain large-deviation approximations for the distribution of the sum of discrete random variables and show that they give sufficiently accurate results in various cases. It is noted that the large-deviation approximation connected with the saddlepoint approximation is widely discussed by Barndorff-Nielsen and Cox (1989) and Jensen (1995).

In sampling from the finite population of size N, there is a problem how to construct a design in which the inclusing probability of k-unit in the population is equal to prescribed π_k and $\sum_{k=1}^{n} \pi_k = n$, and it is a problem whether there exists a sample design of size n which attains the condition. The problem is solved by the Minkowski-Farkas theorem.

References

Akahira, M. (1975a). Asymptotic theory for estimation of location in non-regular cases, I: Order of convergence of consistent estimators. *Rep. Stat. Appl. Res., JUSE* **22**, 8–26.

Akahira, M. (1975b). Asymptotic theory for estimation of location in non-regular cases, II: Bounds of asymptotic distributions of consistent estimators. *Rep. Stat. Appl. Res., JUSE* **22**, 99–115.

Akahira, M. (1986). *The Structure of Asymptotic Deficiency of Estimators.* Queen's Papers in Pure and Applied Mathematics 75, Queen's University Press, Kingston, Canada.

Akahira, M. and Takeuchi, K. (1981). *Asymptotic Efficiency of Statistical Estimators: Concepts and Higher Order Asymptotic Efficiency.* Lecture Notes in Statistics 7, Springer, New York.

Akahira, M. and Takeuchi, K. (1995). *Non-Regular Statistical Estimation.* Lecture Notes in Statistics 107, Springer, New York.

Amari, S. (1985). *Differential-Geometrical Methods in Statistics.* Lecture Notes in Statistics 28, Springer, Berlin.

Bahadur, R. R. (1955). A characterization of sufficiency. *Ann. Math. Statist.* **26**, 286–293.

Barndorff-Nielsen, O. E. and Cox, D. R. (1989). *Asymptotic Techniques for Use in Statistics.* Chapman and Hall, London.

Chapman, D. G. and Robbins, H. (1951). Minimum variance estimation without regularity assumptions. *Ann. Math. Statist.* **22**, 581–586.

Efron, B. (1975). Defining the curvature of a statistical problem (with applications to second order efficiency). *Ann. Statist.* **3**, 1189–1242.

Fisher, R. A. (1925). Theory of statistical estimation. *Proc. Camb. Phil. Soc.* **22**, 700–725.

Fisher, R. A. (1934). Two new properties of mathematical likelihood. *Proc. Roy. Soc. (London) Ser. A* **144**, 285–307.

Fraser, D. A. S. and Guttman, I. (1951). Bhattacharyya bounds without regularity assumptions. *Ann. Math. Statist.* **23**, 629–632.

Ghosh, J. K. (1994). *Higher Order Asymptotics.* NSF-CBMS Regional Conference Series Probab. and Statist., **4**, Inst. of Math. Statist., Hayward, California.

Ghosh, J. K., Sinha, B. K. and Wieand, H. S. (1980). Second order efficiency of the mle with respect to any bounded bowl-shaped loss function. *Ann. Statist.* **8**, 506–521.

Ghosh, J. K. and Subramanyam, K. (1974). Second order efficiency of maximum likelihood estimators. *Sankhyā Ser. A* **36**, 325–358.

Hodges, J. L. and Lehmann, E. L. (1970). Deficiency. *Ann. Math. Statist.* **41**, 783–801.

Ibragimov, I. A. and Has'minskii, R. Z. (1981). *Statistical Estimation: Asymptotic Theory.* Springer, New York.

Jensen, J. L. (1995). *Saddlepoint Approximations.* Clarendon Press, Oxford.

Kiefer, J. (1952). On minimum variance estimators. *Ann. Math. Statist.* **23**, 627–629.

Lehmann, E. L. (1983). *Theory of Point Estimation.* Wiley, New York.

Pfanzagl, J. (1979). First order efficiency implies second order efficiency. In: *Contributions to Statistics. Jaroslav Hájek Memorial Volume.* (J. Jurečková, ed.), 167–196, Academia, Prague.

Pfanzagl, J. and Wefelmeyer, W. (1978). A third order optimum property of the maximum likelihood estimator. *J. Multivariate Anal.* **8**, 1–29.

Pfanzagl, J. and Wefelmeyer, W. (1985). *Asymptotic Expansions for General Statistical Models*. Lecture Notes in Statistics **31**, Springer, Berlin.

Rao, C. R. (1961). Asymptotic efficiency and limiting information. *Proc. Fourth Berkeley Symp. on Math. Statist. and Prob.* **1**, 531–545.

Skibinsky, M. (1967). Adequate subfields and sufficiency. *Ann. Math. Statist.* **38**, 155–161.

Takeuchi, K. (1962). On a fallacy of Gunnar Blom's theorem. *Rep. Stat. Appl. Res., JUSE* **9**, 34–35.

Torgersen, E. N. (1977). Prediction sufficiency when the loss function does not depend on the unknown parameter. *Ann. Statist.* **5**, 155–163.

Weiss, L. and Wolfowitz, J. (1974). *Maximum Probability Estimators and Related Topics*. Lecture Notes in Math. 424, Springer, Berlin.

Zacks, S. (1971). *The Theory of Statistical Inference*. Wiley, New York.

Contents

1990

1991

1992

1993

1997

The Annals of Statistics
1975, Vol. 3, No. 4, 1018–1024

CHARACTERIZATIONS OF PREDICTION SUFFICIENCY (ADEQUACY) IN TERMS OF RISK FUNCTIONS

BY KEI TAKEUCHI AND MASAFUMI AKAHIRA

University of Tokyo and University of Electro-Communications

Prediction sufficiency (adequacy), as it is usually defined in terms of conditional expectations, does imply "real" prediction sufficiency; i.e. sufficiency in terms of risk functions. The converse holds provided we permit the loss to depend on the unknown parameter. This is no longer true if we insist on loss functions which do not involve the unknown parameter. Conditional independence still holds but ordinary sufficiency may fail. If, however, we require equivalence of risk functions, then ordinary sufficiency and, consequently, prediction sufficiency follows.

1. Introduction. It has been shown by Bahadur [2] that sufficiency as defined in terms of conditional expectations, under regularity conditions, implies "real sufficiency" i.e. sufficiency in terms of risk functions. Furthermore it follows from Theorem 11.3 in Bahadur's paper [1] that prediction sufficiency is equivalent to ordinary sufficiency w.r.t. a larger class of probability measures. One may therefore expect similar results to hold for prediction sufficiency (adequacy) as well. In the case of prediction problems it may be of interest to consider loss functions which depend only on the decision to be made and the quantity to be predicted. If we insist on this restriction, then prediction sufficiency in terms of risk functions no longer implies prediction sufficiency as it is defined in terms of conditional expectations. Conditional independence holds but ordinary sufficiency may fail. It will, however, be shown that equivalence of risk functions implies ordinary sufficiency and consequently prediction sufficiency. (One of the authors proved this in an earlier work [7].)

We will, essentially, use the framework of Skibinsky [6]. The notion of adequacy in Skibinsky's paper is, however, replaced by the notion of prediction sufficiency.

2. Theorems. We shall assume that we are given a model consisting of a sample space $(\mathscr{X}, \mathscr{A})$ and a family $\{P_\theta : \theta \in \Theta\}$ of probability measures on $\mathscr{A}$. A sub σ-algebra $\mathscr{B}$ of $\mathscr{A}$ summarizes what can and what can not be observed. Similarily, a sub σ-algebra $\mathscr{C}$ of $\mathscr{A}$ describes what we are interested in predicting. Finally, we are given a sub σ-algebra $\mathscr{B}_0$ of $\mathscr{B}$ and our problem is to decide if anything is lost by basing our predictions on $\mathscr{B}_0$ rather than $\mathscr{B}$.

The prediction problem is assumed to be completely described by a decision space $(T, \mathscr{T})$, i.e. a measurable space and a loss function L from $\Theta \times \mathscr{X} \times T$ to $[0, \infty[$. It will always be assumed that L as a function on $\mathscr{X} \times T$ for given

Received January 1973; revised August 1974.

AMS 1970 *subject classifications.* Primary 62B05; Secondary 62C07.

Key words and phrases. Prediction sufficiency, conditional independence, equality of risk functions.

$\theta \in \Theta$ is $\mathscr{C} \times \mathscr{T}$ measurable. This implies that the loss does not depend on all of $x \in \mathscr{X}$, only on the part of x which is to be predicted.

A decision rule δ will here be defined as a Markov kernel $\delta(S \mid x) \colon S \in \mathscr{T}$, $x \in \mathscr{X}$ which is $\mathscr{B}$ measurable when S is fixed and a probability measure on $\mathscr{T}$ when x is fixed.

If δ is a decision rule then its performance characteristic $\mu_\delta(\cdot \mid \theta)$; $\theta \in \Theta$ may be defined by defining—for each $\theta - \mu_\delta(\cdot \mid \theta)$ as the probability measure on $\mathscr{C} \times \mathscr{T}$ defined by

$$\mu_\delta(C \times S \mid \theta) = \int_C \delta(S \mid x) P_\theta(dx) \,.$$

The risk function $r_\delta(\theta)$; $\theta \in \Theta$ of a decision rule δ is given by:

$$r_\delta(\theta) = \int [\int L_\theta(x, t)\delta(dt \mid x)] P_\theta(dx) \,, \qquad \theta \in \Theta \,.$$

The risk function is determined by the loss function and the performance characteristic through

$$r_\delta(\theta) = \int L_\theta \, d\mu_\delta(\cdot \mid \theta) \,, \qquad \theta \in \Theta \,.$$

A decision rule δ will be called $\mathscr{B}_0$ measurable if $\delta(S \mid \cdot)$ is $\mathscr{B}_0$ measurable for each S.

DEFINITION 1. $\mathscr{B}$ and $\mathscr{C}$ are conditionally independent given $\mathscr{B}_0$ iff

(i) $P_\theta^{\mathscr{C}}(C \mid \mathscr{B}) = P_\theta^{\mathscr{C}}(C \mid \mathscr{B}_0)$ a.e. $[P_\theta]$ for all $C \in \mathscr{C}$ and for all $\theta \in \Theta$.

It is shown in Loève [5], page 351 that (i) and (ii) are equivalent:

(ii) $P_\theta^{\mathscr{X}}(B \cap C \mid \mathscr{B}_0) = P_\theta^{\mathscr{X}}(B \mid \mathscr{B}_0) P_\theta^{\mathscr{X}}(C \mid \mathscr{B}_0)$ a.e. $[P_\theta]$

for all $B \in \mathscr{B}$ and all $C \in \mathscr{C}$ and for all $\theta \in \Theta$.
We define prediction sufficiency (adequacy) and prediction sufficiency in the wide sense as follows:

DEFINITION 2. $\mathscr{B}_0$ is prediction sufficient for $\mathscr{B}$ w.r.t. $\mathscr{C}$ iff $\mathscr{B}_0$ is sufficient for $\mathscr{B}$ and $\mathscr{B}$ and $\mathscr{C}$ are conditionally independent given $\mathscr{B}_0$.

DEFINITION 3. $\mathscr{B}_0$ is prediction sufficient in the wide sense for $\mathscr{B}$ w.r.t. $\mathscr{C}$ iff (a) $\mathscr{B}$ and $\mathscr{C}$ are conditionally independent given $\mathscr{B}_0$ and (b) there exist $\mathscr{B}_0$-measurable sets B_1 and B_2 so that $P_\theta(B_1 \cup B_2) = 1$ for all $\theta \in \Theta$ and $P_{\theta,x}^{\mathscr{B}}(\cdot \mid \mathscr{B}_0)$ is independent of θ if $x \in B_1$ and $P_{\theta,x}^{\mathscr{C}}(\cdot \mid \mathscr{B}_0)$ is independent of θ if $x \in B_2$.

In the following example we shall show that $\mathscr{B}_0$ is prediction sufficient in the wide sense for $\mathscr{B}$ w.r.t. $\mathscr{C}$ but not prediction sufficient for $\mathscr{B}$ w.r.t. $\mathscr{C}$.

We assume that $X_1, X_2, \cdots, X_n$ and Y are random variables such that $X_1, X_2, \cdots, X_n$ are independently and identically distributed as $N(\theta, 1)$ while the conditional distribution of Y given $X_1, X_2, \cdots, X_n$ is $N(0, 1)$ or $N(\theta, 1)$ as $\sum_i X_i > a$ or $\sum_i X_i \leqq a$. Let $\mathscr{B}$, $\mathscr{B}_0$ and $\mathscr{C}$ be the σ-algebras induced by, respectively, $(X_1, X_2, \cdots, X_n)$, min $(a, \sum_i X_i)$ and Y. Then $\mathscr{B}_0$ is prediction sufficient in the wide sense for $\mathscr{B}$ w.r.t. $\mathscr{C}$ but not sufficient for $\mathscr{B}$.

That sufficiency alone is insufficient in prediction problems may be seen by considering, for example, the situation where P_θ does not depend on θ and

1020   KEI TAKEUCHI AND MASAFUMI AKAHIRA

$\mathscr{B}_0 = \{\phi, \mathscr{X}\}$. It is then fairly obvious that prediction of a $\mathscr{C}$ which is not independent of $\mathscr{B}$ should not, in general, be based on $\mathscr{B}_0$.

It follows, as has been pointed out by Skibinsky [6], from Theorem 11.3 in Bahadur [1] that $\mathscr{B}_0$ is prediction sufficient for $\mathscr{B}$ w.r.t. $\mathscr{C}$ if and only if $\mathscr{B}_0$ is sufficient for $\mathscr{B}$ w.r.t. all probability measures on $\mathscr{B}$ of the form:

$$B \cap \longrightarrow P_\theta(B \mid C) \, ,$$

where $C \in \mathscr{C}$, $\theta \in \Theta$ and $P_\theta(C) > 0$.

In analogy with Theorem 10.2 in Bahadur [1] we get:

THEOREM 1. *Suppose $\mathscr{B}_0$ is prediction sufficient for $\mathscr{B}$ w.r.t. $\mathscr{C}$ and there exists a regular conditional probability $P^{\mathscr{B}}(\cdot \mid \mathscr{B}_0)$ of $\mathscr{B}$ given $\mathscr{B}_0$ which does not depend on θ. Let δ be any decision rule from $(\mathscr{X}, \mathscr{B})$ to $(T, \mathscr{T})$ and put $\breve{\delta}(S \mid x) = \int \delta(S \mid x') P_x^{\mathscr{B}}(dx' \mid \mathscr{B}_0); S \in \mathscr{T}, x \in \mathscr{X}$.*

Then $\breve{\delta}$ is $\mathscr{B}_0$ measurable and it has the same performance characteristic as δ. In particular δ and $\breve{\delta}$ have the same risk functions.

PROOF.

$$\mu_\theta(C \times S \mid \breve{\delta}) = \int_C \breve{\delta}(S \mid x) P_\theta(dx) = \int_C E^{\mathscr{B}}(\delta(S \mid \cdot) \mid \mathscr{B}_0) \, dP_\theta$$
$$= \int_C E^{\mathscr{B}}(\delta(S \mid \cdot) \mid \mathscr{B}_0, \mathscr{C}) \, dP_\theta = \int_C \delta(S \mid \cdot) \, dP_\theta = \mu_0(C \times S \mid \delta) \, .$$

REMARK. As is immediately seen from above, we need to allow randomized decision rules. This is not always necessary for the subsequent discussions.

Consequences of "risk prediction sufficiency" for various classes of loss functions. In order to show prediction sufficiency of $\mathscr{B}_0$ we must establish conditional independence and ordinary sufficiency. We will assume that we are given a certain class of loss functions and that to any loss function within that class and to any decision rule δ corresponds a decision rule $\breve{\delta}$ which is $\mathscr{B}_0$ measurable and has uniformly smaller risk than δ. The problem is to decide whether this suffices to establish conditional independence or ordinary sufficiency. It is clear that conditional independence cannot, in general, be established by only considering loss functions which do not depend on x. Similarly loss functions which do not involve θ will, in general, be insufficient to establish ordinary sufficiency.

Conditional independence may, however, be established by considering only loss functions which do not depend on θ.

Similarly, and this follows from corresponding facts for sufficiency (see Bahadur [2], Blackwell [3] and Le Cam [4]), sufficiency of $\mathscr{B}_0$ may be established by considering loss functions which do not depend on x.

Conditional independence may be established by considering the two decision problem with loss functions not depending on θ as follows:

THEOREM 2. *Consider the decision space $T = \{0, 1\}$ and the set of all loss functions, L, of the form*

$$L_\theta(x, 0) = I_C(x) \, , \qquad\qquad x \in \mathscr{X}, \theta \in \Theta \, ,$$
$$L_\theta(x, 1) = p I_{C^c}(x) \, , \qquad\qquad x \in \mathscr{X}, \theta \in \Theta \, ,$$

where $p \in \,]0, 1[$ and $C \in \mathscr{C}$.

Suppose that to each decision rule δ and to each loss function L, of the above form, there corresponds a $\mathscr{B}_o$ measurable decision rule $\tilde{\delta}$ so that

$$r_{\tilde{\delta}}(\theta) \leq r_{\delta}(\theta), \qquad\qquad \theta \in \Theta .$$

Then $\mathscr{B}$ and $\mathscr{C}$ are conditionally independent given $\mathscr{B}_o$.

Before proving the theorem a few remarks may be in order.

REMARK 1. It follows from the proofs that we may restrict attention to non-randomized decision rules.

REMARK 2. The proofs imply also that much smaller sets of loss functions will do. We may, for example, restrict C to a π-system generating $\mathscr{C}$.

REMARK 3. The parameter space Θ does not play any role in this theorem. We may—and shall—in the proof assume that Θ consists of a single point. Conditional independence is, in this situation, equivalent to prediction sufficiency.

PROOF OF THE THEOREM. We may, by Remark 3, omit the subscript θ. Furthermore a decision rule δ may be identified with the critical function $x \curvearrowright \delta(1 \mid x)$. The risk may then be written:

$$r(\delta) = \int L(\cdot, 0)\, dP + \int [L(\cdot, 1) - L(\cdot, 0)]\delta \, dP$$

$$= \int L(\cdot, 0)\, dP + (p + 1) \int \left(\frac{p}{p + 1} - I_C \right) \delta \, dP$$

$$= \int L(\cdot, 0)\, dP + (p + 1) \int \left[\frac{p}{p + 1} - P^{\mathscr{C}}(C \mid \mathscr{B}) \right] \delta \, dP$$

$$\geq \int L(\cdot, 0)\, dP - (p + 1) \int \left[P^{\mathscr{C}}(C \mid \mathscr{B}) - \frac{p}{p + 1} \right]^+ dP ,$$

where "$=$" is obtained iff $\delta = 0$ a.e. or $\delta = 1$ a.e. as $P^{\mathscr{C}}(C \mid \mathscr{B}) < p/(p + 1)$ or $P^{\mathscr{C}}(C \mid \mathscr{B}) > p/(p + 1)$.

The same argument applied to $\mathscr{B}_o$ implies, by the assumption of the theorem, that the minimizing δ may be chosen $\mathscr{B}_o$-measurable and such that $\tilde{\delta} = 0$ a.e. or $\delta = 1$ a.e. as $P^{\mathscr{C}}(C \mid \mathscr{B}_o) < p/(p + 1)$ or $P^{\mathscr{C}}(C \mid \mathscr{B}_o) > p/(p + 1)$. It follows that the event $[P^{\mathscr{C}}(C \mid \mathscr{B}) < p/(p + 1)]$ and the event $[P^{\mathscr{C}}(C \mid \mathscr{B}_o) < p/(p + 1)]$ are equivalent provided $P^{\mathscr{C}}(C \mid \mathscr{B}) \neq p/(p + 1)$ a.e. and $P^{\mathscr{C}}(C \mid \mathscr{B}_o) \neq p/(p + 1)$ a.e. This implies that the random variables $P^{\mathscr{C}}(C \mid \mathscr{B})$ and $P^{\mathscr{C}}(C \mid \mathscr{B}_o)$ have the same distribution. Hence, since $P^{\mathscr{C}}(C \mid \mathscr{B}_o) = E^{\mathscr{C}}(P(C \mid \mathscr{B}) \mid \mathscr{B}_o)$: $P^{\mathscr{C}}(C \mid \mathscr{B}_o) = P^{\mathscr{C}}(C \mid \mathscr{B})$ a.e. It follows that $\mathscr{B}$ and $\mathscr{C}$ are conditionally independent given $\mathscr{B}_o$.

REMARK. This form of the proof was suggested by one of the referees.

A criterion based on least squares prediction theory is, as has been pointed out by one of the referees, even simpler to establish. Consider a sufficiently large class of square integrable and $\mathscr{C}$-measurable random variables g. To a given g we associate the loss function

$$L(x, t) = (g(x) - t)^2, \qquad x \in \mathscr{X}, t \in \,]-\infty, \infty[\, .$$

Then a predictor δ minimizes the risk if and only if it is a version of $E^{\mathscr{C}}(g \,|\, \mathscr{B})$. If $\mathscr{B}_o$ is assumed to be just as good in this situation then $E^{\mathscr{C}}(g \,|\, \mathscr{B}) = E^{\mathscr{C}}(g \,|\, \mathscr{B}_o)$ a.e. This establishes conditional independence if, for example, we admit all functions $g = I_C$ where C runs through a π-system generating $\mathscr{C}$.

For the Lemma and Theorem 3 we assme that $\{P_\theta : \theta \in \Theta\}$ is dominated. Then we get the following lemma.

LEMMA. *If for any $\mathscr{B}$-measurable critical function φ there exists a $\mathscr{B}_o$-measurable cr..ical function ψ such that $E_\theta(\psi) = E_\theta(\varphi)$ for all $\theta \in \Theta$, then $\mathscr{B}_o$ is sufficient for $\mathscr{B}$.*

The proof of the lemma is essentially the same as in Bahadur [2]. The outline is as follows: Let θ_1 and θ_2 be any two points of Θ. Let ϕ be a most powerful test for θ_1 against θ_2. Then for some k

$$\phi(x) = 1 \qquad \text{if} \quad \frac{dP_{\theta_2}}{dP_{\theta_1}} > k \, ,$$

$$= 0 \qquad \text{if} \quad \frac{dP_{\theta_2}}{dP_{\theta_1}} < k \, ,$$

and the set $\{x : (dP_{\theta_2}/dP_{\theta_1})(x) < k\}$ is $\mathscr{B}_o$-measurable. Further for every c (including ∞), the set $\{x : (dP_{\theta_2}/dP_{\theta_1})(x) < c\}$ is $\mathscr{B}_o$-measurable. Hence $\mathscr{B}_o$ is pairwise sufficient for $\mathscr{B}$. Since $\{P_\theta : \theta \in \Theta\}$ is dominated, $\mathscr{B}_o$ is sufficient for $\mathscr{B}$.

The following proposition is an immediate consequence of the lemma:

Suppose that the decision space $T = [0, 1]$ and the loss function L satisfies $L(x, t) = t$, for $0 \leq t \leq 1$. If for any $\mathscr{B}$-measurable decision rule δ, there exists a $\mathscr{B}_o$-measurable decision rule $\tilde{\delta}$ such that $r_{\tilde{\delta}}(\theta) = r_\delta(\theta)$ for all $\theta \in \Theta$, then $\mathscr{B}_o$ is sufficient for $\mathscr{B}$.

From the above, we get the following theorem.

THEOREM 3. *If for any loss function L not depending on θ and for any $\mathscr{B}$-measurable decision rule δ, there exists a $\mathscr{B}_0$-measurable decision rule $\tilde{\delta}$ such that $r_{\tilde{\delta}}(\theta) = r_\delta(\theta)$ for all $\theta \in \Theta$, then $\mathscr{B}_o$ is prediction sufficient for $\mathscr{B}$ w.r.t. $\mathscr{C}$.*

Various criteria for prediction sufficiency may be obtained from these results by considering, in addition to the loss functions described above, loss functions which depend on θ. If we insist on considering only loss functions which do not depend on θ then we will, in general, not be able to conclude prediction sufficiency. This follows by considering the case where $\mathscr{B}_o = \{\phi, \mathscr{X}\}$, $\mathscr{C}$ is ancillary and independent of $\mathscr{B}$ and $\{P_\theta : \theta \in \Theta\}$ is finite.

Let, in this situation, $(T, \mathscr{T})$ be any decision space and L any loss function which does not depend on θ and δ any decision rule from $(\mathscr{X}, \mathscr{B})$ to $(T, \mathscr{T})$.

Choose a $t_o \in T$ such that

$$\int L(x, t_o) P_\theta(dx) \leqq \int \left[\int L(x, t) P_\theta(dx) \right] \nu_\theta(dt) \, , \qquad \theta \in \Theta \, ,$$

where $\nu_\theta(S) = \int \delta(S\,|\,x)P_\theta(dx)$. Then

$$r_\theta(t_o) \leqq r_\theta(\delta)\,, \qquad\qquad \theta \in \Theta\,.$$

The necessary condition for prediction sufficiency in the wide sense is obtained as follows:

THEOREM 4. *Suppose that, for each θ, there are regular conditional probabilities $P_\theta(B\,|\,\mathscr{B}_o): B \in \mathscr{B}$ and $P_J(C\,|\,\mathscr{B}_o): C \in \mathscr{C}$. Suppose further that $\mathscr{B}_o$ is prediction sufficient in the wide sense for $\mathscr{B}$ w.r.t. $\mathscr{C}$.*

Let L be a loss function not depending on θ and assume that there is a $\mathscr{B}_o$-measurable function τ on B_2 and an $\varepsilon \geqq 0$ so that

$$\int L(x',\tau)P(dx'\,|\,\mathscr{B}_o) \leqq \int L(x',t)P(dx'\,|\,\mathscr{B}_o) + \varepsilon\,, \qquad t \in T \ \text{on} \ B_2\,.$$

Then there corresponds to any decision rule δ a $\mathscr{B}_o$-measurable decision rule $\tilde{\delta}$ so that

$$r_\theta(\tilde{\delta}) \leqq r_\theta(\delta) + \varepsilon\,, \qquad\qquad \theta \in \Theta\,.$$

$\tilde{\delta}$ may be defined as τ on B_2 and as $E(\delta(\,\cdot\,|\,\cdot\,)\,|\,\mathscr{B}_o)$ on B_1.

REMARK. There exist, for each $\varepsilon > 0$, a $\mathscr{B}_o$-measurable τ on B_2 satisfying the desired inequality provided

(i) There exists a countable subset $\{t_1, t_2, \cdots, t_n, \cdots\}$ of T such that for all $t \in T$ and for all $x \in B_2$

$$\inf_n \int L(x',t_n)P_x^{\mathscr{C}}(dx'\,|\,\mathscr{B}_o) \leqq \int L(x',t)P_x^{\mathscr{C}}(dx'\,|\,\mathscr{B}_o)\,,$$

and

(ii) For every pair i,j and for any $\varepsilon > 0$ the set

$$M_{ij}(\varepsilon) = \{x: \int L(x',t_i)P_x^{\mathscr{C}}(dx'\,|\,\mathscr{B}_o) < \int L(x',t_j)P_x^{\mathscr{C}}(dx'\,|\,\mathscr{B}_o) + \varepsilon\}$$

is measurable.

PROOF OF THE THEOREM. Put, for each θ,

$$\delta_\theta^*(S\,|\,\cdot) = \int \delta(S\,|\,x')P_\theta(dx'\,|\,\mathscr{B}_o)\,.$$

By the proof of Theorem 1:

$$\begin{aligned}
\varepsilon + r_\theta(\delta) &= \varepsilon + \int [\int L(x,t)\delta_\theta^*(dt\,|\,x)]P_\theta(dx) \\
&= \int_{B_1} [\int L(x,t)\tilde{\delta}(dt\,|\,x)]P_\theta(dx) \\
&\quad + \varepsilon + \int_{B_2} \{\int [\int L(x',t)P_x(dx'\,|\,\mathscr{B}_o)]\delta_\theta^*(dt\,|\,x)\}P_\theta(dx) \\
&\geqq \int_{B_1} [\int L(x,t)\tilde{\delta}(dt\,|\,x)]P_\theta(dx) \\
&\quad + \int_{B_2} \{\int L(x',\tilde{\delta}(x))P_{\theta,x}(dx'\,|\,\mathscr{B}_o)\}P_\theta(dx) \\
&= \int_{B_1} [\int L(x,t)\tilde{\delta}(dt\,|\,x)]P_\theta(dx) + \int_{B_2} L(x,\tilde{\delta}(x))P_\theta(dx) \\
&= r_\theta(\tilde{\delta})\,.
\end{aligned}$$

Acknowledgments. The authors wish to thank Mr. M. Takahashi of Osaka University for valuable suggestions and the referees of the *Annals* for co-operation in completing the final version.

1024   KEI TAKEUCHI AND MASAFUMI AKAHIRA

REFERENCES

[1] BAHADUR, R. R. (1954). Sufficiency and statistical decision functions. *Ann. Math. Statist.* **25** 423–462.
[2] BAHADUR, R. R. (1955). A characterization of sufficiency. *Ann. Math. Statist.* **26** 286–293.
[3] BLACKWELL, D. (1953). Equivalent comparisions of experiments. *Ann. Math. Statist.* **24** 265–272.
[4] LE CAM, L. (1964). Sufficiency and approximate sufficiency. *Ann. Math. Statist.* **35** 1419–1455.
[5] LOÈVE, M. (1963). *Probability Theory*, (3rd. ed.). Van Nostrand, Princeton.
[6] SKIBINSKY, M. (1967). Adequate subfields and sufficiency. *Ann. Math. Statist.* **38** 155–161.
[7] TAKEUCHI, K. (1966). On some statistical prediction procedures. (In Japanese). *Keizaigaku Ronshu* **32** 23–31.

FACULTY OF ECONOMICS   DEPARTMENT OF MATHEMATICS
UNIVERSITY OF TOKYO   UNIVERSITY OF ELECTRO-COMMUNICATIONS
HONGO, BUNKYO-KU   CHOFUGAOKA, CHOFU-SHI
TOKYO, JAPAN   TOKYO, JAPAN

Rep. Univ. Electro-Comm. 26-2, (Sci. & Tech. Sect.), pp. 261—269 February, 1976

On the Second Order Asymptotic Efficiency
of Estimators in Multiparameter Cases[*]

Masafumi AKAHIRA[**] and Kei TAKEUCHI[***]

Abstract

Suppose that X_i's $(i=1, 2, \ldots, n)$ are independently and identically distributed with the density $f(x, \theta, \xi)$, where θ is a real valued parameter and ξ is a real (vector) valued parameter. We consider a (sequence of) estimator(s) which is k-th order asymptotically median unbiased, and define k-th order asymptotic efficiency. We have a formula for the distribution of the second order asymptotically efficient estimator and show that a modified maximum likelihood estimator is second order asymptotically efficient.

1. Introduction

Second order asymptotic efficiency of estimators has been discussed by Pfanzagl [2], Takeuchi and Akahira [3], Takeuchi [4] and Akahira [1] for one dimensional case. In this paper we extend a similar approach to multiparameter cases and obtain a straight-forward generalization of the one-parameter case.

Let $X_1, X_2, \ldots, X_n, \ldots$ be a sequence of independent identically distributed random variables with the density $f(x, \theta, \xi)$, where θ is a real valued parameter and ξ is a real (vector) valued parameter. We assume that ξ is a nuisance parameter. We shall define an estimator of θ to be k-th order asymptotically efficient if the k-th order asymptotic distribution of it attains the bound of the k-th order asymptotic distributions of k-th order asymptotically median unbiased (AMU) estimators of θ. We shall obtain the bound of the second order asymptotic distributions of second order AMU estimators and show that a modified maximum likelihood estimator is second order asymptotically efficient.

2. Notations and definitions

Let $\mathcal{X}$ be an abstract sample space whose generic point is denoted by x, $\mathcal{B}$ a σ-field of subsets of $\mathcal{X}$ and let Θ and Ξ be parameter spaces which are assumed to be open sets in R^1 and R^p respectively. (We denote by R^p a Euclidean p-space with a norm $\|\cdot\|$.) We assume that $\xi=(\xi_1, \ldots, \xi_p)$ $(\in\Xi)$ is a nuisance parameter. We consider a sequence of classes of probability measures $\{P_{\theta,\xi,i} : (\theta, \xi) \in \Theta\times\Xi\}$ $(i=1, 2, \ldots)$ each defined over $(\mathcal{X}, \mathcal{B})$. We shall denote by $(\mathcal{X}^{(n)}, \mathcal{B}^{(n)})$ the n-fold direct products of $(\mathcal{X}, \mathcal{B})$ and the corresponding product measures by $P_{\theta,\xi}^{(n)}=P_{\theta,\xi,1}\times\ldots\times P_{\theta,\xi,n}$. An estimator of θ is defined to be a sequence $\{\hat{\theta}_n\}$

* Received on December 10, 1975
** Statistical Laboratory, University of Electro-Communications
*** University of Tokyo

262   Masfumi AKAHIRA and Kei TAKEUCHI

of $\mathscr{B}^{(n)}$-measurable functions $\hat{\theta}_n$ on $\mathscr{X}^{(n)}$ into Θ $(n=1, 2, \ldots)$.

Definition 1. *For each $k=1, 2, \ldots$, $\{\hat{\theta}_n\}$ is k-th order asymptotically median unbiased (or k-th order AMU) estimator if for any $\vartheta_0=(\theta_0, \xi_0)\epsilon\Theta\times\Xi$, there exists a positive number δ such that*

$$\lim_{n\to\infty} \sup_{\vartheta\epsilon\Theta\times\Xi\,:\,\|\vartheta_0-\vartheta\|<\delta} n^{\frac{k-1}{2}}\left|P_{\vartheta}^{(n)}\{n^{\frac{1}{2}}(\hat{\theta}_n-\theta)\leq 0\}-\frac{1}{2}\right|=0\,;$$

$$\lim_{n\to\infty} \sup_{\vartheta\epsilon\Theta\times\Xi\,:\,\|\vartheta_0-\vartheta\|<\delta} n^{\frac{k-1}{2}}\left|P_{\vartheta}^{(n)}\{n^{\frac{1}{2}}(\hat{\theta}_n-\theta)\geq 0\}-\frac{1}{2}\right|=0.$$

Definition 2. *Suppose that $\{\hat{\theta}_n\}$ is k-th order asymptotically median unbiased, $G_0(t, \theta, \xi)+n^{-\frac{1}{2}}G_1(t, \theta, \xi)+\ldots+n^{-\frac{k-1}{2}}G_{k-1}(t, \theta, \xi)$ is called to be the k-th order asymptotic distribution of it if*

$$\lim_{n\to\infty} n^{\frac{k-1}{2}}\left|P_{\theta,\xi}^{(n)}\{n^{\frac{1}{2}}(\hat{\theta}_n-\theta)\leq t\}-G_0(t, \theta, \xi)-n^{-\frac{1}{2}}G_1(t, \theta, \xi)\right.$$

$$\left.-\ldots-n^{-\frac{k-1}{2}}G_{k-1}(t, \theta, \xi)\right|=0.$$

Consider the problem of testing hypothesis $H^+ : \theta=\theta_0+tn^{-\frac{1}{2}}(t>0)$, against $K : \theta=\theta_0$, $\xi=\xi$. We shall denote by $\beta_n(t, \alpha_n, \theta_0, \xi)$ the power function of the most powerful level α_n test $\left(0<\alpha_n<\frac{1}{2}\right)$. If for $\xi=\xi_0$ and for each $t>0$

$$\sup_{\{\{\alpha_n\}\,:\,\alpha_n=(1/2)+o(n^{-(k-1)/2})\}} \overline{\lim_{n\to\infty}} n^{-\frac{k-1}{2}}\{\beta_n(t, \alpha_n, \theta_0, \xi_0)-H_0^+(t, \theta_0, \xi_0)$$

$$-n^{-\frac{1}{2}}H_1^+(t, \theta_0, \xi_0)-\ldots-n^{-\frac{k-1}{2}}H_{k-1}^+(t, \theta_0, \xi_0)\}=0,$$

then we have

$$G_0(t, \theta_0, \xi_0)\leq H_0^+(t, \theta_0, \xi_0)\,;$$

$$\text{if}\qquad G_i(t, \theta_0, \xi_0)=H_i^+(t, \theta_0, \xi_0)\quad(i=0,\ldots,j-1),$$

$$\text{then}\qquad G_j(t, \theta_0, \xi_0)\leq H_j^+(t, \theta_0, \xi_0)\quad(j=1,\ldots,k)$$

Consider next the problem of testing hypothesis $H^- : \theta=\theta_0+tn^{-\frac{1}{2}}(t<0)$, against $K : \theta=\theta_0, \xi=\xi$. If for $\xi=\xi_0$ and for each $t<0$

$$\inf_{\{\{\alpha_n\}\,:\,\alpha_n=(1/2)+o(n^{-(k-1)/2})\}} \underline{\lim_{n\to\infty}} n^{-\frac{k-1}{2}}\{\beta_n(t, \alpha_n, \theta_0, \xi_0)-H_0^-(t, \theta_0, \xi_0)$$

$$-n^{-\frac{1}{2}}H_1^-(t, \theta_0, \xi_0)-\ldots-n^{-\frac{k-1}{2}}H_{k-1}^-(t, \theta_0, \xi_0)\}=0,$$

then we have

$$G_0(t, \theta_0, \xi_0)\geq H_0^-(t, \theta_0, \xi_0)\,;$$

$$\text{if}\qquad G_i(t, \theta_0, \xi_0)=H_i^-(t, \theta_0, \xi_0)\quad(i=0,\ldots,j-1),$$

$$\text{then}\qquad G_j(t, \theta_0, \xi_0)\geq H_j^-(t, \theta_0, \xi_0)\quad(j=1,\ldots,k).$$

Definition 3. *Suppose that $\{\hat{\theta}_n\}$ is k-th order asymptically median unbiased, it is called to be k-th order asymptotically efficient if for each $\theta\epsilon\Theta$ and each $\xi\epsilon\Xi$*

$$G_i(t, \theta, \xi)=\begin{cases}H_i^+(t, \theta, \xi) & \text{for}\quad t>0,\\ H_i^-(t, \theta, \xi) & \text{for}\quad t<0,\end{cases}$$

$i=0,\ldots,k-1.$

In the subsequent discussions we shall deal only with the second order asymptotic efficiency, but the line of discussions can be extended to general cases.

February, 1976 On the Second Order Asymptotic Efficiency of Estimators in Multiparameter Cases   263

3. Second order asymptotic efficiency in multiparameter cases.

We assume that Ξ is an open set of R^1 because in the subsequent discussions there arises no substantial difference between R^1 and $R^p(p\geq2)$. Let $X_1, X_2, \ldots, X_n, \ldots$ be a sequence of independently and identically distributed random variables having a density function $f(x, \theta, \xi)$.

First we shall obtain the bound of the power functions. Consider the problem of testing hypothesis $H^+: \theta=\theta_0+tn^{-\frac{1}{2}}(t>0)$, $\xi=\xi_0+un^{-\frac{1}{2}}$, against $K: \theta=\theta_0$, $\xi=\xi_0$, where u is an arbitrary but fixed constant. Putting $\theta_1=\theta_0+tn^{-\frac{1}{2}}$ and $\xi_1=\xi_0+un^{-\frac{1}{2}}$, we define Z_{in} as follows:

$$Z_{in}=\log\frac{f(X_i, \theta_0, \xi_0)}{f(X_i, \theta_1, \xi_1)}.$$

The most powerful test is given by the following rejection region

$$\sum_{i=1}^{n} Z_{in}>c,$$

where c is some constant.

For each $i=1, 2, \ldots, Z_{in}$ is expanded as follows:

$$Z_{in}=-\frac{\partial}{\partial\theta}\log f(X_i, \theta_0, \xi_0)(tn^{-\frac{1}{2}})-\frac{\partial}{\partial\xi}\log f(X_i, \theta_0, \xi_0)(un^{-\frac{1}{2}})$$

$$-\frac{1}{2}\frac{\partial^2}{\partial\theta^2}\log f(X_i, \theta_0, \xi_0)(t^2n^{-1})-\frac{1}{2}\frac{\partial^2}{\partial\xi^2}\log f(X_i, \theta_0, \xi_0)(u^2n^{-1})$$

$$-\frac{\partial^2}{\partial\theta\partial\xi}\log f(X_i, \theta_0, \xi_0)(tun^{-1})$$

$$-\frac{1}{6}\frac{\partial^3}{\partial\theta^3}\log f(X_i, \theta_0, \xi_0)(t^3n^{-\frac{3}{2}})-\frac{1}{6}\frac{\partial^3}{\partial\xi^3}\log f(X_i, \theta_0, \xi_0)(u^3n^{-\frac{3}{2}})$$

$$-\frac{1}{2}\frac{\partial^3}{\partial\theta^2\partial\xi}\log f(X_i, \theta_0, \xi_0)(t^2un^{-\frac{3}{2}})-\frac{1}{2}\frac{\partial^3}{\partial\theta\partial\xi^2}\log f(X_i, \theta_0, \xi_0)(tu^2n^{-\frac{3}{2}})$$

$$+o(n^{-\frac{3}{2}}).$$

Hence $\sum_{i=1}^{n} Z_{in}$ is asymptotically normal. If $\theta=\theta_0$ and $\xi=\xi_0$, then the asymptotic mean μ_0 and the asymptotic variance σ_0^2 of $\sum_{i=1}^{n} Z_{in}$ are given by

$$\mu_0=\frac{1}{2}(I_{00}t^2+I_{11}u^2+2I_{01}tu)+o(n^{-\frac{1}{2}});$$

$$\sigma_0^2=I_{00}t^2+I_{11}u^2+2I_{01}tu+o(n^{-\frac{1}{2}}),$$

where

$$I_{00}=E_{\theta_0, \xi_0}\left\{\frac{\partial}{\partial\theta}\log f(X_i, \theta_0, \xi_0)\right\}^2=-E_{\theta_0, \xi_0}\left\{\frac{\partial^2}{\partial\theta^2}\log f(X_i, \theta_0, \xi_0)\right\};$$

$$I_{01}=E_{\theta_0, \xi_0}\left\{\frac{\partial}{\partial\theta}\log f(X_i, \theta_0, \xi_0)\frac{\partial}{\partial\xi}\log f(X_i, \theta_0, \xi_0)\right\}$$

$$=-E_{\theta_0, \xi_0}\left\{\frac{\partial^2}{\partial\theta\partial\xi}\log f(X_i, \theta_0, \xi_0)\right\};$$

$$I_{11}=E_{\theta_0, \xi_0}\left\{\frac{\partial}{\partial\xi}\log f(X_i, \theta_0, \xi_0)\right\}^2=-E_{\theta_0, \xi_0}\left\{\frac{\partial^2}{\partial\xi^2}\log f(X_i, \theta_0, \xi_0)\right\}.$$

If $\theta=\theta_1$ $\xi=\xi_1$, then if follows that

Masfumi AKAHIRA and Kei TAKEUCHI

$$E_{\theta_1,\xi_1}\left\{\frac{\partial}{\partial\theta}\log f(X_i,\theta_0,\xi_0)\right\}$$

$$=E_{\theta_0,\xi_0}\left\{\frac{\partial}{\partial\theta}\log f(X_i,\theta_0,\xi_0)\frac{f(X_i,\theta_1,\xi_1)}{f(X_i,\theta_0,\xi_0)}\right\}$$

$$=E_{\theta_0,\xi_0}\left\{\frac{\partial}{\partial\theta}\log f(X_i,\theta_0,\xi_0)\right\}^2(tn^{-\frac{1}{2}})$$

$$+E_{\theta_0,\xi_0}\left\{\frac{\partial}{\partial\theta}\log f(X_i,\theta_0,\xi_0)\frac{\partial}{\partial\xi}\log f(X_i,\theta_0,\xi_0)\right\}(un^{-\frac{1}{2}})+o(n^{-\frac{1}{2}});$$

$$E_{\theta_1,\xi_1}\left\{\frac{\partial}{\partial\xi}\log f(X_i,\theta_0,\xi_0)\right\}$$

$$=E_{\theta_0,\xi_0}\left\{\frac{\partial}{\partial\theta}\log f(X_i,\theta_0,\xi_0)\frac{\partial}{\partial\xi}\log f(X_i,\theta_0,\xi_0)\right\}(tn^{-\frac{1}{2}})$$

$$+E_{\theta_0,\xi_0}\left\{\frac{\partial}{\partial\xi}\log f(X_i,\theta_0,\xi_0)\right\}^2(un^{-\frac{1}{2}})+o(n^{-\frac{1}{2}}).$$

For $\theta=\theta_1$ and $\xi=\xi_1$, the asymptotic mean μ_1 and the asymptotic variance $\sigma_1{}^2$ of $\sum_{i=1}^{n}Z_{in}$ are given by

$$\mu_1=-\frac{1}{2}(I_{00}t^2+I_{11}u^2+2I_{10}tu)+o(n^{-\frac{1}{2}});$$

$$\sigma_1{}^2=\sigma_0{}^2+o(n^{-\frac{1}{2}}).$$

Hence the asymptotic power of the most powerful test of $\sum_{i=1}^{n}Z_{in}$ is obtained as follows:

$$\Phi\left(\frac{\mu_0+\mu_1}{\sigma_0}\right)=\Phi(\sqrt{I_{00}t^2+I_{11}u^2+2I_{01}tu}) \qquad (3.1)$$

where $\Phi(x)=\int_{-\infty}^{x}\frac{1}{\sqrt{2\pi}}e^{-\frac{u^2}{2}}du$.

Since u can take arbitrary values, then the power function of the tests of the (composite) hypothesis is not larger than the infimum of (3.1) with respect to u. A u minimizing $I_{00}t^2+I_{11}u^2+2I_{01}tu$ is given by $u_0=-(I_{01}/I_{11})t$, and it follows that

$$\Phi(\sqrt{I_{00}t^2+I_{11}u_0{}^2+2I_{01}tu_0})=\sqrt{I^*}\,t,$$

where $I^*=I_{00}-(I_{01}{}^2/I_{11})$.

In order to obtain the expansion of $\sum_{i=1}^{n}Z_{in}$ up to the order of $n^{-\frac{1}{2}}$ we put

$$J_{000}=E_{\theta_0,\xi_0}\left\{\frac{\partial^2}{\partial\theta^2}\log f(X_i,\theta_0,\xi_0)\frac{\partial}{\partial\theta}\log f(X_i,\theta_0,\xi_0)\right\};$$

$$J_{001}=E_{\theta_0,\xi_0}\left\{\frac{\partial^2}{\partial\theta^2}\log f(X_i,\theta_0,\xi_0)\frac{\partial}{\partial\xi}\log f(X_i,\theta_0,\xi_0)\right\};$$

$$J_{010}=E_{\theta_0,\xi_0}\left\{\frac{\partial^2}{\partial\theta\partial\xi}\log f(X_i,\theta_0,\xi_0)\frac{\partial}{\partial\theta}\log f(X_i,\xi_0,\theta_0)\right\};$$

$$J_{011}=E_{\theta_0,\xi_0}\left\{\frac{\partial^2}{\partial\theta\partial\xi}\log f(X_i,\theta_0,\xi_0)\frac{\partial}{\partial\xi}\log f(X_i,\theta_0,\xi_0)\right\};$$

$$J_{110}=E_{\theta_0,\xi_0}\left\{\frac{\partial^2}{\partial\xi^2}\log f(X_i,\theta_0,\xi_0)\frac{\partial}{\partial\theta}\log f(X_i,\theta_0,\xi_0)\right\};$$

$$J_{111}=E_{\theta_0,\xi_0}\left\{\frac{\partial^2}{\partial\xi^2}\log f(X_i,\theta_0,\xi_0)\frac{\partial}{\partial\xi}\log f(X_i,\theta_0,\xi_0)\right\};$$

February, 1976 On the Second Order Asymptotic Efficiency of Estimators in Multiparameter Cases 265

$$K_{000}=E_{\theta_0,\xi_0}\left\{\frac{\partial}{\partial\theta}\log f(X_i,\theta_0,\xi_0)\right\}^3;$$

$$K_{001}=E_{\theta_0,\xi_0}\left[\left\{\frac{\partial}{\partial\theta}\log f(X_i,\theta_0,\xi_0)\right\}^2\left\{\frac{\partial}{\partial\xi}\log f(X_i,\theta_0,\xi_0)\right\}\right];$$

$$K_{011}=E_{\theta_0,\xi_0}\left[\left\{\frac{\partial}{\partial\theta}\log f(X_i,\theta_0,\xi_0)\right\}\left\{\frac{\partial}{\partial\xi}\log f(X_i,\theta_0,\xi_0)\right\}^2\right];$$

$$K_{111}=E_{\theta_0,\xi_0}\left\{\frac{\partial}{\partial\xi}\log f(X_i,\theta_0,\xi_0)\right\}^3.$$

Under suitable regularity conditions the following hold:

$$E_{\theta_0,\xi_0}\left\{\frac{\partial^3}{\partial\theta^3}\log f(X_i,\theta_0,\xi_0)\right\}=-3J_{000}-K_{000};$$

$$E_{\theta_0,\xi_0}\left\{\frac{\partial^2}{\partial\theta^2}\log f(X_i,\theta_0,\xi_0)\frac{\partial}{\partial\xi}\log f(X_i,\theta_0,\xi_0)\right\}=-J_{001}-2J_{010}-K_{001};$$

$$E_{\theta_0,\xi_0}\left\{\frac{\partial}{\partial\xi}\log f(X_i,\theta_0,\xi_0)\frac{\partial^2}{\partial\xi^2}\log f(X_i,\theta_0,\xi_0)\right\}=-J_{110}-2J_{011}-K_{011};$$

$$E_{\theta_0,\xi_0}\left\{\frac{\partial^3}{\partial\xi^3}\log f(X_i,\theta_0,\xi_0)\right\}=-3J_{111}-K_{111}.$$

If $\theta=\theta_0$ and $\xi=\xi_0$, then the asymptotic mean μ_0 of $\sum_{i=1}^{n}Z_{in}$ up to the order of $n^{-\frac{1}{2}}$ is given as follows:

$$\mu_0=\frac{1}{2}(I_{00}t^2+I_{11}u^2+2I_{01}tu)$$

$$+\frac{1}{2}n^{-\frac{1}{2}}\{J_{000}t^2+(J_{001}+2J_{010})t^2u+(J_{110}+2J_{011})tu^2+J_{111}u^3\}$$

$$+\frac{1}{6}n^{-\frac{1}{2}}(K_{000}t^3+3K_{001}t^2u+3K_{011}t^2u+K_{111}u^3)+o(n^{-\frac{1}{2}}).$$

When $u=-(I_{01}/I_{11})t$, μ_0 is written as the following form:

$$\mu_0=\frac{1}{2}I^*t^2+\frac{1}{6}n^{-\frac{1}{2}}(3J^*+K^*)t^3+o(n^{-\frac{1}{2}}).$$

Similarly the asymptotic variance $\sigma_0{}^2$ and the asymptotic third moment γ_0 of $\sum_{i=1}^{n}Z_{in}$ are given

$$\sigma_0{}^2=I^*t^2+J^*t^3n^{-\frac{1}{2}}+o(n^{-\frac{1}{2}});$$

$$\gamma_0=E_{\theta_0,\xi_0}\left(\sum_{i=1}^{n}Z_{in}-\mu_0\right)^3\sim K^*t^3n^{-\frac{1}{2}}+o(n^{-\frac{1}{2}}).$$

If $\theta=\theta_1$ and $\xi=\xi_1$, then if follows that

$$\log\frac{f(X_i,\theta_1,\xi_1)}{f(X_i,\theta_0,\xi_0)}$$

$$=1+\frac{\partial}{\partial\theta}\log f(X_i,\theta_0,\xi_0)(tn^{-\frac{1}{2}})+\frac{\partial}{\partial\xi}\log f(X_i,\theta_0,\xi_0)(un^{-\frac{1}{2}})$$

$$+\frac{1}{2}\left\{\frac{\partial^2}{\partial\theta^2}\log f(X_i,\theta_0,\xi_0)(t^2n^{-1})+\frac{\partial^2}{\partial\xi^2}\log f(X_i,\theta_0,\xi_0)(u^2n^{-1})\right.$$

$$\left.+2\frac{\partial^2}{\partial\theta\partial\xi}\log f(X_i,\theta_0,\xi_0)(tun^{-1})\right\}$$

$$+\frac{1}{2}\left[\left\{\frac{\partial}{\partial\theta}\log f(X_i,\theta_0,\xi_0)\right\}^2t^2n^{-1}+\left\{\frac{\partial}{\partial\xi}\log f(X_i,\theta_0,\xi_0)\right\}^2u^2n^{-1}\right.$$

Masfumi Akahira and Kei Takeuchi

$$+2\left\{\frac{\partial}{\partial\theta}\log f(X_i, \theta_0, \xi_0)\right\}\left\{\frac{\partial}{\partial\xi}\log f(X_i, \theta_0, \xi_0)\right\}(tun^{-1})\right]+o(n^{-1}).$$

Hence we have

$$\mu_1 = \mu_0 + (I_{00}t^2 + I_{11}u^2 + 2I_{01}tu)$$

$$-n^{-\frac{1}{2}}\{J_{000}t^3 + (J_{001} + 2J_{010})t^2u + (J_{110} + 2J_{011})tu^2 + J_{111}u^3\}$$

$$-\frac{1}{2}n^{-\frac{1}{2}}(K_{000}t^3 + 3K_{001}t^2u + 3K_{011}tu^2 + K_{111}u^3).$$

If $u = -(I_{01}/I_{00})t$, then it follows that

$$\mu_1 = -\frac{1}{2}I^*t^2 - \frac{1}{6}n^{-\frac{1}{2}}(3J^* + 2K^*)t^3 + o(n^{-\frac{1}{2}}).$$

Similarly we obtain

$$\sigma_1{}^2 = \sigma_0{}^2 + n^{-\frac{1}{2}}(K_{000}t^3 + 3K_{001}t^2u + 3K_{011}tu^2 + K_{111}u^3) + o(n^{-\frac{1}{2}})$$

$$= I^*t^2 + n^{-\frac{1}{2}}(J^* + K^*)t^3 + o(n^{-\frac{1}{2}});$$

$$\gamma_1 = E_{\theta_1, \xi_1}\left(\sum_{i=1}^{n} Z_{in} - \mu_1\right)^3 = \gamma_0 + o(n^{-\frac{1}{2}}) = -K^*n^{-\frac{1}{2}}t^3 + o(n^{-\frac{1}{2}}).$$

If we choose a c such that

$$P_{\theta_1, \xi_1}{}^{(n)}\left\{\sum_{i=1}^{n} Z_{in} > c\right\} = \frac{1}{2} + o(n^{-\frac{1}{2}}),$$

then $c(=c_1)$ is given by

$$c_1 = -\frac{I^*}{2}t^2 - \frac{3J^* + 2K^*}{6}n^{-\frac{1}{2}}t^3 + \frac{K^*}{6I^*}n^{-1}.$$

Hence we have

$$P_{\theta_0, \xi_0}\left\{\sum_{i=1}^{n} Z_{in} > c_1\right\} = \Phi(t\sqrt{I^*}) + \phi(t\sqrt{I^*})\frac{3J^* + 2K^*}{6\sqrt{I^*}}t^2n^{-\frac{1}{2}} + o(n^{-\frac{1}{2}}) \qquad (3.2)$$

where $\Phi(x) = \int_{-\infty}^{x}\phi(u)du$ with $\phi(u) = \frac{1}{\sqrt{2\pi}}e^{-\frac{u^2}{2}}$.

The bound of the power functions of the problem of testing hypothesis is obtained by (3.2), that is,

$$H_0{}^+(t, \theta_0, \xi_0) = \Phi(t\sqrt{I^*}) \qquad (3.3)$$

$$H_1{}^+(t, \theta_0, \xi_0) = \phi(t\sqrt{I^*})\frac{(3J^* + 2K^*)}{6\sqrt{I^*}}t^2 \qquad (3.4)$$

By a similar way as the case $t > 0$ it is shown that in the case $t < 0$ the bound of the power functions is also given by (3.3) and (3.4). Thus we have

Theorem 1. *The bound for the distributions of second order asymptotically median unbiased estimators is given by*

$$G_0(t, \theta_0, \xi_0) = \Phi(t\sqrt{I^*});$$

$$G_1(t, \theta_0, \xi_0) = \phi(t\sqrt{I^*})\frac{(3J^* + 2K^*)}{6\sqrt{I^*}}t^2,$$

where I^*, J^* and K^* are defined in the context.

Let $\hat{\theta}(=\{\hat{\theta}_n\})$ and $\hat{\xi}(=\{\hat{\xi}_n\})$ be maximum likelihood estimators of θ and ξ, respectively Since

$$\sum_{i=1}^{n}\frac{\partial}{\partial\theta}\log f(X_i, \hat{\theta}, \hat{\xi}) = 0;$$

$$\sum_{i=1}^{n} \frac{\partial}{\partial \xi} \log f(X_i, \hat{\theta}, \hat{\xi}) = 0,$$

expanding them in the neighborhood of $\theta = \theta_0$ and $\xi = \xi_0$ respectively, we have

$$\sum_{i=1}^{n} \frac{\partial}{\partial \theta} \log f(X_i, \theta_0, \xi_0)$$

$$+ \left\{ \sum_{i=1}^{n} \frac{\partial^2}{\partial \theta^2} \log f(X_i, \theta_0, \xi_0)(\hat{\theta}-\theta_0) + \sum_{i=1}^{n} \frac{\partial^2}{\partial \theta \partial \xi} \log f(X_i, \theta_0, \xi_0)(\hat{\xi}-\xi_0) \right\}$$

$$+ \frac{1}{2} \left\{ \sum_{i=1}^{n} \frac{\partial^3}{\partial \theta^3} \log f(X_i, \theta_0, \xi_0)(\hat{\theta}-\theta_0)^2 + \sum_{i=1}^{n} \frac{\partial^3}{\partial \theta \partial \xi^2} \log f(X_i, \theta_0, \xi_0)(\hat{\xi}-\xi_0)^2 \right.$$

$$\left. + 2\sum_{i=1}^{n} \frac{\partial^3}{\partial \theta^2 \partial \xi} \log f(X_i, \theta_0, \xi_0)(\hat{\theta}-\theta_0)(\hat{\xi}-\xi_0) \right\}$$

$$\sim 0\,;$$

$$\sum_{i=1}^{n} \frac{\partial}{\partial \xi} \log f(X_i, \theta_0, \xi_0)$$

$$+ \left\{ \sum_{i=1}^{n} \frac{\partial^2}{\partial \theta \partial \xi} \log f(X_i, \theta_0, \xi_0)(\hat{\theta}-\theta_0) + \sum_{i=1}^{n} \frac{\partial^2}{\partial \xi^2} \log f(X_i, \theta_0, \xi_0)(\hat{\xi}-\xi_0) \right\}$$

$$+ \frac{1}{2} \left\{ \sum_{i=1}^{n} \frac{\partial^3}{\partial \theta^2 \partial \xi} \log f(X_i, \theta_0, \xi_0)(\hat{\theta}-\theta_0)^2 + \sum_{i=1}^{n} \frac{\partial^3}{\partial \xi^3} \log (X_i, \theta_0, \xi_0)(\hat{\xi}-\xi_0)^2 \right.$$

$$\left. + 2\sum_{i=1}^{n} \frac{\partial^3}{\partial \theta \partial \xi^2} \log f(X_i, \theta_0, \xi_0)(\hat{\theta}-\theta_0)(\hat{\xi}-\xi_0) \right\}$$

$$\sim 0.$$

Putting

$$Z_0 = n^{-\frac{1}{2}} \sum_{i=1}^{n} \frac{\partial}{\partial \theta} \log f(X_i, \theta_0, \xi_0)\,;$$

$$Z_1 = n^{-\frac{1}{2}} \sum_{i=1}^{n} \frac{\partial}{\partial \xi} \log f(X_i, \theta_0, \xi_0)\,;$$

$$Z_{00} = n^{-\frac{1}{2}} \sum_{i=1}^{n} \left\{ \frac{\partial^2}{\partial \theta^2} \log f(X_i, \theta_0, \xi_0) + I_{00} \right\}\,;$$

$$Z_{11} = n^{-\frac{1}{2}} \sum_{i=1}^{n} \left\{ \frac{\partial^2}{\partial \xi^2} \log f(X_i, \theta_0, \xi_0) + I_{11} \right\}\,;$$

$$Z_{01} = n^{-\frac{1}{2}} \sum_{i=1}^{n} \left\{ \frac{\partial^2}{\partial \theta \partial \xi} \log f(X_i, \theta_0, \xi_0) + I_{01} \right\},$$

we see that they are asymptotically normal with mean 0.

Since

$$0 = Z_0 + (-I_{00} + n^{-\frac{1}{2}} Z_{00}) n^{\frac{1}{2}} (\hat{\theta}-\theta_0) + (-I_{01} + n^{-\frac{1}{2}} Z_{01}) n^{\frac{1}{2}} (\hat{\xi}-\xi_0)$$

$$+ \frac{1}{2}(-3J_{000} - K_{000}) n^{\frac{1}{2}} (\hat{\theta}-\theta_0)^2 + \frac{1}{2}(-J_{110} - 2J_{011} - K_{011}) n^{\frac{1}{2}} (\hat{\xi}-\xi_0)^2$$

$$+ (-J_{001} - 2J_{010} - K_{001}) n^{\frac{1}{2}} (\hat{\theta}-\theta_0)(\hat{\xi}-\xi_0) + o(n^{-\frac{1}{2}})\,;$$

$$0 = Z_1 + (-I_{01} + n^{-\frac{1}{2}} Z_{01}) n^{\frac{1}{2}} (\hat{\theta}-\theta_0) + (-I_{11} + n^{-\frac{1}{2}} Z_{11}) n^{\frac{1}{2}} (\hat{\xi}-\xi_0)$$

$$+ \frac{1}{2}(-J_{001} - 2J_{010} - K_{001}) n^{\frac{1}{2}} (\hat{\theta}-\theta_0)^2 + \frac{1}{2}(-3J_{111} - K_{111}) n^{\frac{1}{2}} (\hat{\xi}-\xi_0)^2$$

$$+ (-J_{110} - 2J_{011} - K_{011}) n^{\frac{1}{2}} (\hat{\theta}-\theta_0)(\hat{\xi}-\xi_0) + o(n^{-\frac{1}{2}}),$$

it follows that

Masfumi AKAHIRA and Kei TAKEUCHI

$$(I_{00}I_{11}-I_{01}{}^2)n^{\frac{1}{2}}(\hat{\theta}-\theta_0)$$

$$=I_{11}Z_0-I_{01}Z_1+n^{-\frac{1}{2}}(I_{11}Z_{00}-I_{01}Z_{01})n^{\frac{1}{2}}(\hat{\theta}-\theta_0)+n^{-\frac{1}{2}}(I_{11}Z_{01}-I_{01}Z_{11})n^{\frac{1}{2}}$$

$$\cdot(\hat{\xi}-\xi_0)-\frac{1}{2}n^{-\frac{1}{2}}(3J_{000}I_{11}-J_{001}I_{01}-2J_{010}I_{01}+K_{000}I_{11}-K_{001}I_{01})$$

$$\cdot\{n^{\frac{1}{2}}(\hat{\theta}-\theta_0)\}^2-\frac{1}{2}n^{-\frac{1}{2}}(J_{110}I_{11}+2J_{011}I_{11}-3J_{111}I_{01}+K_{011}I_{11}-K_{111}I_{01})$$

$$\cdot\{n^{\frac{1}{2}}(\hat{\xi}-\xi_0)\}^2-n^{-\frac{1}{2}}(J_{001}I_{11}-J_{110}I_{01}+2J_{010}I_{11}-2J_{011}I_{01}+K_{001}I_{11}$$

$$-K_{011}I_{01})\{n^{\frac{1}{2}}(\hat{\theta}-\theta_0)(\hat{\xi}-\xi_0)\}+o(n^{-\frac{1}{2}})$$

Putting $D=I_{00}I_{11}-I_{01}{}^2$, $a=-I_{01}/I_{11}$, $b=-I_{01}/I_{00}$,

we have

$$n^{\frac{1}{2}}(\hat{\theta}-\theta)$$

$$=\frac{I_{11}}{D}(Z_0+aZ_1)+\frac{I_{11}{}^2}{D^2}n^{-\frac{1}{2}}(Z_{00}+aZ_{01})(Z_0+aZ_1)+\frac{I_{11}I_{00}}{D^2}n^{-\frac{1}{2}}(Z_{01}+aZ_{11})(Z_1+bZ_0)$$

$$-\frac{I_{11}{}^3}{2D^3}n^{-\frac{1}{2}}(3J_{000}+J_{01}a+2J_{010}a+K_{000}+K_{001}a)(Z_0+aZ_1)^2$$

$$-\frac{I_{11}I_{00}{}^2}{2D^3}n^{-\frac{1}{2}}(J_{110}+2J_{011}+3J_{111}a+K_{011}+K_{111}a)(Z_1+bZ_0)^2$$

$$-\frac{I_{11}{}^2I_{00}}{D^3}n^{-\frac{1}{2}}(J_{001}+J_{110}a+2J_{010}+2J_{011}a+K_{001}+K_{001}a)$$

$$\cdot(Z_0+aZ_1)(Z_1+bZ_0)+o(n^{-\frac{1}{2}}).$$

Since

$$E_{\theta_0,\xi_0}(Z_0{}^2)=I_{00},\ \ E_{\theta_0,\xi_0}(Z_1{}^2)=I_{11},\ \ E_{\theta_0,\xi_0}(Z_0Z_1)=I_{01},$$

$$E_{\theta_0,\xi_0}(Z_{00}Z_0)=J_{000},\ \ E_{\theta_0,\xi_0}(Z_{00}Z_1)=J_{001},$$

$$E_{\theta_0,\xi_0}(Z_{01}Z_0)=J_{010},\ \ E_{\theta_0,\xi_0}(Z_{01}Z_1)=J_{011},$$

$$E_{\theta_0,\xi_0}(Z_{11}Z_0)=J_{110},\ \ E_{\theta_0,\xi_0}(Z_{11}Z_1)=J_{111},$$

if follows that

$$E_{\theta_0,\xi_0}\{n^{\frac{1}{2}}(\hat{\theta}-\theta_0)\}=-\frac{J^*+K^*}{2I^{*2}}n^{-\frac{1}{2}}+o(n^{-\frac{1}{2}});$$

$$V_{\theta_0,\xi_0}\{n^{\frac{1}{2}}(\hat{\theta}-\theta_0)\}=\frac{1}{I^*}+o(n^{-\frac{1}{2}}),$$

$$E_{\theta_0,\xi_0}\{n^{\frac{1}{2}}(\hat{\theta}-\theta_0)\}^3=-\frac{3J^*+2K^*}{I^{*3}}n^{-\frac{1}{2}}+o(n^{-\frac{1}{2}}).$$

If we define $\hat{\theta}^*$ by

$$\hat{\theta}^*=\hat{\theta}-\frac{K^*}{6nI^*}$$

then $\hat{\theta}^*$ is second order AMU and second order asymptotically efficient at $\theta=\theta_0$ and $\xi=\xi_0$. Since K^* and I^* depend on θ_0 and ξ_0, we may change $\hat{\theta}^*$ into the following form $\hat{\theta}^{**}$:

$$\hat{\theta}^{**}=\hat{\theta}-\frac{K^*(\hat{\theta},\hat{\xi})}{6nI^*(\hat{\theta},\hat{\xi})}.$$

Thus we have

Theorem 2. *$\hat{\theta}^{**}$ is second order AMU and second order asymptotically efficient.*

For any consistent estimator $\{\hat{\theta}_n{}^0\}$ if the asymptotic variance of $\{\hat{\theta}_n{}^0\}$ is given by

$$V_{\theta, \xi}\{n^{\frac{1}{2}}(\hat{\theta}_n{}^0-\theta)\} = \frac{1}{I^*}+o(n^{-\frac{1}{2}}),$$

then we can transform $\{\hat{\theta}_n{}^0\}$ into $\{\hat{\theta}_n{}^{0*}\}$ such that the latter is second order AMU. In a similar way as [3], if there exists the second order asymptotic distribution, it is shown that $\{\hat{\theta}_n{}^{0*}\}$ must be second order asymptotic efficient. We assume that $\{\hat{\theta}_n{}^0\}$ has a second order asymptotic distribution and

$$E_{\theta, \xi}\{n^{\frac{1}{2}}(\hat{\theta}_n{}^0-\theta)\} = c_1(\theta, \xi)n^{-\frac{1}{2}}+o(n^{-\frac{1}{2}}) ;$$

$$V_{\theta, \xi}\{n^{\frac{1}{2}}(\hat{\theta}_n{}^0-\theta)\} = \frac{1}{I^*(\theta, \xi)}+o(n^{-\frac{1}{2}})$$

$$E_{\theta, \xi}\{n^{\frac{1}{2}}(\hat{\theta}_n{}^0-\theta)\}^3 = c_3(\theta, \xi)n^{-\frac{1}{2}}+o(n^{-\frac{1}{2}}).$$

Let $\hat{\theta}^0(=\{\hat{\theta}_n{}^0\})$ and $\hat{\xi}^0(=\{\hat{\xi}_n{}^0\})$ be consistent estimators of θ and ξ, respectively. Putting

$$\hat{\theta}_n{}^{0*} = \hat{\theta}_n{}^0-n^{-1}\left\{c_1(\hat{\theta}^0, \hat{\xi}^0)-\frac{1}{6}I^*(\hat{\theta}^0, \hat{\xi}^0)c_3(\hat{\theta}^0, \hat{\xi}^0)\right\},$$

we see that $\{\hat{\theta}_n{}^{0*}\}$ is second order asymptotically efficient.

References

[1] Akahira, M., "A note on the second order asymptotic efficiency of estimators in an autoregessive process," Rep. Univ. Electro-Comm. 26-1, (1975).

[2] Pfanzagl, J., "On asymptotically complete classes," Proceedings of the Summer Research Institute on Statistical Inference for Stochastic Processes Vol. 2, (1975).

[3] Takeuchi K. and Akahira, M., "On the second order asymptotic efficiencies of estimators," to appear in Proceedings of the Third USSR-Japan Symposium on Probability Theory. Lecture Notes in Mathematics. Springer Verlag.

[4] Takeuchi, K., "Tôkei-teki suitei no Zenkinriron (Asymptotic Theory of Statistical Estimation)," (in Japanese), Kyoiku-Shuppan, Tokyo, (1974).

Rep. Univ. Electro-Comm. 27–1, (Sci. & Tech. Sect.), pp. 95—115 August, 1976

On Gram-Charlier-Edgeworth Type Expansion of the Sums of Random Variables (I)*

Kei TAKEUCHI** and Masafumi AKAHIRA***

Abstract

Gram-Charlier-Edgeworth type expansion of the sum of i.i.d. random variables without higher order moments are given. Exact formulas for the asymptotic expansion of the density for special cases are given and it is shown that the results are generalized to the case when the density function is approximated by a rational function.

1. Introduction

Gram-Charlier-Edgeworth (G-C-E) expansion is applied to the sum of independently and identically distributed (i. i. d.) random variables which satisfy some regularity conditions. More precisely suppose that X_1, X_2, $\cdots$, X_n, $\cdots$ are i. i. d. random variables whose characteristic function $\phi(t)$ $(=E(\exp it\, X_i))$ satisfies the following :

(i) $\phi(t)$ is k-times continuously differentiable at $t=0$, which implies that moments of X_i up to the k-th order exist.

(ii) For any $\rho>0$

$$\sup_{|t|>\rho} |\phi(t)|<1.$$

(iii) For some $m>0$,

$$\int|\phi(t)|^m dt<\infty.$$

Then it can be shown in [4] that the usual Gram-Charlier-Edgeworth expansion ([3]) is applicable both to the density and the distribution functions of $\sum_1^n X_i$.

The above regularity conditions are not the widest possible [1] but similar necessary conditions exclude various cases which are not altogether trivial. For example the condition (ii) excludes discrete distributions, and (i) does the distributions which have infinite moments. The purpose of this paper is to show that expansions similar to the G-C-E expansion can be obtained for cases where the above regularity conditions do not hold true.

The case when X_i's are integer-valued and have moments was discussed by one of the authers in [4] where it was shown that

* Received on April 27, 1976
The results of this paper have first been presented at the Third USSR-Japan Symposium on Probability Theory at Tashkent, August 1975.
** University of Tokyo
*** Statistical Laboratory, University of Electro-Communications

                     Kei Takeuchi and Masafumi Akahira

$$Pr\left\{\frac{\sum\limits_1^n (X_i-\mu)}{\sqrt{n}\,\sigma}\leq z\right\}$$

$$=\Phi(z)-\varphi(z)\left\{\frac{\beta_3}{6\sqrt{n}}\,h_2(z)+\frac{\beta_4}{24n}\,h_3(z)+\frac{\beta_3{}^2}{72n}\,h_5(z)-\frac{1}{24\sigma^2 n}\,h_1(z)\right\}+o\left(\frac{1}{n}\right)$$

where $\mu=E(X_i)$, $\sigma^2=V(X_i)$, $\beta_3=E(X_i-\mu)^3/\sigma^3$, $\beta_4=\{E(X_i-\mu)^4/\sigma^4\}-3$ and $\Phi(z)=\int_{-\infty}^z \varphi(u)du$ with $\varphi(u)=(1/\sqrt{2\pi})e^{-(u^2/2)}$ and $h_i(z)$ is the Hermite's polynomials i.e. $h_i(z)=(-1)^i\,e^{z^2/2}\times\left\{\dfrac{d^i}{dz^i}e^{-(z^2/2)}\right\}$. Further expansions for this case was also given in [4].

In this paper we mainly deal with the case when moments do not exist. If the second order moments exist, but higher order ones do not, then we may obtain the expansion of the type

$$Pr\left\{\frac{\sum\limits_1^n(X_i-\mu)}{\sqrt{n}\,\sigma}\leq z\right\}=\Phi(z)+\frac{1}{\sqrt{n}}g_3(z)+\cdots$$

but $g_3(z),\cdots$ are no longer equal to the derivatives of $\varphi(z)$. When even the second order moments do not exist, but the distribution of X_i belongs to the domain of attraction of the stable law of the exponent α, then we may get an expansion

$$Pr\left\{\frac{\sum\limits_1^n X_i-b_n}{a_n}\leq z\right\}=F(z)+c_{1n}{}^{-1}f_1(z)+c_{2n}{}^{-1}f_2(z)+\cdots$$

where $F(z)$ is the distribution of the stable law, and $\{c_{1n}\}$, $\{c_{2n}\}$, $\cdots$ are sequences of positive numbers, and $f_i(z)$ are some functions.

In this paper we first deal with some specific examples and get results which represent various types of possible solutions to the problem, and then prove that the solutions cover fairly general situations. Here we shall show the results only for integer α (i.e. $\alpha=1$ or 2) but similar results may be obtained for fractional α, which will be given in the subsequent paper [5].

2. Continuous cases

We first note that if the characteristic function of the distribution is integrable, then it is absolutely continuous, and the density function is obtained by the Fourier transform of the characteristic function.

Example 2.1. t-distributions with odd degrees of freedom: suppose that X_i's are i. i. d. random variables having the continuous distribution with the density

$$f(x)=\frac{1}{B\left(\dfrac{1}{2},\ p-\dfrac{1}{2}\right)(1+x^2)^p}$$

where $p=1, 2, \cdots$.

This distribution does not have the moments of higher order than $(2p-2)$-th order.

We shall obtain the characteristic function $\phi_p(t)$ of the distribution.

Put

$$l_p(x)=\frac{1}{2\pi}\int_{-\infty}^{\infty}|t|^{p-1}e^{-|t|-itx}dt.$$

Since

August, 1976 　　　On Gram-Charlier-Edgeworth Type Expansion of the Sums

$$l_p(x)=\frac{1}{2\pi}\int_0^\infty t^{p-1}e^{-(1+it)x}dt+\frac{1}{2\pi}\int_0^\infty t^{p-1}e^{-(1-it)x}dt$$

$$=\frac{1}{2\pi}\Gamma(p)\left\{\frac{1}{(1+ix)^p}+\frac{1}{(1-ix)^p}\right\}$$

$$=\frac{(p-1)!}{\pi(1+x^2)^p}(1-{}_pC_2x^2+{}_pC_4x^4-\cdots)\qquad(2.1)$$

where ${}_pC_i$ $(i=2, 4, \cdots)$ are binomial coefficients, it follows that

$$f_p(x)=a_{p,1}l_1(x)+a_{p,2}l_2(x)+\cdots+a_{p,p}l_p(x),$$

where for each $i=1, \cdots, p$, $a_{p,i}$ is some constant.

Hence we have

$$\phi_p(t)=(a_{p,1}+a_{p,2}|t|+\cdots+a_{p,p}|t|^{p-1})e^{-|t|}.$$

Since $\phi_p(0)=1$, it is seen that $a_{p,1}=1$. From above we obtain

$$\log\phi_p(t)=-|t|+\log(1+a_{p,2}|t|+\cdots+a_{p,p}|t|^{p-1}).$$

(I) $p\geq2$. Since

$$E_p(x)=\int xf_p(x)dx=0\qquad(p\geq2),$$

expanding $\log\phi_p(t)$ we have

$$\log\phi_p(t)=-\frac{1}{2}c_{p,2}t^2+\frac{1}{24}c_{p,4}t^4+\cdots$$

$$+\frac{1}{6}c_{p,3}|t|^3+\frac{1}{120}c_{p,5}|t|^5+\cdots,$$

where for each $i=2, 3, \cdots$, $c_{p,i}$ is some constant.

Since there exist the moments of X_i up to $(2p-2)$-th order, it follows that $c_{p,3}=c_{p,5}=\cdots=c_{p,2p-3}=0$.

Letting $\tilde{\phi}_n(t)=E[\exp(itZ_n)]$ with $Z_n=\sum_1^n(X_i-\mu)/(\sqrt{n}\,\sigma)$, we obtain

$$\log\tilde{\phi}_n(t)=n\log\phi_p\left(\frac{t}{\sqrt{n}\,\sigma}\right)$$

$$=-\frac{t^2}{2}+\frac{\gamma_4}{24}\frac{t^4}{n}+\frac{\gamma_6}{720}\frac{t^6}{n^2}+\cdots$$

$$+\frac{\gamma_3}{6}\frac{|t|^3}{n^{1/2}}+\frac{\gamma_5}{120}\frac{|t|^5}{n^{3/2}}+\cdots$$

$$=e^{-(t^2/2)}\left(1+\frac{\gamma_3}{6n^{1/2}}|t|^3+\frac{\gamma_4}{24n}t^4+\frac{\gamma_3^2}{72n}t^6+\frac{\gamma_5}{120n^{3/2}}|t|^5\right.$$

$$\left.+\frac{\gamma_3\gamma_4}{144n^{3/2}}|t|^7+\frac{\gamma_3^3}{1296n^{3/2}}|t|^9+\cdots\right).$$

where for each $i=3, 4, \cdots$, γ_i is some constant.

For each odd number r we define $g_r(x)$ and $G_r(x)$ as follows:

$$g_r(x)=\frac{1}{2\pi}\int|t|^re^{-(t^2/2)}e^{-itx}dt\,;$$

$$G_r(x)=\int_{-\infty}^x g_r(u)du.$$

Then the density function $\tilde{f}_n(z)$ is given by

$$\tilde{f}_n(z) = \frac{1}{\sqrt{2\pi}} e^{-(z^2/2)} \left\{ 1 + \frac{\gamma_4}{24n} H_4(z) + \frac{\gamma_3^2}{72n} H_6(z) + \cdots \right.$$

$$+ \frac{\gamma_3}{6n^{1/2}} g_3(x) + \frac{\gamma_5}{120n^{3/2}} g_5(x) + \frac{\gamma_3\gamma_4}{144n^{3/2}} g_7(x)$$

$$\left. + \frac{\gamma_3^3}{1296n^{3/2}} g_9(x) + \cdots \right\}.$$

Further the distribution function of Z_n is obtained as follows:

$$F_n(z) = \Phi(z) - \varphi(z) \left\{ \frac{\gamma_4}{24n} H_3(z) + \frac{\gamma_3^2}{72n} H_5(z) + \cdots \right.$$

$$+ \frac{\gamma_3}{6n^{1/2}} G_3(z) + \frac{\gamma_5}{120n^{3/2}} G_5(z) + \frac{\gamma_3\gamma_4}{144n^{3/2}} G_7(z)$$

$$\left. + \frac{\gamma_3^3}{1296n^{3/2}} G_9(z) + \cdots \right\},$$

where $G_i(z) = \int_{-\infty}^{z} g_i(x) dx$.

Since the distribution of Z_n is symmetric, it follows that $G_1(0) = 0$.
By (4.4) of Appendix of section 4 we have

$$G_1(x) = \frac{1}{\pi} e^{-(x^2/2)} \int_0^x e^{t^2/2} dt. \tag{2.2}$$

Hence it is clear that $g_1(x)$ is also calculated.
Since

$$g_3(x) = \frac{1}{2\pi} \int_{-\infty}^{\infty} |t|^3 e^{-(t^2/2) - itx} dt$$

$$= \frac{1}{2\pi} \int_{-\infty}^{\infty} t^2 |t| e^{-(t^2/2) - itx} dt$$

$$= (-1) \frac{d^2}{dx^2} g_1(x)$$

$$= -\frac{d^3}{dx^3} G_1(x),$$

it follows from (2.2) that $g_3(x)$ is calculated.
In general we have for each odd number r

$$g_r(x) = (-1)^{(r-1)/2} \frac{d^r}{dx^r} G_1(x) \, ;$$

$$G_r(x) = (-1)^{(r-1)/2} \frac{d^{r-1}}{dx^{r-1}} G_1(x).$$

Futher since

$$\frac{d}{dx} G_1(x) = g_1(x) = \frac{1}{\pi} - x G_1(x),$$

repeating differentials we obtain

$$\frac{d^2}{dx^2} G_1(x) = -G_1(x) - x g_1(x)$$

$$= (x^2 - 1) G_1(x) - \frac{x}{\pi} \, ;$$

$$\frac{d^3}{dx^3}G_1(x)=2xG_1(x)+(x^2-1)g_1(x)-\frac{1}{\pi}$$

$$=-(x^3-3x)G_1(x)+\frac{1}{\pi}(x^2-2);$$

$$\frac{d^4}{dx^4}G_1(x)=(x^4-6x^2+3)G_1(x)-\frac{1}{\pi}(x^3-5x);$$

$$\cdots.$$

Hence we generally have for each odd number r

$$\frac{d^r}{dx^r}G_1(x)=(-1)^r H_r(x)G_1(x)-\frac{1}{\pi}K_{r-1}(x),$$

where K_{r-1} is the polynomial of $(r-1)$-th degree.

For large x, $G_1(x)$ may be expanded as follows:

$$G_1(x)=\frac{1}{\pi}e^{-(x^2/2)}\int_0^x e^{(x-u)^2/2}\,du$$

$$=\frac{1}{\pi}\int_0^x e^{(u^2/2)-xu}\,du$$

$$=\frac{1}{\pi}\int_0^x\Big(1+\frac{u^2}{2}+\frac{u^4}{8}+\frac{u^6}{48}+\cdots\Big)e^{-xu}\,du$$

$$\sim\frac{1}{\pi}\Big(\frac{1}{x}+\frac{1}{x^3}+\frac{3}{x^5}+\frac{15}{x^7}+\cdots\Big).$$

For example we have

$$G_3(x)\sim(x^3-3x)G_1(x)+\frac{1}{\pi}(x^2-2)$$

$$\sim\frac{1}{\pi}\Big(\frac{6}{x^2}+\frac{60}{x^4}+\cdots\Big).$$

In general $G_1(x)$ and $g_1(x)$ have for large x the order of x^{-r+1} and that of x^{-r}, respectively.

(II) $p=2$. In this case if follows that

$$f_2(x)=\frac{2}{\pi(1+x^2)^2};$$

$$l_1(x)=\frac{1}{\pi(1+x^2)};$$

$$l_2(x)=\frac{1}{\pi(1+x^2)^2}(1-x^2).$$

Since $f_2(x)=l_1(x)+l_2(x)$, the characteristic function is

$$\phi_2(t)=(1+|t|)e^{-|t|}.$$

In a similar way as the case when $p\geq 2$, the characteristic function $\tilde{\phi}_n(t)$ of Z_n is given by

$$\tilde{\phi}_n(t)=e^{-(t^2/2)}\Big(1+\frac{|t|^3}{3\sqrt{n}}-\frac{t^4}{4n}+\frac{t^6}{18n}+\cdots\Big).$$

Hence the distribution function $F_n(z)$ of Z_n is expanded as follows:

$$F_n(z)\sim\varPhi(z)+\frac{1}{3\sqrt{n}}G_3(z)-\varphi(z)\Big\{\frac{1}{4n}H_3(z)-\frac{1}{18n}H_5(z)+\cdots\Big\}.$$

(III) $p=1$. Since $f_1(x)$ is the Cauchy density function, it is easily seen that the distribution of Z_n is also the Cauchy distribution.

Example 2.2. (Non-symmetric case)

Suppose that X_i's are i. i. d. random variables with the density function $f_p(x)$ as follows:

                     Kei Takeuchi and Masafumi Akahira

$$f_p(x)=\begin{cases}\dfrac{p-1}{(1+x)^p}, & x>0;\\[2mm] 0, & x\leq 0,\end{cases}$$

where $p>1$.

We shall first obtain the characteristic function $\phi_p(t)$ $(=E(\exp itX_i))$.

Since

$$\phi_p(t)=\int_0^\infty\frac{(p-1)e^{itx}}{(1+x)^p}dx,\tag{2.3}$$

we have for $p>2$.

$$\begin{aligned}\phi_p{}'(t)&=\frac{d}{dt}\phi_p(t)\\[2mm]&=\int\frac{(p-1)ix\,e^{itx}}{(1+x)^p}dx\\[2mm]&=\frac{p-1}{p-2}i\phi_{p-1}(t)-i\phi_p(t).\end{aligned}\tag{2.4}$$

On the other hand applying integration by parts to the right-hand side of (2.3) we obtain

$$\phi_p(t)=1+\frac{it}{p-2}\phi_{p-1}(t).\tag{2.5}$$

Substituting (2.5) into (2.4) we have

$$\phi_p{}'(t)=\left(\frac{p-1}{t}-i\right)\phi_p(t)-\frac{p-1}{t}.\tag{2.6}$$

In order to solve the differential equation (2.6) we put

$$g_p{}'(t)=\left(\frac{p-1}{t}-i\right)g_p(t).$$

If $t>0$, then it follows that

$$g_p(t)=C_1t^{p-1}e^{-it}\tag{2.7}$$

where C_1 is integration constant.

Putting

$$\phi_p(t)=\psi(t)g_p(t)\tag{2.8}$$

we have

$$\begin{aligned}\phi_p{}'(t)&=\psi_p(t)g_p{}'(t)+\psi_p{}'(t)g_p(t)\\[2mm]&=\left(\frac{p-1}{t}-i\right)\phi_p(t)+\psi_p{}'(t)g_p(t)\end{aligned}$$

Then it follows that

$$\psi_p{}'(t)g_p(t)=-(p-1)/t.$$

Hence it is easily seen from (2.7) that

$$\psi_p{}'(t)=-(p-1)C_1^{-1}t^{-p}e^{it}.\tag{2.9}$$

Considering that $\phi_p(0)=1$, we have from (2.7), (2.8) and (2.9)

$$\phi_p(t)=(p-1)\left(\int_t^\infty u^{-p}e^{iu}du+C_2\right)t^{p-1}e^{-it}\quad\text{for }t>0\tag{2.10}$$

where C_2 is integration constant.

It is seen from (2.5) that the solution (2.10) of (2.6) is also applicable to the case when $1<p\leq 2$.

August, 1976 On Gram-Charlier-Edgeworth Type Expansion of the Sums 101

Since

$$\lim_{t\to\infty} e^{it} t^{-(p-1)} \phi_p(t) = \lim_{t\to\infty} \int_0^\infty \frac{(p-1)e^{it(x+1)}}{t^p(1+x)^p} t\, dx$$

$$= \lim_{t\to\infty} \int_t^\infty \frac{(p-1)e^{iy}}{y^p} dy$$

$$= 0$$

it follows that $C_2 = 0$.

Hence we have for $t > 0$

$$\phi_p(t) = (p-1)\left(\int_t^\infty u^{-p} e^{iu} du\right) t^{p-1} e^{-it}. \tag{2.11}$$

Next we consider the case when $t < 0$. Since $\phi_p(t)$ is the conjugate complex number of $\phi_p(-t)$, it follows that for every $t < 0$

$$\phi_p(t) = (p-1)\left(\int_{|t|}^\infty u^{-p} e^{-iu} du\right) |t|^{p-1} e^{-it}. \tag{2.12}$$

If t is a small and positive number, then $\phi_p(t)$ may be expanded as follows.

Put

$$A_p + iB_p = \int_1^\infty u^{-p} e^{iu} du.$$

Since

$$u^{-p}\left\{e^{iu} - 1 - iu + \frac{u^2}{2} + \cdots - \frac{(iu)^{p-1}}{(p-1)!}\right\} \quad \text{(for integer } p\text{)}$$

is bounded for $0 < u < 1$, it is integrable.

Putting

$$C_p + iD_p = \int_0^1 u^{-p}\left\{e^{iu} - 1 - iu + \frac{u^2}{2} + \cdots - \frac{(iu)^{p-1}}{(p-1)!}\right\} du \quad \text{(for integer } p\text{)}$$

we have

$$\int_t^\infty u^{-p} e^{iu} du = A_p + iB_p + \int_t^1 u^{-p} e^{iu} du$$

$$= A_p + iB_p + \int_t^1 u^{-p}\left\{e^{iu} - 1 - iu - \cdots - \frac{(iu)^{p-1}}{(p-1)!}\right\} du$$

$$+ \int_t^1 u^{-p}\left\{1 + iu + \cdots + \frac{(iu)^{p-1}}{(p-1)!}\right\} du$$

$$= \alpha_p + i\beta_p - \int_0^t u^{-p}\left\{e^{iu} - 1 - iu - \cdots - \frac{(iu)^{p-1}}{(p-1)!}\right\} du$$

$$+ \int_t^1 u^{-p}\left\{1 + iu + \cdots + \frac{(iu)^{p-1}}{(p-1)!}\right\} du$$

$$= \alpha_p + i\beta_p - \int_0^t u^{-p}\left\{\frac{(iu)^p}{p!} + \frac{(iu)^{p+1}}{(p+1)!} + \frac{(iu)^{p+2}}{(p+2)!} + \cdots\right\} du$$

$$+ \int_t^1 \left\{u^{-p} + iu^{1-p} + \cdots + \frac{i^{p-1}u^{-1}}{(p-1)!}\right\} du \tag{2.13}$$

for integer p, where $\alpha_p = A_p + C_p$ and $\beta_p = B_p + D_p$.

If $p(>1)$ is integer, then

Kei Takeuchi and Masafumi Akahira

$$\int_t^\infty u^{-p}e^{iu}du=\alpha_p+i\beta_p-\int_0^t u^{-p}\left\{\frac{(iu)^p}{p!}+\frac{(iu)^{p+1}}{(p+1)!}+\frac{(iu)^{p+2}}{(p+2)!}+\cdots\right\}du$$

$$+\sum_{k=0}^{p-1}\frac{i^k}{k!}\int_t^1 u^{k-p}du$$

$$=\alpha_p+i\beta_p-\frac{i^p}{p!}t-\frac{i^{p+1}}{2(p+1)!}t^2+\frac{i^p}{3(p+2)!}t^3+\cdots$$

$$+\sum_{k=0}^{p-2}\frac{i^k}{k!}\frac{1}{k-p+1}(1-t^{k-p+1})-\frac{i^{p-1}}{(p-1)!}\log t$$

$$=(\alpha_p+\alpha_p')+i(\beta_p+\beta_p')-\sum_{k=0}^{P-2}\frac{i^k}{k!}\frac{1}{k-p+1}t^{k-p+1}-\frac{i^{p-1}}{(p-1)!}\log t-\frac{i^p}{p!}t$$

$$-\frac{i^{p+1}}{2(p+1)!}t^2+\frac{i^p}{3(p+2)!}t^3+\cdots$$

where α_p' and β_p' denote the real part and the imaginary part of

$$\sum_{k=0}^{p-2}\frac{i^k}{k!}\frac{1}{k-p+1}$$

respectively.

Putting $\alpha_p{}^*=\alpha_p+\alpha_p'$ and $\beta_p{}^*=\beta_p+\beta_p'$ we obtain

$$\int_t^\infty u^{-p}e^{iu}du=\alpha_p{}^*+i\beta_p{}^*$$

$$-\sum_{k=0}^{p-2}\frac{i^k}{k!}\frac{1}{k-p+1}t^{k-p+1}-\frac{i^{p-1}}{(p-1)!}\log t-\frac{i^p}{p!}t$$

$$-\frac{i^{p+1}}{2(p+1)!}t^2+\frac{i^p}{3(p+2)!}t^3+\cdots$$

From (2. 11) we have for small $t>0$

$$\phi_p(t)=(p-1)\left(\int_t^\infty u^{-p}e^{-iu}du\right)t^{p-1}e^{-it}$$

$$=(p-1)\left\{\sum_{k=0}^{p-2}i^k a_k t^k-\frac{i^{p-1}}{(p-1)!}t^{p-1}\log|t|+\alpha_p{}^*t^{p-1}+i\beta_p{}^*t^{p-1}-\frac{i^p}{p!}t^p\right.$$

$$+o(|t|^p)\bigg\}\left(1-it-\frac{1}{2}t^2+\frac{i}{6}t^3+\cdots\right)$$

$$=(p-1)\sum_{k=0}^{p-2}i^k a_k t^k-\frac{i^{p-1}}{(p-2)!}t^{p-1}\log|t|+(p-1)\alpha_p{}^*t^{p-1}+i(p-1)\beta_p{}^*t^{p-1}$$

$$-\frac{i^p}{(p-1)!}t^p-(p-1)\sum_{k=0}^{p-2}i^{k+1}a_k t^{k+1}+\frac{i^p}{(p-2)!}t^p\log|t|$$

$$-i(p-1)\alpha_p{}^*t^p+(p-1)\beta_p{}^*t^p-\frac{1}{2}(p-1)\sum_{k=0}^{p-2}i^k a_k t^{k+2}$$

$$+\frac{1}{6}(p-1)\sum_{k=0}^{p-3}i^{k+1}a_k t^{k+3}+\cdots+a_0\frac{(-i)^p}{(p-1)!}t^p$$

$$+o(|t|^p)\tag{2.14}$$

For $t<0$, $\phi_p(t)$ is given by the conjugate complex number of $\phi_p(-t)$.

When $p=2$, we have from (2.13)

$$\int_t^\infty u^{-2}e^{iu}du=\alpha_2+i\beta_2-\int_0^t u^{-2}(e^{iu}-1-iu)du+\int_t^1 u^{-2}(1+iu)du$$

$$=\alpha_2+i\beta_2-\int_0^t u^{-2}\left\{\frac{(iu)^2}{1}+\frac{(iu)^3}{6}+\cdots\right\}du+$$

$$+\frac{1}{t}-1-i\log t$$

$$=\alpha_2-1+i\beta_2+\frac{1}{t}-i\log t+\frac{t}{2}+\cdots.$$

Similarly we have

$$\int_{|t|}^{\infty}u^{-2}e^{iu}du=\alpha_2-1+i\beta_2+\frac{1}{|t|}+i\log|t|+\frac{|t|}{2}+\cdots.$$

From (2. 11) and (2. 12) we obtain

$$\phi_2(t)=1-it\log|t|+(\alpha_2-1)|t|+(\beta_2-1)it$$
$$+t^2\log|t|+\beta_2 t^2-i(\alpha_2-1)t|t|+o(t^2).$$

Since

$$\alpha_2+i\beta_2=\int_0^1 u^{-2}(e^{iu}-1-iu)du+\int_1^{\infty}u^{-2}e^{iu}du,$$

it follows that

$$\alpha_2=\int_0^1 u^{-2}(\cos u-1)du+\int_1^{\infty}u^{-2}\cos u\,du\,;$$

$$\beta_2=\int_0^1 u^{-2}(\sin u-u)du+\int_1^{\infty}u^{-2}\sin u\,du.$$

We can calculate α_2 as follows :

$$\alpha_2=\int_0^{\infty}u^{-2}(\cos u-1)du+1$$

$$=-u^{-1}(\cos u-1)\Big|_0^{\infty}-\int_0^{\infty}\frac{\sin u}{u}du+1$$

$$=1-\frac{\pi}{2}.$$

Since the value of β_2 is not easily obtained, putting $\beta_2=\gamma+1$ we have

$$\phi_2(t)=1-it\log|t|-\frac{\pi}{2}|t|+i\gamma t+t^2\log|t|$$

$$+(\gamma+1)t^2+i\frac{\pi}{2}t|t|+o(t^2) \tag{2. 15}$$

Letting $\tilde{\phi}_n(t)=E[\exp(itZ_n)]$ with $Z_n=\left(\sum_1^n X_i-a_n\right)\Big/b_n$ we obtain

$$\log\tilde{\phi}_n(t)=n\log\phi_2\left(\frac{t}{b_n}\right)-i\frac{a_n}{b_n}t$$

$$=-i\frac{a_n}{b_n}t-i\frac{n}{b_n}t(\log|t|-\log b_n)-\frac{n\pi}{2b_n}|t|+i\frac{n\gamma}{b_n}t$$

$$+\frac{nt^2}{b_n^2}(\log|t|-\log b_n)+(\gamma+1)\frac{nt^2}{b_n^2}+i\frac{n\pi}{2b_n^2}t|t|$$

$$-\frac{nt^2}{2b_n^2}(\log|t|-\log b_n)^2+\frac{nt^2}{8b_n^2}(\pi^2-4\gamma^2)$$

$$-i\frac{n\pi\gamma}{2b_n^2}t|t|+o\left(\frac{n}{b_n^2}\right).$$

If $b_n=n$ and $a_n=n\gamma+n\log n$, then

Kei Takeuchi and Masafumi Akahira

$$\log \tilde{\phi}_n(t) = -it \log|t| - \frac{\pi}{2}|t|$$

$$-\frac{\log n}{n}t^2 + \frac{1}{n}t^2 \log|t| + \frac{\gamma+1}{n}t^2$$

$$+i\frac{\pi}{2n}t|t| - \frac{(\log n)^2}{2n}t^2 + \frac{\log n}{n}t^2 \log|t|$$

$$-\frac{1}{2n}t^2(\log|t|)^2 + \frac{1}{8n}(\pi^2 - 4\gamma^2)t^2 - i\frac{\pi\gamma}{2n}t|t|$$

$$+o\left(\frac{1}{n}\right).$$

Hence it is seen that the asymptotic distribution of $\left(\sum_1^n X_i - n\log n - n\gamma\right)\Big/n$ is the distribution function with the characteristic function $\exp\left\{-\frac{\pi}{2}|t| - it \log|t|\right\}$. This is the spacial case of a stable law with the characteristic exponent $1([1], [2])$. By the expansion of the asymptotic distribution, it follows that the order of the next term is $(\log n)^2/n$.

Next we consider the case when $p=3$. We calculate α_3 and β_3 as follows:

$$\alpha_3 = \int_0^1 u^{-3}\left(\cos u - 1 + \frac{u^2}{2}\right)du + \int_1^\infty u^{-3}\cos u\, du$$

$$= -\frac{u^{-2}}{2}\left(\cos u - 1 + \frac{u^2}{2}\right)\Big|_0^1 - \frac{1}{2}\int_0^1 u^{-2}(\sin u - u)du$$

$$-\frac{u^{-2}}{2}\cos u\Big|_1^\infty - \frac{1}{2}\int_1^\infty u^{-2}\sin u\, du$$

$$= \frac{1}{4} - \frac{\beta_2}{2} = -\frac{\gamma}{2} - \frac{1}{4};$$

$$\beta_3 = \int_0^1 u^{-3}(\sin u - u)du + \int_1^\infty u^{-3}\sin u\, du$$

$$= -\frac{u^{-2}}{2}(\sin u - u)\Big|_0^1 + \frac{1}{2}\int_0^1 u^{-2}(\cos u - 1)du$$

$$-\frac{u^{-2}}{2}\sin u\Big|_1^\infty + \frac{1}{2}\int_0^1 u^{-2}\cos u\, du$$

$$= \frac{1}{2} + \frac{\alpha_2}{2} = 1 - \frac{\pi}{4}.$$

From (2. 13) we have for small $t>0$

$$2\int_t^\infty u^{-3}e^{iu}du = -\gamma - \frac{1}{2} + i\left(2 - \frac{\pi}{2}\right) - 2\int_0^t u^{-3}\left(e^{iu} - 1 - iu + \frac{u^2}{2}\right)du$$

$$+2\int_t^1 u^{-3}\left(1 + iu - \frac{u^2}{2}\right)du$$

$$= -\gamma - \frac{1}{2} + i\left(2 - \frac{\pi}{2}\right) - \frac{it}{3} - \left(1 - \frac{1}{t^2}\right) - 2i\left(1 - \frac{1}{t}\right)$$

$$+\log t + o(t).$$

Hence we have from (2. 11) and (2. 12)

$$\phi_3(t)=t^2\Big(1-it-\frac{t^2}{2}+\frac{it^3}{6}+\cdots\Big)\times$$

$$\times\Big\{\frac{1}{t^2}+\frac{2i}{t}+\log|t|-\gamma-\frac{3}{2}-i\frac{\pi}{2}\frac{t}{|t|}-\frac{i}{3}t+o(|t|)\Big\}$$

$$=1+it+t^2\log|t|-\gamma t^2-i\frac{\pi}{2}t|t|+o(t^2). \tag{2.16}$$

Letting $\tilde{\phi}_n(t)=E[\exp(it\,Z_n)]$ with $Z_n=\sum_{1}^{n}(X_i-1)/\sqrt{n\log n}$ we oktain

$$\log\tilde{\phi}_n(t)=-i\sqrt{\frac{n}{\log n}}t+n\log\phi_3\Big(\frac{t}{\sqrt{n\log n}}\Big)$$

$$=\frac{t^2}{\log n}(\log|t|-\log\sqrt{n\log n})-\frac{\gamma}{\log n}t^2$$

$$-i\frac{\pi t|t|}{2\log n}-\frac{1}{2\log n}t^2+o\Big(\frac{1}{\log n}\Big)$$

$$=-\frac{1}{2}\Big(1+\frac{\log\log n}{\log n}+\frac{2\gamma+1}{\log n}\Big)t^2$$

$$+\frac{1}{\log n}t^2\log|t|-i\frac{\pi}{2\log n}t|t|+o\Big(\frac{1}{\log n}\Big).$$

Putting

$$c_n{}^2=1+\frac{\log\log n}{\log n}+\frac{2\gamma+1}{\log n}$$

we have

$$\tilde{\phi}_n(t)=\Big\{1+\frac{1}{\log n}\Big(t^2\log|t|-i\frac{\pi}{2}t|t|\Big)\Big\}\exp\Big(-\frac{c_n{}^2}{2}t^2\Big)$$

$$+o\Big(\frac{1}{\log n}\Big). \tag{2.17}$$

It is clear that the first term of the right-hand side of (2.17) corresponds to the normal distribution. In order to obtain the density function of Z_n corresponding to the right-hand side of (2.17) we need the Fourier transforms of $(t^2\log|t|)e^{-(t^2/2)}$ and $it|t|e^{-(t^2/2)}$. The discussion will be done in section 4 in detail.

Further we consider the case when $p=4$.

Then we calculate α_4 and β_4 as follows:

$$\alpha_4=\int_0^1 u^{-4}\Big(\cos u-1+\frac{u^2}{2}\Big)du+\int_1^\infty u^{-4}\cos u\,du$$

$$=-\frac{u^{-3}}{3}\Big(\cos u-1+\frac{u^2}{2}\Big)\Big|_0^1-\frac{1}{3}\int_0^1 u^{-3}(\sin u-u)du$$

$$-\frac{u^{-3}}{3}\cos u\Big|_1^\infty-\frac{1}{3}\int_1^\infty u^{-3}\sin u\,du$$

$$=\frac{1}{6}-\frac{\beta_3}{3}$$

$$=-\frac{1}{6}+\frac{\pi}{12};$$

Kei TAKEUCHI and Masafumi AKAHIRA

$$\beta_4 = \int_0^1 u^{-4}\left(\sin u - u + \frac{u^3}{6}\right)du + \int_1^\infty u^{-4}\sin u\, du$$

$$= -\frac{u^{-3}}{3}\left(\sin u - u + \frac{u^3}{6}\right)\Big|_0^1 + \frac{1}{3}\int_1^\infty u^{-3}\left(\cos u - 1 + \frac{u^2}{2}\right)du$$

$$-\frac{u^{-3}}{3}\sin u\Big|_1^\infty + \frac{1}{3}\int_1^\infty u^{-3}\cos u\, du$$

$$= \frac{5}{18} + \frac{\alpha_3}{3}$$

$$= \frac{7}{36} - \frac{\gamma}{6}.$$

From (2.13) we have for small $t > 0$

$$3\int_t^\infty u^{-4}e^{iu}du = 3(\alpha_4 + i\beta_4) - 3\int_0^t u^{-4}\left(e^{iu} - 1 - iu + \frac{u^2}{2} + \frac{iu^3}{6}\right)du$$

$$+ 3\int_t^1 u^{-4}\left(1 + iu - \frac{u^2}{2} - \frac{iu^3}{6}\right)du$$

$$= 3(\alpha_4 + i\beta_4) - \frac{it}{8} - \left(1 - \frac{1}{t^3}\right) - \frac{3}{2}i\left(1 - \frac{1}{t^2}\right)$$

$$+ \frac{3}{2}\left(1 - \frac{1}{t}\right) + \frac{i}{2}\log t + o(t).$$

Hence it follows from (2.11) and (2.12) that

$$\phi_4(t) = \left(1 - it - \frac{t^2}{2} + \frac{it^3}{6} + \cdots\right)$$

$$\cdot\left(1 + \frac{3}{2}it - \frac{3}{2}t^2 + \frac{i}{2}t^3\log|t| + \cdots\right)$$

$$= 1 + \frac{1}{2}it - \frac{t^2}{2} + \frac{i}{2}t^3\log|t| + \cdots. \tag{2.18}$$

Letting $\tilde{\phi}_n(t) = E[\exp(it Z_n)]$ with $Z_n = 2\left(\sum_1^n X_i - \frac{n}{2}\right)\big/\sqrt{3n}$ we obtain

$$\log \tilde{\phi}_n(t) = -\frac{n}{3\sqrt{n}}it + n\log\phi_4\left(\frac{2t}{\sqrt{3n}}\right)$$

$$= -\frac{t^2}{2} - \frac{4\log n}{3\sqrt{3n}}it^3 + o\left(\frac{\log n}{\sqrt{n}}\right).$$

Hence it follows that

$$\tilde{\phi}_n(t) = e^{-(t^2/2)}\left(1 - \frac{4\log n}{3\sqrt{3n}}it^3 + o\left(\frac{\log n}{\sqrt{n}}\right)\right). \tag{2.19}$$

Corresponding to the right-hand side of (2.19) the density function of Z_n is expanded as follows:

$$\frac{1}{\sqrt{2\pi}}e^{-(z^2/2)}\left\{1 - \frac{4\log n}{3\sqrt{3n}}(z^3 - 3z)\right\} + o\left(\frac{\log n}{\sqrt{n}}\right).$$

Example 2.3. (Symmetric case)

Let X_i $(i = 1, 2, \cdots)$ be i.i.d. random variables with the density function $f_p(x)$ as follows:

$$f_p(x) = \frac{p-1}{2(1 + |x|)^p}, \quad -\infty < x < \infty, \tag{2.20}$$

where $p > 1$.

In order to calculate the characteristic function $\phi_p(t)$ $(= E(\exp it X_i))$, we have

$$\phi_p(t)=\int_{-\infty}^{\infty}\frac{(p-1)e^{itx}}{2(1+|x|)^p}dx$$

$$=\int_{-\infty}^{0}\frac{(p-1)e^{itx}}{2(1-x)^p}dx+\int_{0}^{\infty}\frac{(p-1)e^{itx}}{2(1+x)^p}dx$$

$$=\int_{0}^{\infty}\frac{(p-1)e^{-itx}}{2(1+x)^p}dx+\int_{0}^{\infty}\frac{(p-1)e^{itx}}{2(1+x)^p}dx.$$

Since $\int_{0}^{\infty}(p-1)e^{-itx}\big/\{2(1+x)^p\}dx$ is the conjugate complex number of $\int_{0}^{\infty}(p-1)e^{itx}\big/$ $\{2(1+x)^p\}dx$, the characteristic function $\phi_p(t)$ agrees with the real part of the functions (2.11) and (2.12) in the case of Example 2.2.

If p is integer, then it follows from (2.15), (2.16) and (2.14) that for every t

$$\phi_2(t)=1-\frac{\pi}{2}|t|-t^2\log|t|+(\gamma+1)t^2+o(t^2);\tag{2.21}$$

$$\phi_3(t)=1+t^2\log|t|-\gamma t^2-\frac{\pi}{2}|t|^3+o(|t|^3);\tag{2.22}$$

$$\phi_p(t)=1+\left\{-\frac{1}{2}-\frac{p-1}{2(p-3)}+\frac{p-1}{p-2}\right\}t^2+o(t^2),\quad(p\geq4).\tag{2.23}$$

When $p=2$, letting $\tilde{\phi}_n(t)=E[\exp(itZ_n)]$ with $Z_n=\sum_{1}^{n}X_i\big/n$ we obtain

$$\log\tilde{\phi}_n(t)=n\log\phi_2\left(\frac{t}{n}\right)$$

$$=-\frac{\pi}{2}|t|+\frac{t^2}{n}(\log|t|-\log n)+\frac{\gamma+1}{n}t^2,$$

$$-\frac{\pi^2}{8n}t^2+o(t^2).$$

Hence it follows that

$$\tilde{\phi}_n(t)=e^{-(\pi/2)|t|}\left\{1-\frac{\log n}{n}t^2+\frac{1}{n}t^2\log|t|+\frac{8(\gamma+1)-\pi^2}{8n}t^2\right\}$$

$$+o(t^2).\tag{2.24}$$

The distribution function with the characteristic function $e^{-(\pi/2)|t|}$ is the Cauchy distribution having the following density function $f(x)$:

$$f(x)=\frac{2}{\pi^2+4x^2}.$$

In order to obtain the density function of Z_n corresponding to the right-hand side of (2.24), we need the Fourier transforms of $t^2e^{-(\pi/2)|t|}$ and $(t^2\log|t|)e^{-(\pi/2)|t|}$. The discussion will be done in section 4 in detail.

When $p=3$, putting $Z_n=\sum_{1}^{n}X_i\big/\sqrt{n\log n}$, we have

$$\log\tilde{\phi}_n(t)=\frac{t^2}{\log n}\left(\log|t|-\frac{1}{2}\log n-\frac{1}{2}\log\log n\right)$$

$$-\frac{\gamma t^2}{\log n}-\frac{\pi}{2\sqrt{n}(\log n)^{3/2}}|t|^3+o\left(\frac{1}{n}\right).$$

Hence it follows that

$$\tilde{\phi}_n(t)=e^{-t^2/2}\left\{1+\frac{1}{\log n}t^2\log|t|+\frac{1}{2(\log n)^2}t^4(\log t)^2+\cdots\right\}.\tag{2.25}$$

                     Kei Takeuchi and Masafumi Akahira

Then we note that the expansion of the power of $1/\log n$ is obtained. In section 3 we shall get the density function corresponding to the second term of the right-hand side of (2.25). But it is difficult to obtain the density function corresponding to the third term of the right-hand side of (2.25).

The asymptotic expansions given as above are also applicable to the other distribution cases.

Example 2.4. Let X_i's $(i=1, 2, \cdots)$ be i.i.d. random variables with the density function $f(x)$ as follows:

$$f(x)=\frac{1}{2(1+x^2)^{3/2}}.$$

Putting

$$g(x)=f(x)-\frac{1}{2(1+|x|)^3}$$

we have

$$\int_0^x y^2 g(y)dy=\frac{1}{2}\log(\sqrt{x^2+1}+x)-\frac{x}{2\sqrt{1+x^2}}$$

$$-\frac{1}{2}\log(1+x)-\left(\frac{1}{1+x}-1\right)+\left\{\frac{1}{(1+x)^2}-1\right\}$$

$$=\frac{1}{2}\log\{(\sqrt{x^2+1}+x)/(x+1)\}-\frac{x}{2\sqrt{1+x^2}}-\frac{x}{(1+x)^2}.$$

Hence is follows that

$$\int_{-\infty}^{\infty}x^2 g(x)dx=\log 2.$$

Since

$$\int_{-\infty}^{\infty}g(x)dx=\frac{1}{2};\quad \int_{-\infty}^{\infty}xg(x)dx=0,$$

it is seen that

$$\int_{-\infty}^{\infty}e^{itx}g(x)dx=\frac{1}{2}+\frac{\log 2}{2}t^2+o(t^2).$$

By Example 2.3 we obtain

$$\int_{-\infty}^{\infty}e^{itx}f(x)dx=\frac{1}{2}\int_{-\infty}^{\infty}\frac{e^{itx}}{(1+|x|)^3}dx+\int_{-\infty}^{\infty}e^{itx}g(x)dx$$

$$=1+\frac{1}{2}t^2\log|t|-\frac{\log 2+\gamma}{2}t^2+o(t^2).$$

Putting $Z_n=\sum_1^n X_i\big/\sqrt{n\log n}$, we see that Z_n is asymptotically normal and the next order of the expansion is given by $(\log n)^{-1}$.

Example 2.5. Let U_i $(i=1, 2, \cdots)$ be i.i.d. random variables with the uniform distribution on the open interval $(0, 1)$.

Put

$$X_i=\frac{1}{2}\left(\frac{1}{1-U_i}-\frac{1}{U_i}\right)\ (i=1, 2, \cdots).$$

Since

$$x=\frac{1}{2}\left(\frac{1}{1-u}-\frac{1}{u}\right),$$

it follows that

$$\frac{dx}{du}=\frac{1}{2}\left\{\frac{1}{u^2}+\frac{1}{(1-u)^2}\right\}$$
$$=\frac{1}{2}\left\{\frac{1}{u^2(1-u)^2}-\frac{2}{u(1-u)}\right\}.$$

On the other hand since

$$x^2=\frac{(1-2u)^2}{4u^2(1-u)^2}=\frac{1}{4u^2(1-u)^2}-\frac{1}{u(1-u)},$$

we have

$$\sqrt{x^2+1}=\frac{1}{2u(1-u)}-1.$$

Hence

$$\frac{dx}{du}=2(x^2+1+\sqrt{x^2+1}\,).$$

Therefore the density function $f(x)$ of X_i is given by

$$f(x)=\frac{1}{2(x^2+1+\sqrt{x^2+1})}.$$

Putting

$$g(x)=\frac{1}{2(1+x^2)(1+x^2+\sqrt{1+x^2})}$$

we obtain

$$f(x)=\frac{1}{2(1+x^2)}-\frac{1}{2(1+x^2)^{3/2}}+g(x).$$

Then

$$\int f(x)dx=1\,;$$

$$\int g(x)dx=1-\frac{\pi}{2}+1=2-\frac{\pi}{2}\,;$$

$$\int x^2g(x)dx=\int\{f(x)-g(x)\}\,dx=\frac{\pi}{2}-1.$$

It follows by Example 2.4 that the characteristic function $\phi(t)$ of X_i is given as follows:

$$\phi(t)=1+\frac{\pi}{2}e^{-|t|}-\frac{t^2}{2}\log|t|+\gamma t^2-\frac{1}{2}\left(\frac{\pi}{2}-1\right)t^2+o(t^2)$$
$$=1-\frac{\pi}{2}|t|-\frac{t^2}{2}\log|t|+\left(\gamma+\frac{1}{2}\right)t^2+o(t^2).$$

Then the asymptotic expansion of the distribution of $\sum\limits_{1}^{n}X_i\big/n$ agrees essentially with the case when the density function is $1/\{2(1+|x|^2)^2\}$.

The above results obtained may be generalized into more general case.

Suppose that X_i's are i.i.d. random variables with a density $f(x)$ such that

(i) $f(x)=0$ for $x<0\,;$

(ii) there exists a bounded rational function $g(x)$ such that for some positive integer m

$$\int_0^\infty x^k\{f(x)-g(x)\}\,dx=M_k,\quad k=0,1,\cdots,m$$

exist.

                     Kei TAKEUCHI and Masafumi AKAHIRA

Then the characteristic function of X_i is expressed as

$$E(e^{itx_i})=\int_0^\infty e^{itx}g(x)dx+\int_0^\infty e^{itx}\{f(x)-g(x)\}dx$$

$$=\int_0^\infty e^{itx}g(x)dx$$

$$+\left\{M_0+M_1it+\cdots+\frac{M_k}{k!}(it)^k\right\}+o(|t|^k).$$

On the other hand $g(x)$ may be decomposed as

$$g(x)=\sum_j\sum_{k=1}^{\nu_j}\frac{c_{jk}}{(x-\alpha_j)^k}+\sum_j\sum_{s=1}^{\nu_{j}'}\frac{d_{jk}}{\{\gamma_j+(x-\beta_j)^2\}^s}$$

Since $g(x)$ is bounded, $\alpha_j<0$ for all j, hence $\int_0^\infty\frac{c_{jk}e^{itx}}{(x-\alpha_i)^k}dx$ can be expanded around $t=0$,

as was done in Example 2. 2., and

$$\int_0^\infty\frac{d_{jk}e^{itx}}{\{\gamma_j+(x-\beta_j)^2\}^s}dx=\int_0^\infty\frac{d_{jk}e^{it\beta_j}e^{itx}}{(\gamma_j+x^2)^s}dx+\int_0^{\beta_j}\frac{d_{jk}e^{itx}}{\{\gamma_j+(x-\beta_j)^2\}^s}dx \qquad (2.26)$$

Then the second term of (2. 26) is expanded as in Example 2. 1 and the third one is easily
shown to be expanded into a polynomial of t. This establishes the expansion of $E(e^{itx})$ and

we can get the asymptotic expansion of the distribution of $\sum_1^n X_i$ similarly as in Examples 2. 4

and 2. 5.

   For more general cases if the density $f(x)$ satisfies the assumption that $f^+(x)$ and $f^-(x)$
defined by

$$f^+(x)=\begin{cases}0 & \text{for}\quad x<0;\\ f(x) & \text{for}\quad x>0;\end{cases}$$

$$f^-(x)=\begin{cases}0 & \text{for}\quad x<0;\\ f(-x) & \text{for}\quad x>0;\end{cases}$$

satisfy the condition (ii), then the above arguments are applicable to f^+ and f^- separately and
we get a similar expansion.

3. Lattice cases

   Let X_i $(i=1, 2, \cdots)$ be non-negative integer-valued random variables with

$$Pr\{X_i=x\}=p_k(x)=\frac{(k-1)!}{(x+1)\cdots(x+k)}, \quad x=0, 1, \cdots, \ k\geq2.$$

Putting.

$$\psi_k(\theta)=\sum_{x=0}^\infty\theta^x p_k(x)$$

we have

$$\frac{d^k}{d\theta^k}\{\psi_k(\theta)\theta^k\}=(k-1)!\sum_{x=0}^\infty\theta^x=\frac{(k-1)!}{1-\theta}.$$

Then

$$\psi_k(\theta)=\theta^{-k}\int_0^\theta\frac{(\theta-\eta)^{k-1}}{1-\eta}d\eta+\chi(\theta),$$

where $\chi(\theta)$ is some polynomial of $(k-1)$th degree.
Hence the characteristic function of X_i is given by $\psi_k(e^{it})$ and

$$Pr\left\{\sum_1^n X_i = x\right\} = \frac{1}{2\pi}\int_{-\pi}^{\pi}\{\psi_k(e^{it})\}^n\, e^{-itx}\, dt.$$

For example we consider the case when $k=2$.

Then

$$\psi_2(\theta) = \theta^{-2}\int_0^\theta \frac{\theta-\eta}{1-\eta}d\eta + a\theta + b$$

$$= \frac{(1-\theta)\log(1-\theta)}{\theta^2} + \frac{1}{\theta} + a\theta + b. \tag{3.1}$$

It follows from (3.1) that $a=b=0$.

Hence we have

$$\phi(t) = \psi_2(e^{it}) = e^{-2it}(1-e^{it})\log(1-e^{it}) + e^{-it}.$$

Since

$$\log(1-e^{it}) = \log(-it) + \log\{(1-e^{it})/(-it)\}$$

$$= \log|t| - i\frac{\pi}{2}\frac{t}{|t|} + \log\left(1 + \frac{it}{2} + \cdots\right)$$

$$= \log|t| - i\frac{\pi}{2}\frac{t}{|t|} + \frac{1}{2}it + o(|t|),$$

it follows that

$$\phi(t) = -it\left(\log|t| - i\frac{\pi}{2}\frac{t}{|t|} - \frac{1}{2}it\right) + 1 - it + o(|t|)$$

$$= 1 - it\log|t| - \frac{\pi}{2}|t| - it + o(|t|).$$

Putting $z = (x - n\log n + n)/n$, we have

$$Pr\left\{\sum_1^n X_i = x\right\} \sim \frac{1}{2n\pi}\int_{-\infty}^{\infty}\exp\left(-\frac{\pi}{2}|t| - it\log|t| - itz\right)dt.$$

In a similar way as the continuous case we may be able to expand up to the higher orders.

Similarly we may expand the probability of $\sum_1^n X_i$ for any integer k. Also as was done in section 2, we can expand the characteristic function if $Pr\{X_i = x\} = p(x)$ satisfies the following:

(i) $p(x) = 0$ for $x < 0$;

(ii) there are constants $a_2, a_3, \cdots, a_m$ such that

$$\sum_{x=0}^{\infty} x^k\{p(x) - a_2 p_2(x) - \cdots - a_m p_m(x)\} = M_k$$

for $k = 0, 1, \cdots, l$ exist and is finite.

4. Appendix

In order to get the density functions corresponding to the characteristic functions $\tilde{\phi}_n(t)$ given in sections 2 and 3 we need the Fourier transforms of $it|t|e^{-t^2/2}$, $(\log|t|)e^{-t^2/2}$, $t^2 e^{-(\pi/2)|t|}$ and $t^2(\log|t|)e^{-|t|}$.

For the purpose we shall calculate

$$g_1(x) = \frac{1}{\pi}\int_0^{\infty} t(\cos tx)e^{-t^2/2}\, dt \tag{4.1}$$

Hence we have

$$G_1(x)=\int_0^x g_1(t)dt=\frac{1}{\pi}\int_0^\infty (\sin tx)e^{-t^2/2}dt.$$

Applying integration by parts to (4.1) we obtain

$$g_1(x)=\frac{1}{\pi}(-\cos tx)e^{-t^2/2}\Big|_0^\infty -\frac{1}{\pi}\int_0^\infty x(\sin tx)e^{-t^2/2}dt$$

$$=\frac{1}{\pi}-xG_1(x).$$

Putting

$$G_1(x)=u(x)e^{-x^2/2} \tag{4.2}$$

we have

$$g_1(x)=G_1{}'(x)$$
$$=u'(x)e^{-x^2/2}-xu(x)e^{-x^2/2}$$
$$=u'(x)e^{-x^2/2}-xG_1(x).$$

Then it follows that

$$u(x)=\frac{1}{\pi}\int_0^x e^{t^2/2}dt+C \tag{4.3}$$

where C is an integration constant.

If $G_1(0)=0$, it is seen from (4.2) and (4.3) that $C=0$. From (4.2) and (4.3) we have

$$G_1(x)=\frac{1}{\pi}e^{-x^2/2}\int_0^x e^{t^2/2}dt. \tag{4.4}$$

Next we calculate th Fourier transform of $it|t|e^{-t^2/2}$. By (4.4) we have

$$\frac{1}{2\pi}\int it|t|e^{-(t^2/2)-itx}dt$$

$$=-\frac{d}{dx}\frac{1}{2\pi}\int |t|e^{-(t^2/2)-itx}dt$$

$$=-g_1{}'(x)$$

$$=-\frac{d^2}{dx^2}G_1(x)$$

$$=\frac{x}{\pi}-(x^2-1)G_1(x). \tag{4.5}$$

Further we shall calculate the Fourier transforms of $(\log|t|)e^{-t^2/2}$ and $(t^2\log|t|)e^{-t^2/2}$.
Put

$$K(x)=\frac{1}{2\pi}\int_{-\infty}^\infty (\log|t|)e^{-(t^2/2)-itx}dt;$$

$$K^*(x)=\frac{1}{2\pi}\int_0^\infty (\log t)e^{-(t^2/2)-itx}dt.$$

Then it follows that $K(x)=K^*(x)+K^*(-x)$. The first and second derivatives of $K^*(x)$ are given as follows:

$$K^{*\prime}(x)=-\frac{i}{2\pi}\int_0^\infty (t\log t)e^{-(t^2/2)-itx}dt$$

$$K^{*\prime\prime}(x)=-\frac{1}{2\pi}\int_0^\infty (t^2\log t)e^{-(t^2/2)-itx}dt.$$

Applying integration by parts we have

$$K^{*\prime\prime}(x)=-\frac{1}{2\pi}(1+\log t-ixt\log t)e^{-(t^2/2)-itx}dt$$

$$=-\frac{1}{2\sqrt{2\pi}}e^{-(x^2/2)}-K^*(x)-xK^{*\prime}(x).$$

Hence $K^*(x)$ is the solution of the second order differential equation

$$y^{\prime\prime}+xy^{\prime}+y=-\frac{1}{2}\varphi(x), \tag{4.6}$$

where $\varphi(x)=\frac{1}{\sqrt{2\pi}}e^{-x^2/2}.$

In order to solve (4.6), we put

$$y^{\prime\prime}+xy^{\prime}+y=0. \tag{4.7}$$

Since $y^{\prime\prime}=-(xy^{\prime}+y)=-(xy)^{\prime}$, it follows that $y=c_1e^{-x^2/2}+c_2$, where c_1 and c_2 are integration constants.   Hence $y=\varphi(x)$ is the solution of (4.7).   Putting $y=h(x)\varphi(x)$, we substitute it into (4.6).   Then it follows that

$$h^{\prime\prime}\varphi+2h^{\prime}\varphi^{\prime}+h\varphi^{\prime\prime}+xh^{\prime}\varphi+xh\varphi^{\prime}+h\varphi=\varphi.$$

Since $\varphi^{\prime}=-x\varphi$ and $\varphi^{\prime\prime}=(x^2-1)\varphi$, we have

$$h^{\prime\prime}-xh^{\prime}=-\frac{1}{2}.$$

If $h^{\prime\prime}-xh^{\prime}=0$, then $h^{\prime}=e^{x^2/2}$.   Putting $h^{\prime}=g(x)e^{x^2/2}$ we obtain

$$g^{\prime}e^{x^2/2}+gxe^{x^2/2}-gxe^{x^2/2}=-\frac{1}{2}.$$

Since $g^{\prime}=-\frac{1}{2}e^{-x^2/2}$, it follows that

$$g(x)=-\frac{1}{2}\int e^{-x^2/2}dx+c_1;$$

$$h(x)=-\frac{1}{2}\int\left(\int e^{-x^2/2}dx+c_1\right)e^{x^2/2}dx+c_2;$$

$$K^*(x)=-\frac{1}{2}\varphi(x)\left\{\int\left(\int e^{-x^2/2}dx+c_1\right)e^{x^2/2}dx+c_2\right\}.$$

Since $K(x)$ is an odd function, it is seen that

$$K(x)=-\varphi(x)\left\{\int_0^x\left(\int_0^x e^{-x^2/2}dx\right)e^{x^2/2}dx+C\right\}, \tag{4.8}$$

where

$$C=-\frac{K(0)}{\varphi(0)}=-\frac{1}{\sqrt{2\pi}}\int(\log|t|)e^{-t^2/2}dt.$$

Further it follows from (2.1) that the Fourier transform of $t^2e^{-(\pi/2)|t|}$ is given by

$$\frac{16\pi^2}{(4x^2+\pi^2)^3}\left(1-\frac{12}{\pi^2}x^2\right). \tag{4.9}$$

Next we shall calculate the Fourier transform of $t^2(\log|t|)e^{-|t|}$.   Putting

$$h(x)=\frac{1}{2\pi}\int_{-\infty}^{\infty}t^2(\log|t|)e^{-itx-|t|}dt;$$

$$H(x)=\frac{1}{2\pi}\int_{-\infty}^{\infty}(\log|t|)e^{-itx-|t|}dt,$$

114   Kei TAKEUCHI and Masafumi AKAHIRA

we obtain

$$h(x) = -H''(x), \qquad (4.10)$$

On the other hand putting

$$H^*(x) = \frac{1}{2\pi} \int_0^\infty (\log t) e^{-itx-t} dt,$$

we have

$$H(x) = \frac{1}{2\pi} \int_0^\infty (\log t) e^{-itx-t} dt + \frac{1}{2\pi} \int_{-\infty}^0 (\log|t|) e^{-itx+t} dt$$

$$= \frac{1}{2\pi} \int_0^\infty (\log t) e^{-itx-t} dt + \frac{1}{2\pi} \int_0^\infty (\log|t|) e^{itx-t} dt$$

$$= H^*(x) + H^*(-x).$$

Since

$$H^{*\prime}(x) = \frac{1}{2\pi} \int_0^\infty -(it \log t) e^{-itx-t} dt,$$

applying integration by parts we obtain

$$H^{*\prime}(x) = -\frac{i}{2\pi(1+ix)} \int_0^\infty (\log t + 1) e^{-itx-t} dt$$

$$= -\frac{i}{1+ix} \left\{ H^*(x) + \frac{1}{2\pi} \int_0^\infty e^{-itx-t} dt \right\}$$

$$= -\frac{1}{x-i} H^*(x) + \frac{i}{2\pi(x-i)^2}. \qquad (4.11)$$

In order to solve (4.11) putting

$$H^*(x) = \frac{L(x)}{x-i}$$

we have

$$H^{*\prime}(x) = -\frac{L(x)}{(x-i)^2} + \frac{L'(x)}{x-i}$$

$$= -\frac{H^*(x)}{x-i} + \frac{L'(x)}{x-i}.$$

Hence it follows that

$$L'(x) = \frac{i}{2\pi(x-i)} = -\frac{1}{2\pi(1+ix)}.$$

Since

$$L(x) = -\frac{1}{2\pi} \log(1+ix) + C$$

$$= \frac{i}{2\pi} \left\{ \frac{1}{2} \log(1+x^2) + i \tan^{-1} x + C \right\},$$

where C is an integration constant, it is seen that

$$H^*(x) = \frac{x+i}{2\pi(1+x^2)} \left\{ \frac{i}{2} \log(1+x^2) - \tan^{-1} x + C \right\};$$

$$H(x) = \frac{1}{2\pi(1+x^2)} \left\{ -\log(1+x^2) - 2x \tan^{-1} x + C \right\}. \qquad (4.12)$$

By the value of $H(0)$ we may be able to decide that of C. From (4.10) and (4.12) $h(x)$ is also obtained.

For example we shall get the density function corresponding to the right-hand side of (2. 17). Since from (4. 5) and (4. 8)

$$\frac{1}{2\pi}\int_{-\infty}^{\infty} it\,|t|\,e^{-(c_n^2/2)t^2 - itx}\,dt = -\frac{1}{c_n^3}G_1{}''\left(\frac{x}{c_n}\right);$$

$$\frac{1}{2\pi}\int_{-\infty}^{\infty} (t^2 \log|t|)e^{-(c_n^2/2)t^2 - itx}\,dt = \frac{1}{c_n^3}K''\left(\frac{x}{c_n}\right) + \frac{1}{c_n}\varphi''\left(\frac{x}{c_n}\right),$$

the density function corresponding to (2. 17) is given as follows:

$$\frac{1}{c_n}\varphi\left(\frac{x}{c_n}\right) + \frac{1}{\log n}\left\{\frac{1}{c_n^3}G_1{}''\left(\frac{x}{c_n}\right) + \frac{1}{c_n}\varphi''\left(\frac{x}{c_n}\right) + \frac{1}{c_n^3}K''\left(\frac{x}{c_n}\right)\right\} + o\left(\frac{1}{\log n}\right)$$

References

[1] Gnedenko, B. V. and Kolmogorov, A. N., "Limit Distributions for Sums of Independent Random Variables," Addison-Wesley, Cambridge, Massachussets, 1954 (Translated from Russian).

[2] Ibragimov, I. A. and Linnik, Yu. V., "Independent and Stationary Sequences of Random Variables," Wolters-Noordhoff, the Netherland (1971) (Translated from Russian).

[3] Kendall, M. G. and Stuart, A., "The Advanced Theory of Statistics Volume 1" Charles Griffin, (1969).

[4] Takeuchi, K, "Kakuritsu-Bunpu no Kinji (Approximations of Probability Distributions)," (In Japanese), Kyoiku-Shuppan, Tokyo, (1975).

[5] Takeuchi, K. and Akahira, M., "On Gram-Charlier-Edgeworth type expansion of the sums of random variables (II)," Rep. Univ. Electro-Comm. 27-1, (1976).

Rep. Univ. Electro-Comm. 27-1, (Sci. & Tech. Sect.), pp. 117—123 August, 1976

On Gram-Charlier-Edgeworth Type Expansion of the Sums of Random Variables (II)*

Kei TAKEUCHI** and Masafumi AKAHIRA***

Abstract

Gram-Charlier-Edgeworth type expansions of the sums of independent random variables without higher order moments are given. It is shown that the asymptotic distribution functions for some cases are stable laws with fractional characteristic exponents, and the asymptotic expansion of the density function for one case is also given.

1. Introduction

In the previous paper [4], we have obtained Gram-Charlier-Edgeworth (G.-C.-E.) type expansions of the sums of independent random variables with density functions which are approximated by rational functions. Here in this paper we shall obtain G.-C.-E. expansions for the case when the density function is of the type

$$f(x) \propto (1+x)^{-p}, \ x>0 \, ; \ =0, \ x \leq 0$$

or

$$f(x) \propto (1+|x|)^{-p}$$

which p is not an integer.

2. Non-symmetric case

Let $X_1, X_2, \cdots, X_n, \cdots$ be i. i. d. random variables having the density $f_p(x)$ as follows:

$$f_p(x) = \begin{cases} \dfrac{p-1}{(1+x)^p}, & x>0 \, ; \\[2mm] 0, & x \leq 0, \end{cases}$$

where $p>1$ and p is a non-integer.

We shall first obtain the characteristic function $\phi_p(t) \ (=E(\exp it X_i))$. By Example 2.2 of [4] we have

$$\phi_p(t) = \begin{cases} (p-1)\left(\displaystyle\int_t^\infty u^{-p} e^{iu}\, du\right) t^{p-1} e^{-it}, & t>0 \, ; \\[3mm] (p-1)\left(\displaystyle\int_{|t|}^\infty u^{-p} e^{-iu}\, du\right) |t|^{p-1} e^{-it}, & t<0. \end{cases} \tag{2.1}$$

If t is a small and positive number, the $\phi_p(t)$ may be expanded as follows.

  * received on June 9, 1976
    The results of this paper have first been presented at the Third USSR-Japan Symposium on Probability Theory at Tashkent, August 1975.
 ** University of Tokyo
*** Statistical Laboratory, University of Electro-Communications

118   Kei TAKEUCHI and Masafumi AKAHIRA

Put

$$A_p + iB_p = \int_1^\infty u^{-p} e^{iu} du. \tag{2.2}$$

Since

$$u^{-p}\left\{ e^{iu} - 1 - iu + \frac{u^2}{2} + \cdots - \frac{(iu)^{[p]}}{[p]!} \right\}$$

is bounded for $0 < u < 1$ and non-integer p, it is integrable.
Putting

$$C_p + iD_p = \int_0^1 u^{-p}\left\{ e^{iu} - 1 - iu + \frac{u^2}{2} + \cdots - \frac{(iu)^{[p]}}{[p]!} \right\} du \tag{2.3}$$

we have for non-integer $p(>1)$,

$$\int_t^\infty u^{-p} e^{iu} du = A_p + iB_p + \int_t^1 u^{-p} e^{iu} du$$

$$= A_p + iB_p$$
$$+ \int_t^1 u^{-p}\left\{ e^{iu} - 1 - iu - \cdots - \frac{(iu)^{[p]}}{[p]!} \right\} du$$
$$+ \int_t^1 u^{-p}\left\{ 1 + iu + \cdots + \frac{(iu)^{[p]}}{[p]!} \right\} du$$

$$= \alpha_p + i\beta_p$$
$$- \int_0^t u^{-p}\left\{ e^{iu} - 1 - iu - \cdots - \frac{(iu)^{[p]}}{[p]!} \right\} du$$
$$+ \int_t^1 u^{-p}\left\{ 1 + iu + \cdots + \frac{(iu)^{[p]}}{[p]!} \right\} du$$

$$= \alpha_p + i\beta_p$$
$$- \int_0^t u^{-p}\left\{ \frac{(iu)^{[p]+1}}{([p]+1)!} + \frac{(iu)^{[p]+2}}{([p]+2)!} + \cdots \right\} du$$
$$+ \int_t^1 \left(u^{-p} + iu^{1-p} + \cdots + \frac{i^{[p]}u^{[p]-p}}{[p]!} \right) du$$

$$= \alpha_p + i\beta_p$$
$$- \int_0^t \left\{ \frac{i^{[p]+1}}{([p]+1)!} u^{-p+[p]+1} + \frac{i^{[p]+2}}{([p]+2)} u^{-p+[p]+2} + \cdots \right\} du$$
$$+ \sum_{k=0}^{[p]} \frac{i^k}{k!} \int_t^1 u^{k-p}$$

$$= \alpha_p + i\beta_p$$
$$- \frac{i^{[p]+1}}{([p]+1)!(-p+[p]+2)} t^{-p+[p]+2} - \frac{i^{[p]+2}}{([p]+2)!(-p+[p]+3)} t^{-p+[p]+3}$$
$$- \cdots$$
$$+ \sum_{k=0}^{[p]} \frac{i^k}{k!} \frac{1}{k-p+1} - \sum_{k=0}^{[p]} \frac{i^k}{k!} \frac{t^{k-p+1}}{k-p+1}$$

$$= \alpha_p + \alpha_p' + i(\beta_p + \beta_p') - \sum_{k=0}^{[p]} \frac{i^k}{k!} \frac{t^{k-p+1}}{k-p+1}$$
$$- \frac{i^{[p]+1}}{([p]+1)!(-p+[p]+2)} t^{-p+[p]+2}$$
$$- \frac{i^{[p]+2}}{([p]+2)!(-p+[p]+3)} t^{-p+[p]+3} - \cdots, \tag{2.4}$$

where $\alpha_p{}'$ and $\beta_p{}'$ denote the real part and the imaginary part of

$$\sum_{k=0}^{[p]}\frac{i^k}{k!}\frac{1}{k-p+1}$$

respectively, and $\alpha_p = A_p + C_p$ and $\beta_p = B_p + D_p$.

Put $\alpha_p{}^* = \alpha_p + \alpha_p{}'$ and $\beta_p{}^* = \beta_p + \beta_p{}'$. From (2.1) we have for small $t>0$

$$\phi_p(t) = e^{-it}(p-1)t^{p-1}\int_t^\infty u^{-p}e^{-iu}du$$

$$=\left(1-it-\frac{t^2}{2}+\frac{i}{6}t^3+\frac{t^4}{24}+\cdots\right)$$

$$\cdot\left\{(p-1)\alpha_p{}^*t^{p-1}+i(p-1)\beta_p{}^*t^{p-1}-\sum_{k=0}^{[p]}\frac{(p-1)i^k k^k}{k!(k-p+1)}\right.$$

$$-\frac{(p-1)i^{[p]+1}}{([p]+1)!(-p+[p]+2)}t^{[p]+1}$$

$$\left.-\frac{(p-1)i^{[p]+2}}{([p]+2)!(-p+[p]+3)}t^{[p]+2}-\cdots\right\} \tag{2.5}$$

For example we consider the case when $2<p<3$.

Then it follows from (2.4) that for small $t>0$

$$\int_t^\infty u^{-p}e^{iu}du = \alpha_p + i\beta_p - \int_0^t u^{-p}\left(e^{iu}-1-iu+\frac{u^2}{2}\right)du$$

$$+\int_t^1 u^{-p}\left(1+iu-\frac{u^2}{2}\right)du$$

$$=\alpha_p + i\beta_p - \frac{i}{6(4-p)}t^{4-p} - \frac{1}{p-1}(1-t^{1-p})$$

$$-\frac{i}{p-2}(1-t^{2-p}) - \frac{1}{2(3-p)}(1-t^{3-p}) + o(t^2).$$

From (2.1) we have for small $t>0$

$$\phi_p(t) = \left(1-it-\frac{t^2}{2}+\frac{it^3}{6}-\cdots\right)$$

$$\cdot\left\{1+\frac{p-1}{p-2}it+(\gamma_p+i\delta_p)t^{p-1}+\frac{p-1}{2(3-p)}t^2+\cdots\right\}$$

$$=1+\frac{1}{p-2}it+(\gamma_p+i\delta_p)t^{p-1}+\frac{p-1}{(p-2)(3-p)}t^2+\cdots,$$

where γ_p is some negative constant and δ_p is some real constant.

Hence it follows from (2.1) that

$$\phi_p(t)=1+\frac{1}{p-2}it+\left(\gamma_p+i\delta_p\frac{t}{|t|}\right)|t|^{p-1}+\frac{p-1}{(p-2)(3-p)}t^2+\cdots.$$

Letting $\tilde\phi_n(t)=E[\exp(itZ_n)]$ with $Z_n=n^{-1/(p-1)}\sum_1^n\left(X_i-\frac{1}{p-2}\right)$,

we obtain

$$\log\tilde\phi_n(t)=\left(\gamma_p+i\delta_p\frac{t}{|t|}\right)|t|^{p-1}+\left\{\frac{p-1}{(p-2)(3-p)}-\frac{1}{2}\right\}n^{-(3-p)/(p-1)}t^2$$

$$+\cdots. \tag{2.6}$$

As immediately seen from above, the asymptotic distribution of Z_n is a stable law with the characteristic exponent $p-1$ ([2], [3]) and the order of the next term is $n^{-(3-p)/(p-1)}\ (>n^{-1})$. Next we shall obtain the asymptotic expansion of the density function $g(x:p-1)$ corresponding to $\tilde\phi_n(t)$ given by (2.6). From (2.6) we have for $2<p<3$

$$\tilde{\phi}_n(t)=\left[\exp\left\{\left(\gamma_p+i\delta_p\frac{t}{|t|}\right)|t|^{p-1}\right\}\right]\cdot\left\{1+c_p n^{-(3-p)/(p-1)}t^2+\frac{1}{2}c_p^2 n^{-2(3-p)/(p-1)}t^4+\cdots\right\},$$

where $\quad c_p=\dfrac{p-1}{(p-2)(3-p)}-\dfrac{1}{2}.$

It follows by Theorem 2.4.2 of [3] (page 55) that for $2<p<3$

$$\frac{1}{2\pi}\int_{-\infty}^{\infty}e^{-itx}\exp\left\{\left(\gamma_p+i\delta_p\frac{t}{|t|}\right)|t|^{p-1}\right\}dt$$

$$\sim\frac{1}{\pi x}\sum_{n=0}^{\infty}(-1)^{n+1}\frac{\Gamma(n(p-1)+1)}{n!}\sin\left[\frac{1}{2}\pi n\left\{p-1-(3-p)\frac{\delta_p}{\gamma_p\tan\frac{1}{2}\pi(p-1)}\right\}\right]x^{-n(p-1)} \quad (2.7)$$

holds as $x\to\infty$.

Hence for $2<p<3$

$$g(x:\ p-1)\sim\frac{1}{\pi x}\sum_{n=0}^{\infty}(-1)^{n+1}\frac{\Gamma(n(p-1)+1)}{n!}\sin\left[\frac{1}{2}\pi n\left\{p-1-(3-p)\frac{\delta_p}{\gamma_p\tan\frac{\pi}{2}(p-1)}\right\}\right]x^{-n(p-1)}$$

holds as $x\to\infty$.

Next we shall calculate explicitely the value of (2.7) for $2<p<3$. We first have

$$\frac{1}{2\pi}\int_{-\infty}^{\infty}e^{-itx}\exp\left\{\left(\gamma_p+i\delta_p\frac{t}{|t|}\right)|t|^{p-1}\right\}dt$$

$$=\frac{1}{2\pi}\int_{-\infty}^{\infty}(\cos tx-i\sin tx)\exp\left\{\left(\gamma_p+i\delta_p\frac{t}{|t|}\right)|t|^{p-1}\right\}dt$$

$$=\frac{1}{2\pi}\int_{0}^{\infty}(\cos tx)\exp\{(\gamma_p+i\delta_p)t^{p-1}\}dt$$

$$+\frac{1}{2\pi}\int_{-\infty}^{0}(\cos tx)\exp\{(\gamma_p-i\delta_p)(-t)^{p-1}\}dt$$

$$=\frac{1}{2\pi}\int_{0}^{\infty}(\cos tx)_p\exp\{(\gamma+i\delta_p)t^{p-1}\}dt$$

$$+\frac{1}{2\pi}\int_{0}^{\infty}(\cos tx)\exp\{(\gamma_p-i\delta_p)t^{p-1}\}dt \quad (2.8)$$

Since for non-negative integer k and for a complex number α

$$\int_{0}^{\infty}t^{2k}e^{-\alpha t^{p-1}}dt$$

$$=\frac{1}{p-1}\alpha^{-(2k+1)/(p-1)}\Gamma\left(\frac{2k+1}{p-1}\right),$$

it follows that

$$\frac{1}{2\pi}\int_{0}^{\infty}(\cos tx)\exp\{(\gamma_p+i\delta_p)t^{p-1}\}dt$$

$$=\frac{1}{2\pi}\int_{0}^{\infty}\sum_{k=0}^{\infty}\frac{(-1)^k}{(2k)!}(tx)^{2k}\exp\{(\gamma_p+i\delta_p)t^{p-1}\}dt$$

$$=\frac{1}{2\pi}\sum_{k=0}^{\infty}\frac{(-1)^k x^{2k}}{(2k)!}\int_{0}^{\infty}t^{2k}\exp\{(\gamma_p+i\delta_p)t^{p-1}\}dt$$

$$=\frac{1}{2\pi}\sum_{k=0}^{\infty}\frac{(-1)^k x^{2k}}{(2k)!}\frac{1}{p-1}(-\gamma_p-i\delta_p)^{-(2k+1)/(p-1)}\Gamma\left(\frac{2k+1}{p-1}\right)$$

$$=\frac{1}{2\pi(p-1)}\sum_{k=0}^{\infty}(-1)^k\frac{\Gamma\left(\dfrac{2k+1}{p-1}\right)}{\Gamma(2k+1)}(-\gamma_p-i\delta_p)^{-(2k+1)/(p-1)}x^{2k}. \quad (2.9)$$

Putting

$$a_{p,k}e^{ib_{p,k}}=(-\gamma_p-i\delta_p)^{-(2k+1)/(p-1)},$$

we have from (2.8) and (2.9)

$$\frac{1}{2\pi}\int_{-\infty}^{\infty}e^{-itx}\exp\left\{\left(\gamma_p+i\delta_p\frac{t}{|t|}\right)|t|^{p-1}\right\}dt$$

$$=\frac{1}{\pi(p-1)}\sum_{k=0}^{\infty}(-1)^k\frac{\Gamma\left(\dfrac{2k+1}{p-1}\right)}{\Gamma(2k+1)}a_{p,k}(\cos b_{p,k})x^{2k} \qquad (2.10)$$

for $2<p<3$.

We put

$$g_p(x)=\frac{1}{\pi(p-1)}\sum_{k=0}^{\infty}(-1)^k\frac{\Gamma\left(\dfrac{2k+1}{p-1}\right)}{\Gamma(2k+1)}a_{p,k}(\cos b_{p,k})x^{2k}. \qquad (2.11)$$

The result (2.10) is essentially same as that of Feller [1], page 549.

3.　Symmetric case

Let X_i $(i=1, 2, \cdots)$ be i. i. d. random variables with the density function $f_p(x)$ as follows:

$$f_p(x)=\frac{p-1}{2(1+|x|)^p}, \quad -\infty<x<\infty,$$

where $p>1$ and p is a non-integer.

In order to calculate the characteristic function $\phi_p(t)$ $(=(\exp itX_i))$, we have

$$\phi_p(t)=\int_{-\infty}^{\infty}\frac{(p-1)e^{itx}}{2(1+|x|)^p}dx$$

$$=\int_{-\infty}^{0}\frac{(p-1)e^{itx}}{2(1-x)^p}dx+\int_{0}^{\infty}\frac{(p-1)e^{itx}}{2(1+x)^p}dx$$

$$=\int_{0}^{\infty}\frac{(p-1)e^{-itx}}{2(1+x)^p}dx+\int_{0}^{\infty}\frac{(p-1)e^{itx}}{2(1+x)^p}dx$$

Since $\int_{0}^{\infty}(p-1)e^{itx}\big/\{2(1+x)^p\}\,dx$ is the conjugate complex number of $\int_{0}^{\infty}(p-1)e^{itx}\big/\{2(1+x)^p\}$ $\cdot dx$, the characteristic function $\phi_p(t)$ agrees with the real part of the characteristic function (2.1) in the case of section 2.

If p is non-integer, then it follows from (2.5) that for every t

$$\phi_p(t)=\begin{cases}1+(p-1)\alpha_p^*|t|^{p-1}+\lambda_p t^2+o(t^2) & \text{for } 1<p<3;\\ 1+\lambda_p t^2+(p-1)\alpha_p^*|t|^{p-1}+o(|t|^{p-1}) & \text{for } 3<p,\end{cases} \qquad (3.1)$$

where $\lambda_p=\dfrac{1}{(2-p)(p-3)}$.

Note that $\lambda_p<0$ for $p>3$.

Since

$$\alpha_p^*=\alpha_p+\alpha_p'=A_p+C_p+\alpha_p',$$

it follows from (2.2) and (2.3) that α_p^* is the real part of

$$\int_{1}^{\infty}u^{-p}e^{iu}du+\int_{0}^{1}u^{-p}\left\{e^{iu}-1-iu+\frac{u^2}{2}+\cdots-\frac{(iu)^{[p]}}{[p]!}\right\}du+\sum_{k=0}^{[p]}\frac{i}{k!}\frac{1}{k-p+1}.$$

Hence it is also seen that for $1<p<3$,

$$\alpha_p^*<0.$$

                      Kei TAKEUCHI and Masatumi AKAHIRA

If $1<p<3$, then letting $\tilde{\phi}_n(t)=E[\exp(itZ_n)]$ with $Z_n=n^{-1/(p-1)}\sum_1^n X_i$ we obtain

$$\log\tilde{\phi}_n(t)=n\log\phi_p(t)$$

$$=\begin{cases}(p-1)\alpha_p^*|t|^{p-1}+\lambda_p n^{(p-3)/(p-1)}t^2-\dfrac{(p-1)^2}{2}\alpha_p^{*2}\dfrac{|t|^{2(p-1)}}{n}+o(n^{-1}) & \text{for } 1<p<2\,;\\[2mm](p-1)\alpha_p^*|t|^{p-1}-\dfrac{(p-1)^2}{2}\alpha_p^{*2}\dfrac{|t|^{2(p-1)}}{n}+\lambda_p n^{(p-3)/(p-1)}t^2+o(n^{(p-3)/(p-1)}) & \text{for } 2<p\end{cases}$$

$<3.$

Hence it follows that

$$\tilde{\phi}_n(t)=\begin{cases}e^{(p-1)\alpha_p^*|t|^{p-1}}\left\{1+\lambda_p n^{(p-3)/(p-1)}t^2-\dfrac{(p-1)^2}{2}\alpha_p^{*2}\dfrac{|t|^{2(p-1)}}{n}\right\}+o(n^{-1}) & \text{for } 1<p<2\,;\\[3mm]e^{(p-1)\alpha_p^*|t|^{p-1}}\left\{1-\dfrac{(p-1)^2}{2}\alpha_p^{*2}\dfrac{|t|^{2(p-1)}}{n}+\lambda_p n^{(p-3)/(p-1)}t^2\right\}+o(n^{(p-3)/(p-1)}) & \text{for } 2<p<3.\end{cases}$$

If $p>3$, then letting $\tilde{\phi}_n(t)=E[\exp(itZ_n)]$ with $Z_n=\sum_1^n X_i\big/\sqrt{n}$ we have

$$\log\tilde{\phi}_n(t)=n\log\phi_p\left(\frac{t}{\sqrt{n}}\right)$$

$$=\begin{cases}\lambda_p t^2+(p-1)\alpha_p^* n^{(3-p)/2}|t|^{p-1}+o(n^{(3-p)/2}) & \text{for } 3<p<5\,;\\[2mm]\lambda_p t^2-\dfrac{\lambda_p^2 t^4}{2n}+o(n^{-1}) & \text{for } 5<p,\end{cases}$$

Hence it follows that

$$\tilde{\phi}_n(t)=\begin{cases}e^{\lambda_p t^2}\{1+(p-1)\alpha_p^* n^{(3-p)/2}|t|^{p-1}\}+o(n^{(3-p)/2}) & \text{for } 3<p<5\,;\\[2mm]e^{\lambda_p t^2}\left(1-\dfrac{\lambda_p^2}{2n}t^4\right)+o(n^{-1}) & \text{for } 5<p.\end{cases}$$

Considering the case when $p=5/2$, we have from (3.1)

$$\phi_{5/2}(t)=1-\gamma'|t|^{3/2}+5t^2+o(t^2)$$

where $\gamma'=-\dfrac{1}{2}\alpha_{5/2}^*>0.$

Letting $\tilde{\phi}_n(t)=E[\exp(itZ_n)]$ with $Z_n=n^{-2/3}\sum_1^n X_i$, we obtain

$$\log\tilde{\phi}_n(t)=-\gamma'|t|^{3/2}+5n^{-1/3}t^2-\frac{\gamma'^2}{2}n^{-1}|t|^3+\cdots.$$

Hence it follows that

$$\tilde{\phi}_n(t)=e^{-\lambda'|t|^{3/2}}\left(1+5n^{-1/3}t^2+\frac{25}{2}n^{-2/3}t^4-\frac{\gamma'^2}{2}n^{-1}|t|^3+\frac{125}{6}n^{-1}t^6+\cdots\right).$$

As is immediately seen from above, the expansion of the power of $n^{-1/3}$ is obtained and the distribution function having the characteristic function $e^{-\gamma'|t|^{3/2}}$ is a stable law with characteristic exponent $3/2$ ([2], [3]).

It is easily seen that terms in the expansion of the density function corresponding to terms $e^{-\gamma'|t|^{3/2}t^{2k}}$ is

$$i^k\frac{d^{2k}}{dx^{2k}}g_{3/2}(x),$$

where $g_{3/2}(x)$ is the density function of the stable law given by (2.11).

Finally it is remarked that when $p>3$, the leading term corresponds to the normal distribution but in the subsequent terms in the expansion of the characteristic function there appear terms like $|t|^k e^{-t^2/2}$ where k is not an integer, and the density function corresponding to this is

$$\frac{1}{\pi}\int_0^\infty t^k (\cos tx) e^{-t^2/2} dt$$

$$=\frac{1}{\pi}\int_0^\infty \sum_{j=0}^\infty \frac{t^{k+2j}(-x^2)^j}{(2j)!} e^{-t^2/2} dt$$

$$=\frac{1}{\pi}\sum_{j=0}^\infty 2^{(k+2j-1)/2} \frac{\Gamma\left(\frac{k+2j+1}{2}\right)}{\Gamma(2j+1)}(-x^2)^j.$$

References

[1] Feller, W., "An Introduction to Probability Theory and Its Applications, Vol. II. John Wiley and Sons, New York, (1966).

[2] Gnedenko, B. V. and Kolmogorov, A. N., "Limit Distributions for Sums of Independent Random Variables," Addision-Wesley, Cambridge, Massachussets, 1954 (Translated from Russian).

[3] Ibragimov, I. A. and Linnik, Yu. V., "Independent and Stationary sequences of Random Variables," Wolters-Noordhoff, the Netherlands (1971) (Translated from Russian).

[4] Takeuchi, K. and Akahira, M., "On Gram-Charlier-Edgeworth type expansion of the sums of random variables (I)," Rep. Univ. Electro-Comm. 27–1, (1976).

ON THE SECOND ORDER ASYMPTOTIC EFFICIENCIES OF ESTIMATORS *

Kei TAKEUCHI and Masafumi AKAHIRA
University of Tokyo and University of Electro-Communications

1. Introduction

Second order efficiency of asymptotically efficient estimators has been discussed by R.A.Fisher [6], C.R.Rao [9], [10] and others in terms of the loss of information. Recently Chibisov ([4], [5]) has shown that a ML (maximum likelihood) estimator is second order asymptotically efficient in some sense. Pfanzagl ([7], [8]) obtained similar results. One of the authors established similar results in a book written in Japanese [11] in terms of the asymptotic distribution of the estimators. In this paper we shall present the outline of the discussion given in [11] and proceed further to the third order asymptotic efficiency. Further it is shown that the results can be extended to non-regular situations.

2. Notations and definitions

Let $\mathcal{X}$ be an abstract sample space whose generic point is denoted by x, $\mathcal{B}$ a σ-field of subsets of $\mathcal{X}$, and let Θ be a parameter space, which is assumed to be an open set in a Euclidean 1-space $\mathbf{R}^1$. We shall denote by $(\mathcal{X}^{(n)}, \mathcal{B}^{(n)})$ the n-fold direct products of $(\mathcal{X}, \mathcal{B})$. For each $n = 1, 2, \ldots$, the points of $\mathcal{X}^{(n)}$ will be denoted by $\tilde{x}_n = (x_1, \ldots, x_n)$. We consider a sequence of classes of probability measures $\{P_{n,\theta} : \theta \in \Theta\}$ $(n = 1, 2, \ldots)$ each defined on $(\mathcal{X}^{(n)}, \mathcal{B}^{(n)})$ such that for each $n = 1, 2, \ldots$ and each $\theta \in \Theta$ the following holds:

$$P_{n,\theta}\left(B^{(n)}\right) = P_{n+1,\theta}\left(B^{(n)} \times \mathcal{X}\right)$$

for all $B^{(n)} \in \mathcal{B}^{(n)}$.

An estimator of θ is defined to be a sequence $\{\hat{\theta}_n\}$ of $\mathcal{B}^{(n)}$-measurable functions $\hat{\theta}_n$ on $\mathcal{X}^{(n)}$ into Θ $(n = 1, 2, \ldots)$. For simplicity we denote an estimator as $\hat{\theta}_n$ instead of $\{\hat{\theta}_n\}$. For an increasing sequence of positive numbers $\{c_n\}$ $(\lim_{n \to \infty} c_n = \infty)$ an estimator $\hat{\theta}_n$ is called consistent with order $\{c_n\}$ (or $\{c_n\}$-consistent for short) if for every $\varepsilon > 0$ and every $\vartheta \in \Theta$, there exist a

---

*This paper is retyped with the correction of typographical errors.

sufficiently small positive number δ and a sufficiently large number L satisfying the following :

$$\overline{\lim_{n\to\infty}} \sup_{\theta:|\theta-\vartheta|<\delta} P_{n,\theta}\left\{c_n|\hat{\theta}_n - \theta| \geq L\right\} < \varepsilon.$$

The order $\{c_n\}$ of convergence of consistent estimators and its bound are discussed in [1] and [11]. In the subsequent discussions we shall deal only with the case when $c_n = \sqrt{n}$. Let $\hat{\theta}_n$ be a $\{\sqrt{n}\}$-consistent estimator.

<u>Definition 1.</u> $\hat{\theta}_n$ is asymptotically median unbiased (or AMU for short) if for any $\vartheta \in \Theta$ there exists a positive number δ such that

$$\lim_{n\to\infty} \sup_{\theta:|\theta-\vartheta|<\delta} \left|P_{n,\theta}\left\{\sqrt{n}(\hat{\theta}_n - \theta) \leq 0\right\} - \frac{1}{2}\right| = 0;$$

$$\lim_{n\to\infty} \sup_{\theta:|\theta-\vartheta|<\delta} \left|P_{n,\theta}\left\{\sqrt{n}(\hat{\theta}_n - \theta) \geq 0\right\} - \frac{1}{2}\right| = 0.$$

<u>Definition 2.</u> For $\hat{\theta}_n$ asymptotically median unbiased $F_\theta(t)$ is called an asymptotic distribution of it if

$$\lim_{n\to\infty} \left|P_{n,\theta}\left\{\sqrt{n}(\hat{\theta}_n - \theta) \leq t\right\} - F_\theta(t)\right| = 0.$$

Since $\hat{\theta}_n$ is a $\{\sqrt{n}\}$-consistent estimator, it follows that $F_\theta(-\infty) = 0$ and $F_\theta(\infty) = 1$. Let $\hat{\theta}_n$ be a AMU estimator. Then it follows that $F_\theta(0) = 1/2$ for all $\theta \in \Theta$. Let θ_0 be arbitrary and fixed in Θ. Putting

$$A_n(t) = \left\{\tilde{x}_n : \hat{\theta}_n(\tilde{x}_n) - \theta_0 \leq t/\sqrt{n}\right\}$$

we have

$$\lim_{n\to\infty} P_{n,\theta_0}\left\{A_n(t)\right\} = F_{\theta_0}(t);$$

$$\lim_{n\to\infty} P_{n,\theta_0+(t/\sqrt{n})}\left\{A_n(t)\right\} = \frac{1}{2}.$$

Consider the problem of testing hypothesis $H^+ : \theta = \theta_0 + (t/\sqrt{n})$ $(t > 0)$ against $K : \theta = \theta_0$. We shall denote by $\beta_n(t, \alpha_n, \theta_0)$ the power of the most powerful level α_n test $(0 < \alpha_n < 1)$. Then we obtain for each $t > 0$

$$(2.1) \qquad F_{\theta_0}(t) \leq \sup_{\{\{\alpha_n\}:\lim_{n\to\infty}\alpha_n=1/2\}} \overline{\lim_{n\to\infty}} \beta_n(t, \alpha_n, \theta_0).$$

Denote by $\bar{\beta}_{\theta_0}(t)$ the right-hand side of (2.1).

Consider next the problem of testing hypothesis $H^- : \theta = \theta_0 + (t/\sqrt{n})$ $(t < 0)$ against alternative $K : \theta = \theta_0$. In a similar way as the case $t > 0$ we define $\beta_n(t, \alpha_n, \theta_0)$ and $\bar{\beta}_{\theta_0}(t)$. Then we have for each $t < 0$

606

$$(2.2) \qquad F_{\theta_0}(t) \geq 1 - \bar{\beta}_{\theta_0}(t).$$

Since θ_0 is arbitrary, the bounds of the asymptotic distributions of AMU estimators are obtained as follows :

$$F_\theta(t) \leq \overline{\beta}_\theta(t) \qquad \text{for all } t > 0;$$
$$F_\theta(t) \geq 1 - \overline{\beta}_\theta(t) \qquad \text{for all } t < 0.$$

For any $\theta \in \Theta$ letting $\overline{\beta}_\theta(0) = 1/2$ we make the following definition.

<u>Definition 3.</u> An asymptotically median unbiased estimator $\hat{\theta}_n$ is called asymptotically efficient if for each $\theta \in \Theta$

$$F_\theta(t) = \begin{cases} \overline{\beta}_\theta(t) & \text{for all } t \geq 0; \\ 1 - \overline{\beta}_\theta(t) & \text{for all } t < 0. \end{cases}$$

<u>Definition 4.</u> $\hat{\theta}_n$ is second order asymptotically median unbiased (or second order AMU for short) if for any $\vartheta \in \Theta$, there exists a positive number δ such that

$$\lim_{n \to \infty} \sup_{\theta:|\theta-\vartheta|<\delta} \sqrt{n} \left| P_{n,\theta} \left\{ \sqrt{n} \left(\hat{\theta}_n - \theta \right) \leq 0 \right\} - \frac{1}{2} \right| = 0;$$

$$\lim_{n \to \infty} \sup_{\theta:|\theta-\vartheta|<\delta} \sqrt{n} \left| P_{n,\theta} \left\{ \sqrt{n} \left(\hat{\theta}_n - \theta \right) \geq 0 \right\} - \frac{1}{2} \right| = 0.$$

<u>Definition 5.</u> For $\hat{\theta}_n$ second order asymptotically median unbiased $F_\theta(t) + (1/\sqrt{n})G_\theta(t)$ is called a second order asymptotic distribution of it if

$$\lim_{n \to \infty} \sqrt{n} \left| P_{n,\theta} \left\{ \sqrt{n} \left(\hat{\theta}_n - \theta \right) \leq t \right\} - F_\theta(t) - (1/\sqrt{n})G_\theta(t) \right| = 0.$$

Let θ_0 be arbitrary but fixed in Θ. Then we have

$$P_{n,\theta_0} \{A_n(t)\} = F_{\theta_0}(t) + \frac{1}{\sqrt{n}} G_{\theta_0}(t) + o\left(\frac{1}{\sqrt{n}} \right);$$

$$P_{n,\theta_0+(t/\sqrt{n})} \{A_n(t)\} = \frac{1}{2} + o\left(\frac{1}{\sqrt{n}} \right).$$

Consider the problem of testing hypothesis $H^+ : \theta = \theta_0 + (t/\sqrt{n})$ $(t > 0)$ against alternative $K : \theta = \theta_0$. Let $\beta_n(t, \alpha_n, \theta_0)$ denote the power of the most powerful level α_n test $(0 < \alpha_n < 1)$. We assume that

$$\sup_{\left\{ \{\alpha_n\}: \alpha_n = \frac{1}{2} + o\left(\frac{1}{\sqrt{n}}\right) \right\}} \beta_n(t, \alpha_n, \theta_0)$$

$$607$$

$$= \beta_{\theta_0}(t) + \frac{1}{\sqrt{n}} \gamma_{\theta_0}(t) + o\left(\frac{1}{\sqrt{n}}\right).$$

Considering the case $t < 0$ in a similar way as the case $t > 0$, we make the following definition.

<u>Definition 6.</u> A second order asymptotically median unbiased estimator $\hat{\theta}_n$ is called second order asymptotically efficient if for each $\theta \in \Theta$

$$F_\theta(t) = \begin{cases} \beta_\theta(t) & \text{for all } t \geq 0, \\ 1 - \beta_\theta(t) & \text{for all } t < 0, \end{cases}$$

$$G_\theta(t) = \begin{cases} \gamma_\theta(t) & \text{for all } t \geq 0, \\ -\gamma_\theta(t) & \text{for all } t < 0, \end{cases}$$

where for each $\theta \in \Theta$ $\beta_\theta(0) = 1/2$ and $\gamma_\theta(0) = 0$.

In a similar and obvious way we may define the third or k-th order asymptotic efficiency of estimators.

3. Asymptotic efficiency and second order asymptotic efficiency

Let $X_1, X_2, \ldots, X_n, \ldots$ be independently and identically distributed random variables with a density function $f(x, \theta)$ satisfying (i), (ii) and (iii).

(i) $\{x : f(x, \theta) > 0\}$ does not depent on θ.

(ii) For almost all $x[\mu]$, $f(x, \theta)$ is twice continuously differentiable in θ.

(iii) For each $\theta \in \Theta$

$$0 < I_\theta$$

$$= \int \left\{ \frac{\partial}{\partial \theta} \log f(x, \theta) \right\}^2 f(x, \theta) d\mu(x)$$

$$= -\int \left\{ \frac{\partial^2}{\partial \theta^2} \log f(x, \theta) \right\} f(x, \theta) d\mu(x)$$

$$< \infty.$$

608

Let θ_0 be arbitrary but fixed in Θ. Consider the problem of testing hypothesis $H^+ : \theta = \theta_0 + (t/\sqrt{n})$ $(t > 0)$ against alternative $K : \theta = \theta_0$. Then the rejection region of the most powerful test is given by

$$T_n = \sum_{i=1}^{n} Z_{ni} > c,$$

where $Z_{ni} = \log\{f(X_i, \theta_0)/f(X_i, \theta_0 + \frac{t}{\sqrt{n}})\}$. Since

$$T_n = \sum_{i=1}^{n} Z_{ni} \sim -\frac{t}{\sqrt{n}} \sum_{i=1}^{n} \frac{\partial}{\partial\theta} \log f(X_i, \theta)$$
$$-\frac{t^2}{2n} \sum_{i=1}^{n} \frac{\partial^2}{\partial\theta^2} \log f(X_i, \theta),$$

if $\theta = \theta_0$, then T_n is asymptotically normal with mean $t^2 I_{\theta_0}/2$ and variance $t^2 I_{\theta_0}$ and if $\theta = \theta_0 + (t/\sqrt{n})$, then T_n is asymptotically normal with mean $-t^2 I_{\theta_0}/2$ and variance $t^2 I_{\theta_0}$. Hence it follows that

$$\overline{\beta}_{\theta_0}(t) = \Phi\left(t\sqrt{I_{\theta_0}}\right),$$

where $\Phi(u) = \int_{-\infty}^{u} \frac{1}{\sqrt{2\pi}} e^{-x^2/2} dx$. From (2.1) we have

$$F_{\theta_0}(t) \leq \Phi\left(t\sqrt{I_{\theta_0}}\right) \text{ for all } t > 0.$$

In a similar way as the case $t > 0$, we obtain from (2.2)

$$F_{\theta_0}(t) \geq 1 - \Phi\left(|t|\sqrt{I_{\theta_0}}\right) = \Phi\left(t\sqrt{I_{\theta_0}}\right).$$

Since θ_0 is arbitrary we have now established the following well known theorem.

<u>Theorem 3.1.</u> Under conditions (i), (ii) and (iii), if $\sqrt{n}(\hat{\theta}_n - \theta)$ is asymptotically normal with mean 0 and variance $1/I_\theta$, then $\hat{\theta}_n$ is asymptotically efficient.

It has been well established that under some regularity conditions the maximum likelihood estimator $\hat{\theta}_{ML}$ has the same asymptotic distributions as above, hence $\hat{\theta}_{ML}$ is asymptotically efficient.

Using Gram-Charlier expansion of the distribution of $\sum_{i=1}^{n} Z_{ni}$ we get the asymptotic series of the power of the most powerful test.

609

We further assume the following :

(iv) $f(x, \theta)$ is three times continuously differentiable in θ.

(v) There exist

$$J_\theta = E_\theta \left[\left\{ \frac{\partial^2}{\partial \theta^2} \log f(X, \theta) \right\} \left\{ \frac{\partial}{\partial \theta} \log f(X, \theta) \right\} \right]$$

and

$$K_\theta = E_\theta \left[\left\{ \frac{\partial}{\partial \theta} \log f(X, \theta) \right\}^3 \right]$$

and the following hold :

$$E_\theta \left[\frac{\partial^3}{\partial \theta^3} \log f(X, \theta) \right] = -3 J_\theta - K_\theta.$$

We denote $(\partial/\partial\theta) f(x, \theta)$, $(\partial^2/\partial\theta^2) f(x, \theta)$ and $(\partial^3/\partial\theta^3) f(x, \theta)$ by f_θ, $f_{\theta\theta}$ and $f_{\theta\theta\theta}$, respectively.
Since

$$\frac{\partial^3}{\partial \theta^3} \log f(x, \theta) = \frac{f_{\theta\theta}(x, \theta)}{f(x, \theta)} - \frac{3 f_{\theta\theta}(x, \theta) f_\theta(x, \theta)}{\{f(x, \theta)\}^2} + \frac{2 \{f_\theta(x, \theta)\}^3}{\{f(x, \theta)\}^3}$$

$$= \frac{f_{\theta\theta}(x, \theta)}{f(x, \theta)} - 3 \left\{ \frac{\partial}{\partial \theta} \log f(x, \theta) \right\} \left\{ \frac{\partial^2}{\partial \theta^2} \log f(x, \theta) \right\}$$

$$- \left\{ \frac{\partial}{\partial \theta} \log f(x, \theta) \right\}^3,$$

it follows by the last condition of (v) that

$$\int f_{\theta\theta\theta}(x, \theta) d\mu(x) = 0.$$

Let $t > 0$. If $\theta = \theta_0$, then

610

$$T_n = \sum_{i=1}^{n} Z_{ni} \sim - \frac{t}{\sqrt{n}} \sum_{1}^{n} \frac{\partial}{\partial \theta} \log f(X_i, \theta_0)$$

$$- \frac{t^2}{2n} \sum_{1}^{n} \frac{\partial^2}{\partial \theta^2} \log f(X_i, \theta_0)$$

$$- \frac{t^3}{6n\sqrt{n}} \sum_{1}^{n} \frac{\partial^3}{\partial \theta^3} \log f(X_i, \theta_0).$$

Hence it follows that

$$E_{\theta_0}(T_n) \sim \frac{t^2}{2} I + \frac{t^3}{6\sqrt{n}}(3J + K),$$

$$V_{\theta_0}(T_n) \sim n \left(\frac{t^2}{n} I + \frac{t^3}{n\sqrt{n}} J \right) = t^2 I + \frac{t^3}{\sqrt{n}} J,$$

$$E_{\theta_0} \left[\{T_n - E_{\theta_0}(T_n)\}^3 \right] \sim - \frac{t^3}{\sqrt{n}} K,$$

where I, J and K denote I_{θ_0}, J_{θ_0} and K_{θ_0}, respectively. Put $\theta_1 = \theta_0 + (t/\sqrt{n})$. If $\theta = \theta_1$, then

$$T_n \sim - \frac{t}{\sqrt{n}} \sum_{1}^{n} \frac{\partial}{\partial \theta} \log f(X_i, \theta_1)$$

$$+ \frac{t^2}{2n} \sum_{1}^{n} \frac{\partial^2}{\partial \theta^2} \log f(X_i, \theta_1)$$

$$- \frac{t^3}{6n\sqrt{n}} \sum_{1}^{n} \frac{\partial^3}{\partial \theta^3} \log f(X_i, \theta_1).$$

Hence it follows that

$$E_{\theta_1}(T_n) \sim - \frac{t^2}{2} I' + \frac{t^3}{6\sqrt{n}}(3J' + K'),$$

where I', J' and K' denote I_{θ_1}, J_{θ_1} and K_{θ_1}, respectively. On the other hand we have

611

$$I' \sim I + \frac{t}{\sqrt{n}}\frac{\partial}{\partial\theta}I_{\theta_0}$$

$$=I + \frac{t}{\sqrt{n}}\frac{\partial}{\partial\theta}\int\left\{\frac{\partial}{\partial\theta}\log f(x,\theta_0)\right\}^2 f(x,\theta_0)d\mu$$

$$=I + \frac{t}{\sqrt{n}}\int 2\left\{\frac{\partial^2}{\partial\theta^2}\log f(x,\theta_0)\right\}\left\{\frac{\partial}{\partial\theta}\log f(x,\theta_0)\right\}f(x,\theta_0)d\mu$$

$$+ \frac{t}{\sqrt{n}}\int\left\{\frac{\partial}{\partial\theta}\log f(x,\theta_0)\right\}^3 f(x,\theta_0)d\mu$$

$$=I + \frac{t}{\sqrt{n}}(2J+K).$$

Hence we obtain

$$E_{\theta_1}(T_n) \sim -\frac{t^2}{2}I - \frac{t^3}{6\sqrt{n}}(3J+2K).$$

Since

$$J' \sim J + \frac{t}{\sqrt{n}}\frac{\partial}{\partial\theta}J_{\theta_0};$$

$$K' \sim K + \frac{t}{\sqrt{n}}\frac{\partial}{\partial\theta}K_{\theta_0},$$

it follows by a similar way as above that

$$V_{\theta_1}(T_n) \sim t^2 I + \frac{t^3}{\sqrt{n}}(J+K);$$

$$E_{\theta_1}\left[\{T_n - E_{\theta_1}(T_n)\}^3\right] \sim -\frac{t^3}{\sqrt{n}}K.$$

Letting a_n be a rejection bound, we have

$$P_{n,\theta_1}\{T_n < a_n\} = P_{n,\theta_1}\left\{\frac{T_n + (t^2 I/2)}{t\sqrt{I}} < \frac{a_n + (t^2 I/2)}{t\sqrt{I}}\right\}.$$

Putting $c_n = \{a_n + (t^2 I/2)\}/(t\sqrt{I})$, we obtain

$$P_{n,\theta_1}\{T_n < a_n\}$$

$$= \Phi(c_n) - \phi(c_n)\left\{-\frac{t^2}{6\sqrt{n}I}(3J+2K) + \frac{t}{2\sqrt{n}I}(J+K)c_n\right.$$

$$\left. - \frac{1}{6\sqrt{n}I I}K(c_n^2 - 1)\right\} + o\left(\frac{1}{\sqrt{n}}\right),$$

where $\phi(u) = \Phi'(u) = \frac{1}{\sqrt{2\pi}}e^{-u^2/2}$. If $P_{n,\theta_1}\{T_n < a_n\} = 1/2$, then it follows that $c_n = O(1/\sqrt{n})$ and $\Phi(c_n) = (1/2) + c_n\phi(c_n)$. Since

612

$$c_n = -\frac{t^2}{6\sqrt{n}I}(3J + 2K) + \frac{K}{6\sqrt{n}II} + o\left(\frac{1}{\sqrt{n}}\right),$$

we have

$$a_n = -\frac{t^2 I}{2} - \frac{t^3}{6\sqrt{n}I}(3J + 2K) + \frac{tK}{6\sqrt{n}I} + o\left(\frac{1}{\sqrt{n}}\right).$$

Then we have

$$P_{n,\theta_0}\{T_n \geq a_n\}$$

$$= 1 - P_{n,\theta_0}\{T_n < a_n\}$$

$$= 1 - P_{n,\theta_0}\left\{\frac{T_n - (t^2 I/2)}{t\sqrt{I}} - c_n < -t\sqrt{I}\right\}$$

$$= 1 - \Phi(-t\sqrt{I}) + \phi(-t\sqrt{I})\left\{\frac{t^2}{6\sqrt{n}I}(3J + K) - c_n + \frac{t}{2\sqrt{n}I}J(-t\sqrt{I})\right.$$

$$\left. - \frac{1}{6\sqrt{n}II}K(t^2 I - 1)\right\} + o\left(\frac{1}{\sqrt{n}}\right)$$

$$= \Phi(t\sqrt{I}) + \frac{t^2}{6\sqrt{n}I}(3J + 2K)\phi(t\sqrt{I}) + o\left(\frac{1}{\sqrt{n}}\right).$$

Since θ_0 is arbitrary, it follows that for each $t > 0$

$$G_\theta(t) = \Phi\left(t\sqrt{I_\theta}\right),$$

$$\gamma_\theta(t) = \frac{t^2}{6\sqrt{I_\theta}}(3J_\theta + 2K_\theta)\phi\left(t\sqrt{I_\theta}\right).$$

In a similar way as the case $t > 0$, we have for each $t < 0$,

$$G_\theta(t) = \Phi\left(-t\sqrt{I}\right),$$

$$\gamma_\theta(t) = -\frac{t^2}{6\sqrt{I_\theta}}(3J_\theta + 2K_\theta)\phi\left(t\sqrt{I_\theta}\right).$$

Therefore we have now established the following which is analogous to the result of Pfanzagl [7].

Theorem 3.2. Under conditions (i)~(v), if

613

$$(3.1) \qquad P_{n,\theta}\left\{\sqrt{nI_\theta}(\hat\theta_n - \theta) \le t\right\} = \Phi(t) + \frac{3J_\theta + 2K_\theta}{6\sqrt{n}I_\theta^{3/2}}t^2\phi(t) + o\left(\frac{1}{\sqrt{n}}\right),$$

then $\hat\theta_n$ is second order asymptotically efficient.

(3.1) means thet $\sqrt{nI_\theta}(\hat\theta_n - \theta)$ has an asymptotic distribution with mean $-\{(3J_\theta + 2K_\theta)/(6\sqrt{n}I_\theta^{1/2})\} + o(1/\sqrt{n})$ and variance $1 + o(1/\sqrt{n})$ and third moment $-(3J_\theta + 2K_\theta)/(\sqrt{n}I_\theta^{3/2})$.

Let $X_1, X_2, \ldots, X_n, \ldots$ be independently and identically distributed random variables with an exponential distribution having the following density function $f(x,\theta)$:

$$(3.2) \qquad f(x,\theta) = \begin{cases} \frac{1}{\theta}e^{-x/\theta}, & x > 0, \\ 0, & x \le 0. \end{cases}$$

Since

$$\log f(X,\theta) = -\frac{X}{\theta} - \log\theta,$$

$$\frac{\partial}{\partial\theta}\log f(X,\theta) = \frac{X}{\theta^2} - \frac{1}{\theta},$$

$$\frac{\partial^2}{\partial\theta^2}\log f(X,\theta) = -\frac{2X}{\theta^3} + \frac{1}{\theta^2},$$

it follows that

$$I_\theta = \frac{1}{\theta^4}E_\theta\left[(X-\theta)^2\right] = \frac{1}{\theta^2},$$

$$J_\theta = -\frac{1}{\theta^5}E_\theta\left[(X-\theta)(2X-\theta)\right] = -\frac{2}{\theta^5}E_\theta\left[(X-\theta)^2\right] = -\frac{2}{\theta^3},$$

$$K_\theta = \frac{1}{\theta^6}E_\theta\left[(X-\theta)^3\right] = \frac{2}{\theta^3}.$$

If $\sqrt{n}(\hat\theta_n - \theta)/\theta$ has an asymptotic distribution with mean $1/(3\sqrt{n})$ and variance $1 + o(1/\sqrt{n})$ and third moment $2/\sqrt{n}$, then $\hat\theta_n$ is second order asymptotically efficient.

The maximum likelihood estimator of θ is given by $\hat\theta_{ML} = \bar X = \sum_1^n X_i/n$. Putting

$$\hat\theta^*_{ML} = \left(1 + \frac{1}{3n}\right)\bar X,$$

we have

$$E_\theta\left[\frac{\sqrt{n}(\hat\theta^*_{ML} - \theta)}{\theta}\right] = \frac{1}{3\sqrt{n}},$$

614

$$V_\theta\left[\frac{\sqrt{n}\hat{\theta}^*_{ML}}{\theta}\right] = \left(1+\frac{1}{3n}\right)^2 = 1 + o\left(\frac{1}{\sqrt{n}}\right),$$

$$E_\theta\left[\left\{\frac{\sqrt{n}(\hat{\theta}^*_{ML}-\theta)}{\theta}\right\}^3\right] = \frac{2}{\sqrt{n}}\left(1+\frac{1}{3n}\right)^3 = \frac{2}{\sqrt{n}} + o\left(\frac{1}{\sqrt{n}}\right).$$

Hence $\hat{\theta}^*_{ML}$ is second order asymptotically efficient. In this case $\overline{X}$ is sufficient statistic and the distribution has monotone likelihood ratio. Consider the problem of testing hypothesis $\theta = \theta_1$ against alternative $\theta = \theta_0$, where $\theta_1 = \theta_0 + (t/\sqrt{n})$ $(t > 0)$. Then the rejection region of the most powerful test is given by the following form:

$$\overline{X} < c.$$

On the other hand since the cumulants of $\overline{X}$ are given as follows:

$$V_\theta(\overline{X}) = \frac{\theta^2}{n},$$

$$E_\theta\left[(\overline{X}-\theta)^3\right] = \frac{2\theta^3}{n^2}.$$

$\sqrt{n}(\overline{X}-\theta)/\theta$ has an asymptotic distribution with mean 0 and variance 1 and third cumulant $2/\sqrt{n}$.
Using Gram-Charlier expansion we have

$$(3.3) \qquad P_{n,\theta}\left\{\frac{\sqrt{n}(\overline{X}-\theta)}{\theta} < c\right\}$$
$$= \Phi(c) - \frac{1}{3\sqrt{n}}\phi(c)(c^2-1).$$

Let $\theta = \theta_1(= \theta_0 + t/\sqrt{n})$ $(t > 0)$. If (3.3) agrees with $1/2$ up to the order of $1/\sqrt{n}$, then the following must hold:

$$c\phi(c) - \frac{1}{3\sqrt{n}}\phi(c)(c^2-1) \doteqdot 0.$$

Hence it follows that $c = -1/(3\sqrt{n})$. The rejection region of the level $1/2$ test of hypothesis $\theta = \theta_1$ is given by

$$\overline{X} < \left(1 - \frac{1}{3n}\right)\theta_1.$$

615

This agrees asymptotically with

$$\hat{\theta}^*_{ML} = \left(1 + \frac{1}{3n}\right)\overline{X}(< \theta_1)$$

up to the order of $1/\sqrt{n}$.

Assume that $\sqrt{n}(\hat{\theta}_n - \theta)$ has the asymptotic normal distribution with mean 0 and variance $1/I_\theta$, and that the cumulants of the asymptotic distribution are given as follows:

$$\sqrt{n}E_\theta(\hat{\theta}_n - \theta) = \frac{1}{\sqrt{n}}c_1(\theta) + o\left(\frac{1}{\sqrt{n}}\right);$$

$$nV_\theta(\hat{\theta}_n) = \frac{1}{I_\theta} + \frac{1}{\sqrt{n}}c_2(\theta) + o\left(\frac{1}{\sqrt{n}}\right);$$

$$E_\theta\left[\left\{\sqrt{n}\left(\hat{\theta}_n - E_\theta(\hat{\theta}_n)\right)\right\}^3\right] = \frac{1}{\sqrt{n}}c_3(\theta) + o\left(\frac{1}{\sqrt{n}}\right).$$

Further the m-th $(m \geq 4)$ cumulants of $\sqrt{n}\hat{\theta}_n$ is assumed to be less than order $1/\sqrt{n}$. We assume that $c_1(\theta)$, $c_2(\theta)$ and $c_3(\theta)$ are continuous in θ. Put $\hat{\theta}^*_n = \hat{\theta}_n - (1/n)k(\hat{\theta}_n)$. Since $\hat{\theta}^*_n$ agrees with $\hat{\theta}_n - (1/n)k(\theta)$ up to the order of $1/n$, the cumulants of the asymptotic distribution of $\sqrt{n}I_\theta(\hat{\theta}^*_n - \theta)$ are given as follows:

$$\sqrt{n}I_\theta E_\theta(\hat{\theta}^*_n - \theta) = \sqrt{\frac{I_\theta}{n}}\{c_1(\theta) - k(\theta)\} + o\left(\frac{1}{\sqrt{n}}\right);$$

$$nI_\theta V_\theta(\hat{\theta}^*_n) = 1 + \frac{I_\theta c_2(\theta)}{\sqrt{n}} + o\left(\frac{1}{\sqrt{n}}\right);$$

$$E_\theta\left[\left\{\sqrt{nI_\theta}\left(\hat{\theta}^*_n - E_\theta(\hat{\theta}^*_n)\right)\right\}^3\right] = \frac{I_\theta^{3/2}c_3(\theta)}{\sqrt{n}} + o\left(\frac{1}{\sqrt{n}}\right).$$

If

$$k(\theta) = c_1(\theta) - \frac{1}{6}I_\theta c_3(\theta),$$

then in the Gram-Charlier expansion the following holds:

$$P_{n,\theta}\left\{\sqrt{n}\left(\hat{\theta}^*_n - \theta\right) \leq 0\right\} = \frac{1}{2} + o\left(\frac{1}{\sqrt{n}}\right).$$

It also follows that

(3.4)
$$P_{n,\theta}\left\{\sqrt{nI_\theta}\left(\hat{\theta}^*_n - \theta\right) \leq t\right\}$$

$$= \Phi(t) - \frac{1}{\sqrt{n}}\phi(t)\left[\frac{c_2(\theta)}{2I_\theta}t + \frac{c_3(\theta)}{I_\theta^{3/2}}t^2\right] + o\left(\frac{1}{\sqrt{n}}\right).$$

Hence it is seen that (3.4) agrees with the bound (3.1) if and only if $c_2(\theta) = 0$ and $c_3(\theta) = (3J_\theta + 2K_\theta)/I_\theta^3$.

If $c_2(\theta) = 0$, then $c_3(\theta)$ must automatically be equal to $-(3J_\theta + 2K_\theta)/I_\theta^3$. Indeed since $\phi(t)t^2 > 0$ for all $t \neq 0$, it follows that (3.4) $\gtrless$ (3.1) uniformly in θ if $c_3(\theta) \gtrless -(3J_\theta + 2K_\theta)/I_\theta^3$. This means that the bound (3.1) fails to hold at either positive or negative t. Hence if $c_2(\theta) = 0$, we must have $c_3(\theta) = -(3J_\theta + 2K_\theta)/I_\theta^3$. This apparently seems mysterious. However it will be naturally understood from different examples that if the second moment is decided up to order of $1/\sqrt{n}$, then the third moment is done. We shall verify the above fact by the maximum likelihood estimator. The second order asymptotic efficiency of the maximum likelihood estimator will be verified using the above fact. The continuous differentiability of the likelihood function is assumed up to the necessary order. Let $\hat{\theta}_{ML}$ be a maximum likelihood estimator. Since

$$0 = \sum_1^n \frac{\partial}{\partial\theta}\log f(X_i, \hat{\theta}_{ML})$$

$$= \sum_1^n \frac{\partial}{\partial\theta}\log f(X_i, \theta) + \sum_1^n \left\{\frac{\partial^2}{\partial\theta^2}\log f(X_i, \theta)\right\}\left(\hat{\theta}_{ML} - \theta\right)$$

$$+ \frac{1}{2}\sum_1^n \left\{\frac{\partial^3}{\partial\theta^3}\log f(X_i, \theta^*)\right\}\left(\hat{\theta}_{ML} - \theta\right)^2,$$

putting $T_n = n(\hat{\theta}_{ML} - \theta)$ we have

$$0 = \frac{1}{\sqrt{n}}\sum_1^n \frac{\partial}{\partial\theta}\log f(X_i, \theta) + \frac{1}{n}\left\{\sum_1^n \frac{\partial^2}{\partial\theta^2}\log f(X_i, \theta)\right\}T_n$$

$$+ \frac{1}{2n\sqrt{n}}\left\{\sum_1^n \frac{\partial^3}{\partial\theta^3}\log f(X_i, \theta^*)\right\}T_n^2.$$

Put

$$Z_1 = \frac{1}{\sqrt{n}}\sum_1^n \frac{\partial}{\partial\theta}\log f(X_i, \theta);$$

$$Z_2 = \frac{1}{\sqrt{n}}\sum_1^n \left\{\frac{\partial^2}{\partial\theta^2}\log f(X_i, \theta) + I_\theta\right\};$$

$$W = \frac{1}{n}\sum_1^n \frac{\partial^3}{\partial\theta^3}\log f(X_i, \theta^*).$$

Then Z_1 and Z_2 have the asymptotic normal distributions with mean 0 and variance I_θ and $L_\theta (= E_\theta \{ (\partial^2/\partial\theta^2) \log f(X,\theta) + I_\theta \}^2)$ and covariance J_θ, respectively and also W converges in probability to $-3J_\theta - K_\theta$. Hence it follows that

$$Z_1 + \left(-I_\theta + \frac{1}{\sqrt{n}} Z_2 \right) T_n - \frac{3J_\theta + K_\theta}{2\sqrt{n}} T_n^2 \sim 0,$$

$$T_n \sim \frac{1}{I_\theta} Z_1 + \frac{1}{\sqrt{n} I_\theta^2} Z_1 Z_2 - \frac{3J_\theta + K_\theta}{2\sqrt{n} I_\theta^3} Z_1^2.$$

Since covariance of Z_1 and the term of the order is 0, it follows that

$$V_\theta(T_n) = \frac{1}{I_\theta} + o\left(\frac{1}{\sqrt{n}} \right).$$

Hence it is seen that the maximum likelihood estimator is second order asymptotically efficient. If we indeed calculate the asymptotic cumulants, they are obtained as follows:

$$E_\theta(T_n) \sim -\frac{J_\theta + K_\theta}{2\sqrt{n} I_\theta^2} + o\left(\frac{1}{\sqrt{n}} \right);$$

$$V_\theta(T_n) \sim \frac{1}{I_\theta} + o\left(\frac{1}{\sqrt{n}} \right);$$

$$E_\theta \left[\{ T_n - E_\theta(T_n) \}^3 \right] \sim -\frac{3J_\theta + 2K_\theta}{\sqrt{n} I_\theta^3} + o\left(\frac{1}{\sqrt{n}} \right).$$

Hence it is shown that if the second moment is decided up to the order of $1/\sqrt{n}$, the third moment is equal to what it should be.

As another example, we consider the location parameter case. Let $X_1, X_2, \ldots, X_n, \ldots$ be a sequence of independently and identically distributed random variables with a density function $f(x - \theta)$. It is well known that under appropriate regularity conditions the best linear estimator $\hat{\theta}_n = \sum_{i=1}^n c_{in} X_{(i|n)} + c_{0n}$ is asymptotically efficient, where $X_{(1|n)} < \cdots < X_{(n|n)}$ and c_{in} are optimal constants as was established by Blom and others. Then the estimator $\hat{\theta}_n$ is second order asymptotically efficient. Indeed, if $U_{(i|n)}$ are order statistics from the uniform distribution, then it follows that

618

$$F^{-1}(X_{(i|n)}) = \theta + F^{-1}\left(\frac{i}{n+1}\right)$$
$$+ \left\{F^{-1}\left(\frac{i}{n+1}\right)\right\}'\left(U_{(i|n)} - \frac{i}{n+1}\right)$$
$$+ \frac{1}{2}\left\{F^{-1}\left(\frac{i}{n+1}\right)\right\}''\left(U_{(i|n)} - \frac{i}{n+1}\right)^2 + R.$$

Let $c_{0n} = -\sum_{i=1}^{n} c_{in} F^{-1}(i/(n+1))$ and $\sum_{i=1}^{n} c_{in} = 1$.
Putting

$$a_{in} = \left\{F^{-1}\left(\frac{i}{n+1}\right)\right\}',$$
$$b_{in} = \left\{F^{-1}\left(\frac{i}{n+1}\right)\right\}'',$$

we have

$$\sqrt{n}\left(\hat{\theta}_n - \theta\right) = \sqrt{n}\sum_{i=1}^{n} c_{in} a_{in}\left(U_{(i|n)} - \frac{i}{n+1}\right)$$
$$+ \frac{\sqrt{n}}{2}\sum_{i=1}^{n} c_{in} b_{in}\left(U_{(i|n)} - \frac{i}{n+1}\right)^2 + \sqrt{n} R_{in}.$$

Then the asymptotic variance is given by

$$V(\sqrt{n}(\hat{\theta}_n - \theta))$$
$$\sim n\sum_{i=1}^{n} c_{in}^2 a_{in}^2 \frac{i(n+1-i)}{(n+1)(n+2)} + 2n\sum\sum_{i<j} c_{in} c_{jn} a_{in} a_{jn} \frac{i(n+1-j)}{(n+1)^2(n+2)}$$
$$+ n\sum\sum_{i\leq j} c_{in} c_{jn} a_{in} b_{jn} \frac{i(n+1-j)(n+1-2j)}{(n+1)^3(n+2)(n+3)}$$
$$+ n\sum\sum_{i>j} c_{in} c_{jn} a_{in} b_{jn} \frac{j(n+1-i)(n+1-2j)}{(n+1)^3(n+2)(n+3)}.$$

Let $J(u)$ be a function sufficiently smooth and put $c_{in} + (1/n)J(i/(n+1))$. Putting $h(u) = f(F^{-1}(u))$, we have

$$\left(F^{-1}(u)\right)' = h(u)^{-1}$$
$$\left(F^{-1}(u)\right)'' = -h'(u)h(u)^{-2}.$$

Since

$$V(\sqrt{n}(\hat{\theta}_n - \theta))$$

619

$$\sim 2 \iint\limits_{u<v} u(1-v)J(u)J(v)h(u)^{-1}h(v)^{-1}dudv$$

$$+ \frac{1}{n} \iint\limits_{u<v} u(1-v)(1-2v)J(u)J(v)h(u)^{-1}h'(v)h(v)^{-2}dudv$$

$$+ \frac{1}{n} \iint\limits_{u>v} v(1-u)(1-2v)J(u)J(v)h(u)^{-1}h'(v)h(v)^{-2}dudv$$

$$+ o\left(\frac{1}{n}\right),$$

it follows that

$$J(u) = -h(u)h'(u)/I,$$

where $I = \int \{f'(x)\}^2 / f(x)dx = \int \{h'(u)\}^2 du$. Then we have

$$V(\sqrt{n}(\hat{\theta}_n - \theta)) \sim \frac{1}{I} + o\left(\frac{1}{\sqrt{n}}\right).$$

Let $\hat{\theta}_n^*$ be a modified estimator which is second order AMU. Then it will be shown that $\hat{\theta}_n^*$ is second order asymptotically efficient. Since

$$J = -\int \left\{ \frac{d^2}{dx^2} \log f(x) \right\} \left\{ \frac{d}{dx} \log f(x) \right\} f(x)dx$$

$$= \frac{1}{2} \int \left\{ \frac{d}{dx} \log f(x) \right\}^2 f'(x)dx$$

$$= \frac{1}{2} \int \left\{ \frac{d}{dx} \log f(x) \right\}^3 f(x)dx$$

$$= \frac{1}{2} \int \{h'(u)\}^3 du$$

$$= -\frac{K}{2},$$

the third asymptotic moment of the second order asymptotically efficient estimator must be $-K/(2n^2 I^3)$. Indeed, since

$$\text{cov}\left[\left(U_{(i|n)} - \frac{i}{n+1} \right) \left(U_{(j|n)} - \frac{j}{n+1} \right), \ \left(U_{(k|n)} - \frac{k}{n+1} \right)^2 \right]$$

620

$$= 2cov(U_{(i|n)}, U_{(k|n)})cov(U_{(j|n)}, U_{(k|n)}) + o\left(\frac{1}{n}\right);$$

$$E\left[\left(U_{(i|n)} - \frac{i}{n+1}\right)\left(U_{(j|n)} - \frac{j}{n+1}\right)\left(U_{(k|n)} - \frac{k}{n+1}\right)\right]$$

$$= \frac{i(n+1-j)(n+1-k)}{(n+1)^2(n+2)(n+3)}, \quad i \le j \le k,$$

it follows that

$$E\left[\left\{\hat{\theta}_n - E(\hat{\theta}_n)\right\}^3\right]$$

$$\sim \frac{6}{n^2} \iiint\limits_{u<v<w} u(1-2v)(1-w)J(u)J(v)J(w)h(u)^{-1}h(v)^{-1}h(w)^{-1}dudvdw$$

$$+ \frac{3}{n^2} \int \left\{\int (\min(u,v) - uv)\, J(u)h(u)du\right\}^2 J(v)h(v)^{-2}h'(v)dv + o\left(\frac{1}{n^2}\right).$$

Putting $J(u) = -h(u)h''(u)/I$, we have

$$E\left[\left\{\hat{\theta}_n - E(\hat{\theta}_n)\right\}^3\right]$$

$$\sim -\frac{1}{n^2 I^3}\int \{h'(u)\}^3\, du - \frac{1}{n^2 I^3}\int h(u)h'(u)h''(u)du$$

$$= -\frac{1}{n^2 I^3}\int \{h'(u)\}^3\, du + \frac{3}{2n^2 I^3}\int \{h'(u)\}^3\, du$$

$$= -\frac{K}{2n^2 I^3}.$$

Hence it is shown that the third asymptotic moment is $-K/(2n^2 I^3)$.

The second order asymptotic efficiency of ML estimators depends on the assumption of differentiability of the density function. The following example is to make this point clear. Suppose that X_i's are independently and identically distributed according to the double (bilateral) exponential distribution with the density function

$$f(x - \theta) = \frac{1}{2}e^{-|x-\theta|},$$

where θ is unknown location parameter. In this case it can be shown that the first order asymptotic bound is obtained with $I = 1$, and it can be easily shown that the ML estimator

621

$$\hat{\theta}_{ML} = \operatorname*{med}_{1 \le i \le n} X_i$$

is asymptotically efficient. The second order asymptotic bound is obtained after some algebraic manipulations (the detail is given in [11]) as follows:

(3.5)
$$\Phi(t) - \frac{t^2}{6\sqrt{n}}\phi(t)\operatorname{sgn} t.$$

But the asymptotic distribution of $\hat{\theta}_{ML}$ is given by

$$P_{n,\theta}\left\{\sqrt{n}\left(\hat{\theta}_{ML} - \theta\right) \le t\right\} = \Phi(t) - \frac{t^2}{2\sqrt{n}}\phi(t)\operatorname{sgn} t + o\left(\frac{1}{\sqrt{n}}\right),$$

which differs in the second term from (3.5).

4. Third order asymptotic efficiency

We proceed to the problem of third order asymptotic efficiency. $\hat{\theta}_n$ is called third order asymptotically median unbiased if for any $\vartheta \in \Theta$ there exists a positive number δ such that

$$\lim_{n\to\infty} \sup_{\theta:|\theta-\vartheta|<\delta} n\left| P_{n,\theta}\left\{\sqrt{n}\left(\hat{\theta}_n - \theta\right) \le 0\right\} - \frac{1}{2}\right| = 0;$$

$$\lim_{n\to\infty} \sup_{\theta:|\theta-\vartheta|<\delta} n\left| P_{n,\theta}\left\{\sqrt{n}\left(\hat{\theta}_n - \theta\right) \ge 0\right\} - \frac{1}{2}\right| = 0.$$

Let $\hat{\Theta}_3$ be the class of all third order asymptotically median unbiased estimators. For all $\hat{\theta}_n \in \hat{\Theta}_3$ we shall obtain the bounds of G_θ and H_θ such that

$$\lim_{n\to\infty} n\left| P_{n,\theta}\left\{\sqrt{n}\left(\hat{\theta}_n - \theta\right) \le t\right\} - F_\theta(t) - \frac{1}{\sqrt{n}}G_\theta(t) - \frac{1}{n}H_\theta(t)\right| = 0.$$

In a similar way as section 3 we consider the problem of testing hypothesis $\theta = \theta_0 + (t/\sqrt{n})$ against alternative $\theta = \theta_0$ and the asymptotic expansion of the power of the most powerful test of it.

The continuous differentiability of the likelihood function is assumed up to the necessary order. Putting $\psi' = (\partial/\partial\theta)\log f(x,\theta)$,

622

$\psi'' = (\partial^2/\partial\theta^2)\log f(x,\theta)$ and $\psi^{(i)} = (\partial^i/\partial\theta^i)\log f(x,\theta)$ $(i=3,4)$, we have

$$T_n = \sum_{i=1}^{n} Z_{ni}$$

$$= -\frac{t}{\sqrt{n}}\sum_{1}^{n}\psi'(X_i) - \frac{t^2}{2n}\sum_{1}^{n}\psi''(X_i) - \frac{t^3}{6n\sqrt{n}}\sum_{1}^{n}\psi^{(3)}(X_i)$$

$$- \frac{t^4}{24n^2}\sum_{1}^{n}\psi^{(4)}(X_i) + o\left(\frac{1}{n}\right),$$

where $Z_{ni} = \log\{f(X_i,\theta_0)/f(X_i,\theta_0 + (t/\sqrt{n}))\}$. Define

$$L = E_{\theta_0}\left\{\psi^{(3)}(X)\psi'(X)\right\},$$

$$M = E_{\theta_0}\left\{\psi''(X)^2\right\},$$

$$N = E_{\theta_0}\left\{\psi''(X)\psi'(X)^2\right\},$$

$$H = E_{\theta_0}\left\{\psi'(X)^4\right\}.$$

Since under appropriate conditions the following holds:

$$E_{\theta_0}\left\{\psi^{(4)}(X)\right\} = -4L - 3M - 6N - H,$$

the asymptotic cumulants are given as follows:

$$E_{\theta_0}(T_n) \sim \frac{t^2 I}{2} + \frac{t^3}{6\sqrt{n}}(3J + K) + \frac{t^4}{24n}(4L + 3M + 6N + H),$$

$$V_{\theta_0}(T_n) \sim t^2 I + \frac{t^3}{\sqrt{n}}J + \frac{t^4}{4n}(M - I^2) + \frac{t^4}{3n}L,$$

$$E_{\theta_0}\left[\{T_n - E_{\theta_0}(T_n)\}^3\right] \equiv \gamma_{\theta_0}(T_n) \sim -\frac{t^3}{\sqrt{n}}K - \frac{3t^4}{2n}(N + I^2),$$

$$E_{\theta_0}\left[\{T_n - E_{\theta_0}(T_n)\}^4\right] - 3\left\{V_{\theta_0}(T_n)\right\}^2 \equiv \delta_{\theta_0}(T_n) \sim \frac{t^4}{n}(H - 3I^2).$$

If $\theta = \theta_1(= \theta_0 + (t/\sqrt{n}))$, then for any measurable function g

$$E_{\theta_1}\left\{g(X)\right\}$$

623

$$= E_{\theta_0}\left[g(X)\left\{1 + \frac{t}{\sqrt{n}}\frac{f'_{\theta_0}(X)}{f_{\theta_0}(X)} + \frac{t^2}{2n}\frac{f''_{\theta_0}(X)}{f_{\theta_0}(X)} + \frac{t^3 f^{(3)}_{\theta_0}(X)}{6n\sqrt{n}f_{\theta_0}(X)} + \cdots\right\}\right]$$

$$= E_{\theta_0}\left[g(X)\left\{1 + \frac{t}{\sqrt{n}}\psi'(X) + \frac{t^2}{2n}\left(\psi''(X) + \psi'(X)^2\right)\right.\right.$$

$$+ \frac{t^3}{6n\sqrt{n}}\left(\psi^{(3)}(X) + 3\psi''(X)\psi'(X) + \psi'(X)^3\right)$$

$$\left.\left.+ \cdots\right\}\right].$$

Hence we have

$$E_{\theta_1}(T_n) \sim E_{\theta_0}(T_n) - E_{\theta_0}\left[\left\{\frac{t}{\sqrt{n}}\sum_1^n\psi'(X_i) + \frac{t^2}{2n}\sum_1^n\psi''(X_i) + \frac{t^3}{6n\sqrt{n}}\sum_1^n\psi^{(3)}(X_i)\right\}\right.$$

$$\left\{\frac{t}{\sqrt{n}}\sum_1^n\psi'(X_i) + \frac{t^2}{2n}\sum_1^n\left(\psi''(X_i) + \psi'(X_i)^2\right)\right.$$

$$\left.\left.+ \frac{t^3}{6n\sqrt{n}}\sum_1^n\left(\psi^{(3)}(X_i) + 3\psi''(X_i)\psi'(X_i) + \psi'(X_i)^3\right)\right\}\right]$$

$$+ o\left(\frac{1}{n}\right)$$

$$= -\frac{t^2}{2}I - \frac{t^2}{6\sqrt{n}}(3J + 2K) - \frac{t^4}{24n}(4L + 3M + 12N + 3H).$$

Similarly we obtain

$$V_{\theta_1}(T_n) \sim t^2I + \frac{t^3}{\sqrt{n}}(J + K) + \frac{t^4}{12n}(4L + 3M + 18N + 6H - 3I^2),$$

$$\gamma_{\theta_1}(T_n) \sim -\frac{t^3}{\sqrt{n}}K - \frac{t^4}{2n}(3N + 2H - 3I^2),$$

$$\delta_{\theta_1}(T_n) \sim \frac{t^4}{n}(H - 3I^2).$$

Put $V_n = \{T_n - E_{\theta_1}(T_n)\} / \left(t\sqrt{I}\right)$. If

$$V_{\theta_1}(V_n) = \frac{1}{t^2I}V_{\theta_1}(T_n)$$

$$= 1 + \frac{1}{\sqrt{n}}\beta_1 + \frac{1}{n}\beta_2 + o\left(\frac{1}{n}\right),$$

624

$$\gamma_{\theta_1}(V_n) = \frac{1}{t^3 I \sqrt{I}} \gamma_{\theta_1}(T_n)$$

$$= \frac{1}{\sqrt{n}} \gamma_1 + \frac{1}{n} \gamma_2 + o\left(\frac{1}{n}\right),$$

$$\delta_{\theta_1}(V_n) = \frac{1}{t^4 I} \delta_{\theta_1}(T_n)$$

$$= \frac{1}{n} \delta + o\left(\frac{1}{n}\right),$$

then we have

$$P_{n,\theta_1}\{V_n \leq a\}$$
$$= \Phi(a) - \phi(a)\left\{\frac{1}{2}\left(\frac{\beta_1}{\sqrt{n}} + \frac{\beta_2}{n}\right)a + \frac{1}{6}\left(\frac{\gamma_1}{\sqrt{n}} + \frac{\gamma_2}{n}\right)(a^2 - 1)\right.$$
$$+ \left(\frac{\delta}{24n} + \frac{\beta_1^2}{8n}\right)(a^3 - 3a) + \frac{\beta_1\gamma_1}{12n}(a^4 + 6a^2 + 3)$$
$$\left. + \frac{\gamma_1^2}{72n}(a^5 - 10a^3 - 15a)\right\} + o\left(\frac{1}{n}\right).$$

Choose a such that

$$P_{n,\theta_1}\{V_n \leq a\} = \frac{1}{2} + o\left(\frac{1}{n}\right).$$

Since

$$\Phi(a) = \frac{1}{2} + a\phi(a) - \frac{1}{2}a^3\phi'(a) + \cdots,$$

it follows that

$$a = \frac{\beta_1}{2\sqrt{n}}a - \left(\frac{\gamma_1}{6\sqrt{n}} + \frac{\gamma_2}{n}\right) + \frac{1}{4n}\beta_1\gamma_1 + o\left(\frac{1}{n}\right)$$
$$= -\frac{\gamma_1}{6\sqrt{n}} - \frac{\gamma_2}{n} + \frac{\beta_1\gamma_1}{6n} + o\left(\frac{1}{n}\right).$$

If for $\theta = \theta_0$

$$E_{\theta_1}(V_n) - E_{\theta_0}(V_n) - a = -t\sqrt{I} + \frac{1}{\sqrt{n}}\alpha_1' + \frac{1}{n}\alpha_2' + o\left(\frac{1}{n}\right),$$

$$V_{\theta_0}(V_n) = \frac{1}{\sqrt{n}}\beta_1' + \frac{1}{n}\beta_2' + o\left(\frac{1}{n}\right),$$

625

$$\gamma_{\theta_0}(V_n) = \frac{1}{\sqrt{n}}\gamma_1' + \frac{1}{n}\gamma_2' + o\left(\frac{1}{n}\right),$$

$$\delta_{\theta_0}(V_n) = \frac{1}{n}\delta_1' + o\left(\frac{1}{n}\right),$$

then we have

$$(4.1) \qquad P_{n,\theta_0}\{V_n \geq a\}$$

$$= 1 - P_{n,\theta_0}\{V_n < a\}$$

$$= \Phi(t') + \phi(t')\left\{\frac{1}{\sqrt{n}}\alpha_1' + \frac{1}{n}\alpha_2' - \left(\frac{\beta_1'}{2\sqrt{n}} + \frac{\beta_2'}{2n} + \frac{\alpha_1'^2}{2n}\right)t'\right.$$

$$+ \left(\frac{\gamma_1'}{6\sqrt{n}} + \frac{\gamma_2'}{2n} + \frac{\alpha_1'\beta_1'}{2n}\right)\left(t'^2 - 1\right)$$

$$- \left(\frac{\delta_1'}{24n} + \frac{\alpha_1'\gamma_1'}{6n} + \frac{\beta_1'^2}{8n}\right)\left(t'^3 - 3t'\right)$$

$$+ \frac{\beta_1'\gamma_1'}{12n}\left(t'^4 - 6t'^2 + 3\right)$$

$$\left. - \frac{\gamma_1'^2}{72n}\left(t'^5 - 10t'^3 + 15t'\right)\right\} + o\left(\frac{1}{n}\right),$$

where $t' = t\sqrt{I}$. Substituting in (4.1)

$$\alpha_1' = \frac{t'^2}{2I^{3/2}}(2J + K) - \frac{K}{6I^{3/2}},$$

$$\alpha_2' = \frac{t'^3}{12I^2}(4L + 3M + 6N + 3I^2) + \frac{t'}{6I^3}K(J + K),$$

$$\beta_1' = \frac{t'J}{I^{3/2}},$$

$$\beta_2' = \frac{t'^2}{4I^2}(M - I^2) + \frac{t'^2}{3I^2}L,$$

$$\gamma_1' = -\frac{K}{I^{3/2}},$$

626

$$\gamma_2' = -\frac{3t'}{2I^2}(N + I^2),$$

$$\delta = \frac{1}{I^2}(H - 3I^2),$$

we obtain

$$P_{n,\theta_0}\{V_n \geq a\}$$
$$= \Phi(t') + \phi(t')\left[\frac{t'^2}{6\sqrt{n}I^{3/2}}(3J + 2K) - \frac{t'^5}{72nI^{3/2}}(3J + 2K)^2\right.$$
$$- \frac{t'^3}{72nI^3}\left\{(3J + 2K)^2 - 3I(4L + 3M + 6N - H + 6I^2)\right\}$$
$$\left. + \frac{t'}{72nI^3}\left\{2K^2 + 9I(2N + H - I^2)\right\}\right]$$
$$+ o\left(\frac{1}{n}\right).$$

Hence it follows that for $t > 0$

$$P_{n,\theta}\left\{\sqrt{nI}\left(\hat{\theta}_n - \theta\right) \leq t\right\}$$
$$\leq \Phi(t) - \phi(t)\left\{\frac{\beta_3''}{6\sqrt{n}} + \frac{\beta_3''}{6\sqrt{n}}(t^2 - 1) + \frac{\beta_3''^2}{72n}(t^5 - 10t^3 + 15t)\right.$$
$$+ \left(\frac{\beta_4''}{24n} + \frac{\beta_3''^2}{36n}\right)(t^3 - 3t)$$
$$\left. + \left(\frac{\beta_2''}{2n} + \frac{\beta_3''^2}{72n}\right)t\right\} + o\left(\frac{1}{n}\right),$$

where $\beta_3'' = -(3J + 2K)/I^{3/2}$,

$$\beta_4'' = \frac{1}{I^3}\left\{3(3J + 2K)^2 - I(4L + 3M + 6N - H + 6I^2)\right\},$$

$$\beta_2'' = \frac{1}{36I^3}\left\{17(3J + 2K)^2 - 2K^2 - 9I(4L + 3M + 8N + 5I^2)\right\}.$$

Further the moments of the third order asymptotically efficient estimator must be as follows:

627

$$(4.2) \qquad E_\theta(\hat{\theta}_n) = \theta + \frac{\beta_3''}{6\sqrt{n}I^{3/2}} + o\left(\frac{1}{n\sqrt{n}}\right),$$

$$(4.3) \qquad V_\theta(\sqrt{n}\hat{\theta}_n) = I + \frac{\beta_2''}{nI} + o\left(\frac{1}{n}\right),$$

$$(4.4) \qquad \gamma_\theta(\sqrt{n}\hat{\theta}_n) = \frac{\beta_3''}{\sqrt{n}I^{3/2}} + o\left(\frac{1}{n}\right),$$

$$(4.5) \qquad \delta_\theta(\sqrt{n}\hat{\theta}_n) = \frac{\beta_4''}{nI^2} + o\left(\frac{1}{n}\right).$$

Next we shall consider the example of the exponential distribution with the density (3.2) given in section 3. Since

$$\hat{\theta}_n = \left(1 + \frac{1}{3n}\right)\overline{X}$$

it follows that

$$(4.6) \qquad E_\theta(\hat{\theta}_n) = \theta + \frac{\theta}{3n},$$

$$(4.7) \qquad V_\theta(\sqrt{n}\hat{\theta}_n) = \left(1 + \frac{2}{3n}\right)\theta^2,$$

$$(4.8) \qquad \gamma_\theta(\sqrt{n}\hat{\theta}_n) = \frac{2}{n}\theta^3 + o\left(\frac{1}{n}\right),$$

$$(4.9) \qquad \delta_\theta(\sqrt{n}\hat{\theta}_n) = \frac{6}{n}\theta^4 + o\left(\frac{1}{n}\right).$$

Since $I = 1/\theta^2$, $J = -2/\theta^3$, $K = 2/\theta^3$, $L = 6/\theta^4$, $M = 5/\theta^4$, $N = -5/\theta^4$ and $H = 9/\theta^4$, it follows from (4.2)$\sim$(4.5) that $\beta_3'' = 2$, $\beta_4'' = 6$ and $\beta_2'' = 2/3$. Then the moments of (4.2)$\sim$(4.5) are equal to those of (4.6)$\sim$(4.9), respectively. This fact should hold because $\overline{X}$ is a sufficient statistic.

We next consider the maximum likelihood estimator $\hat{\theta}_{ML}$. Putting $T_n = \sqrt{n}(\hat{\theta}_{ML} - \theta)$, we have

$$Z_1 + \left(-1 + \frac{1}{\sqrt{n}}Z_2\right)T_n + \frac{1}{2\sqrt{n}}\left\{-(3J + K) + \frac{1}{\sqrt{n}}Z_3\right\}T_n^2$$

$$- \frac{4L + 3M + 6N + H}{6n}T_n^3 \sim 0,$$

628

where $Z_3 = (1/\sqrt{n}) \sum_1^n \left\{ \psi^{(3)}(X_i) + 3J + K \right\}$. We also obtain

$$T_n \sim \frac{1}{I} Z_1 + \frac{1}{\sqrt{n} I^2} \left(Z_2 Z_1 - \frac{3J+K}{2I} \right) Z_1^2$$

$$+ \frac{1}{nI^3} \left\{ Z_1 Z_2^2 + \frac{1}{2} Z_1^2 Z_3 - \frac{3(3J+K)}{2I} Z_1^2 Z_2 \right.$$

$$\left. + \frac{(3J+K)^2}{2I^2} Z_1^3 - \frac{4L + 3M + 6N + H}{6} Z_1^3 \right\}.$$

Since

$$E(Z_1^2 Z_2) \sim \frac{1}{\sqrt{n}} (N + I^2),$$

$$E(Z_1^3 Z_2) \sim \frac{3}{n} IJ,$$

$$E(Z_1^4 Z_2^2) \sim \frac{3}{n} (M - I^2) + 12J^2,$$

it follows that

$$E(T_n) \sim -\frac{J+K}{2\sqrt{n} I^2} + o\left(\frac{1}{n} \right),$$

$$V(T_n) \sim \frac{1}{I} + \frac{7J^2 + 14JK + 5K^2}{24I^4} - \frac{L + 4N + H + I^2}{nI^3} + o\left(\frac{1}{n} \right),$$

$$\gamma(T_n) \sim -\frac{3J + 2K}{nI} + o\left(\frac{1}{n} \right),$$

$$\delta(T_n) \sim \frac{12(J+K)(2J+K)}{nI^5} - \frac{4L + 12N + 3H - 3I}{nI^4} + o\left(\frac{1}{n} \right).$$

We define a modified maximum likelihood estimator $\hat{\theta}^*_{ML}$ as follows:

$$\hat{\theta}^*_{ML} = \hat{\theta}_{ML} + \frac{K_{\hat{\theta}_{ML}}}{6nI^2_{\hat{\theta}_{ML}}}.$$

Then $\hat{\theta}^*_{ML}$ is third order AMU, that is,

$$P_{n,\theta} \left\{ \hat{\theta}^*_{ML} - \theta \le 0 \right\} = \frac{1}{2} + o\left(\frac{1}{n} \right).$$

Further it follows that

$$nV(\hat{\theta}^*_{ML}) = n\left(1 + \frac{1}{6n}\frac{\partial}{\partial\theta}\frac{K_\theta}{I_\theta}\right)^2 V(\hat{\theta}_{ML})$$

$$\sim V(T_n) + \frac{(3N+H)I - 2K(2J+K)}{3nI^4},$$

where

$$\frac{\partial}{\partial\theta}K_\theta = \frac{\partial}{\partial\theta}\int \{\psi'(x)\}^3 f(x)dx$$

$$= 3\int \psi''(x)\psi'(x)^2 f(x)dx + \int \{\psi'(x)\}^4 f(x)dx$$

$$= 3N + H,$$

$$\frac{\partial}{\partial\theta}I_\theta = 2J + K.$$

In general the maximum likelihood estimator is shown not to be third order asymptotically efficient. It is enough to consider a special case, and we take the Cauchy distribution as an example which has the following density function $f(x,\theta)$:

$$f(x,\theta) = \frac{1}{\pi(1 + (x-\theta)^2)}.$$

Let $X_1, X_2, \ldots, X_n, \ldots$ be independent identically distributed random variables with above density.
Since

$$\frac{\partial}{\partial\theta}\log f(x,0) = -\frac{\partial}{\partial x}\log f(x,0) = \frac{2x}{1+x^2},$$

$$\frac{\partial^2}{\partial\theta^2}\log f(x,0) = \frac{\partial^2}{\partial x^2}\log f(x,0) = -\frac{2(1-x^2)}{(1+x^2)^2},$$

$$\frac{\partial^3}{\partial\theta^3}\log f(x,0) = -\frac{\partial^3}{\partial x^3}\log f(x,0) = -\frac{4x(3-x^2)}{(1+x^2)^3},$$

it follows that $I = 1/2$, $J = K = 0$, $L = -3/4$, $M = 7/8$, $N = -1/8$ and $H = 3/8$. Since the density is symmetric, the maximum likelihood estimator $\hat{\theta}_{ML}$ is third order AMU. The second and fourth cumulants are given as follows:

630

$$(4.10) \qquad V(\sqrt{n}\hat{\theta}_{ML}) \sim 2 + \frac{5}{n},$$

$$(4.11) \qquad \gamma(\sqrt{n}\hat{\theta}_{ML}) \sim \frac{66}{n}.$$

On the other hand since the values of the bound of the asymptotic distribution is given by

$$\beta_2'' = 1/4, \quad \beta_4'' = 0,$$

they are larger than the above moments (4.10) and (4.11). Hence $\hat{\theta}_{ML}$ is not third order asymptotically efficient.

5. Some further remarks

The above results can be extended to the cases (I), (II), (III), (IV).

(I) When the distribution is discrete, some modification is necessary to apply the above argument.

(II) Let X_i's be independent but not identically distributed random variables with density functions $f_i(x_i, \theta)$. Let $\hat{\theta}_n$ be a second order AMU estimator. When $c_n(\hat{\theta}_n - \theta)$ has an asymptotic distribution, we obtain the bound of the distributions of $\hat{\theta}_n$. Let θ_0 be arbitrary but fixed in Θ. Consider the most powerful test of the problem of testing hypothesis $\theta = \theta_1 \, (= \theta_0 + c_n^{-1}a, a > 0)$ against alternative $\theta = \theta_0$. Then the test statistic of it is given by

$$(5.1) \qquad T_n = \sum_1^n \log \frac{f_i(X_i, \theta_1)}{f_i(X_i, \theta_0)}$$

$$= -\sum_1^n \left\{ \frac{\partial}{\partial\theta} \log f_i(X_i, \theta_0) \right\} \cdot (c_n^{-1}a)$$

$$-\frac{1}{2}\sum_1^n \left\{ \frac{\partial^2}{\partial\theta^2} \log f_i(X_i, \theta_0) \right\} \cdot (c_n^{-1}a)^2$$

$$-\frac{1}{6}\sum_1^n \left\{ \frac{\partial^3}{\partial\theta^3} \log f_i(X_i, \theta_0) \right\} \cdot (c_n^{-1}a)^3$$

$$+\sum_{i=1}^n R_{in}.$$

Put

$$I_i = E_{\theta_0}\left\{\frac{\partial}{\partial\theta}\log f_i(X_i,\theta_0)\right\}^2 = -E_{\theta_0}\left\{\frac{\partial^2}{\partial\theta^2}\log f_i(X_i,\theta_0)\right\}.$$

It follows that the variance of the first term of the right-hand side of (5.1) is given by $(c_n^{-1}a)^2\sum_1^n I_i$. When $c_n = \sqrt{\tilde{I}_n}$, where $\tilde{I}_n = \sum_1^n I_i$, it's variance converges to some finite value. If the following "Lindeberg type" condition holds: for every $\varepsilon > 0$,

$$\lim_{n\to\infty}\frac{1}{\tilde{I}_n}\sum_1^n E_{\theta_0}\left[\left\{\frac{\partial}{\partial\theta}\log f_i(X_i,\theta_0)\right\}^2 \chi_{in}(\varepsilon)\right] = 0,$$

where

$$\chi_{in}(\varepsilon) = \begin{cases} 1, & \text{if } \frac{\partial}{\partial\theta}\log f_i(X_i,\theta_0) > \varepsilon\sqrt{\tilde{I}_n}, \\ 0, & \text{otherwise}, \end{cases}$$

then the first term of the right-hand side of (5.1) is asymptotically normal with mean 0 and variance a^2. The necessary condition for this to hold is given by $\lim_{n\to\infty}\max_{1\le i\le n} I_i/\tilde{I}_n = 0$. If a similar condition as above holds: for every $\varepsilon > 0$,

$$\lim_{n\to\infty}\frac{1}{\tilde{I}_n}\sum_1^n E_{\theta_0}\left[\left\{-\frac{\partial^2}{\partial\theta^2}\log f_i(X_i,\theta_0)\right\}\chi_{in}^*(\varepsilon)\right] = 0,$$

where

$$\chi_{in}^*(\varepsilon) = \begin{cases} 1, & \text{if } -\frac{\partial^2}{\partial\theta^2}\log f_i(X_i,\theta_0) > \varepsilon\tilde{I}_n, \\ 0, & \text{otherwise}, \end{cases}$$

then the second term of the right-hand side of (5.1) converges in probability to $a^2/2$.

If under alternative $\theta = \theta_0$, similar regularity conditions as above, then the test statistic T_n is asymptotically normal with mean $-a^2/2$ and variance a^2. Put $c_n = \sqrt{\tilde{I}_n}$. For $a > 0$, the supremum of the power functions is given by $\Phi(a)$. Hence if $\sqrt{\tilde{I}_n}(\hat{\theta}_n - \theta)$ is asymptotically normal with mean 0 and variance 1, then $\hat{\theta}_n$ is an asymptotically efficient estimator.

Let $\hat{\theta}_{ML}$ be a maximum likelihood estimator. Then

$$0 = \sum_1^n\left\{\frac{\partial}{\partial\theta}\log f_i(X_i,\theta)\right\}\left(\hat{\theta}_{ML}-\theta\right)$$

$$+ \sum_1^n\left\{\frac{\partial^2}{\partial\theta^2}\log f_i(X_i,\theta)\right\}\left(\hat{\theta}_{ML}-\theta\right)^2$$

$$+ \sum_1^n R'_{in}.$$

632

Under some regularity conditions $\sqrt{\tilde{I}_n}(\hat{\theta}_{ML} - \theta)$ is asymptotically normal with mean 0 and variance 1. <u>Hence the maximum likelihood estimator $\hat{\theta}_{ML}$ is asymptotically efficient.</u>

In order to consider the higher order of the bound of the power functions, we must decide the orders of

$$\tilde{K}_n = \sum_1^n K_i = \sum_1^n E_{\theta_0}\left[\left\{\frac{\partial}{\partial\theta}\log f_i(X_i,\theta_0)\right\}^3\right]$$

and

$$\tilde{J}_n = \sum_1^n J_i = \sum_1^n E_{\theta_0}\left[\left\{\frac{\partial^2}{\partial\theta^2}\log f_i(X_i,\theta_0)\right\}\left\{\frac{\partial}{\partial\theta}\log f_i(X_i,\theta_0)\right\}\right].$$

Then the asymptotic moments of the test statistic T_n are formally obtained as follows:

$$E_{\theta_0}(T_n) \sim \frac{a^2}{2} - \frac{3\tilde{J}_n + \tilde{K}_n}{6\tilde{I}_n^{3/2}}a^3,$$

$$V_{\theta_0}(T_n) \sim a^2 + \frac{\tilde{J}_n}{\tilde{I}_n^{3/2}}a^3,$$

$$E_{\theta_0}\{T_n - E_{\theta_0}(T_n)\}^3 \sim -\frac{\tilde{K}_n}{\tilde{I}_n^{3/2}}a^3.$$

Similarly we have for $\theta = \theta_1$

$$E_{\theta_1}(T_n) \sim -\frac{a}{2} - \frac{3\tilde{J}_n + 2\tilde{K}_n}{6\tilde{I}_n^{3/2}}a^3,$$

$$V_{\theta_1}(T_n) \sim a^2 + \frac{\tilde{J}_n + \tilde{K}_n}{\tilde{I}_n^{3/2}}a^3,$$

$$E_{\theta_1}\{T_n - E_{\theta_1}(T_n)\}^3 \sim -\left(\tilde{K}_n/\tilde{I}_n^{3/2}\right)a^3.$$

If $\tilde{J}_n$ and $\tilde{K}_n$ have the same order which is less than $\tilde{I}_n^{3/2}$ and the residual terms are negligible, we can apply Gram-Charlier expansion to the asymptotic distribution of the statistic T_n. Put $d_n^{-1} = \tilde{K}_n/\tilde{I}_n^{3/2}$. If

$$\lim_{n\to\infty} d_n\left|P_{n,\theta}\left\{\sqrt{\tilde{I}_n}\left(\hat{\theta}_n - \theta\right) \leq 0\right\} - 1/2\right| = 0$$

uniformly in any neighborhood of θ_0, then for $a > 0$

633

$$(5.2) \quad \varlimsup_{n\to\infty} d_n \left[P_{n,\theta_0} \left\{ \sqrt{\tilde{I}_n} \left(\hat{\theta}_n - \theta \right) \le a \right\} - \Phi(a) - \frac{3\tilde{J}_n + 2\tilde{K}_n}{6\tilde{I}_n^{3/2}} a^2 \phi(a) \right] \le 0.$$

For $a < 0$, a similar inequality as above holds. If the equality "$=$" of (5.2) holds, then $\hat{\theta}_n$ may be called second order asymptotically efficient. The necessary condition for second order asymptotic efficiency is that $\hat{\theta}_n$ has the following asymptotic moments:

$$E_\theta \left\{ \sqrt{\tilde{I}_n} \left(\hat{\theta}_n - \theta \right) \right\} \sim -\frac{3\tilde{J}_n + 2\tilde{K}_n}{3\tilde{I}_n^{3/2}} + o\left(\frac{1}{d_n}\right);$$

$$V_\theta \left\{ \sqrt{\tilde{I}_n} \left(\hat{\theta}_n - \theta \right) \right\} \sim 1 + o\left(\frac{1}{d_n}\right);$$

$$E_\theta \left\{ \sqrt{\tilde{I}_n} \left(\hat{\theta}_n - \theta \right) \right\}^3 \sim -\frac{3\tilde{J}_n + 2\tilde{K}_n}{\tilde{I}_n^{3/2}} + o\left(\frac{1}{d_n}\right).$$

If $\hat{\theta}_n$ has a smooth asymptotic distribution and $V_\theta \left\{ \sqrt{\tilde{I}_n} \left(\hat{\theta}_n - \theta \right) \right\}$ $\sim 1 + o(1/d_n)$, then the modified estimator $\hat{\theta}_n^*$ which is second order asymptotically median unbiased is second order asymptotically efficient. Let

$$\hat{\theta}_{ML}^* = \hat{\theta}_{ML} + \frac{\tilde{K}_n(\hat{\theta}_{ML})}{6n\tilde{I}_n(\hat{\theta}_{ML})^2}.$$

Under regularity conditions $\hat{\theta}_{ML}^*$ is second order asymptotically efficient. If the following generally hold:

$$\tilde{I}_n/n \to \overline{I} \quad (n \to \infty),$$
$$\tilde{J}_n/n \to \overline{J} \quad (n \to \infty),$$
$$\tilde{K}_n/n \to \overline{K} \quad (n \to \infty),$$

then it follows that $d_n = \sqrt{n}$. Since the second term of the asymptotic distribution has order of $1/\sqrt{n}$, a similar result as the i.i.d. case holds. But the fact does not always hold. For the purpose we consider the following examples.

Let $X_i = \theta z_i + U_i$, $i = 1, 2, \ldots$, where z_i's are constants and U_i's are independent identically distributed ramdom variables with a density $f(u)$. Putting

634

$$I = \int \left\{ \frac{d}{du} \log f(u) \right\}^2 f(u)\,du = \int \frac{\{f'(u)\}^2}{f(u)}\,du,$$

$$J = \int \left\{ \frac{d^2}{du^2} \log f(u) \right\} \left\{ \frac{d}{du} \log f(u) \right\} f(u)\,du$$

$$= \int \left[\frac{f''(u)f'(u)}{f(u)} - \frac{\{f'(u)\}^3}{\{f(u)\}^2} \right] du,$$

$$K = \int \left\{ \frac{d}{du} \log f(u) \right\}^3 f(u)\,du$$

$$= \int \frac{\{f'(u)\}^3}{\{f(u)\}^2}\,du,$$

we have for each i

$$I_i = z_i^2 I, \quad J_i = -z_i^3 J, \quad K_i = -z_i^3 K.$$

Then it follows that

$$\tilde{I}_n = \left(\sum_1^n z_i^2 \right) I, \quad \tilde{J} = -\left(\sum_1^n z_i^3 \right) J, \quad \tilde{K} = -\left(\sum_1^n z_i^3 \right) K.$$

By Lindeberg's condition we obtain

$$\max_{1 \le i \le n} z_i^2 \Big/ \sum_1^n z_i^2 \to 0 \quad (n \to \infty).$$

Since

$$\frac{|\sum_1^n z_i^3|}{(\sum_1^n z_i^2)^{3/2}} \le \frac{\max |z_i|}{(\sum_1^n z_i^2)^{1/2}} \to 0 \quad (n \to \infty),$$

it follows that

$$\tilde{J}_n \Big/ \tilde{I}_n^{3/2} \to 0 \quad (n \to \infty),$$

$$\tilde{K}_n \Big/ \tilde{I}_n^{3/2} \to 0 \quad (n \to \infty).$$

Then we can take $d_n = \left(\sum_1^n z_i^2 \right)^{3/2} \Big/ \left(\sum_1^n z_i^3 \right)$. If $\sum_1^n z_i^2/n \to m_2$ $(n \to \infty)$ and $\sum_1^n z_i^3/n \to m_3(\ne 0)$, then it follows that $d_n = O(\sqrt{n})$. Then the second term of the asymptotic distribution has order of 1.

635

But letting $z_i = i$, we have

$$\tilde{I}_n \doteqdot n(n+1)(2n+1)I/6 \sim n^3 I/3.$$

Hence if $\sqrt{n^3 I/3}(\hat{\theta}_n - \theta)$ is asymptotically normal with mean 0 and variance 1, then $\hat{\theta}_n$ is an asymptotically efficiency estimator. Since

$$\tilde{J}_n = -\frac{n^2(n+1)^2}{4}J \sim -\frac{n^4}{4}J,$$
$$\tilde{K}_n = -\frac{n^2(n+1)^2}{4}K \sim -\frac{n^4}{4}K,$$

it follows that $d_n = O(\sqrt{n})$. Hence in this case the second term of the asymptotic distribution has also order of $1/\sqrt{n}$.

If $z_i \to 0$ $(i \to \infty)$, then it follows by Lindeberg's conditions that the following must hold:

$$\tilde{I}_n = \left(\sum_1^n z_i^2\right) I \to \infty \quad (n \to \infty).$$

But it is possible that $\sum_1^n z_i^2 \to \infty$ $(n \to \infty)$ and $\sum_1^n z_i^3 < \infty$. Since $d_n = O(\tilde{I}_n^{1/2})$, letting $z_i = 1/\sqrt{i}$ we obtain $\tilde{I}_n = O(\log n)$. Hence $\sqrt{\log n}(\hat{\theta}_n - \theta)$ is asymptotically normal and the second term of the asymptotic distribution has order of $(\log n)^{-1/2}$.

Let X_i's be independently distributed random variables with the following density:

$$f(x_i, \theta) = \frac{1}{\theta^{p_i}\Gamma(p_i)}x_i^{p_i-1}e^{-x_i/\theta}, \quad x > 0,$$

where p_i's are known positive integer and θ is an unknown parameter. Since

$$\log f_i(X_i, \theta) = (p_i - 1)\log X_i - X_i/\theta - p_i \log \theta - \log \Gamma(p_i),$$
$$\frac{\partial}{\partial \theta}\log f_i(X_i, \theta) = \frac{X_i}{\theta^2} - \frac{p_i}{\theta},$$
$$\frac{\partial^2}{\partial \theta^2}\log f_i(X_i, \theta) = -\frac{2X_i}{\theta^3} + \frac{p_i}{\theta^2},$$

it follows that

636

$$I_i = E_\theta\left[\left(\frac{X_i}{\theta^2} - \frac{p_i}{\theta}\right)^2\right] = \frac{p_i}{\theta^2},$$

$$J_i = -E_\theta\left[\left(\frac{X_i}{\theta^2} - \frac{p_i}{\theta}\right)\left(\frac{2X_i}{\theta^3} - \frac{p_i}{\theta^2}\right)\right] = -\frac{2p_i}{\theta^3},$$

$$K_i = E_\theta\left[\left(\frac{X_i}{\theta^2} - \frac{p_i}{\theta}\right)^3\right] = \frac{2p_i}{\theta^3}.$$

Then we have

$$\tilde{I}_n = \sum_1^n p_i/\theta^2, \quad \tilde{J}_n = -2\sum_1^n p_i/\theta^3, \quad \tilde{K}_n = 2\sum_1^n p_i/\theta^3.$$

Since $d_n = (\sum_1^n p_i)^{1/2}$, the asymptotic distribution of the second order asymptotically efficient estimator $\hat{\theta}_n$ is given by

$$P_{n,\theta}\left\{\sqrt{\sum_1^n p_i}\left(\hat{\theta}_n - \theta\right)/\theta < a\right\} \sim \Phi(a) + \frac{1}{3\sqrt{\sum_1^n p_i}}a^2\phi(a).$$

The maximum likelihood estimator $\hat{\theta}_{ML}$ of θ is also given by $\sum_1^n X_i/\sum_1^n p_i$. Let

$$\hat{\theta}_{ML}^* = \left(1 + \frac{1}{3\sum_1^n p_i}\right)\hat{\theta}_{ML}.$$

Then $\hat{\theta}_{ML}^*$ is a second order asymptotically efficient estimator. Since in this case $\sum_1^n X_i$ is a sufficient statistic, the second order AMU estimator based on it must be second order asymptotically efficient. Indeed it is easily shown that the asymptotic distribution of $\hat{\theta}_{ML}^*$ agrees with the above bound of the asymptotic distributions.

(III) Let $\{X_i\}$ be defined recursively by

$$X_i = \theta X_{i-1} + U_i, \quad i = 1, 2, \ldots,$$

where $X_0 = 0$ and $\{U_i\}$ is a sequence of independent identically distributed real random variables having density f with mean 0 and variance σ^2. Let θ_0 be arbitrary but fixed in Θ. We consider the problem of testing hypothesis $\theta = \theta_1$ $(= \theta_0 + c_n^{-1}a)$ against alternative $\theta = \theta_0$. Then the test statistic of the most powerful test is given by

$$\sum_{i=1}^n Z_{ni} = \sum_{i=1}^n \log\frac{f(X_i - \theta_0 X_{i-1})}{f(X_i - \theta_1 X_{i-1})}.$$

637

By Taylor expansion we have

$$\sum_{i=1}^{n} Z_{ni} = \sum_{i=1}^{n} c_n^{-1} a X_{i-1} \frac{f'(U_i)}{f(U_i)}$$
$$- \frac{1}{2} \sum_{i=1}^{n} c_n^{-2} a^2 X_{i-1}^2 \left\{ \frac{\partial^2}{\partial U_i^2} \log f(U_i) \right\}$$
$$+ \sum_{i=1}^{n} R_{in}.$$

Since for $|\theta| > 1$, X_i is stochastically order of θ_i, the asymptotic distribution of $\sum_{i=1}^{n} Z_{ni}$ is not always written as an easy type. If $|\theta| < 1$, then at least X_i is asymptotically stationary. If $c_n = \sqrt{n}$, then $\sum_{i=1}^{n} Z_{ni}$ is asymptotically normal ([11]). From above it is shown that $\sqrt{n}(\hat{\theta}_n - \theta)$ is asymptotically normal with mean 0 and variance $(1 - \theta^2)/I\sigma^2$, where $I = \int \{f'(u)\}^2 / f(u) du$ if $\hat{\theta}_n$ is an asymptotically efficient estimator.

When $|\theta| < 1$, using asymptotic moments we can formally obtain the bound of the second order asymptotic distributions of second order AMU estimators of θ and show that a modified least squares estimator of θ is second order asymptotically efficient (Akahira [2]).

(IV) When the parameter θ is vector valued, we may decompose it to $\theta = \binom{\xi}{\eta}$ where ξ is a real number and η is a real vector, and we consider estimating ξ, while η is treated as a "nuisance" parameter. Then the second order asymptotic bound for any estimator $\hat{\xi}$ of ξ is given in a similar but somewhat complicated algebraic expression, and it can be shown that a modified ML estimator attains this bound under some regularity conditions ([3]).

References

[1] Akahira, M., "Asymptotic theory for estimation of location in non-regular cases, I: Order of convergence of consistent estimators," Rep. Stat. Appl. Res., JUSE, 22, (1975).

[2] Akahira, M., "A note on the second asymptotic efficiency of estimators in an autoregressive process," Rep. Univ. Electro-Comm. 26–1, (1975).

638

[3] Akahira, M. and Takeuchi, K., "On the second order asymptotic efficiency of estimators in multiparameter cases," Rep. Univ. Electro-Comm. 26–2, (1976).

[4] Chibisov, D. M., "On the normal approximation for a certain class of statistics," Proc. Sixth Berkeley Symp. on Math. Statist. and Prob. 1, (1972).

[5] Chibisov, D. M., "Asymptotic expansions for Neyman's $C(\alpha)$ tests," Proc. Second Japan-USSR Symp. on Probability Theory. Lecture Notes in Mathematics. Springer Verlag, Berlin, (1973).

[6] Fisher, R. A., "Theory of statistical estimation," Proc. Camb. Phil. Soc. 22, (1925).

[7] Pfanzagl, J., "Asymptotic expansions related to minimum contrast estimators," Ann. Statist. 1, (1973).

[8] Pfanzagl, J., "On asymptotically complete classes," Proc. Summer Research Inst. of Statistical Inference for Stochastic Processes, 2, (1975).

[9] Rao, C. R., "Asymptotic efficiency and limiting information," Proc. Fourth Berkeley Symp. on Math. Statist. and Prob. 1, (1961).

[10] Rao, C. R., "Efficient estimates and optimum inference procedures in large samples," J. Roy. Statist. Soc. 24 (B), (1962).

[11] Takeuchi, K., "Tokei-teki suitei no Zenkinriron (Asymptotic Theory of Statistical Estimation)," (In Japanese) Kyoiku-Shuppan, Tokyo, (1974).

K. Takeuchi
Faculty of Economics
University of Tokyo
Hongo, Bunkyo-ku
Tokyo, Japan

M. Akahira
Department of Mathematics
University of Electro-Communications
Chofugaoka, Chofu-shi
Tokyo, Japan

Ann. Inst. Statist. Math.
29 (1977), Part A, 397–406

EXTENSION OF EDGEWORTH TYPE EXPANSION OF THE DISTRIBUTION OF THE SUMS OF I.I.D. RANDOM VARIABLES IN NON-REGULAR CASES

Kei Takeuchi and Masafumi Akahira*

(Received Oct. 21, 1976; revised July 7, 1977)

Abstract

The asymptotic expansions of the distributions of the sums of independent identically distributed random variables are given by Edgeworth type expansions when moments do not necessarily exist, but when the density can be approximated by rational functions.

1. Introduction

Asymptotic distributions for the sums of independent and identically distributed (i.i.d.) random variables has been extensively studied for many years ([1]). Suppose that $X_1, X_2, \cdots, X_n$ is a sequence of i.i.d. random variables with mean μ and the distribution function $F(x)$. Then the asymptotic expansion for the distribution of $Y_n = \left(\sum_1^n X_i - n\mu \right) \Big/ \sqrt{n}$ up to the order $n^{-m/2+1}$ is given by the classical Edgeworth expansion when F has the moments up to the mth order as well as the continuous density ([2], [3]). The purpose of the present paper is to extend the result to the case when the moments do not necessarily exist assuming only that the density function $f(x)$ of F is approximated by some appropriate rational functions as $|x| \to \infty$. Put, for some real constants M_n and $V_n \ (>0)$, $Z_n = \left(\sum_1^n X_i - M_n \right) \Big/ V_n$. Then, as is well known, if a $G(x)$ is a limiting distribution of a sequence of distributions of Z's, it is necessarily a stable distribution. We shall show in the next section that if this is the case then under the assumption stated above V_n must be either n or $\sqrt{n \log n}$ or $\sqrt{n}$ and the corresponding leading term of the asymptotic expansion is the Cauchy, or the unsymmetric stable distribution with characteristic exponent 1, or the normal distribution, respectively. The second term is shown to be of order either

---

* Supported in part by the Sakkokai Foundation.

                 KEI TAKEUCHI AND MASAFUMI AKAHIRA

$1/n$, $(\log n)/n$, $1/\log n$ or $1/\sqrt{n}$ and the density function corresponding to the second term are also given. Generally the relative magnitudes of the successive terms are complicated and is difficult to write down explicitly but may be obtained for each particular case. The details of the computation are given in [4] and may be referred to for further examples. Further the similar results for fractional characteristic exponents are given in [5] (see also Cramér [6] for this case).

2. Results

We shall obtain the Edgeworth type expansions for the sums of random variables not necessarily with finite moments but with the density $f(x)$ which satisfies the following condition:

There exist two rational functions $g^+(x)$ and $g^-(x)$ such that for a positive integer m $(\geqq 2)$ and some $0<\delta<1$

(a) $\lim\limits_{x\to\infty} x^{m+1+\delta}|f(x)-g^+(x)|=0$;

(b) $\lim\limits_{x\to-\infty} |x|^{m+1+\delta}|f(x)-g^-(x)|=0$.

Remark 1. m depends on the order of the expansion and should not be smaller than 2 in order that we have meaningful expansions.

Remark 2. It should be remarked that since the distribution has a density $f(x)$, we have

$$\sup_{|t|>\rho} |\phi(t)|<1$$

for any $\rho>0$.

We shall first show that the following proposition holds.

PROPOSITION. $g^+(x)$ and $g^-(x)$ may be taken of the form

$$g^+(x)=\sum_{j=2}^{m+1} \frac{\alpha_j}{(x+1)^j} \; ; \qquad g^-(x)=\sum_{j=2}^{m+1} \frac{\beta_j}{(x-1)^j} ,$$

where α_j and β_j are real constants.

PROOF. We first decompose $g^+(x)$ into the sums of partial fractions such as

$$g^+(x)=\sum_i \sum_j \frac{B_{ij}}{(x+D_i)^j}+\sum_i \sum_j \frac{E_{ij}}{(x^2+F_i x+G_i)^j} .$$

But each term in above expression can be approximated by a sum of the terms of the type $\beta_j/(x+1)^j$ up to order x^{-m-1} when x is large. Therefore $g^+(x)$ may be replaced by a sum of such terms. Similar is true of $g^-(x)$.

Now let us denote that

$$\phi_j(t)=\int_0^\infty \frac{(j-1)e^{itx}}{(1+x)^j}\,dx\ ,$$

then we have

$$\int_{-\infty}^0 \frac{(j-1)e^{itx}}{(1-x)^j}\,dx=\phi_j(-t)=\overline{\phi_j(t)}\ .$$

Thus if the density $f(x)$ satisfies the condition above, the characteristic function can be expressed as

$$(1)\qquad \phi(t)=\int e^{itx}f(x)dx=\sum_j \alpha_j\phi_j(t)+\sum_j \beta_j\phi_j(-t)+\psi(t)$$

and since

$$\int_0^\infty x^m\,|f(x)-g^+(x)|\,dx<\infty\ ;\qquad \int_{-\infty}^0 |x|^m\,|f(x)-g^-(x)|\,dx<\infty\ ,$$

the remainder term $\psi(t)$ can be differentiated m times at $t=0$, so that we have

$$(2)\qquad \psi(t)=\sum_{p=0}^m c_p t^p+o(|t|^m)\ .$$

It is shown by Takeuchi and Akahira [4] that for small $|t|>0$ and $j\geqq 2$

$$(3)\qquad \phi_j(t)=(j-1)\left(\int_{|t|}^\infty u^{-j}e^{-iu}du\right)|t|^{j-1}e^{-it}$$

$$=(j-1)\left\{\sum_{k=0}^{j-2} i^k a_k |t|^k-\frac{i^{j-1}}{(j-1)!}|t|^{j-1}\log|t|+\alpha_j^*|t|^{j-1}\right.$$

$$\left.+i\beta_j^*|t|^{j-1}-\frac{i^j}{j}|t|^j\right\}\left(1-it-\frac{1}{2}t^2+\frac{i}{6}t^3+\cdots\right)$$

$$=1+\sum_{k=1}^m a_{jk}|t|^k+\sum_{k=1}^m b_{jk}t|t|^{k-1}+\sum_{k=j}^{m+1} c_{jk}|t|^{k-1}\log|t|$$

$$+\sum_{k=j}^m d_{jk}t|t|^{k-1}\log|t|+o(|t|^m)\ ,$$

where $a_k=-1/k!(k-j+1)$ and α_j^* and β_j^* are real constants equal to the real part and the imaginary part of

$$\int_1^\infty u^{-j}e^{iu}du+\int_0^1 u^{-j}\left\{e^{iu}-1-iu+\frac{u^2}{2}+\cdots-\frac{(iu)^{j-1}}{(j-1)!}\right\}du+\sum_{k=0}^{j-2}\frac{i^k}{k!(k-j+1)}$$

respectively, and a_{jk}, b_{jk}, c_{jk} and d_{jk} are certain complex numbers. Hence we have from (1), (2) and (3)

$$(4) \qquad \phi(t)=1+\sum_{k=1}^{m} A_k |t|^k+\sum_{k=1}^{m} B_k t|t|^{k-1}+\sum_{j}\sum_{k=j}^{m+1} c_{jk}|t|^{k-1}\log|t|$$

$$+\sum_{j}\sum_{k=j}^{m} d_{jk}t|t|^{k-1}\log|t|+o(|t|^m) \, ,$$

where A_k and B_k are certain complex numbers.

Let $X_1, X_2,\cdots, X_n,\cdots$ be a sequence of i.i.d. random variables with the characteristic function $\phi(t)$. We define $z_n=\left(\sum_{i=1}^{n} X_i-M_n\right)\Big/V_n$ whose characteristic function $\phi_n(t)$ is expressed as

$$\phi_n(t)=\mathrm{E}\,(e^{itZ_n})=\left\{\phi\Big(\frac{t}{V_n}\Big)\right\}^n\exp\Big(-i\frac{M_n}{V_n}t\Big) \, .$$

Since

$$\log\phi_n(t)=n\log\phi\Big(\frac{t}{V_n}\Big)-i\frac{M_n}{V_n}t \, ,$$

it follows from (4) that

$$(5)\quad \log\phi_n(t)=-i\frac{M_n}{V_n}t+n\left[\left\{\sum_{k=1}^{m} A_k\frac{|t|^k}{V_n^k}+\sum_{k=1}^{m} B_k\frac{t|t|^{k-1}}{V_n^k}+\sum_{k=1}^{m} C_k\frac{|t|^k}{V_n^k}\right.\right.$$

$$\cdot(\log|t|-\log V_n)+\sum_{k=1}^{m} D_k\frac{t|t|^{k-1}}{V_n^k}(\log|t|-\log V_n)$$

$$\left.\left.+\cdots\right\}-\frac{1}{2}\{\ \ \}^2+\frac{1}{3}\{\ \ \}^3-\cdots\right] \, .$$

It is noted that since $\phi_n(t)$ is a characteristic function B_1 is a pure imaginary if not equal to zero, and that term can be cancelled out by an appropriate choice of M_n. Without loss of generality we put $B_1=0$.

In order to obtain V_n and the leading term of $\phi_n(t)$ it is enough to consider the following cases:

(i) $A_1\neq0$,

(ii) $A_1=0$, $A_2\neq0$,

(iii) $A_1=A_2=0$, $C_1=D_1=0$, $C_2>0$, $D_2=0$,

(iv) $A_1=A_2=0$, $C_1>0$, $D_1=0$.

Since the asymptotic distribution of Z_n is stable, it is seen from (5) that each case is as follows:

Case (i). $V_n=n$ necessarily holds and then $C_1=0$. If $D_1=0$, then the leading term of $\phi_n(t)$ is given by $e^{A_1|t|}$. If $D_1\neq0$, then M_n must be $O(n\log n)$. When $D_1\neq0$, we have the characteristic function $\phi_n(t)$ with the leading term of the type $\exp\{A_1|t|+D_1 t\log|t|\}$ which is the unsymmetric stable law of characteristic exponent 1.

Case (ii). $V_n=n$ necessarily holds and then $B_2=C_1=C_2=D_1=D_2=0$.

Hence the leading term of $\phi_n(t)$ is given by $e^{A_2 t^2}$.

Case (iii). $V_n = \sqrt{n \log n}$ necessarily holds. Hence the leading term of $\phi_n(t)$ is given by $e^{-C_2 t^2}$.

Case (iv). $V_n = n \log n$ and the leading term of $\phi_n(t)$ must be $e^{-C_1 |t|}$. If $C_1 > 0$, then F belongs to the domain of attraction of the Cauchy distribution, and the density $f(x)$ is expressed as

$$f(x) = \begin{cases} \alpha(1+x)^{-2} + o(x^{-2}) & \text{as } x \to \infty , \\ \beta(1-x)^{-2} + o(x^{-2}) & \text{as } x \to -\infty , \end{cases}$$

with $\alpha + \beta > 0$. But this implies that F belongs to the domain of normal attraction of the stable law with characteristic exponent 1 (see [1], p. 181), which contradicts the assumption $V_n = n \log n$. Thus the case (iv) can be excluded from our consideration, except for the trivial case $C_1 = 0$.

In other cases the leading term of $\phi_n(t)$ is equal to 1 which is the degenerate case.

When the leading term is obtained in each case, the remaining terms may be arranged in the order of magnitude. A few examples are in order to show the point.

Example 1. Let X_i's $(i = 1, 2, \cdots)$ be i.i.d. random variables with the density function $f(x)$ given by

$$f(x) = \frac{1}{2(x^2 + 1 + \sqrt{x^2 + 1})} .$$

Since $f(x)$ is a symmetric function and $f(x) \sim 1/2x^2$ as $x \to \infty$, $F(y)$ belongs to the domain of normal attraction of a stable law with characteristic exponent 1 ([1]).

Note that X_i can be expressed as $X_i = (1/(1 - U_i) - 1/U_i)/2$, where U_i is distributed uniformly in $(0, 1)$. Putting

$$g(x) = \frac{1}{2(1 + x^2)(1 + x^2 + \sqrt{1 + x^2})} ;$$

$$h(x) = \frac{1}{2(1 + x^2)^{3/2}} - \frac{1}{2(1 + |x|)^3}$$

we have

$$(6) \qquad f(x) = \frac{1}{2(1 + x^2)} - \frac{1}{2(1 + x^2)^{3/2}} + g(x)$$

$$= \frac{1}{2(1 + x^2)} - \frac{1}{2(1 + |x|)^3} - h(x) + g(x) .$$

 KEI TAKEUCHI AND MASAFUMI AKAHIRA

If we have $m=2$, $0<\delta<1$ and $g^{\pm}(x)=1/2(1+x^2)-1/2(1+|x|)^3$, then (a) and (b) are satisfied. First we obtain

$$\int_0^x y^2 h(y)dy = \frac{1}{2}\log(\sqrt{x^2+1}+x) - \frac{x}{2\sqrt{1+x^2}} - \frac{1}{2}\log(1+x)$$

$$- \left(\frac{1}{1+x}-1\right) + \frac{1}{(1+x)^2} - 1$$

$$= \frac{1}{2}\log\{(\sqrt{x^2+1}+x)/(x+1)\} - \frac{x}{2\sqrt{1+x^2}} - \frac{x}{(1+x)^2} \;.$$

Hence it follows that

$$\int_{-\infty}^{\infty} x^2 h(x)dx = \log 2 \;.$$

Since

$$\int_{-\infty}^{\infty} h(x)dx = 1/2 \;; \qquad \int_{-\infty}^{\infty} xh(x)dx = 0 \;,$$

it is seen that

$$(7) \qquad \int_{-\infty}^{\infty} e^{itx}h(x)dx = \frac{1}{2} + \frac{\log 2}{2}t^2 + o(t^2) \;.$$

Since

$$\int_{-\infty}^{\infty} g(x)dx = 2 - \frac{\pi}{2} \;; \qquad \int_{-\infty}^{\infty} x^2 g(x)dx = \frac{\pi}{2} - 1 \;,$$

we have

$$(8) \qquad \int_{-\infty}^{\infty} e^{itx}g(x)dx = 2 - \frac{\pi}{2} - \frac{1}{2}\left(\frac{\pi}{2}-1\right)t^2 + o(t^2) \;.$$

It follows by Takeuchi and Akahira [4] that

$$(9) \qquad \int_{-\infty}^{\infty} \frac{e^{itx}}{(1+|x|)^3}dx = 1 + t^2\log|t| - \gamma t^2 + o(t^2) \;,$$

where γ is some constant. From (6)–(9) we obtain

$$\phi(t) = \int_{-\infty}^{\infty} e^{itx}f(x)dx$$

$$= \frac{\pi}{2}e^{-|t|} - \frac{1}{2} - \frac{t^2}{2}\log|t| + \frac{\gamma}{2}t^2 - \frac{1}{2} - \frac{\log 2}{2}t^2 + 2 - \frac{\pi}{2}$$

$$- \frac{1}{2}\left(\frac{\pi}{2}-1\right)t^2 + o(t^2)$$

$$= 1 - \frac{\pi}{2}|t| - \frac{t^2}{2}\log|t| + \frac{1}{2}(1-\gamma-\log 2)t^2 + o(t^2) \;.$$

Letting $\phi_n(t)=\mathrm{E}\,[\exp\,itZ_n]$ with $Z_n=\sum_1^n X_i/n$, we obtain

$$\log\,\phi_n(t)=n\,\log\,\phi\left(\frac{t}{n}\right)$$

$$=-\frac{\pi}{2}|t|-\frac{t^2}{2n}(\log\,|t|+\log\,n)+\frac{1}{2}(1-\gamma-\log\,2)\frac{t^2}{n}$$

$$-\frac{\pi^2}{8n}t^2+o\left(\frac{1}{n}\right)\,.$$

Since

$$\phi_n(t)=e^{-(\pi/2)|t|}\left[1-\frac{\log\,n}{2n}t^2-\frac{1}{2n}t^2\,\log\,|t|\right.$$

$$\left.+\frac{1}{8n}\{4(1-\gamma-\log\,2)-\pi^2\}t^2\right]+o\left(\frac{1}{n}\right)\,,$$

the asymptotic expansion of the distribution of Z_n agrees essentially with the Cauchy distribution. The second term in the expansion has the form

$$\frac{1}{n}(a\,\log\,n+b)f_1(x)+cf_2(x)\,,$$

where $f_1(x)$ and $f_2(x)$ are the functions that are Fourier transforms of

$$t^2e^{-(\pi/2)|t|}\qquad\text{and}\qquad(t^2\,\log\,|t|)e^{-(\pi/2)|t|}$$

respectively. Hence $f_1(x)$ and $f_2(x)$ are equal to $-g''(x)$ and $-h''(x)$, respectively, where

$$g(x)=\frac{1}{\pi(c'^2+x^2)}\,;$$

$$h(x)=\frac{1}{2\pi(1+x^2)}\{-\log\,(1+x^2)-2x\,\tan^{-1}\,x+C''\}$$

with certain constants c' and C''.

Example 2. Let X_i's $(i=1, 2, \cdots)$ be i.i.d. random variables with the normal density with mean 0 and variance 1. Let $Y_i=1/X_i$ $(i=1, 2, \cdots)$. Then the density function $f(y)$ of Y_i is given by

$$f(y)=\frac{1}{\sqrt{2\pi}}\frac{1}{y^2}e^{-1/2y^2}\,.$$

Since $f(y)$ is a symmetric function and $f(y)\sim1/(\sqrt{2\pi}y^2)$ as $y\to\infty$, $F(x)$ belongs to the domain of normal attraction of a stable law with the characteristic exponent 1 ([1]). Put

$$(10) \qquad g(y) = f(y) - \frac{\sqrt{2}}{\sqrt{\pi}\,(1+2y^2)}\,.$$

If we have $m=2$, $0<\delta<1$ and $g^{\pm}(x)=2/(\sqrt{\pi}\,(1+2x^2))$, then (a) and (b) are satisfied. First we have

$$(11) \qquad \int_{-\infty}^{\infty} g(y)dy = 1-\sqrt{\pi}\;;$$

$$(12) \qquad \int_{-\infty}^{\infty} yg(y)dy = 0\,.$$

Further we obtain

$$(13) \qquad \int_{-\infty}^{\infty} y^2 g(y)dy = \frac{1}{\sqrt{2\pi}} \int_{-\infty}^{\infty} \left(e^{-1/2y^2} - \frac{2y^2}{1+2y^2} \right) dy$$

$$= \sqrt{\frac{2}{\pi}} \int_{0}^{\infty} \left\{ \frac{1}{x^2} e^{-x^2/2} - \frac{2}{x^2(2+x^2)} \right\} dx$$

$$= \sqrt{\frac{2}{\pi}} \int_{0}^{\infty} (2u)^{-3/2} \left(e^{-u} - \frac{1}{1+u} \right) du$$

$$= \frac{1}{2\sqrt{\pi}} \left\{ -2\Gamma\left(\frac{1}{2}\right) + 2 \int_{0}^{\infty} \frac{u^{-1/2}}{(1+u)^2} du \right\}$$

$$= -1 + \frac{1}{\sqrt{\pi}} \int_{0}^{\infty} v^{1/2}(1-v)^{-1/2}dv$$

$$= -1 + \frac{1}{\sqrt{\pi}} B\left(\frac{3}{2},\frac{1}{2}\right)$$

$$= -1 + \frac{\sqrt{\pi}}{2}\,.$$

From (10)–(13) we have

$$\phi(t) = \int_{-\infty}^{\infty} e^{ity} f(y)dy$$

$$= \int_{-\infty}^{\infty} \frac{\sqrt{2}\,e^{ity}}{\sqrt{\pi}\,(1+2y^2)} dy + \int_{-\infty}^{\infty} e^{ity} g(y)dy$$

$$= \sqrt{\pi} \int_{-\infty}^{\infty} \frac{\exp{(i(t/\sqrt{2}\,)x)}}{\pi(1+x^2)} dx + \int_{-\infty}^{\infty} e^{ity} g(y)dy$$

$$= \sqrt{\pi}\,\exp\left(-\frac{|t|}{\sqrt{2}}\right) + 1 - \sqrt{\pi} - \frac{1}{2}\left(-1+\frac{\sqrt{\pi}}{2}\right)t^2 + o(t^2)$$

$$= 1 - \sqrt{\frac{\pi}{2}}\,|t| + \frac{t^2}{2} + o(t^2)\,.$$

Letting $\phi_n(t) = \mathrm{E}\,[\exp{(itZ_n)}]$ with $Z_n = \sum_{1}^{n} X_i/n$, we obtain

$$\log \phi_n(t) = n \log \phi\left(\frac{t}{n}\right) = -\sqrt{\frac{\pi}{2}}|t| + \left(\frac{\pi}{4} + \frac{1}{2}\right)\frac{t^2}{n} + o\left(\frac{1}{n}\right) .$$

Since

$$\phi_n(t) = \left(\exp\left(-\sqrt{\frac{\pi}{2}}|t|\right)\right)\left\{1 + \left(\frac{\pi}{4} + \frac{1}{2}\right)\frac{t^2}{n}\right\} + o\left(\frac{1}{n}\right) ,$$

the asymptotic expansion of the distribution of Z_n agrees essentially with the case when the density function is $\sqrt{2}/\{\sqrt{\pi}\,(\pi + 2x^2)\}$ and the next term is $(a + bx^2 + cx^4)/\{\sqrt{2\pi}(1 + x^2)\}^3$.

Example 3. X_i's $(i = 1, 2, \cdots)$ are distributed according to t-distribution with 2-degrees of freedom i.e. with the density

$$f(x) = \frac{1}{2(1 + x^2)^{3/2}} .$$

Note that the distribution has no finite variance, still it belongs to the domain of attraction of the normal.   First we have

$$f(x) = h(x) + \frac{1}{2(1 + |x|)^3} ,$$

where $h(x)$ is given in Example 1.

The conditions (a) and (b) are satisfied with $m = 2$, $0 < \delta < 1$ and

$$g^{\pm}(x) = \frac{1}{2(1 + |x|)^3} .$$

From Example 1 we obtain

$$\phi(t) = \int_{-\infty}^{\infty} e^{itx} f(x) dx = 1 + \frac{1}{2} t^2 \log |t| - \frac{\log 2 + \gamma}{2} t^2 + o(t^2) ,$$

where γ is some constant.   Letting $\phi_n(t) = \mathrm{E}\,[\exp\,(itZ_n)]$ with $Z_n = \sum_1^n X_i$ $/\sqrt{n \log n}$, we have

$$\log \phi_n(t) = n \log \phi\left(\frac{t}{\sqrt{n \log n}}\right)$$
$$= \frac{1}{2}\frac{t^2 \log |t|}{\log n} - \frac{1}{4}\left\{1 + \frac{\log \log n}{\log n} + \frac{2(\log 2 + \gamma)}{\log n}\right\} t^2 + o\left(\frac{1}{\log n}\right)$$
$$= e^{-(c_n/4)t^2}\left\{1 + \frac{1}{2}\frac{t^2 \log |t|}{\log n} + o\left(\frac{1}{n \log n}\right)\right\} ,$$

where

$$c_n = 1 + \frac{\log \log n}{\log n} + \frac{2(\log 2 + \gamma)}{\log n} .$$

                KEI TAKEUCHI AND MASAFUMI AKAHIRA

The leading term of the characteristic function $\phi_n(t)$ is equal to $e^{-t^2/4}$ which corresponds to the normal distribution. And the second term $g(x)$ is the Fourier transform of

$$\frac{1}{2}(t^2 \log |t|)e^{-t^2/4} \; ;$$

that is,

$$g(x) = -\pi G''(x) \; ,$$

where

$$G(x) = -\frac{2}{\sqrt{\pi}} e^{-x^2} \left\{ \int_0^x \left(\int_0^x e^{-x^2} dx \right) e^{x^2} dx + C \right\}$$

with

$$C = -\frac{1}{4\sqrt{\pi}} \int_{-\infty}^{\infty} (\log |t|)e^{-t^2/4} dt \; .$$

Acknowledgement

The authors wish to thank the referee for his kind comments.

UNIVERSITY OF TOKYO
UNIVERSITY OF ELECTRO-COMMUNICATIONS

REFERENCES

[1] Gnedenko, B. V. and Kolmogorov, A. N. (1954). *Limit Distributions for Sums of Independent Random Variables*, Addison-Wesley, Cambridge, Massachussets, (Translated from Russian).

[2] Ibragimov, I. A. and Linnik, Yu. V. (1971). *Independent and Stationary Sequences of Random Variables*, Wolters-Noordhoff, the Netherlands, (Translated from Russian).

[3] Petrov, V. V. (1975). *Sums of Independent Random Variables*, Springer-Verlag, (Translated from Russian).

[4] Takeuchi, K. and Akahira, M. (1976). On Gram-Charlier-Edgeworth type expansion of the sums of random variables (I), *Rep. Univ. Electro-Comm.* **27**-1, 95–115.

[5] Takeuchi, K. and Akahira, M. (1976). On Gram-Charlier-Edgeworth type expansion of the sums of random variables (II), *Rep. Univ. Electro-Comm.* **27**-1, 117–123.

[6] Cramér, H. (1963). On asymptotic expansions for sums of independent random variables with a limiting stable distribution, *Sankhyā*, Ser. A, **25**, 13–24.

Rep. Univ. Electro-Comm. 28-2, (Sci. & Tech. Sect.), pp. 259−269, February, 1978

On Gram-Charlier-Edgeworth Type Expansion of the Sums of Random Variables (III): Multivariate Cases*

Kei TAKEUCHI** and Masafumi AKAHIRA***

Abstract

Gram-Charlier-Edgeworth type expansion of the sums of i.i.d. random variables without higher order moments was discussed in the previous papers. In this paper it is shown that the expansion is extended to the multidimensional cases.

1. Introduction

The authors discussed in the previous papers ([1],[2],[3]) Gram-Charlier-Edgeworth (G-C-E) type expansions for the sums of i.i.d. random variables which do not necessarily have moments up to the required order. Now it is extended to the multidimensional case. Suppose that X_1, X_2, ..., X_n, ... is a sequence of i.i.d. p-dimensional continuous random vectors. Now suppose that the density function $f(x)$ is of the following form:

$$f(x) = \frac{g(\tau)}{(1+r)^{\beta+p+1}}$$

where $r = (x'x)^{1/2}$, τ is vector of $p-1$ angles of the x-vector and β is a positive constant. Then the characteristic function of X is computed as

$$\phi(t) = \int e^{it'x} f(x)\,dx$$

$$= \int\int \frac{r^p g(\tau)}{(1+r)^{\beta+p+1}} \exp\left\{ir\rho h(\theta,\tau)\right\} dr d\tau,$$

where $\rho = (t't)^{1/2}$ and θ is the vector of angles of t. If we denote

$$\widetilde{\phi}_{p,\beta}(u) = \int \frac{r^p}{(1+r)^{\beta+p+1}} \exp(iur)\,dr,$$

we get

$$\phi(t) = \int \widetilde{\phi}_{p,\beta}(\rho h(\theta,\tau))\,d\tau.$$

* Received on October 6, 1977
** University of Tokyo
*** Statistical Laboratory, University of Electro-Communications

Therefore the characteristic function of $Z_n = a_n^{-1} \sum_1^n X_i - b_n$ is given by

$$\left\{ \int \widetilde{\phi}_{p,\beta} \left(a_n^{-1} \rho h \left(\theta, \tau \right) \right) d\tau \right\}^n \exp \left(-i b_n' t \right).$$

Then expanding $\widetilde{\phi}_{p,\beta} \left(\cdot \right)$ near the origin, as was done in the previous paper, we get the asymptotic expansion of the characteristic function of Z_n and thence its density. Explicit form of the asymptotic density for some special cases are given.

Then the results are generalized to the case when

$$f \left(x \right) = g_1 \left(x \right) + \ldots + g_p \left(x \right) + h \left(x \right),$$

where $g_i \left(x \right)$ are all above type and $h \left(x \right)$ satisfies the condition

$$\lim_{|x'x| \to \infty} |x'x|^{\frac{r+p}{2}} h \left(x \right) = 0$$

for some $r > 0$.

2. Results

Suppose that the density is of the form

$$f \left(x \right) = c_1 f_1 \left(x \right) + c_2 f_2 \left(x \right) + \ldots + c_k f_k \left(x \right) + g \left(x \right),$$

where $f_i \left(i = 1, 2, \ldots, k \right)$ are functions of the form given in section 1. If for some r

$$| x'x |^{\frac{r+2}{2}} g \left(x \right) \to 0 \quad \left(x'x \to \infty \right),$$

then it is possible to extend the characteristic function

$$\phi \left(t \right) = \int f \left(x \right) e^{it'x} dx$$

around $t = 0$ up to appropriate order. Hence it is seen that the asymptotic expansion of the distribution of Z_n holds. In the subsequent discussion we deal with multivariate t-distribution cases and more general cases.

Let $X_1, X_2, \ldots, X_n, \ldots$ be a sequence of i.i.d. random vectors having a multivariate t-distribution with a density

$$f \left(x \right) = c \left(1 + x'Ax \right)^{-\frac{\nu+p}{2}}, \quad \nu > 0,$$

where A is a positive definite matrix.

Suppose that Y is a random vector having p-dimensional normal distribution with mean vector 0 and covariance matrix A^{-1}, and νS^2 is distributed like χ^2 with ν degree of freedom, and νS^2 and Y are mutually independent. Then it is seen that the t-distribution is equal to the distribution of Y/S. Since the conditional distribution of Y/S given S is normal with mean vector 0 and covariance matrix A^{-1}/S^2, it follows that the characteristic function of the t-distribution is given by

$$\phi(t) = E(\exp it'Y/S)$$

$$= E\left[\exp\left\{-\frac{1}{2}t'A^{-1}t\right)/S^2\right\}\right]$$

$$= \frac{1}{2\Gamma(\frac{\nu}{2})}\int_0^\infty \left(\frac{S^2}{2}\right)^{\frac{\nu}{2}-1}\exp\left\{-\frac{1}{2}\left(S^2 + \frac{t'A^{-1}t}{S^2}\right)\right\}dS^2 .$$

Hence it is easily seen that $\phi(t)$ is a function only through and $t'A^{-1}t$, Put $\phi(t) = \psi_\nu(\tau)$, where $\tau = (t'A^{-1}t)$. Since the form of $\phi(t)$ does not depend on the dimension p, it follows that in the case $p = 1$, $\psi_\nu(\sqrt{\nu}|\tau|)$ is the characteristic function of the t-distribution with ν degree of freedom. When ν is odd number, ψ_ν is easily obtained as follows: for $\tau > 0$

$$\psi_1(\tau) = \frac{1}{\pi}\int \frac{e^{i\tau x}}{1+x^2}\,dx = e^{-\tau} ;$$

$$\psi_3(\tau) = (1+\tau)e^{-\tau} ;$$

$$\psi_5(\tau) = \left\{1 + \tau + (\tau^2/3)\right\}e^{-\tau} ;$$

$$\psi_7(\tau) = \left\{1 + \tau + (2\tau^2/5) + (\tau^3/15)\right\}e^{-\tau} ;$$

When $\nu = 1$, $Z_n = \sum_i X_i/n = \overline{X}$ is identically multivariate Cauchy.

When $\nu = 3$, the characteristic function of $Z_n = \sum_i X_i/n = \overline{X}$ is expanded as follows:

$$\left\{\psi_3\left(\frac{\tau}{\sqrt{n}}\right)\right\}^n \sim e^{-\frac{\tau^2}{2}}\left\{(1 + \frac{\tau^3}{3\sqrt{n}} - \frac{\tau^4}{4n} + \frac{\tau^6}{18n} + o\left(\frac{1}{n}\right))\right\}$$

$$= e^{-\frac{t'A^{-1}t}{2}}\left\{1 + \frac{(t'A^{-1}t)}{3n} - \frac{(t'A^{-1}t)^2}{4n} + \frac{(t'A^{-1}t)^3}{18n} + o\left(\frac{1}{n}\right)\right\} \qquad (2.1)$$

Hence the asymptotic distribution of Z_n is normal with covariance matrix A^{-1}. In order to get the below the second term of (2.1) we consider the case when $A = I$. This does not lose the generality since the general case is reduced to the above by appropriate linear transformation. Then it follows that

$$\left\{\psi_3\left(\frac{\tau}{\sqrt{n}}\right)\right\}^n \sim e^{-\frac{1}{2}\sum_i t_i^2}\left\{1 + \frac{1}{3\sqrt{n}}(\sum_i t_i^2)^{1/2} - \frac{(\sum_i t_i^2)}{4n} + \frac{(\sum_i t_i^2)^3}{18n}\right.$$

$$\left. + o\left(\frac{1}{n}\right)\right\}$$

In the asymptotic expansion of the density corresponding to the characteristic function, the corresponding term to $(t_i^2\, t_j^2 \cdots)\exp(-\frac{1}{2}\sum_i t_i^2)$ is given by

Kei TAKEUCHI and Masafumi AKAHIRA

$$\left(- \frac{\partial^2}{\partial x_i^2} \right) \left(- \frac{\partial^2}{\partial x_j^2} \right) \cdots \left(\frac{1}{\sqrt{2n}} \right)^p \exp\left(-\frac{1}{2} \sum_i x_i^2 \right)$$

For the term of order $n^{-\frac{1}{2}}$ putting

$$G(x) = \frac{1}{(2\pi)^p} \int \cdots \int \left(\sum_j t_j^2 \right)^{1/2} \left\{ \exp\left(-\frac{1}{2} \sum_j t_j^2 - \sum_j it_j x_j \right) \right\} \prod_j dt_j$$

the corresponding term to $\left(\sum_i t_i^2 \right)^{3/2} \exp\left(-\frac{1}{2} \sum_i t_i^2 \right)$ is equal to $- \sum_i \frac{\partial^2}{\partial x_i^2} G(x)$.

In order to obtain $G(x)$ we consider the case when $p = 2$. Making a transformation $t_1 = \tau \cos\theta$, $t_2 = \tau \sin\theta$, $x_1 = r \cos\alpha$, $x_2 = r \sin\alpha$, we get

$$G(x_1, x_2) = \frac{1}{(2\pi)^2} \int_0^{2\pi} \int_0^{\pi} \tau^2 \exp\left\{ -\frac{\tau^2}{2} - i\tau r \cos(\theta - \alpha) \right\} d\tau d\theta$$

$$= \frac{1}{(2\pi)^2} \int_0^{\infty} \int_{-\pi}^{\pi} \tau^2 \exp\left(-\frac{\tau^2}{2} - i\tau r \cos\theta \right) d\theta d\tau.$$

Since $\cos\left(\theta + \frac{\pi}{2} \right) = -\cos\theta$, it follows that

$$G(x_1, x_2) = \frac{1}{(2\pi)^2} \int_0^{\infty} \int_{-\frac{\pi}{2}}^{\frac{\pi}{2}} \tau^2 \left\{ \exp\left(-\frac{\tau^2}{2} - i\tau r \cos\theta \right) \right.$$

$$\left. + \exp\left(-\frac{\tau^2}{2} + i\tau r \cos\theta \right) \right\} d\theta d\tau$$

$$= \frac{1}{(2\pi)^2} \int_{-\frac{\pi}{2}}^{\frac{\pi}{2}} \int_{-\infty}^{\infty} \tau^2 \exp\left(-\frac{\tau^2}{2} - i\tau r \cos\theta \right) d\tau d\theta.$$

Since for any real number y

$$\frac{1}{2\pi} \int_{-\pi}^{\pi} \tau^2 e^{-\frac{\tau^2}{2} - i\tau y} d\tau = \left(-\frac{d^2}{dy^2} \right) \frac{1}{\sqrt{2\pi}} e^{-\frac{y^2}{2}} = (1 - y^2) \frac{1}{\sqrt{2\pi}} e^{-\frac{y^2}{2}},$$

it is seen that

$$G(x_1, x_2) = \frac{1}{(2\pi)^{3/2}} \int_{-\frac{\pi}{2}}^{\frac{\pi}{2}} (1 - r^2 \cos^2\theta) e^{-\frac{r^2 \cos^2\theta}{2}} d\theta.$$

Putting $g(r^2) = G(x_1, x_2)$, we have

$$\left(\frac{\partial^2}{\partial x_1^2} + \frac{\partial^2}{\partial x_2^2} \right) G = \left(\frac{\partial^2 r^2}{\partial x_1^2} + \frac{\partial^2 r^2}{\partial x_2^2} \right) \frac{dG}{dr^2} + \left\{ \left(\frac{\partial r}{\partial x_1} \right)^2 + \left(\frac{\partial r}{\partial x_2} \right)^2 \right\} \frac{d^2 G}{d(r^2)^2}$$

$$= 4g'(r^2) + 4r^2 g''(r^2).$$

Consequently each term is expressed as a function of r^2.

When $p \geqq 3$, putting $r^2 = \sum_i x_i^2$ and making an orthogonal transformation $t_1' = \sum_i x_i t_i / r$, $t_2' = \sum_j c_j t_j, \ldots, t_j' = \sum_j c_{pj} t_j$ we obtain

$$G(x) = \frac{1}{(2\pi)^p} \int \cdots \int \tau \left\{ \exp\left(-\frac{1}{2} \sum t_i'^2 - it_1' r\right) \right\} \prod_i dt_i' .$$

Further transforming $\tau \cos\theta_p = t_1'$, $\tau \cos\theta_p \cos\theta_{p-1} = t_2'$, $\ldots$, we have

$$G(x) = \frac{1}{(2\pi)^p} \int \cdots \int \tau^p \left| \prod_{j=2}^{p-1} \sin^{p-1}\theta_j \right| \left\{ \exp\left(-\frac{\tau^2}{2} - i\tau r \cos\theta_1\right) \right\} d\tau \prod_j d\theta_j$$

$$= c_p \int_0^\infty \int_0^{2\pi} \tau^p \left\{ \exp\left(-\frac{\tau^2}{2} - i\tau r \cos\theta_1\right) \right\} d\tau d\theta_1 ,$$

where c_p is some constant. For even number p, the following generally holds:

$$G(x) = c_p \int_{-\frac{\pi}{2}}^{\frac{\pi}{2}} (-1)^{\frac{p}{2}} \varphi^{(p)}(r \cos\theta_1) d\theta_1 ,$$

where $\varphi(x) = \dfrac{1}{\sqrt{2\pi}} e^{-\frac{x^2}{2}}$.

Since for odd number p

$$\frac{2}{\pi} \int_0^\infty \tau e^{-\frac{t^2}{2} - i\tau x} dx = \frac{1}{\pi} \left\{ \int_0^x \varphi(t)^{-1} dt \right\} \varphi(x) = G_1(x) \text{ (say)},$$

it follows that

$$G(x) = c_p' \int_{-\frac{\pi}{2}}^{\frac{\pi}{2}} (-1)^{\frac{p-1}{2}} G_1^{(p-1)}(r \cos\theta_1) d\theta_1 = g_1(r^2) .$$

Since

$$\left(\sum_i \frac{\partial^2}{\partial x_i^2} \right) G(x) = 2p g_p'(r^2) + 4r^2 g_p''(r^2),$$

the asymptotic expansion of the density may be possible.

For odd number ν ($\geqq 5$) the asymptotic expansion of the density may be similarly calculated.

When ν is even number, it may be impossible to express the characteristic function in a simple form. Putting $\nu = 2m$ we have

$$\psi_\nu(t) = 1 - \frac{\tau^2}{2(\nu-2)} + \frac{\tau^4}{4(\nu-2)(\nu-4)} + \cdots + c_m \tau^{2m} \log \tau + o(\tau^{2m} \log \tau).$$

For $\nu = 2$ we obtain

$$\psi_2(\tau) = 1 + \frac{1}{2}\tau^2 \log \tau - \frac{c}{2}\tau^2 + o(\tau^2).$$

Then the logarism of the characteristic function of $Z_n = \sum_i X_i / \sqrt{n \log n}$ is given by

$$n \log \psi_2\left(\frac{\tau}{\sqrt{n \log n}}\right) = \frac{\tau^2}{2 \log n}\left(\log \tau - \frac{1}{2}\log n - \frac{1}{2}\log n\right)$$

$$- \frac{c\tau^2}{2 \log n} + O\left(\frac{1}{n}\right).$$

For an appropriate number c_n, the characteristic function of $c_n Z_n$ is expanded as

$$e^{-\frac{\tau^2}{2}}\left(1 + \frac{1}{\log n}\tau^2 \log \tau + o\left(\frac{1}{\log n}\right)\right).$$

The corresponding density to $e^{-\frac{\tau^2}{2}}$ is normal and that to $(\tau^2 \log \tau)e^{-\frac{\tau^2}{2}}$ may be calculated as follows: Putting

$$K_1(x) = \frac{1}{\pi}\int_0^\infty (\log t)\, e^{-\frac{t^2}{2}} (\cos tx)\, dt\ ;$$

$$K_2(x) = \frac{1}{\pi}\int_0^\infty (t \log t)\, e^{-\frac{t^2}{2}} (\cos tx)\, dt,$$

we have

$$c_p \int_{-\frac{\pi}{2}}^{\frac{\pi}{2}} K_1^{(p)}(r \cos \theta)\, d\theta = K_1(r) \text{ for even number } p\ ;$$

$$c_p \int_{-\frac{\pi}{2}}^{\frac{\pi}{2}} K_2^{(p)}(r \cos \theta)\, d\theta = K_2(r) \text{ for odd number } p\ ;$$

It follows from Takeuchi and Akahira [1] that

$$K_1(x) = -\varphi(x)\left[\int_0^x\left\{\int_0^y e^{-\frac{t^2}{2}} dt\right\} e^{\frac{y^2}{2}} dy + c_1\right]\ ;$$

$$K_2(x) = K_1^{*\prime}(x)$$

$$K_1^*(x) = \int_0^x e^{-\frac{u^2}{2}} K_1(u)\, du + c_2,$$

where c_1 and c_2 are certain constants.

Hence it is seen from the above that the asymptotic expansion of the density may be possible.

We shall deal with more general densities. In the case $p = 2$, we consider that the density of X_i is expressed as

$$f(x_1, x_2) = \frac{c}{r(1 + r^2)^{\frac{v+1}{2}}} g(\alpha)$$

, where $r = \sqrt{x_1{}^2 + x_2{}^2}$, $\alpha = \tan^{-1} x_2 / x_1$ and c is some constant, and $g(\alpha + \pi) = g(\alpha)$. Putting $\tau = t_1{}^2 + t_2{}^2$ and $\theta = \tan^{-1} t_2 / t_1$, we obtain the following characteristic function:

$$\phi(t_1, t_2) = \int\int f(x_1, x_2) e^{it_1 x_1 + it_2 x_2} dx_1 dx_2$$

$$= \int_{-\frac{\pi}{2}}^{\frac{\pi}{2}} \int_{-\infty}^{\infty} \frac{1}{(1 + r^2)^{\frac{v+1}{2}}} e^{i\tau t \cos(\theta - \alpha)} g(\alpha) dr d\alpha$$

$$= \int_{-\frac{\pi}{2}}^{\frac{\pi}{2}} \psi_v(\tau \cos(\theta - \alpha)) g(\alpha) d\alpha.$$

For example a density $f(x_1, x_2)$ is given by

$$f(x_1, x_2) = \frac{c}{1 + x_1{}^4 + x_2{}^4}$$

where c is some constant.

Since

$$\frac{c}{1 + x_1{}^4 + x_2{}^4} \doteqdot \frac{K}{\{1 + (x_1{}^2 + x_2{}^2)\}^2} \cdot \frac{(x_2{}^2 + x_2{}^2)^2}{x_1{}^4 + x_2{}^4}$$

where K is some constant, we have

$$f(x_1, x_2) \doteqdot \frac{K}{(1 + r^2)^2} \cdot \frac{1}{\cos^4 \alpha + \sin^4 \alpha}.$$

On the other hand it follows that

$$\psi_2(t) = (1 + |t|) e^{-|t|}$$

(See [1], page 99).

Hence

$$\phi(t_1, t_2) = \int_{-\frac{\pi}{2}}^{\frac{\pi}{2}} \frac{1 + |\tau \cos(\theta - \alpha)|}{\cos^4 \alpha + \sin^4 \alpha} e^{-|\tau \cos(\theta - \alpha)| d\alpha} . |$$

                    Kei TAKEUCHI and Masafumi AKAHIRA

If $\nu = 1$, then

$$\phi(t_1, t_2) = \int_{-\frac{\pi}{2}}^{\frac{\pi}{2}} e^{-|\tau \cos(\theta-\alpha)|} g(\alpha)\, d\alpha$$

$$= 1 - \tau \int_{-\frac{\pi}{2}}^{\frac{\pi}{2}} |\cos(\theta - \alpha)|\, g(\alpha)\, d\alpha$$

$$+ \frac{\tau^2}{2} \int_{-\frac{\pi}{2}}^{\frac{\pi}{2}} \cos^2(\theta - \alpha) g(\alpha)\, d\alpha + o(\tau^2). \tag{2.2}$$

Though the last term of (2.2) is expressed as

$$\tau^2 \int_{-\frac{\pi}{2}}^{\frac{\pi}{2}} \cos^2(\theta - \alpha) g(\alpha)\, d\alpha$$

$$= \tau^2 \int_{-\frac{\pi}{2}}^{\frac{\pi}{2}} (\cos^2\theta \cos^2\alpha + 2\cos\theta s\ \theta \text{ in } \cos\alpha \sin\alpha$$

$$+ \sin^2\theta \sin^2\alpha) g(\alpha)\, d\alpha$$

$$= A_{11} t_1^2 + 2 A_{12} t_1 t_2 + A_{22} t_2^2 ,$$

$$\tau \int_{-\frac{\pi}{2}}^{\frac{\pi}{2}} |\cos(\theta - \alpha)|\, g(\alpha)\, d\alpha = \tau h(\theta) \quad \text{(say)}$$

is not done as a simple formula of t_1 and t_2. Hence the characteristic function of $\sum_i X_i/n$ is the following form:

$$\left\{ \phi\left(\frac{t_1}{n}, \frac{t_2}{n}\right) \right\}^n \sim e^{-\tau h(\theta)} \left\{ 1 + \frac{1}{n} (A_{11} t_1^2 + 2 A_{12} t_1 t_2 + A_{22} t_2^2 + \cdots) + o\left(\frac{1}{n}\right) \right\}$$

$$\tag{2.3}$$

The density corresponding to the first term of the right-hand side of (2.3) is given by

$$f^*(x_1, x_2) = \frac{1}{(2\pi)^2} \int_{-\infty}^{\infty} \int_{0}^{\infty} e^{-\tau h(\theta)} e^{-i\tau r \cos(\theta-\alpha)}\, dt_1 dt_2$$

$$= \frac{1}{2\pi^2} \int_{-\frac{\pi}{2}}^{\frac{\pi}{2}} \int_{0}^{\infty} \tau e^{-\tau h(\theta)} \cos\ \tau r \cos(\theta - \alpha) \} \, d\tau d\theta$$

$$= \frac{1}{2\pi^2} \int_{-\frac{\pi}{2}}^{\frac{\pi}{2}} \frac{1 - r^4 \cos^4(\theta - \alpha)}{\{h(\theta)^2 + r^2 \cos^2(\theta - \alpha)\}^2}\, d\theta . \tag{2.4}$$

In general the distribution of the density (2.4) is not Cauchy. Since the asymptotic distribution is a stable law, it is shown from the above that in two-dimensional case the stable law with characteristic exponent 1 is not always two-variate Cauchy. But for every a_1 and a_2 the distribution of $a_1 X_1 + a_2 X_2$ is always Cauchy. The corresponding density to the second term of the right-hand side of (2.4) is given by

$$- \frac{1}{n} (A_{11} \frac{\partial^2}{\partial x_1^2} + 2A_{12} \frac{\partial^2}{\partial x_1 \partial x_2} + A_{22} \frac{\partial^2}{\partial x_2^2}) f^* (x_1, x_2) .$$

If $\nu = 3$, then

$$\phi (t_1, t_2) = 1 - \frac{1}{2} (A_{11} t_1^2 + 2A_{12} t_1 t_2 + A_{22} t^2)$$

$$+ \frac{\tau^3}{3} \int | \cos (\theta - \alpha) |^3 g (\alpha) d\alpha + o (\tau^3) .$$

Since

$$\{ \phi (\frac{t_1}{\sqrt{n}} , \frac{t_2}{\sqrt{n}}) \} \sim e^{-\frac{1}{2} t' A t} \{ 1 + \frac{\tau^3}{3\sqrt{n}} h_3 (\theta) + o (\frac{1}{\sqrt{n}}) \} ,$$

it follows that the first term corresponds to a normal distribution. Making an appropriate linear transformation we may have $A = I$. Then the corresponding term of the density to $\tau^3 e^{-\frac{\tau^2}{2}} h_3(\theta)$ is expressed by

$$- (\frac{\partial^2}{\partial x_1^2} + \frac{\partial^2}{\partial x_2^2}) \frac{1}{2\pi} \int_{-\frac{\pi}{2}}^{\frac{\pi}{2}} \frac{1}{\sqrt{2\pi}} \{ 1 - r^2 \cos^2 (\theta - \alpha) \} h (\theta) e^{-\frac{r^2 \cos^2 (\theta - \alpha)}{2}} d\theta .$$

If ν is odd number, then similar expansions hold.

If $\nu = 2$, then

$$\phi(t_1, t_2) = 1 + \frac{1}{2} \tau^2 (\log \tau) \int \cos^2 (\theta - \alpha) g (\alpha) d\alpha$$

$$+ \frac{1}{2} \int \{ \log | \cos (\theta - \alpha) | \} g (\alpha) d\alpha$$

$$- \frac{c}{2} \tau^2 \int \cos^2 (\theta - \alpha) g (\alpha) d\alpha + o (\tau^2) .$$

By a suitable linear transformation we have

$$\phi (t_1, t_2) = 1 + \frac{1}{2} \tau^2 \log \tau - \frac{1}{2} \{ c + h_4 (\theta) \} \tau^2 + o (\tau^2) .$$

It is seen that the asymptotic distribution of $\sum_i X_i / \sqrt{n \log n}$ is normal. The term of order $1/\log n$ is calculated by the same way as the multivariate t-distribution case.

268   Kei TAKEUCHI and Masafumi AKAHIRA

The above discussion is extended to a general dimensional case. Indeed, we consider the following density:

$$f(x) = \frac{c}{r^{p-1}(1 + r^2)^{\nu/2}} g(\alpha) ,$$

where $r^2 = x'x$ and α is a vector of $(p-1)$ angles and c is some constant. If $g(-\alpha) = g(\alpha)$, then the characteristic function of X is given by

$$p(t) = \int \psi_\nu(\tau, h(\theta, \alpha)) g(\alpha) d\alpha.$$

Hence the same discussions as two-dimensional case may be possible in the multidimensional case.

If $\nu = 1$, then the asymptotic distribution of $\sum_i X_i/n$ is a stable law with characteristic exponent 1 but it is not always multivariate Cauchy. For $\nu = 2$ and $\nu \geq 2$, the asymptotic distributions of $\sum_i X_i / \sqrt{n \log n}$ and $\sum_i X_i / \sqrt{n}$ are normal, respectively.
The asymptotic expansion of the density is essentially similar to the case $p = 2$.

It is also seen that the symmetricity of the density is not essential. Indeed, if in case $p = 2$ a density is given by

$$f(x_1, x_2) = \frac{q}{r(1 + r)^{q+1}} g(\alpha) ,$$

Then the characteristic function is expressed as

$$\phi(t_1, t_2) = \int_{-\pi}^{\pi} \psi_q(\tau \cos(\theta - \alpha)) g(\alpha) d\alpha ,$$

where $\quad \overline{\psi}_q(t) = \int_0^\infty \frac{q e^{itr}}{(1 + r)^{q+1}} dr .$

Then we may obtain the asymptotic expansion of the characteristic function of Z_n by the same way as the above symmetric case. When $p \geq 3$, then similar discussions hold.

In the discussion of this paper it is essential that the order of $f(x)$ as $x'x \to \infty$ is independent of directions. But we may avoid the point by the following way. Indeed, in the case $p = 2$ we consider a density

$$f(x_1, x_2) = \frac{c}{1 + x_1^2 + x_2^4} .$$

Then the characteristic function is obtained by

$$\phi(t_1, t_2) = 2 \int_{-\frac{\pi}{2}}^{\frac{\pi}{2}} \int_0^\infty \frac{cr e^{i\tau r(\theta - \alpha)}}{1 + r^2 \cos^2\alpha + r^4 \sin^4\alpha} dr d\alpha .$$

If

$$\frac{r}{1 + r^2 \cos^2\alpha + r^4 \sin^4\alpha} = \frac{1}{(\sin^4\alpha)(1+r)^3} + g(r) \ ,$$

$g(r)$ is order of r^{-4}. Since we may make such expansion in detail, we obtain the asymptotic expansion of $\phi(t_1, t_2)$ around the origin. But in such cases it may difficult that standardized constants are common. In this case we have to consider the asymptotic distribution of $(\sum_i X_{1i}/n, \sum_i X_{2i}/n)$.

References

[1] Takeuchi, K. and Akahira, M., "On Gram-Charlier-Edgeworth type expansion of the sums of random variables (I)," Rep. Univ. Electro-Comm. **27**-1, (1976), 95—115.

[2] Takeuchi, K. and Akahira, M., "On Gram-Charlier-Edgeworth type expansion of the sums of random variables (II)," Rep. Univ. Electro-Comm. **27**-1, (1976), 117—123.

[3] Takeuchi, K. and Akahira, M., "Extension of Edgeworth type expansion of the distribution of the sums of i.i.d. random variables in non-regular cases," Ann. Inst. Statist. Math. **29**, Part A, (1977), 397—406.

Rep. Univ. Electro-Comm. 28 − 2 , (Sci. & Tech. Sect.), pp. 271−293 February, 1978

Third Order Asymptotic Efficiency of Maximum Likelihood Estimator for Multiparameter Exponential Case*

Kei TAKEUCHI** and Masafumi AKAHIRA***

Abstract

For multiparameter exponential family distributions the asymptotic expansion of the distribution of extended regular (ER) estimators is obtained up to the order n^{-1}. And it is shown that the (modified) maximum likelihood estimator attains the third order asymptotic efficiency among ER best asymptotically normal (BAN) estimators.

1. Introduction

Third order asymptotic efficiency of maximum likelihood (ML) estimators have been discussed by J. Pfanzagl and W. Wefelmeyer ([2]), (who adopted the terminology) and also by J.K. Ghosh and K. Subramanyam ([1]), for cases when sufficient statistics exist. In this paper we shall establish more general results for the multiparameter exponential family, introducing a differential operator, and show that (modified) ML estimator is always optimal up to the order n^{-1} among similar modified class of estimators.

The method is not restricted to the exponential family and will be applied to more general cases in a subsequent paper ([3]).

2. Formulation of the problem

Suppose that $X_1, X_2, \dots, X_n$ are independently and identically distributed according to an exponential type distribution, i.e. with the density

$$f(x, \theta) = h(x)\, c(\theta)\, \exp\Big\{ \sum_{i=1}^{m} s_i(\theta)\, t_i(x) \Big\} ,$$

where θ is a p-dimensional real vector-valued parameter which is supposed to be in an open set $\Theta \subset R^p$, and $s_i(\theta)$'s are continuous real valued functions and $t_i(x)$'s are real valued measurable functions. Define

$$c^*(s_1, \cdots, s_m) = \int h(x) \exp\Big\{ \sum_{i=1}^{m} s_i t_i(x) \Big\} d\mu(x)$$

---

  * Received on December 7, 1977
 ** University of Tokyo
*** Statistical Laboratory, University of Electro-Communications

as a real valued function of $s_1, \ldots, s_m$, and denote $S^* \subset R^m$ be the set of $(s_1, \ldots, s_m)$ for which c^* is finite. We assume that for every $\theta \in \Theta$, $(s_1(\theta), \ldots, s_m(\theta))$ is an inner point of S^*. We have, obviously that $c = c^{*-1}$.

We assume that $s_1(\theta), \ldots, s_m(\theta)$ are linearly independent when θ varies in Θ, and also that $t_1(x), \ldots, t_m(x)$ are linearly independent functions of x. (The domain of x does not need be specified)

For the sample $(X_1, \ldots, X_n)$, a sufficient statistic is given by

$$T = (T_1, \ldots, T_m);$$

$$T_i = \sum_{j=1}^{n} t_i(X_j) / n .$$

From the assumption T has the moment generating function given by

$$E \left(\exp \sum_{i=1}^{m} \xi_i T_i \right)$$

$$= c^*(s_1(\theta) + \xi_1, \ldots, s_m(\theta) + \xi_m) / c^*(s_1(\theta), \ldots, s_m(\theta)).$$

And the joint cumulants of T_i's are given by

$$\kappa_{ijk \cdots} = \frac{\partial^r}{\partial s_i \partial s_j \partial s_k \cdots} \log c^* \bigg|_{s_1 = s_1(\theta), \ldots, s_m = s_m(\theta)}.$$

We assume that T_i's are continuously (w. r. t. the Lebesgue measure) distributed and that their joint density admits Edgeworth expansion.

We also assume that $s_i(\theta)$'s are continuously differential four times.

Now we shall define a class of estimators called extended regular (ER) estimators which are expressed as

$$\hat{\theta} = g(T_1, \ldots, T_m) + \frac{1}{n} h(T_1, \ldots, T_m).$$

where g is three times continuously differentiable and h is continuously differentiable, and both are independent of n.

An ER estimator is consistent iff

$$g(\mu_1(\theta), \ldots, \mu_m(\theta)) \equiv \theta \tag{2.1}$$

where $\mu_i(\theta) = E_\theta(T_i)$ $(i = 1, \ldots, m)$.

and if $\hat{\theta}$ is a consistent ER estimator, $\sqrt{n}(\hat{\theta} - \theta)$ is asymptotically distributed according to the normal distribution with mean vector 0 and the variance-convariance matrix

$$\Sigma = \sum_i \sum_j \sigma_{ij} a_i a_j' , \quad \sigma_{ij} = V_\theta(T_i, T_j)$$

where a_i's are vectors defined by

$$a_i = \frac{\partial g}{\partial T_i} \bigg|_{T_i = \mu_i(\theta)} \tag{2.2}$$

It is well known that

$$\Sigma - I^{-1} \geq 0 \quad (\text{positive semi-definite})$$

where I is the Fisher Information matrix given by

$$\begin{aligned}
I &= -E_\theta \left[\frac{\partial^2}{\partial\theta\,\partial\theta'} \log f(X, \theta) \right] \\[2mm]
&= E_\theta \left[\frac{\partial}{\partial\theta} \log f(X, \theta) \cdot \frac{\partial}{\partial\theta'} \log f(X, \theta) \right] \\[2mm]
&= E_\theta \left[\sum_i \frac{\partial s_i}{\partial\theta} (T_i - \mu) \cdot \sum_i \frac{\partial s_i}{\partial\theta'} (T_i - \mu) \right] \\[2mm]
&= \sum_i \sum_j \sigma_{ij} \frac{\partial s_i}{\partial\theta} \frac{\partial s_j}{\partial\theta'}
\end{aligned}$$

And also it is easily seen that the equality is attained iff

$$a_i = I^{-1} \frac{\partial s_i}{\partial\theta} \tag{2.3}$$

We call an ER estimator which satisfies (2.1) and (2.3) an ER best asymptotically normal (BAN) estimator.

3. ER BAN estimators

Now we expand each component of $\hat{\theta}$ into Taylor's series as

$$\begin{aligned}
\hat{\theta}_\alpha = \theta_\alpha &+ \sum_i a_i^\alpha (T_i - \mu_i) + \frac{1}{2} \sum_i \sum_j b_{ij}^\alpha (T_i - \mu_i)(T_j - \mu_j) \\[2mm]
&+ \frac{1}{6} \sum_i \sum_j \sum_k c_{ijk}^\alpha (T_i - \mu_i)(T_j - \mu_j)(T_k - \mu_k) \\[2mm]
&+ \frac{1}{n} \Delta_\alpha + \frac{1}{n} \sum_i d_i^\alpha (T_i - \mu_i) + R_\alpha ,
\end{aligned} \tag{3.1}$$

where $a_i^\alpha = \dfrac{\partial g_\alpha}{\partial T_i}$, $b_{ij}^\alpha = \dfrac{\partial^2 g_\alpha}{\partial T_i \partial T_j}$, $c_{ijk}^\alpha = \dfrac{\partial^3 g_\alpha}{\partial T_i \partial T_j \partial T_k}$,

$$\Delta_\alpha = h(\mu_1, \ldots, \mu_m), \quad d_i^\alpha = \frac{\partial h}{\partial T_i}$$

and all the derivatives are evaluated at $T_i \equiv \mu_i(\theta)$.

By putting $T_i' = \sqrt{n}\,(T_i - \mu_i)$, (3.1) can be rewritten as

$$\sqrt{n}\,(\theta_\alpha - \theta_\alpha) = \sum_i a_i^\alpha T_i' + \frac{1}{2\sqrt{n}}\sum_i \sum_j b_{ij}^\alpha T_i'T_j'$$

$$+ \frac{1}{6n} \sum_i \sum_j \sum_k c_{ijk}^\alpha T_i'T_j'T_k'$$

$$+ \frac{1}{\sqrt{n}}\,\Delta_\alpha + \frac{1}{n}\sum_i d_i^\alpha T_i' + R_\alpha'\,.$$

It is shown that R_α' is stochastically of order smaller than n^{-1}.

We shall denote that

$$Y_\alpha = \sqrt{n}\,(\hat{\theta}_\alpha - \theta_\alpha) - R_\alpha\,.$$

Then we have $\sqrt{n}\,(\hat{\theta}_\alpha - \theta_\alpha) - Y_\alpha = o_p\,(1/n)$, moreover, we have

Theorem 3.1 The joint distribution of $\sqrt{n}\,(\hat{\theta}_\alpha - \theta_\alpha)$ and Y_α coincide up to the order n^{-1}, more precisely, for any bounded continuous function ϕ, we have

$$\varlimsup_{n \to \infty} n\,|\,E_\theta\,[\phi\,(Y_1,\,\ldots,\,Y_p)\,] - E_\theta\,[\phi\,(\sqrt{n}\,(\hat{\theta}_1 - \theta),\ldots,$$

$$\sqrt{n}\,(\hat{\theta}_p - \theta))\,]\,| = 0\,.$$

Proof. Observing that $Pr\,\{\,n\,|\,R'\,|\,>\,\epsilon\,\} = o\,(1/n)$ for every $\epsilon > 0$, we can show that

$$\lim_{n \to \infty} n\,|\,E\,[\exp \sum_\alpha it_\alpha\,(\hat{\theta}_\alpha - \theta_\alpha)\,] - E\,[\exp \sum_\alpha it_\alpha\,y_\alpha\,]\,| = 0$$

uniformly in $(t_1,\ldots,\,t_p)$ in any compact subset of R^p, which establishes the theorem. (QED)

Corollary 3.1. For any measurable set $C \subset R^p$ whose boundary is of Lebesgue measure 0, we have

$$\varlimsup_{n \to \infty} n\,|\,P_\theta\,\{Y \in C\} - P_\theta\,\{\sqrt{n}\,(\hat{\theta} - \theta) \in C\}\,| = 0.$$

From Theorem 3.1 we have

Theorem 3.2. The joint distribution of $\sqrt{n}\,(\hat{\theta}_\alpha - \theta_\alpha)$ $(\alpha = 1,\cdots, P\,)$ is asymptotically equal to a continuous distribution up to the order n^{-1}.

And for the distribution of Y_α's we have

Theorem 3.3. The joint distribution of Y_α's admits Gram-Charlier-Edgeworth type expansion up to the order n^{-1}.

Proof. The theorem is established by algebraic evaluation of the joint characteristic function of Y_α's, the detail of which is obmitted.

Consequently the joint density of $\sqrt{n}\,(\hat{\theta}_\alpha - \theta_\alpha)$ $(\alpha = 1,\cdots, P\,)$ can be expanded up to the order n^{-1} if we have joint cumulants of Y_α's up to the fourth.

4. Some preliminary results

In order to evaluate the joint cumulants, we introduce the following differential operator.

$$D_\alpha = \sum_\beta I^{\alpha\beta} \frac{\partial}{\partial\theta_\beta} \, , \text{ where } I^{\alpha\beta} \text{ denotes the element of } I^{-1} \text{ and the random variable}$$

$$U_\alpha = \sqrt{n} \, \sum_\alpha \sum_\beta I^{\alpha\beta} \frac{\partial s_i}{\partial\theta_\beta} (T_i - \mu_i)$$

$$= \sum_i (D_\alpha s_i) \, T'_i$$

$$= D_\alpha \{ \sqrt{n} \, \log f(X, \theta) \} \, .$$

The following lemma will be frequently applied in the subsequent discussions.

Lemma 4.1. Suppose that Z_θ is a function of $T_1, \dots, T_m$ and of θ, differentiable in θ and that

$$\frac{\partial}{\partial\theta} E_\theta (Z_\theta) = \int \frac{\partial}{\partial\theta} \{ Z_\theta \prod_i f(x_i, \theta) \} \prod_i d\mu(x_i)$$

holds. Then we have

$$E_\theta (U_\alpha Z_\theta) = \frac{1}{\sqrt{n}} \, D_\alpha E_\theta (Z_\theta) - \frac{1}{\sqrt{n}} \, E_\theta (D_\alpha Z_\theta) \, .$$

Proof.

$$D_\alpha E_\theta (Z_\theta)$$

$$= \int (D_\alpha Z_\theta) \prod_i f(x_i, \theta) \prod_i d\mu(x_i) + \int Z_\theta D_\alpha \prod_i f(x_i, \theta) \prod_i d\mu(x_i)$$

$$= E_\theta (D_\alpha D_\theta) + \int Z_\theta D_\alpha (\sum_i \log f(x_i, \theta)) \prod_i f(x_i, \theta) \prod_i d\mu(x_i)$$

$$= E_\theta (D_\alpha Z_\theta) + \sqrt{n} \, E_\theta (Z_\theta U_\alpha)$$

from which the theorem is directly derived. (QED)

We shall use the following symbols.

$$J_{\alpha\beta\cdot\gamma} = E_\theta [\frac{\partial^2}{\partial\theta_\alpha \partial\theta_\beta} \log f(X, \theta) \cdot \frac{\partial}{\partial\theta_\gamma} \log f(X, \theta)]$$

$$= \sum_i \sum_j \sigma_{ij} (\frac{\partial^2 s_i}{\partial\theta_\alpha \partial\theta_\beta}) (\frac{\partial s_j}{\partial\theta_\gamma}) \, ;$$

$$K_{\alpha\beta\gamma} = E_\theta [\frac{\partial}{\partial\theta_\alpha} \log f(X, \theta) \cdot \frac{\partial}{\partial\theta_\beta} \log f(X, \theta) \cdot \frac{\partial}{\partial\theta_\gamma} \log f(X, \theta)]$$

$$= \sum_i \sum_j \sum_k \kappa_{ijk} (\frac{\partial s_i}{\partial\theta_\alpha}) (\frac{\partial s_j}{\partial\theta_\beta}) (\frac{\partial s_k}{\partial\theta_\gamma}) \, ;$$

Kei TAKEUCHI and Masafumi AKAHIRA

$$J^{\alpha\beta\cdot\gamma} = \sum_{\alpha'} \sum_{\beta'} \sum_{\gamma'} I^{\alpha\alpha'} I^{\beta\beta'} I^{\gamma\gamma'} J_{\alpha'\beta'\cdot\gamma'} \;;$$

$$K^{\alpha\beta\gamma} = \sum_{\alpha'} \sum_{\beta'} \sum_{\gamma'} I^{\alpha\alpha'} I^{\beta\beta'} I^{\gamma\gamma'} K_{\alpha'\beta'\gamma'} \;;$$

$$L_{\alpha\beta\gamma\cdot\delta} = E_\theta \left[\frac{\partial^3}{\partial\theta_\alpha \partial\theta_\beta \partial\theta_\gamma} \log f(X,\theta) \cdot \frac{\partial}{\partial\theta_\delta} \log f(X,\theta) \right]$$

$$= \sum_i \sum_j \sigma_{ij} \left(\frac{\partial^3 s_i}{\partial\theta_\alpha \partial\theta_\beta \partial\theta_\gamma} \right) \left(\frac{\partial s_j}{\partial\theta_\delta} \right) \;;$$

$$M_{\alpha\beta\cdot\gamma\delta} = E_\theta \left[\frac{\partial^2}{\partial\theta_\alpha \partial\theta_\beta} \log f(X,\theta) \cdot \frac{\partial^2}{\partial\theta_\gamma \partial\theta_\delta} \log f(X,\theta) \right] - I_{\alpha\beta} I_{\gamma\delta}$$

$$= \sum_i \sum_j \sigma_{ij} \left(\frac{\partial^2 s_i}{\partial\theta_\alpha \partial\theta_\beta} \right) \left(\frac{\partial^2 s_i}{\partial\theta_\gamma \partial\theta_\delta} \right) \;;$$

$$N_{\alpha\beta\cdot\gamma\delta} = E_\theta \left[\frac{\partial^2}{\partial\theta_\alpha \partial\theta_\beta} \log f(X,\theta) \cdot \frac{\partial}{\partial\theta_\gamma} \log f(X,\theta) \cdot \frac{\partial}{\partial\theta_\delta} \log f(X,\theta) \right]$$

$$+ I_{\alpha\beta} I_{\gamma\delta}$$

$$= \sum_i \sum_j \sum_k \kappa_{ijk} \left(\frac{\partial^2 s_i}{\partial\theta_\alpha \partial\theta_\beta} \right) \left(\frac{\partial s_i}{\partial\theta_\gamma} \right) \left(\frac{\partial s_k}{\partial\theta_\delta} \right) \;;$$

$$H_{\alpha\beta\gamma\delta} = E_\theta \left[\frac{\partial}{\partial\theta_\alpha} \log f(X,\theta) \cdot \frac{\partial}{\partial\theta_\beta} \log f(X,\theta) \cdot \frac{\partial}{\partial\theta_\gamma} \log f(X,\theta) \right.$$

$$\left. \cdot \frac{\partial}{\partial\theta_\delta} \log f(X,\theta) \right] - (I_{\alpha\beta} I_{\gamma\delta} + I_{\alpha\gamma} I_{\beta\delta} + I_{\alpha\delta} I_{\beta\gamma})$$

$$= \sum_i \sum_j \sum_k \sum_l \kappa_{ijkl} \left(\frac{\partial s_i}{\partial\theta_\alpha} \right) \left(\frac{\partial s_j}{\partial\theta_\beta} \right) \left(\frac{\partial s_k}{\partial\theta_\gamma} \right) \left(\frac{\partial s_l}{\partial\theta_\delta} \right) \;;$$

$$L^{\alpha\beta\gamma\cdot\delta} = \sum_{\alpha'} \sum_{\beta'} \sum_{\gamma'} \sum_{\delta'} I^{\alpha\alpha'} I^{\beta\beta'} I^{\gamma\gamma'} I^{\delta\delta'} L_{\alpha'\beta'\gamma'\cdot\delta'} \;;$$

$$M^{\alpha\beta\cdot\gamma\delta} = \sum_{\alpha'} \sum_{\beta'} \sum_{\gamma'} \sum_{\delta'} I^{\alpha\alpha'} I^{\beta\beta'} I^{\gamma\gamma'} I^{\delta\delta'} M_{\alpha'\beta'\cdot\gamma'\delta'} \;;$$

$$N^{\alpha\beta\cdot\gamma\delta} = \sum_{\alpha'} \sum_{\beta'} \sum_{\gamma'} \sum_{\delta'} I^{\alpha\alpha'} I^{\beta\beta'} I^{\gamma\gamma'} I^{\delta\delta'} N_{\alpha'\beta'\cdot\gamma'\delta'} \;;$$

$$H^{\alpha\beta\gamma\delta} = \sum_{\alpha'} \sum_{\beta'} \sum_{\gamma'} \sum_{\delta'} I^{\alpha\alpha'} I^{\beta\beta'} I^{\gamma\gamma'} I^{\delta\delta'} H_{\alpha'\beta'\gamma'\delta'} \;.$$

Using these symbols we have

$$D_\alpha I_{\beta\gamma} = \sum_i \sum_j D_\alpha \left(\sigma_{ij} \frac{\partial s_i}{\partial \theta_\beta} \frac{\partial s_j}{\partial \theta_\gamma} \right)$$

$$= \sum_i \sum_j (D_\alpha \sigma_{ij}) \left(\frac{\partial s_i}{\partial \theta_\beta}\right) \left(\frac{\partial sj}{\partial \theta_\gamma}\right) + \sum_i \sum_j \sigma_{ij} \left(D_\alpha \frac{\partial s_i}{\partial \theta_\beta}\right) \left(\frac{\partial s_j}{\partial \theta_\gamma}\right)$$

$$+ \sum_i \sum_j \sigma_{ij} \left(\frac{\partial s_i}{\partial \theta_\beta}\right) \left(D_\alpha \frac{\partial s_j}{\partial \theta_\gamma}\right)$$

$$= \sum_{\alpha'} I^{\alpha\alpha'} \sum_i \sum_j \sum_k \kappa_{ijk} \left(\frac{\partial s_i}{\partial \theta_\beta}\right) \left(\frac{\partial s_j}{\partial \theta_\gamma}\right) \left(\frac{\partial s_k}{\partial \theta_{\alpha'}}\right)$$

$$+ \sum_{\alpha'} I^{\alpha\alpha'} \sum_i \sum_j \sigma_{ij} \left(\frac{\partial^2 s_i}{\partial \theta_{\alpha'} \partial \theta_\beta}\right) \left(\frac{\partial s_j}{\partial \theta_\gamma}\right)$$

$$+ \sum_{\alpha'} I^{\alpha\alpha'} \sum_i \sum_j \sigma_{ij} \left(\frac{\partial s_i}{\partial \theta_\beta}\right) \left(\frac{\partial^2 s_j}{\partial \theta_{\alpha'} \partial \theta_\gamma}\right)$$

$$= \sum_{\alpha'} I^{\alpha\alpha'} K_{\alpha'\beta'\gamma} + \sum_{\alpha'} I^{\alpha\alpha'} J_{\alpha'\beta\cdot\gamma} + \sum_{\alpha'} I^{\alpha\alpha'} J_{\alpha'\gamma\cdot\beta} \ ;$$

$$D_\alpha I^{\beta\gamma} = -\sum_{\beta'} \sum_{\gamma'} I^{\beta\beta'} I^{\gamma\gamma'} D_\alpha I_{\beta'\gamma'}$$

$$= -\sum_{\beta'} \sum_{\gamma'} \sum_{\alpha'} I^{\beta\beta'} I^{\gamma\gamma'} I^{\alpha\alpha'} \left(K_{\alpha'\beta'\gamma'} + J_{\alpha'\beta'\cdot\gamma'} + J_{\alpha'\gamma'\cdot\beta'}\right)$$

$$= -K^{\alpha\beta\gamma} - J^{\alpha\beta\cdot\gamma} - J^{\alpha\gamma\cdot\beta} \ ;$$

$$D_\alpha J_{\beta\gamma\cdot\delta} = \sum_i \sum_j (D_\alpha \sigma_{ij}) \left(\frac{\partial^2 s_i}{\partial \theta_\beta \partial \theta_\gamma}\right) \left(\frac{\partial^2 s_k}{\partial \theta_\delta^2}\right)$$

$$+ \sum_i \sum_j \sigma_{ij} \left(D_\alpha \frac{\partial^2 s_i}{\partial \theta_\beta \partial \theta_\gamma}\right) \left(\frac{\partial^2 s_i}{\partial \theta_\gamma^2}\right)$$

$$+ \sum_i \sum_j \sigma_{ij} \left(\frac{\partial^2 s_i}{\partial \theta_\beta \partial \theta_\gamma}\right) \left(D_\alpha \frac{\partial^2 s_j}{\partial \theta_\delta^2}\right)$$

$$= \sum_{\alpha'} I^{\alpha\alpha'} \sum_i \sum_j \sum_k \kappa_{ijk} \left(\frac{\partial^2 s_i}{\partial \theta_\beta \partial \theta_\gamma}\right) \left(\frac{\partial s_j}{\partial \theta_\delta}\right) \left(\frac{\partial s_k}{\partial \theta_{\alpha'}}\right)$$

$$+ \sum_{\alpha'} I^{\alpha\alpha'} \sum_i \sum_j \left(\frac{\partial^3 s_i}{\partial \theta_{\alpha'} \partial \theta_\beta \partial \theta_\gamma}\right) \left(\frac{\partial s_j}{\partial \theta_\delta}\right)$$

$$+ \sum_{\alpha'} I^{\alpha\alpha'} \sum_i \sum_j \left(\frac{\partial^2 s_i}{\partial \theta_\beta \partial \theta_\gamma}\right) \left(\frac{\partial^2 s_j}{\partial \theta_{\alpha'} \partial \theta_\delta}\right)$$

 Kei TAKEUCHI and Masafumi AKAHIRA

$$= \sum_{\alpha'} I^{\alpha\alpha'} \left(N_{\beta\gamma\cdot\alpha'\delta} + L_{\alpha'\beta\gamma\cdot\delta} + M_{\beta\gamma\cdot\alpha'\delta} \right) \; ;$$

$$D_\alpha K_{\beta\gamma\delta} = \sum_i \sum_j \sum_k (D_\alpha \kappa_{ijk}) \left(\frac{\partial s_i}{\partial\theta_\beta}\right) \left(\frac{\partial s_j}{\partial\theta_\gamma}\right) \left(\frac{\partial s_k}{\partial\theta_\delta}\right)$$

$$+ \sum_i \sum_j \sigma_{ij} \left(D_\alpha \frac{\partial s_i}{\partial\theta_\beta}\right) \left(\frac{\partial s_j}{\partial\theta_\gamma}\right) \left(\frac{\partial s_k}{\partial\theta_\delta}\right)$$

$$+ \sum_i \sum_j \sigma_{ij} \left(\frac{\partial s_i}{\partial\theta_\beta}\right) \left(D_\alpha \frac{\partial s_j}{\partial\theta_\gamma}\right) \left(\frac{\partial s_k}{\partial\theta_\delta}\right)$$

$$+ \sum_i \sum_j \sigma_{ij} \left(\frac{\partial s_i}{\partial\theta_\beta}\right) \left(\frac{\partial s_j}{\partial\theta_\gamma}\right) \left(D_\alpha \frac{\partial s_k}{\partial\theta_\delta}\right)$$

$$= \sum_{\alpha'} I^{\alpha\alpha'} \left(H_{\alpha'\beta\gamma\delta} + N_{\alpha'\beta\cdot\gamma\delta} + N_{\alpha'\gamma\cdot\beta\delta} + N_{\alpha'\delta\cdot\beta\gamma} \right) \; ;$$

$$D_\alpha J^{\beta\gamma\cdot\delta} = \sum_{\beta'} \sum_{\gamma'} \sum_{\delta'} (D_\alpha I^{\beta\beta'}) I^{\gamma\gamma'} I^{\delta\delta'} J_{\beta'\gamma'\cdot\delta'}$$

$$+ \sum_{\beta'} \sum_{\gamma'} \sum_{\delta'} I^{\beta\beta'} (D_\alpha I^{\gamma\gamma'}) I^{\delta\delta'} J_{\beta'\gamma'\cdot\delta'}$$

$$+ \sum_{\beta'} \sum_{\gamma'} \sum_{\delta'} I^{\beta\beta'} I^{\gamma\gamma'} (D_\alpha I^{\delta\delta'}) J_{\beta'\gamma'\cdot\delta'}$$

$$+ \sum_{\beta'} \sum_{\gamma'} \sum_{\delta'} I^{\beta\beta'} I^{\gamma\gamma'} I^{\delta\delta'} D_\alpha J_{\beta'\gamma'\cdot\delta'}$$

$$= - \sum_{\beta'} \sum_{\gamma'} \sum_{\delta'} (K^{\alpha\beta\beta'} + J^{\alpha\beta\cdot\beta'} + J^{\alpha\beta'\cdot\beta}) I^{\gamma\gamma'} I^{\delta\delta'} J_{\beta'\gamma'\cdot\delta'}$$

$$- \sum_{\beta'} \sum_{\gamma'} \sum_{\delta'} (K^{\alpha\gamma\gamma'} + J^{\alpha\gamma\cdot\gamma'} + J^{\alpha\gamma'\cdot\gamma}) I^{\beta\beta'} I^{\delta\delta'} J_{\beta'\gamma'\cdot\delta'}$$

$$- \sum_{\beta'} \sum_{\gamma'} \sum_{\delta'} (K^{\alpha\delta\delta'} + J^{\alpha\delta\cdot\delta'} + J^{\alpha\delta'\cdot\delta}) I^{\beta\beta'} I^{\gamma\gamma'} J_{\beta'\gamma'\cdot\delta'}$$

$$+ \sum_{\alpha'} \sum_{\beta'} \sum_{\gamma'} \sum_{\delta'} I^{\alpha\alpha'} I^{\beta\beta'} I^{\gamma\gamma'} I^{\delta\delta'} \left(N_{\beta'\gamma'\cdot\alpha'\delta'} + L_{\alpha'\beta'\gamma'\cdot\delta'} + M_{\beta'\gamma'\cdot\alpha'\delta'} \right) .$$

Introducing still other notations as

$$J^{\alpha\beta}_{\cdot\gamma'} = \sum_{\alpha'} \sum_{\beta'} I^{\alpha\alpha'} I^{\beta\beta'} J_{\alpha'\beta'\cdot\gamma'} \; ;$$

$$J^{\alpha\cdot\gamma}_{\beta} = \sum_{\alpha'} \sum_{\gamma'} I^{\alpha\alpha'} I^{\gamma\gamma'} J_{\alpha'\beta'\cdot\gamma'} \; ;$$

$$K^{\alpha\beta}_{\gamma'} = \sum_{\alpha'} \sum_{\beta'} I^{\alpha\alpha'} I^{\beta\beta'} K_{\alpha'\beta'\gamma'}$$

the above can be abreviated as

$$D_\alpha J^{\beta\gamma\cdot\delta} = - \sum_{\beta'} (K^{\alpha\beta\beta'} + J^{\alpha\beta\cdot\beta'} + J^{\alpha\beta'\cdot\beta}) J^{\gamma\cdot\delta}_{\beta'}$$

$$- \sum_{\gamma'} (K^{\alpha\gamma\gamma'} + J^{\alpha\gamma\cdot\gamma'} + J^{\alpha\gamma'\cdot\gamma}) J^{\beta\cdot\delta}_{\gamma'}$$

$$- \sum_{\delta'} (K^{\alpha\delta\delta'} + J^{\alpha\delta\cdot\delta'} + J^{\alpha\delta'\cdot\delta}) J^{\beta\gamma}_{\cdot\delta'}$$

$$+ N^{\beta\gamma\cdot\alpha\delta} + L^{\alpha\beta\gamma\cdot\delta} + M^{\alpha\delta\cdot\beta\gamma} \ ; \tag{4.1}$$

$$D_\alpha K^{\beta\gamma\delta} = \sum_{\beta'} \sum_{\gamma'} \sum_{\delta'} (D_\alpha I^{\beta\beta'}) I^{\gamma\gamma'} I^{\delta\delta'} K_{\beta'\gamma'\delta'}$$

$$+ \sum_{\beta'} \sum_{\gamma'} \sum_{\delta'} I^{\beta\beta'} (D_\alpha I^{\gamma\gamma'}) I^{\delta\delta'} K_{\beta'\gamma'\delta'}$$

$$+ \sum_{\beta'} \sum_{\gamma'} \sum_{\delta'} I^{\beta\beta'} I^{\alpha\alpha'} (D_\alpha I^{\delta\delta'}) K_{\beta'\gamma'\delta'}$$

$$+ \sum_{\beta'} \sum_{\gamma'} \sum_{\delta'} I^{\beta\beta'} I^{\gamma\gamma'} I^{\delta\delta'} (D_\alpha K_{\beta'\gamma'\delta'})$$

$$= - \sum_{\beta'} (K^{\alpha\beta\beta'} + J^{\alpha\beta\cdot\beta'} + J^{\alpha\beta'\cdot\beta}) K^{\gamma\delta}_{\beta'}$$

$$- \sum_{\gamma'} (K^{\alpha\gamma\gamma'} + J^{\alpha\gamma\cdot\gamma'} + J^{\alpha\gamma'\cdot\gamma}) K^{\beta\delta}_{\gamma'}$$

$$- \sum_{\delta'} (K^{\alpha\delta\delta'} + J^{\alpha\delta\cdot\delta'} + J^{\alpha\delta'\cdot\delta}) K^{\beta\gamma}_{\delta'}$$

$$+ H^{\alpha\beta\gamma\delta} + N^{\alpha\beta\cdot\gamma\delta} + N^{\alpha\gamma\cdot\beta\delta} + N^{\alpha\delta\cdot\beta\gamma} \ . \tag{4.2}$$

We have

$$D_\alpha U_\beta = D_\alpha D_\beta \sum_i s_i T'_i$$

$$= \sum_i (D_\alpha D_\beta s_i) T'_i - \sqrt{n} \sum_i D_\beta s_i (D_\alpha \mu_i)$$

$$= \sum_i (D_\alpha D_\beta s_i) T'_i - \sqrt{n} \sum_i \sum_j (D_\beta s_i) \sigma_{ij} (D_\alpha s_j)$$

$$= \sum_i (D_\alpha D_\beta s_i) T'_i$$

$$- \sqrt{n} \sum_{\alpha'} \sum_{\beta'} \sum_i \sum_j I^{\alpha\alpha'} I^{\beta\beta'} \sigma_{ij} \left(\frac{\partial s_i}{\partial \theta_{\alpha'}}\right) \left(\frac{\partial s_j}{\partial \theta_{\beta'}}\right)$$

$$= \sum_i (D_\alpha D_\beta s_i) T'_i - \sqrt{n} \sum_{\alpha'} \sum_{\beta'} I^{\alpha\alpha'} I^{\beta\beta'} I_{\alpha'\beta'}$$

$$= \sum_i (D_\alpha D_\beta s_i) T'_i - \sqrt{n} I^{\alpha\beta} .$$

we shall denote

$$V_{\alpha\beta} = \sum_i (D_\alpha D_\beta s_i) T'_i .$$

280   Kei TAKEUCHI and Masafumi AKAHIRA

Then we have the second lemma:

Lemma 4.2   If $U_\beta Z_\theta$ and Z_θ both satisfy the condition of the lemma 4.1 we have

$$E_\theta(U_\alpha U_\beta Z_\theta) = \frac{1}{\sqrt{n}} D_\alpha E_\theta(U_\beta Z_\theta) + I^{\alpha\beta} E_\theta(Z_\theta)$$

$$- \frac{1}{\sqrt{n}} E_\theta(V_{\alpha\beta} Z_\theta) - \frac{1}{\sqrt{n}} E_\theta(U_\alpha D_\alpha Z_\theta)$$

$$= I^{\alpha\beta} E_\theta(Z_\theta) - \frac{1}{\sqrt{n}} E_\theta(V_{\alpha\beta} Z_\theta) + \frac{1}{n} D_\alpha D_\beta E_\theta(Z_\theta)$$

$$- \frac{1}{n} D_\alpha E_\theta(D_\beta Z_\theta) - \frac{1}{n} D_\beta E_\theta(D_\alpha Z_\theta) + \frac{1}{n} E_\theta(D_\alpha D_\beta Z_\theta)$$

Proof.   By lemma 4.1, we have

$$E_\theta(U_\alpha U_\beta Z_\theta) = \frac{1}{\sqrt{n}} D_\alpha E_\theta(U_\beta Z_\theta) - \frac{1}{\sqrt{n}} E_\theta(D_\alpha(U_\beta Z_\theta))$$

$$= \frac{1}{\sqrt{n}} D_\alpha E_\theta(U_\beta Z_\theta) - \frac{1}{\sqrt{n}} E_\theta\left\{ (V_{\alpha\beta} - \sqrt{n}\, I^{\alpha\beta}) Z_\theta \right\}$$

$$- \frac{1}{\sqrt{n}} E_\theta(U_\beta D_\alpha Z_\theta)$$

which establishes the first half of the lemma. The second half is derived by applying the lemma
4.1, to terms, $E_\theta(U_\beta Z_\theta)$ and $E_\theta(U_\beta D_\alpha Z_\theta)$ in the above. (QED)

We shall derive the moments of U_α and $V_{\beta\gamma}$. Obviously it is shown that

$$E(U_\alpha) = 0 \;\; ; \;\; E(V_{\alpha\beta}) = 0 \;\; ; \;\; E(U_\alpha U_\beta) = I^{\alpha\beta}.$$

We have

$$V_{\alpha\beta} = \sum_i (D_\alpha D_\beta s_i)\, T'_i$$

$$= \sum_i \sum_{\beta'} D_\alpha \left(I^{\beta\beta'} \frac{\partial s_i}{\partial \theta_\beta} \right) T'_i$$

$$= \sum_i \sum_{\beta'} (D_\alpha I^{\beta\beta'}) \left(\frac{\partial s_i}{\partial \theta_{\beta'}} \right) T'_i + \sum_i \sum_{\alpha'} \sum_{\beta'} I^{\alpha\alpha'} I^{\beta\beta'} \left(\frac{\partial^2 s_i}{\partial \theta_{\alpha'} \partial \theta_{\beta'}} \right) T'_i$$

$$= - \sum_i \sum_{\beta'} (K^{\alpha\beta\beta'} + J^{\alpha\beta\cdot\beta'} + J^{\alpha\beta'\cdot\beta}) \left(\frac{\partial s_i}{\partial \theta_{\beta'}} \right) T'_i$$

$$+ \sum_i \sum_{\alpha'} \sum_{\beta'} I^{\alpha\alpha'} I^{\beta\beta'} \left(\frac{\partial^2 s_i}{\partial \theta_{\alpha'} \partial \theta_{\beta'}} \right) T'_i.$$

Hence it follows that

$$E(V_{\alpha\beta}U_\gamma) = -\sum_{\beta'}\sum_{\gamma'} (K^{\alpha\beta\beta'} + J^{\alpha\beta\cdot\beta'} + J^{\alpha\beta'\cdot\beta}) \, I^{\gamma\gamma'} I_{\beta'\gamma'}$$

$$+ \sum_{\alpha'}\sum_{\beta'}\sum_{\gamma'} I^{\alpha\alpha'} I^{\beta\beta'} I^{\gamma\gamma'} J_{\alpha'\beta'\cdot\gamma'}$$

$$= -K^{\alpha\beta\gamma} - J^{\alpha\beta\cdot\gamma} - J^{\alpha\gamma\cdot\beta} + J^{\alpha\beta\cdot\gamma}$$

$$= -K^{\alpha\beta\gamma} - J^{\alpha\gamma\cdot\beta} \;;$$

$$E(V_{\alpha\beta}V_{\gamma\delta}) = \sum_{\beta'}\sum_{\delta'} (K^{\alpha\beta\beta'} + J^{\alpha\beta\cdot\beta'} + J^{\alpha\beta'\delta'}) I_{\beta'\gamma'} \cdot (K^{\gamma\delta\delta'} + J^{\gamma\delta\cdot\delta'} + J^{\gamma\delta'\cdot\delta})$$

$$- \sum_{\beta'}\sum_{\delta'} (K^{\alpha\beta\beta'} + J^{\alpha\beta\cdot\beta'} + J^{\alpha\beta'\cdot\beta}) \, I^{\gamma\gamma'} I^{\delta\delta'} J_{\gamma'\delta'\cdot\beta'}$$

$$- \sum_{\beta'}\sum_{\gamma'} (K^{\gamma\delta\delta'} + J^{\gamma\delta\cdot\delta'} + J^{\gamma\delta'\cdot\delta}) I^{\alpha\alpha'} I^{\beta\beta'} J_{\alpha'\beta'\cdot\delta'}$$

$$+ \sum_{\alpha'}\sum_{\beta'}\sum_{\gamma'}\sum_{\delta'} I^{\alpha\alpha'} I^{\beta\beta'} I^{\gamma\gamma'} I^{\delta\delta'} M_{\alpha'\beta'\cdot\gamma'\delta'}$$

$$= \sum_{\delta'} (K^{\alpha\beta}_{\delta'} + J^{\alpha\beta}_{\cdot\delta'} + J^{\alpha\cdot\beta}_{\delta'}) (K^{\gamma\delta\delta'} + J^{\gamma\delta\cdot\delta'} J^{\gamma\delta'\cdot\delta})$$

$$- \sum_{\beta'} (K^{\alpha\beta\beta'} + J^{\alpha\beta\cdot\beta'} + J^{\alpha\beta'\cdot\beta}) J^{\gamma\delta}_{\beta'}$$

$$- \sum_{\delta'} (K^{\gamma\delta\delta'} + J^{\gamma\delta\cdot\delta'} + J^{\gamma\delta'\cdot\delta}) J^{\alpha\beta}_{\cdot\delta'}$$

$$+ M^{\alpha\beta\cdot\gamma\delta}$$

$$= \sum_{\delta'} (K^{\alpha\beta}_{\delta'} + J^{\alpha\cdot\beta}_{\delta'}) (K^{\gamma\delta\cdot\delta'} + J^{\gamma\delta'\cdot\delta})$$

$$- \sum_{\delta'} J^{\alpha\beta}_{\cdot\delta'} J^{\gamma\delta\cdot\delta'} + M^{\alpha\beta\cdot\gamma\delta} \;; \tag{4.3}$$

$$\sqrt{n}\, E(U_\alpha U_\beta U_\gamma) = K^{\alpha\beta\gamma} \;;$$

$$\sqrt{n}\, E(V_{\alpha\beta} U_\gamma U_\delta) = -\sum_{\beta'}\sum_{\gamma'}\sum_{\delta'} (K^{\alpha\beta\beta'} + J^{\alpha\beta\cdot\beta'} + J^{\alpha\beta'\cdot\beta}) \, I^{\gamma\gamma'} I^{\delta\delta'} K_{\beta'\gamma'\delta'}$$

$$+ \sum_{\alpha'}\sum_{\beta'}\sum_{\gamma'}\sum_{\delta'} I^{\alpha\alpha'} I^{\beta\beta'} I^{\gamma\gamma'} I^{\delta\delta'} N_{\alpha'\beta'\cdot\gamma'\delta'}$$

$$= -\sum_{\beta'} (K^{\alpha\beta\beta'} + J^{\alpha\beta\cdot\beta'} + J^{\alpha\beta'\cdot\beta}) K^{\gamma\delta}_{\beta'} + N^{\alpha\beta\cdot\gamma\delta} \;; \tag{4.4}$$

$$E(U_\alpha U_\beta U_\gamma U_\delta) = \frac{1}{n} H^{\alpha\beta\gamma\delta} + I^{\alpha\beta} I^{\gamma\delta} + I^{\alpha\gamma} I^{\beta\delta} + I^{\alpha\delta} I^{\beta\gamma}.$$

5. Asymptotic cumulants of ER BAN estimators

We shall obtain asymptotic joint cumulants of Z_α's making use of the results of the previous section. Note that the asymptotic cumulants of Z_α's are equal to those of $\sqrt{n}\,(\hat{\theta}_\alpha - \theta_\alpha)$ up to the order n^{-1}.

We can write

$$Y_\alpha = U_\alpha + \frac{1}{2\sqrt{n}}\, Q_\alpha + \frac{1}{n}\, W_\alpha \,.$$

And we have
$$E(Y_\alpha) = \frac{1}{2\sqrt{n}} E(Q_\alpha) + o\left(\frac{1}{n}\right)$$

We shall denote $\frac{1}{2} E(Q_\alpha) = \mu_\alpha$, and

$$Z_\alpha = Y_\alpha - \frac{1}{\sqrt{n}}\, \mu_\alpha \,.$$

Then we have

$$D_\alpha Z_\beta = -\sqrt{n}\, D_\alpha \theta_\beta - \frac{1}{\sqrt{n}} D_\alpha \mu_\beta = -\sqrt{n}\, I^{\alpha\beta} - \frac{1}{\sqrt{n}} D_\alpha \mu_\beta \,.$$

First we shall obtain the variance of Z_α up to the order n^{-1}.
We have

$$E(Z_\alpha - U_\alpha)^2 = \frac{1}{4n} V(Q_\alpha) + o\left(\frac{1}{n}\right)$$

and we have

$$Cov(Z_\alpha, Z_\beta) = E(Z_\alpha Q_\beta) + o\left(\frac{1}{n}\right)$$

$$= E(U_\alpha Z_\beta) + E(U_\beta Z_\alpha) - E(U_\alpha U_\beta) + E(Z_\alpha - U_\alpha)^2$$

$$+ o\left(\frac{1}{n}\right) \,.$$

Applying the lemma 4.1 of the previous section we get

$$E(U_\alpha Z_\beta) = \frac{1}{\sqrt{n}} D_\alpha E(Z_\beta) - \frac{1}{\sqrt{n}} E(D_\alpha Z_\beta)$$

$$= I^{\alpha\beta} + \frac{1}{n} D_\alpha \mu_\beta \,;$$

$$E(U_\beta Z_\alpha) = I^{\alpha\beta} + \frac{1}{n} D_\beta \mu_\alpha \,.$$

Hence we have

$$Cov\,(Z_\alpha, Z_\beta) \;=\; Cov\,(Y_\alpha, Y_\beta)$$

$$= I^{\alpha\beta} + \frac{1}{n}\,D_\alpha\,\mu_\beta + \frac{1}{n}\,D_\beta\,\mu_\alpha + \frac{1}{4n}\,V(Q_\alpha) + o\!\left(\frac{1}{n}\right).$$

As to the third cumulants, the following equation is referred to

$$Z_\alpha Z_\beta Z_\gamma = \frac{1}{2}\,(\,U_\alpha Z_\beta Z_\gamma + U_\beta Z_\alpha Z_\gamma + U_\gamma Z_\alpha Z_\beta)$$

$$-\frac{1}{2}\,U_\alpha U_\beta U_\gamma + \frac{1}{2}\,U_\alpha\,(Z_\beta - U_\beta)\,(Z_\alpha - U_\gamma)$$

$$+\frac{1}{2}\,U_\beta\,(Z_\alpha - U_\alpha)\,(Z_\gamma - U_\gamma) + \frac{1}{2}\,U_\gamma\,(Z_\alpha - U_\alpha)\,(Z_\beta - U_\beta)$$

$$+\,(Z_\alpha - U_\alpha)\,(Z_\beta - U_\beta)\,(Z_\gamma - U_\gamma)\,.$$

Since the last four terms are of magnitude of order smaller than n^{-1}, we have

$$E\,(Z_\alpha Z_\beta Z_\gamma) = \frac{1}{2}\,E\,(U_\alpha Z_\beta Z_\gamma + U_\beta Z_\alpha Z_\gamma + U_\gamma Z_\alpha Z_\beta)$$

$$-\frac{1}{2}\,E\,(U_\alpha U_\beta U_\gamma) + o\!\left(\frac{1}{n}\right).$$

Furthermore we drive that

$$E\,(U_\alpha Z_\beta Z_\gamma) = \frac{1}{\sqrt{n}}\,D_\alpha\,E\,(Z_\beta Z_\gamma) - \frac{1}{\sqrt{n}}\,E\,\{\,(D_\alpha Z_\beta)\,Z_\gamma\,\}$$

$$-\frac{1}{\sqrt{n}}\,E\{Z_\beta\,(D_\alpha Z_\gamma)\}$$

$$= \frac{1}{\sqrt{n}}\,D_\alpha I^{\beta\gamma} + o\!\left(\frac{1}{n}\right)$$

$$= -\frac{1}{\sqrt{n}}\,(K^{\alpha\beta\gamma} + J^{\alpha\beta\cdot\gamma} + J^{\alpha\gamma\cdot\beta}) + o\!\left(\frac{1}{n}\right).$$

Thus the third cumulant is given as

$$E\,(Z_\alpha Z_\beta Z_\gamma) = -\frac{1}{\sqrt{n}}\,(2K^{\alpha\beta\gamma} + J^{\alpha\beta\cdot\gamma} + J^{\alpha\gamma\cdot\beta} + J^{\beta\gamma\cdot\alpha}) + o\!\left(\frac{1}{n}\right)$$

$$= \frac{1}{\sqrt{n}}\,\beta_{\alpha\beta\gamma} + o\!\left(\frac{1}{n}\right).$$

As to the fourth cumulants, we shall need the following equation,

Kei TAKEUCHI and Masafumi AKAHIRA

$$Z_\alpha Z_\beta Z_\gamma Z_\delta = \frac{2}{3} (U_\alpha Z_\beta Z_\gamma Z_\delta + U_\beta Z_\alpha Z_\gamma Z_\delta + U_\gamma Z_\alpha Z_\beta Z_\delta + U_\delta Z_\alpha Z_\beta Z_\gamma)$$

$$- \frac{1}{3} (U_\alpha U_\beta Z_\gamma Z_\delta + U_\alpha U_\gamma Z_\beta Z_\delta + U_\alpha U_\delta Z_\beta Z_\gamma + U_\beta U_\gamma Z_\alpha Z_\delta + U_\beta U_\delta Z_\alpha Z_\gamma$$

$$+ U_\alpha U_\delta Z_\alpha Z_\beta) + \frac{1}{3} U_\alpha U_\beta U_\gamma U_\delta + \frac{1}{3} U_\alpha (Z_\beta - U_\beta)(Z_\gamma - U_\gamma)(Z_\delta - U_\delta)$$

$$+ \frac{1}{3} U_\beta (Z_\alpha - U_\alpha)(Z_\gamma - U_\gamma)(Z_\delta - U_\delta) + \frac{1}{3} U_\gamma (Z_\alpha - U_\alpha)(Z_\beta - U_\beta)$$

$$\cdot (Z_\delta - U_\delta) + \frac{1}{3} U_\delta (Z_\alpha - U_\alpha)(Z_\beta - U_\beta)(Z_\gamma - U_\gamma)$$

$$+ (Z_\alpha - U_\alpha)(Z_\beta - U_\beta)(Z_\gamma - U_\gamma)(Z_\delta - U_\delta) .$$

The last four terms in the above are shown to be of magnitude of order less than n^{-1}, hence we have

$$E (Z_\alpha Z_\beta Z_\gamma Z_\delta) = \frac{2}{3} E (U_\alpha Z_\beta Z_\gamma Z_\delta + U_\beta Z_\alpha Z_\gamma Z_\delta + U_\gamma Z_\alpha Z_\beta Z_\delta + U_\delta Z_\alpha Z_\beta Z_\gamma)$$

$$- \frac{1}{3} E (U_\alpha U_\beta Z_\gamma Z_\delta + U_\alpha U_\gamma Z_\beta Z_\delta + U_\alpha U_\delta Z_\beta Z_\gamma + U_\beta U_\gamma Z_\alpha Z_\delta$$

$$+ U_\beta U_\delta Z_\alpha Z_\gamma + U_\gamma U_\delta Z_\alpha Z_\beta) + \frac{1}{3} E (U_\alpha U_\beta U_\gamma U_\delta) + o (\frac{1}{n}) \qquad (5.1)$$

Applying the lemma 1 in the previous section, it follows that

$$E (U_\alpha Z_\beta Z_\gamma Z_\delta) = \frac{1}{\sqrt{n}} D_\alpha E (Z_\beta Z_\gamma Z_\delta) - \frac{1}{\sqrt{n}} E ((D_\alpha Z_\beta) Z_\gamma Z_\delta)$$

$$- \frac{1}{\sqrt{n}} E ((D_\alpha Z_\gamma) Z_\beta Z_\delta) - \frac{1}{\sqrt{n}} E ((D_\alpha Z_\delta) Z_\beta Z_\gamma)$$

$$= - \frac{1}{n} D_\alpha (2K^{\beta\gamma\delta} + J^{\beta\gamma\cdot\delta} + J^{\beta\delta\cdot\gamma} + J^{\gamma\delta\cdot\beta})$$

$$+ (I^{\alpha\beta} + \frac{1}{n} D_\alpha \mu_\beta) E (Z_\gamma Z_\delta) + (I^{\alpha\gamma} + \frac{1}{n} D_\alpha \mu_\gamma) E (Z_\beta Z_\delta)$$

$$+ (I^{\alpha\delta} + \frac{1}{n} D_\alpha \mu_\delta) E (Z_\beta Z_\gamma) + o (\frac{1}{n})$$

$$= - \frac{1}{n} D_\alpha (2K^{\beta\gamma\delta} + J^{\beta\gamma\cdot\delta} + J^{\beta\delta\cdot\gamma} + J^{\gamma\delta\beta})$$

$$+ I^{\alpha\beta} E (Z_\gamma Z_\delta) + I^{\alpha\gamma} E (Z_\beta Z_\delta) + I^{\alpha\delta} E (Z_\beta Z_\gamma)$$

$$+ \frac{1}{n} \left(I^{\gamma\delta} D_\alpha \mu_\beta + I^{\beta\delta} D_\alpha \mu_\gamma + I^{\beta\gamma} D_\alpha \mu_\delta \right) + o\left(\frac{1}{n}\right).$$

It also follows that

$$E(U_\alpha U_\beta Z_\gamma Z_\delta) = I^{\alpha\beta} E(Z_\gamma Z_\delta) - \frac{1}{\sqrt{n}} E(V_{\alpha\beta} Z_\gamma Z_\delta)$$

$$+ \frac{1}{n} D_\alpha D_\beta E(Z_\gamma Z_\delta) - \frac{1}{n} D_\alpha E((D_\beta Z_\gamma) Z_\delta) - \frac{1}{n} D_\alpha E(Z_\gamma D_\beta Z_\delta)$$

$$- \frac{1}{n} D_\beta E((D_\alpha Z_\gamma) Z_\delta) - \frac{1}{n} D_\beta E(Z_\gamma D_\alpha Z_\delta)$$

$$+ \frac{1}{n} E((D_\alpha D_\beta Z_\gamma) Z_\delta) + \frac{1}{n} E((D_\alpha Z_\gamma)(D_\beta Z_\delta))$$

$$+ \frac{1}{n} E((D_\beta Z_\gamma)(D_\alpha Z_\delta)) + \frac{1}{n} E((D_\alpha D_\beta Z_\delta) Z_\gamma)$$

$$= I^{\alpha\beta} E(Z_\gamma Z_\delta) - \frac{1}{\sqrt{n}} E(V_{\alpha\beta} Z_\gamma Z_\delta) + \frac{1}{n} D_\alpha D_\beta I^{\gamma\delta}$$

$$+ \left(I^{\alpha\gamma} + \frac{1}{n} D_\alpha \mu_\gamma\right)\left(I^{\beta\delta} + \frac{1}{n} D_\beta \mu_\delta\right)$$

$$+ \left(I^{\beta\gamma} + \frac{1}{n} D_\beta \mu_\gamma\right)\left(I^{\alpha\delta} + \frac{1}{n} D_\alpha \mu_\delta\right) + o\left(\frac{1}{n}\right).$$

We shall show that

$$E(V_{\alpha\beta} Z_\gamma Z_\delta) = E(V_{\alpha\beta} U_\gamma U_\delta) + \frac{1}{\sqrt{n}} E(V_{\alpha\beta} V_{\gamma\delta}) + \frac{1}{\sqrt{n}} E(V_{\alpha\beta} V_{\delta\gamma})$$

$$+ o\left(\frac{1}{\sqrt{n}}\right).$$

In order to prove this, it is first noted that

$$E(V_{\alpha\beta} Z_\beta Z_\delta) = E(V_{\alpha\beta} U_\gamma U_\delta) + E(V_{\alpha\beta} U_\delta Z_\gamma)$$

$$- E(V_{\alpha\beta} U_\gamma U_\delta) + o\left(\frac{1}{\sqrt{n}}\right).$$

And we have

$$E(U_\gamma V_{\alpha\beta} Z_\delta) = \frac{1}{\sqrt{n}} D_\gamma E(V_{\alpha\beta} Z_\delta) - \frac{1}{\sqrt{n}} E((D_\gamma V_{\alpha\beta}) Z_\delta)$$

$$- \frac{1}{\sqrt{n}} E(V_{\alpha\beta} D_\gamma Z_\delta)$$

$$= \frac{1}{\sqrt{n}} \, D_\gamma E(V_{\alpha\beta} Z_\delta) - \frac{1}{\sqrt{n}} \, E((D_\gamma V_{\alpha\beta}))Z_\delta) + o(\frac{1}{\sqrt{n}})$$

$$= \frac{1}{\sqrt{n}} \, D_\gamma E(V_{\alpha\beta} U_\delta) - \frac{1}{\sqrt{n}} \, E((D_\gamma V_{\alpha\beta}))U_\delta) + o(\frac{1}{\sqrt{n}}) \, .$$

On the other hand, it holds that

$$E(U_\gamma V_{\alpha\beta} U_\delta) = \frac{1}{\sqrt{n}} \, D_\gamma E(V_{\alpha\beta} U_\delta) - \frac{1}{\sqrt{n}} \, E((D_\gamma V_{\alpha\beta})U_\delta)$$

$$- \frac{1}{\sqrt{n}} \, E(V_{\alpha\beta} V_{\gamma\delta}) + o(\frac{1}{\sqrt{n}}) \, .$$

It follows that

$$E(U_\gamma V_{\alpha\beta} Z_\delta) = E(U_\gamma V_{\alpha\beta} U_\delta) + E(V_{\alpha\beta} V_{\gamma\delta}) + o(\frac{1}{\sqrt{n}})$$

from which the desired relation is obtained.

Thus we get

$$E(U_\alpha U_\beta Z_\gamma Z_\delta) = I^{\alpha\beta} E(Z_\gamma Z_\delta) - \frac{1}{\sqrt{n}} \, E(V_{\alpha\beta} U_\gamma U_\delta)$$

$$- \frac{1}{n} E(V_{\alpha\beta} V_{\gamma\delta}) - \frac{1}{n} E(V_{\alpha\beta} V_{\delta\gamma})$$

$$+ I^{\alpha\gamma} I^{\beta\delta} + I^{\alpha\delta} I^{\beta\gamma} + \frac{1}{n} D_\alpha D_\beta I^{\gamma\delta}$$

$$+ \frac{1}{n} (I^{\alpha\gamma} D_\beta \mu_\delta + I^{\beta\gamma} D_\alpha \mu_\delta + I^{\alpha\delta} D_\beta \mu_\gamma + I^{\beta\gamma} D_\alpha \mu_\gamma) + o(\frac{1}{n}) \, .$$

Substituting (4.1), (4.2), (4.3) and (4.4) into the terms $E(V_{\alpha\beta} U_\gamma U_\delta)$, $E(V_{\alpha\beta} V_{\gamma\delta})$, $E(V_{\alpha\beta} V_{\delta\gamma})$ and $D_\alpha D_\beta I^{\gamma\delta} = -D_\alpha(K^{\beta\gamma\delta} + J^{\beta\gamma\cdot\delta} + J^{\beta\delta\cdot\gamma})$, we obtain after some algebraic manipulations,

$$E(U_\alpha U_\beta Z_\gamma Z_\delta) = I^{\alpha\beta} E(Z_\gamma Z_\delta) + I^{\alpha\gamma} I^{\beta\delta} + I^{\alpha\delta} I^{\beta\gamma}$$

$$+ \frac{1}{n} (I^{\alpha\gamma} D_\beta \mu_\delta + I^{\alpha\beta} D_\beta \mu_\gamma + I^{\beta\gamma} D_\alpha \mu_\delta + I^{\beta\delta} D_\alpha \mu_\delta)$$

$$- \frac{1}{n} (H^{\alpha\beta\gamma\delta} + 2N^{\alpha\beta\cdot\gamma\delta} + N^{\alpha\gamma\cdot\beta\delta} + N^{\alpha\delta\cdot\beta\gamma}$$

$$+ N^{\beta\gamma\cdot\alpha\delta} + N^{\beta\delta\cdot\alpha\gamma} + L^{\alpha\beta\gamma\cdot\delta} + L^{\alpha\beta\delta\cdot\gamma} + 2M^{\alpha\beta\cdot\gamma\delta}$$

$$+ M^{\alpha\gamma\cdot\beta\delta} + M^{\alpha\delta\cdot\beta\gamma})$$

$$+ \frac{1}{n} \sum_{\gamma'} (K^{\alpha\gamma\gamma'} + J^{\alpha\gamma\cdot\gamma'} + J^{\alpha\gamma'\cdot\gamma}) (K^{\beta\delta}_{\gamma'} + J^{\beta\delta}_{\gamma'} + J^{\beta\cdot\delta}_{\gamma'})$$

$$+ \frac{1}{n} \sum_{\delta'} (K^{\alpha\delta\delta'} + J^{\alpha\delta\cdot\delta'} + J^{\alpha\delta'\cdot\delta}) (K^{\beta\gamma}_{\delta'} + J^{\beta\gamma}_{\delta'} + J^{\beta\cdot\gamma}_{\delta'})$$

$$+ \frac{1}{n} \sum_{\gamma'} J^{\alpha\beta\cdot\gamma'} (2K^{\beta\delta}_{\gamma'} + 2J^{\gamma\delta\cdot\gamma'} + J^{\gamma\gamma'\cdot\delta} + J^{\gamma'\delta\cdot\gamma}) + o\left(\frac{1}{n}\right).$$

Calculating similarly we have

$$E(U_\alpha Z_\beta Z_\gamma Z_\delta) = I^{\alpha\beta} E(Z_\gamma Z_\delta) + I^{\alpha\gamma} E(Z_\beta Z_\delta) + I^{\alpha\delta} E(Z_\beta Z_\gamma)$$

$$+ \frac{1}{n} (I^{\gamma\delta} D_\alpha \mu_\beta + I^{\beta\delta} D_\alpha \mu_\gamma + I^{\beta\gamma} D_\alpha \mu_\beta)$$

$$- \frac{1}{n} (2H^{\alpha\beta\gamma\delta} + 2N^{\alpha\beta\cdot\gamma\delta} + 2N^{\alpha\gamma\cdot\beta\delta} + 2N^{\alpha\delta\cdot\beta\gamma} + N^{\beta\gamma\cdot\alpha\delta}$$

$$+ N^{\beta\delta\cdot\alpha\gamma} + N^{\gamma\delta\cdot\alpha\beta} + L^{\alpha\beta\gamma\cdot\delta} + L^{\alpha\beta\delta\cdot\gamma} + L^{\alpha\gamma\delta\cdot\beta}$$

$$+ M^{\alpha\delta\cdot\beta\gamma} + M^{\alpha\gamma\cdot\beta\delta} + M^{\alpha\beta\cdot\gamma\delta})$$

$$+ \frac{1}{n} \sum_{\beta'} (K^{\alpha\beta\beta'} + J^{\alpha\beta\cdot\beta'} + J^{\alpha\beta'\cdot\beta}) (2K^{\gamma\delta}_{\beta'} + J^{\gamma\delta}_{\beta'} + J^{\gamma\cdot\delta}_{\beta'} + J^{\delta\cdot\gamma}_{\beta'})$$

$$+ \frac{1}{n} \sum_{\gamma'} (K^{\alpha\gamma\gamma'} + J^{\alpha\gamma\cdot\gamma'} + J^{\alpha\gamma'\cdot\gamma}) (2K^{\beta\delta}_{\gamma'} + J^{\beta\delta}_{\gamma'} + J^{\beta\cdot\delta}_{\gamma'} + J^{\delta\cdot\beta}_{\gamma'})$$

$$+ \frac{1}{n} \sum_{\delta'} (K^{\alpha\delta\delta'} + J^{\alpha\delta\cdot\delta'} + J^{\alpha\delta'\cdot\delta}) (2K^{\beta\gamma}_{\delta'} + J^{\beta\gamma}_{\delta'} + J^{\beta\cdot\gamma}_{\delta'} + J^{\gamma\cdot\beta}_{\delta'})$$

$$+ o\left(\frac{1}{n}\right).$$

Substituting all those forms together with

$$E(U_\alpha U_\beta U_\gamma U_\delta) = \frac{1}{n} H^{\alpha\beta\gamma\delta} + I^{\alpha\beta} I^{\gamma\delta} + I^{\alpha\gamma} I^{\beta\delta} + I^{\alpha\delta\cdot\beta\gamma}$$

into (5.1) we have the fourth cumulants as

$$E(Z_\alpha Z_\beta Z_\gamma Z_\delta) - E(Z_\alpha Z_\beta) E(Z_\gamma Z_\delta) - E(Z_\alpha Z_\gamma) E(Z_\beta Z_\delta) - E(Z_\alpha Z_\delta)$$

$$\cdot E(Z_\beta Z_\gamma)$$

$$= (I^{\alpha\beta} - E(Z_\alpha Z_\beta)) (I^{\gamma\delta} - E(Z_\gamma Z_\delta)) + (I^{\alpha\gamma} - E(Z_\alpha Z_\gamma))$$

$$\cdot (I^{\beta\delta} - E(Z_\beta Z_\delta)) + (I^{\alpha\delta} - E(Z_\alpha Z_\delta)) (I^{\beta\gamma} - E(Z_\beta Z_\gamma))$$

$$- \frac{1}{n} (H^{\alpha\beta\gamma\delta} + 2N^{\alpha\beta\cdot\gamma\delta} + 2N^{\alpha\gamma\cdot\beta\delta} + 2N^{\alpha\delta\cdot\beta\gamma} + 2N^{\beta\gamma\cdot\alpha\delta} + 2N^{\beta\delta\cdot\alpha\gamma}$$

$$+ 2N^{\gamma\delta\cdot\gamma\beta} + L^{\alpha\beta\gamma\cdot\delta} + L^{\alpha\beta\delta\cdot\gamma} + L^{\alpha\gamma\delta\cdot\beta} + L^{\beta\gamma\delta\cdot\alpha})$$

$$+ \frac{1}{n} \sum_{\beta'} (K^{\alpha\beta\cdot\beta'} + J^{\alpha\beta\cdot\beta'} + J^{\alpha\beta'\cdot\beta}) (2K^{\gamma\delta}_{\beta'} + J^{\gamma\cdot\delta}_{\beta'} + J^{\delta\cdot\gamma}_{\beta'})$$

$$+ \frac{1}{n} \sum_{\gamma'} (K^{\alpha\gamma\cdot\gamma'} + J^{\alpha\gamma\cdot\gamma'} + J^{\alpha\gamma'\cdot\gamma}) (2K^{\beta\delta}_{\gamma'} + J^{\beta\cdot\delta}_{\gamma'} + J^{\delta\cdot\beta}_{\gamma'})$$

$$+ \frac{1}{n} \sum_{\delta'} (K^{\alpha\delta\delta'} + J^{\alpha\delta\cdot\delta'} + J^{\alpha\delta'\cdot\delta}) (2K^{\beta\gamma}_{\delta'} + J^{\beta\cdot\gamma}_{\delta'} + J^{\gamma\cdot\beta}_{\delta'})$$

$$+ \frac{1}{n} \sum_{\gamma'} (K^{\beta\gamma\cdot\gamma'} + J^{\beta\gamma\cdot\gamma'} + J^{\beta\gamma'\cdot\gamma}) (2K^{\alpha\delta}_{\gamma'} + J^{\alpha\cdot\delta}_{\gamma'} + J^{\delta\cdot\alpha}_{\gamma'})$$

$$+ \frac{1}{n} \sum_{\delta'} (K^{\beta\delta\cdot\delta'} + J^{\beta\delta\cdot\delta'} + J^{\beta\delta'\cdot\delta}) (2K^{\alpha\gamma}_{\delta'} + J^{\alpha\cdot\gamma}_{\delta'} + J^{\gamma\cdot\alpha}_{\delta'})$$

$$+ \frac{1}{n} \sum_{\delta'} (K^{\gamma\delta\delta'} + J^{\gamma\delta\cdot\delta'} + J^{\gamma\delta'\cdot\delta}) (2K^{\alpha\beta}_{\delta'} + J^{\alpha\cdot\beta}_{\delta'} + J^{\beta\cdot\alpha}_{\delta'}) + o(\frac{1}{n}).$$

The first three terms in the above are of order smaller than n^{-1}

We denote

$$\Delta^{\alpha\beta\cdot\gamma\delta} = \sum_{\beta'} (K^{\alpha\beta\cdot\beta'} + J^{\alpha\beta\cdot\beta'} + J^{\alpha\beta'\cdot\beta}) (2K^{\gamma\delta}_{\beta'} + J^{\gamma\cdot\delta}_{\beta'} + J^{\delta\cdot\gamma}_{\beta'})$$

$$= \sum_{\xi} \sum_{\alpha'} \sum_{\beta'} \sum_{\gamma'} \sum_{\delta'} \sum_{\xi'} I^{\alpha\alpha'} I^{\beta\beta'} I^{\gamma\gamma'} I^{\delta\delta'} I^{\xi\xi'} (K_{\alpha'\beta'\xi'} + J_{\alpha'\beta'\cdot\xi'} + J_{\alpha'\xi'\cdot\beta'})$$

$$\cdot (2K_{\gamma'\delta'\xi'} + J_{\gamma'\xi'\cdot\delta'} + J_{\delta'\xi'\cdot\gamma'})$$

Then the 4th cumulant is given by

$$-\frac{1}{n} H^{\alpha\beta\gamma\delta} - \frac{2}{n} (N^{\alpha\beta\cdot\gamma\delta} + N^{\alpha\gamma\cdot\beta\delta} + N^{\alpha\delta\cdot\beta\gamma} + N^{\beta\gamma\cdot\alpha\delta} + N^{\beta\delta\cdot\alpha\gamma} + N^{\alpha\delta\cdot\alpha\beta})$$

$$-\frac{1}{n} (L^{\alpha\beta\gamma\cdot\delta} + L^{\alpha\beta\delta\cdot\gamma} + L^{\alpha\gamma\delta\cdot\beta} + L^{\beta\gamma\delta\cdot\alpha})$$

$$+\frac{1}{n} (\Delta^{\alpha\beta\cdot\gamma\delta} + \Delta^{\alpha\gamma\cdot\beta\delta} + \Delta^{\alpha\delta\cdot\beta\gamma} + \Delta^{\beta\gamma\cdot\alpha\delta} + \Delta^{\beta\delta\cdot\alpha\gamma} + \Delta^{\gamma\delta\cdot\alpha\beta})$$

$$+ o(\frac{1}{n}) = \frac{1}{n} \beta_{\alpha\beta\gamma\delta} + o(\frac{1}{n}) \tag{5.2}$$

Summarizing the computations we have

Theorem 5.1. The asymptotic cumulants of $\sqrt{n} \, (\hat{\theta}_\alpha - \theta_\alpha) \, (\alpha = 1, \cdots, P)$ up to the order n^{-1}

are given by

$$E \, (\sqrt{n} \, (\hat{\theta}_\alpha - \theta_\alpha)) = \frac{1}{\sqrt{n}} \, \mu_\alpha + o(\frac{1}{n}) \, ;$$

$$Cov \, (\sqrt{n} \, (\hat{\theta}_\alpha - \theta_\alpha), \sqrt{n} \, (\hat{\theta}_\beta - \theta_\beta)) = I^{\alpha\beta} + \frac{1}{n} D_\alpha \mu_\beta + \frac{1}{n} D_\beta \mu_\alpha$$

$$+ \frac{1}{4n} Cov(Q_\alpha, Q_\beta) + o\left(\frac{1}{n}\right) ;$$

$$\kappa_3\left(\sqrt{n}\,(\hat{\theta}_\alpha - \theta_\alpha),\ \sqrt{n}\,(\hat{\theta}_\beta - \theta_\beta),\ \sqrt{n}\,(\hat{\theta}_\gamma - \theta_\gamma)\right) = \frac{1}{\sqrt{n}} \beta_{\alpha\beta\gamma} + o\left(\frac{1}{n}\right) ;$$

$$\kappa_4\left(\sqrt{n}\,(\hat{\theta}_\alpha - \theta_\alpha),\ \sqrt{n}\,(\hat{\theta}_\beta - \theta_\beta),\ \sqrt{n}\,(\hat{\theta}_\gamma - \theta_\gamma),\ \sqrt{n}\,(\hat{\theta}_\delta - \theta_\delta)\right)$$

$$= \frac{1}{n} \beta_{\alpha\beta\gamma\delta} - o\left(\frac{1}{n}\right),$$

where $\beta_{\alpha\beta\gamma}$ and $\beta_{\alpha\beta\gamma\delta}$ are given above.

It is also shown that

Theorem 5.2. The joint distribution of $\sqrt{n}\,(\hat{\theta}_\alpha - \theta_\alpha)$ $(\alpha = 1, \cdots, P)$ is equivalent up to the order n^{-1} to a continuous distribution, whose density is given as follows

$$f(y_1, \cdots, y_p) = \frac{|I|^{\frac{1}{2}}}{(\sqrt{2\pi})^p} \left[\exp -\frac{1}{2}\left\{ \sum_\alpha \sum_\beta I_{\alpha\beta}\, y_\alpha\, y_\beta \right\}\right]$$

$$\left\{ 1 + \frac{1}{6\sqrt{n}} \sum_\alpha \sum_\beta \sum_\gamma \beta_{\alpha\beta\gamma} H_{\alpha\beta\gamma} + \frac{1}{\sqrt{n}} \sum_\alpha \mu_\alpha H_\alpha \right.$$

$$+ \frac{1}{72n} \sum_\alpha \sum_\beta \sum_\gamma \sum_{\alpha'} \sum_{\beta'} \sum_{\gamma'} \beta_{\alpha\beta\gamma}\beta_{\alpha'\beta'\gamma'}\, H_{\alpha\beta\gamma\,\alpha'\beta'\gamma'}$$

$$+ \frac{1}{24n} \sum_\alpha \sum_\beta \sum_\gamma \sum_\delta \beta_{\alpha\beta\gamma\delta}\, H_{\alpha\beta\gamma\delta}$$

$$+ \frac{1}{2n} \sum_\alpha \sum_\beta (\mu_\alpha\mu_\beta + D_\alpha\mu_\beta + D_\beta\mu_\alpha)\, H_\alpha H_\beta$$

$$\left. + \frac{1}{4n} \sum_\alpha \sum_\beta Cov(Q_\alpha, Q_\beta)\, H_\alpha H_\beta \right\} + o\left(\frac{1}{n}\right)$$

where $H_{\alpha\beta}\ldots$ are (mlutivariate) Hermite polynomials defined by

$$H_\alpha = \sum_\beta I_{\alpha\beta}\, y_\beta ;$$

$$H_{\alpha\beta} = H_\alpha H_\beta - I_{\alpha\beta} ;$$

$$H_{\alpha\beta\gamma} = H_\alpha H_\beta H_\gamma - I^{\alpha\beta} H_\gamma - I^{\alpha\gamma} H_\beta - I^{\beta\gamma} H_\alpha ,$$

etc..

It should be remarked that the third and the fourth cumulants are equal for all ER BAN estimators up to the order n^{-1}.

6. Asymptotic higher order efficiency of ML estimators

Now we impose some higher order conditions on ER estimators. Either of the three conditions may be taken into consideration.

a) higher order asymptotic unbiasedness : $\mu_\alpha = 0$,

b) higher order asymptotic coordinate-wise median unbiasedness:

$$\lim_{n \to \infty} n \quad |\ Pr\{\ \sqrt{n}\ (\theta_\alpha - \theta_\alpha) < 0\} - \frac{1}{2}\ | = 0$$

Then μ_α should be expressed in terms of 3rd cumulants, i.e. $\eta_\alpha = \frac{1}{6}\beta_{\alpha\alpha\alpha}$.

c) higher order asymptotic mode unbiasedness :

the asymptotic density has its mode at the origin ;

again μ_α should be expressed in terms of 3rd cumulants.

Now we can consider the class S of estimators which are ER BAN estimators satisfying one of the above conditions.

It should be remarked that for given g which satisfies the condition

$$\frac{\partial}{\partial T_i}\ g\ \Big|_{\ T_i = \mu_i} = D_\alpha s_i$$

we can determine h so that

$$\mu_\alpha = \frac{1}{2}\ \sum_i \sum_j\ b_{ij}\ \sigma_{ij}\ +\ h(\mu_1, \ldots, \mu_m)$$

where $b_{ij} = \dfrac{\partial^2}{\partial T_i \partial T_j}\ g\ \Big|_{\ T_i = \mu_i}$

satisfies the condition stated above. Therefore for any g there is an ER estimator which belongs to the class S.

Now for the estimators in S, the asymptotic distribution up to the order n^{-1} is determined except for the term

$$\frac{1}{4n}\ Cov\ (Q_\alpha, Q_\beta)$$

Therefore we shall minimize the matrix

$$\Lambda = \{\ \tau_{\alpha\beta}\ \}$$

$$\tau_{\alpha\beta} = Cov(Q_\alpha, Q_\beta) = nE(Z_\alpha - U_\alpha)(Z_\beta - U_\beta) + o(1).$$

$\sqrt{n}\ (Z_\alpha - U_\alpha)$ must satisfy the following conditions :

$$\sqrt{n}\ E(U_\beta U_\gamma(Z_\alpha - U_\alpha)) = D_\beta E(U_\gamma(Z_\alpha - U_\alpha)) - E((D_\beta U_\gamma)(Z_\alpha - U_\alpha))$$

$$- E(U_\gamma D_\beta (Z_\alpha - U_\alpha))$$

$$= E(U_\alpha V_{\beta\alpha}) + o(1) = -K^{\alpha\beta\gamma} - J^{\beta\gamma\cdot\alpha} + o(1) ;$$

$$\sqrt{n}\ E(U_\beta V_{\gamma\delta} (Z_\alpha - U_\alpha)) = D_\beta E(V_{\gamma\delta} (Z_\alpha - U_\alpha))$$

$$- E((D_\beta V_{\gamma\delta})(Z_\alpha - U_\alpha)) - E(V_{\gamma\delta} D_\beta (Z_\alpha - U_\alpha))$$

$$= E(V_{\gamma\delta} V_{\beta\alpha}) + o(1) .$$

Under these conditions $nCov\,(Z_\alpha - U_\alpha, Z_\beta - V_\beta)$ is minimized (asymptotically) when it is expressed as

$$\sqrt{n}\,(Z_\alpha - U_\alpha) = \sum_\beta \sum_\gamma \lambda_{\alpha\beta\gamma}\, U_\beta\, U_\gamma + \sum_\alpha \sum_\beta \sum_\gamma \sum_\delta \lambda_{\alpha\beta\gamma\delta}\, U_\beta\, V_{\gamma\delta} + o(1) ,$$

where $\lambda_{\alpha\beta\gamma}$ and $\lambda_{\alpha\beta\gamma\delta}$ are Lagrangean multipliers determined so that the conditions are satisfied.

Now instead of determining explicitly $\lambda_{\alpha\beta\gamma}$ and $\lambda_{\alpha\beta\gamma\delta}$, we shall show that the above equation is actually satisfied when g is equal to the ML estimator.

The ML estimator is determined by the equations

$$\sum_i \frac{\partial s_i}{\partial \theta_\alpha} (T_i - \mu_i) = 0, \quad \alpha = 1, \ldots, p . \tag{6.1}$$

Since $(s_1 \ldots, s_m)$ and $(\mu_1, \ldots, \mu_m)$ correspond one to one and the mapping is continuous, $\hat\theta_0$ is consistent provided that $(s_1(\theta_1) \ldots s_m(\theta_1)) \neq (s_1(\theta_2) \ldots s_m(\theta_2))$ if $\theta_1 \neq \theta_2$. Moreover if for all θ and for any t_α, $(\Sigma t_\alpha \frac{\partial s_1}{\partial \theta_\alpha}, \ldots, \Sigma t_\alpha \frac{\partial s_m}{\partial \theta_2}) \neq (0, \ldots 0)$, it is shown that

$$Pr \left\{ \| \hat\theta - \theta \| > \epsilon \right\} = o(n^{-k}) \quad \text{for any } k .$$

Consequently we can expand the estimator around the true value of θ.

Differentiating (6.1) with respect to T_i, we have

$$\sum_\alpha \frac{\partial \hat\theta_\alpha}{\partial T_i} \left\{ \sum_j \frac{\partial^2 s}{\partial \theta_\alpha \partial \theta_\beta} (T_j - \mu_j) - \sum_j \frac{\partial s_j}{\partial \theta_\beta} \frac{\partial \mu_j}{\partial \theta_\alpha} \right\} + \frac{\partial s_i}{\partial \theta_\beta} = 0 ,$$

$$\beta = 1, \ldots, p. \tag{6.2}$$

Putting $T_i = \mu_i$ we have

$$-\sum_\alpha I_{\alpha\beta} \frac{\partial \hat\theta_\alpha}{\partial T_i} + \frac{\partial s_i}{\partial \theta_\beta} = 0, \quad \beta = 1, \ldots, p ;$$

$$\frac{\partial \hat\theta_\alpha}{\partial T_i} = \Sigma I^{\alpha\beta} \frac{\partial s_i}{\partial \theta_\beta}$$

which is to be expected since ML estimator is a BAN estimator. Differentiating (6.2) again with respect to T_i, we have

$$\sum_\alpha \frac{\partial^2 \hat\theta_\alpha}{\partial T_i \partial T_j} \left\{ \sum_k \frac{\partial^2 s_k}{\partial \theta_\alpha \partial \theta_\beta} (T_k - \mu_k) - \sum_k \frac{\partial s_k}{\partial \theta_\alpha} \frac{\partial \mu_k}{\partial \theta_\beta} \right\}$$

$$+ \sum_\alpha \sum_\gamma \left(\frac{\partial \hat\theta_\alpha}{\partial T_i}\right) \left(\frac{\partial \hat\theta_\gamma}{\partial T_j}\right) \left\{ \sum_k \frac{\partial^2 s_k}{\partial \theta_\alpha \partial \theta_\beta \partial \theta_\gamma} (T_k - \mu_k) - \sum_k \frac{\partial^2 s_k}{\partial \theta_\alpha \partial \theta_\beta} \frac{\partial \mu_k}{\partial \theta_\gamma} \right.$$

$$\left. - \sum_k \frac{\partial^2 s_k}{\partial \theta_\beta \partial \theta_\gamma} \frac{\partial \mu_k}{\partial \theta_\alpha} - \sum_k \frac{\partial s_k}{\partial \theta_\beta} \frac{\partial^2 \mu_k}{\partial \theta_\alpha \partial \theta_\gamma} \right\}$$

$$+ \sum_\alpha \frac{\partial \hat\theta_\alpha}{\partial T_i} \frac{\partial^2 s_i}{\partial \theta_\alpha \partial \theta_\beta} + \sum_\gamma \frac{\partial \hat\theta_\gamma}{\partial T_j} \frac{\partial^2 s_j}{\partial \theta_\gamma \partial \theta_\beta} = 0, \quad \beta = 1, \ldots, p$$

Substituting $T_i = \mu_i$, $\dfrac{\partial \hat\theta_\alpha}{\partial T_i} = D_\alpha s_i$, $\sum \dfrac{\partial^2 s_k}{\partial \theta_\alpha \partial \theta_\beta} \dfrac{\partial \mu_k}{\partial \theta_\gamma} = J_{\alpha\beta\cdot\gamma}$,

$$\sum \frac{\partial s_k}{\partial \theta_\beta} \frac{\partial^2 \mu_k}{\partial \theta_\alpha \partial \theta_\gamma} = J_{\alpha\gamma\cdot\beta} + \kappa_{\alpha\beta\gamma},$$

it follows that

$$- \sum_\alpha \left(\frac{\partial^2 \hat\theta_\alpha}{\partial T_i \partial T_j}\right) I^{\alpha\beta} - \sum_\alpha \sum_\gamma (D_\alpha s_i)(D_\gamma s_j)(J_{\alpha\beta\cdot\gamma} + J_{\alpha\gamma\cdot\beta} + J_{\beta\gamma\cdot\alpha} + K_{\alpha\beta\gamma})$$

$$+ \sum_\alpha (D_\alpha s_i) \frac{\partial^2 s_j}{\partial \theta_\alpha \partial \theta_\beta} + \sum_\gamma (D_\alpha s_j) \frac{\partial^2 s_i}{\partial \theta_\alpha \partial \theta_\beta} = 0$$

Hence we have

$$\frac{\partial^2 \hat\theta_\alpha}{\partial T_i \partial T_j} = - \sum_\beta \sum_\gamma \sum_\delta I^{\alpha\beta} (J_{\beta\gamma\cdot\delta} + J_{\beta\delta\cdot\gamma} + J_{\gamma\delta\cdot\beta} + K_{\beta\gamma\delta})(D_\gamma s_i)(D_\delta s_j)$$

$$+ \sum_\beta \sum_\gamma I^{\alpha\beta} \left(\frac{\partial^2 s_j}{\partial \theta_\beta \partial \theta_\gamma}\right)(D_\gamma s_i) + \sum_\beta \sum_\gamma I^{\alpha\beta} \left(\frac{\partial^2 s_i}{\partial \theta_\beta \partial \theta_\gamma}\right)(D_\gamma s_j)$$

We have

$$\sum_j \sum_\beta I^{\alpha\beta} \left(\frac{\partial^2 s_j}{\partial \theta_\beta \partial \theta_\gamma}\right) T_j' = \sum_{\gamma'} I_{\gamma\gamma'} V_{\alpha\gamma'}$$

$$+ \sum_{\beta'} \sum_{\gamma'} I_{\gamma\gamma'} (K^{\alpha\beta'\gamma'} + J^{\alpha\beta'\cdot\gamma'} + J^{\alpha\gamma'\cdot\beta'})$$

$$\cdot \sum_i \left(\frac{\partial s_i}{\partial \theta_{\beta'}}\right) T_i'$$

$$= \sum_{\delta} I_{\gamma\delta} \; V_{\alpha\delta} + \sum_{\alpha'} \sum_{\beta} I^{\alpha\alpha'} \; (K_{\alpha'\beta\gamma} + J_{\alpha'\beta\cdot\gamma} + J_{\alpha'\gamma\cdot\beta}) \; U_{\beta} \; .$$

Consequently,

$$\sum_{i} \sum_{j} \left(\frac{\partial^2 \hat{\theta}_{\alpha}}{\partial T_i \partial T_j} \right) T_i' \, T_j' \; = \; Q_{\alpha} \; = \; \sum_{\alpha'} \sum_{\beta'} \sum_{\gamma} I^{\alpha\alpha'} \; (K_{\alpha'\beta\gamma} + J_{\beta\gamma\cdot\alpha'}) \, U_{\beta} \, U_{\gamma}$$

$$+ \; 2 \sum_{\delta} \sum_{\gamma} I_{\gamma\delta} \; V_{\alpha\delta} \; U_{\gamma}$$

which has the form required for minimizing solution of Λ.

Thus if Λ^* denotes the variance-covariance matrix of Q_{α} corresponding to the (extended) ML estimator and Λ be any other estimator in the set S, we have

$$\Lambda - \Lambda^* \geq 0 \quad \text{(non-negative definite)}.$$

From this the following theorem is established

Theorem 6.1. Let $\hat{\theta}^*$ be the extended ML estimator in S and $\hat{\theta}$ be any other estimator in S and C by ane convex set in R^p containing the origin. Then it holds that

$$\lim_{n \to \infty} n \, [\, Pr \, \{ \, \sqrt{n} \, (\hat{\theta}^* - \theta) \, \epsilon \; C \, \} - Pr \, \{ \, \sqrt{n} \, (\hat{\theta} - \theta) \, \epsilon \; C \, \} \,] \geq 0$$

We also have

Theorem 6.2. Suppose that L is bounded, negatively unimodal function around the origin, i.e., for any c the set $\{ u : L(u) \leq c \} \subset R^p$ is convex and contains the origin. Let $\hat{\theta}^*$ be the extended ML estimator in S and $\hat{\theta}$ be any other estimator in S. Then

$$\overline{\lim_{n \to \infty}} \; n \, [\, E \, \{ \, L \, (\sqrt{n} \, (\hat{\theta}^* - \theta)) \, \} - E \, \{ \, L \, (\sqrt{n} \, (\hat{\theta} - \theta)) \, \} \,] \leq 0$$

References

[1] Ghosh, J.K. and Subramanyam, K., "Second order efficiency of maximum likelihood estimators," Sankhyā Ser. A, 36, (1974).

[2] Pfanzagl, J. and Wefelmeyer, W., "A third order optimum property of the maximum likelihood estimator," Preprints in Statistics 26, University of Cologne (1977).

[3] Takeuchi, K. and Akahira, M., "Third order asymptotic efficiency of maximum likelihood estimator in general case," (to appear).

Rep. Univ. Electro-Comm. 29–1, (Sci. & Tech. Sect.), pp. 37–45 August, 1978

Asymptotic Optimality of the Generalized Bayes Estimator[*]

Kei TAKEUCHI[**] and Masafumi AKAHIRA[***]

Abstract

For all symmetric loss functions, the generalized Bayes estimator is second order asymptotically efficient in the class A_2 of the all second order asymptotically median unbiased (AMU) estimators and third order asymptotically efficient in the restricted class D of estimators. When the loss function is not symmetric, the generalized Bayes estimator is second order asymptotically efficient in the class A_2 but not third order asymptotically efficient in the restricted wider class C than the class D.

1. Introduction

The expansion of a generalized Bayes estimator with respect to a loss function of the type $L(\theta)=|\theta|^a (a\geq 1)$ is obtained by Gusev ([4]). His result can be extended to all symmetric loss functions. Strasser ([6]) also obtained asymptotic expansions of the distribution of the generalized Bayes estimator.

It is shown in this paper that for all symmetric loss function the generalized Bayes estimator $\hat{\theta}_n$ is asymptotically expanded as

$$\sqrt{n}\,(\hat{\theta}_n-\theta)=\frac{Z_1(\theta)}{I(\theta)}+\frac{1}{\sqrt{n}}\left[\alpha_\theta+\frac{\beta_\theta\{3J(\theta)+K(\theta)\}}{6I(\theta)}+\frac{Z_1(\theta)Z_2(\theta)}{I(\theta)^2}-\frac{3J(\theta)+K(\theta)}{2I(\theta)^3}Z_1(\theta)^2\right]+o_p\left(\frac{1}{\sqrt{n}}\right),$$

where the symbols of the right-hand side are defined in the contents. And the asymptotic distributions of the estimators are the same up to the order n^{-1} except for constant location shift. Therefore if it is properly adjusted to be second order AMU and third order AMU, it is second order asymptotically efficient in the class A_2 of the all second order AMU estinators and also third order asymptotically efficient among the estimators belonging to the class D for $c_n=\sqrt{n}$, where D is the class whose element $\hat{\theta}_n$ is third order AMU and asymptotically expanded as

$$c_n(\hat{\theta}_n-\theta)=\frac{Z_1(\theta)}{I(\theta)}+c_n^{-1}Q(\theta)+o_p(c_n^{-1})$$

and $Q(\theta)=O_p(1)$, $E[Z_1(\theta)Q(\theta)^i]=o(1)$ $(i=1,2)$, where E denotes asymptotic expectation and the distribution of $c_n(\hat{\theta}_n-\theta)$ admits Edgeworth expansion ([9], [10]).

But when the loss function is not symmetric, the generalized Bayes estimator $\hat{\theta}_n$ has the asymptotic form:

$$\sqrt{n}\,(\hat{\theta}_n-\theta)=\frac{Z_1(\theta)}{I(\theta)}+\frac{1}{\sqrt{n}}\left[\alpha_\theta+\frac{\beta_\theta\{3J(\theta)+K(\theta)\}}{6I(\theta)}+\frac{Z_1(\theta)Z_2(\theta)}{I(\theta)^2}-\frac{3J(\theta)+K(\theta)}{2I(\theta)^3}Z_1(\theta)^2\right]$$

   * Received on June 9, 1978
  ** University of Tokyo
*** Statistical Laboratory, University of Electro-Communications

38   Kei Takeuchi and Masafumi Akahira

$$+\frac{\gamma_\theta}{\sqrt{n}}\left\{Z_2(\theta)-\frac{3J(\theta)+K(\theta)}{I(\theta)}Z_1(\theta)\right\}+o_p\left(\frac{1}{\sqrt{n}}\right)$$

and the existence of the last term excludes it from the class D, and it belongs to only the wider class C whose element satisfies all but $E[Z_1(\theta)Q(\theta)^i]=o(1)$ $(i=1,2)$ in the conditions of the class D ([9], [10]), and since the last term depends on the specific choice of the loss function, no uniform third order asymptotic efficiency property holds. Note that $A_2 \supset A_3 \supset C \supset D$.

2.   Results

Let $(\mathcal{X}, \mathcal{B})$ be a sample space.   We consider a family of probability measures on $\mathcal{B}$, $\mathcal{P}=\{P_\theta : \theta\in\Theta\}$, where the index set Θ is called a parameter space.   We assume that Θ is an open set in a Euclidean 1-space R^1.   Consider n-fold direct products $(\mathcal{X}^{(n)}, \mathcal{B}^{(n)})$ of $(\mathcal{X}, \mathcal{B})$ and corresponding product measures $P_\theta^{(n)}$ of P_θ.   An estimator of θ is defined to be a sequence $\{\hat{\theta}_n\}$ of $\mathcal{B}^{(n)}$-measurable functions $\hat{\theta}_n$ on $\mathcal{X}^{(n)}$ into Θ $(n=1, 2, \cdots)$.   For simplicity we denote an estimator as $\hat{\theta}_n$ instead of $\{\hat{\theta}_n\}$.   For increasing sequence of positive numbers $\{c_n\}$ (c_n tending to infinity) an estimator $\hat{\theta}_n$ is called consistent with order $\{c_n\}$ (or $\{c_n\}$-consistent for short) if for every $\varepsilon>0$ and every $\vartheta\in\Theta$ there exist a sufficiently small positive number δ and a sufficiently large positive number L satisfying the following :

$$\lim_{n\to\infty}\sup_{\theta:|\theta-\vartheta|<\delta} P_\theta^{(n)}\{c_n|\hat{\theta}_n-\theta|\geq L\}<\varepsilon \quad ([1]).$$

For each $k=1, 2, \cdots$, a $\{c_n\}$-consistent estimator $\hat{\theta}_n$ is k-th order asymptotically median unbiased (or k-th order AMU) estimator if for any $\vartheta\in\Theta$, there exists a positive number δ such that

$$\lim_{n\to\infty}\sup_{\theta:|\theta-\vartheta|<\delta} c_n^{k-1}\left|P_\theta^{(n)}\{\hat{\theta}_n\leq\theta\}-\frac{1}{2}\right|=0 ;$$

$$\lim_{n\to\infty}\sup_{\theta:|\theta-\vartheta|<\delta} c_n^{k-1}\left|P_\theta^{(n)}\{\hat{\theta}_n\geq\theta\}-\frac{1}{2}\right|=0.$$

We have defined a k-th order AMU estimator to be k-th order asymptotically efficient in the class A_k of the all k-th order AMU estimators if the k-th order asymptotic distribution of it attains uniformly the bound of the k-th order asymptotic distributions of the estimators in the class A_k ([2], [7]).   If the class A_k is changed for the class C and the class D in the definition in the case when $k=3$, third order asymptotically efficient estmators in the class C and in the class D are defined, respectively ([9], [10]).   It is genearally shown by Pfanzagl and Wefelmeyer ([5]) and Akahira and Takeuchi ([3], [8], [9], [10]) that there exist second order asymptotically efficient estimators in the class A_2 but not third order asymptotically efficient estimators in the class A_3.   But it was also shown in [9] and [10] that if we restrict the class of estimators, appropriately, we have higher order asymptotically efficient estimators among the restricted class of estimators, and that the maximum likelihood estimator belongs to the class of higher order asymptotically efficient estimators.

We assume that for each $\theta\in\Theta$ P_θ is absolutely continuous with respect to σ-finite measure μ.   We denote a density $dP_\theta/d\mu$ by $f(x, \theta)$.   Then the joint density is given by $\prod_{i=1}^{n} f(x_i, \theta)$.

In the subsequent discussion we shall deal with the case when $c_n=\sqrt{n}$.   Let $\Theta=R^1$.   Let $L_n(u)$ be a bounded non-negative and monotone increasing function of $|u|$ and $\pi(\theta)$ be a non-negative function. Define a posterior density $p_n(\theta\,|\,\tilde{x}_n)$ and a posterior risk $r_n(d\,|\,\tilde{x}_n)$ by

$$p_n(\theta\,|\,\tilde{x}_n)=\left\{\prod_{i=1}^{n} f(x_i,\,\theta)\right\}\pi(\theta)\left[\int_{\Theta}\left(\prod_{i=1}^{n} f(x_i,\,\theta)\right)\pi(\theta)d\theta\right]^{-1},$$

and

$$r_n(d\,|\,\tilde{x}_n)=\int_{\Theta} L_n(d-\theta)p_n(\theta\,|\,\tilde{x}_n)d\theta$$

respectively, where $\tilde{x}_n=(x_1,\,x_2,\,\cdots,\,x_n)$. Now suppose that $\lim\limits_{n\to\infty} L_n(u/\sqrt{n}\,)=L^*(u)$ for all real number u. We define

$$r_n{}^*(d\,|\,\tilde{x}_n)=\int_{\Theta} L^*(\sqrt{n}\,(d-\theta))p_n(\theta\,|\,\tilde{x}_n)d\theta.$$

An estimator $\hat{\theta}_n$ is called a generalized Bayes estimator with respect to a loss function L^* and a prior density π if

$$r_n{}^*(\hat{\theta}_n\,|\,\tilde{x}_n)=\inf_{d\in\Theta} r_n{}^*(d\,|\,\tilde{x}_n).$$

Then $\hat{t}_n=\sqrt{n}\,(\hat{\theta}_n-\theta)$ may be also called a generalized Bayes estimator w. r. t. L^* and π. Since

$$\lim_{n\to\infty}\left|\inf_{d\in\Theta}\int_{\Theta} L_n(d-\theta)\tilde{p}(\theta)d\theta-\inf_{d\in\Theta}\int_{\Theta} L^*(\sqrt{n}\,(d-\theta))\tilde{p}(\theta)d\theta\right|=0$$

uniformly in every posterior density $\tilde{p}(\theta)$, it follows that for a generalized Bayes estimator $\hat{\theta}_n$

$$\lim_{n\to\infty}\left|\inf_{d\in\Theta} r_n(d\,|\,\tilde{x}_n)-r_n{}^*(\hat{\theta}_n\,|\,\tilde{x}_n)\right|=0.$$

Suppose that $X_1,\,X_2,\,\cdots,\,X_n,\,\cdots$ is a sequence of i. i. d. random variables with a density $f(x,\,\theta)$ satisfying (i)$\sim$(iv).

   (i) $\{x:f(x,\,\theta)>0\}$ does not depend on θ;

   (ii) For almost all $x[\mu]$, $f(x,\,\theta)$ is three times continuously differentiable in θ;

   (iii) For each $\theta\in\Theta$

$$0<I(\theta)=E_\theta\left[\left\{\frac{\partial}{\partial\theta}\log f(x,\,\theta)\right\}^2\right]=-E_\theta\left[\frac{\partial^2}{\partial\theta^2}\log f(x,\,\theta)\right]<\infty;$$

   (iv) There exist

$$J(\theta)=E_\theta\left[\left\{\frac{\partial^2}{\partial\theta^2}\log f(X,\,\theta)\right\}\left\{\frac{\partial}{\partial\theta}\log f(X,\,\theta)\right\}\right]$$

and

$$K(\theta)=E_\theta\left[\left\{\frac{\partial}{\partial\theta}\log f(X,\,\theta)\right\}^3\right]$$

and the following holds:

$$E_\theta\left[\frac{\partial^3}{\partial\theta^3}\log f(X,\,\theta)\right]=-3J(\theta)-K(\theta).$$

By the following way we have shown in [7] that a maximum likelihood estimator (MLE) is second order asymptotically efficient. Let $\hat{\theta}_{ML}$ be an MLE. By Taylor expansion we have

$$0=\sum_{i=1}^{n}\frac{\partial}{\partial\theta}\log f(X_i,\,\hat{\theta}_{ML})$$

$$=\sum_{i=1}^{n}\frac{\partial}{\partial\theta}\log(X_i,\,\theta)+\sum_{i=1}^{n}\left\{\frac{\partial^2}{\partial\theta^2}\log f(X_i,\,\theta)\right\}(\hat{\theta}_{ML}-\theta)$$

$$+\frac{1}{2}\sum_{i=1}^{n}\left\{\frac{\partial^3}{\partial\theta^3}\log f(X_i,\theta^*)\right\}(\hat{\theta}_{ML}-\theta)^2,$$

where $\quad |\theta^*-\theta|\leq|\hat{\theta}_{ML}-\theta|$.

Putting $T_n=\sqrt{n}\,(\hat{\theta}_{ML}-\theta)$ we obtain

$$0=\frac{1}{\sqrt{n}}\sum_{i=1}^{n}\frac{\partial}{\partial\theta}\log f(X_i,\theta)+\frac{1}{n}\left\{\sum_{i=1}^{n}\frac{\partial^2}{\partial\theta^2}\log f(X_i,\theta)\right\}T_n$$

$$+\frac{1}{2n\sqrt{n}}\left\{\sum_{i=1}^{n}\frac{\partial^3}{\partial\theta^3}\log f(X_i,\theta^*)\right\}T_n^2.$$

Set

$$Z_1(\theta)=\frac{1}{\sqrt{n}}\sum_{i=1}^{n}\frac{\partial}{\partial\theta}\log f(X_i,\theta)\,;$$

$$Z_2(\theta)=\frac{1}{\sqrt{n}}\sum_{i=1}^{n}\left\{\frac{\partial^2}{\partial\theta^2}\log f(X_i,\theta)+I(\theta)\right\}\,;$$

$$W(\theta)=\frac{1}{n}\sum_{i=1}^{n}\frac{\partial^3}{\partial\theta^3}\log f(X_i,\theta),$$

where
$$I(\theta)=E_\theta\left[\left\{\frac{\partial}{\partial\theta}\log f(X,\theta)\right\}^2\right].$$

Then it follows that $W(\theta)$ converges in probability to $-3J(\theta)-K(\theta)$. Hence the following theorem holds :

Theorem 1. ([4]).   Under the conditions (i)$\sim$(iv)

$$\sqrt{n}\,(\hat{\theta}_{ML}-\theta)=\frac{Z_1(\theta)}{I(\theta)}+\frac{Z_1(\theta)Z_2(\theta)}{\sqrt{n}\,I(\theta)^2}-\frac{3J(\theta)+K(\theta)}{2\sqrt{n}\,I(\theta)^3}Z_1(\theta)^2+o_P\left(\frac{1}{\sqrt{n}}\right)$$

up to order $n^{-1/2}$ as $n\to\infty$.

Put

$$\hat{\theta}_{ML}^*=\hat{\theta}_{ML}+\frac{K(\hat{\theta}_{ML})}{6nI(\hat{\theta}_{ML})^2}$$

Then $\hat{\theta}_{ML}^*$ is second order AMU. From Theorem 1 we have established the following :

Theorem 2. ([4]).   Under the conditions (i)$\sim$(iv), $\hat{\theta}_{ML}^*$ is second order asymptotically efficient in the class A_2.

It will be shown that the generalized Bayes estimator w. r. t. a loss function and a prior density is second order asymptotically efficient in the class A_2.

Let θ_0 be a true parameter of $\theta(\epsilon\Theta)$.

Further we assume the following :

(v) $\pi(\theta)$ is twice differentiable in θ.

Then we have

$$p_n(\theta\,|\,\tilde{X}_n)/p_n(\theta_0\,|\,\tilde{X}_n)$$

$$=\exp[\log\{p_n(\theta\,|\,\tilde{X}_n)/p_n(\theta_0\,|\,\tilde{X}_n)\}]$$

$$=\exp\left[\sum_{i=1}^{n}\log\{f(X_i,\theta)/f(X_i,\theta_0)\}+\log\{\pi(\theta)/\pi(\theta_0)\}\right]$$

$$=\exp\left[\sum_{i=1}^{n}\log f(X_i,\theta)-\sum_{i=1}^{n}\log f(X_i,\theta_0)+\log\pi(\theta)-\log\pi(\theta_0)\right]$$

$$=\exp\left[\left\{\sum_{i=1}^{n}\frac{\partial}{\partial\theta}\log f(X_i,\theta_0)\right\}(\theta-\theta_0)+\frac{1}{2}\left\{\sum_{i=1}^{n}\frac{\partial^2}{\partial\theta^2}\log f(X_i,\theta_0)\right\}(\theta-\theta_0)^2\right.$$

$$+\frac{1}{6}\left\{\sum_{i=1}^{n}\frac{\partial^3}{\partial\theta^3}\log(X_i,\theta*)\right\}(\theta-\theta_0)^3+\frac{\pi'(\theta_0)}{\pi(\theta_0)}(\theta-\theta_0)+o\left(\frac{1}{\sqrt{n}}\right)\Bigg]$$

$$=\exp\Bigg[\sqrt{n}\{Z_1(\theta_0)\}(\theta-\theta_0)+\frac{1}{2}\{\sqrt{n}Z_2(\theta_0)-nI(\theta_0)\}(\theta-\theta_0)^2$$

$$+\frac{n}{6}W(\theta*)(\theta-\theta_0)^3+\frac{\pi'(\theta_0)}{\pi(\theta_0)}(\theta-\theta_0)+o\left(\frac{1}{\sqrt{n}}\right)\Bigg],$$

where $\tilde{X}=(X_1,X_2,\cdots,X_n)$ and $|\theta*-\theta_0|\leq|\theta-\theta_0|$.

Since $W(\theta)$ converges in probability to $-3J(\theta)-K(\theta)$, it follows that

$$p_n(\theta|\tilde{X}_n)/p_n(\theta_0|\tilde{X}_n)$$

$$=\exp\Bigg[\{Z_1(\theta_0)\}\sqrt{n}(\theta-\theta_0)+\frac{1}{2}\left\{\frac{Z_2(\theta_0)}{\sqrt{n}}-I(\theta_0)\right\}\{\sqrt{n}(\theta-\theta_0)\}^2$$

$$-\frac{1}{6\sqrt{n}}\{3J(\theta_0)+K(\theta_0)\}\{\sqrt{n}(\theta-\theta_0)\}^3+\frac{\pi'(\theta_0)}{\sqrt{n}\,\pi(\theta_0)}\{\sqrt{n}(\theta-\theta_0)\}+o_p\left(\frac{1}{\sqrt{n}}\right)\Bigg]$$

Putting $t=\sqrt{n}(\theta-\theta_0)$ we obtain

(1) $p_n(\theta|\tilde{X}_n)/p(\theta_0|\tilde{X}_n)$

$$=\exp\Bigg[Z_1(\theta_0)t+\frac{1}{2}\left\{\frac{Z_2(\theta_0)}{\sqrt{n}}-I(\theta_0)\right\}t^2-\frac{3J(\theta_0)+K(\theta_0)}{6\sqrt{n}}t^3+\frac{\pi'(\theta_0)}{\sqrt{n}\,(\theta_0)\pi}t+o_p\left(\frac{1}{\sqrt{n}}\right)\Bigg]$$

$$=\exp\Bigg[-\frac{I(\theta_0)}{2}\left\{t-\frac{Z_1(\theta_1)}{I(\theta_0)}\right\}^2+\frac{Z_1(\theta_0)^2}{2I(\theta_0)}+\frac{\pi'(\theta_0)}{\sqrt{n}\,\pi(\theta_0)}t+\frac{Z_2(\theta_0)}{2\sqrt{n}}t^2+\frac{3J(\theta_0)+K(\theta_0)}{6\sqrt{n}}t^3+o_p\left(\frac{1}{\sqrt{n}}\right)\Bigg]$$

$$=\left\{\exp\frac{Z_1(\theta_0)^2}{2I(\theta_0)}\right\}\left[\exp-\frac{I(\theta_0)}{2}\left\{t-\frac{Z_1(\theta_0)}{I(\theta_0)}\right\}^2\right]$$

$$\cdot\exp\left\{\frac{\pi'(\theta_0)}{\sqrt{n}\,\pi(\theta_0)}t+\frac{Z_2(\theta_0)}{2\sqrt{n}}t^2-\frac{3J(\theta_0)+K(\theta_0)}{6\sqrt{n}}t^3+o_p\left(\frac{1}{\sqrt{n}}\right)\right\}.$$

$$=\left\{\exp\frac{Z_1(\theta_0)^2}{2I(\theta_0)}\right\}\left[\exp-\frac{I(\theta_0)}{2}\left\{t-\frac{Z_1(\theta_0)}{I(\theta_0)}\right\}^2\right]$$

$$\cdot\left\{1+\frac{\pi'(\theta_0)}{\sqrt{n}\,\pi(\theta_0)}t+\frac{Z_2(\theta_0)}{2\sqrt{n}}t^2-\frac{3J(\theta_0)+K(\theta_0)}{6\sqrt{n}}t^3+o_p\left(\frac{1}{\sqrt{n}}\right)\right\}$$

$$=q_n(t,\theta_0|\tilde{X}_n)\quad\text{(say)}.$$

Let $\hat{t}_n=\sqrt{n}(\hat{\theta}_n-\theta_0)$. Then the posterior risk is given by

(2) $$r_n*(\hat{\theta}_n|\tilde{x}_n)=\frac{1}{\sqrt{n}}p_n(\theta_0|\tilde{x}_n)\int_{-\infty}^{\infty}L*(\hat{t}_n-t)q_n(t,\theta_0|\tilde{x}_n)dt.$$

Further we assume the following :

(vi) $L*(u)$ is a convex function ;

(vii) $\displaystyle\int_{-\infty}^{\infty}L*(-u)q_n(u+t,\theta_0|\tilde{x}_n)du$ is partially differentiable with respect to t under the

integral sign.

By (2) and the assumption (vi) it is shown that the generalized Bayes estimator $\hat{t}_n$ w. r. t. $L*(\cdot)$ and $\pi(\cdot)$ is given by a solution u of the equation

$$(d/du)\int_{-\infty}^{\infty}L*(u-t)q_n(t,\theta_0|\tilde{x}_n)dt=0.$$

Since by (vii)

$$\frac{d}{du}\int_{-\infty}^{\infty}L*(u-t)q_n(t,\theta_0|\tilde{x}_n)dt$$

Kei TAKEUCHI and Masafumi AKAHIRA

42

$$=\frac{d}{du}\int_{-\infty}^{\infty} L^*(-t)q_n(u+t,\theta_0\,|\,\tilde{x}_n)dt$$

$$=\frac{d}{dt}\int_{-\infty}^{\infty} L^*(-u)q_n(t+u,\theta_0\,|\,\tilde{x}_n)du$$

$$=\int_{-\infty}^{\infty} L^*(-u)\Big\{\frac{d}{dt}q_n(t+u,\theta_0\,|\,\tilde{x}_n)\Big\}du,$$

the generalized Bayes estimator $\hat{t}_n$ is obtained by a solution of the equation

$$(3)\quad \int_{-\infty}^{\infty} L^*(-u)\Big\{\frac{d}{dt}q_n(t_n+u,\theta_0\,|\,\tilde{x}_n)\Big\}du=0.$$

Since $t=\sqrt{n}(\theta-\theta_0)$ and $\hat{t}_n=\sqrt{n}(\hat{\theta}_n-\theta_0)$, the solution $\hat{\theta}_n$ of (3) may be called to be the generalized Bayes estimator.

From (1) and (3) we have

$$0=\int_{-\infty}^{\infty} L^*(-u)e^{-\frac{I(\theta_0)}{2}\left\{\hat{t}_n+u-\frac{Z_1(\theta_0)}{I(\theta_0)}\right\}^2}\Bigg[-I(\theta_0)\Big\{\hat{t}_n+u-\frac{Z_1(\theta_0)}{I(\theta_0)}\Big\}$$

$$\cdot\Big\{1+\frac{\pi'(\theta_0)}{\sqrt{n}\,\pi(\theta_0)}(\hat{t}_n+u)+\frac{Z_2(\theta_0)}{2\sqrt{n}}(\hat{t}_n+u)^2-\frac{3J(\theta_0)+K(\theta_0)}{6\sqrt{n}}$$

$$\cdot(t_n+u)^3+o_p\Big(\frac{1}{\sqrt{n}}\Big)\Big\}+\frac{\pi'(\theta_0)}{\sqrt{n}\,\pi(\theta_0)}+\frac{Z_2(\theta_0)}{\sqrt{n}}(\hat{t}_n+u)$$

$$-\frac{3J(\theta_0)+K(\theta_0)}{2\sqrt{n}}(\hat{t}_n+u)^2+o_p\Big(\frac{1}{\sqrt{n}}\Big)\Bigg]du$$

Putting $\hat{u}=\hat{t}_n-\dfrac{Z_1(\theta_0)}{I(\theta_0)}$, we obtain

$$(4)\quad 0=\int_{-\infty}^{\infty} L^*(-u)e^{-\frac{I(\theta_0)}{2}(u+\hat{u})^2}\Bigg[-I(\theta_0)(u+\hat{u})\Big\{1+\frac{\pi'(\theta_0)}{\sqrt{n}\,\pi(\theta_0)}\Big(u+\hat{u}+\frac{Z_1(\theta_0)}{I(\theta_0)}\Big)$$

$$+\frac{Z_2(\theta_0)}{2\sqrt{n}}\Big(u+\hat{u}+\frac{Z_1(\theta_0)}{I(\theta_0)}\Big)^2-\frac{3J(\theta_0)+K(\theta_0)}{6\sqrt{n}}\Big(u+\hat{u}+\frac{Z_1(\theta_0)}{I(\theta_0)}\Big)^3$$

$$+o_p\Big(\frac{1}{\sqrt{n}}\Big)\Big\}+\frac{\pi'(\theta_0)}{\sqrt{n}\,\pi(\theta_0)}+\frac{Z_2(\theta_0)}{\sqrt{n}}\Big(u+\hat{u}+\frac{Z_1(\theta_0)}{I(\theta_0)}\Big)-\frac{3J(\theta_0)+K(\theta_0)}{2\sqrt{n}}$$

$$\cdot\Big(u+\hat{u}+\frac{Z_1(\theta_0)}{I(\theta_0)}\Big)^2+o_p\Big(\frac{1}{\sqrt{n}}\Big)\Bigg]du$$

$$=\int_{-\infty}^{\infty} L^*(-u)e^{-\frac{I(\theta_0)}{2}u^2}\Big\{1-I(\theta_0)u\hat{u}+o\Big(\frac{1}{\sqrt{n}}\Big)\Big\}$$

$$\Bigg[-I(\theta_0)u-I(\theta_0)\hat{u}-\frac{I(\theta_0)\pi'(\theta_0)}{\sqrt{n}\,\pi(\theta_0)}(u+\hat{u})\Big\{u+\hat{u}+\frac{Z_1(\theta_0)}{I(\theta_0)}\Big\}$$

$$-\frac{I(\theta_0)Z_2(\theta_0)}{2\sqrt{n}}(u+\hat{u})\Big\{u^2+\hat{u}^2+\frac{Z_1(\theta_0)^2}{I(\theta_0)^2}+2u\hat{u}+\frac{2Z_1(\theta_0)}{I(\theta_0)}\hat{u}+\frac{2Z_1(\theta_0)}{I(\theta_0)}u\Big\}$$

$$+\frac{I(\theta_0)\{3J(\theta_0)+K(\theta_0)\}}{6\sqrt{n}}(u+\hat{u})\Big\{u^3+\hat{u}^3+\frac{Z_1(\theta_0)^3}{I(\theta_0)^3}+3u^2\hat{u}$$

$$+3u^2\frac{Z_1(\theta_0)}{I(\theta_0)}+3\hat{u}^2\frac{Z_1(\theta_0)}{I(\theta_0)}+3\hat{u}^2u+3\frac{Z_1(\theta_0)^2}{I(\theta_0)^2}u+3\frac{Z_1(\theta_0)^2}{I(\theta_0)^2}\hat{u}$$

$$+6u\hat{u}\frac{Z_1(\theta_0)}{I(\theta_0)}\Big\}+\frac{\pi'(\theta_0)}{\sqrt{n}\,\pi(\theta_0)}+\frac{Z_2(\theta_0)}{\sqrt{n}}u+\frac{Z_2(\theta_0)}{\sqrt{n}}\hat{u}+\frac{Z_1(\theta_0)Z_2(\theta_0)}{I(\theta_0)\sqrt{n}}$$

$$-\frac{3J(\theta_0)+K(\theta_0)}{2\sqrt{n}}\Big\{u^2+\hat{u}^2+\frac{Z_1(\theta_0)^2}{I(\theta_0)^2}+2u\hat{u}+\frac{2Z_1(\theta_0)}{I(\theta_0)}\hat{u}+\frac{2Z_1(\theta_0)}{I(\theta_0)}u\Big\}$$

$$+o_p\!\left(\frac{1}{\sqrt{n}}\right)\!\Big]du.$$

(I) Symmetric loss function.

We assume the following :

(viii) $L^*(u)$ is a symmetric loss function about zero.

We define

$$(5)\quad \varphi_k=\int_{-\infty}^{\infty}L^*(u)u^k e^{-\frac{I(\theta_0)}{2}u^2}du.\quad (k=2j\ ;\ j=0,1,2,\cdots).$$

Note that $\displaystyle\int_{-\infty}^{\infty}L^*(u)u^{2j+1}e^{-\frac{I(\theta_0)}{2}u^2}du=0\quad (j=0,1,2,\cdots).$

It follows from (4), (5) and the assumption (ix) that

$$\begin{aligned}
0=&-I\varphi_0\hat{u}-\frac{I(\theta_0)\pi'(\theta_0)}{\sqrt{n}\,\pi(\theta_0)}\varphi_2-\frac{I(\theta_0)Z_2(\theta_0)}{2\sqrt{n}}\Big\{2\varphi_2\hat{u}+\frac{2Z_1(\theta_0)}{I(\theta_0)}\varphi_2+\varphi_2\hat{u}\\
&+\frac{Z_1(\theta_0)^2}{I(\theta_0)^2}\varphi_0\hat{u}\Big\}+\frac{I(\theta_0)\{3J(\theta_0)+K(\theta_0)\}}{6\sqrt{n}}\Big\{\varphi_4+3\varphi_2\frac{Z_1(\theta_0)^2}{I(\theta_0)^2}+6\varphi_2\hat{u}\frac{Z_1(\theta_0)}{I(\theta_0)}+o_p(1)\Big\}\\
&+\frac{\pi'(\theta_0)}{\sqrt{n}\,\pi(\theta_0)}\varphi_0+\frac{Z_2(\theta_0)}{\sqrt{n}}\varphi_0\hat{u}+\frac{Z_1(\theta_0)Z_2(\theta_0)}{I(\theta_0)\sqrt{n}}\varphi_0\\
&-\frac{3J(\theta_0)+K(\theta_0)}{2\sqrt{n}}\Big\{\varphi_2+\frac{Z_1(\theta_0)^2}{I(\theta_0)^2}\varphi_0+\frac{2Z_1(\theta_0)}{I(\theta_0)}\varphi_0\hat{u}\Big\}+I(\theta_0)^2\varphi_2\hat{u}+o_p\!\left(\frac{1}{\sqrt{n}}\right).
\end{aligned}$$

Since

$$\begin{aligned}
I(\theta_0)\{\varphi_0-I(\theta_0)\varphi_2\}\hat{u}=&\frac{\pi'(\theta_0)}{\pi(\theta_0)\sqrt{n}}\{\varphi_0-I(\theta_0)\varphi_2\}+\frac{3J(\theta_0)+K(\theta_0)}{6\sqrt{n}}\{I(\theta_0)\varphi_4-3\varphi_2\}\\
&+\frac{Z_1(\theta_0)Z_2(\theta_0)}{I(\theta_0)\sqrt{n}}\{\varphi_0-I(\theta_0)\varphi_2\}\\
&-\frac{3J(\theta_0)+K(\theta_0)}{2I(\theta_0)^2\sqrt{n}}Z_1(\theta_0)^2\{\varphi_0-I(\theta_0)\varphi_2\}+o_p\!\left(\frac{1}{\sqrt{n}}\right)
\end{aligned}$$

and $\varphi_0-I(\theta_0)\varphi_2\neq0,$

it follows that

$$\begin{aligned}
(6)\quad \hat{u}=&\frac{\pi'(\theta_0)}{I(\theta_0)\pi(\theta^3)\sqrt{n}}+\frac{I(\theta_0)\varphi_4-3\varphi_2}{\varphi_0-I(\theta_0)\varphi_2}\cdot\frac{3J(\theta_0)+K(\theta_0)}{6I(\theta_0)\sqrt{n}}+\frac{Z_1(\theta_0)Z_2(\theta_0)}{I(\theta_0)^2\sqrt{n}}\\
&-\frac{3J(\theta_0)+K(\theta_0)}{2I(\theta_0)^3\sqrt{n}}Z_1(\theta_0)^2+o_p\!\left(\frac{1}{\sqrt{n}}\right).
\end{aligned}$$

Hence we have

$$(7)\quad \hat{t}_n=\frac{Z_1(\theta_0)}{I(\theta_0)}+\hat{u},$$

where $\hat{u}$ is given by (6). Since $\hat{t}_n=\sqrt{n}\,(\hat{\theta}_n-\theta_0)$, we modify $\hat{\theta}_n$ to be second order AMU and denote it by $\hat{\theta}_n{}^*$.

From Theorem 1, (6) and (7) it follows that the MLE $\hat{\theta}_{ML}{}^*$ is asymptotically equivalent to the generalized Bayes estimator $\hat{\theta}_n{}^*$ up to order $n^{-1/2}$. By Theorem 2 it is seen that $\hat{\theta}_n{}^*$ is second order asymptotically efficient in the class A_2. It follows from (6) and (7) that the generalized Bayes estimator $\hat{\theta}_n{}^*$ belongs to the class D. Then the asymptotic distribution of $\hat{\theta}_n{}^*$ is equivalent to that of the MLE $\hat{\theta}_{ML}{}^*$ up to order n^{-1}. Since $\hat{\theta}_{ML}{}^*$ is third order asymptotically efficient in the class $D([9],[10])$, $\hat{\theta}_n{}^*$ is also so.

Hence we have established the following:

Theorem 3. Under the conditions (i)$\sim$(viii), the generalized Bayes estimator $\hat{\theta}_n{}^*$ is second order asymptotically efficient in the class A_2 and also third order asymptotically efficient in the class D.

Remark: In the location parameter case, $L^*(u)=u^2$, $\pi(\theta)\equiv1$, the generalized Bayes estimator is reduced to the Pitman estimator. It follows by Theorem 3 that the Pitman estimator is second order asymptotically efficient in the class A_2 if it is properly adjusted to be second order asymptotically median unbiased.

(II) Asymmetric loss function.

In the case when the loss function $L^*(\cdot)$ is asymmetric we shall obtain a similar result as the symmetric case. Further we assume the following:

(ix) $L^*(u)$ is an asymmetric loss function;

(x)   There exists a c satisfying

$$\int_{-\infty}^{\infty} L^*(c-u)ue^{-\frac{I(\theta_0)}{2}u^2}\,du=0.$$

Then we define

$$\psi_k=\int_{-\infty}^{\infty} L^*(c-u)u^k e^{-\frac{I(\theta_0)}{2}u^2}\,du \quad (k=0, 1, 2, \cdots).$$

Put $\hat{u}=\hat{t}_n-\dfrac{Z_1(\theta_0)}{I(\theta_0)}$ and $\hat{v}=\hat{u}+c$.

In a similar way as the symmetric loss function we have from (4) and the assumptions (ix) and (x)

$$(8)\quad \hat{v}=\frac{\pi'(\theta_0)}{I(\theta_0)\pi(\theta_0)\sqrt{n}}+\frac{I(\theta_0)\psi_4-3\psi_2}{\psi_0-I(\theta_0)\psi_2}\cdot\frac{3J(\theta_0)+K(\theta_0)}{6I(\theta_0)\sqrt{n}}+\frac{Z_1(\theta_0)Z_2(\theta_0)}{I(\theta_0)^2\sqrt{n}}$$
$$-\frac{3J(\theta_0)+K(\theta_0)}{2I(\theta_0)^3\sqrt{n}}Z_1(\theta_0)^2-\frac{\psi_3}{2\{\psi_0-I(\theta_0)\psi_2\}\sqrt{n}}\Big\{Z_2(\theta_0)$$
$$-\frac{3J(\theta_0)+K(\theta_0)}{I(\theta_0)}Z_1(\theta_0)\Big\}+o_p\Big(\frac{1}{\sqrt{n}}\Big).$$

Hence we have

$$(9)\quad \hat{t}_n=\frac{Z_1(\theta_0)}{I(\theta_0)}-c+\hat{v},$$

where $\hat{v}$ is given by (8). Since $t_n=\sqrt{n}\,(\hat{\theta}_n-\theta_0)$, we modify $\hat{\theta}_n$ to be second order AMU and denote it by $\hat{\theta}_n{}^*$. Since there exists the fifth term of the right-hand side of (8), the generalized Bayes estimator $\hat{\theta}_n{}^*$ does not belong to D but to C ([9], [10]). However it follows from Theorem 1, (8) and (9) that the asymptotic distribution of $\hat{\theta}_n{}^*$ is equivalent to that of the MLE $\hat{\theta}_{ML}$ up to order $n^{-1/2}$ ([9], [10]). Since the fifth term of the right-hand side of (8) depends on the specific choice of the loss function, the modified generalized Bayes estimator $\hat{\theta}^{**}(\epsilon A_3)$ is not third order asymptotically efficient estimator in the class C ([9], [10]). Hence we have established the following:

Theorem 4. Under the conditions (i)$\sim$(vii), (ix) and (x), the generalized Bayes estimator is second order asymptotically efficient in the class A_2 but $\hat{\theta}_n{}^{**}(\epsilon A_3)$ is not third order asymptotically efficient in the class C.

References

[1] Akahira, M.: "Asymptotic theory for estimation of location in non-regular cases, I: Order of convergence of consistent estimators, Rep. Stat. Appl. Res., JUSE, **22**, 8–26 (1975).

[2] Akahira, M. and Takeuchi, K.: "On the second order asymptotic efficiency of estimators in multiparameter cases," Rep. Univ. Electro-Comm., **26**, 261–269 (1976).

[3] Akahira, M. and Takeuchi, K.: "Discretized likelihood methods——Asymptotic properties of discretized likelihood estimators (DLE's)," (to appear).

[4] Gusev, S. I.: "Asymptotic expansions associated with some statistical estimators in the smooth case I. Expansions of random variables," Theory Prob. Applications, **20**, 470–498 (1975).

[5] Pfanzagl, J. and Wefelmeyer, W.: "A third order optimum property of maximum likelihood estimator," J. Multivariate Anal. **8**, 1–29 (1978).

[6] Strasser, H.: "Asymptotic expansions for Bayes procedures," In Recent Development in Statistics (J. R. Barra *et al.*, Ed.) North-Holland 9–35, (1977).

[7] Takeuchi, K. and Akahira, M.: "On the second order asymptotic efficiencies of estimators," Proc. of the Third Japan-USSR Symp. on Prob. Theory, Lecture Notes in Math. **550**, Springer-Verlag, 604–638 (1976).

[8] Takeuchi, K. and Akahira, M.: "Third order asymptotic efficiency of maximum likelihood estimator for multiparameter exponential case," Rep. Univ. Electro-Comm. **28**, 271–293 (1978).

[9] Takeuchi, K. and Akahira, M.: "On the asymptotic efficiency of estimators," (in Japanese), A report of the Symposium on Some Problems of Asymptotic Theory, Annual Meeting of the Mathematical Society of Japan, 1–24 (1978).

[10] Takeuchi, K. and Akahira, M.: "Third order asymptotic efficiency of maximum likelihood estimator in general case," (to appear).

Rep. Stat. Appl. Res., JUSE
Vol. 26, No. 3, September, 1979

__A Section__

On the Second Order Asymptotic Efficiency of Unbiased Confidence Intervals

MASAFUMI AKAHIRA* and KEI TAKEUCHI**

1. Introduction

It has been known in [4] that the confidence interval based on maximum likelihood (or asymptotically efficient) estimator $\hat{\theta}_n$ given as $[\hat{\theta}_n - \{u_{\alpha/2}/\sqrt{I(\hat{\theta}_n)}\}, \hat{\theta}_n + \{u_{\alpha/2}/\sqrt{I(\hat{\theta}_n)}\}]$ is asymptotically unbiased, and asymptotically efficient.

The purpose of this paper is to discuss the problem of the second order asymptotic efficiency of confidence intervals, that is, to consider the asymptotic power of the confidence interval up to the order $n^{-1/2}$ in the neighborhood of the true value of the parameter.

It will be shown that asymptotically unbiased, and second order asymptotically (most powerful) confidence interval at $\theta' = \theta + tn^{-1/2}$ depends on the specified value of t, thus proving that there exists no uniformly second order asymptotically most powerful unbiased confidence interval. The situation is in contrast to the case of the point estimation where maximum likelihood (or usually BAN) estimators are uniformly second order asymptotically efficient ([5], [7]), and is related to the curvature of the parameter space as was discussed by Efron [3].

2. Notations and Definitions

Let $\mathfrak{X}$ be an abstract sample space whose generic point is denoted by x, $\mathfrak{B}$ a σ-field of subsets of $\mathfrak{X}$, and let Θ be a parameter space, which is assumed to be an open set in a Euclidean 1-space R^1. We shall denote by $(\mathfrak{X}^{(n)}, \mathfrak{B}^{(n)})$ the n-fold direct products of $(\mathfrak{X}, \mathfrak{B})$. For each $n = 1, 2, \ldots,$ the points of $\mathfrak{X}^{(n)}$ will be denoted by $\tilde{x}_n = (x_1, \ldots, x_n)$ and the corresponding random variables by $\tilde{X}_n = (X_1, \ldots, X_n)$. We consider a sequence of classes of probability measures $\{P_{n,\theta} : \theta \in \Theta\}$ $(n = 1, 2, \ldots)$ each defined on $(\mathfrak{X}^{(n)}, \mathfrak{B}^{(n)})$ such that for each $n = 1, 2, \ldots$ and each $\theta \in \Theta$ the following holds:

$$P_{n,\theta}(B^{(n)}) = P_{n+1,\theta}(B^{(n)} \times \mathfrak{X})$$

for all $B^{(n)} \in \mathfrak{B}^{(n)}$. Let c_n be the maximum order of convergence of consistent estimators ([1]).

A subinterval of Θ, $[\underline{\theta}_n(\tilde{X}_n), \bar{\theta}_n(\tilde{X}_n)]$ whose limit depends on the observed random variables $\tilde{X}_n$ (properly measurable statistics) is called consistent of order c_n if for every $t(\neq 0)$

$$\lim_{|t| \to \infty} \varliminf_{n \to \infty} P_{n,\theta}\{\theta + tc_n^{-1} \in [\underline{\theta}_n(\tilde{X}_n), \bar{\theta}_n(\tilde{X}_n)]\} = 1.$$

A consistent interval $[\underline{\theta}_n(\tilde{X}_n), \bar{\theta}_n(\tilde{X}_n)]$ is called an α-asymptotically unbiased confidence (α-AUC) interval, $0 < \alpha < 1$, if for every $\vartheta \in \Theta$ and every $t(\neq 0)$

$$\varlimsup_{n \to \infty} P_{n,\theta}\{\theta + tc_n^{-1} \in [\underline{\theta}_n(\tilde{X}_n), \bar{\theta}_n(\tilde{X}_n)]\} \leq 1 - \alpha,$$

and for every $\vartheta \in \Theta$ there exists a positive number δ such that

Received March, 1979.
* Statistical Laboratory, University of Electro-Communications, Chofu, Tokyo 182, Japan
** Faculty of Economics, University of Tokyo, Hongô, Bunkyo-ku, Tokyo 113, Japan

$$\varliminf_{n\to\infty} \inf_{\theta:\,|\theta-\vartheta|<\delta} P_{n,\theta}\{\theta \in [\underline{\theta}_n(\tilde{X}_n),\, \bar{\theta}_n(\tilde{X}_n)]\} \geq 1 - \alpha.$$

Further an α-AUC interval $[\underline{\theta}_n(\tilde{X}_n),\, \bar{\theta}(\tilde{X}_n)]$ is called to be second order α-AUC if for every $\theta \in \Theta$ and every $t(\neq 0)$

$$\varlimsup_{n\to\infty} c_n[P_{n,\theta}\{\theta + tc_n^{-1} \in [\underline{\theta}_n(\tilde{X}_n),\, \bar{\theta}_n(\tilde{X}_n)]\} - (1 - \alpha)] \leq 0,$$

and for every $\vartheta \in \Theta$ there exists a positive number δ such that

$$\varliminf_{n\to\infty} \inf_{\theta:\,|\theta-\vartheta|<\delta} c_n[P_{n,\theta}\{\theta \in [\underline{\theta}_n(\tilde{X}_n),\, \bar{\theta}_n(\tilde{X}_n)]\} - (1 - \alpha)] \geq 0.$$

An α-AUC interval $[\underline{\theta}_n^*(\tilde{X}_n),\, \bar{\theta}_n^*(\tilde{X}_n)]$ is called asymptotically efficient if for every α-AUC interval $[\underline{\theta}_n(\tilde{X}_n),\, \bar{\theta}_n(\tilde{X}_n)]$, every $\theta \in \Theta$ and every $t(\neq 0)$

$$\varlimsup_{n\to\infty} [P_{n,\theta}\{\theta + tc_n^{-1} \in [\underline{\theta}_n^*(\tilde{X}_n),\, \bar{\theta}_n^*(\tilde{X}_n)]\} - P_{n,\theta}\{\theta + tc_n^{-1} \in [\underline{\theta}_n(\tilde{X}_n),\, \bar{\theta}_n(\tilde{X}_n)]\}] \leq 0$$

A second order α-AUC interval $[\underline{\theta}_n^*(\tilde{X}_n),\, \bar{\theta}_n^*(\tilde{X}_n)]$ is called second order asymptotically efficient if for every second order α-AUC interval $[\underline{\theta}_n(\tilde{X}_n),\, \bar{\theta}_n(\tilde{X}_n)]$, every $\theta \in \Theta$ and every $t(\neq 0)$

$$\varlimsup_{n\to\infty} c_n[P_{n,\theta}\{\theta + tc_n^{-1} \in [\underline{\theta}_n^*(\tilde{X}_n),\, \bar{\theta}_n^*(\tilde{X}_n)]\} - P_{n,\theta}\{\theta + tc_n^{-1} \in [\underline{\theta}_n(\tilde{X}_n),\, \bar{\theta}_n(\tilde{X}_n)]\}] \leq 0.$$

Note that our definition of the asymptotically efficient confidence interval is different from that of Wald [10] (asymptotically shortest confidence interval in his terminology) in that a) he considered only exact confidence intervals, and b) he defined efficiency in terms of the supremum of the difference of powers of two intervals over all pairs of parameter values of θ, θ' which implies uniformity of convergence of the power function, which in turn implies the boundedness from below of the Fisher information; and both aspects limit the applicability of the theory.

3. Second Order Asymptotic Efficiency of Unbiased Confidence Intervals.

We assume that for each $\theta \in \Theta$ P_θ is absolutely continuous with respect to a σ-finite measure μ. We denote a density $dP_\theta/d\mu$ by $f(x, \theta)$. Then the joint density is given by $\prod_{i=1}^{n} f(x_i, \theta)$. In the subsequent discussion we shall deal with the case when $c_n = \sqrt{n}$.

Suppose that $X_1, X_2, \ldots, X_n, \ldots$ is a sequence of independently and identically distributed (i.i.d.) random variables with the density $f(x, \theta)$ satisfying (i) $\sim$ (iv).

(i) $\{x: f(x, \theta) > 0\}$ does not depend on θ;

(ii)   For almost all $x[\mu]$, $f(x, \theta)$ is three times continuously differentiable in θ;

(iii)   For each $\theta \in \Theta$

$$0 < I(\theta) = E_\theta\left[\left\{\frac{\partial}{\partial\theta} \log f(X, \theta)\right\}^2\right] = -E_\theta\left[\frac{\partial^2}{\partial\theta^2} \log f(X, \theta)\right] < \infty;$$

(iv)   There exist

$$J(\theta) = E_\theta\left[\left\{\frac{\partial^2}{\partial\theta^2} \log f(X, \theta)\right\}\left\{\frac{\partial}{\partial\theta} \log f(X, \theta)\right\}\right]$$

and

$$K(\theta) = E_\theta\left[\left\{\frac{\partial}{\partial\theta} \log f(X, \theta)\right\}^3\right]$$

and the following holds:

$$E_\theta\left[\frac{\partial^3}{\partial\theta^3} \log f(X, \theta)\right] = -3J(\theta) - K(\theta).$$

We put

$$Z_1(\theta) = \frac{1}{\sqrt{n}} \sum_{i=1}^{n} \frac{\partial}{\partial\theta} \log f(X_i, \theta);$$

$$Z_2(\theta) = \frac{1}{\sqrt{n}} \sum_{i=1}^{n} \left\{\frac{\partial^2}{\partial\theta^2} \log f(X_i, \theta) + I(\theta);\right\};$$

$$W(\theta) = \frac{1}{n} \sum_{i=1}^{n} \frac{\partial^3}{\partial \theta^3} \log f(X_i, \theta).$$

Let θ_0 be arbitrary but fixed in Θ. Consider the problem of testing hypothesis $H: \theta = \theta_0$ against alternative $K: \theta = \theta_0 + tn^{-1/2}$ ($t \neq 0$). By the fundamental lemma of Neyman and Pearson it follows that the most powerful locally unbiased test function has a rejection region of the following type:

$$\frac{\prod_{i=1}^{n} f(X_i, \theta_0 + tn^{-1/2})}{\sum_{i=1}^{n} f(X_i, \theta_0)} > \lambda_0 + \lambda_1 tn^{-1/2} \sum_{i=1}^{n} \frac{\partial}{\partial \theta} \log f(X_i, \theta), \tag{3.1}$$

where λ_0 and λ_1 are certain real numbers, and if it is actually (grobally) unbiased it is the most powerful unbiased test function. By Taylor expansion we obtain

$$\prod_{i=1}^{n} f(X_i, \theta_0 + tn^{-1/2}) / \prod_{i=1}^{n} f(X_i, \theta_0)$$

$$= \exp\left[\sum_{i=1}^{n} \left\{ \log f(X_i, \theta_0 + tn^{-1/2}) - \log f(X_i, \theta_0) \right\} \right]$$

$$= \exp\left\{ \frac{t}{\sqrt{n}} \sum_{i=1}^{n} \frac{\partial}{\partial \theta} \log f(X_i, \theta_0) + \frac{t^2}{2n} \sum_{i=1}^{n} \frac{\partial^2}{\partial \theta^2} \log f(X_i, \theta_0) \right.$$

$$\left. + \frac{t^3}{6n\sqrt{n}} \sum_{i=1}^{n} \frac{\partial^3}{\partial \theta^3} \log f(X_i, \theta^*) \right\},$$

$$= \exp\left[tZ_1(\theta_0) + \frac{t^2}{2\sqrt{n}} \{Z_2(\theta_0) - \sqrt{n}\, I(\theta_0)\} + \frac{t^3}{6\sqrt{n}} W(\theta^*) \right]$$

$$= \exp\left\{ -\frac{t^2}{2} I(\theta_0) + tZ_1(\theta_0) + \frac{t^2}{2\sqrt{n}} Z_2(\theta) + \frac{t^3}{6\sqrt{n}} W(\theta^*) \right\},$$

where $|\theta^* - \theta| < tn^{-1/2}$. Since $W(\theta)$ converges in probability to $-3J(\theta) - K(\theta)$, it follows that

$$\prod_{i=1}^{n} f(X_i, \theta_0 + tn^{-1/2}) / \prod_{i=1}^{n} f(X_i, \theta_0) \tag{3.2}$$

$$= \exp\left[-\frac{t^2}{2} I(\theta_0) + tZ_1(\theta_0) + \frac{t^2}{2\sqrt{n}} Z_2(\theta_0) - \frac{t^3}{6\sqrt{n}} \{3J(\theta_0) + K(\theta_0)\} + o_p\left(\frac{1}{\sqrt{n}}\right) \right]$$

$$= e^{-(t^2/2)I(\theta_0)} \cdot e^{tZ_1(\theta_0)} \left[1 + \frac{t^2}{2\sqrt{n}} Z_2(\theta_0) - \frac{t^3}{6\sqrt{n}} \{3J(\theta_0) + K(\theta_0)\} + o_p\left(\frac{1}{\sqrt{n}}\right) \right].$$

First we consider the problem up to constant order. It follows by (3.1) and (3.2) that there exists the solutions $\underline{Z}_1$ and $\bar{Z}_1 (\underline{Z}_1 < \bar{Z}_1)$ of the following equation of $Z_1(\theta_0)$:

$$e^{-(t^2/2)I(\theta_0)} e^{tZ_1(\theta_0)} = \lambda_0 + \lambda_1 tZ_1(\theta_0). \tag{3.3}$$

Then it follows that

$$e^{-(t^2/2)I(\theta_0)} \cdot e^{tZ_1(\theta_0)} > \lambda_0 + \lambda_1 tZ_1(\theta_0)$$

if and only if

$$Z_1(\theta_0) < \underline{Z}_1, Z_1(\theta_0) > \bar{Z}_1,$$

It can be easily seen that in order to make the α-AUC interval we have to take $\underline{Z}_1 = -u_{\alpha/2}\sqrt{I(\theta_2)}$ and $\bar{Z}_1 = u_{\alpha/2}\sqrt{I(\theta_0)}$ (See [4]), where $u_{\alpha/2}$ denotes upper $100\alpha/2$-percent point of the standard normal distribution. In order to determine λ_0 and λ_1 we substitute $\underline{Z}_1 = -u_{\alpha/2}\sqrt{I(\theta_0)}$ and $\bar{Z}_1 = u_{\alpha/2}\sqrt{I(\theta_0)}$ in the equations

$$e^{-(t^2/2)I(\theta_0)} \cdot e^{t\underline{Z}_1} = \lambda_0 + \lambda_1 t\underline{Z}_1 :$$
$$e^{-(t^2/2)I(\theta_0)} \cdot e^{t\bar{Z}_1} = \lambda_0 + \lambda_1 t\bar{Z}_1.$$

Then we have

$$e^{-(t^2/2)I(\theta_0)} \cdot e^{-tu_{\alpha/2}\sqrt{I(\theta_0)}} = \lambda_0 + \lambda_1 tu_{\alpha/2}\sqrt{I(\theta_0)}.$$
$$e^{-(t^2/2)I(\theta_0)} \cdot e^{tu_{\alpha/2}\sqrt{I(\theta_0)}} = \lambda_0 + \lambda_1 tu_{\alpha/2}\sqrt{I(\theta_0)}.$$

Hence we obtain

$$\lambda_0 = e^{-(t^2/2)I(\theta_0)} \cosh(tu_{\alpha/2}\sqrt{I(\theta_0)});$$

$$\lambda_1 = \frac{1}{u_{\alpha/2}\sqrt{I(\theta_0)}} e^{-(t^2/2)I(\theta_0)} \sinh\left(tu_{\alpha/2}\sqrt{I(\theta_0)}\right).$$

If there exist $\underline{\theta}_n{}^*(\tilde{X}_n)$ and $\bar{\theta}_n{}^*(\tilde{X}_n)$ such that

$$Z_1(\underline{\theta}_n^*(\tilde{X}_n)) = u_{\alpha/2}\sqrt{I(\underline{\theta}_n^*\tilde{X}_n))}$$

$$Z_1(\bar{\theta}_n^*(\tilde{X}_n)) = -u_{\alpha/2}\sqrt{I(\bar{\theta}_n^*(\tilde{X}_n))},$$

then the α-*AUC* interval $[\underline{\theta}_n^*(\tilde{X}_n), \bar{\theta}_n^*(\tilde{X}_n)]$ is asymptotically efficient. Hence we have established the following:

Theorem 1. Under the conditions (i)~(iii), the α-*AUC* interval $[\underline{\theta}_n^*(\tilde{X}_n), \bar{\theta}_n^*(\tilde{X}_n)]$ is asymptotically efficient.

We can actually construct the asymptotically efficient α-*AUC* interval using the maximum likelihood estimator (MLE) $\hat{\theta}_{ML}$. We put

$$\underline{\theta}_n^*(\tilde{X}_n) = \hat{\theta}_{ML} + \frac{\gamma'}{\sqrt{n}};$$

$$\bar{\theta}_n^*(\tilde{X}_n) = \hat{\theta}_{ML} + \frac{\gamma}{\sqrt{n}}.$$

We have

$$Z_1(\underline{\theta}_n^*) = Z_1\left(\hat{\theta}_{ML} + \frac{\gamma'}{\sqrt{n}}\right)$$

$$= \frac{\gamma'}{\sqrt{n}}Z_1'(\hat{\theta}_{ML}) + o_p\left(\frac{1}{\sqrt{n}}\right)$$

$$= \frac{\gamma'}{\sqrt{n}}\{Z_2(\hat{\theta}_{ML}) - I(\hat{\theta}_{ML})\sqrt{n}\} + o_p\left(\frac{1}{\sqrt{n}}\right)$$

$$= -\gamma' I(\hat{\theta}_{ML}) + o_p\left(\frac{1}{\sqrt{n}}\right)$$

Since

$$Z_1(\underline{\theta}_n^*) = u_{1/2}\sqrt{I(\underline{\theta}_n^*)},$$

it follows that

$$-\gamma' I(\hat{\theta}_{ML}) + o_p\left(\frac{1}{\sqrt{n}}\right)$$

$$= u_{\alpha/2}\sqrt{I(\hat{\theta}_{ML})} + o_p\left(\frac{1}{\sqrt{n}}\right).$$

Hence we have

$$\gamma' = -\frac{u_{\alpha/2}}{\sqrt{I(\hat{\theta}_{ML})}}.$$

By a way similar to the above we obtain

$$\gamma = \frac{u_{\alpha/2}}{\sqrt{I(\hat{\theta}_{ML})}}.$$

Hence we have established the following:

Theorem 2. Under the conditions (i)~(iii), the α-*AUC* interval

$$\left[\hat{\theta}_{ML} - \frac{u_{\alpha/2}}{\sqrt{I(\hat{\theta}_{ML})n}}, \hat{\theta}_{ML} + \frac{u_{\alpha/2}}{\sqrt{I(\hat{\theta}_{ML})n}}\right]$$

is asymptotically efficient.

Next we consider the case up to the order of $n^{-1/2}$, i.e., second order case ([6]). We put $Z_1(\theta_0) = \bar{Z}_1 + \frac{\Delta}{\sqrt{n}}$ in the following equation:

$$e^{-(t^2/2)I(\theta_0)} \cdot e^{tZ_1(\theta_0)}\left[1 + \frac{t^2}{2\sqrt{n}}Z_2(\theta_0) - \frac{t^3}{6\sqrt{n}}\{3J(\theta_0) + K(\theta_0)\}\right] = \lambda_0 + \lambda_1 t Z_1(\theta_0) \quad (3.4)$$

We have

ON THE SECOND ORDER ASYMPTOTIC EFFICIENCY $\qquad$ *103/33*

$$\lambda_0 + \lambda_1 t\left(\bar{Z}_1 + \frac{\Delta}{\sqrt{n}}\right)$$

$$= e^{-(t^2/2)I} \cdot e^{t(\bar{Z}_1 + (\Delta/\sqrt{n}))}\left\{1 + \frac{t^2}{2\sqrt{n}}Z_2 - \frac{t^3}{6\sqrt{n}}3J + K\right\}$$

$$= e^{-(t^2/2)I} \cdot e^{t\bar{Z}_1}\left(1 + \frac{t}{\sqrt{n}}\Delta\right)\left\{1 + \frac{t^2}{2\sqrt{n}}Z_2 - \frac{t^3}{6\sqrt{n}}(3J + K)\right\}.$$

It follows by (3.3) and (3.4) that

$$1 + \lambda_1 t\frac{\Delta}{\sqrt{n}}e^{-t\bar{Z}_1 + (t^2/2)I} = 1 + \frac{t\Delta}{\sqrt{n}} + \frac{t^2}{2\sqrt{n}}Z_2 - \frac{t^3}{6\sqrt{n}}(3J + K).$$

Hence we obtain

$$\Delta = -\frac{1}{t(1 - \lambda_1 e^{-t\bar{Z}_1 + (t^2/2)I})}\left\{\frac{t^2}{2}Z_2 - \frac{t^3}{6}(3J + K)\right\}. \tag{3.5}$$

Putting $Z_1(\theta_0) = \underline{Z}_1 + \dfrac{\Delta'}{\sqrt{n}}$ in the equation (3.4), we similarly have

$$\Delta' = -\frac{1}{t(1 - \lambda_1 e^{-t\underline{Z}_1 + (t^2/2)I})}\left\{\frac{t^2}{2}Z_2 - \frac{t^3}{6}(3J + K)\right\}. \tag{3.6}$$

The rejection region like the type of (1) is given by

$$Z_1(\theta_0) - \frac{\Delta'}{\sqrt{n}} < \underline{Z}_1, \quad Z_1(\theta_0) - \frac{\Delta}{\sqrt{n}} > \bar{Z}_1, \tag{3.7}$$

where Δ and Δ' are given by (3.5) and (3.6), respectively. We have

$$Z_1(\theta_0) - \frac{\Delta}{\sqrt{n}} = Z_1(\theta_0) + \frac{c_1}{\sqrt{n}}Z_2(\theta_0) - \frac{c_2}{\sqrt{n}},$$

$$Z_1(\theta_0) - \frac{\Delta'}{\sqrt{n}} = Z_1(\theta_0) + \frac{c_1'}{\sqrt{n}}Z_2(\theta^0) - \frac{c_2'}{\sqrt{n}};$$

where

$$c_1 = \frac{t}{2(1 - \lambda_1 e^{-t\bar{Z}_1 + (t^2/2)I})}, \quad c_2 = \frac{t^2(3J + K)}{6(1 - \lambda_1 e^{-t\bar{Z}_1 + (t^2/2)I})},$$

$$c_1' = \frac{t}{2(1 - \lambda_1 e^{-t\underline{Z}_1 + (t^2/2)I})}, \quad c_2' = \frac{t^2(3J + K)}{6(1 - \lambda_1 e^{-t\bar{Z}_1 + (t^2/2)I})}.$$

It follows from (3.7) that the rejection region like the type (3.1) is given by

$$Z_1(\theta_0) + \frac{c_1'}{\sqrt{n}}Z_2(\theta_0) < \underline{Z}_1 + \frac{c_2'}{\sqrt{n}}, \quad Z_1(\theta_0) + \frac{c_1}{\sqrt{n}}Z_2(\theta_0) > \bar{Z}_1 + \frac{c_2}{\sqrt{n}},$$

where c_1, c_1', c_2 and c_2' are given above. We put

$$A_1 = \left\{\tilde{x}_n : Z_1(\theta_0) + \frac{c_1}{\sqrt{n}}Z_2(\theta_0) > \bar{Z}_1 + \frac{c_2}{\sqrt{n}}\right\},$$

$$A_2 = \left\{\tilde{x}_n : Z_1(\theta_0) + \frac{c_1'}{\sqrt{n}}Z_2(\theta_0) < \underline{Z}_1 + \frac{c_2'}{\sqrt{n}}\right\}.$$

In order to construct the second order α-AUC interval we have to determine $\underline{Z}_1$ and $\bar{Z}_1$ such that

$$P_{n,\theta_0}(A_1) + P_{n,\theta_0}(A_2) = \alpha + o\left(\frac{1}{n}\right); \tag{3.8}$$

$$\frac{\partial}{\partial\theta}P_{n,\theta_0}(A_1) + \frac{\partial}{\partial\theta}P_{n,\theta_0}(A_2) = o\left(\frac{1}{n}\right). \tag{3.9}$$

If χ_A denotes the indicator function of the set A, then (3.8) is given by

$$E_{n,\theta_0}(\chi_{A_1}) + E_{n,\theta_0}(\chi_{A_2}) = \alpha + o\left(\frac{1}{n}\right). \tag{3.10}$$

Since

$$\frac{\partial}{\partial\theta}P_{n,\theta}(A) = \int \chi_A(\tilde{x}_n)\left\{\frac{\partial}{\partial\theta}\prod_{i=1}^{n} f(x_i, \theta)\right\}d\mu^{(n)}$$

$$= \int \chi_A(\tilde{x}_n)\left\{\frac{\partial}{\partial\theta}\log\prod_{i=1}^{n} f(x_i, \theta)\right\}\prod_{i=1}^{n} f(x_i, \theta)\,d\mu^{(n)}$$

$$= \int \chi_A(\tilde{x}_n) \Big\{ \sum_{i=1}^{n} \frac{\partial}{\partial \theta} \log f(x_i, \theta) \Big\} \prod_{i=1}^{n} f(x_i, \theta) \, d\mu^{(n)}$$

$$= E_{n, \theta} \Big[\chi_A(\tilde{X}_n) \sum_{i=1}^{n} \frac{\partial}{\partial \theta} \log f(x_i, \theta) \Big],$$

where $\mu^{(n)}$ denotes the n-fold direct product measure of μ, it follows that (3.9) is given by

$$E_{n, \theta_0}[\chi_{A_1}(\tilde{X}_n) Z_1(\theta_0)] + E_{n, \theta_0}[\chi_{A_2}(\tilde{X}_n) Z_1(\theta_0)] = o\Big(\frac{1}{n}\Big). \tag{3.11}$$

Putting

$$Z_1^*(\theta_0) = Z_1(\theta_0) + \frac{c_1}{\sqrt{n}} Z_2(\theta_0); \quad Z_1^{**}(\theta_0) = Z_1(\theta_0) + \frac{c_1'}{\sqrt{n}} Z_2(\theta_0),$$

we have from (3.10)

$$o\Big(\frac{1}{n}\Big) = E_{n, \theta_1} \Big[\chi_{A_1}(\tilde{x}_n) \Big\{ Z_1^*(\theta_0) - \frac{c_1}{\sqrt{n}} Z_2(\theta_0) \Big\} \Big] \tag{3.12}$$

$$+ E_{n, \theta_0} \Big[\chi_{A_2}(\tilde{x}_n) \Big\{ Z_1^{**}(\theta_0) - \frac{c_1'}{\sqrt{n}} Z_2(\theta_0) \Big\} \Big]$$

$$= E_{n, \theta_0}[\chi_{A_1}(\tilde{X}_n) Z_1^*(\theta_0)] - \frac{c_1}{\sqrt{n}} E_{n, \theta_0}[\chi_{A_1}(\tilde{x}_n) Z_2(\theta_0)]$$

$$+ E_{n, \theta_0}[\chi_{A_2}(\tilde{X}_n) Z_1^{**}(\theta_0)] - \frac{c_1'}{\sqrt{n}} E_{n, \theta_0}[\chi_{A_2}(\tilde{X}_n) Z_2(\theta_0)].$$

Since

$$E_{n, \theta_0}[Z_1^2(\theta_0)] = I(\theta_0); \quad E_{n, \theta_0}[Z_1(\theta_0) Z_2(\theta_0)] = J(\theta_0);$$
$$E_{n, \theta_0}[Z_2^2(\theta_0)] = M(\theta_0) - I^2(\theta_0),$$

where

$$M(\theta) = E_\theta \Big[\Big\{ \frac{\partial^2}{\partial \theta^2} \log f(x, \theta) \Big\}^2 \Big],$$

it follows that

$$V_{n, \theta_0}(Z_1^*(\theta_0)) = E_{n, \theta_0} \Big[\Big\{ Z_1(\theta_0) + \frac{c_1}{\sqrt{n}} Z_2(\theta_0) \Big\}^2 \Big] \tag{3.13}$$

$$= E_{n, \theta_0}[Z_1^2(\theta_0)] + \frac{2c_1}{\sqrt{n}} E_{n, \theta_0}[Z_1(\theta_0) Z_2(\theta_0)] + o\Big(\frac{1}{\sqrt{n}}\Big)$$

$$= I(\theta_0) + \frac{2c_1}{\sqrt{n}} J(\theta_0) + o\Big(\frac{1}{\sqrt{n}}\Big);$$

$$\mathrm{Cov}_{n, \theta_0}(Z_1^*(\theta_0), Z_2(\theta_0)) = E_{n, \theta_0}[Z_1^*(\theta_0) Z_2(\theta_0)] - E_{n, \theta_0}[Z_1^*(\theta_0)] E_{n, \theta_0}[Z_2(\theta_0)] \tag{3.14}$$

$$= E_{n, \theta_0} \Big[\Big\{ Z_1(\theta_0) + \frac{c_1}{\sqrt{n}} Z_2(\theta_0) \Big\} Z_2(\theta_0) \Big]$$

$$= E_{n, \theta_0}[Z_1(\theta_0) Z_2(\theta_0)] + \frac{c_1}{\sqrt{n}} E_{n, \theta_0}[Z_2^2(\theta_0)]$$

$$= J(\theta_0) + \frac{c_1}{\sqrt{n}} \{ M(\theta_0) - I^2(\theta_0) \} + o\Big(\frac{1}{\sqrt{n}}\Big).$$

From (3.13) and (3.14) we have

$$E_{n, \theta_0}[Z_2(\theta_0) \mid Z_1^*(\theta_0)] = \frac{\mathrm{Cov}_{n, \theta_0}(Z_1^*(\theta_0), Z_2(\theta_0))}{V_{n, \theta_0}(Z_1^*(\theta_0))} Z_1^*(\theta_0)$$

$$= \frac{J(\theta_0)}{I(\theta_0)} Z_1^*(\theta_0) + o\Big(\frac{1}{\sqrt{n}}\Big).$$

Hence we obtain

$$E_{n, \theta_0}[\chi_{A_1}(\tilde{X}_n) Z_2(\theta_0)] = E_{n, \theta_0}[\chi_{A_1}(\tilde{X}_n) E_{n, \theta_0}(Z_2(\theta_0)) \mid Z_1^*(\theta_0))] \tag{3.15}$$

$$= \frac{J(\theta_0)}{I(\theta_0)} E_{n, \theta_0}[\chi_{A_1}(\tilde{X}_n) Z_1^*(\theta_0)] + o\Big(\frac{1}{\sqrt{n}}\Big).$$

By a way similar to the above we obtain

ON THE SECOND ORDER ASYMPTOTIC EFFICIENCY   *105/35*

$$E_{n,\theta_0}[\chi_{A_2}(\tilde{X}_n)Z_2(\theta_0)] = \frac{J(\theta_0)}{I(\theta_0)} E_{n,\theta_0}[\chi_{A_2}(\tilde{X}_n)Z_1^{**}(\theta_0)] + o\left(\frac{1}{\sqrt{n}}\right). \tag{3.16}$$

Form (3.12), (3.15) and (3.16) it follows that

$$\left\{1 - \frac{c_1 J(\theta_0)}{I(\theta_0)\sqrt{n}}\right\} E_{n,\theta_0}[\chi_{A_1}(\tilde{X}_n)Z_1^*(\theta_0)] + \left\{1 - \frac{c_1' J(\theta_0)}{I(\theta_0)\sqrt{n}}\right\} E_{n,\theta_0}[\chi_{A_2}(\tilde{X}_n)Z_1^{**}(\theta_0)] = o\left(\frac{1}{n}\right). \tag{3.17}$$

It follows by Edgeworth expansion that the probability density function $f(x)$ of $Z_1^*(\theta_0)/\sigma$, where $\sigma^2 = V_{n,I_0}(Z_1^*(\theta_0))((3.13))$, is given by

$$f(x) = \phi(x) + \frac{\beta_3}{6\sqrt{n}} H_3(x)\phi(x) + o\left(\frac{1}{\sqrt{n}}\right),$$

where $\phi(x) = \frac{1}{\sqrt{2\pi}} e^{-x^2/2}$ and β_3 is the third order cumulant of Z_1^*/σ and $H_3(x)$ is a Hermite polynomial, i.e. $H_3(x) = x^3 - 3x$. Since

$$\int xH_3(x)\phi(x)\,dx = \int x(x^3 - 3x)\phi(x)\,dx$$

$$= \int (x^4 - 3x^2)\phi(x)\,dx$$

$$= \int \{(x^4 - 6x^2 + 3) + 3(x^2 - 1)\}\phi(x)\,dx$$

$$= \int \{H_4(x) + 3H_2(x)\}\phi(x)\,dx$$

$$= \int H_4(x)\phi(x)\,dx + 3\int H_2(x)\phi(x)\,dx$$

$$= \{-H_3(x) - 3H_1(x)\}\phi(x)$$

$$= (-x^3 + 3x - 3x)\phi(x)$$

$$= -x^3\phi(x),$$

where $H_i(x)$ ($i = 1, 2, 3, 4$) are Hermite polynomials, it follows that

$$\int xf(x)\,dx = \int x\left\{\phi(x) + \frac{\beta_3}{6\sqrt{n}} H_3(x)\phi(x)\right\}dx + o\left(\frac{1}{\sqrt{n}}\right) \tag{3.18}$$

$$= \int x\phi(x)\,dx - \frac{\beta_3}{6\sqrt{n}} \int xH_3(x)\phi(x)\,dx + o\left(\frac{1}{\sqrt{n}}\right)$$

$$= -\phi(x) - \frac{\beta_3}{6\sqrt{n}} x^3\phi(x) + o\left(\frac{1}{\sqrt{n}}\right).$$

Since by a way similar to the above, the probability density function $g(x)$ of $Z_1^{**}(\theta_0)/\sigma'$, where $\sigma'^2 = V_{n,\theta_0}(Z_1^{**}(\theta_0)) = I(\theta_0) + \frac{2c_1'}{\sqrt{n}} J(\theta_0) + o\left(\frac{1}{\sqrt{n}}\right)$, is given by

$$g(x) = \phi(x) + \frac{\beta_3'}{6\sqrt{n}} H_3(x)\phi(x) + o\left(\frac{1}{\sqrt{n}}\right),$$

where β_3' is the third order cumulant of $Z_1^{**}(\theta_0)/\sigma'$, it follows that

$$\int xg(x)\,dx = -\phi(x) - \frac{\beta_3'}{6\sqrt{n}} x^3\phi(x) + o\left(\frac{1}{\sqrt{n}}\right). \tag{3.19}$$

Putting

$$k_1 = \bar{Z}_1 + \frac{c_2}{\sqrt{n}}; \quad k_1' = \underline{Z}_1 + \frac{c_2'}{\sqrt{n}}$$

we have

$$A_1 = \{Z_1^* > k_1\}; \quad A_2 = \{Z_1^{**} < k_1'\}.$$

By (3.18) we obtain

$$E_{n,\theta_0}[\chi_{A_1}(\tilde{X}_n)Z_1^*(\theta_0)] = \left[-\phi\left(\frac{x}{\sigma}\right) - \frac{\beta_3}{6\sqrt{n}} \left(\frac{x}{\sigma}\right)^3 \phi\left(\frac{x}{\sigma}\right)\right]_{k_1}^{\infty} \tag{3.20}$$

$$= \phi\left(\frac{k_1}{\sigma}\right) + \frac{\beta_3}{6\sqrt{n}} \left(\frac{k_1}{\sigma}\right)^3 \phi\left(\frac{k_1}{\sigma}\right).$$

By (3.19) we also have

$$E_{n,\theta_0}[\chi_{A_2}(\tilde{X}_n)Z_1^{**}(\theta_0)] = \left[-\phi\left(\frac{x}{\sigma'}\right) - \frac{\beta_3'}{6\sqrt{n}}\left(\frac{x}{\sigma'}\right)^3\phi\left(\frac{x}{\sigma}\right)\right]_{-\infty}^{k_1'} \tag{3.21}$$

$$= -\phi\left(\frac{k_1'}{\sigma'}\right) - \frac{\beta_3'}{6\sqrt{n}}\left(\frac{k_1'}{\sigma'}\right)^3\phi\left(\frac{k_1'}{\sigma'}\right).$$

Since $\beta_3 = \beta_3' = K(\theta_0)$, it follows by (3.17), (3.20) and (3.21) that

$$o\left(\frac{1}{n}\right) = \left(1 - \frac{c_1 J}{I\sqrt{n}}\right)E_{n,\theta_0}(\chi_{A_1}Z_1^*) + \left(1 - \frac{c_1' J}{I\sqrt{n}}\right)E_{n,\theta_0}(\chi_{A_1}Z_1^{**}) \tag{3.22}$$

$$= \left(1 - \frac{c_1 J}{I\sqrt{n}}\right)\left\{\phi\left(\frac{k_1}{\sigma}\right) + \frac{K}{6\sqrt{n}}\left(\frac{k_1}{\sigma}\right)^3\phi\left(\frac{k_1}{\sigma}\right)\right\}$$

$$- \left(1 - \frac{c_1' J}{I\sqrt{n}}\right)\left\{\phi\left(\frac{k_1'}{\sigma'}\right) + \frac{K}{6\sqrt{n}}\left(\frac{k_1'}{\sigma'}\right)^3\phi\left(\frac{k_1'}{\sigma'}\right)\right\}$$

$$= \left(1 - \frac{c_1 J}{I\sqrt{n}}\right)\left\{1 + \frac{K}{6\sqrt{n}}\left(\frac{k_1}{\sigma}\right)^3\right\}\phi\left(\frac{\bar{Z}_1 + \dfrac{c_2}{\sqrt{n}}}{\sigma}\right)$$

$$- \left(1 - \frac{c_1' J}{I\sqrt{n}}\right)\left\{1 + \frac{K}{6\sqrt{n}}\left(\frac{k_1'}{\sigma'}\right)^3\right\}\phi\left(\frac{\underline{Z}_1 + \dfrac{c_2'}{\sqrt{n}}}{\sigma'}\right).$$

From (3.22) we have

$$\phi\left(\frac{\bar{Z}_1 + \dfrac{c_2}{\sqrt{n}}}{\sigma}\right)\bigg/\phi\left(\frac{\underline{Z}_1 + \dfrac{c_2'}{\sqrt{n}}}{\sigma'}\right) = \frac{\left(1 - \dfrac{c_1' J}{I\sqrt{n}}\right)\left\{1 + \dfrac{K}{6\sqrt{n}}\left(\dfrac{k_1'}{\sigma'}\right)^3\right\}}{\left(1 - \dfrac{c_1 J}{I\sqrt{n}}\right)\left\{1 + \dfrac{K}{6\sqrt{n}}\left(\dfrac{k_1}{\sigma}\right)^3\right\}} \tag{3.23}$$

$$= \left\{1 - \frac{c_1' J}{I\sqrt{n}} + \frac{K}{6\sqrt{n}}\left(\frac{k_1'}{\sigma'}\right)^3\right\}\left\{1 + \frac{c_1 J}{I\sqrt{n}} - \frac{K}{6\sqrt{n}}\left(\frac{k_1}{\sigma}\right)^3\right\}$$

$$= 1 + \frac{c_1 - c_1'}{I\sqrt{n}}J + \frac{K}{6\sqrt{n}}\left\{\left(\frac{k_1'}{\sigma'}\right)^3 - \left(\frac{k_1}{\sigma}\right)^3\right\} + o\left(\frac{1}{\sqrt{n}}\right).$$

Since $\sigma^2 = \sigma'^2 = I(\theta_0) + o\left(\dfrac{1}{\sqrt{n}}\right)$, it follows from (3.23) that

$$-\frac{1}{2I}\left\{\left(\bar{Z}_1 + \frac{c_2}{\sqrt{n}}\right)^2 - \left(\underline{Z}_1 + \frac{c_2'}{\sqrt{n}}\right)^2\right\} \tag{3.24}$$

$$= \log\left\{1 + \frac{(c_1 - c_1')J}{I\sqrt{n}} + \frac{K}{6I^{3/2}\sqrt{n}}(\underline{Z}_1^3 - \bar{Z}_1^3)\right\},$$

$$= \frac{(c_1 - c_1')J}{I\sqrt{n}} + \frac{K}{6I^{3/2}\sqrt{n}}(\underline{Z}_1^3 - \bar{Z}_1^3) + o\left(\frac{1}{\sqrt{n}}\right).$$

Putting $\bar{Z}_1 = u_{\alpha/2}\sqrt{I(\theta_0)} + \dfrac{\eta}{\sqrt{n}}$ and $\underline{Z}_1 = -u_{\alpha/2}\sqrt{I(\theta_0)} + \dfrac{\eta'}{\sqrt{n}}$, we have from (3.24)

$$-\frac{u_{\alpha/2}}{\sqrt{I}}\frac{\eta + \eta' + c_2 + c_2'}{\sqrt{n}}$$

$$= \frac{(c_1 - c_1')J}{I\sqrt{n}} + \frac{K}{6I^{3/2}\sqrt{n}}(I^{3/2}u_{\alpha/2}^3 + I^{3/2}u_{\alpha/2}^3).$$

Hence we have

$$\eta + \eta' = -\frac{(c_1 - c_1')J}{u_{\alpha/2}\sqrt{I}} - \frac{K\sqrt{I}}{3}u_{\alpha/2}^2 - c_2 - c_2'. \tag{3.25}$$

Since

$$\int f(x)\,dx = \int \left\{\phi(x) + \frac{K}{6\sqrt{n}}H_3(x)\phi(x)\right\}dx$$

$$= \Phi(x) - \frac{K}{6\sqrt{n}}\int H_3(x)\phi(x)\,dx$$

$$= \Phi(x) - \frac{K}{6\sqrt{n}} H_2(x)\phi(x)$$

$$= \Phi(x) - \frac{K}{6\sqrt{n}}(x^2 - 1)\phi(x);$$

$$\int g(x)\,dx = \Phi(x) - \frac{K}{6\sqrt{n}}(x^2 - 1)\phi(x),$$

where $\Phi(x)$ is a standard normal distribution function, it follows from (3.10) that

$$\alpha + o\left(\frac{1}{n}\right) = E_{n,\theta_0}(\chi_{A_1}) + E_{n,\theta_0}(\chi_{A_2})$$

$$= 1 - \Phi\left(\frac{k_1}{\sigma}\right) + \frac{K}{6\sqrt{n}}\left(\frac{k_1^2}{\sigma^2} - 1\right)\phi\left(\frac{k_1}{\sigma}\right)$$

$$+ \Phi\left(\frac{k_1'}{\sigma'}\right) - \frac{K}{6\sqrt{n}}\left(\frac{k_1'^2}{\sigma'^2} - 1\right)\phi\left(\frac{k_1'}{\sigma'}\right).$$

Hence we have

$$1 - \alpha + o\left(\frac{1}{n}\right) = \Phi\left(\frac{\bar{Z}_1 + \frac{c_2}{\sqrt{n}}}{\sigma}\right) - \Phi\left(\frac{\underline{Z}_1 + \frac{c_1'}{\sqrt{n}}}{\sigma'}\right)$$

$$+ \frac{K}{6\sqrt{n}}\left\{\left(\frac{\underline{Z}_1^2}{\sigma'^2} - 1\right)\phi\left(\frac{\underline{Z}_1}{\sigma'}\right) - \left(\frac{\bar{Z}_1^2}{\sigma^2} - 1\right)\phi\left(\frac{\bar{Z}_1}{\sigma}\right)\right\}. \tag{3.26}$$

Since we put

$$\bar{Z}_1 = u_{\alpha/2}\sqrt{I} + \frac{\eta}{\sqrt{n}}; \quad \underline{Z}_1 = -u_{\alpha/2}\sqrt{I} + \frac{\eta'}{\sqrt{n}},$$

we have from (3.26)

$$\Phi\left(\frac{u_{\alpha/2}\sqrt{I} + \frac{c_2 + \eta}{\sqrt{n}}}{\sigma}\right) - \Phi\left(\frac{-u_{\alpha/2}\sqrt{I} + \frac{c_2' + \eta'}{\sqrt{n}}}{\sigma'}\right) = 1 - \alpha + o\left(\frac{1}{n}\right). \tag{3.27}$$

Since

$$\sigma = \sqrt{I(\theta_0)} + \frac{c_1 J(\theta_0)}{\sqrt{I(\theta_0)n}} + o\left(\frac{1}{\sqrt{n}}\right);$$

$$\sigma' = \sqrt{I(\theta_0)} + \frac{c_1' J(\theta_0)}{\sqrt{I(\theta_0)n}} + o\left(\frac{1}{\sqrt{n}}\right),$$

it follows that

$$\frac{u_{\alpha/2}\sqrt{I} + \frac{c_2 + \eta}{\sqrt{n}}}{\sigma} = \left(u_{\alpha/2} + \frac{c_2 + \eta}{\sqrt{In}}\right)\left\{1 - \frac{c_1 J}{I\sqrt{n}} + o\left(\frac{1}{\sqrt{n}}\right)\right\}$$

$$= u_{\alpha/2} - u_{\alpha/2}\frac{c_1 J}{I\sqrt{n}} + \frac{c_2 + \eta}{\sqrt{In}} + o\left(\frac{1}{\sqrt{n}}\right);$$

$$\frac{-u_{\alpha/2}\sqrt{I} + \frac{c_2' + \eta'}{\sqrt{n}}}{\sigma'} = -u_{\alpha/2} + u_{\alpha/2}\frac{c_1' J}{I\sqrt{n}} + \frac{c_2' + \eta'}{\sqrt{In}} + o\left(\frac{1}{\sqrt{n}}\right).$$

Hence we have

$$\Phi\left(\frac{u_{\alpha/2}\sqrt{I} + \frac{c_2 + \eta}{\sqrt{n}}}{\sigma}\right) = 1 - \frac{\alpha}{2} + \frac{1}{\sqrt{n}}\phi(u_{\alpha/2})\left(-\frac{u_{\alpha/2}c_1 J}{\sqrt{I}} + \frac{c_2 + \eta}{\sqrt{I}}\right) + o\left(\frac{1}{\sqrt{n}}\right)$$

$$\Phi\left(\frac{-u_{\alpha/2}\sqrt{I} + \frac{c_2' + \eta'}{\sqrt{n}}}{\sigma'}\right) = \frac{\alpha}{2} + \frac{1}{\sqrt{n}}\phi(u_{\alpha/2})\left(\frac{u_{\alpha/2}c_1' J}{I} + \frac{c_2' + \eta'}{\sqrt{I}}\right) + o\left(\frac{1}{\sqrt{n}}\right).$$

From (3.27) we obtain

$$\frac{1}{\sqrt{n}}\phi(u_{\alpha/2})\left(-\frac{u_{\alpha/2}(c_1 + c_1')J}{I} + \frac{c_2 - c_2' + \eta - \eta'}{\sqrt{I}}\right) = o\left(\frac{1}{n}\right). \tag{3.28}$$

Since by (3.28)

$$\frac{c_2 - c_2' + \eta - \eta'}{\sqrt{I}} = \frac{u_{\alpha/2}(c_1 + c_1')J}{I},$$

it follows that

$$\eta - \eta' = \frac{u_{\alpha/2}(c_1 + c_1')J}{\sqrt{I}} - c_2 + c_2'. \tag{3.29}$$

From (3.25) and (3.29) we obtain

$$\eta = \frac{u_{\alpha/2}(c_1 + c_1')J}{2\sqrt{I}} - \frac{(c_1 - c_1')J\sqrt{I}}{2u_{\alpha/2}} - \frac{K\sqrt{I}}{6}u_{\alpha/2}^2 - c_2; \tag{3.30}$$

$$\eta' = -\frac{u_{\alpha/2}(c_1 + c_1')J}{2\sqrt{I}} - \frac{(c_1 - c_1')J\sqrt{I}}{2u_{\alpha/2}} - \frac{K\sqrt{I}}{6}u_{\alpha/2}^2 - c_2'. \tag{3.31}$$

where

$$c_1 = \frac{t}{2(1 - \lambda_1 e^{-tu_{\alpha/2}\sqrt{I} + (t^2/2)I})}; \quad c_2 = \frac{t^2(3J + K)}{6(1 - \lambda_1 e^{-tu_{\alpha/2}\sqrt{I} + (t^2/2)I})};$$

$$c_1' = \frac{t}{2(1 - \lambda_1 e^{tu_{\alpha/2}\sqrt{I} + (t^2/2)I})}; \quad c_2' = \frac{t^2(3J + K)}{6(1 - \lambda_1 e^{tu_{\alpha/2}\sqrt{I} + (t^2/2)I})};$$

with

$$\lambda_1 = \frac{1}{u_{\alpha/2}\sqrt{I(\theta_0)}}e^{-(t^2/2)I}\sinh(tu_{\alpha/2}\sqrt{I}) + O(1/\sqrt{n}).$$

Hence we have

$$\bar{Z}_1 = u_{\alpha/2}\sqrt{I} + \frac{\eta}{\sqrt{n}}; \quad \underline{Z}_1 = -u_{\alpha/2}\sqrt{I} + \frac{\eta'}{\sqrt{n}},$$

where η and η' are given by (3.30) and (3.31), respectively. We also obtain

$$k_1 = u_{\alpha/2}\sqrt{I} + \frac{c_2 + \eta}{\sqrt{n}}; \quad k' = -u_{\alpha/2}\sqrt{I} + \frac{c_2' + \eta'}{\sqrt{n}}.$$

We put

$$u(\theta; t) = c_2 + \eta; \quad 1(\theta; t) = c_2' + \eta'.$$

If there exist $\underline{\theta}_n^{**}(\tilde{x}_n)$ and $\bar{\theta}_n^{**}(\tilde{x}_n)$ in the neighborhood of maximum likelihood estimator (MLE) θ_{ML} such that

$$Z_1(\underline{\theta}_n^{**}(\tilde{x}_n)) + \frac{c_1}{\sqrt{n}}Z_2(\underline{\theta}_n^{**}(\tilde{x}_n)) = u_{\alpha/2}\sqrt{I(\underline{\theta}_n^{**}(\tilde{x}_n))} + \frac{u(\underline{\theta}_n^{**}(\tilde{x}_n); t)}{\sqrt{n}} \tag{3.32}$$

$$Z_1(\bar{\theta}_n^{**}(\tilde{x}_n)) + \frac{c_1'}{\sqrt{n}}Z_2(\bar{\theta}_n^{**}(\tilde{x}_n)) = -u_{\alpha/2}\sqrt{I(\bar{\theta}_n^{**}(\tilde{x}_n))} + \frac{1(\bar{\theta}_n^{**}(\tilde{x}_n); t)}{\sqrt{n}} \tag{3.33}$$

([2]), then the second order α-AUC interval $[\underline{\theta}_n^{**}(\tilde{x}_n), \bar{\theta}_n^{**}(\tilde{x}_n)]$ is second order asymptotically efficient. Hence we have established the following:

Theorem 3. Under conditions (i)$\sim$(iv), the second order α-AUC interval $[\underline{\theta}_n^{**}(\tilde{x}_n), \bar{\theta}_n^{**}(\tilde{x}_n)]$ is second order asymptotically efficient.

We shall actually construct the second order asymptotically efficient α-AUC interval using the MLE $\hat{\theta}_{ML}$. We put

$$\underline{\theta}_n^{**}(\tilde{x}_n) = \hat{\theta}_{ML} - \frac{u_{\alpha/2}}{\sqrt{I(\hat{\theta}_{ML})n}} + \frac{\delta'}{n};$$

$$\bar{\theta}_n^{**}(\tilde{x}_n) = \hat{\theta}_{ML} + \frac{u_{\alpha/2}}{\sqrt{I(\hat{\theta}_{ML})n}} + \frac{\delta}{n}.$$

we have

$$Z_1(\underline{\theta}_n^{**}) + \frac{c_1}{\sqrt{n}}Z_2(\underline{\theta}_n^{**})$$

$$= Z_1\left(\hat{\theta}_{ML} - \frac{u_{\alpha/2}}{\sqrt{I(\hat{\theta}_{ML})n}} + \frac{\delta'}{n}\right) + \frac{c_1}{\sqrt{n}}Z_2\left(\hat{\theta}_{ML} - \frac{u_{\alpha/2}}{\sqrt{I(\hat{\theta}_{ML})n}} + \frac{\delta'}{n}\right) + o_p\left(\frac{1}{n}\right)$$

ON THE SECOND ORDER ASYMPTOTIC EFFICIENCY $\quad 109/39$

$$= \left(-\frac{u_{\alpha/2}}{\sqrt{I(\hat{\theta}_{ML})n}} + \frac{\delta'}{n}\right) Z_1'(\hat{\theta}_{ML}) + \frac{1}{2}\left(-\frac{u_{\alpha/2}}{\sqrt{I(\hat{\theta}_{ML})n}} + \frac{\delta'}{n}\right)^2 Z_1''(\hat{\theta}_{ML})$$
$$+ \frac{c_1}{\sqrt{n}} Z_2(\hat{\theta}_{ML}) + \frac{c_1}{\sqrt{n}}\left(-\frac{u_{\alpha/2}}{\sqrt{I(\hat{\theta}_{ML})n}} + \frac{\delta'}{n}\right) Z_2'(\hat{\theta}_{ML}) + o_p\left(\frac{1}{n}\right).$$

Since

$$Z_1'(\theta_0) = Z_2(\theta_0) - \sqrt{n}\, I(\theta_0);$$
$$I'(\theta_0) = 2J(\theta_0) + K(\theta_0),$$

it follows that

$$Z_1''(\theta_0) = Z_2'(\theta_0) - \sqrt{n}\, I'(\theta_0)$$
$$\sim -\sqrt{n}\, J(\theta_0) - \sqrt{n}\{2J(\theta_0) + K(\theta_0)\}$$
$$= -\sqrt{n}\{3J(\theta_0) + K(\theta_0)\}.$$

We also obtain

$$Z_2'(\theta_0) = \frac{1}{\sqrt{n}} \sum_{i=1}^{n} \frac{\partial^3}{\partial\theta^3} \log f(x_i, \theta_0) + \sqrt{n}\, I'(\theta_0)$$
$$= \frac{1}{\sqrt{n}} \sum_{i=1}^{n} \frac{\partial^3}{\partial\theta^3} \log f(x_i, \theta_0) + \sqrt{n}\{2J(\theta_0) + K(\theta_0)\}$$
$$\sim \sqrt{n}\{-3J(\theta_0) - K(\theta_0)\} + \sqrt{n}\{2J(\theta_0) + K(\theta_0)\}$$
$$= -J(\theta_0)\sqrt{n}.$$

Hence we have

$$Z_1(\underline{\theta}_n^{**}) + \frac{c_1}{\sqrt{n}} Z_2(\underline{\theta}_n^{**})$$
$$= \left(-\frac{u_{\alpha/2}}{\sqrt{I(\hat{\theta}_{ML})n}} + \frac{\delta'}{n}\right)\{Z_2(\hat{\theta}_{ML}) - \sqrt{n}\, I(\hat{\theta}_{ML})\}$$
$$+ \frac{u_{\alpha/2}^2}{\sqrt{2I(\hat{\theta}_{ML})n}}\{-\sqrt{n}\,(3J(\hat{\theta}_{ML}) + K(\hat{\theta}_{ML}))\} + \frac{c_1}{\sqrt{n}} Z_2(\hat{\theta}_{ML})$$
$$+ \frac{u_{\alpha/2}}{\sqrt{I(\hat{\theta}_{ML})n}}(\sqrt{n}\, J(\hat{\theta}_{ML})) + o_p\left(\frac{1}{\sqrt{n}}\right)$$
$$= u_{\alpha/2}\sqrt{I(\hat{\theta}_{ML})} - \frac{u_{\alpha/2}}{\sqrt{I(\hat{\theta}_{ML})n}} Z_2(\hat{\theta}_{ML}) + \frac{c_1}{\sqrt{n}} Z_2(\hat{\theta}_{ML})$$
$$- \frac{u_{\alpha/2}^2}{2I(\theta_{ML})\sqrt{n}}\{3J(\hat{\theta}_{ML}) + K(\hat{\theta}_{ML})\} + \frac{c_1 u_{\alpha/2} J(\hat{\theta}_{ML})}{\sqrt{I(\hat{\theta}_{ML})}} + o_p\left(\frac{1}{\sqrt{n}}\right).$$

Since

$$\sqrt{I(\underline{\theta}_n^{**})} = \sqrt{I(\hat{\theta}_{ML})} + o_p\left(\frac{1}{\sqrt{n}}\right)$$
$$u(\underline{\theta}_n^{**}, t) = u(\hat{\theta}_{ML}; t) + o_p\left(\frac{1}{\sqrt{n}}\right),$$

it follows by (3.32) that

$$-\frac{u_{\alpha/2}}{\sqrt{I(\hat{\theta}_{ML})n}} Z_2(\hat{\theta}_{ML}) - \frac{\delta'}{\sqrt{n}} I(\hat{\theta}_{ML}) + \frac{c_1}{\sqrt{n}} Z_2(\hat{\theta}_{ML}) - \frac{u_{\alpha/2}^2}{2I(\hat{\theta}_{ML})\sqrt{n}}\{3J(\hat{\theta}_{ML}) + K(\hat{\theta}_{ML})\}$$
$$+ \frac{c_1 u_{\alpha/2} J(\hat{\theta}_{ML})}{\sqrt{I(\hat{\theta}_{ML})n}} = -\frac{u_{\alpha/2}^2\{2J(\hat{\theta}_{ML}) + K(\hat{\theta}_{ML})\}}{2\sqrt{I(\hat{\theta}_{ML})n}}.$$

Then we have

$$\delta' = \left(c_1 - \frac{u_{\alpha/2}}{\sqrt{I(\hat{\theta}_{ML})}}\right) \frac{Z_2(\hat{\theta}_{ML})}{I(\hat{\theta}_{ML})} + c_1 u_{\alpha/2} \frac{J(\hat{\theta}_{ML})}{I(\hat{\theta}_{ML})^{3/2}} \tag{3.34}$$

$$-\frac{u_{\alpha/2}^2}{2I(\hat{\theta}_{ML})}\left\{\frac{3J(\hat{\theta}_{ML})+K(\hat{\theta}_{ML})}{I(\hat{\theta}_{ML})}+\frac{2J(\hat{\theta}_{ML})+K(\hat{\theta}_{ML})}{\sqrt{I(\hat{\theta}_{ML})}}\right\}.$$

In a similar way to the above we obtain

$$\delta=\left(c_1+\frac{u_{\alpha/2}}{\sqrt{I(\hat{\theta}_{ML})}}\right)\frac{Z_2(\hat{\theta}_{ML})}{I(\hat{\theta}_{ML})}-c_1 u_{\alpha/2}\frac{J(\hat{\theta}_{ML})}{I(\hat{\theta}_{ML})^{3/2}} \tag{3.35}$$

$$-\frac{u_{\alpha/2}^2}{2I(\hat{\theta}_{ML})}\left\{\frac{3J(\hat{\theta}_{ML})+K(\hat{\theta}_{ML})}{I(\hat{\theta}_{ML})}+\frac{2J(\hat{\theta}_{ML})+K(_{ML})}{\sqrt{I(\hat{\theta}_{ML})}}\right\}.$$

Hence we have established the following:

Theorem 4. Under conditions (i) $\sim$ (iv), the second order α-*AUC* interval

$$\left[\hat{\theta}_{ML}-\frac{u_{\alpha/2}}{\sqrt{I(\hat{\theta}_{ML})n}}+\frac{\delta'}{n},\hat{\theta}_{ML}+\frac{u_{\alpha/2}}{\sqrt{I(\hat{\theta}_{ML})n}}+\frac{\delta}{n}\right] \tag{3.36}$$

is second order asymptotically efficient, where δ' and δ are given by (3.34) and (3.35), respectively.

Remark. It should be noted that both of δ' and δ depend on t and $Z_2(\cdot)$. The second order α-*AUC* interval (3.36) is most powerful but not uniformly most powerful because of the dependence on t. This testifies the fact that the asymptotically sufficient statistic up to the order $n^{-1/2}$ is given by the pair $(\hat{\theta}_{ML}, Z_2(\hat{\theta}_{ML}))$ ([9], also [8]), while in the case of point estimation $\hat{\theta}_{ML}$ is alone asymptotically efficien up to the order $n^{-1/2}$, i.e. second order.

References

[1]   Akahira, M.: Asymptotic theory for estimation of location in non-regular cases, I: Order of convergence of consistent estimators, *Rep. Stat. Appl. Res., JUSE,* **22** (1975) 8–26.

[2]   Akahira, M. and Takeuchi, K.: Discretized likelihood methods—Asymptotic properties of discretized likelihood estimator (DLE's), *Ann. Inst. Statist. Math.,* **31** (1979), Part A, 39–56.

[3]   Efron, B.: Defining the curvature of a statistical problem (with applications to second order efficiency), *Ann. Statist.,* **3** (1975) 1189–1242.

[4]   Lehmann, E. L.: *Testing Statistical Hypotheses,* John Wiley (1959).

[5]   Pfanzagl, J.: Asymptotic expansions related to maximum contrast estimators, *Ann. Statist.,* **1** (1973) 993–1026.

[6]   Takeuchi, K.: *Asymptotic expansions for the critical limits of unbiased tests,* Discussion paper No. 131, Kyoto Institute of Economics Research, Kyoto University. (1978).

[7]   Takeuchi, K. and Akahira, M.: On the second order asymptotic efficiencies of estimators, *In Proceedings of the Third Japan-USSR Symposium on Probability Theory (G. Maruyama and J. V. Prokhorov Eds.) Lecture Notes in Mathematics* **550** (1976) Springer-Verlag, Berlin, 604–638.

[8]   Michel, R.: *Higher* order asymptotic sufficiency, *Sankhya,* **40** (1978) 76–84.

[9]   Suzuki, T.: Asymptotic sufficiency up to higher orders and its applications to statistical tests and estimates, *Osaka J, Math.,* **15** (1978) 575–588.

[10] Wald, A.: Asymptotically shortest confidence intervals, *Ann. Math. Statist.,* **13** (1942) 127–137.

Rep. Stat. Appl. Res., JUSE.
Vol. 26, No. 4, December, 1979

A Section

Remarks on the Asymptotic Efficiency and Inefficiency of Maximum Probability Estimators

MASAFUMI AKAHIRA* AND KEI TAKEUCHI**

ABSTRACT

Suppose that X_i ($i = 1, 2, \ldots, n$) are distributed according to a two-sided truncated normal distribution with unknown mean θ. It is shown that the maximum probability estimator (MPE) of Weiss and Wolfowitz is asymptotically inadmissible and has smaller concentration of probability than the asymptotically efficient estimator $\hat{\theta}_{PT} = (\min X_i + \max X_i)/2$.

1. Introduction

Weiss and Wolfowitz ([7]) defined the concept of the maximum probability estimator (MPE) which $\hat{\theta}^r_{MP}$ is defined as the value of which maximizes

$$\int_{\theta-rc_n^{-1}}^{\theta+rc_n^{-1}} \prod_{i=1}^{n} f(x_i, \theta) d\theta$$

with appropriate choice of the sequence of constants $\{c_n\}$. And they proved that asymptotically $P_{n,\theta}\{c_n|\hat{\theta}_n - \theta| < r\}$ is maximized by the MPE thus defined ([9]). However the estimator $\hat{\theta}^r_{MP}$ depends on the choice of r, and the theorem does not say anything about

$$P_{n,\theta}\{c_n|\hat{\theta}^r_{MP} - \theta| < r'\} \text{ for } r' \neq r$$

and it is not ruled out that $\hat{\theta}^r_{MP}$ may behave badly outside the fixed interval $(-r, r)$. In regular cases, $\hat{\theta}^r_{MP}$ is asymptotically equivalent to the maximum likelihood estimator (MLE), and is asymptotically independent of r. But in non-regular cases $\hat{\theta}^r_{MP}$ does depend on r (which is also emphasized by Weiss and Wolfowitz themselves), the asymptotic behavior of $\hat{\theta}^r_{MP}$ is not well investigated. In [2], [3], [4] and [5] we defined one-sided and two-sided asymptotically efficient estimator in non-regular cases, as the one which maximizes

$$\lim_{n\to\infty} P_{n,\theta}\{c_n(\hat{\theta}_n - \theta) < a\} \qquad \text{for } a > 0$$

or

$$\lim_{n\to\infty} P_{n,\theta}\{c_n(\hat{\theta}_n - \theta) > -b\} \qquad \text{for } b > 0$$

and

$$\lim_{n\to\infty} P_{n,\theta}\{c_n|\hat{\theta}_n - \theta| < c\} \qquad \text{for } c > 0$$

Received 4 July, 1979
* Statistical Laboratory, University of Electro-Communications, Chofu, Tokyo 182, Japan
** Faculty of Economics, University of Tokyo, Hongo, Bunkyo-ku, Tokyo 113, Japan

uniformly in θ and a or b or c among asymptotically median unbiased (AMU) estimators which satisfy the condition

$$\lim_{n\to\infty} P_{n,\theta}\{\hat{\theta}_n \leq \theta\} = \lim_{n\to\infty} P_{n,\theta}\{\hat{\theta}_n \geq \theta\} = \frac{1}{2}$$

uniformly in θ.

Then the question is: if $\hat{\theta}^r_{MP}$ is AMU, then is it asymptotically efficient in the sense above? The purpose of this paper is to show that the answer is not always yes, even in the case when asymptotically efficient estimator does exist, implying that the MPE may asymptotically inadmissible.

The case discussed below is also covered by Weiss and Wolfowitz [8] and their Z_n is basically asymptotically equivalent to $\hat{\theta}_{PT}$ and at least as good as $\hat{\theta}^r_{MP}$ but they did not show explicitly the inadmissibility of the latter.

2. Results

Let $(\mathfrak{X}, \mathfrak{B})$ be a sample space. We consider a family of probability measures on $\mathfrak{B}$, $\mathcal{P} = \{P_\theta : \theta \in \Theta\}$, where the index set Θ is called the parameter space. We assume that Θ is an open set in a Euclidean 1-space R^1. Consider n-fold direct products $(\mathfrak{X}^{(n)}, \mathfrak{B}^{(n)})$ of $(\mathfrak{X}, \mathfrak{B})$ and the corresponding product measures $P_{n,\theta}$ of P_θ. An estimator of θ is defined to be a sequence $\{\hat{\theta}_n\}$ of $\mathfrak{B}^{(n)}$-measurable functions $\hat{\theta}_n$ on $\mathfrak{X}^{(n)}$ into Θ $(n = 1, 2, \ldots)$. For simplicity we denote an estimator by $\hat{\theta}_n$ instead of $\{\hat{\theta}_n\}$. For an increasing sequence of positive numbers $\{c_n\}$ (c_n tending to infinity) an estimator $\hat{\theta}_n$ is called consistent with order $\{c_n\}$ (or $\{c_n\}$-consistent for short) if for every $\epsilon > 0$ and every $\vartheta \in \Theta$ there exist a sufficiently small positive number δ and a sufficiently large positive number L satisfying the following:

$$\varliminf_{n\to\infty} \sup_{\theta: |\theta-\vartheta|<\delta} P_{n,\theta}\{c_n|\hat{\theta}_n - \theta| \geq L\} < \epsilon \qquad ([1]).$$

A $\{c_n\}$-consistent estimator $\hat{\theta}_n$ is called asymptotically median unbiased (AMU) estimator if for any $\vartheta \in \Theta$ there exists a positive number δ such that

$$\lim_{n\to\infty} \sup_{\theta:|\theta-\vartheta|<\delta} \left| P_{n,\theta}\{\hat{\theta}_n \leq \theta\} - \frac{1}{2} \right| = 0;$$

$$\lim_{n\to\infty} \sup_{\theta:|\theta-\vartheta|<\delta} \left| P_{n,\theta}\{\hat{\theta}_n \geq \theta\} - \frac{1}{2} \right| = 0.$$

A distribution function $F_\theta(t)$ is called to be the asymptotic distribution of $c_n(\hat{\theta}_n - \theta)$ (or $\hat{\theta}_n$ for short) if for each real number t, $F_\theta(t)$ is continuous in θ and for any $\vartheta \in \Theta$ there exists a positive number d such that at any continuity point t of $F_\theta(t)$

$$\lim_{n\to\infty} \sup_{\theta:|\theta-\vartheta|<d} |P_{n,\theta}\{c_n(\hat{\theta}_n - \theta) \leq t\} - F_\theta(t)| = 0.$$

Let $\mathfrak{X} = \Theta = R^1$. Suppose that $X_1, X_2, \ldots, X_n, \ldots$ is a sequence of independently and identically distributed random variables with a density $f(x:\theta)$. Suppose that θ is a location parameter, i.e., $f(x:\theta) = f(x - \theta)$. Further we assume that

$$f(x) = \begin{cases} ce^{-(x^2/2)} & \text{for } |x| < 1; \\ 0 & \text{for } |x| \geq 1, \end{cases}$$

where c is some constant.

In this case it is shown by Akahira [1] that there exists a consistent estimator with order $\{n\}$, which is the largest possible order.

Let $R = (-r, r)$. The maximum probability estimator $\hat{\theta}^r_{MP}$ of θ is defined as d maximizing

$$\int_{d-r/n}^{d+r/n} \prod_{i=1}^{n} f(x_i - \theta)d\theta$$

(Weiss and Wolfowitz [7], [9]), and in this case $\hat{\theta}^r_{MP}$ is shown to be

$$
\hat{\theta}^r_{MP} =
\begin{cases}
\dfrac{1}{2}(\min X_i + \max X_i) \text{ for } \max X_i - \min X_i > 2 - \dfrac{2r}{n}; \\[2ex]
\min X_i + 1 - \dfrac{r}{n} \text{ for } \bar{X} > \min X_i + 1 - \dfrac{r}{n}, \max X_i - \min X_i \leq 2 - \dfrac{2r}{n}; \\[2ex]
\max X_i - 1 + \dfrac{r}{n} \text{ for } \bar{X} < \max X_i - 1 + \dfrac{r}{n}, \max X_i - \min X_i \leq 2 - \dfrac{2r}{n}; \\[2ex]
\bar{X} \text{ for } \max X_i - 1 + \dfrac{r}{n} \leq \bar{X} \leq \min X_i + 1 - \dfrac{r}{n};
\end{cases}
$$

where $\bar{X} = \dfrac{1}{n} \sum\limits_{i=1}^{n} X_i$.

Remark. The definition of $\hat{\theta}^r_{MP}$ in the first case is not unique, but it does not affect the subsequent conclusions.

First we note that the densities of the asymptotic distributions (asymptotic densities) of min $X_i + 1$ and max $X_i - 1$ are given by

$$
g_1(u) = \begin{cases} ke^{-ku} & \text{for } u > 0; \\ 0 & \text{for } u \leq 0; \end{cases}
\tag{2.1}
$$

$$
g_2(u) = \begin{cases} ke^{ku} & \text{for } u < 0; \\ 0 & \text{for } u \geq 0; \end{cases}
\tag{2.2}
$$

respectively, where $k = ce^{-1/2}$. Now we consider another estimator $\hat{\theta}_{PT} = (\min X_i + \max X_i)/2$ which is asymptotically equivalent to the Pitman (best location invariant) estimator. Since min $X_i + 1$ and max $X_i - 1$ are asymptotically independent, the density of the asymptotic distribution of $\hat{\theta}_{PT} = \frac{1}{2}(\min X_i + \max X_i)$ is given by

$$
g(u) = ke^{-2k|u|}.
\tag{2.3}
$$

Next we shall obtain the asymptotic density of $\hat{\theta}^r_{MP}$. The asymptotic conditional density of $\frac{1}{2}$ $(\min X_i + \max X_i)$ given $\max X_i - \min X_i > 2 - 2r/n$ is obtained by

$$
h_1(u) = \begin{cases} \dfrac{k}{1 - K(r)} (e^{-2k|u|} - e^{-2kr}) & \text{for } |u| < r; \\[2ex] 0 & \text{for } |u| \geq r; \end{cases}
$$

where $K(r) = (1 + 2kr)e^{-2kr}$.

It can be shown that $\bar{X}$ is asymptotically independent of min X_i and max X_i, or more precisely $\sqrt{n}\,\bar{X}$ is asymptotically normally distributed independently of $n(\min X_i + 1)$ and $n(\max X_i - 1)$. Hence it follows that the asymptotic conditional density of min $X_i + 1 - r/n$ given $\bar{X} > \min X_i + 1 - r/n$ and max $X_i - \min X_i \leq 2 - 2r/n$ is obtained by

$$
h_2(u) = \begin{cases} \dfrac{k}{K(r)} e^{-2kr} & \text{for } |u| < r; \\[2ex] \dfrac{k}{K(r)} e^{-k(u+r)} & \text{for } |u| \geq r. \end{cases}
$$

Further the asymptotic conditional density of max $X_i - 1 + r/n$ given $\bar{X} < \max X_i - 1 + r/n$ and max $X_i - \min X_i \leq 2 - 2r/n$ is obtained by

$$
h_3(u) = \begin{cases} \dfrac{k}{K(r)} e^{-2kr} & \text{for } |u| < r; \\[2ex] \dfrac{k}{K(r)} e^{-k(r-u)} & \text{for } u \leq -r. \end{cases}
$$

Since the order of the probability of max $X_i - 1 + r/n < \bar{X} < \min X_i + 1 - r/n$ is $1/n$, the probability of $\bar{X}$ given max $X_i - 1 + r/n < \bar{X} < \min X_i + 1 - r/n$ is negligible in this case.

Hence the asymptotic density of $\hat{\theta}^r_{MP}$ is given by

$$
g^*(u) = \begin{cases} ke^{-2k|u|} & \text{for } |u| < r; \\[2ex] \dfrac{k}{2} e^{-k(|u|+r)} & \text{for } |u| \geq r. \end{cases}
\tag{2.4}
$$

Let $P_{n,\theta}$ be the probability measures with the density $\prod_{i=1}^{n} f(x_i - \theta)$.

Next we shall get the lower bound of

$$
\begin{aligned}
P_{n,\theta}\{n\,|\theta_n - \theta| \geq u\} \\
= P_{n,\theta}\{n(\hat{\theta}_n - \theta) \leq -u\} + P_{n,\theta}\{n(\hat{\theta}_n - \theta) \geq u\}
\end{aligned}
\tag{2.5}
$$

for all AMU estimators. Let u be arbitrarily positive but fixed and $\hat{\theta}_n$ be AMU. Let θ_0 be any fixed in Θ and put $\theta_1 = \theta_0 - (u/n)$ and $\theta_2 = \theta_0 + (u/n)$.

Since for each real number t $P_{n,\theta}\{n(\hat{\theta}_n - \theta) \leq t\}$ converges uniformly in a neighborhood of θ_0, it follows that for sufficiently large n

$$
P_{n,\theta_1}\{\hat{\theta}_n \leq \theta_0\} = P_{n,\theta_1}\{n(\hat{\theta}_n - \theta_1) \leq u\} \sim P_{n,\theta_0}\{n(\hat{\theta}_n - \theta_0) \leq u\};
$$
$$
P_{n,\theta_2}\{\hat{\theta}_n \leq \theta_0\} = P_{n,\theta_2}\{n(\hat{\theta}_n - \theta_2) \leq -u\} \sim P_{n,\theta_0}\{n(\hat{\theta}_n - \theta_0) \leq -u\}.
$$

In order to get the lower bound of (2.5) it is sufficient to find the maximum value of

$$
P_{n,\theta_2}\{\hat{\theta}_n \leq \theta_0\} - P_{n,\theta_1}\{\hat{\theta}_n \leq \theta_0\}
\tag{2.6}
$$

for AMU estimators $\hat{\theta}_n$. Let $B_n = \{\tilde{x}_n : \hat{\theta}_n(\tilde{x}_n) \leq \theta_0\}$ and ϕ_n be an indicator function of B_n.

Then it follows that

$$
\int_{-\infty}^{\infty} \cdots \int_{-\infty}^{\infty} \phi_n(\tilde{x}_n) \prod_{i=1}^{n} f(x_i - \theta_0) \prod_{i=1}^{n} dx_i = E_{n,\theta_0}(\phi_n) \longrightarrow \frac{1}{2}\,(n \longrightarrow \infty).
$$

We also have

$$
\int_{-\infty}^{\infty} \cdots \int_{-\infty}^{\infty} \phi_n(\tilde{x}_n) \left\{ \prod_{i=1}^{n} f(x_i, \theta_2) - \prod_{i=1}^{n} f(x_i, \theta_1) \right\} \prod_{i=1}^{n} dx_i
$$
$$
= E_{n,\theta_2}(\phi_n) - E_{n,\theta_1}(\phi_n).
\tag{2.7}
$$

If the maximum value of (2.7) is obtained, then it is that of (2.6). In a similar way as the proof of the fundamental lemma of Neyman and Person, it is shown that ϕ_n maximizing (2.7) is given by

$$
\phi_n^*(\tilde{x}_n) = \begin{cases} 1, \prod\limits_{i=1}^{n} f(x_i, \theta_2) - \prod\limits_{i=1}^{n} f(x_i, \theta_1) > \lambda \prod\limits_{i=1}^{n} f(x_i, \theta_0); \\[2ex] 0, \prod\limits_{i=1}^{n} f(x_i, \theta_2) - \prod\limits_{i=1}^{n} f(x_i, \theta_1) < \lambda \prod\limits_{i=1}^{n} f(x_i, \theta_0), \end{cases}
$$

for some λ.

Since θ_1 and θ_2 are symmetric with respect to θ_0, it is clear that $\lambda = 0$.

Hence

$$
\phi_n^*(\tilde{x}_n) = \begin{cases} 1, \text{ if } \prod\limits_{i=1}^{n} f(x_i, \theta_2) > \prod\limits_{i=1}^{n} f(x_i, \theta_1), \text{ that is,} \\[2ex] \quad \text{(a)}\ \prod\limits_{i=1}^{n} f(x_i, \theta_2) > 0,\ \prod\limits_{i=1}^{n} f(x_i, \theta_1) = 0; \\[2ex] \quad \text{or} \\[2ex] \quad \text{(b)}\ \prod\limits_{i=1}^{n} f(x_i, \theta_2) > 0,\ \prod\limits_{i=1}^{n} f(x_i, \theta_1) > 0 \text{ and } \bar{x} < c'; \\[2ex] 0, \text{ if } \prod\limits_{i=1}^{n} f(x_i, \theta_2) < \prod\limits_{i=1}^{n} f(x_i, \theta_1), \text{ that is,} \\[2ex] \quad \text{(a')}\ \prod\limits_{i=1}^{n} f(x_i, \theta_2) = 0,\ \prod\limits_{i=1}^{n} f(x_i, \theta_1) > 0; \\[2ex] \quad \text{or} \\[2ex] \quad \text{(b')}\ \prod\limits_{i=1}^{n} f(x_i, \theta_2) > 0,\ \prod\limits_{i=1}^{n} f(x_i, \theta_1) > 0 \text{ and } \bar{x} > c', \end{cases}
$$

where $\bar{x} = \dfrac{1}{n} \sum\limits_{i=1}^{n} x_i$ and c' is some constant

Putting

$$A = \left\{ \tilde{x}_n \,\middle|\, \max x_i - 1 > \theta_0 - \frac{u}{n},\ \min x_i + 1 > \theta_0 + \frac{u}{n} \right\},$$

$$B = \left\{ \tilde{x}_n \,\middle|\, \max x_i - 1 < \theta_0 - \frac{u}{n},\ \min x_i + 1 < \theta_0 + \frac{u}{n} \right\},$$

$$C = \left\{ \tilde{x}_n \,\middle|\, \max x_i - 1 < \theta_0 - \frac{u}{n},\ \min x_i + 1 > \theta_0 + \frac{u}{n} \right\},$$

$$D = \{ \tilde{x}_n \,|\, \bar{x} < c' \},$$
$$D' = \{ \tilde{x}_n \,|\, \bar{x} > c' \}$$

we have

$$\phi_n^*(\tilde{x}_n) = \begin{cases} 1, & \text{if } \tilde{x}_n \in A \cup (C \cap D); \\ 0, & \text{if } \tilde{x}_n \in B \cup (C \cap D'). \end{cases} \tag{2.8}$$

Since $\bar{X}$, $\max X_i - 1$ and $\min X_i + 1$ are mutually asymptotically independent, it is seen that the conditional probability $P_{n,\theta}\{D\,|\,C\}$ of D given C is given by for sufficiently large n

$$P_{n,\theta}\{D\,|\,C\} = P_{n,\theta}\{D\}$$

In order that we get

$$E_{n,\theta_0}(\phi_n^*) \longrightarrow \frac{1}{2} \ (n \longrightarrow \infty)$$

we must have

$$P_{n,\theta_0}\{D\} \longrightarrow \frac{1}{2} \ (n \longrightarrow \infty).$$

It follows from (2.1), (2.2) and (2.8) that for sufficiently large n

$$E_{n,\theta_2}(\phi_n^*) - E_{n,\theta_1}(\phi_n^*)$$

$$\sim P_{n,\theta_2}(A) + \frac{1}{2} P_{n,\theta_2}(C) - P_{n,\theta_1}(A) - \frac{1}{2} P_{n,\theta_1}(C) \tag{2.9}$$

$$\sim P_{n,\theta_2}(A)$$
$$\sim 1 - e^{-2ku}$$

From (2.5), (2.6), (2.7) and (2.9) it is shown that

$$\varliminf_{n\to\infty} P_{n,\theta}\{n\,|\,\hat{\theta}_n - \theta\,| \geq u\} \geq e^{-2ku} \tag{2.10}$$

for all AMU estimators $\hat{\theta}_n$.

Hence the right-hand side of (10) is the desired lower bound. Therefore it is shown that the bound of the asymptotic distribution of $n\,|\,\hat{\theta}_n - \theta\,|$ for AMU estimators $\hat{\theta}_n$ is given by

$$\varlimsup_{n\to\infty} P_{n,\theta}\{n\,|\,\hat{\theta}_n - \theta\,| < u\} \leq 1 - e^{-2ku}. \tag{2.11}$$

Since both of the asymptotic densities of $\hat{\theta}_{PT}$ and $\hat{\theta}^r_{MP}$ are symmetric, it is easily seen that these estimator are AMU.
From (2.3) and (2.4) we have

$$\lim_{n\to\infty} P_{n,\theta}\{n\,|\,\hat{\theta}_{PT} - \theta\,| < u\} = 1 - e^{-2ku}; \tag{2.12}$$

$$\lim_{n\to\infty} P_{n,\theta}\{n\,|\,\hat{\theta}_{MP} - \theta\,| < u\} = \begin{cases} 1 - e^{-2ku} & \text{for } 0 < u < r; \\ 1 - e^{-k(u+r)} & \text{for } r \leq u. \end{cases} \tag{2.13}$$

Hence it is shown by (2.11), (2.12) and (2.13) that $\hat{\theta}_{PT}$ is two-sided asymptotically efficient in the sense that the asymptotic distribution of $\hat{\theta}_{PT}$ attains uniformly the bound given by the right-hand side of (2.11) but $\hat{\theta}^r_{MP}$ is not so.

Remark that in this case the MLE is asymptotically (also in small sample) uniformly (in u) inadmissible which fact can be established in the following way. It is easily seen that

$$\hat{\theta}_{ML} = \begin{cases} \min X_i + 1 & \text{if } \bar{X} > \min X_i + 1, \\ \max X_i - 1 & \text{if } \bar{X} < \max X_i - 1, \\ \bar{X} & \text{if } \max X_i - 1 < \bar{X} < \min X_i + 1 \end{cases}$$

and asymptotically, it is equivalent to

$$\tilde{\theta}_n = \begin{cases} \min X_i + 1 & \text{with probability } 1/2, \\ \max X_i - 1 & \text{with probability } 1/2. \end{cases}$$

Since the asymptotic density of $\tilde{\theta}_n$ is symmetric, it is easily shown that $\hat{\theta}_{ML}$ is AMU. Thus the asymptotic distribution of $n|\hat{\theta}_{ML} - \theta|$ is easily seen to be

$$\lim_{n \to \infty} P_{n,\theta}\{n|\hat{\theta}_{ML} - \theta| < u\} = 1 - e^{-ku}$$

which has twice as large as large scale as the asymptotic distribution of $\hat{\theta}_{PT}$, the asymptotic efficiency of $\hat{\theta}_{ML}$ compared with θ_{PT} is exactly 50%!

It should be also noted that $\hat{\theta}_{ML}$ is the limit of $\hat{\theta}^r_{MP}$ as $r \to 0$, and that $\hat{\theta}^r_{MP}$ is inadmissible outside the interval $-r < u < r$, and that the interval converges to the null set when r goes to zero. On the order hand when r becomes large, the interval of u in which $\hat{\theta}^r_{MP}$ is admissible extends and the probability outside the interval also concentrates to the center, and as $r \to \infty$, $\hat{\theta}^r_{MP}$ approaches to $\hat{\theta}_{PT}$ which is uniformly efficient. Thus in this case the choice of larger r is *always* better than the choice of smaller r, which, although remarkable enough, does not hold in the other cases.

It is also noted that the Pitman estimator is the generalized Bayes estimator with uniform prior and quadratic loss, but $\hat{\theta}_{PT}$ is asymptotically equivalent not only to the Pitman estimator but also to any generalized Bayes estimators with symmetric loss function ([6]).

We have defined an AMU estimator $\hat{\theta}_n^*$ to be one-sided asymptotically efficient in the sense that the asymptotic distribution of $n(\hat{\theta}^* - \theta)$ uniformly attains the bound of the asymptotic distributions of $n(\hat{\theta}_n - \theta)$ for AMU estimators $\hat{\theta}_n$([3], [4], [5]). We have showned that the bound of the asymptotic distribution of $n(\hat{\theta}_n - \theta)$ for AMU estimators $\hat{\theta}_n$ is given the following density:

$$g^{**}(u) = \begin{cases} ke^{-k|u|} & \text{for } |u| \leq \dfrac{1}{k}\log 2; \\ 0 & \text{for } |u| > \dfrac{1}{k}\log 2, \end{cases}$$

(See the proof of Theorem 3.1 of [2]). Comparing (2.3) and (2.4), we may also conclude that $\hat{\theta}_{PT}$ and $\hat{\theta}^r_{MP}$ are asymptotically equivalent for $|u| < r$ but for $|u| \geq r$ $\hat{\theta}_{PT}$ is asymptotically better than $\hat{\theta}^r_{MP}$ in the sense that the asymptotic distribution of $\hat{\theta}_{PT}$ is uniformly nearer to the bound than that of $\hat{\theta}^r_{MP}$. Though $\hat{\theta}^r_{MP}$ and $\hat{\theta}_{PT}$ are not one-sided asymptotically efficient in the sense that their asymptotic distributions uniformly attains the above bound ([3], [4], [5]), it is seen that $\hat{\theta}_{PT}$ is asymptotically better than $\hat{\theta}_{MP}$.

REFERENCES

[1] Akahira, M.: Asymptotic theory for estimation of location in non-regular cases, I:Order of convergence of consistent estimators. *Rep. Stat. Appl. Res., JUSE* **22**, (1975) 8–26.

[2] Akahira, M.: Asymptotic theory for estimation of location in non-regular cases, II:Bounds of asymptotic distributions of consistent estimators. *Rep. Stat. Appl. Res., JUSE* **22**, (1975) 99–115.

[3] Akahira, M.: On the asymptotic efficiency of estimators in an autoregressive process. *Ann. Inst. Statist. Math.* **28**, (1976) 35–48.

[4] Takeuchi, K.: *Tokei-teki Suitei no Zenkinriron (Asymptotic Theory of Statistical Estimation) (in Japanese)*, Kyoiku Shuppan, Tokyo. (1974)

[5] Takeuchi, K. and Akahira, M.: On the second order asymptotic efficiencies of estimators. *Proc. of the 3rd Japan-USSR Symp, on Prob. Theory, Lecture Notes in Math.* **550**, (1976) 604–638, Springer Verlag.

[6] Takeuchi, K. and Akahira, M.: Asymptotic optimality of the generalized Bayes estimator. *Rep. Univ. Electro-Comm.,* **29**, (1978) 37–45.

[7] Weiss, L. and Wolfowitz, J.: Maximum probability estimators. *Ann. Inst. Statist. Math.,* **19**, (1967) 193–206.

[8] Weiss, L. and Wolfowitz, J.: Generalized maximum likelihood estimators in a particular case. *Theor. Probability Appl.,* **13**, (1968) 622–627.

[9] Weiss, L. and Wolfowitz, J.: Maximum Probability Estimators and Related Topics. *Lecture Notes in Math.* **424**, (1974) Springer Verlag.

Ann. Inst. Statist. Math.
31 (1979), Part A, 39-56

DISCRETIZED LIKELIHOOD METHODS—ASYMPTOTIC PROPERTIES OF DISCRETIZED LIKELIHOOD ESTIMATORS (DLE'S)[1]

MASAFUMI AKAHIRA[2] AND KEI TAKEUCHI

(Received Feb. 8, 1978; revised Feb. 13, 1979)

Abstract

Suppose that $X_1, X_2, \cdots, X_n, \cdots$ is a sequence of i.i.d. random variables with a density $f(x, \theta)$. Let c_n be a maximum order of consistency. We consider a solution $\hat{\theta}_n$ of the discretized likelihood equation

$$\sum_{i=1}^{n} \log f(X_i, \hat{\theta}_n + rc_n^{-1}) - \sum_{i=1}^{n} \log f(X_i, \hat{\theta}_n) = a_n(\hat{\theta}_n, r)$$

where $a_n(\theta, r)$ is chosen so that $\hat{\theta}_n$ is asymptotically median unbiased (AMU). Then the solution $\hat{\theta}_n$ is called a discretized likelihood estimator (DLE). In this paper it is shown in comparison with DLE that a maximum likelihood estimator (MLE) is second order asymptotically efficient but not third order asymptotically efficient in the regular case. Further it is seen that the asymptotic efficiency (including higher order cases) may be systematically discussed by the discretized likelihood methods.

1. Introduction

Recently second order asymptotic efficiency has been studied by Chibisov [4], [5], Pfanzagl [8], [9], Takeuchi and Akahira [3], [11], Efron [6], Ghosh and Subramanyam [7] and others. Furthermore third order asymptotic efficiency was discussed in Takeuchi and Akahira [11], [12], [13] and Pfanzagl and Wefelmeyer [10]. In this paper using the discretized likelihood method we consider the asymptotic efficiency of estimators including higher order cases.

Suppose that $X_1, X_2, \cdots, X_n, \cdots$ is a sequence of i.i.d. random variables with a density $f(x, \theta)$. Let c_n be a maximum order of consistent estimator of θ. We have proposed a solution $\hat{\theta}_n$ of the discretized likeli-

---

[1] The results of this paper have been presented by the first author at the meeting on "Asymptotic Methods of Statistics" at the Mathematical Institute in Oberwolfach of West Germany, November 1977.

[2] Partially supported by Sakkokai Foundation.

40   MASAFUMI AKAHIRA AND KEI TAKEUCHI

hood equation

$$(1.1) \qquad \sum_{i=1}^{n} \log f(X_i, \hat{\theta}_n + rc_n^{-1}) - \sum_{i=1}^{n} \log f(X_i, \hat{\theta}_n) = a_n(\hat{\theta}_n, r) \ ,$$

where $a_n(\theta, r)$ is chosen so that $\hat{\theta}_n$ is asymptotically median unbiased (AMU) (the possibility of which will be shown in the context) [3]. Then the solution $\hat{\theta}_n$ is called a discretized likelihood estimator (DLE). If for each real number r,

$$\sum_{i=1}^{n} \log f(X_i, \theta + rc_n^{-1}) - \sum_{i=1}^{n} \log f(X_i, \theta)$$

is locally monotone in θ, then the asymptotic distribution of the DLE $\hat{\theta}_n$ attains the bound of the asymptotic distributions (discussed below) of AMU estimators of θ at r. It is easily seen that there is at least one estimator which attain the bound. In regular cases with $c_n = \sqrt{n}$ the left-hand side of (1.1) is expanded as

$$(1.2) \qquad \frac{\partial}{\partial \theta} \sum_{i=1}^{n} \log f(X_i, \hat{\theta}_n) + \frac{r}{2\sqrt{n}} \frac{\partial^2}{\partial \theta^2} \sum_{i=1}^{n} \log f(X_i, \hat{\theta}_n) + \cdots = a_n(\hat{\theta}_n, r) \ .$$

We derive from (1.2) the order to which the maximum likelihood estimator (MLE) is asymptotically efficient. In this paper it is shown that an MLE is second order asymptotically efficient but not third order asymptotically efficient. The motivation for the definition of the DLE is that; when we test the hypothesis $\theta = \theta_0 + rc_n^{-1}$ against $\theta = \theta_0$, the most powerful test is given by rejecting the hypothesis if

$$\sum_{i=1}^{n} \log f(X_i, \theta_0 + rc_n^{-1}) - \sum_{i=1}^{n} \log f(X_i, \theta_0) < k_n$$

hence if an estimator $\hat{\theta}_r$ is defined so that the event $\hat{\theta}_r > \theta_0$ is equivalent to the above inequality (at least asymptotically up to some order), then $\hat{\theta}_r$ is efficient (asymptotically up to some order) for specified choice of r. Therefore if $\hat{\theta}_r$ can be defined independently of r, then it is shown to be efficient (asymptotically up to the above mentioned order), and if not, we can establish that there does not exist any efficient (in the same sense) estimator. It is also seen that the asymptotic efficiency (including higher order cases) may be systematically discussed by the discretized likelihood method.

2.   Notations and definitions

Let $(\mathscr{X}, \mathscr{B})$ be a sample space. We consider a family of probability measures on $\mathscr{B}$, $\mathscr{P} = \{P_\theta : \theta \in \Theta\}$, where the index set Θ is called the parameter space. We assume that Θ is an open set in a Euclidean 1-space R^1. Consider n-fold direct products $(\mathscr{X}^{(n)}, \mathscr{B}^{(n)})$ of $(\mathscr{X}, \mathscr{B})$ and the corresponding product measures $P_{n,\theta}$ of P_θ. An estimator of θ is defined to be a sequence $\{\hat{\theta}_n\}$ of $\mathscr{B}^{(n)}$-measurable functions $\hat{\theta}_n$ on $\mathscr{X}^{(n)}$ into Θ $(n=1, 2, \cdots)$. For simplicity we denote an estimator as $\hat{\theta}_n$ instead of $\{\hat{\theta}_n\}$. For increasing sequence of positive numbers $\{c_n\}$ (c_n tending to infinity) an estimator $\hat{\theta}_n$ is called consistent with order $\{c_n\}$ (or $\{c_n\}$-consistent for short) if for every $\varepsilon>0$ and every $\vartheta \in \Theta$ there exist a sufficiently small positive number δ and a sufficiently large number L satisfying the following:

$$\varlimsup_{n\to\infty} \sup_{\theta:|\theta-\vartheta|<\delta} P_{n,\theta}\{c_n|\hat{\theta}_n-\theta|\geqq L\}<\varepsilon \qquad ([1]).$$

For each $k=1, 2, \cdots$, a $\{c_n\}$-consistent estimator $\hat{\theta}_n$ is kth order asymptotically median unbiased (or kth order AMU) estimator if for any $\vartheta \in \Theta$, there exists a positive number δ such that

$$\lim_{n\to\infty} \sup_{\theta:|\theta-\vartheta|<\delta} c_n^{k-1}\left| P_{n,\theta}\{\hat{\theta}_n\leqq\theta\} - \frac{1}{2}\right|=0 ;$$

$$\lim_{n\to\infty} \sup_{\theta:|\theta-\vartheta|<\delta} c_n^{k-1}\left| P_{n,\theta}\{\hat{\theta}_n\geqq\theta\} - \frac{1}{2}\right|=0 .$$

For $\hat{\theta}_n$ kth order AMU, $G_0(t, \theta)+c_n^{-1}G_1(t, \theta)+\cdots+c_n^{-(k-1)}G_{k-1}(t, \theta)$ is called to be the kth order asymptotic distribution of $c_n(\hat{\theta}_n-\theta)$ (or $\hat{\theta}_n$ for short) if

$$\lim_{n\to\infty} c_n^{k-1}|P_{n,\theta}\{c_n(\hat{\theta}_n-\theta)<t\} -G_0(t, \theta)-c_n^{-1}(t, \theta)-\cdots-c_n^{-(k-1)}G_{k-1}(t, \theta)|=0 .$$

We note that $G_i(t, \theta)$ $(i=1, \cdots, k-1)$ may be generally absolute continuous functions, hence the asymptotic distributions for any fixed n may not be a distribution function.

Suppose that $\hat{\theta}_n$ is AMU and has the kth order asymptotic distribution $G_0(t, \theta)+c_n^{-1}G_1(t, \theta)+\cdots+c_n^{-(k-1)}G_{k-1}(t, \theta)$. Letting θ_0 ($\in \Theta$) be arbitrary but fixed we consider the problem of testing hypothesis $H^+: \theta= \theta_0+tc_n^{-1}$ $(t>0)$ against $K: \theta=\theta_0$. Put $\Phi_{1/2}=\{\{\phi_n\}: E_{n,\theta_0+tc_n^{-1}}(\phi_n)=1/2+ o(c_n^{-(k-1)}),\ 0\leqq\phi_n(\tilde{x}_n)\leqq 1$ for all $\tilde{x}_n \in \mathscr{X}^{(n)}$ $(n=1, 2,\cdots)\}$. Putting $A\hat{\theta}_{n,\theta_0}= \{c_n(\hat{\theta}_n-\theta_0)\leqq t\}$, we have

$$\lim_{n\to\infty} P_{n,\theta_0+tc_n^{-1}}(A\hat{\theta}_{n,\theta_0})=\lim_{n\to\infty} P_{n,\theta_0+tc_n^{-1}}\{\hat{\theta}_n\leqq\theta_0+tc_n^{-1}\} =\frac{1}{2} .$$

Hence it is seen that a sequence $\{\chi_{A\hat{\theta}_{n,\theta_0}}\}$ of the indicators (or characteristic functions) of $A\hat{\theta}_{n,\theta_0}$ $(n=1, 2,\cdots)$ belongs to $\Phi_{1/2}$. If

42   MASAFUMI AKAHIRA AND KEI TAKEUCHI

$$\sup_{\{\phi_n\}\,\in\,\Phi_{1/2}}\ \varlimsup_{n\to\infty}\ c_n^{k-1}\{E_{n,\,\theta_0}(\phi_n)-H_0^+(t,\,\theta_0)-c_n^{-1}H_1^+(t,\,\theta_0)-$$

$$\cdots-c_n^{-(k-1)}H_{k-1}^+(t,\,\theta_0)\}=0\ ,$$

then we have

$$G_0(t,\,\theta_0)\leqq H_0^+(t,\,\theta_0)\ ;$$

and for any positive integer $j\ (\leqq k)$ if $G_i(t,\,\theta_0)=H_i^+(t,\,\theta_0)\ (i=1,\cdots,j-1)$, then

$$G_j(t,\,\theta_0)=H_j^+(t,\,\theta_0)\ .$$

Consider next the problem of testing hypothesis $H^-:\theta=\theta_0+tc_n^{-1}\ (t<0)$ against $K:\theta=\theta_0$. If

$$\inf_{\{\phi_n\}\,\in\,\Phi_{1/2}}\ \varliminf_{n\to\infty}\ c_n^{k-1}\{E_{n,\,\theta_0}(\phi_n)-H_0^-(t,\,\theta_0)-c_n^{-1}H_1^-(t,\,\theta_0)-$$

$$\cdots-c_n^{-(k-1)}H_{k-1}^-(t,\,\theta_0)\}=0\ ,$$

then we have

$$G_0(t,\,\theta_0)\geqq H_0^-(t,\,\theta_0)\ ;$$

and for any positive integer $j\ (\leqq k)$ if $G_i(t,\,\theta_0)=H_i^-(t,\,\theta_0)\ (i=0,\cdots,j-1)$, then $G_j(t,\,\theta_0)\geqq H_j^-(t,\,\theta_0)$.

$\hat{\theta}_n$ is called to be kth order asymptotically efficient if the kth order asymptotic distribution of it attains uniformly the bound of the kth order asymptotic distributions of kth order AMU estimators, that is, for each $\theta\in\Theta$

$$G_i(t,\,\theta)=\begin{cases}H_i^+(t,\,\theta)&\text{for }t>0\ ,\\[2mm]H_i^-(t,\,\theta)&\text{for }t<0\ ,\end{cases}$$

$i=0,\cdots,k-1$ ([2], [11]). (Note that for $t=0$ we have $G_i(0,\,\theta)=H_i^+(0,\,\theta)=H_i^-(0,\,\theta)\ (i=0,\cdots,k-1)$ from the condition of kth order asymptotically median unbiasedness.)

We assume that for each $\theta\in\Theta$ P_θ is absolutely continuous with respect to σ-finite measure μ.

We denote a density $dP_\theta/d\mu$ by $f(x,\,\theta)$. Let $L(\theta;\tilde{x}_n)$ be a likelihood function, that is, $L(\theta;\tilde{x}_n)=\prod_{i=1}^{n}f(x_i,\,\theta)$, where $\tilde{x}_n=(x_1,\,x_2,\cdots,x_n)$. For each $k=1,2,\cdots$, a $\{c_n\}$-consistent estimator $\hat{\theta}_n$ is called discretized likelihood estimator (DLE) if for each real number r, $\hat{\theta}_n$ satisfies the discretized likelihood equation

$$(2.1)\qquad \log L(\hat{\theta}_n+rc_n^{-1};\tilde{x}_n)-\log L(\hat{\theta}_n;\tilde{x}_n)=a_n(\hat{\theta}_n,\,r)\ ,$$

where $a_n(\theta, r)$ is a function in θ and r and it also depends on n. The function $a_n(\theta, r)$ is not defined for the moment but will be determined in the sequel so that the solution obtained from the above equation be asymptotically median unbiased up to kth order. It should be noted that DLE $\hat{\theta}_n$ is required to be $\{c_n\}$-consistent and we do not claim that the solution of the equation (2.1) be $\{c_n\}$-consistent. We implicitely claim that there exists $\underline{a}$ solution of the equation in the $O(c_n^{-1})$-neighborhood of the true value. In the practical situation we have to obtain the DLE first by finding a $\{c_n\}$-consistent estimator $\tilde{\theta}_n$ in some way or another then find a solution of the equation in the neighborhood of $\tilde{\theta}_n$. Suppose that for given function $a_n(\theta, r)$,

$$(2.2) \qquad \log L(\theta + rc_n^{-1}; \tilde{x}_n) - \log L(\theta; \tilde{x}_n) - a_n(\theta, r)$$

is locally monotone in θ with probability larger than $1 - o(c_n^{-(k-1)})$. For regular case the particular form of $a_n(\theta, r)$ will be given later (e.g. page 47 etc.). For the present it is only necessary to remark that $a_n(\theta, r)$ is of the magnitude of order smaller than the previous terms of (2.2). Then the kth order asymptotic distribution of the DLE $\hat{\theta}_{\mathrm{DL}}$ attains the bound of the kth order asymptotic distributions of kth order AMU estimators of θ at r. Indeed, it follows by the monotone of (2.2) that for any $\vartheta \in \Theta$, there exists a positive number δ such that

$$(2.3) \qquad \lim_{n \to \infty} \sup_{\theta:\, |\theta - \vartheta| < \delta} c_n^{k-1} |P_{n,\theta}\{\hat{\theta}_n > \theta - rc_n^{-1}\} - P_{n,\theta}\{\log L(\theta, \tilde{x}_n)$$
$$- \log L(\theta - rc_n^{-1}, \tilde{x}_n) > a_n(\theta - rc_n^{-1}, r)\}| = 0 \ .$$

Letting $\theta_0\ (\in \Theta)$ be arbitrary but fixed we consider the problem of testing hypothesis $H: \theta = \theta_0 - rc_n^{-1}\ (r > 0)$ against alternative $K: \theta = \theta_0$. Putting $A\hat{\theta}_{n,\theta} = \{c_n(\hat{\theta}_n - \theta) > -r\}$ we have $P_{n, \theta_0 - rc_n^{-1}}\ (A\hat{\theta}_{n,\theta_0}) = 1/2 + o(c_n^{-(k-1)})$. Let $\mathcal{U}_k$ be the class of the all kth order AMU estimators. Set $\Phi_{1/2} = \{\{\phi_n\}: \mathrm{E}_{n, \theta_0 - rc_n^{-1}}(\phi_n) = 1/2 + o(c_n^{-(k-1)}),\ 0 \leq \phi_n(\tilde{x}_n) \leq 1$ for all $\tilde{x}_n \in \mathcal{X}^{(n)}\ (n = 1, 2, \cdots)\}$. It is noted that every sequence $\{\chi_{A\hat{\theta}_{n,\theta}}(\tilde{x}_n)\}$ of the indicators of the sets $A\hat{\theta}_{n,\theta}$ with the estimators $\hat{\theta}_n$ in $\mathcal{U}_k$ is contained in $\Phi_{1/2}$. In order to obtain the upper bound of $\varlimsup_{n \to \infty} P_{n, \theta_0}(A\hat{\theta}_{n,\theta_0})$ in $\mathcal{U}_k$, it is sufficient to find a sequence $\{\phi_n^*\}$ of the tests which maximize $\varlimsup_{n \to \infty} \mathrm{E}_{n, \theta_0}(\phi_n)$ in $\Phi_{1/2}$. It is shown by the Neyman-Pearson fundamental lemma that ϕ_n^* has the rejection S_n satisfying

$$\sum_{i=1}^{n} \log \frac{f(X_i, \theta_0 - rc_n^{-1})}{f(X_i, \theta_0)} < k_n \ ,$$

where k_n is some constant. Then it follows from (2.3) that the upper

44   MASAFUMI AKAHIRA AND KEI TAKEUCHI

bound of $\varliminf\limits_{n\to\infty} P_{n,\theta_0}(A_{\hat{\theta}_n,\theta_0})$ in $\mathcal{U}_k$ is given by $\varliminf\limits_{n\to\infty} \mathrm{E}_{n,\theta_0}(\phi_n^*)$. Hence the kth

order asymptotic distribution of the DLE $\hat{\theta}_{\mathrm{DL}}$ attains the bound of the kth order asymptotic distributions of kth order AMU estimators at $-r$. In a similar way as the case when $r>0$, we also obtain for $r<0$ the upper bound of $\varliminf\limits_{n\to\infty} P_{n,\theta_0}(A_{\hat{\theta}_n,\theta_0}^c)$ in $\mathcal{U}_k$ of the same form. Hence the desired result also holds for the case $r<0$. In later sections it will be seen that $\hat{\theta}_{\mathrm{DL}}$ is asymptotically efficient up to second order. Note that the DLE usually depends on r in cases more than third order.

In the subsequent discussion we shall deal with the case when $c_n = \sqrt{n}$.

3. Second order asymptotic efficiency

Suppose that $X_1, X_2, \cdots, X_n, \cdots$ is a sequence of i.i.d. random variables with a density $f(x, \theta)$ satisfying (i)~(iv).

(i) $\{x: f(x, \theta)>0\}$ does not depend on θ ;

(ii)   For almost all $x[\mu]$, $f(x, \theta)$ is three times continuously differentiable in θ ;

(iii)   For each $\theta \in \Theta$

$$0<I(\theta)=\mathrm{E}_\theta\left[\left\{\frac{\partial}{\partial\theta}\log f(X, \theta)\right\}^2\right]=-\mathrm{E}\left[\frac{\partial^2}{\partial\theta^2}\log f(X, \theta)\right]<\infty ;$$

(iv)   There exist

$$J(\theta)=\mathrm{E}_\theta\left[\left\{\frac{\partial^2}{\partial\theta^2}\log f(X, \theta)\right\}\left\{\frac{\partial}{\partial\theta}\log f(X, \theta)\right\}\right]$$

and

$$K(\theta)=\mathrm{E}_\theta\left[\left\{\frac{\partial}{\partial\theta}\log f(X, \theta)\right\}^3\right]$$

and the following holds:

$$\mathrm{E}_\theta\left[\frac{\partial^3}{\partial\theta^3}\log f(X, \theta)\right]=-3J(\theta)-K(\theta) .$$

By the following way we have shown in [11] that an MLE is second order asymptotically efficient. Let $\hat{\theta}_{\mathrm{ML}}$ be a maximum likelihood estimator. By Taylor expansion we have

$$0=\sum_{i=1}^n \frac{\partial}{\partial\theta}\log f(X_i, \hat{\theta}_{\mathrm{ML}})$$

$$=\sum_{i=1}^n \frac{\partial}{\partial\theta}\log f(X_i, \theta)+\sum_{i=1}^n \left\{\frac{\partial^2}{\partial\theta^2}\log f(X_i, \theta)\right\}(\hat{\theta}_{\mathrm{ML}}-\theta)$$

$$+\frac{1}{2}\sum_{i=1}^{n}\left\{\frac{\partial^3}{\partial\theta^3}\log f(X_i,\theta^*)\right\}(\hat{\theta}_{\mathrm{ML}}-\theta)^2\,,$$

where $|\theta^*-\theta|\leqq|\hat{\theta}_{\mathrm{ML}}-\theta|$. Putting $T_n=\sqrt{n}\,(\hat{\theta}_{\mathrm{ML}}-\theta)$ we obtain

$$0=\frac{1}{\sqrt{n}}\sum_{i=1}^{n}\frac{\partial}{\partial\theta}\log f(X_i,\theta)+\frac{1}{n}\left\{\sum_{i=1}^{n}\frac{\partial^2}{\partial\theta^2}\log f(X_i,\theta)\right\}T_n$$

$$+\frac{1}{2n\sqrt{n}}\left\{\sum_{i=1}^{n}\frac{\partial^3}{\partial\theta^3}\log f(X_i,\theta^*)\right\}T_n^2\,.$$

Set

$$Z_1(\theta)=\frac{1}{\sqrt{n}}\sum_{i=1}^{n}\frac{\partial}{\partial\theta}\log f(X_i,\theta)\;;$$

$$Z_2(\theta)=\frac{1}{\sqrt{n}}\sum_{i=1}^{n}\left\{\frac{\partial^2}{\partial\theta^2}\log f(X_i,\theta)+I(\theta)\right\}\;;$$

$$W(\theta)=\frac{1}{n}\sum_{i=1}^{n}\frac{\partial^3}{\partial\theta^3}\log f(X_i,\theta)\,.$$

Then it follows that $W(\theta)$ converges in probability to $-3J(\theta)-K(\theta)$. Hence the following theorem holds:

THEOREM 1 ([11]). *Under conditions* (i)$\sim$(iv)

$$(3.1)\quad \sqrt{n}\,(\hat{\theta}_{\mathrm{ML}}-\theta)=\frac{Z_1(\theta)}{I(\theta)}+\frac{Z_1(\theta)Z_2(\theta)}{\sqrt{n}\,I(\theta)^2}-\frac{3J(\theta)+K(\theta)}{2\sqrt{n}\,I(\theta)^3}Z_1(\theta)^2+o_p\left(\frac{1}{\sqrt{n}}\right)$$

up to order $n^{-1/2}$ *as* $n\to\infty$.

Put

$$\hat{\theta}_{\mathrm{ML}}^*=\hat{\theta}_{\mathrm{ML}}+\frac{K(\hat{\theta}_{\mathrm{ML}})}{6nI(\hat{\theta}_{\mathrm{ML}})^2}\,.$$

Then $\hat{\theta}_{\mathrm{ML}}^*$ is second order AMU. From Theorem 1 we have established the following:

THEOREM 2 ([11]). *Under conditions* (i)$\sim$(iv), $\hat{\theta}_{\mathrm{ML}}$ *is second order asymptotically efficient.*

In the sequel we obtain the same result for DLE. We further assume the following:

(v) For given function $a_n(\theta,r)$

$$\log L(\theta+rn^{-1/2},\tilde{x}_n)-\log L(\theta,\tilde{x}_n)-a_n(\theta,r)$$

is locally monotone in θ with probability larger than $1-o(n^{-1})$.

Remark. In usual situation it is generally true since $(1/n)(\partial^2/\partial\theta^2)\cdot$

$\log L(\theta, \tilde{x}_n)$ is asymptotically equal to $-I(\theta)$ (<0) and $a_n(\theta, r)$ is smaller order than n^{-1}, and is usually of constant order $(O(1))$ as is shown below. Let $\hat{\theta}_n$ be an DLE. Since

$$\sum_{i=1}^{n} \log f(X_i, \hat{\theta}_n + rn^{-1/2}) - \sum_{i=1}^{n} \log f(X_i, \hat{\theta}_n) = a_n \ ,$$

it follows by Taylor expansion that

$$\frac{1}{\sqrt{n}} \sum_{i=1}^{n} \frac{\partial}{\partial \theta} \log f(X_i, \hat{\theta}_n) + \frac{r}{2n} \sum_{i=1}^{n} \frac{\partial^2}{\partial \theta^2} \log f(X_i, \hat{\theta}_n)$$

$$+ \frac{r^2}{6n\sqrt{n}} \sum_{i=1}^{n} \frac{\partial^3}{\partial \theta^3} \log f(X_i, \theta_n^*) = \frac{a_n}{r} \ ,$$

where

$$|\theta_n^* - \hat{\theta}_n| < \frac{r}{\sqrt{n}} \ .$$

Since $(1/n) \sum_{i=1}^{n} \{(\partial^3/\partial \theta^3) \log f(X_i, \theta)\}$ converges in probability to $-3J(\theta) - K(\theta)$, it is seen that

$$(3.2) \qquad Z_1(\hat{\theta}_n) + \frac{r}{2}\left\{ -I(\hat{\theta}_n) + \frac{1}{\sqrt{n}} Z_2(\hat{\theta}_n)\right\}$$

$$- \frac{r^2}{6\sqrt{n}} \{3J(\hat{\theta}_n) + K(\hat{\theta}_n)\} + o_p\left(\frac{1}{\sqrt{n}}\right) = \frac{a_n}{r} \ .$$

On the other hand we have

$$(3.3) \quad Z_1(\hat{\theta}_n) = Z_1(\theta) + \frac{1}{\sqrt{n}} \{Z_2(\theta) - \sqrt{n} I(\theta)\} T_n - \frac{3J(\theta) + K(\theta)}{2\sqrt{n}} T_n^2 + o_p\left(\frac{1}{\sqrt{n}}\right) \ ,$$

where $T_n = \sqrt{n}(\hat{\theta}_n - \theta)$. Since

$$Z_1(\hat{\theta}_n) = Z_1(\theta) + \frac{1}{\sqrt{n}} Z_2(\theta) T_n - I(\theta) T_n - \frac{3J(\theta) + K(\theta)}{2\sqrt{n}} T_n^2 + o_p\left(\frac{1}{\sqrt{n}}\right) \ ,$$

it follows that

$$(3.4) \qquad T_n = \frac{1}{I(\theta)} \left\{ -Z_1(\hat{\theta}_n) + Z_1(\theta) + \frac{1}{\sqrt{n}} Z_2(\theta) T_n \right.$$

$$\left. - \frac{3J(\theta) + K(\theta)}{2\sqrt{n}} T_n^2 \right\} + o_p\left(\frac{1}{\sqrt{n}}\right) \ .$$

Since

$$I(\hat{\theta}_n) = I(\theta) + \frac{1}{\sqrt{n}} \{2J(\theta) + K(\theta)\} T_n + o_p\left(\frac{1}{\sqrt{n}}\right) \ ;$$

$$J(\hat{\theta}_n)=J(\theta)+\frac{1}{\sqrt{n}}J'(\theta)T_n+o_p\!\left(\frac{1}{\sqrt{n}}\right);$$

$$K(\hat{\theta}_n)=K(\theta)+\frac{1}{\sqrt{n}}K'(\theta)T_n+o_p\!\left(\frac{1}{\sqrt{n}}\right);$$

$$Z_2(\hat{\theta}_n)=Z_2(\theta)-J(\theta)T_n+o_p(1),$$

it follows from (3.2) that

$$Z_1(\hat{\theta}_n)=\frac{a_n}{r}+\frac{r}{2}I(\theta)+\frac{r}{6\sqrt{n}}\{3rJ(\theta)+rK(\theta)-3Z_2(\theta)\}$$

$$+\frac{r}{2\sqrt{n}}\{3J(\theta)+K(\theta)\}T_n+o_p\!\left(\frac{1}{\sqrt{n}}\right)$$

up to order $n^{-1/2}$. From (3.4) we have

$$T_n=-\frac{a_n'}{rI}-\frac{r^2(3J+K)}{6I\sqrt{n}}+\frac{Z_1}{I}+\frac{Z_1Z_2}{I^2\sqrt{n}}-\frac{(3J+K)Z_1^2}{2I^3\sqrt{n}}+o_p\!\left(\frac{1}{\sqrt{n}}\right)$$

up to order $n^{-1/2}$, where $a_n=-(rI/2)+a_n'$ with $a_n'=o(1/\sqrt{n})$ so that $\hat{\theta}_n$ be AMU. In order to have second order asymptotic median unbiasedness of $\hat{\theta}_n$ we put

$$a_n'=-\frac{r^3(3J+K)}{6\sqrt{n}}-\frac{rK}{6I\sqrt{n}}+o\!\left(\frac{1}{\sqrt{n}}\right)$$

and we denote by T_n^* the corresponding T_n to this particular value of a_n. Then we may also denote $T_n^*=\sqrt{n}(\hat{\theta}_n^*-\theta)$. Then it is shown that $\hat{\theta}_n^*$ is second order AMU. Hence we have established the following theorem.

THEOREM 3. *Under conditions* (i)$\sim$(v), *the DLE $\hat{\theta}_n^*$ with a_n defined above satisfies the following*:

$$T_n^*=\sqrt{n}(\hat{\theta}_n^*-\theta)$$

$$=\frac{K}{6I^2\sqrt{n}}+\frac{Z_1(\theta)}{I(\theta)}+\frac{Z_1(\theta)Z_2(\theta)}{I(\theta)^2\sqrt{n}}-\frac{3J(\theta)+K(\theta)}{2I(\theta)^3\sqrt{n}}Z_1(\theta)^2$$

$$+\frac{r}{2I(\theta)\sqrt{n}}\left\{Z_2(\theta)-\frac{3J(\theta)+K(\theta)}{I(\theta)}Z_1(\theta)\right\}+o_p\!\left(\frac{1}{\sqrt{n}}\right)$$

up to order $n^{-1/2}$ as $n\to\infty$.

Remark. Since we have

$$\mathrm{E}_\theta\left[Z_1(\theta)\left\{Z_2(\theta)-\frac{3J(\theta)+K(\theta)}{I(\theta)}Z_1(\theta)\right\}\right]=0;$$

$$V_\theta(T_n^*)=\frac{1}{I(\theta)}+o\left(\frac{1}{\sqrt{n}}\right),$$

and the third order cumulant is equal to that of the MLE up to the order $n^{-1/2}$. Hence the asymptotic expansion of the distribution of T_n^* is equal to that of $\sqrt{n}(\hat{\theta}_{\mathrm{ML}}^*-\theta)$ up to the order $n^{-1/2}$. Therefore the DLE $\hat{\theta}_n^*$ is asymptotically equivalent to the MLE $\hat{\theta}_{\mathrm{ML}}^*$ up to that order.

4. Third order asymptotic efficiency

We proceed to the problem of third order asymptotic efficiency. We further assume the following:

(vi) For almost all $x[\mu]$, $f(x,\theta)$ is four times continuously differentiable in θ;

(vii) there exist

$$L(\theta)=\mathrm{E}_\theta\left[\left\{\frac{\partial^3}{\partial\theta^3}\log f(X,\theta)\right\}\left\{\frac{\partial}{\partial\theta}\log f(X,\theta)\right\}\right];$$

$$M(\theta)=\mathrm{E}_\theta\left[\left\{\frac{\partial^2}{\partial\theta^2}\log f(X,\theta)\right\}^2\right];$$

$$N(\theta)=\mathrm{E}_\theta\left[\left\{\frac{\partial^2}{\partial\theta^2}\log f(X,\theta)\right\}\left\{\frac{\partial}{\partial\theta}\log f(X,\theta)\right\}^2\right]$$

and

$$H(\theta)=\mathrm{E}_\theta\left[\left\{\frac{\partial}{\partial\theta}\log f(X,\theta)\right\}^4\right]$$

and the following holds:

$$\mathrm{E}_\theta\left[\frac{\partial^4}{\partial\theta^4}\log f(X,\theta)\right]=-4L(\theta)-3M(\theta)-6N(\theta)-H(\theta).$$

Let $\hat{\theta}_n$ be an DLE. Since

$$\sum_{i=1}^n\log f\left(X_i,\hat{\theta}_n+\frac{r}{\sqrt{n}}\right)-\sum_{i=1}^n\log f(X_i,\hat{\theta}_n)=a_n,$$

it follows that

$$\frac{1}{\sqrt{n}}\sum_{i=1}^n\frac{\partial}{\partial\theta}\log f(X_i,\hat{\theta}_n)+\frac{r}{2n}\sum_{i=1}^n\frac{\partial^2}{\partial\theta^2}\log f(X_i,\hat{\theta}_n)$$

$$+\frac{r^2}{6n\sqrt{n}}\sum_{i=1}^n\frac{\partial^3}{\partial\theta^3}\log f(X_i,\hat{\theta}_n)+\frac{r^3}{24n^2}\sum_{i=1}^n\frac{\partial^4}{\partial\theta^4}\log f(X_i,\theta^*)$$

$$=\frac{a_n}{r},$$

where $|\theta^*-\theta|<r/\sqrt{n}$. Since $(1/n)\sum_{i=1}^{n}(\partial^4/\partial\theta^4)\log f(X_i,\theta)$ converges in probability to $-4L(\theta)-3M(\theta)-6N(\theta)-H(\theta)$, it is seen that

$$Z_1(\hat\theta_n)+\frac{r}{2}\left\{-I(\hat\theta_n)+\frac{1}{\sqrt{n}}Z_2(\hat\theta_n)\right\}-\frac{r^2}{6\sqrt{n}}\{3J(\hat\theta_n)+K(\hat\theta_n)\}$$

$$+\frac{r^2}{6n}Z_3(\hat\theta_n)-\frac{r^3}{24n}\{4L(\hat\theta_n)+3M(\hat\theta_n)+6N(\hat\theta_n)+H(\hat\theta_n)\}\sim\frac{a_n}{r}\ ,$$

where

$$Z_3(\theta)=\frac{1}{\sqrt{n}}\sum_{i=1}^{n}\left\{\frac{\partial^3}{\partial\theta^3}\log f(X_i,\theta)+3J(\theta)+K(\theta)\right\}\ .$$

Hence

$$(4.1)\qquad Z_1(\hat\theta_n)\sim\frac{a_n}{r}+\frac{1}{2}I(\hat\theta_n)-\frac{r}{2\sqrt{n}}Z_2(\hat\theta_n)+\frac{r^2}{6\sqrt{n}}\{3J(\hat\theta_n)+K(\hat\theta_n)\}$$

$$-\frac{r^2}{6n}Z_3(\hat\theta_n)+\frac{r^3}{24n}\{4L(\hat\theta_n)+3M(\hat\theta_n)+6N(\hat\theta_n)+H(\hat\theta_n)\}\ .$$

On the other hand we have

$$Z_1(\hat\theta_n)=Z_1(\theta)+\frac{1}{\sqrt{n}}\{Z_2(\hat\theta_n)-\sqrt{n}\,I(\theta)\}T_n-\frac{3J(\theta)+K(\theta)}{2\sqrt{n}}T_n^2$$

$$-\frac{1}{6n}\{4L(\theta)+3M(\theta)+6N(\theta)+H(\theta)\}T_n^3+o_p\left(\frac{1}{n}\right)\ ,$$

where $T_n=\sqrt{n}(\hat\theta_n-\theta)$. We obtain

$$(4.2)\qquad T_n=\frac{1}{I(\theta)}\left[-Z_1(\hat\theta_n)+Z_1(\theta)+\frac{1}{\sqrt{n}}Z_2(\theta)T_n-\frac{3J(\theta)+K(\theta)}{2\sqrt{n}}T_n^2\right.$$

$$\left.-\frac{1}{6n}\{4L(\theta)+3M(\theta)+6N(\theta)+H(\theta)\}T_n^3\right]+o_p\left(\frac{1}{n}\right)\ .$$

Since

$$I(\hat\theta_n)=I(\theta)+\frac{1}{\sqrt{n}}\{2J(\theta)+K(\theta)\}T_n$$

$$+\frac{1}{2n}\{2L(\theta)+2M(\theta)+5N(\theta)+H(\theta)\}T_n^2+o_p\left(\frac{1}{n}\right)\ ;$$

$$J(\hat\theta_n)=J(\theta)+\frac{1}{\sqrt{n}}\{L(\theta)+M(\theta)+N(\theta)\}T_n+o_p\left(\frac{1}{\sqrt{n}}\right)\ ;$$

$$K(\hat\theta_n)=K(\theta)+\frac{1}{\sqrt{n}}\{3N(\theta)+H(\theta)\}T_n+o_p\left(\frac{1}{\sqrt{n}}\right)\ ;$$

$$L(\hat{\theta}_n)=L(\theta)+\frac{1}{\sqrt{n}}L'(\theta)T_n+o_p\left(\frac{1}{\sqrt{n}}\right)\ ;$$

$$M(\hat{\theta}_n)=M(\theta)+\frac{1}{\sqrt{n}}M'(\theta)T_n+o_p\left(\frac{1}{\sqrt{n}}\right)\ ;$$

$$N(\hat{\theta}_n)=N(\theta)+\frac{1}{\sqrt{n}}N'(\theta)T_n+o_p\left(\frac{1}{\sqrt{n}}\right)\ ;$$

$$H(\hat{\theta}_n)=H(\theta)+\frac{1}{\sqrt{n}}H'(\theta)T_n+o_p\left(\frac{1}{\sqrt{n}}\right)\ ;$$

$$Z_2(\hat{\theta}_n)=Z_2(\theta)-J(\theta)T_n-\frac{1}{2\sqrt{n}}\{2L(\theta)+M(\theta)+N(\theta)\}T_n^2+o_p\left(\frac{1}{\sqrt{n}}\right)\ ;$$

and

$$Z_3(\hat{\theta}_n)=\frac{1}{\sqrt{n}}Z_3'(\theta)T_n+o_p\left(\frac{1}{\sqrt{n}}\right)\ ,$$

it follows from (4.1) that

$$\begin{aligned}
Z_1(\hat{\theta}_n)=&\frac{a_n}{r}+\frac{r}{2}I-\frac{r}{2\sqrt{n}}Z_2+\frac{r^2(3J+K)}{6\sqrt{n}}-\frac{r^2}{6n}Z_3+\frac{r}{2\sqrt{n}}(3J+K)T_n\\
&+\frac{r^3}{24n}(4L+3M+6N+H)+\frac{r}{4n}(4L+3M+6N+H)T_n^2\\
&+\frac{r^2}{6n}(3L+3M+6N+H)T_n+o_p\left(\frac{1}{n}\right)
\end{aligned}$$

up to order n^{-1} as $n\to\infty$. Under conditions (i)$\sim$(vii) we have from (4.2)

$$\begin{aligned}
(4.3)\quad T_n=&-\frac{a_n}{rI}-\frac{r}{2}-\frac{r^2(3J+K)}{6I\sqrt{n}}+\frac{Z_1}{I}+\frac{rZ_2}{2I\sqrt{n}}+\frac{r^2Z_3}{6In}-\frac{r^3}{24In}\\
&-\frac{r}{2I\sqrt{n}}(3J+K)T_n-\frac{r^2\mathcal{L}'}{2In}T_n+\frac{Z_2}{I\sqrt{n}}T_n-\frac{3J+K}{2I\sqrt{n}}T_n^2\\
&-\frac{r}{4In}\mathcal{L}T_n^2-\frac{1}{6In}\mathcal{L}T_n^3+o_p\left(\frac{1}{n}\right)\\
=&-\frac{a_n''}{rI}+\frac{K}{6I^2\sqrt{n}}+\frac{Z_1}{I}+\frac{r}{2I\sqrt{n}}\left(Z_2-\frac{3J+K}{I}Z_1\right)+\frac{Z_1Z_2}{I^2\sqrt{n}}\\
&-\frac{(3J+K)Z_1^2}{2I^3\sqrt{n}}+\frac{1}{I^3n}\left\{Z_1Z_2^2-(3J+K)Z_1^2Z_2-\frac{\mathcal{L}}{6}Z_1^3\right.\\
&+\frac{(3J+K)^2Z_1^2}{2I^2}-\frac{(3J+K)Z_1Z_2}{2I}-\frac{(3J+K)KZ_1}{6I}+\frac{KZ_2}{6}\\
&-\frac{r(3J+K)}{12}-\frac{r^2I(3J+K)Z_2}{4}+\frac{r^2(3J+K)Z_1}{4}
\end{aligned}$$

$$-\frac{3r(3J+K)Z_1Z_2}{2}+\frac{r(3J+K)^2Z_1^2}{4I}+\frac{rIZ_2^2}{2}-\frac{r^2I\mathcal{L}'Z_1}{2}$$

$$+\frac{r(3J+K)^2Z_1^2}{2I}-\frac{r}{4}\mathcal{L}Z_1^2\Big\}+o_p\Big(\frac{1}{n}\Big)$$

up to order n^{-1} as $n\to\infty$, where $a_n=-(rI/2)-r^3(3J+K)/6I\sqrt{n}+rK/6I\sqrt{n}+a_n''$ with $a_n''=O(1/n)$ and $\mathcal{L}=4L+3M+6N+H$ and $\mathcal{L}'=3L+3M+6N+H$. We denote by T_n^* the corresponding T_n to this particular value of a_n. Then we may also denote $T_n^*=\sqrt{n}(\hat{\theta}_n^*-\theta)$. On the other hand under conditions (i)$\sim$(vii) we have obtained in [11]

$$(4.4)\quad \sqrt{n}(\hat{\theta}_{\mathrm{ML}}-\theta)=\frac{Z_1}{I}+\frac{1}{I^2\sqrt{n}}\Big\{Z_1Z_2-\frac{(3J+K)Z_1^2}{2I}\Big\}$$

$$+\frac{1}{I^3n}\Big\{Z_1Z_2^2+\frac{1}{2}Z_1^2Z_3-\frac{3J+K}{2I}Z_1^2Z_2$$

$$+\frac{(3J+K)^2}{2I^2}Z_1^3-\frac{4L+3M+6N+H}{6I}Z_1^3\Big\}+o_p\Big(\frac{1}{n}\Big)\ .$$

Let $\hat{\theta}_{\mathrm{ML}}^*$ be a modified MLE so that it is third order AMU. Comparing (4.3) with (4.4), we see that $\sqrt{n}(\hat{\theta}_{\mathrm{ML}}^*-\theta)$ and T_n^* are essentially different in the order n^{-1}. We put $T_{\mathrm{ML}}^*=\sqrt{n}(\hat{\theta}_{\mathrm{ML}}^*-\theta)$. Note that the difference of T_{ML}^* and T_n^* appears in the linear term of order $n^{-1/2}$ and other terms of order n^{-1}. It is shown in [12], [13] and [14] that the asymptotic cumulants are determined up to order n^{-1} by the terms of order up to $n^{-1/2}$ if the first term is equal to $Z_1(\theta)/I(\theta)$; and the fourth order cumulant is identical in the first term of order n^{-1} for all asymptotically efficient estimators. In the third order cumulants we have

$$\kappa_3(T_{\mathrm{ML}}^*-\mathrm{E}_\theta(T_{\mathrm{ML}}^*))=\frac{\beta_3(\theta)}{\sqrt{n}}+o\Big(\frac{1}{n}\Big)\ ;$$

$$\kappa_3(T_n^*-\mathrm{E}_\theta(T_n^*))=\frac{\beta_3(\theta)}{\sqrt{n}}+\frac{\gamma_3(\theta)}{n}+o\Big(\frac{1}{n}\Big)\ .$$

Therefore there is a difference of asymptotic distributions of $\sqrt{n}(\hat{\theta}_{\mathrm{ML}}^*-\theta)$ and T_n^* in the term of n^{-1} if $\gamma_3(\theta)\neq0$. If we denote by Theorem 3 and (4.3)

$$T_n^*=\frac{Z_1}{I}-\frac{J+K}{6I^2\sqrt{n}}+\frac{r}{\sqrt{n}}L^*+\frac{1}{\sqrt{n}}(Q-c)+\frac{1}{n}R+o_p\Big(\frac{1}{n}\Big)\ ,$$

where

$$L^*=\frac{1}{2I}\Big(Z_2-\frac{3J+K}{I}Z_1\Big)\ ;$$

$$Q = \frac{1}{I^2}\left(Z_1 Z_2 - \frac{3J+K}{2I} Z_1^2 \right) ;$$

$$c = -\frac{J+K}{2I^2} .$$

We decompose that

$$(4.5) \quad \frac{\beta_3(\theta)}{\sqrt{n}} + \frac{\gamma_3(\theta)}{n} + o\left(\frac{1}{n}\right) = -\frac{1}{2I^3} \, \mathrm{E}_\theta \,(Z_1^3) + \frac{3}{2I} \, \mathrm{E}_\theta \,[Z_1(T_n^* - \mathrm{E}_\theta \,(T_n^*))^2]$$

$$+ \frac{3}{2In} \, \mathrm{E}_\theta \,[Z_1(rL^* + Q - c)^2] + o\left(\frac{1}{n}\right) .$$

Note that $\mathrm{E}_\theta \,(Q) = c$. We have $\mathrm{E}_\theta \,(Z_1^3) = K/\sqrt{n}$. For the second term of the right-hand side of (4.5) we use the following lemma (see [12], Lemma 4.1).

LEMMA. *Suppose that $X_1, X_2, \cdots, X_n, \cdots$ is a sequence of i.i.d. random variables with a density $f(x, \theta)$ which is differentiable in θ for almost all $x[\mu]$. Suppose further that T_θ is a $\mathscr{B}^{(n)}$-measurable function on $\mathscr{X}^{(n)}$ into R^1 and differentiable function of θ and that*

$$\frac{\partial}{\partial \theta} \, \mathrm{E}_\theta \,(T_\theta) = \int_{\mathscr{X}^{(n)}} \frac{\partial}{\partial \theta} \left\{ T_\theta \prod_{i=1}^{n} f(x_i, \theta) \right\} \prod_{i=1}^{n} d\mu(x_i)$$

holds. Then we have

$$\mathrm{E}_\theta \,(Z_1 T_\theta) = \frac{1}{\sqrt{n}} \frac{\partial}{\partial \theta} \, \mathrm{E}_\theta \,(T_\theta) - \frac{1}{\sqrt{n}} \, \mathrm{E}_\theta \left(\frac{\partial T_\theta}{\partial \theta} \right) .$$

By the lemma we obtain

$$(4.6) \quad \mathrm{E}_\theta \,[Z_1(T_n^* - \mathrm{E}_\theta \,(T_n^*))^2]$$

$$= \frac{1}{\sqrt{n}} \frac{\partial}{\partial \theta} \, \mathrm{V}_\theta \,(T_n^*) - \frac{2}{\sqrt{n}} \, \mathrm{E}_\theta \left[(T_n^* - \mathrm{E}_\theta \,(T_n^*)) \right.$$

$$\left. \cdot \left\{ \frac{\partial}{\partial \theta} (T_n^* - \mathrm{E}_\theta \,(T_n^*)) \right\} \right]$$

$$= \frac{1}{\sqrt{n}} \frac{\partial}{\partial \theta} \left(\frac{1}{I} \right) + o\left(\frac{1}{n}\right)$$

$$= -\frac{2J+K}{I^2\sqrt{n}} + o\left(\frac{1}{n}\right) .$$

We also obtain

$$(4.7) \quad \mathrm{E}_\theta \,[Z_1(rL^* + Q - c)^2]$$

$$= 2r \, \mathrm{E}_\theta \,[Z_1 L^*(Q - c)] + o(1) = \frac{r}{I} \left\{ \frac{M}{I} - \frac{(3J+K)K}{2I^2} \right\} + o(1) .$$

From (4.5), (4.6) and (4.7) we have

$$\gamma_3(\theta)=\frac{3r}{2I(\theta)^2}\left[\frac{M(\theta)}{I(\theta)}-\frac{\{3J(\theta)+K(\theta)\}K(\theta)}{2I(\theta)^2}\right].$$

Since the third order asymptotic distribution of T_n^* attains the bound of the third order asymptotic distributions of third order AMU estimators at r, there is no third order AMU estimator which uniformly attains the bound, if $\gamma_3(\theta)$ is not equal to zero. Hence we have established:

THEOREM 4. *Under conditions* (i)$\sim$(vii), $\hat{\theta}_{\mathrm{ML}}^*$ *is not third order asymptotically efficient if* $\gamma_3(\theta)\neq 0$.

5. Maximum log-likelihood estimator

Instead of the equation (1.1) we may take a solution $\hat{\theta}_n$ of the discretized likelihood equation:

$$(5.1)\qquad \sum_{i=1}^{n}\log f(X_i,\hat{\theta}_n+rc_n^{-1})-\sum_{i=1}^{n}\log f(X_i,\hat{\theta}_n-rc_n^{-1})=0.$$

The solution $\hat{\theta}_n$ is θ which maximizes

$$\int_{-rc_n^{-1}}^{rc_n^{-1}}\sum_{i=1}^{n}\log f(X_i,\theta+t)dt=\int_{-rc_n^{-1}}^{rc_n^{-1}}\log L(\theta+t)dt$$

where $L(\theta)$ denotes the likelihood function. Then $\hat{\theta}_n$ is called maximum log-likelihood estimator (MLLE) ([3]). If $\log L(\hat{\theta}_n+t)$ is locally linearized, $\hat{\theta}_n$ agrees with maximum probability estimator by Weiss and Wolfowitz ([15]).

Now we shall consider a location parameter case when the density function f satisfies the following assumption:

(viii) $f(x,\theta)=f(x-\theta)$ and $f(x)$ is symmetric w.r.t. the origin.

It follows by the symmetricity of f that the solution $\hat{\theta}_n$ of (5.1) is AMU. We also have

$$J(\theta)=K(\theta)=0.$$

Let $\hat{\theta}_n$ be an MLLE. Then

$$\sum_{i=1}^{n}\log f(X_i-\hat{\theta}_n-(r/\sqrt{n}))-\sum_{i=1}^{n}\log f(X_i-\hat{\theta}_n+(r/\sqrt{n}))=0.$$

Without loss of generality we assume that $\theta_0=0$. Since

$$\sum_{i=1}^{n}\log f(X_i-\hat{\theta}_n-(r/\sqrt{n}))-\sum_{i=1}^{n}\log f(X_i)$$

         MASAFUMI AKAHIRA AND KEI TAKEUCHI

$$-\left\{\sum_{i=1}^{n}\log f(X_i-\hat{\theta}_n+(r/\sqrt{n}))-\sum_{i=1}^{n}\log f(X_i)\right\}=0\ ,$$

it follows by Taylor expansion around $\theta=0$ that

$$\frac{2r}{\sqrt{n}}\left[\frac{\partial}{\partial\theta}\sum_{i=1}^{n}\log f(X_i-\theta)\right]_{\theta=0}+\frac{2r\hat{\theta}_n}{\sqrt{n}}\left[\frac{\partial^2}{\partial\theta^2}\sum_{i=1}^{n}\log f(X_i-\theta)\right]_{\theta=0}$$

$$+\left(\frac{r\hat{\theta}_n^2}{\sqrt{n}}+\frac{r^3}{3n\sqrt{n}}\right)\left[\frac{\partial^3}{\partial\theta^3}\sum_{i=1}^{n}\log f(X_i-\theta)\right]_{\theta=0}$$

$$+\frac{1}{3}\left(\frac{r\hat{\theta}_n}{\sqrt{n}}+\frac{r^3\hat{\theta}_n}{n\sqrt{n}}\right)\left[\frac{\partial^4}{\partial\theta^4}\log f(X_i-\theta)\right]_{\theta=0}\sim0\ ,$$

where $|\theta^*|\leqq|\theta|$. Putting $T_n=\sqrt{n}\,\hat{\theta}_n$, we obtain

$$2rZ_1+2r\left(\frac{Z_2}{\sqrt{n}}-I(0)\right)T_n+\{Z_3-\sqrt{n}(3J(0)+K(0))\}$$

$$\cdot\left(\frac{rT_n^2}{n}+\frac{r^3}{3n}\right)-\frac{1}{3}\{4L(0)+3M(0)+6N(0)+H(0)\}$$

$$\cdot\left(\frac{r}{n}T_n^3+\frac{r^3}{n}T_n\right)\sim0\ .$$

Hence it follows that

$$T_n=\frac{Z_1}{I}+\frac{Z_2}{I\sqrt{n}}\left(\frac{Z_1}{I}+\frac{Z_1Z_2}{I^2n}\right)+\frac{r^2}{6In}Z_3+\frac{1}{2In}Z_3\frac{Z_1^2}{I^2}$$

$$-\frac{1}{6In}(4L+3M+6N+H)\left(\frac{Z_1^3}{I^3}+r^2\frac{Z_1}{I}\right)+o_p\left(\frac{1}{n}\right)\ .$$

Under conditions (i)$\sim$(vii) we have

$$(5.2)\quad T_n=\frac{Z_1}{I}+\frac{Z_1Z_2}{I^2\sqrt{n}}+\frac{1}{I^3n}\left\{Z_1Z_2^2+\frac{1}{2}Z_1^2Z_3-\frac{1}{6I}(4L+3M+6N+H)Z_1^3\right\}$$

$$+\frac{r^2Z_3}{6In}-\frac{r^2(4L+3M+6N+H)}{6I^2n}Z_1+o_p\left(\frac{1}{n}\right)$$

up to order n^{-1} as $n\to\infty$. As was stated preciously the difference in the order n^{-1} term between (4.4) and (5.2) does not affect the asymptotic distribution up to the order n^{-1}. Hence we have established:

THEOREM 5. *Under conditions* (i)$\sim$(iv) *and* (vi)$\sim$(viii), *the MLLE* $\hat{\theta}_n$ *is asymptotically equivalent to the MLE* $\hat{\theta}_{\mathrm{ML}}$ *up to order* n^{-1}.

It is shown in [12], [13] and [14] that $\hat{\theta}_{\mathrm{ML}}$ maximizes the symmetric probability

$$P_{n,\theta}\{\sqrt{n}\,|\hat{\theta}_{\mathrm{ML}}-\theta|<u\}$$

up to the order n^{-1} among all regular AMU estimators. Therefore it is seen that

$$\lim_{n\to\infty} n[P_{n,\theta}\{\sqrt{n}\,|\hat{\theta}_{\mathrm{ML}}-\theta|<u\}-P_{n,\theta}\{\sqrt{n}\,|\hat{\theta}_{\mathrm{DL}}-\theta|<u\}]\geqq 0$$

for all u, where $\hat{\theta}_{\mathrm{DL}}$ denotes DLE. The asymptotic distribution of the MLLE $\hat{\theta}_{\mathrm{MLL}}$ is equal to that of $\hat{\theta}_{\mathrm{ML}}$ up to the order n^{-1}, but not in the $n^{-3/2}$ and we can show that

$$\lim_{n\to\infty} n^{3/2}[P_{n,\theta}\{\sqrt{n}\,|\hat{\theta}_{\mathrm{MLL}}-\theta|<r\}-P_{n,\theta}\{\sqrt{n}\,|\hat{\theta}_{\mathrm{ML}}-\theta|<r\}]\geqq 0$$

in general situations. Hence for symmetric intervals $\hat{\theta}_{\mathrm{MLL}}$ is not fourth order asymptotically efficient.

6. Conclusion remarks on discretized likelihood methods[1]

If we also define as asymptotic efficient estimator as $\hat{\theta}_n$ which maximize $\lim_{n\to\infty} P_{n,\theta}\{c_n|\hat{\theta}_n-\theta|<r\}$ among AMU estimators $\hat{\theta}_n$, then $\hat{\theta}_n^*$ satisfying the following equation (6.1) is asymptotically efficient:

$$(6.1)\qquad \prod_{i=1}^{n} \frac{f(X_i,\hat{\theta}_n^*+rc_n^{-1})}{f(X_i,\hat{\theta}_n^*)} - \prod_{i=1}^{n} \frac{f(X_i,\hat{\theta}_n^*-rc_n^{-1})}{f(X_i,\hat{\theta}^*)} = a_n\,,$$

where a_n is chosen so that $\hat{\theta}_n^*$ is AMU. Then (6.1) is also expressed as

$$(6.2)\qquad \exp\{\log L(\hat{\theta}_n^*+rc_n^{-1})-\log L(\hat{\theta}_n^*)\}$$
$$-\exp\{\log L(\hat{\theta}_n^*-rc_n^{-1})-\log L(\hat{\theta}_n^*)\}=a_n\,.$$

If $\exp(\log L)$ is linearize, then (6.2) is reduced to the type of (5.1), where in this case the right-hand side of (5.1) is not always zero.

Hence it is seen from the above that the asymptotic efficiency (including higher order cases) may be systematically discussed by the discretized likelihood methods.

Further the discretized likelihood methods may be also applied to the statistical estimation theory of the fixed sample size.

University of Electro-Communications
University of Tokyo

References

[1] Akahira, M. (1975). Asymptotic theory for estimation of location in non-regular cases, I: Order of convergence of consistent estimators, *Rep. Statist. Appl. Res., JUSE*, **22**, 8–26.

[2] Akahira, M. and Takeuchi, K. (1976). On the second order asymptotic efficiency of

---

[1] This section is based on Akahira and Takeuchi [3].

estimators in multiparameter cases, *Rep. Univ. Electro-Comm.*, **26**, 261–269.

[3] Akahira, M. and Takeuchi, K. (1977). Asymptotic properties of estimators obtained from discretized likelihood equations, Annual Meeting of the Mathematical Society of Japan.

[4] Chibisov, D. M. (1972). On the normal approximation for a certain class of statistics, *Proc. 6th Berkeley Symp. Math. Statist. Prob.*, **1**, 153–174.

[5] Chibisov, D. M. (1973). Asymptotic expansions for Neyman's $C(\alpha)$ tests, *Proc. 2nd Japan-USSR Symp. on Prob. Theory*, (Lecture Notes in Math. 330) Springer-Verlag, 16–45.

[6] Efron, B. (1975). Defining the curvature of a statistical problem (with applications to second order efficiency), *Ann. Statist.*, **3**, 1189–1242.

[7] Ghosh, J. K. and Subramanyam, K. (1974). Second order efficiency of maximum likelihood estimators, *Sankhyā*, A, **36**, 325–358.

[8] Pfanzagl, J. (1973). Asymptotic expansions related to minimum contrast estimators, *Ann. Statist.*, **1**, 993–1026.

[9] Pfanzagl, J. (1975). On asymptotically complete classes, *Statistical Inference and Related Topics, Proc. of the Summer Research Institute on Statistical Inference for Stochastic Processes*, **2**, 1–43.

[10] Pfanzagl, J. and Wefelmeyer, W. (1978). A third order optimum property of the maximum likelihood estimator, *J. Multivariate Anal.*, **8**, 1–29.

[11] Takeuchi, K. and Akahira, M. (1976). On the second order asymptotic efficiencies of estimators, *Proc. of the 3rd Japan-USSR Symp. on Prob. Theory*, (Lecture Notes in Math. 550) Springer-Verlag, 604–638.

[12] Takeuchi, K. and Akahira, M. (1978). Third order asymptotic efficiency of maximum likelihood estimator for multiparameter exponential case, *Rep. Univ. Electro-Comm.*, **28**, 271–293.

[13] Takeuchi, K. and Akahira, M. (1978). On the asymptotic efficiency of estimators, (in Japanese), A report of the Symposium on Some Problems of Asymptotic Theory, Annual Meeting of the Mathematical Society of Japan, 1–24.

[14] Takeuchi, K. and Akahira, M. (1979). Third order asymptotic efficiency of maximum likelihood estimator in general case, to appear.

[15] Weiss, L. and Wolfowitz, J. (1967). Maximum probability estimators, *Ann. Inst. Statist. Math.*, **19**, 193–206.

Ann. Inst. Statist. Math.
31 (1979), Part A, 403–415

ASYMPTOTIC OPTIMALITY OF THE GENERALIZED BAYES ESTIMATOR IN MULTIPARAMETER CASES

KEI TAKEUCHI AND MASAFUMI AKAHIRA

(Received July 28, 1978; revised Oct. 24, 1979)

Abstract

The higher order asymptotic efficiency of the generalized Bayes estimator is discussed in multiparameter cases.

For all symmetric loss functions, the generalized Bayes estimator is second order asymptotically efficient in the class A_2 of the all second order asymptotically median unbiased (AMU) estimators and third order asymptotically efficient in the restricted class D of estimators.

1. Introduction

The expansion of a generalized Bayes estimator with respect to a loss function of the type $L(\theta)=|\theta|^a$ $(a\geq 1)$ is obtained by Gusev [5]. His result can be extended to all symmetric loss functions. Strasser [8] also obtained asymptotic expansions of the distribution of the generalized Bayes estimator.

In one parameter case the second order (or third order) asymptotic efficiency of the generalized Bayes estimator has been discussed by Takeuchi and Akahira [12].

It is shown in this paper that in multiparameter case for all symmetric loss function the generalized Bayes estimator $\hat{\theta}$ is asymptotically expanded as

$$\sqrt{n}\,(\hat{\theta}-\theta)=U-\frac{1}{2\sqrt{n}}I^{-1}V+\frac{1}{\sqrt{n}}I^{-1}W+o_p\!\left(\frac{1}{\sqrt{n}}\right),$$

where the symbols of the right-hand side are defined in the contexts. And the asymptotic distributions of the estimators are the same up to the order n^{-1} except for constant location shift. Therefore if it is properly adjusted to be asymptotically median unbiased, it is third order asymptotically efficient among the estimators belonging to the class D ([3], [4], [11]) whose element $\hat{\theta}$ is third order AMU and is asymptotically expanded as

$$\sqrt{n}\,(\hat{\theta}-\theta)=U+\frac{1}{\sqrt{n}}Q+o_p\!\left(\frac{1}{\sqrt{n}}\right)$$

and $Q_\alpha=O_p(1)$ $(\alpha=1,\cdots,p)$ and $\mathrm{E}\,[U_\alpha Q_\beta^k]=o(1)$ $(k=1,2)$ for all $\alpha,\beta=1,\cdots,p$, where E denotes asymptotic expectation and $U=(U_1,\cdots,U_p)'$ and $Q=(Q_1,\cdots,Q_p)'$.

2. Results

Let $(\mathcal{X},\mathcal{B})$ be a sample space. We consider a family of probability measures on $\mathcal{B}$, $\mathcal{P}=\{P_\theta:\theta\in\Theta\}$, where the index set Θ is called a parameter space. We assume that Θ is an open set in a Euclidean p-space R^p with a norm denoted by $\|\cdot\|$. Then an element θ of Θ may be denoted by $(\theta_1,\cdots,\theta_p)$. Consider n-fold direct products $(\mathcal{X}^{(n)},\mathcal{B}^{(n)})$ of $(\mathcal{X},\mathcal{B})$ and the corresponding product measures $P_\theta^{(n)}$ of P_θ. An estimator of θ is defined to be a sequence $\{\hat{\theta}_n\}$ of $\mathcal{B}^{(n)}$-measurable functions $\hat{\theta}_n$ on $\mathcal{X}^{(n)}$ into Θ $(n=1,2,\cdots)$. For simplicity we denote an estimator as $\hat{\theta}$ instead of $\{\hat{\theta}_n\}$. Then $\hat{\theta}$ may be denoted by $(\hat{\theta}_1,\cdots,\hat{\theta}_p)$. For an increasing sequence of positive numbers $\{c_n\}$ $(c_n$ tending to infinity) an estimator is called consistent with order $\{c_n\}$ (or $\{c_n\}$-consistent for short) if for every $\varepsilon>0$ any every $\vartheta\in\Theta$ there exist a sufficiently small positive number δ and a sufficiently large number L satisfying the following:

$$\varlimsup_{n\to\infty}\ \sup_{\theta:\,\|\theta-\vartheta\|<\delta}\ P_\theta^{(n)}\{c_n\|\hat{\theta}-\theta\|\geqq L\}<\varepsilon\qquad([1])\ .$$

For each $k=1,2,\cdots$, a $\{c_n\}$-consistent estimator $\hat{\theta}$ is kth order asymptotically median unbiased (or kth order AMU) estimator if for each $\vartheta\in\Theta$ and each $\alpha=1,\cdots,p$, there exists a positive number δ such that

$$\lim_{n\to\infty}\ \sup_{\theta:\,\|\theta-\vartheta\|<\delta}\ c_n^{k-1}\left|P_\theta^{(n)}\{\hat{\theta}_\alpha\leqq\theta_\alpha\}-\frac{1}{2}\right|=0\ ;$$

$$\lim_{n\to\infty}\ \sup_{\theta:\,\|\theta-\vartheta\|<\delta}\ c_n^{k-1}\left|P_\theta^{(n)}\{\hat{\theta}_\alpha\geqq\theta_\alpha\}-\frac{1}{2}\right|=0\ .$$

For $\hat{\theta}$ kth order AMU, $G_0(t,\theta_\alpha)+c_n^{-1}G_1(t,\theta_\alpha)+\cdots+c_n^{-(k-1)}G_{k-1}(t,\theta_\alpha)$ $(\alpha=1,\cdots,p)$ is called to be the kth order asymptotic marginal distribution of $c_n(\hat{\theta}-\theta)$ (or $\hat{\theta}$ for short) if

$$\lim_{n\to\infty}c_n^{k-1}|P_\theta^{(n)}\{c_n(\hat{\theta}_\alpha-\theta_\alpha)<t\}-G_0(t,\theta_\alpha)$$
$$-c_n^{-1}G_1(t,\theta_\alpha)-\cdots-c_n^{-(k-1)}G_{k-1}(t,\theta_\alpha)|=0\ .$$

We note that $G_i(t,\theta_\alpha)$ $(i=1,\cdots,k-1;\ \alpha=1,\cdots,p)$ may be generally absolutely continuous functions, hence the asymptotic marginal distributions for any fixed n may not be a distribution function.

Suppose that $\hat{\theta}$ is kth order AMU and has the kth order marginal asymptotic distribution $G_0(t, \theta_\alpha)+c_n^{-1}G_1(t, \theta_\alpha)+\cdots+c_n^{-(k-1)}G_{k-1}(t, \theta_\alpha)$ $(\alpha=1, \cdots, p)$ and the joint distribution of $\hat{\theta}$ admits asymptotic expansion up to kth order, i.e., the order of $c_n^{-(k-1)}$. Letting θ_0 ($\in \Theta$) be arbitrary but fixed. Denote θ_0 by $(\theta_{01}, \cdots, \theta_{0p})$. Let α be arbitrary but fixed in $1, \cdots, p$. We consider the problem of testing hypothesis $H^+: \theta_\alpha=\theta_{0\alpha}+tc_n^{-1}$ ($t>0$) against $K: \theta_\alpha=\theta_{0\alpha}$. Put $\Phi_{1/2}=\{\{\phi_n\}; \mathrm{E}_{\theta_{0\alpha}+tc_n^{-1}}^{(n)}(\phi_n)=1/2+o(c_n^{-(k-1)}), 0\leq \phi_n(\tilde{x}_n)\leq 1$ for all $\tilde{x}_n \in \mathscr{X}^{(n)}$ $(n=1, 2, \cdots)\}$. Putting $A_{\hat{\theta}_\alpha, \theta_{0\alpha}}=\{c_n(\hat{\theta}_\alpha-\theta_{0\alpha})\leq t\}$, we have

$$\lim_{n\to\infty} P_{\theta_{0\alpha}+tc_n^{-1}}^{(n)}(A_{\hat{\theta}_\alpha, \theta_{0\alpha}})=\lim_{n\to\infty} P_{\theta_{0\alpha}+tc_n^{-1}}^{(n)}\{\hat{\theta}_\alpha\leq\theta_{0\alpha}+tc_n^{-1}\}=\frac{1}{2} \ .$$

Hence it is seen that a sequence $\{\chi_{A_{\hat{\theta}_\alpha, \theta_{0\alpha}}}\}$ of the indicators (or characteristic functions) of $A_{\hat{\theta}_\alpha, \theta_{0\alpha}}$ $(n=1, 2, \cdots)$ belongs to $\Phi_{1/2}$. If

$$\sup_{\{\phi_n\}\in\Phi_{1/2}} \varlimsup_{n\to\infty} c_n^{k-1}\{\mathrm{E}_{\theta_{0\alpha}}^{(n)}(\phi_n)-H_0^+(t, \theta_{0\alpha})-c_n^{-1}H_1^+(t, \theta_{0\alpha})-\cdots$$
$$-c_n^{-(k-1)}H_{k-1}^+(t, \theta_{0\alpha})\}=0 \ ,$$

then we have

$$G_0(t, \theta_{0\alpha})\leq H_0^+(t, \theta_{0\alpha}) \ ;$$

and for any positive integer j $(\leq k)$ if $G_i(t, \theta_{0\alpha})=H_i^+(t, \theta_{0\alpha})$ $(i=1, \cdots, j-1)$ then

$$G_j(t, \theta_{0\alpha})=H_j^+(t, \theta_{0\alpha}) \ .$$

Consider next the problem of testing hypothesis $H^-: \theta_\alpha=\theta_{0\alpha}+tc_n^{-1}$ $(t<0)$ against $K: \theta_\alpha=\theta_{0\alpha}$. If

$$\sup_{\{\phi_n\}\in\Phi_{1/2}} \lim_{n\to\infty} c_n^{k-1}\{\mathrm{E}_{\theta_{0\alpha}}^{(n)}(\phi_n)-H_0^-(t, \theta_{0\alpha})-c_n^{-1}H_1^-(t, \theta_{0\alpha})-\cdots$$
$$-c_n^{-(k-1)}H_{k-1}^-(t, \theta_{0\alpha})\}=0 \ ,$$

then we have

$$G_0(t, \theta_{0\alpha})\geq H_0^-(t, \theta_{0\alpha}) \ ;$$

and for any positive integer j $(\leq k)$ if $G_i(t, \theta_{0\alpha})=H_i^-(t, \theta_{0\alpha})$ $(i=1, \cdots, j-1)$, then $G_j(t, \theta_{0\alpha})\geq H_j^-(t, \theta_{0\alpha})$.

$\hat{\theta}$ is called to be kth order asymptotically efficient in the class $\boldsymbol{A}_k$ of the all kth order AMU estimators if the kth order asymptotic marginal distribution of it attains uniformly the bound of the kth order asymptotic marginal distributions of kth order AMU estimators, that

          KEI TAKEUCHI AND MASAFUMI AKAHIRA

is, for each $\alpha=1,\cdots,p$

$$G_i(t,\theta_\alpha)=\begin{cases} H_i^+(t,\theta_\alpha) & \text{for } t>0, \\ H_i^-(t,\theta_\alpha) & \text{for } t<0, \end{cases}$$

$i=0,\cdots,k-1$ ([2], [4], [9]). [Note that for $t=0$ and each $\alpha=1,\cdots,p$ we have $G_i(0,\theta_\alpha)=H_i^+(0,\theta_\alpha)=H_i^-(0,\theta_\alpha)$ $(i=0,\cdots,k-1)$ from the condition of kth order asymptotically median unbiasedness.]

$\hat{\theta}$ is called to be third order asymptotically efficient in the class D if the third order asymptotic marginal distribution of it attains uniformly the bound of the third order asymptotic marginal distributions of estimators in D. It is generally shown by Pfanzagl and Wefelmeyer [6], [7] and Akahira and Takeuchi [3], [4], [10], [11], [14] that there exist second order asymptotically efficient estimators but not third order asymptotically efficient estimators in the class A_3. But it was also shown in [4], [10] and [11] that if we restrict the class of estimators appropriately, we have asymptotically efficient estimators among the restricted class of estimators and that the maximum likelihood estimator belongs to the class of higher order asymptotically efficient estimators.

We assume that for each $\theta \in \Theta$ P_θ is absolutely continuous with respect to σ-finite measure μ. We denote a density $dP_\theta/d\mu$ by $f(x,\theta)$. Then the joint density is given by $\prod_{i=1}^{n} f(x_i,\theta)$.

In the subsequent discussion we shall deal with the case when $c_n=\sqrt{n}$. Let $\Theta=R^p$. Let $L_n(u)$ $(u=(u_1,\cdots,u_p)\in R^p)$ be a bounded nonnegative and quasi-convex function around the origin, i.e. for any c the set $\{u: L(u)\leq c\}$ $(\subset R^p)$ is convex and contains the origin and $\pi(\theta)$ be a non-negative function. Define a posterior density $p_n(\theta|\tilde{x}_n)$ and a posterior risk $r_n(d|\tilde{x}_n)$ by

$$p_n(\theta|\tilde{x}_n)=\left\{\prod_{i=1}^{n} f(x_i,\theta)\right\}\pi(\theta)\left[\int_\Theta \left(\prod_{i=1}^{n} f(x_i,\theta)\right)\pi(\theta)d\theta\right]^{-1}$$

and

$$r_n(d|\tilde{x}_n)=\int_\Theta L_n(d-\theta)p_n(\theta|\tilde{x}_n)d\theta ,$$

respectively, where $\tilde{x}_n=(x_1,\cdots,x_n)$. Now suppose that $\lim_{n\to\infty} L_n(u/\sqrt{n})=L^*(u)$ for all $u\in R^p$. We define

$$r_n^*(d|\tilde{x}_n)=\int_\Theta L^*(\sqrt{n}(d-\theta))p_n(\theta|\tilde{x}_n)d\theta .$$

An estimator $\hat{\theta}$ is called a generalized Bayes estimator with respect to a loss function L^* and a prior density π if

$$r_n^*(\hat{\theta}\,|\,\tilde{x}_n)=\inf_{d\in\theta} r_n^*(d\,|\,\tilde{x}_n)\ .$$

Then $\hat{t}=\sqrt{n}(\hat{\theta}-\theta)$ may also be called a generalized Bayes estimator w.r.t. L^* and π. Since

$$\lim_{n\to\infty}\left|\inf_{d\in\theta}\int_\theta L_n(d-\theta)\tilde{p}(\theta)d\theta-\inf_{d\in\theta}\int_\theta L^*(\sqrt{n}(d-\theta))\tilde{p}(\theta)d\theta\right|=0$$

uniformly in every posterior density $\tilde{p}(\theta)$, it follows that for a generalized Bayes estimator

$$\lim_{n\to\infty}|\inf_{d\in\theta} r_n(d\,|\,\tilde{x}_n)-r_n^*(\hat{\theta}\,|\,\tilde{x}_n)|=0\ .$$

Suppose that $X_1, X_2,\cdots, X_n,\cdots$ is a sequence of i.i.d. random variables with a density $f(x,\theta)$ satisfying (i)–(iv).

(i) $\{x\colon f(x,\theta)>0\}$ does not depend on θ.

(ii) For almost all $x[\mu]$, $f(x,\theta)$ is three times continuously differentiable in θ_α $(\alpha=1,\cdots, p)$.

(iii) For each α, β $(\alpha, \beta=1,\cdots, p)$

$$0<I_{\alpha\beta}(\theta)=\mathrm{E}_\theta\left[\left\{\frac{\partial}{\partial\theta_\alpha}\log f(x,\theta)\right\}\left\{\frac{\partial}{\partial\theta_\beta}\log f(x,\theta)\right\}\right]$$

$$=-\mathrm{E}_\theta\left[\frac{\partial^2}{\partial\theta_\alpha\partial\theta_\beta}\log f(x,\theta)\right]<\infty\ .$$

(iv) There exist

$$J_{\alpha\beta\cdot\gamma}(\theta)=\mathrm{E}_\theta\left[\left\{\frac{\partial^2}{\partial\theta_\alpha\partial\theta_\beta}\log f(x,\theta)\right\}\left\{\frac{\partial}{\partial\theta_\gamma}\log f(x,\theta)\right\}\right]\,,$$

$$K_{\alpha\beta\gamma}(\theta)=\mathrm{E}_\theta\left[\left\{\frac{\partial}{\partial\theta_\alpha}\log f(x,\theta)\right\}\left\{\frac{\partial}{\partial\theta_\beta}\log f(x,\theta)\right\}\right.$$

$$\left.\cdot\left\{\frac{\partial}{\partial\theta_\gamma}\log f(x,\theta)\right\}\right]$$

and

$$W_{\alpha\beta\gamma}(\theta)=\mathrm{E}_\theta\left[\frac{\partial^3}{\partial\theta_\alpha\partial\theta_\beta\partial\theta_\gamma}\log f(x,\theta)\right]$$

and the following holds:

$$\mathrm{E}_\theta\left[\frac{\partial^3}{\partial\theta_\alpha\partial\theta_\beta\partial\theta_\gamma}\log f(x,\theta)\right]=-J_{\alpha\beta\cdot\gamma}(\theta)-J_{\alpha\gamma\cdot\beta}(\theta)-J_{\beta\gamma\cdot\alpha}(\theta)-K_{\alpha\beta\gamma}(\theta)\ .$$

It was shown by the same way in [9] that a maximum likelihood estimator (MLE) is second order asymptotically efficient. Let $\hat{\theta}$ be an MLE. By Taylor expansion we have

408   KEI TAKEUCHI AND MASAFUMI AKAHIRA

$$0 = \sum_\alpha \sum_i \frac{\partial}{\partial \theta_\alpha} \log f(X_i, \hat{\theta})$$

$$= \sum_\alpha \left\{ \sum_i \frac{\partial}{\partial \theta_\alpha} \log f(X_i, \theta) \right\} (\hat{\theta}_\alpha - \theta_\alpha) + \sum_\alpha \sum_\beta \left\{ \sum_i \frac{\partial^2}{\partial \theta_\alpha \partial \theta_\beta} \log f(X_i, \theta) \right\}$$

$$\cdot (\hat{\theta}_\alpha - \theta_\alpha)(\hat{\theta}_\beta - \theta_\beta) + \frac{1}{2} \sum_\alpha \sum_\beta \sum_\gamma \left\{ \sum_i \frac{\partial^3}{\partial \theta_\alpha \partial \theta_\beta \partial \theta_\gamma} \log f(X_i, \theta^*) \right\}$$

$$\cdot (\hat{\theta}_\alpha - \theta_\alpha)(\hat{\theta}_\beta - \theta_\beta)(\hat{\theta}_\gamma - \theta_\gamma) \ ,$$

where $\|\theta^* - \theta\| \leq \|\hat{\theta} - \theta\|$.   Putting $T = \sqrt{n}(\hat{\theta} - \theta)$ we obtain

$$0 = \frac{1}{\sqrt{n}} \sum_\alpha \left\{ \sum_i \frac{\partial}{\partial \theta_\alpha} \log f(X_i, \theta) \right\} T_\alpha + \frac{1}{n} \sum_\alpha \sum_\beta \left\{ \sum_i \frac{\partial^2}{\partial \theta_\alpha \partial \theta_\beta} \log f(X_i, \theta) \right\} T_\alpha T_\beta$$

$$+ \frac{1}{2n\sqrt{n}} \sum_\alpha \sum_\beta \sum_\gamma \left\{ \sum_i \frac{\partial^3}{\partial \theta_\alpha \partial \theta_\beta \partial \theta_\gamma} \log f(X_i, \theta^*) \right\} T_\alpha T_\beta T_\gamma \ ,$$

where $T = (T_1, \cdots, T_p)'$.   Set

$$Z_\alpha(\theta) = \frac{1}{\sqrt{n}} \sum_{i=1}^n \frac{\partial}{\partial \theta_\alpha} \log f(X_i, \theta) \ ;$$

$$Z_{\alpha\beta}(\theta) = \frac{1}{\sqrt{n}} \sum_{i=1}^n \frac{\partial^2}{\partial \theta_\alpha \partial \theta_\beta} \log f(X_i, \theta) + I_{\alpha\beta}(\theta) \ .$$

Then it follows that $W_{\alpha\beta\gamma}(\theta)$ converges in probability to $-\{J_{\alpha\beta\cdot\gamma}(\theta) + J_{\alpha\gamma\cdot\beta}(\theta) + J_{\beta\gamma\cdot\alpha}(\theta) + K_{\alpha\beta\gamma}(\theta)\}$.   We put $\rho_{\alpha\beta\gamma}(\theta) = J_{\alpha\beta\cdot\gamma}(\theta) + J_{\alpha\gamma\cdot\beta}(\theta) + J_{\beta\gamma\cdot\alpha}(\theta) + K_{\alpha\beta\gamma}(\theta)$. Hence the following theorem holds:

THEOREM 1.   *Under conditions* (i)–(iv)

$$\sqrt{n}(\hat{\theta} - \theta) = U - \frac{1}{2\sqrt{n}} I^{-1}V + \frac{1}{\sqrt{n}} I^{-1}W + o_p\left(\frac{1}{\sqrt{n}}\right) \ ,$$

where $I = (I_{\alpha\beta})$ *and* $P = (P_{\alpha\beta})$ *are matrices and* $V = (\sum_\beta \sum_\gamma \rho_{\alpha\beta\gamma} U_\beta U_\gamma)$, $W = (\sum_\beta U_\beta Z_{\beta\gamma})$ *and* $U = (U_\alpha)$ *are p-dimensional column vectors and* $U_\alpha = \sum_\beta I^{\alpha\beta} Z_\beta$ *and* $I^{\alpha\beta}$ *denotes the* (α, β)-*element of the inverse matrix of the information matrix* I.

Since the proof of the theorem is essentially same as that of one parameter case ([9]), it is omitted.

Put

$$\hat{\theta}^* = \hat{\theta} + \frac{1}{6n} Y \ ,$$

where $Y = (\sum_\beta \sum_\gamma U_\beta W_{\alpha\beta\gamma})$ is a column vector.   Then $\hat{\theta}^*$ is second order

AMU. From Theorem 1 we have established the following:

THEOREM 2. *Under conditions* (i)–(iv) $\hat{\theta}^*$ *is second order asymptotically efficient in the class* $\boldsymbol{A_2}$.

Since the proof of the theorem is essentially same as that of one parameter case ([9]), it is omitted.

It will be shown that the generalized Bayes estimator w.r.t. a loss function and a prior density is second order asymptotically efficient. Let θ_0 be a true parameter of θ ($\in \Theta$). Denote θ and θ_0 by $(\theta_1, \cdots, \theta_p)'$ and $(\theta_{01}, \cdots, \theta_{0p})'$ respectively. Further we assume the following:

(v) For each $\alpha = 1, \cdots, p$, $\pi(\theta)$ is twice partially differentiable in θ_α. Then we have

$$
p_n(\theta \mid \tilde{x}_n)/p_n(\theta_0 \mid \tilde{x}_n)
$$

$$
= \exp\left[\log\{p_n(\theta \mid \tilde{x}_n)/p_n(\theta_0 \mid \tilde{x}_n)\}\right]
$$

$$
= \exp\left[\sum_{i=1}^{n} \log\{f(x_i, \theta)/f(x_i, \theta_0)\} + \log\{\pi(\theta)/\pi(\theta_0)\}\right]
$$

$$
= \exp\left[\sum_{i=1}^{n} \log f(x_i, \theta) - \sum_{i=1}^{n} \log f(x_i, \theta_0) + \log \pi(\theta) - \log \pi(\theta_0)\right]
$$

$$
= \exp\left[\sum_{\alpha} \left\{\sum_{i=1}^{n} \frac{\partial}{\partial \theta_\alpha} \log f(x_i, \theta_0)\right\}(\theta_\alpha - \theta_{0\alpha})\right.
$$

$$
+ \frac{1}{2} \sum_{\alpha} \sum_{\beta} \left\{\sum_{i=1}^{n} \frac{\partial^2}{\partial \theta_\alpha \partial \theta_\beta} \log f(x_i, \theta_0)\right\}(\theta_\alpha - \theta_{0\alpha})(\theta_\beta - \theta_{0\beta})
$$

$$
+ \frac{1}{6} \sum_{\alpha} \sum_{\beta} \sum_{\gamma} \left\{\sum_{i=1}^{n} \frac{\partial^3}{\partial \theta_\alpha \partial \theta_\beta \partial \theta_\gamma} \log f(x_i, \theta^*)\right\}
$$

$$
\cdot (\theta_\alpha - \theta_{0\alpha})(\theta_\beta - \theta_{0\beta})(\theta_\gamma - \theta_{0\gamma}) + \sum_{\alpha} \frac{\pi'_\alpha(\theta_0)}{\pi(\theta_0)}(\theta_\alpha - \theta_{0\alpha}) + o\left(\frac{1}{\sqrt{n}}\right)\bigg]
$$

$$
= \exp\left[\sqrt{n} \sum_{\alpha} Z_\alpha(\theta_0)(\theta_\alpha - \theta_{0\alpha}) + \frac{1}{2} \sum_{\alpha} \sum_{\beta} \{\sqrt{n} Z_{\alpha\beta}(\theta_0) - n I_{\alpha\beta}(\theta_0)\}\right.
$$

$$
\cdot (\theta_\alpha - \theta_{0\alpha})(\theta_\beta - \theta_{0\beta}) + \frac{n}{6} \sum_{\alpha} \sum_{\beta} \sum_{\gamma} W_{\alpha\beta\gamma}(\theta^*)(\theta_\alpha - \theta_{0\alpha})(\theta_\beta - \theta_{0\beta})
$$

$$
\cdot (\theta_\gamma - \theta_{0\gamma}) + \sum_{\alpha} \frac{\pi'_\alpha(\theta_0)}{\pi(\theta_0)}(\theta_\alpha - \theta_{0\alpha}) + o\left(\frac{1}{\sqrt{n}}\right)\bigg] ,
$$

where $\pi'_\alpha(\theta) = \partial \pi(\theta)/\partial \theta_\alpha$ ($\alpha = 1, \cdots, p$). It follows that

$$
p_n(\theta \mid \tilde{x}_n)/p_n(\theta_0 \mid \tilde{x}_n)
$$

$$
= \exp\left[\sum_{\alpha} Z_\alpha(\theta_0)\{\sqrt{n}(\theta_\alpha - \theta_{0\alpha})\} + \frac{1}{2} \sum_{\alpha} \sum_{\beta} \left\{\frac{Z_{\alpha\beta}(\theta_0)}{\sqrt{n}} - I_{\alpha\beta}(\theta_0)\right\}\right.
$$

$$
\cdot \{n(\theta_\alpha - \theta_{0\alpha})(\theta_\beta - \theta_{0\beta})\} - \frac{1}{6\sqrt{n}} \sum_{\alpha} \sum_{\beta} \sum_{\gamma} \rho_{\alpha\beta\gamma}(\theta_0)
$$

$$
\cdot \{n\sqrt{n}(\theta_\alpha - \theta_{0\alpha})(\theta_\beta - \theta_{0\beta})(\theta_\gamma - \theta_{0\gamma})\}
$$

410   KEI TAKEUCHI AND MASAFUMI AKAHIRA

$$+ \frac{1}{\sqrt{n}\,\pi(\theta_0)} \sum_\alpha \pi'_\alpha(\theta_0)\{\sqrt{n}\,(\theta_\alpha - \theta_{0\alpha})\} + o_p\!\left(\frac{1}{\sqrt{n}}\right)\Big] .$$

Putting $t_\alpha = \sqrt{n}\,(\theta_\alpha - \theta_{0\alpha})$ $(\alpha = 1, \cdots, p)$ we obtain

(1) $p_n(\theta \,|\, \tilde{x}_n)/p_n(\theta_0 \,|\, \tilde{x}_n)$

$$= \exp\left[\sum_\alpha Z_\alpha(\theta) t_\alpha + \frac{1}{2} \sum_\alpha \sum_\beta \left\{ \frac{Z_{\alpha\beta}(\theta_0)}{\sqrt{n}} - I_{\alpha\beta}(\theta_0) \right\} t_\alpha t_\beta \right.$$

$$- \frac{1}{6\sqrt{n}} \sum_\alpha \sum_\beta \sum_\gamma \rho_{\alpha\beta\gamma}(\theta_0) t_\alpha t_\beta t_\gamma$$

$$\left. + \frac{1}{\sqrt{n}\,\pi(\theta_0)} \sum_\alpha \pi'_\alpha(\theta_0) t_\alpha + o_p\!\left(\frac{1}{\sqrt{n}}\right)\right]$$

$$= \exp\left[-\frac{1}{2} \sum_\alpha \sum_\beta I_{\alpha\beta}(\theta_0)(t_\alpha - U_\alpha)(t_\beta - U_\beta) + \frac{1}{2} \sum_\alpha \sum_\beta I_{\alpha\beta}(\theta_0) U_\alpha U_\beta \right.$$

$$+ \frac{1}{\sqrt{n}\,\pi(\theta_0)} \sum_\alpha \pi'_\alpha(\theta_0) t_\alpha + \frac{1}{2\sqrt{n}} \sum_\alpha \sum_\beta Z_{\alpha\beta}(\theta_0) t_\alpha t_\beta$$

$$\left. - \frac{1}{6\sqrt{n}} \sum_\alpha \sum_\beta \sum_\gamma \rho_{\alpha\beta\gamma}(\theta_0) t_\alpha t_\beta t_\gamma + o_p\!\left(\frac{1}{\sqrt{n}}\right)\right]$$

$$= \left(\exp \frac{1}{2} \sum_\alpha \sum_\beta I_{\alpha\beta}(\theta_0) U_\alpha U_\beta \right)$$

$$\cdot \left[\exp\left\{ -\frac{1}{2} \sum_\alpha \sum_\beta I_{\alpha\beta}(\theta_0)(t_\alpha - U_\alpha)(t_\beta - U_\beta) \right\} \right]$$

$$\cdot \exp\left\{ \frac{1}{\sqrt{n}\,\pi(\theta_0)} \sum_\alpha \pi'_\alpha(\theta_0) t_\alpha + \frac{1}{2\sqrt{n}} \sum_\alpha \sum_\beta Z_{\alpha\beta}(\theta_0) t_\alpha t_\beta \right.$$

$$\left. - \frac{1}{6\sqrt{n}} \sum_\alpha \sum_\beta \sum_\gamma \rho_{\alpha\beta\gamma}(\theta_0) t_\alpha t_\beta t_\gamma + o_p\!\left(\frac{1}{\sqrt{n}}\right)\right\}$$

$$= \left(\exp \frac{1}{2} \sum_\alpha \sum_\beta I_{\alpha\beta}(\theta_0) U_\alpha U_\beta \right)$$

$$\cdot \left[\exp\left\{ -\frac{1}{2} \sum_\alpha \sum_\beta I_{\alpha\beta}(\theta_0)(t_\alpha - U_\alpha)(t_\beta - U_\beta) \right\} \right]$$

$$\cdot \left\{ 1 + \frac{1}{\sqrt{n}\,\pi(\theta_0)} \sum_\alpha \pi'_\alpha(\theta_0) t_\alpha + \frac{1}{2\sqrt{n}} \sum_\alpha \sum_\beta Z_{\alpha\beta}(\theta_0) t_\alpha t_\beta \right.$$

$$\left. - \frac{1}{6\sqrt{n}} \sum_\alpha \sum_\beta \sum_\gamma \rho_{\alpha\beta\gamma}(\theta_0) t_\alpha t_\beta t_\gamma + o_p\!\left(\frac{1}{\sqrt{n}}\right)\right\}$$

$$= q_n(t, \theta_0 \,|\, \tilde{x}_n) \qquad \text{(say)},$$

where $t = (t_1, \cdots, t_p)'$. Let $\hat{t} = \sqrt{n}\,(\hat{\theta} - \theta_0)$. Then the posterior risk is given by

(2) $r_n^*(\hat{t} \,|\, \tilde{x}_n) = \dfrac{1}{\sqrt{n}} \, p_n(\theta_0 \,|\, \tilde{x}_n) \displaystyle\int L^*(\hat{t} - t) q_n(t, \theta_0 \,|\, \tilde{x}_n) dt .$

ASYMPTOTIC OPTIMALITY OF THE GENERALIZED BAYES ESTIMATOR   411

Further we assume the following:

(vi) $L^*(u)$ is a convex function;

(vii)   For each $\alpha=1,\cdots,p,$ $\int L^*(-u)q_n(u+t,\theta_0|\tilde{x}_n)du$ is continuously partially differentiable with respect to t_α under the integral sign.

By (2) and the assumption (vi) it is shown that the generalized Bayes estimator $\hat{t}$ w.r.t. $L^*(\cdot)$ and $\pi(\cdot)$ is given as a solution u of the equation

$$(d/du_\alpha)\int L^*(u-t)q_n(t,\theta_0|\tilde{x}_n)dt=0 \qquad (\alpha=1,\cdots,p)\,.$$

Since by (vii)

$$\frac{d}{du_\alpha}\int L^*(u-t)q_n(t,\theta_0|\tilde{x}_n)dt$$

$$=\frac{d}{du_\alpha}\int L^*(-t)q_n(u+t,\theta_0|\tilde{x}_n)dt$$

$$=\frac{d}{du_\alpha}\int L^*(-u)q_n(t+u,\theta_0|\tilde{x}_n)du$$

$$=\int L^*(-u)\left\{\frac{d}{dt_\alpha}q_n(t+u,\theta_0|\tilde{x}_n)\right\}du \qquad (\alpha=1,\cdots,p)\,,$$

the generalized Bayes estimator $\hat{t}$ is obtained by a solution of the equation

$$(3) \qquad \int L^*(-u)\left\{\frac{d}{dt_\alpha}q_n(t+u,\theta_0|\tilde{x}_n)\right\}du=0 \qquad (\alpha=1,\cdots,p)\,.$$

Since $t=\sqrt{n}(\theta-\theta_0)$ and $\hat{t}=\sqrt{n}(\hat{\theta}-\theta_0),$ $\hat{t}$ may be called to be the generalized Bayes estimator. From (1) and (3) we have

$$0=\int L^*(-u)\exp\left[-\frac{1}{2}\sum_\alpha\sum_\beta I_{\alpha\beta}(\theta_0)(\hat{t}_\alpha+u_\alpha-U_\alpha)(\hat{t}_\beta+u_\beta-U_\beta)\right]$$

$$\cdot\left[-\sum_\beta I_{\alpha\beta}(\theta_0)(\hat{t}_\beta+u_\beta-U_\beta)\left\{1+\frac{1}{\sqrt{n}\,\pi(\theta_0)}\sum_\alpha\pi'_\alpha(\theta_0)(\hat{t}_\alpha+u_\alpha)\right.\right.$$

$$+\frac{1}{2\sqrt{n}}\sum_\alpha\sum_\beta Z_{\alpha\beta}(\theta_0)(\hat{t}_\alpha+u_\alpha)(\hat{t}_\beta+u_\beta)$$

$$\left.-\frac{1}{6\sqrt{n}}\sum_\alpha\sum_\beta\sum_\gamma\rho_{\alpha\beta\gamma}(\theta_0)(\hat{t}_\alpha+u_\alpha)(\hat{t}_\beta+u_\beta)(\hat{t}_\gamma+u_\gamma)\right\}$$

$$+\frac{1}{\sqrt{n}\,\pi(\theta_0)}\pi'_\alpha(\theta_0)+\frac{1}{\sqrt{n}}\sum_\beta Z_{\alpha\beta}(\theta_0)(\hat{t}_\beta+u_\beta)$$

$$\left.-\frac{1}{2\sqrt{n}}\sum_\beta\sum_\gamma\rho_{\alpha\beta\gamma}(\theta_0)(\hat{t}_\beta+u_\beta)(\hat{t}_\gamma+u_\gamma)\right]du+o_p\left(\frac{1}{\sqrt{n}}\right)\,.$$

412   KEI TAKEUCHI AND MASAFUMI AKAHIRA

Putting $\hat{u}_\alpha = \hat{t}_\alpha - U_\alpha$ $(\alpha = 1, \cdots, p)$ we obtain

$$
\begin{aligned}
0 = \int L^*(-u) & \left[\exp\left\{ -\frac{1}{2} \sum_\alpha \sum_\beta I_{\alpha\beta}(\theta_0)(u_\alpha + \hat{u}_\alpha)(u_\beta + \hat{u}_\beta) \right\} \right] \\
& \cdot \left[-\sum_\beta I_{\alpha\beta}(u_\beta + \hat{u}_\beta) \left\{ 1 + \frac{1}{\sqrt{n}\,\pi(\theta_0)} \sum_\alpha \pi_\alpha'(\theta_0)(u_\alpha + \hat{u}_\alpha + U_\alpha) \right. \right. \\
& \quad + \frac{1}{2\sqrt{n}} \sum_\alpha \sum_\beta Z_{\alpha\beta}(\theta_0)(u_\alpha + \hat{u}_\alpha + U_\alpha)(u_\beta + \hat{u}_\beta + U_\beta) \\
& \quad \left. - \frac{1}{6\sqrt{n}} \sum_\alpha \sum_\beta \sum_\gamma \rho_{\alpha\beta\gamma}(\theta_0)(u_\alpha + \hat{u}_\alpha + U_\alpha)(u_\beta + \hat{u}_\beta + U_\beta)(u_\gamma + \hat{u}_\gamma + U_\gamma) \right\} \\
& \quad + \frac{1}{\sqrt{n}\,\pi(\theta_0)} \pi_\alpha'(\theta_0) + \frac{1}{\sqrt{n}} \sum_\beta Z_{\alpha\beta}(\theta_0)(\hat{u}_\beta + u_\beta + U_\beta) \\
& \quad \left. - \frac{1}{2\sqrt{n}} \sum_\beta \sum_\gamma \rho_{\alpha\beta\gamma}(\theta_0)(\hat{u}_\beta + u_\beta + U_\beta)(\hat{u}_\gamma + u_\gamma + U_\gamma) \right] du + o_p\left(\frac{1}{\sqrt{n}} \right) .
\end{aligned}
$$

Further we assume the following.

(viii) $L^*(u)$ is a symmetric loss function about the origin.
We define

$$
M = \int L^*(-u) \exp\left[-\frac{1}{2} \sum_\alpha \sum_\beta I_{\alpha\beta}(\theta_0)u_\alpha u_\beta \right] du ;
$$

$$
P_{\alpha\beta} = \int L^*(-u)u_\alpha u_\beta \exp\left[-\frac{1}{2} \sum_\alpha \sum_\beta I_{\alpha\beta}(\theta_0)u_\alpha u_\beta \right] du
$$
$$
(\alpha, \beta = 1, \cdots, p) ;
$$

$$
Q_{\alpha\beta\gamma\delta} = \int L^*(-u)u_\alpha u_\beta u_\gamma u_\delta \exp\left[-\frac{1}{2} \sum_\alpha \sum_\beta I_{\alpha\beta}(\theta_0)u_\alpha u_\beta \right] du
$$
$$
(\alpha, \beta, \gamma, \delta = 1, \cdots, p) .
$$

Note that by (viii)

$$
\int L^*(-u)u_\alpha \exp\left(-\frac{1}{2} \sum_\alpha \sum_\beta I_{\alpha\beta}(\theta_0)u_\alpha u_\beta \right) du
$$
$$
= \int L^*(-u)u_\alpha u_\beta u_\gamma \exp\left(-\frac{1}{2} \sum_\alpha \sum_\beta I_{\alpha\beta}(\theta_0)u_\alpha u_\beta \right) du = 0
$$
$$
(\alpha, \beta, \gamma = 1, \cdots, p) .
$$

Then we have

$$
\begin{aligned}
0 = \int L^*(-u) & \left[\exp\left\{ -\frac{1}{2} \sum_\alpha \sum_\beta I_{\alpha\beta}(\theta_0)u_\alpha u_\beta \right\} \right] \left\{ 1 - \sum_\gamma \sum_\delta I_{\gamma\delta}(\theta_0)u_\gamma u_\delta \right\} \\
& \cdot \left[-\sum_\beta I_{\alpha\beta}(\theta_0)u_\beta - \sum_\beta I_{\alpha\beta}(\theta_0)\hat{u}_\beta - \sum_\beta \sum_\gamma \frac{I_{\alpha\beta}(\theta_0)\pi_\gamma'(\theta_0)}{\sqrt{n}\,\pi(\theta_0)} \right.
\end{aligned}
$$

$$\cdot (u_\beta+\hat{u}_\beta)(u_\gamma+\hat{u}_\gamma+U_\gamma)-\frac{1}{2\sqrt{n}}\sum_\beta\sum_\gamma\sum_\delta I_{\alpha\beta}(\theta_0)Z_{\gamma\delta}(\theta_0)$$

$$\cdot (u_\beta+\hat{u}_\beta)(u_\gamma u_\delta+U_\gamma U_\delta+2u_\gamma\hat{u}_\delta+2U_\gamma\hat{u}_\delta+2U_\gamma u_\delta)$$

$$+\frac{1}{6\sqrt{n}}\sum_\beta\sum_\gamma\sum_\delta\sum_\xi I_{\alpha\beta}(\theta_0)\rho_{\gamma\delta\xi}(\theta_0)(u_\beta+\hat{u}_\beta)(u_\gamma u_\delta u_\xi+U_\gamma U_\delta U_\xi$$

$$+3u_\gamma u_\delta\hat{u}_\xi+3u_\beta u_\gamma U_\xi+3U_\gamma U_\delta u_\xi+3U_\gamma U_\delta\hat{u}_\xi+6u_\gamma\hat{u}_\delta U_\xi)$$

$$+\frac{1}{\sqrt{n}}\frac{\pi'_\alpha(\theta_0)}{\pi(\theta_0)}+\frac{1}{\sqrt{n}}\sum_\beta Z_{\alpha\beta}(\theta_0)u_\beta+\frac{1}{\sqrt{n}}\sum_\beta Z_{\alpha\beta}(\theta_0)U_\beta-\frac{1}{2\sqrt{n}}$$

$$\cdot \sum_\beta\sum_\gamma \rho_{\alpha\beta\gamma}(\theta_0)(u_\beta u_\gamma+U_\beta U_\gamma+2u_\beta\hat{u}_\gamma+2U_\beta u_\gamma+2U_\beta\hat{u}_\gamma)\Big]du+o_p\Big(\frac{1}{\sqrt{n}}\Big)$$

$$=-M\sum_\beta I_{\alpha\beta}(\theta_0)\hat{u}_\beta-\frac{1}{\sqrt{n}}\sum_\beta\sum_\gamma I_{\alpha\beta}(\theta_0)\frac{\pi'_\gamma(\theta_0)}{\pi(\theta_0)}P_{\beta\gamma}-\frac{1}{\sqrt{n}}\sum_\beta\sum_\gamma\sum_\delta I_{\alpha\beta}(\theta_0)$$

$$\cdot Z_{\gamma\delta}(\theta_0)U_\gamma P_{\beta\delta}+\frac{1}{6\sqrt{n}}\sum_\beta\sum_\gamma\sum_\delta\sum_\xi I_{\alpha\beta}(\theta_0)\rho_{\gamma\delta\xi}(\theta_0)(Q_{\beta\gamma\delta\xi}+3U_\gamma U_\delta P_{\beta\xi})$$

$$+\frac{\pi'_\alpha(\theta_0)}{\sqrt{n}\,\pi(\theta_0)}M+\frac{1}{\sqrt{n}}\sum_\beta U_\beta Z_{\alpha\beta}(\theta_0)M-\frac{1}{2\sqrt{n}}\sum_\beta\sum_\gamma \rho_{\alpha\beta\gamma}(\theta_0)$$

$$\cdot (P_{\beta\gamma}+U_\beta U_\gamma M)+\sum_\beta\sum_\gamma\sum_\delta I_{\gamma\beta}(\theta_0)I_{\alpha\delta}(\theta_0)P_{\delta\gamma}\hat{u}_\beta+o_p\Big(\frac{1}{\sqrt{n}}\Big)$$

$$(\alpha=1,\cdots,p)\,.$$

Using a matrix representation we obtain

$$(IPI-MI)\hat{u}=\frac{1}{\sqrt{n}}(PI-ME)\pi^*+\frac{1}{\sqrt{n}}L-\frac{1}{2\sqrt{n}}(PI-ME)V$$

$$+\frac{1}{\sqrt{n}}(PI-ME)W+o_p\Big(\frac{1}{\sqrt{n}}\Big)\,,$$

where E is an unit matrix, and L, V and W are column vectors with

$$\pi^*=\Big(\frac{\pi'_\alpha(\theta_0)}{\pi(\theta_0)}\Big)\,;$$

$$L=\Big(-\frac{1}{6}\sum_\beta\sum_\gamma\sum_\delta\sum_\xi I_{\alpha\beta}(\theta_0)\rho_{\gamma\delta\xi}(\theta_0)Q_{\beta\gamma\delta\xi}+\frac{1}{2}\sum_\beta\sum_\gamma \rho_{\alpha\beta\gamma}(\theta_0)P_{\beta\gamma}\Big)\,;$$

$$V=(\sum_\beta\sum_\gamma \rho_{\alpha\beta\gamma}(\theta_0)U_\beta U_\gamma)\,;$$

$$W=(\sum_\beta U_\beta Z_{\beta\gamma}(\theta_0))\,.$$

Since it is derived from (vi) that the matrix $IPI-MI$ is positive definite, it follows that

$$(4)\qquad \hat{u}=\frac{1}{\sqrt{n}}I^{-1}\pi^*+\frac{1}{\sqrt{n}}(IPI-MI)^{-1}L-\frac{1}{2\sqrt{n}}I^{-1}V$$

$$+\frac{1}{\sqrt{n}}I^{-1}W+o_p\left(\frac{1}{\sqrt{n}}\right).$$

Hence we have

$$(5) \qquad \hat{t}_\alpha=\hat{u}_\alpha+U_\alpha \qquad (\alpha=1,\cdots,p),$$

where $\hat{u}=(\hat{u}_1,\cdots,\hat{u}_p)'$ is given by (4). Since $\hat{t}=\sqrt{n}(\hat{\theta}-\theta_0)$, we modify $\hat{\theta}$ to be second order AMU and denote by $\hat{\theta}^*$. From Theorem 1, (4) and (5) it follows that the MLE $\hat{\theta}^*_{ML}$ is asymptotically equivalent to the generalized Bayes estimator $\hat{\theta}^*$ up to order $n^{-1/2}$. By Theorem 2 it is seen that $\hat{\theta}^*$ is second order asymptotically efficient in the class $\boldsymbol{A}_2$.

We have defined in [4], [11] and [12] the class $\boldsymbol{D}$ as the set of the all third order AMU estimators $\hat{\theta}$ satisfying the following:

(a) $\hat{\theta}$ is asymptotically expanded as

$$\sqrt{n}(\hat{\theta}-\theta)=U+\frac{1}{\sqrt{n}}Q+o_p\left(\frac{1}{\sqrt{n}}\right)$$

and $Q_\alpha=O_p(1)$ $(\alpha=1,\cdots,p)$ and $\mathrm{E}(U_\alpha Q_\beta^k)=o(1)$ $(k=1,2)$ for all $\alpha,\beta=1,\cdots,p$, where E denotes the asymptotic expectation of $U_\alpha Q_\beta^k$ with

$$U_\alpha=\frac{1}{\sqrt{n}}\sum_\beta I^{\alpha\beta}\frac{\partial}{\partial\theta_\beta}\sum_{i=1}^{n}\log f(X_i,\theta) \qquad (\alpha=1,\cdots,p);$$

(b) The joint distribution of $\hat{\theta}$ admits Edgeworth expansion.
It follows from (4) and (5) that the generalized Bayes estimator $\hat{\theta}^*$ belongs to the class $\boldsymbol{D}$. Then the asymptotic marginal distribution of $\hat{\theta}^*$ is equivalent to that of the MLE $\hat{\theta}^*_{ML}$ up to order n^{-1}. Since $\hat{\theta}^*_{ML}$ is third order asymptotically efficient in the class $\boldsymbol{D}$ ([4], [10], [11]), $\hat{\theta}^*$ is also so. Hence we have established:

THEOREM 3. *Under the assumptions* (i)–(viii), *the generalized Bayes estimator $\hat{\theta}^*$ is second order asymptotically efficient in the class $\boldsymbol{A}_2$ and also third order asymptotically efficient in the class $\boldsymbol{D}$.*

UNIVERSITY OF TOKYO
UNIVERSITY OF ELECTRO-COMMUNICATIONS

REFERENCES

[1] Akahira, M. (1975). Asymptotic theory for estimation of location in non-regular cases, I: Order of convergence of consistent estimators, *Rep. Statist. Appl. Res., JUSE*, **22**, 8–26.

[2] Akahira, M. and Takeuchi, K. (1976). On the second order asymptotic efficiency of estimators in multiparameter cases, *Rep. Univ. Electro-Comm.*, **26**, 261–269.

ASYMPTOTIC OPTIMALITY OF THE GENERALIZED BAYES ESTIMATOR   415

[3] Akahira, M. and Takeuchi, K. (1979). Discretized likelihood methods—Asymptotic properties of discretized likelihood estimators (DLE's), *Ann. Inst. Statist. Math.*, **31**, A, 39–56.

[4] Akahira, M. and Takeuchi, K. (1979). The Concept of Asymptotic Efficiency and Higher Order Asymptotic Efficiency in Statistical Estimation Theory, Lecture Note.

[5] Gusev, S. I. (1975). Asymptotic expansions associated with some statistical estimators in the smooth case 1. Expansions of random variables, *Theory Prob. Appl.*, **20**, 470–498.

[6] Pfanzagl, J. and Wefelmeyer, W. (1978). A third-order optimum property of maximum likelihood estimator, *J. Multivariate Anal.*, 8, 1–29.

[7] Pfanzagl, J. and Wefelmeyer, W. (1979). Addendum to "A third-order optimum property of the maximum likelihood estimator", *J. Multivariate Anal.*, **9**, 179–182.

[8] Strasser, H. (1977). Asymptotic expansions for Bayes procedures, *Recent Development in Statistics* (ed. J. R. Barra et al.), North-Holland, 9–35.

[9] Takeuchi, K. and Akahira, M. (1976). On the second order asymptotic efficiencies of estimators, *Proceedings of the Third Japan-USSR Symposium on Probability Theory* (eds. G. Maruyama and J. V. Prokhorov), Lecture Notes in Mathematics 550, Springer-Verlag, Berlin, 604–638.

[10] Takeuchi, K. and Akahira, M. (1978). Third order asymptotic efficiency of maximum likelihood estimator for multiparameter exponential case, *Rep. Univ. Electro-Comm.*, **28**, 271–293.

[11] Takeuchi, K. and Akahira, M. (1978). On the asymptotic efficiency of estimators, (in Japanese), A report of the Symposium on Various Problems of Asymptotic Theory, Annual Meeting of the Mathematical Society of Japan, 1–24.

[12] Takeuchi, K. and Akahira, M. (1978). Asymptotic optimality of the generalized Bayes estimator, *Rep. Univ. Electro-Comm.*, **29**, 37–45.

[13] Takeuchi, K. and Akahira, M. (1979). Note on non-regular asymptotic estimation—What "non-regularity" implies, *Rep. Univ. Electro-Comm.*, **30**, 63–66.

[14] Takeuchi, K. and Akahira, M. (1979). Third order asymptotic efficiency of maximum likelihood estimator in general case, (to appear).

Rep. Univ. Electro-Comm. 30–1, (Sci. & Tech. Sect.), pp. 63—66 August, 1979

Note on Non-Regular Asymptotic Estimation
—— What "Non-Regularity" Implies[*]

Kei TAKEUCHI[**] and Masafumi AKAHIRA[***]

Abstract

In the asymptotic theory of statistical estimation the authors tried to compare the regular versus non-regular situation to clarify the significance and implication of each of the regularity conditions. This paper summarizes the results far obtained.

1. Introduction

Asymptotic theory of estimation is usually based on a set of regularity conditions. Regular case has been studied since Cramér [5] to Huber [6], and Pfanzagl [7] among others. Non-regular case has been studied by various authors at various occasions and Weiss and Wolfowitz [9] gave general results covering non-regular cases. The authors tried to compare the regular versus non-regular situation to clarify the significance and implication of each of the regularity conditions ([3]). In this paper we summarize the results thus far obtained.

2. Results

Suppose that $X_1, X_2, \cdots, X_n, \cdots$ are a sequence of i.i.d. random variables whose distribution F_ξ depends on an unknown parameter $\xi \in \Xi \subset R^p$. A parameter to be estimated is considered to be a real-valued function $\theta = \theta(\xi)$ of ξ.

In the regular case it is assumed that

(i) The distribution F_ξ is dominated by some measure μ with the density f_ξ;

(ii) The support, i.e. the set $A_\xi = \{x : f_\xi(x) > 0\}$ is independent of ξ;

(iii) $f_\xi(x)$ is continuous in ξ for almost all x $[\mu]$;

(iv) $f_\xi(x)$ is continuously differentiable (up to specified order) in ξ for almost all x $[\mu]$;

(v) The information matrix $I_\xi = E\left[\left\{\frac{\partial}{\partial \xi} \log f_\xi(x)\right\}\left\{\frac{\partial}{\partial \xi} \log f_\xi(x)\right\}'\right]$ exists.

Moreover it is usually assumed that (vi) $\theta(\xi)$ is continuously differentiable (up to specified order) in ξ. Then with some other set of regularity conditions it is shown that if the maximum likelihood estimator (m.l.e.) is $\hat{\theta} = \theta(\hat{\xi})$ whose $\hat{\xi}$ is the m.l.e. of ξ, $\sqrt{n}(\hat{\theta} - \theta)$ is asymptotically normal with mean 0 and variance $\left(\frac{\partial \theta}{\partial \xi}\right)' I^{-1}\left(\frac{\partial \theta}{\partial \xi}\right)$ and $\hat{\theta}$ is asymptotically efficient, and it is also shown to be second order asymptotically efficient.

[*] Received on May 30, 1979.
  This paper was submitted to the Second Vilnius Conference on Probability Theory and Mathematical Statistics, 1977 to which the authors could not attend nor present it.
[**] University of Tokyo
[***] Statistical Laboratory, University of Electro-Communications

64   Kei TAKEUCHI and Masafumi AKAHIRA

The conclusion involves following statements :

(a) The order of convergence of estimators is $n^{1/2}$;

(b) Asymptotic bound for the asymptotically efficient estimators is given by the normal distribution.

(c) The m.l.e. asymptotically attains uniformly that bound ;

(d) It also attains the bound up to the order $n^{-1/2}$ (Pfanzagl [7], Takeuchi and Akahira [8]).

If either of the above assumptions (i)$\sim$(v) fails to hold, some of the conclusions above may not necessarily hold true.

(1) In the undominated case, if the family of the probability distributions is decomposed into dominated subclasses such that

$$\varXi = \bigcup_\alpha \varXi_\alpha \ , \ \ \varXi_\alpha \frown \varXi_\beta = \phi \ \ (\alpha \neq \beta)$$

and correspondingly the sample space $\mathscr{X}$ is also decomposed.

$$\mathscr{X} = \bigcup_\alpha \mathscr{X}_\alpha$$

and $\quad Pr\{X \in \mathscr{X}_\alpha | \theta\} = 1 \quad$ if $\quad \theta \in \varXi_\alpha$;

$\qquad Pr\{X \in \mathscr{X}_\alpha | \theta\} = 0 \quad$ if $\quad \theta \notin \varXi_\alpha,$

then each subclass $\varXi_\alpha$ can be treated separately. For other undominated case little, if any, has been known.

(2) If the support depends on ξ, the order of convergence of estimators is not necessarily to be $n^{1/2}$. Assuming smoothness of $\theta(\xi)$ the order usually depends on the power α defined by

$$1 - P_{\xi'}\{A_\xi\} = O(\|\xi - \xi'\|^\alpha)$$

as ξ' approaches to ξ, where $P_{\xi'}$ is the probability measure with the density $f_{\xi'}$. If $f_\xi(x)$ is sufficiently regular, the maximum order of convergence is

$$n^{1/\alpha} \qquad \text{when} \qquad 0 < \alpha < 2 \ ;$$

$$(n \log n)^{1/2} \qquad \text{when} \qquad \alpha = 2 \qquad ;$$

$$n^{1/2} \qquad \text{when} \qquad \alpha > 2 \qquad ;$$

(Akahira [1]).

(3) Discontinuity of $f_\xi(x)$ in ξ plays similar role as the non-identical support (actually the later can be dealt with as the case when $f_\xi(x) = 0$ for some x).

(4) When the support depends on $\varXi$, an (uniformly) asymptotically efficient estimator may not exist, and m.l.e. may not be asymptotically good.

In the case of convergence of order $n^{1/\alpha}(\alpha < 2)$ m.l.e. is not asymptotically efficient, nor even asymptotically admissible: Suppose that

$$f(x-\theta) = \begin{cases} c \exp\left\{-\dfrac{1}{2}(x-\theta)^2\right\} & \text{if} \ \ |x-\theta| < 1 \ ; \\ 0 & \text{otherwise,} \end{cases}$$

where c is some constant.

Then the m.l.e. is asymptotically equivalent to

$$\min_i X_i + 1 \quad \text{with probability } 1/2 \ ,$$

$$\max_i X_i - 1 \quad \text{with probability } 1/2 ,$$

while the estimator $\hat{\theta}^* = (\min X_i + \max X_i)/2$ is asymptotically uniformly more concentrated. (The asymptotic distributions of the two are equal except for the scale being twice as large for the m. l. e. as for $\hat{\theta}^*$). In this case there exists one-sided asymptotically efficient estimator, but no estimator is asymptotically uniformly efficient. If we define two-sided efficiency by

$$\inf_{\hat{\theta}} \lim_{n \to \infty} Pr\{|\hat{\theta} - \theta| < t\}$$

together with the asymptotically median unbiased (a. m. u.) condition of $\hat{\theta}$ (Takeuchi and Akahira [8]), then $\hat{\theta}^*$ above is uniformly (in t) asymptotically two-side efficient.

Also Weiss and Wolfowitz's maximum probability estimator is uniformly inferior to $\hat{\theta}^*$ (Akahira and Takeuchi [3], [4]).

When the order of convergence is $(n \log n)^{1/2}$ and the bound of the asymptotic distributions is normal, then m. l. e. is asymptotically efficient.

(5) If the density is not smooth enough, while Fisher information is well defined, then m. l. e. is asymptotically efficient but not second order (that is in the order $n^{-1/2}$) asymptotically efficient as in the regular case : Suppose that

$$f(x - \theta) = \frac{1}{2} \exp\{-|x - \theta|\}$$

The asymptotic distribution of the m. l. e. $\hat{\theta}$ is given by

$$\lim_{n \to \infty} \sqrt{n} \left| Pr\{\sqrt{n}(\hat{\theta} - \theta) \leq t\} - \Phi(t) - \frac{t^2}{2\sqrt{n}} \phi(t) sgn(t) \right| = 0,$$

where $\Phi(u) = \int_{-\infty}^{u} \phi(t) \, dt$ with $\phi(t) = \frac{1}{\sqrt{2\pi}} e^{-t^2/2}$, while the bound of the asymptotic distributions of a. m. u. estimators is given by

$$\Phi(t) + \frac{t^2}{6\sqrt{n}} \phi(t) sgn(t).$$

which can not be uniformly attained (Takeuchi and Akahira [8]) but is attained for a specified t_0 by an estimator $\tilde{\theta}$ satisfying

$$-\sum_i |X_i - \tilde{\theta}| + \sum_i \left| X_i - \tilde{\theta} + \frac{t_0}{n} \right| - c = 0,$$

where c is determined so that $\tilde{\theta}$ be a. m. u. (Akahira and Takeuchi [2], [3]).

(6) When the Fisher information is infinity, the asymptotic distribution can be obtained if the distribution of $(\partial/\partial\theta)\log f(X, \theta)$ belongs to the attraction of some stable law, and m. l. e. may or may not be asymptotically efficient : Suppose that

$$X_i = (X_{1i}, X_{2i}) \quad (i = 1, 2, \cdots) ;$$
$$X_{2i} = \theta X_{1i} + U_i \quad (i = 1, 2, \cdots) ,$$

where U_i's are i. i. d. with the density $f(u)$ for which $I = \int \{f'(u)\}^2/f(u) du < \infty$ and $E(U_i) = 0$ $(i = 1, 2, \cdots)$. The Fisher information is $E(X_{1i}^2)I$, hence is infinity $E(X_{1i}^2) = \infty$. If X_i belongs to the non-normal attraction domain of normal law, then the order of maximum convergence is $(n \log n)^{1/2}$ and m. l. e. is asymptotically efficient, and when X_i belongs to the normal attraction domain of the stable law with characteristic exponent $\alpha(<2)$, then the maximum of convergence is $n^{1/\alpha}$, and the m. l. e. attains the order but not asymptotic efficiency.

The above examples are chosen to illustrate the situation which usually happens in more general cases.

References

[1] Akahira, M.: "Asymptotic theory for estimation of location in non-regular cases, I: Order of convergence of consistent estimators, Rep. Stat. Appl. Res., JUSE, **22**, 8–26 (1975).

[2] Akahira, M. and Takeuchi, K.: "Discretized likelihood methods——Asymptotic properties of discretized likelihood estimators (DLE's)," Ann. Inst. Statist. Math. **31**, Part A, 39–56 (1979).

[3] Akahira, M. and Takeuchi, K.: "The Concept of Asymptotic Efficiency and Higher Order Asymptotic Efficiency in Statistical Estimation Theory," Lecture Note (1979).

[4] Akahira, M. and Takeuchi, K.: "Remarks on the asymptotic efficiency and inefficiency of maximum probability estimators," (to appear).

[5] Cramér, H.: "Mathematical Methods of Statistics," Princeton University Press (1946).

[6] Huber, P.: "The behavior of maximum likelihood estimates under non-standard conditions," Proc. Fifth Berkeley Symp. on Math. Statist. Prob. **1**, 221–233 (1967).

[7] Pfanzagl, J.: "Asymptotic expansions related to minimum contrast estimators," Ann. Statist., **1**, 993–1026 (1973).

[8] Takeuchi, K. and Akahira, M.: "On the second order asymptotic efficiencies of estimators," Proc. Third Japan–USSR Symp. on Probability Theory. Lecture Notes in Mathematics **550**, Springer Verlag, 604–638 (1976).

[9] Weiss, L. and Wolfowitz, J.: "Maximum Probability Estimators and Related Topics," Lecture Notes in Mathematics **424**, Springer Verlag (1974).

Austral. J. Statist., **22** (3), 1980, 332–335

A NOTE ON PREDICTION SUFFICIENCY (ADEQUACY) AND SUFFICIENCY[1]

MASAFUMI AKAHIRA AND KEI TAKEUCHI

University of Electro-Communications and University of Tokyo

Summary

Let $(\mathcal{X}, \mathcal{A})$ be a measurable space and $\mathcal{P}$ a family of probability measures on $\mathcal{A}$. Let $\mathcal{B}$ and $\mathcal{C}$ be sub σ-algebras of $\mathcal{A}$ and $\mathcal{B}_0$ a sub σ-algebra of $\mathcal{B}$. It is shown that if $\mathcal{B}_0$ is prediction sufficient (adequate) for $\mathcal{B}$ with respect to $\mathcal{C}$ and $\mathcal{P}$, and $\mathcal{S}$ is sufficient for $\mathcal{B}_0^{\vee}\mathcal{C}$ with respect to $\mathcal{P}$ then $\mathcal{S}$ is sufficient for $\mathcal{B}^{\vee}\mathcal{C}$ with respect to $\mathcal{P}$; that if $\mathcal{P}$ is homogeneous and $(\mathcal{B}_0; \mathcal{B}, \mathcal{C})$ is Markov for $\mathcal{P}$, and $\mathcal{B}_0^{\vee}\mathcal{C}$ is sufficient for $\mathcal{B}^{\vee}\mathcal{C}$ with respect to $\mathcal{P}$, then $\mathcal{B}_0$ is sufficient for $\mathcal{B}$ with respect to $\mathcal{P}$; and by example that the Markov property is necessary for the latter proposition to hold.

1. Introduction

Prediction sufficiency (adequacy) was defined and discussed by, among others, Skibinsky (1967), Takeuchi & Akahira (1975), and Torgersen (1977). The purpose of this note is to show the relation with prediction sufficiency and the "usual" concept of sufficiency in the following situation. Suppose that the observations are given sequentially in time in two sets. Let (X, Y) be a set of observations where X is observed first and Y comes later. In such a situation one may be interested to know how much information in X should be reserved to be later combined with Y without any loss of information in the combined sample. Technically it amounts to giving the condition for the statistic $T = t(X)$, that the combined statistic (T, Y) be sufficient for (X, Y). It is shown that it is sufficient that T be prediction sufficient for Y, and it is also necessary under further conditions. We may further add that in this formulation the distribution of Y can be dependent on X in a general manner so that we can include the sequential design case when the design (that is observation scheme) of Y is determined depending on X.

We will, essentially, use the framework of Skibinsky (1967). The notion of adequacy in Skibinsky's paper is, however, replaced by the notion of prediction sufficiency.

[1] Manuscript received September 13, 1978; revised January 16, 1979.

2. Results

Let $(\mathcal{X}, \mathcal{A})$ be a measurable space and $\mathcal{P}$ a family of probability measures on $\mathcal{A}$. We denote by $\mathcal{A}_1 \vee \mathcal{A}_2$ the smallest σ-algebra which contains each member of two subclasses $\mathcal{A}_1$, $\mathcal{A}_2$ of $\mathcal{A}$. Let χ_A denote the indicator function of a set A. $\mathcal{B}$ and $\mathcal{C}$ denote sub σ-algebras of $\mathcal{A}$, and $\mathcal{B}_0$ denotes a sub σ-algebra of $\mathcal{B}$.

Definition 1. $\mathcal{B}_0$ is said to be sufficient for $\mathcal{B}$ with respect to $\mathcal{P}$, symbolically

$$\mathcal{B}_0 \text{ suf } (\mathcal{P}; \mathcal{B}),$$

if for each $B \in \mathcal{B}$ there exists a $\mathcal{B}_0$-measurable function $\phi_B^{\mathcal{B}_0}$ such that for every $p \in \mathcal{P}$

$$\phi_B^{\mathcal{B}_0} = E_p^{\mathcal{B}_0} \chi_B \quad \text{a.e. } [p],$$

where $E_p^{\mathcal{B}_0} \chi_B$ denotes the conditional expectation under p of χ_B given $\mathcal{B}_0$.

Definition 2. (Skibinsky, 1967; Takeuchi & Akahira, 1975). $\mathcal{B}_0$ is said to be prediction sufficient (adequate) for $\mathcal{B}$ with respect to $\mathcal{C}$ and $\mathcal{P}$, symbolically

$$\mathcal{B}_0 \text{ pred. suf } (\mathcal{P}; \mathcal{B}, \mathcal{C})$$

if $\mathcal{B}_0 \text{ suf } (\mathcal{P}; \mathcal{B})$ and $(\mathcal{B}_0; \mathcal{B}, \mathcal{C})$ is Markov for $\mathcal{P}$, that is, $\mathcal{B}$ and $\mathcal{C}$ are conditionally independent given $\mathcal{B}_0$ for each $p \in \mathcal{P}$.

Lemma. *Let $(\mathcal{X}, \mathcal{A}, p)$ be a probability space. $(\mathcal{B}_0; \mathcal{B}, \mathcal{C})$ is Markov for p if and only if for every $B \in \mathcal{B}$ and every $C \in \mathcal{C}$*

$$E_p^{\mathcal{B}_0 \vee \mathcal{C}}(\chi_{B \cap C}) = \chi_C E_p^{\mathcal{B}_0}(\chi_B) \quad \text{a.e. } [p].$$

Proof. Suppose that $(\mathcal{B}_0, \mathcal{B}, \mathcal{C})$ is Markov for p. For every $B \in \mathcal{B}$, every $C \in \mathcal{C}$, every $B_0 \in \mathcal{B}_0$ and every $C' \in \mathcal{C}$

$$\int_{B_0 \cap C'} E_p^{\mathcal{B}_0 \vee \mathcal{C}} \chi_{B \cap C} \, dp = \int_{B_0 \cap C'} \chi_{B \cap C} \, dp$$
$$= E_p(\chi_{B_0 \cap B} \chi_{C' \cap C}) = E_p[E_p^{\mathcal{B}_0}(\chi_{B_0 \cap B} \chi_{C' \cap C})]$$
$$= E_p[(E_p^{\mathcal{B}_0} \chi_{B_0 \cap B})(E_p^{\mathcal{B}_0} \chi_{C' \cap C})]$$
$$= E_p[E_p^{\mathcal{B}_0}(\chi_{C' \cap C} E_p^{\mathcal{B}_0} \chi_{B_0 \cap B})]$$
$$= E_p[\chi_{C' \cap C} E_p^{\mathcal{B}_0} \chi_{B_0 \cap B}] = E_p[\chi_{B_0 \cap C' \cap C} E_p^{\mathcal{B}_0} \chi_B]$$
$$= \int_{B_0 \cap C'} \chi_C E_p^{\mathcal{B}_0} \chi_B \, dp.$$

Since $\{B_0 \cap C \mid B_0 \in \mathcal{B}_0, \ C \in \mathcal{C}\}$ generates the σ-algebra $\mathcal{B}_0 \vee \mathcal{C}$, it follows that for every $B \in \mathcal{B}$ and every $C \in \mathcal{C}$

$$E_p^{\mathcal{B}_0 \vee \mathcal{C}} \chi_B = \chi_C E_p^{\mathcal{B}_0} \chi_B \quad \text{a.e. } [p].$$

334   MASAFUMI AKAHIRA AND KEI TAKEUCHI

Conversely, if for every $B \in \mathscr{B}$ and every $C \in \mathscr{C}$

$$E_p^{\mathscr{B}_0^{\vee}\mathscr{C}} \chi_B = \chi_C E_p^{\mathscr{B}_0} \chi_B \quad \text{a.e. } [p],$$

then for every $B \in \mathscr{B}$ and every $C \in \mathscr{C}$

$$E_p^{\mathscr{B}_0}(\chi_{B \cap C}) = E_p^{\mathscr{B}_0}(E_p^{\mathscr{B}_0^{\vee}\mathscr{C}} \chi_{B \cap C})$$
$$= E_p^{\mathscr{B}_0}(\chi_C E_p^{\mathscr{B}_0} \chi_B)$$
$$= E_p^{\mathscr{B}_0} \chi_B E_p^{\mathscr{B}_0} \chi_C \quad \text{a.e. } [p].$$

Hence $(\mathscr{B}_0; \mathscr{B}, \mathscr{C})$ is Markov for p.

Theorem 1. *Let $\mathscr{S}$ be a sub σ-algebra of $\mathscr{B}_0^{\vee}\mathscr{C}$. Suppose that $\mathscr{B}_0$ pred. suf $(\mathscr{P}; \mathscr{B}, \mathscr{C})$ and $\mathscr{S}$ suf $(\mathscr{P}; \mathscr{B}_0^{\vee}\mathscr{C})$. Then $\mathscr{S}$ suf $(\mathscr{P}; \mathscr{B}^{\vee}\mathscr{C})$.*

Proof. For every $B \in \mathscr{B}$, every $C \in \mathscr{C}$ and every $p \in \mathscr{P}$

$$E_p^{\mathscr{S}} \chi_{B \cap C} = E_p^{\mathscr{S}}(E_p^{\mathscr{B}_0^{\vee}\mathscr{C}} \chi_{B \cap C})$$
$$= E_p^{\mathscr{S}}(\chi_C E_p^{\mathscr{B}_0} \chi_B) \quad \text{(by Lemma)}$$
$$= E_p^{\mathscr{S}}(\chi_C \phi_B^{\mathscr{B}_0}) \text{ (by } \mathscr{B}_0 \text{ suf } (\mathscr{P}; \mathscr{B})) \text{ a.e. } [p].$$

Since $\chi_C \phi_B^{\mathscr{B}_0}$ is a $\mathscr{B}_0^{\vee}\mathscr{C}$-measurable function and $\mathscr{S}$ suf $(\mathscr{P}; \mathscr{B}_0^{\vee}\mathscr{C})$, it follows that there exists a $\mathscr{S}$-measurable function $\phi_{B \cap C}^{\mathscr{S}}$ such that for every $p \in \mathscr{P}$

$$\phi_{B \cap C}^{\mathscr{S}} = E_p^{\mathscr{S}}(\chi_C \phi_B^{\mathscr{B}_0}) = E_p^{\mathscr{S}} \chi_{B \cap C} \quad \text{a.e. } [p].$$

Since $\{B \cap C \mid B \in \mathscr{B}, C \in \mathscr{C}\}$ generates $\mathscr{B}^{\vee}\mathscr{C}$, it is seen that $\mathscr{S}$ suf $(\mathscr{P}; \mathscr{B}^{\vee}\mathscr{C})$.

For Theorem 2 we assume that $\mathscr{P}$ is homogeneous, that is, p and q are mutually absolutely continuous for every p and q in $\mathscr{P}$.

Theorem 2. *Suppose that $(\mathscr{B}_0; \mathscr{B}, \mathscr{C})$ is Markov for $\mathscr{P}$ and $\mathscr{B}_0^{\vee}\mathscr{C}$ suf $(\mathscr{P}; \mathscr{B}^{\vee}\mathscr{C})$. Then $\mathscr{B}_0$ suf $(\mathscr{P}; \mathscr{B})$.*

Proof. Since $\mathscr{P}$ is homogeneous and $\mathscr{B}_0^{\vee}\mathscr{C}$ suf $(\mathscr{P}; \mathscr{B}^{\vee}\mathscr{C})$ and $(\mathscr{B}_0; \mathscr{B}, \mathscr{C})$ is Markov for $\mathscr{P}$, it follows by the Lemma that for every $B \in \mathscr{B}$ there exists a $\mathscr{B}_0$-measurable function $\phi_B^{\mathscr{B}_0}$ such that for every $p \in \mathscr{P}$

$$E_p^{\mathscr{B}_0}(\chi_B) = E_p^{\mathscr{B}_0}(E_p^{\mathscr{B}_0^{\vee}\mathscr{C}} \chi_B) = E_p^{\mathscr{B}_0}(\phi_B^{\mathscr{B}_0}) = \phi_B^{\mathscr{B}_0} \quad \text{a.e. } [p].$$

Hence $\mathscr{B}_0$ suf $(\mathscr{P}; \mathscr{B})$.

Remark. It is shown by the following counter-example that the homogeneity of $\mathscr{P}$ is necessary for Theorem 2. Suppose that $\mathscr{B} = \mathscr{C} = \{\phi, B, B^c, \mathscr{X}\}$, $\mathscr{B}_0 = \{\phi, \mathscr{X}\}$ and $\mathscr{P} = \{p, q\}$ with $p(B) = q(B^c) = 1$. Then $(\mathscr{B}_0; \mathscr{B}, \mathscr{C})$ is Markov for $\mathscr{P}$ and $\mathscr{B}_0^{\vee}\mathscr{C}$ suf $(\mathscr{P}; \mathscr{B}^{\vee}\mathscr{C})$ but $\mathscr{B}_0$ is not sufficient for $\mathscr{B}$ with respect to $\mathscr{P}$.

The following example shows that the Markov property is necessary for Theorem 2 to hold. Suppose that (X, Y, Z) are jointly normally distributed with means $(\theta, 0, \theta)$ where θ is the unknown parameter and with known variances and convariances to be later specified. Let $\mathcal{B}_0$, $\mathcal{B}$ and $\mathcal{C}$ be the σ-algebras induced by, respectively $X, (X, Y)$ and Z. Let us assume that Y can be expressed as $Y = X - Z + U$, where U is independent of X and Z. We further assume that $\text{Cov}(X, Z) = 0$ and $V(X) = V(Z) = V(U) = 1$, where Cov and V denote covariance and variance, respectively. Then it follows that $E(U) = 0$, $V(Y) = 3$, $\text{Cov}(Y, X) = 1$ and $\text{Cov}(Y, Z) = -1$. And given (X, Z), Y is conditionally normally distributed with mean $X - Z$ and variance 1, hence its conditional distribution is independent of θ. Therefore (X, Z) is sufficient for (X, Y, Z) or equivalently $\mathcal{B}_0^{\vee}\mathcal{C}$ is sufficient for $\mathcal{B}^{\vee}\mathcal{C}$. However, X is not sufficient for (X, Y), that is, $\mathcal{B}_0$ is not sufficient for $\mathcal{B}$, because given X, Y is conditionally normally distributed with mean $X - \theta$ and variance 2, which is dependent on θ. Nor are Y and Z conditionally independent given X, hence $(\mathcal{B}_0, \mathcal{B}, \mathcal{C})$ is not Markov.

References

Skibinsky, M. (1967). Adequate subfields and sufficiency. *Ann. Math. Statist.* **38,** 155–161.

Takeuchi, K. & Akahira, M. (1975). Characterizations of prediction sufficiency (adequacy) in terms of risk functions. *Ann. Statist.* **3,** 1018–1024.

Torgersen, E. N. (1977). Prediction sufficiency when the loss function does not depend on the unknown parameter. *Ann. Statist.* **5,** 155–163.

Rep. Univ. Electro-Comm. 31–1, (Sci. & Tech. Sect.), pp. 89—96   August, 1980

Third Order Asymptotic Efficiency and Asymptotic Completeness of Estimators*

Kei TAKEUCHI** and Masafumi AKAHIRA***

Abstract

In the previous paper it was shown that the maximum likelihood estimator (MLE) was third order asymptotically efficient in multiparameter exponential cases. In this paper it is shown that the result is extended to more general cases. The concept of asymptotic completeness of an estimator is introduced and it is shown that the MLE is higher order asymptotically complete in the appropiate classes.

1. Introduction

In the previous papers [6], [7] and [3] Takeuchi and Akahira have discussed the third order asymptotic efficiency of the maximum likelihood estimator (MLE) and the generalized Bayes estimator in multiparameter exponential cases. In this parer we show that it is possible to extend the result on the MLE to more general cases in a very straightforward manner by introducing appropriate classes of estimators. Further we introduce the concept of an estimator in the sense that it can beat asymptotically any other "regular" estimators by making appropriate transformations if necessary and show that the MLE is higher order asymptotically complete in the appropriate classes.

2. Third order asymptotic efficiency of maximum likelihood estimator in general cases

In order to generalize the results in the previous paper [6], it is necessary to define appropriate classes of estimators among which asymptotic distribution is to be compared. As was shown in Akahira and Takeuchi [3] it is possible to establish the second order asymptotic efficiency of the modified MLE within the class of second order AMU estimators without any further restriction of the class of estimators to be considered. On the contrary in the previous paper [6] we restricted the class of estimators to be of a special type of functions of the sufficient statistic which we called the extended regular estimators. As will be shown later we can not usually have third order asymptotically efficient estimators without such restriction. In more general cases where sufficient statistics of finite dimension do not exist, the argument in the previous paper [6] can not be applied directly. We shall establish the third order asymptotic efficiency of the modified MLE and also of the modified generalized Bayes estimator within the classes **C** and **D** of estimators which are defined below. For the simplicity's sake

* Received on June 10, 1980
** University of Tokyo
*** Statistical Laboratory, University of Electro-Communications

we shall first consider the real parameter (one dimensional) case, and that the sample is independently and identically distributed with all necessary regularity conditions to be specified later.

Let $(\mathcal{X}, \mathcal{B})$ be a sample space. We consider a family of probability measures on $\mathcal{B}$, $\mathcal{P}= \{P_\theta : \theta \in \Theta\}$, where the index set Θ is called a parameter space. We assume that Θ is an open set in a Euclidean 1-space R^1. Consider n-fold direct products $(\mathcal{X}^{(n)}, \mathcal{B}^{(n)})$ of $(\mathcal{X}, \mathcal{B})$ and corresponding product measures $P_{\theta,n}$ of P_θ. For each $n=1,,\ldots$, the points of $\mathcal{X}^{(n)}$ will be denoted by $\tilde{x}_n = (x_1, \ldots, x_n)$. An estimator of θ is defined to be a sequence $\{\hat{\theta}_n\}$ of $\mathcal{B}^{(n)}$-measurable functions $\hat{\theta}_n$ on $\mathcal{X}^{(n)}$ into Θ $(n=1, 2, \ldots)$. For simplicity we denote an estimator as $\hat{\theta}_n$ instead of $\{\hat{\theta}_n\}$. For an increasing sequence of positive numbers $\{c_n\}$ (c_n tending to infinity) an estimator $\hat{\theta}_n$ is called consistent with order $\{c_n\}$ (or $\{c_n\}$-consistent for short) if for every $\varepsilon>0$ and every $\vartheta \in \Theta$ there exists a sufficiently small positive number δ and a sufficiently large positive number L satisfying the following:

$$\varlimsup_{n\to\infty} \sup_{\theta\,:\,|\theta-\vartheta|<\delta} P_{\theta,n}\{c_n|\hat{\theta}_n-\theta|\geq L\} < \varepsilon \quad (\,[\,1\,]\,).$$

For each $k=1, 2, \ldots, a$ $\{c_n\}$-consistent estimator $\hat{\theta}_n$ is called to be k-th order asymptotically median unbiased (or k-th order AMU) if for any $\vartheta \in \Theta$ there exists a positive number δ such that

$$\lim_{n\to\infty} \sup_{\theta\,:\,|\theta-\vartheta|<\delta} c_n^{k-1}\left| P_{\theta,n}\{\hat{\theta}_n \leq \theta\} - \frac{1}{2}\right| = 0\,;$$

$$\lim_{n\to\infty} \sup_{\theta\,:\,|\theta-\vartheta|<\delta} c_n^{k-1}\left| P_{\theta,n}\{\hat{\theta}_n \geq \theta\} - \frac{1}{2}\right| = 0.$$

For $\hat{\theta}_n$ k-th order AMU, $G_0(t, \theta)+c_n^{-1}G_1(t, \theta)+ \ldots +c_n^{-(k-1)}G_{k-1}(t, \theta)$ is called to be the k-th order asymptotic distribution of $c_n(\hat{\theta}_n-\theta)$ (or θ_n for short) if

$$\lim_{n\to\infty} c_n^{k-1}| P_{\theta,n}\{c_n(\hat{\theta}_n-\theta)<t\} -G_0(t, \theta)-c_n^{-1}G_1(t, \theta)- \ldots -c_n^{-(k-1)}G_{k-1}(t, \theta)| = 0.$$

We denote that $G_i(t, \theta)$ $(i=1, \ldots, k-1)$ may be general absolutely continuous function, hence the asymptotic distribution for any fixed n may not be a distribution function.

Suppose that $\hat{\theta}_n$ is k-th order AMU and has the k-th order asymptotic distribution $G_0(t, \theta)+c_n^{-1}G_1(t, \theta)+ \ldots +c_n^{-(k-1)}G_{k-1}(t, \theta)$.

Letting $\theta_0(\in\Theta)$ be arbitrary but fixed we consider the problem of testing hypothesis $H^+: \theta =\theta_0+tc_n^{-1}(t>0)$ against $K: \theta=\theta_0$.

Put $\Phi_{1/2} = \{ \{\phi_n\} : E_{\theta_0+tc_n^{-1}, n}(\phi_n)=1/2+o(c_n^{-(k-1)}), 0\leq \phi_n(\tilde{x}_n)\leq 1$ for all $\tilde{x}_n\in\mathcal{X}^{(n)}$ $(n=1, 2, \ldots)\}$.

Putting $A_{\hat{\theta}_n, \theta_0} = \{c_n(\hat{\theta}_n-\theta_0)\leq t\}$, we have

$$\lim_{n\to\infty} P_{\theta_0+tc_n^{-1},\, n}(A_{\hat{\theta}_n, \theta_0}) = \lim_{n\to\infty} P_{\theta_0+tc_n^{-1}, n}\{\hat{\theta}_n\leq\theta_0+tc_n^{-1}\} = \frac{1}{2}.$$

Hence it is seen that a sequence $\{\chi_{A_{\hat{\theta}_n, \theta_0}}\}$ of the indicators of $A_{\hat{\theta}_n, \theta_0}$ $(n=1, 2, \ldots)$ belongs to $\Phi_{1/2}$. If

$$\sup_{\{\phi_n\}\in\Phi_{1/2}} \varlimsup_{n\to\infty} c_n^{k-1}\{E_{\theta_0,n}(\phi_n)-H_0^+(t, \theta_0)-c_n^{-1}H_1^+(t, \theta_0)- \ldots -c_n^{-(k-1)}H_{k-1}^+(t, \theta_0)\} = 0,$$

then we have

$$G_0(t, \theta_0)\leq H_0^+(t, \theta_0)\,;$$

and for any positive integer $j(\leq k)$ if $G_i(t, \theta_0)=H_i^+(t, \theta_0)$ $(i=0, \ldots, j-1)$, then

$$G_j(t, \theta_0)\leq H_j^+(t, \theta_0).$$

August, 1980   Third Order Asymptotic Efficiency and Asymptotic Completeness of Estimators   91

Consider next problem of testing hypothesis $H^-:\ \theta=\theta_0+tc_n^{-1}\ (t<0)$ against $K:\ \theta=\theta_0$. If

$$\inf_{\{\phi_n\}\in\Phi_{1/2}}\ \lim_{n\to\infty} c_n{}^{k-1}\{E_{\theta_0,n}(\phi_n)-H_0{}^-(t,\theta_0)-c_n{}^{-1}H_1{}^-(t,\theta_0)-\ldots-c_n{}^{-(k-1)}H_{k-1}{}^-(t,\theta_0)\}=0,$$

then we have

$$G_0(t,\theta_0)\geqq H_0{}^-(t,\theta_0)\ ;$$

and for any poistive integer $j(\leqq k)$ if $G_i(t,\theta_0)=H_i{}^-(t,\theta_0)\ (i=0,\ldots,j-1)$ then $G_j(t,\theta_0)\geqq H_j{}^-(t,\theta_0)$.

$\hat\theta_n$ is called to be k-th order asympotically efficient if the k-th order asymptotic distribution of it attains uniformly the bound of the k-th order asymptotic distributions of k-th order AMU estimators, that is, for each $\theta\in\Theta$

$$G_i(t,\theta)=\begin{cases}H_i{}^+(t,\theta) & \text{for}\quad t>0\ ;\\ H_j{}^-(t,\theta) & \text{for}\quad t<0,\end{cases}$$

$i=,\ldots,$ ([2], [3], [4], [5]). [Note that for $t=0$ we have $G_i(0,\theta)=H_i{}^+(0,\theta)=H_i{}^-(0,\theta)\ (i=0,\ldots,k-1)$ from the condition of k-th order asymptotically median unbiasedness.]

Thus if $\hat\theta_n{}^*$ is k-th order asymptotically efficient in the above sense, then for any k-th order AMU estimator $\hat\theta_n$ we have

$$\lim_{n\to\infty} c_n{}^{k-1}[P_{\theta,n}\{-a<c_n(\hat\theta_n{}^*-\theta)<b\}-P_{\theta,n}\{-a<c_n(\hat\theta_n-\theta)<b\}]\geqq0. \qquad (2.1)$$

for all $a,b>0$ and $\theta\in\Theta$. Note that the k-th order asymptotic efficiency defined in the above gives a sufficient condition but not a necessary condition (2.1).

We assume that for each $\theta\in\Theta$ P_θ is absolutely continuous with respect to a σ-finite measure μ. We denote a density $dP_\theta/d\mu$ by $f(x,\theta)$. Then the joint density of $(X_1,\ldots,X_n)$ is given by $\prod_{i=1}^{n} f(x_i,\theta)$. In this section we shall deal with the case when $c_n=\sqrt{n}$.

First we shall restrict our attention to the class of (first order) asymptotic efficient estimators which we shall call the class **A**. It has been well established that any estimator $\hat\theta_n$ in the class **A** can be expressed as

$$\sqrt{n}\,(\hat\theta_n-\theta)=\frac{1}{I(\theta)\sqrt{n}}\frac{\partial L_n}{\partial\theta}+o_p(1),$$

where $L_n=\sum_{i=1}^{n}\log f(X_i,\theta)$.

Secondly, we call the class of estimators which are second AMU and second order asymptotically efficient as class **B**. Then any estimator $\hat\theta_n$ which is second order AMU and for which the distribution of $\sqrt{n}\,(\hat\theta_n-\theta)$ admits the Edgeworth expansion up to the order $n^{-1/2}$ belongs to the class **B**.

Now we generalize the concept into one step further. We call the class **C** the class of estimators $\hat\theta_n$ which are third order AMU and for which the distribution of $\sqrt{n}\,(\hat\theta_n-\theta)$ admits the Edgeworth expansion up to the order n^{-1} and

$$\sqrt{n}\,(\hat\theta_n-\theta)=\frac{1}{I(\theta)\sqrt{n}}\frac{\partial L_n}{\partial\theta}+\frac{1}{\sqrt{n}}Q(\theta)+O_p\left(\frac{1}{n}\right),$$

where $Q(\theta)$ is a quantity of stochastic order 1.

We call the estimator $\hat\theta_n$ belongs to the class **D** if in the above we have

$$E\left[\frac{\partial L_n}{\partial\theta}Q(\theta)^2\right]=0,$$

                         Kei Takeuchi and Masafumi Akahira

where E stands for the asymptotic mean.

It is to be noted that Lemma 4.1 in the previous paper [6] can be generalized straight-forwardly as follows.

Lemma 2.1. Suppose that Z_θ is a function of $X_1, \ldots, X_n$ and θ and is differentiable in θ. Then

$$E_\theta(UZ_\theta) = \frac{1}{\sqrt{n}} \frac{\partial}{\partial \theta} E_\theta(Z_\theta) - \frac{1}{\sqrt{n}} E_\theta\left(\frac{\partial Z_\theta}{\partial \theta}\right),$$

where $U = \frac{\sqrt{n}}{I(\theta)} \frac{\partial}{\partial \theta} \log f(X, \theta)$, provided that the differentiation under the integral sign of E_θ (Z_θ) is allowed.

In the following discussions we assume that expectation of the relevant quantities always exist and the asymptotic value of the expectation and the asymptotic mean are the same at least up to the required order.

Remark: The problem of non-existence of the relevant moments vis-à-vis asymptotic moments can be evaded in the following way. Suppose that we have a smooth and montone increasing transformation $\eta = \psi(\theta)$ so that η is bounded. Instead of θ we may consider η as the parameter to be estimated. Since η is bounded we may assume that all the estimators are also bounded hence have moments of any order, and we can easily formulate the condition for the asymptotic moments and the moments of the asymptotic distribution being equal. If we consider the estimation of η instead of θ the asymptotic theory based on moments is applicable in a straightforward way. Transforming back to the estimator of θ by $\hat{\theta} = \psi^{-1}(\hat{\eta})$ and consider-ing the monotoniticy of the transformation of the distribution, the results are also applicable to the estimation of θ.

Let us first consider the class **C**. Suppose that $\sqrt{n}(\hat{\theta}_n - \theta)$ is expanded as

$$\sqrt{n}(\hat{\theta}_n - \theta) = \frac{Z_1(\theta)}{I(\theta)} + \frac{1}{\sqrt{n}} Q(\theta) + O_p\left(\frac{1}{n}\right),$$

where $Q(\theta) = O_p(1)$ and $Z_1(\theta) = \frac{1}{\sqrt{n}} \sum_{i=1}^{n} \frac{\partial}{\partial \theta} \log f(X_i, \theta)$.

Then we have the following lemma.

Lemma 2.2.

$$E_\theta[Z_1(\theta)Q(\theta)] = o(1)$$

Proof.

Define

$$T_\theta = \sqrt{n}(\hat{\theta}_n - \theta) - \frac{Z_1(\theta)}{I(\theta)} = \frac{1}{\sqrt{n}} Q(\theta) + o_p\left(\frac{1}{\sqrt{n}}\right).$$

Then we have

$$\frac{1}{I(\theta)} E_\theta[Z_1(\theta)T_\theta] = \frac{1}{\sqrt{n}} \frac{\partial}{\partial \theta} E_\theta(T_\theta) - \frac{1}{\sqrt{n}} E_\theta\left(\frac{\partial T_\theta}{\partial \theta}\right)$$

$$= \frac{1}{n} \frac{\partial}{\partial \theta} E_\theta[Q(\theta)] + 1 - \frac{1}{\sqrt{n}} E_\theta\left[\frac{\partial}{\partial \theta}\left\{\frac{Z_1(\theta)}{I(\theta)}\right\}\right] + o_p\left(\frac{1}{n}\right);$$

$$= o_p\left(\frac{1}{\sqrt{n}}\right).$$

From which we have

$$E_\theta[Z_1(\theta)Q(\theta)]=\sqrt{n}\,E_\theta[Z_1(\theta)T_\theta]+o_p(1)$$
$$=o_p(1).$$

Thus we have the asymptotic cumulants of $\sqrt{n}\,(\hat{\theta}_n-\theta)$ up to the order n^{-1} as

$$E_\theta[\sqrt{n}\,(\hat{\theta}_n-\theta)]=\frac{1}{\sqrt{n}}\mu_1(\theta)+\frac{1}{n}\mu_2(\theta)+o\left(\frac{1}{n}\right);$$

$$V_\theta[\sqrt{n}\,(\hat{\theta}_n-\theta)]=\frac{1}{I(\theta)}+\frac{1}{n}v_2(\theta)+o\left(\frac{1}{n}\right);$$

$$\kappa_3[\sqrt{n}\,(\hat{\theta}_n-\theta)]=\frac{1}{\sqrt{n}}\beta_3(\theta)+\frac{1}{n}\gamma_3(\theta)+o\left(\frac{1}{n}\right);$$

$$\kappa_4[\sqrt{n}\,(\hat{\theta}_n-\theta)]=\frac{1}{n}\beta_4(\theta)+o\left(\frac{1}{n}\right).$$

We can show in exactly similar way as in the previous papers [5] and [6] that

$$\beta_3(\theta)=-\frac{3J(\theta)+2K(\theta)}{I(\theta)^3};$$

$$\beta_4(\theta)=-\frac{1}{I(\theta)^4}\{3H^*(\theta)+12N^*(\theta)+4L(\theta)\}+\frac{12}{I(\theta)^5}\{2J(\theta)+K(\theta)\}\{J(\theta)+K(\theta)\}$$

where

$$L(\theta)=E_\theta\left[\left\{\frac{\partial^3}{\partial\theta^3}\log f(X,\,\theta)\right\}\left\{\frac{\partial}{\partial\theta}\log f(X,\,\theta)\right\}\right];$$

$$N^*(\theta)=E_\theta\left[\left\{\frac{\partial^2}{\partial\theta^2}\log f(X,\,\theta)+I(\theta)\right\}\left\{\frac{\partial}{\partial\theta}\log f(X,\,\theta)\right\}^2\right];$$

$$H^*(\theta)=E_\theta\left[\left\{\frac{\partial}{\partial\theta}\log f(X,\,\theta)\right\}^4\right]-3I(\theta)^2.$$

Since the odd order cumulants affects only the asymmetric aspects of the asymptotic distribution, the asymptotic expansion of $P_{\theta,n}\{\sqrt{n}\,|\hat{\theta}_n-\theta|<a\}$ is independent of them. Hence we have the following theorem.

Theorem 2.1. Let $\hat{\theta}_{ML}{}^*$ be the modified MLE in the class **C** and $\hat{\theta}_n$ be any other estimator in the class **C**. Then it holds that

$$\lim_{n\to\infty}n[P_{\theta,n}\{\sqrt{n}\,|\hat{\theta}_{ML}{}^*-\theta|<a\}-P_{\theta,n}\{\sqrt{n}\,|\hat{\theta}_n-\theta|<a\}]\geqq0$$

for all $a>0$ and all $\theta\in\Theta$.

For the class **D** we further have that $\gamma_3(\theta)=0$ and $\mu_1(\theta)=\beta_3(\theta)/6$, hence all the asymptotic moments are determined up to the order n^{-1} except v_2. Since v_2 is minimized for the modified MLE we have the following theorem.

Theorem 2.2. Let $\hat{\theta}_{ML}{}^{**}$ be the modified MLE in the class **D** and $\hat{\theta}_n$ be any other estimator in the class **D**. Then it holds that

$$\lim_{n\to\infty}n[P_{\theta,n}\{-a<\sqrt{n}\,(\hat{\theta}_{ML}{}^{**}-\theta)<b\}-P_{\theta,n}\{-a<\sqrt{n}\,(\hat{\theta}_n-\theta)<b\}]\geqq0$$

for all $a>0$, all $b>0$ and all $\theta\in\Theta$.

3. The concept of asymptotic completeness of an estimator

We consider first the case of a real-valued parameter. The concept of asymptotic efficiency and higher order asymptotic efficiency gives a very strong condition for the "goodness" of an estimator in that it uniformly maximizes asymptotically the probability of its lying within a

                    Kei TAKEUCHI and Masafumi AKAHIRA

neighborhood of the value irrespective of the parameter value and the choice of neighborhood. But only one somewhat artificial restriction is that the estimators considered should be (higher order) asymptotically median unbiased. It is, however, easily seen from the context that the only requirement is that the class of estimators to be compared must have the same asymptotic "location" whatever the exact meaning of the word may be, without which nothing can be derived. Therefore one may consider different types of the definition of "location", and accordingly different type of "adjustment" of the estimator to satisfy the condition. Thus we may consider in the following way. Suppose that we want to establish asymptotic optimality of an estimator $\hat{\theta}_n{}^0$, e. g. MLE, in some class of estimators.

What has been virtually established is that for any estimator $\hat{\theta}_n$ in the class if both $\hat{\theta}_n$ and $\hat{\theta}_n{}^0$ are adjusted like

$$\hat{\theta}_n{}^* = \hat{\theta}_n + \frac{1}{n} g(\hat{\theta}_n) ;$$

$$\hat{\theta}_n{}^{0*} = \hat{\theta}_n{}^0 + \frac{1}{n} h(\hat{\theta}_n{}^0) ;$$

so that $\hat{\theta}_n{}^*$ and $\hat{\theta}_n{}^{0*}$ are both (higher order) asymptotically median unbiased then $\hat{\theta}_n{}^{0*}$ is (higher order) asymptotically uniformly "better" than $\hat{\theta}_n{}^*$. If we change the definition of "location" the way of "adjustment" will be changed, but the same results hold for adjusted estimators $\hat{\theta}_n{}^{0*}$ and $\hat{\theta}_n{}^*$. This fact will lead to the following question: Instead of comparing two adjusted estimators $\hat{\theta}_n{}^*$ and $\hat{\theta}_n{}^{0*}$, isn't it possible to compare $\hat{\theta}_n$ and the adjusted $\hat{\theta}_n{}^*$? Naturally, the way $\hat{\theta}_n{}^*$ is adjusted depends on $\hat{\theta}_n$ to be compared, but is independent of the parameter. And we introduce the following concept.

An estimator $\hat{\theta}_n{}^*$ is called to be asymptotically complete with respect to a class $\mathbf{M}$ of estimators if for any $\hat{\theta}_n \in \mathbf{M}$ there is an "adjusted" estimator

$$\hat{\theta}_n{}^{**} = \hat{\theta}_n{}^* + g_n(\hat{\theta}_n{}^*)$$

so that for any $a > 0$, $b > 0$ and θ we have

$$\lim_{n \to \infty} [P_{\theta, n}\{-a < c_n(\hat{\theta}_n{}^{**} - \theta) < b\} - P_{\theta, n}\{-a < c_n(\hat{\theta}_n - \theta) < b\}] \geqq 0. \qquad (3.1)$$

Now let L be a loss function such that $L(0) = 0$ and $L(u)$ is an increasing function for $u > 0$ and a decreasing function for $u < 0$ and it is bounded. If $\hat{\theta}_n{}^*$ is asymptotically complete in the above sense, then for any estimator $\hat{\theta}_n$ in $\mathbf{M}$ there is $\hat{\theta}_n{}^{**}$ which is a function of $\hat{\theta}_n{}^*$ for which

$$\lim_{n \to \infty} [E_{\theta, n}(L(c_n(\hat{\theta}_n{}^{**} - \theta))) - E_{\theta, n}(L(c_n(\hat{\theta}_n - \theta)))] \leqq 0. \qquad (3.2)$$

Higher order asymptoic completeness can be defined in a similar way. For each $k = 2, 3, \ldots$, an estimator $\hat{\theta}_n{}^*$ is called to be k-th order asymptotically complete with respect to a class $\mathbf{M}$ of estimators if for any $\hat{\theta}_n \in \mathbf{M}$ there is an "adjusted" estimator

$$\hat{\theta}_n{}^{**} = \hat{\theta}_n{}^* + g_n(\hat{\theta}_n{}^*)$$

so that for any $a > 0$, $b > 0$ and θ we have

$$\lim_{n \to \infty} c_n{}^{k-1}[P_{\theta, n}\{-a < c_n(\hat{\theta}_n{}^{**} - \theta) < b\} - P_{\theta, n}\{-a < c_n(\hat{\theta}_n - \theta) < b\}] \geqq 0. \qquad (3.3)$$

Further $\hat{\theta}_n{}^*$ is called to be k-th order asymptotically symmetrically complete with respect to a class $\mathbf{M}$ of estimators when (3.3) holds when $a = b$. If $\hat{\theta}_n{}^*$ is k-th order asymptotically complete with respect to $\mathbf{M}$, then for any estimator $\hat{\theta}_n$ in $\mathbf{M}$ there is $\hat{\theta}_n{}^{**}$ which is a function of $\hat{\theta}_n{}^*$

August, 1980 Third Order Asymptotic Efficiency and Asymptotic Completeness of Estimators 95

for which

$$\lim_{n\to\infty} c_n{}^{k-1}[E_{\theta,n}(L(c_n(\hat{\theta}_n{}^{**}-\theta)))-E_{\theta,n}(L(c_n(\hat{\theta}_n-\theta)))]\leqq 0. \tag{3.4}$$

If $\hat{\theta}_n{}^*$ is k-th order asymptotically symmetrically complete with respect to $\mathbf{M}$, then (3.4) holds for all symmetric loss function around the origin.

Unfortunately the above definition is meaningless as such, since no estimator can asymptotically complete, which can be seen below. Let $\hat{\theta}_n{}^*$ be an asymptotically efficient estimator (e. g. MLE) and $\sqrt{n}(\hat{\theta}_n{}^*-\theta)$ be asymptotically distributed with mean 0 and variance $1/I$, where I is the Fisher information. Let $\hat{\theta}_n$ be an AMU estimator with asymptotic variance larger than $1/I$. Obviously $\hat{\theta}_n$ is not asymptotically efficient and inferior to $\hat{\theta}_n{}^*$. A naturally adjusted estimator of $\hat{\theta}_n{}^*$ corresponding to $\hat{\theta}_n$ will be $\hat{\theta}_n{}^{**}=\hat{\theta}_n{}^*+c/\sqrt{n}$ which has the same asymptotic mean with $\hat{\theta}_n$. Now take $(0<)b(<c)$ small enough and $a(>0)$ large enough. Then we have

$$\overline{\lim_{n\to\infty}}[P_{\theta,n}\{-a<\sqrt{n}(\hat{\theta}_n{}^{**}-\theta)<b\}-P_{\theta,n}\{-a<\sqrt{n}(\hat{\theta}_n{}^*-\theta)<b\}]\leqq 0$$

which contradicts the condition of asymptotic completeness.
However we may adjusted $\hat{\theta}_n{}^*$ we can not get uniformity irrespective of a and b. It is, of course, possible to get the reverse inequality if we adjust $\hat{\theta}_n{}^*$ depending on a and b, but the nice uniformity would be lost.

There is nothing unnatural in this problem: If there is a bias in the estimators, the estimator with large variance may hit upon the true value by chance but one with small variance has little chance to get close to the true value. Therefore we must modify the concept of the asymptotic completeness and call $\hat{\theta}_n{}^*$ asymptotically symmetrically complete if (3.1) holds when $a=b$, then (3.2) holds for all symmetric loss function.

The class $\mathbf{F}$ of estimators here is more specifically defined as the class of estimators $\hat{\theta}_n$ such that $c_n(\hat{\theta}_n-\theta)$ has the asymptotic distribution in the sense that

$$\lim_{n\to\infty} P_{\theta,n}\{c_n(\hat{\theta}_n-\theta)<a\}=F_{\theta}(a),$$

where the convergence is locally uniform in θ and $F_{\theta}(a)$ is absolutely continuous in a and continuous in θ, and we further assume that c_n is equal to the maximum order of consistency. Now we have the following theorem, the proof of which is omitted.

Theorem 3.1. Let $\hat{\theta}_n{}^*$ be asymptotically efficient and the density of the asymptotic distribution of $c_n(\hat{\theta}_n-\theta)$ is symmetric around the origin. Then $\hat{\theta}_n{}^*$ is asymptotically symmetrically complete with respect to the class $\mathbf{F}$.

In this case it is seen that $\hat{\theta}_n{}^*$ need not be modified.

Higher order asymptotic compleness has, however, more meaningful implication. Now we consider the regular case with $c_n=\sqrt{n}$ and we define the classes $\mathbf{B}^*$, $\mathbf{C}^*$ and $\mathbf{D}^*$ which are defined in the same way as the class $\mathbf{B}$, $\mathbf{C}$ and $\mathbf{D}$, respectively but omitting the condition of higher order asymptotic median unbiasedness.
Then the following theorems hold:

Theorem 3.2. The MLE is second order asymptotically complete with respect to the class $\mathbf{B}^*$.

Theorem 3.3. The MLE is third order asymptotically symmetrically complete with respect to the class $\mathbf{C}^*$.

Theorem 3.4. The MLE is third order asymptotically complete with respect to the class $\mathbf{D}^*$.

The outline of the proof of Theorem 3.4 is as follows:

Let $\hat{\theta}_n \in \mathbf{D}^*$ and the asymptotic bias of $\hat{\theta}_n$ be

$$E_{\theta,n}(\sqrt{n}\,(\hat{\theta}_n - \theta)) = \frac{1}{\sqrt{n}}\mu_1(\theta) + o\left(\frac{1}{n}\right)$$

and also of $\hat{\theta}_n^*$

$$E_{\theta,n}(\sqrt{n}\,(\hat{\theta}_n^* - \theta)) = \frac{1}{\sqrt{n}}\mu_1^*(\theta) + o\left(\frac{1}{n}\right).$$

Let the modified estimator $\hat{\theta}_n^{**}$ be defined as

$$\hat{\theta}_n^{**} = \hat{\theta}_n^* + \frac{1}{n}\{\mu_1^*(\hat{\theta}_n^*) - \mu_1^*(\hat{\theta}_n^*)\}.$$

Then we can prove exactly as of Theorem 6.1 in the previous paper [6] that

$$\lim_{n\to\infty} n[P_{\theta,n}\{-a < \sqrt{n}\,(\hat{\theta}_n^{**} - \theta) < b\} - P_{\theta,n}\{-a < \sqrt{n}\,(\hat{\theta}_n - \theta) < b\}] \geq 0$$

for all $a, b > 0$ and θ, because the asymptotic moments of $\sqrt{n}\,(\hat{\theta}_n - \theta)$ and $\sqrt{n}\,(\hat{\theta}_n^{**} - \theta)$ are the same up to the order n^{-1} except the second term in the variance which is smaller for $\hat{\theta}_n^*$.

The proofs of Theorem 3.2 and 3.3 are similar and even simpler.

The classes $\mathbf{C}^*$ and $\mathbf{D}^*$ are much natural classes than classes $\mathbf{C}$ and $\mathbf{D}$ and Theorems 3.2, 3.3 and 3.4 may be nearly the final conclusion of the asymptotic derivability of the MLE because we have the problem of the choice of the way of modification which should be based on the consideration of the asymptotic bias versus asymptotic variance, which can not be decided once and for all, but what is established in the above implies that we only need to consider the estimators which are modified MLE but it is not necessary to take any other (regular) estimators into consideration.

4. Conclusion remark

The conclusion of this paper can be easily generalized to the multiparameter cases and the discussion is very much similar to that in the multiparameter exponential case. So we omit the details of the derivation.

References

[1]   Akahira, M.: Asymptotic theory for estimation of location in non-regular cases, I: Order of convergence of consistent estimators, Rep. Stat. Appl. Res., JUSE, **22**, 8–26 (1975).

[2]   Akahira, M.: A note on the order asymptotic efficiency of estimators in an autoregressive process, Rep. Univ. Electro-Comm., **26**, 143–149 (1975).

[3]   Akahira, M. and Takeuchi, K.: The Concept of Asymptotic Efficiency and Higher Order Asymptotic Efficiency in Statistical Estimation Theory, Lecture Note (1979); Revised (1980).

[4]   Akahira, M. and Takeuchi, K.: Discretized likelihood methods: Asymptotic properties of discretized likelihood estimators (DLE's), Ann. Inst. Statist. Math., **31**, Part A, 39–56 (1979).

[5]   Takeuchi, K. and Akahira, M.: On the second order asymptotic efficiencies of estimators, Proceedings of the Third Japan–USSR Symposium on Probability Theory, Lecture Notes in Mathematics, **550**, Springer-Verlag, 604–638 (1976).

[6]   Takeuchi, K. and Akahira, M.: Third order asymptotic efficiency of maximum likelihood estimator for multiparameter exponential case, Rep. Univ. Electro-Comm., **28**, 271–293 (1978).

[7]   Takeuchi, K. and Akahira, M.: Asymptotic optimality of the generalized Bayes estimator in multiparameter cases, Ann. Inst. Statist. Math., **31**, Part A, 403–415 (1979).

Statistics & Decisions 1, 17-38 (1982)
© Akademische Verlagsgesellschaft 1982

ON ASYMPTOTIC DEFICIENCY OF ESTIMATORS IN POOLED SAMPLES IN THE PRESENCE
OF NUISANCE PARAMETERS

Masafumi Akahira and Kei Takeuchi

Received: revised version: May 4, 1982

Abstract
The concept of asymptotic deficiency is extended to the case when a
common parameter θ is estimated from m sets of independent samples
each of size n , and the asymptotic deficiencies of some asymptotically
efficient estimators relative to the maximum likelihood estimator based
on the pooled sample are discussed in the presence of nuisance parameters.

1. Introduction.
The third order asymptotic efficiency of the maximum likelihood estimator
(MLE) in the case of independently and identically distributed (i.i.d.)
samples has been established by Pfanzagl and Wefelmeyer (1978), Akahira
and Takeuchi (1978, 1981a) and Ghosh, Sinha and Wieand (1980), and it can
be formulated in terms of asymptotic deficiency introduced by Hodges and
Lehmann (1970) (Akahira, 1981; Akahira and Takeuchi 1981a, 1981b).

Let X_1, X_2, ..., X_n, ... be i.i.d. random variables according to
the distribution with a density function $f(x,\theta)$ where θ is a real

---

AMS Subject Classification (1980): 62F12, 62F10.
Key words and phrases: Asymptotic deficiency, asymptotically efficient
   estimator, maximum likelihood estimator, pooled samples, nuisance parameter, loss of information, asymptotic unbiasedness, ancillary statistics.

parameter. Under a suitable set of regularity conditions it can be proved that the MLE $\hat{\theta}_n^*$ of θ can be asymptotically expanded into the form

$$\sqrt{n}(\hat{\theta}_n^* - \theta) = \frac{1}{I} Z + \frac{1}{\sqrt{n}} Q^* + o_p\left(\frac{1}{\sqrt{n}}\right) ,$$

where $Z = \frac{1}{\sqrt{n}} \sum_{i=1}^{n} \frac{\partial}{\partial\theta} \log f(x_i,\theta)$ and $I = E_\theta[\{\frac{\partial}{\partial\theta} \log f(x,\theta)\}^2]$ and that for any asymptotically efficient estimator $\hat{\theta}_n^0$ which admits the same type of expansion

$$\sqrt{n}(\hat{\theta}_n^0 - \theta) = \frac{1}{I} Z + \frac{1}{\sqrt{n}} Q_0 + o_p\left(\frac{1}{\sqrt{n}}\right)$$

we have

$$V(Q_0) \geqq V(Q^*) ,$$

where V designates variance. This implies that we can construct a "modified" estimator (relative to $\hat{\theta}_n^0$) of the form

$$\hat{\theta}_n^{**} = \hat{\theta}_n^* + \frac{1}{n} h(\hat{\theta}_n^*)$$

so that $\hat{\theta}_n^{**}$ has the same asymptotic bias as $\hat{\theta}_n^*$ up to the order n^{-1}, and that

$$\lim_{n\to\infty} n[P_{\theta,n}\{\sqrt{n}|\hat{\theta}_n^{**}-\theta| < a\} - P_{\theta,n}\{\sqrt{n}|\hat{\theta}_n^0-\theta| < a\}] \geqq 0$$

for all θ and all $a > 0$. Moreover, if n_1 is defined so that the modified estimator $\hat{\theta}_{n_1}^{**}$ (with Q^{**}) based on the sample of size n_1 has the same asymptotic distribution as that of the estimator $\hat{\theta}_n^0$ (with Q_0) based on the sample of size $n(>n_1)$ up to the order n^{-1}, we have

$$\lim_{n\to\infty}(n-n_1) = I\{V(Q_0) - V(Q^{**})\}$$

(Akahira, 1981). We call the above limit the asymptotic deficiency of $\hat{\theta}_n^0$ relative to the MLE. Note that the right-hand side of the above

AKAHIRA-TAKEUCHI   1 9

equation does not necessarily mean the difference of the variances of estimators but of the asymptotic variances, hence we do not need to bother about the remainder terms (Akahira and Takeuchi, 1981a).

In this paper we extend this concept to the case when a common parameter θ is estimated from m independent sets of samples and discuss the deficiencies of some asymptotically efficient estimators relative to the MLE based on the pooled sample from the distribution with different nuisance parameters.

2. Asymptotic Deficiencies Of Estimators For Samples From The Distributions With Different Nuisance Parameters.

Suppose that it is required to estimate an unknown real-valued parameter θ based on m samples of size n whose values are X_{ij} $(i = 1,\ldots,m;$ $j = 1,\ldots,n)$ from the distributions with different nuisance parameters ξ_i $(i = 1,\ldots,m)$. Now we assume that for each i X_{i1}, X_{i2}, $\ldots$, X_{in} are independently and identically distributed according to a density function $f(x,\theta,\xi_i)$ with respect to a σ-finite measure μ.

We assume further regularity conditions:

(A.1) For each i the set $\{x\,|\,f(x,\theta,\xi_i) > 0\}$ does not depend on θ and ξ_i.

(A.2) For each i , for almost all $x[\mu]$, $f(x,\theta,\xi_i)$ is three times continuously partially differentiable in θ and ξ_i.

(A.3) For each $\theta \in \Theta$ and each i

$$0 < I_{\theta\theta}^{i} = E[\{\frac{\partial}{\partial\theta} \log f(X_{ij},\theta,\xi_i)\}^2]$$

$$= -E[\frac{\partial^2}{\partial\theta^2} \log f(X_{ij},\theta,\xi_i)]$$

$$< -\infty \;;$$

$$0 < I_{\xi\xi}^{i} = E[\{\frac{\partial}{\partial\xi_i} \log f(X_{ij},\theta,\xi_i)\}^2]$$

$$= -E[\frac{\partial^2}{\partial \xi_i^2} \log f(X_{ij}, \theta, \xi_i)]$$

$$< \infty .$$

(A.4) The parameters are defined to be "orthogonal" in the sense that

$$E[\frac{\partial^2}{\partial \theta \partial \xi_i} \log f(X_{ij}, \theta, \xi_i)] = 0 \quad (i = 1,\ldots,m) .$$

Remark that this assumption is not necessarily restrictive, since otherwise we can redefine the parameters $\xi_i' = g(\theta, \xi_i)$ $(i = 1,\ldots,m)$ so that they have the above orthogonality.

(A.5) There exist

$$J_{\theta\theta\theta}^i = E[\{\frac{\partial^2}{\partial \theta^2} \log f(X_{ij}, \theta, \xi_i)\}\{\frac{\partial}{\partial \theta} \log f(X_{ij}, \theta, \xi_i)\}] ;$$

$$J_{\theta\theta\xi}^i = E[\{\frac{\partial^2}{\partial \theta^2} \log f(X_{ij}, \theta, \xi_i)\}\{\frac{\partial}{\partial \xi_i} \log f(X_{ij}, \theta, \xi_i)\}] ;$$

$$J_{\theta\xi\theta}^i = E[\{\frac{\partial^2}{\partial \theta \partial \xi_i} \log f(X_{ij}, \theta, \xi_i)\}\{\frac{\partial}{\partial \theta} \log f(X_{ij}, \theta, \xi_i)\}] ;$$

$$J_{\theta\xi\xi}^i = E[\{\frac{\partial^2}{\partial \theta \partial \xi_i} \log f(X_{ij}, \theta, \xi_i)\}\{\frac{\partial}{\partial \xi_i} \log f(X_{ij}, \theta, \xi_i)\}] ;$$

$$J_{\xi\xi\theta}^i = E[\{\frac{\partial^2}{\partial \xi_i^2} \log f(X_{ij}, \theta, \xi_i)\}\{\frac{\partial}{\partial \theta} \log f(X_{ij}, \theta, \xi_i)\}] ;$$

$$K_{\theta\theta\theta}^i = E[\{\frac{\partial}{\partial \theta} \log f(X_{ij}, \theta, \xi_i)\}^3] ,$$

$$K_{\theta\theta\xi}^i = E[\{\frac{\partial}{\partial \theta} \log f(X_{ij}, \theta, \xi_i)\}^2 \{\frac{\partial}{\partial \xi_i} \log f(X_{ij}, \theta, \xi_i)\}] ;$$

$$M_{\theta\theta\theta\theta}^i = E[\{\frac{\partial^2}{\partial \theta^2} \log f(X_{ij}, \theta, \xi_i)\}^2] - I_{\theta\theta}^{i^2} ;$$

$$M_{\theta\xi\theta\xi}^i = E[\{\frac{\partial^2}{\partial \theta \partial \xi_i} \log f(X_{ij}, \theta, \xi_i)\}^2] ,$$

and the following hold:

$$E[\frac{\partial^3}{\partial\theta^3} \log f(X_{ij},\theta,\xi_i)] = -3J^i_{\theta\theta\theta}-K^i_{\theta\theta\theta} \ ;$$

$$E[\frac{\partial^3}{\partial\theta^2\partial\xi_i} \log f(X_{ij},\theta,\xi_i)] = -J^i_{\theta\xi\theta} \ ;$$

$$E[\frac{\partial^3}{\partial\theta\partial\xi_i^2} \log f(X_{ij},\theta,\xi_i)] = -J^i_{\theta\xi\xi} \ ,$$

for $i = 1, \ldots, m$.

From the assumption (A.4) it is noted that $K^i_{\theta\theta\xi} = J^i_{\theta\xi\theta} - J^i_{\theta\theta\xi}$. Denote the log likelihood functions $L_i(\theta,\xi_i)$ $(i = 1,\ldots,m)$ and $L(\theta,\xi_1,\ldots,\xi_m)$ by

$$L_i(\theta,\xi_i) = \sum_{j=1}^{n} \log f(x_{ij},\theta,\xi_i) \quad (i = 1,\ldots,m) \ ;$$

$$L(\theta,\xi_1,\ldots,\xi_m) = \sum_{i=1}^{m} L_i(\theta,\xi_i) = \sum_{i=1}^{m} \sum_{j=1}^{n} \log f(x_{ij},\theta,\xi_i) \ .$$

When $\xi_1, \ldots, \xi_m$ are known, we denote by $\hat{\theta}_0$ the MLE of θ based on $L(\theta,\xi_1,\ldots,\xi_m)$. For each $i = 1, \ldots, m$ let $\hat{\theta}_i$ and $\hat{\xi}_i$ be the MLE's of θ and ξ_i based on $L_i(\theta,\xi_i)$, respectively. Let $\hat{\theta}^*, \hat{\xi}^*_1, \ldots, \hat{\xi}^*_m$ be the MLE's of $\theta, \xi_1, \ldots, \xi_m$ based on $L(\theta,\xi_1,\ldots,\xi_m)$, respectively.

(i) <u>The MLE $\hat{\theta}_0$ when $\xi_1, \ldots, \xi_m$ are known.</u>

Since $\hat{\theta}_0$ is the MLE of θ for known parameters $\xi_1, \ldots, \xi_m$, it follows that

$$\frac{\partial}{\partial\theta} L(\hat{\theta}_0,\xi_1,\ldots,\xi_m) = 0 \ .$$

Then it is seen by a similar way as Akahira and Takeuchi (1981a, page 92) that the expansion of $\hat{\theta}_0$ is given by

$$\sqrt{n}(\hat{\theta}_0-\theta) = \frac{1}{I^*} Z^*_\theta + \frac{1}{\sqrt{n}\ I^{*2}} Z^*_{\theta\theta}Z^*_\theta - \frac{3J^*+K^*}{2\sqrt{n}\ I^{*3}} Z^{*2}_\theta + o_p\left(\frac{1}{\sqrt{n}}\right)$$

$$= \frac{1}{I^*} Z^*_\theta + \frac{1}{\sqrt{n}} Q_0 + o_p\left(\frac{1}{\sqrt{n}}\right) \text{ (say)};$$

where

$$Z_\theta^* = \sum_{i=1}^m Z_\theta^i, \quad Z_{\theta\theta}^* = \sum_{i=1}^m Z_{\theta\theta}^i \ , \quad I^* = \sum_{i=1}^m I_{\theta\theta}^i, \quad J^* = \sum_{i=1}^m J_{\theta\theta\theta}^i \ ,$$

and $\ K^* = \sum_{i=1}^m K_{\theta\theta\theta}^i \ $ with

$$Z_\theta^i = \frac{1}{\sqrt{n}} \frac{\partial}{\partial\theta} L_i(\theta,\xi_i) \ ;$$

$$Z_{\theta\theta}^i = \frac{1}{\sqrt{n}} \{ \frac{\partial^2}{\partial\theta^2} L_i(\theta,\xi_i) + nI_{\theta\theta}^i \} \ .$$

Hence it is easily seen that

$$(2.1) \qquad V(Q_0) = \frac{1}{I^{*4}} (M^* I^* - J^{*2}) + \frac{(J^*+K^*)^2}{2I^{*4}} \ ,$$

where $\ M^* = \sum_{i=1}^m M_{\theta\theta\theta\theta}^i .$

    (ii) <u>The MLE's $\hat{\theta}^*$, $\hat{\xi}_1^*$, ..., $\hat{\xi}_m^*$.</u>

Since

$$\frac{\partial}{\partial\theta} L(\hat{\theta}^*,\hat{\xi}_1^*,\ldots,\hat{\xi}_m^*) = 0 \ ,$$

$$\frac{\partial}{\partial\xi_i} L(\hat{\theta}^*,\hat{\xi}_1^*,\ldots,\hat{\xi}_m^*) = 0 \quad (i = 1,\ldots,m) \ ;$$

it follows by a similar way as the case (i) that the stochastic expansion of $\hat{\theta}^*$ is given by

$$\sqrt{n}(\hat{\theta}^*-\theta) = \frac{1}{I^*} Z_\theta^* + \frac{1}{\sqrt{n} \ I^*} Z_{\theta\theta}^* Z_\theta^* - \frac{3J^*+K^*}{2\sqrt{n} \ I^{*3}} Z_\theta^{*2}$$

$$+ \frac{1}{\sqrt{n} \ I^*} \sum_{i=1}^m \frac{1}{I_{\xi\xi}^i} Z_{\theta\xi}^i Z_\xi^i - \frac{1}{\sqrt{n} \ I^{*2}} Z_\theta^* \sum_{i=1}^m \frac{J_{\theta\xi\theta}^i}{I_{\xi\xi}^i} Z_\xi^i$$

$$- \frac{1}{2\sqrt{n} \ I^*} \sum_{i=1}^m \frac{J_{\theta\xi\xi}^i}{I_{\xi\xi}^i} Z_\xi^{i2} + o_p\left(\frac{1}{\sqrt{n}}\right)$$

$$= \frac{1}{I^*} Z_\theta^* + \frac{1}{\sqrt{n}} Q_0 + \frac{1}{\sqrt{n}} Q_1 + o_p\left(\frac{1}{\sqrt{n}}\right) \quad \text{(say)}$$

$$= \frac{1}{I^*} Z_\theta^* + \frac{1}{\sqrt{n}} Q^* + o_p\left(\frac{1}{\sqrt{n}}\right) \quad \text{(say)} \ ,$$

where

$$Z_\xi^i = \frac{1}{\sqrt{n}} \frac{\partial}{\partial \xi_i} L_i(\theta, \xi_i) \; ;$$

$$Z_{\theta\xi}^i = \frac{1}{\sqrt{n}} \frac{\partial^2}{\partial\theta\partial\xi_i} L_i(\theta, \xi_i) \; .$$

Since $\text{Cov}(Q_0, Q_1) = 0$, it follows that

$$(2.2) \qquad V(Q^*) = V(Q_0) + V(Q_1)$$

with

$$V(Q_1) = \frac{1}{I^{*2}} \sum_{i=1}^{m} \frac{1}{I_{\xi\xi}^i} \left(M_{\theta\xi\theta\xi}^i - \frac{J_{\theta\xi\theta}^{i2}}{I_{\theta\theta}^i} - \frac{J_{\theta\xi\xi}^{i2}}{I_{\xi\xi}^i} \right)$$

$$+ \frac{1}{I^{*2}} \sum_{i=1}^{m} \left(\frac{1}{I_{\theta\theta}^i} - \frac{1}{I^*} \right) \frac{J_{\theta\xi\theta}^{i2}}{I_{\xi\xi}^i} + \frac{1}{2I^{*2}} \sum_{i=1}^{m} \frac{J_{\theta\xi\xi}^{i2}}{I_{\xi\xi}^{i2}} \; .$$

(iii) <u>The weighted estimator by $\mathcal{J}_{\theta\theta}^j = -(1/n)(\partial^2/\partial\theta^2) L_i(\hat{\theta}_i, \hat{\xi}_i)$</u>
<u>($i = 1, \ldots, m$)</u>.

We consider the estimator

$$\tilde{\theta} = \frac{\displaystyle\sum_{i=1}^{m} \mathcal{J}_{\theta\theta}^i \, \hat{\theta}_i}{\displaystyle\sum_{i=1}^{m} \mathcal{J}_{\theta\theta}^i} \; .$$

According to the remark by R. A. Fisher (1925) that the second order derivative of the loglikelihood function gives the "actual" information of the MLE, we may compute.

By a similar way as in case (i) we have

$$(2.3) \qquad \sqrt{n}(\tilde{\theta}-\theta) = \frac{1}{I^*} Z_\theta^* + \frac{1}{\sqrt{n}\, I^{*2}} Z_{\theta\theta}^* Z_\theta^* - \frac{1}{\sqrt{n}\, I^{*2}} \sum_{i=1}^{m} \frac{3J_{\theta\theta\theta}^i + K_{\theta\theta\theta}^i}{I_{\theta\theta}^i} Z_\theta^i Z_\theta^*$$

$$+ \frac{1}{2\sqrt{n}\, I^{*2}} \sum_{i=1}^{m} \frac{3J_{\theta\theta\theta}^i + K_{\theta\theta\theta}^i}{I_{\theta\theta}^i} Z_\theta^{i2}$$

$$+ \frac{1}{\sqrt{n}\, I^*} \sum_{i=1}^{m} \frac{1}{I_{\xi\xi}^i} Z_{\theta\xi}^i Z_\xi^i - \frac{1}{\sqrt{n}\, I^{*2}} \left(\sum_{i=1}^{m} \frac{J_{\theta\xi\theta}^i}{I_{\xi\xi}^i} Z_\xi^i \right) Z_\theta^*$$

$$- \frac{1}{2\sqrt{n}\, I^*} \sum_{i=1}^{m} \frac{J_{\theta\xi\xi}^i}{I_{\xi\xi}^i} Z_\xi^{i2} + o_p\!\left(\frac{1}{\sqrt{n}}\right)$$

$$= \frac{1}{I^*} Z_\theta^* + \frac{1}{\sqrt{n}} Q_0 + \frac{1}{\sqrt{n}} Q_1 + \frac{1}{\sqrt{n}} Q_2 + o_p\!\left(\frac{1}{\sqrt{n}}\right)$$

$$= \frac{1}{I^*} Z_\theta^* + \frac{1}{\sqrt{n}} \tilde{Q} + o_p\!\left(\frac{1}{\sqrt{n}}\right) \quad \text{(say)}$$

where

$$Q_2 = \frac{1}{2I^{*2}} \sum_{i=1}^{m} (3J_{\theta\theta\theta}^i + K_{\theta\theta\theta}^i)\left(\frac{1}{I_{\theta\theta}^i} Z_\theta^i - \frac{1}{I^*} Z_\theta^* \right)^2 .$$

Since

$$\mathrm{Cov}(Q_0, Q_2) = \mathrm{Cov}(Q_1, Q_2) = 0 \ ,$$

it follows that

$$(2.4) \qquad V(\tilde{Q}) = V(Q_0) + V(Q_1) + V(Q_2)$$

with

$$V(Q_2) = \frac{1}{2I^{*2}} \sum_{i=1}^{m} \frac{(3J_{\theta\theta\theta}^i + K_{\theta\theta\theta}^i)^2}{I_{\theta\theta}^{i2}} - \frac{(3J^* + K^*)^2}{2I^{*4}} .$$

(iv) <u>The weighted estimator by $\hat{I}_{\theta\theta}^i = I_{\theta\theta}^i(\hat{\theta}_i, \hat{\xi}_i)$ $(i = 1,\ldots,m)$.</u>

We consider the estimator

$$\hat{\bar{\theta}} = \frac{\displaystyle\sum_{i=1}^{m} \hat{I}_{\theta\theta}^i \, \hat{\theta}_i}{\displaystyle\sum_{i=1}^{m} \hat{I}_{\theta\theta}^i}$$

Then the stochastic expansion of $\hat{\bar{\theta}}$ is given by

AKAHIRA-TAKEUCHI

$$\sqrt{n}(\hat{\bar{\theta}}-\theta) = \frac{1}{I^*} z^*_\theta + \frac{1}{\sqrt{n}\, I^*} \sum_{i=1}^{m} \frac{1}{I^i_{\theta\theta}} z^i_{\theta\theta} z^i_\theta + \frac{1}{2\sqrt{n}\, I^*} \sum_{i=1}^{m} \frac{J^i_{\theta\theta\theta}+K^i_{\theta\theta\theta}}{I^{i2}_{\theta\theta}} z^{i2}_\theta$$

$$- \frac{1}{\sqrt{n}\, I^{*2}} \sum_{i=1}^{m} \frac{2J^i_{\theta\theta\theta}+K^i_{\theta\theta\theta}}{I^i_{\theta\theta}} z^i_\theta z^*_\theta$$

$$+ \frac{1}{\sqrt{n}\, I^*} \sum_{i=1}^{m} \frac{1}{I^i_{\xi\xi}} z^i_{\theta\xi} z^i_\xi + \frac{1}{\sqrt{n}\, I^*} \sum_{i=1}^{m} \frac{J^i_{\theta\xi\theta}+K^i_{\theta\theta\xi}}{I^i_{\theta\theta} I^i_{\xi\xi}} z^i_\theta z^i_\xi$$

$$- \frac{1}{2\sqrt{n}\, I^*} \sum_{i=1}^{m} \frac{J^i_{\theta\xi\xi}}{I^{i2}_{\xi\xi}} z^{i2}_\xi - \frac{1}{\sqrt{n}\, I^{*2}} \sum_{i=1}^{m} \frac{2J^i_{\theta\xi\theta}+K^i_{\theta\theta\xi}}{I^i_{\xi\xi}} z^i_\xi z^*_\theta + o_p\!\left(\frac{1}{\sqrt{n}}\right)$$

$$= \frac{1}{I^*} z^*_\theta + \frac{1}{\sqrt{n}} Q_0 + \frac{1}{\sqrt{n}} Q_1 + \frac{1}{\sqrt{n}} Q_3 + \frac{1}{\sqrt{n}} Q_4 + o_p\!\left(\frac{1}{\sqrt{n}}\right)$$

$$= \frac{1}{I^*} z^*_\theta + \frac{1}{\sqrt{n}} \bar{Q} + o_p\!\left(\frac{1}{\sqrt{n}}\right) \quad \text{(say)},$$

where

$$Q_3 = \frac{1}{I^*} \sum_{i=1}^{m} \frac{1}{I^i_{\theta\theta}} z^i_{\theta\theta} z^i_\theta + \frac{1}{2I^*} \sum_{i=1}^{m} \frac{J^i_{\theta\theta\theta}+K^i_{\theta\theta\theta}}{I^i_{\theta\theta}} z^{i2}_\theta$$

$$- \frac{1}{I^{*2}} \sum_{i=1}^{m} \frac{2J^i_{\theta\theta\theta}+K^i_{\theta\theta\theta}}{I^i_{\theta\theta}} z^i_\theta z^*_\theta - \frac{1}{I^{*2}} z^*_{\theta\theta} z^*_\theta + \frac{3J^*+K^*}{2I^{*3}} z^{*2}_\theta \;;$$

$$Q_4 = \frac{1}{I^*} \sum_{i=1}^{m} \left(\frac{J^i_{\theta\xi\theta}+K^i_{\theta\theta\xi}}{I^i_{\xi\xi}}\right)\!\left(\frac{z^i_\theta}{I^i_{\theta\theta}} - \frac{z^*_\theta}{I^*}\right) z^i_\xi$$

We divide Q_3 into two parts Q_5 and Q_6 as follows:

$$Q_3 = \frac{1}{I^*}\Bigg(\sum_{i=1}^{m} \frac{1}{I^i_{\theta\theta}} z^i_{\theta\theta} z^i_\theta - \frac{1}{I^*} z^*_{\theta\theta} z^*_\theta - \frac{1}{I^*} \sum_{i=1}^{m} \frac{J^i_{\theta\theta\theta}}{I^i_{\theta\theta}} z^i_\theta z^*_\theta$$

$$+ \frac{1}{I^*} \sum_{i=1}^{m} J^i_{\theta\theta\theta} z^{*2}_\theta\Bigg) + \frac{1}{2\, I^*} \sum_{i=1}^{m} (J^i_{\theta\theta\theta}+K^i_{\theta\theta\theta})\!\left(\frac{z^i_\theta}{I^i_{\theta\theta}} - \frac{z^*_\theta}{I^*}\right)^2$$

$$= \frac{1}{I^*} \sum_{i=1}^{m} \left(\frac{z^i_\theta}{I^i_{\theta\theta}} - \frac{z^*_\theta}{I^*}\right)\!\left(z^i_{\theta\theta} - \frac{J^i_{\theta\theta\theta}}{I^*} z^*_\theta\right)$$

$$+ \frac{1}{2I^*} \sum_{i=1}^{m} (J^i_{\theta\theta\theta} + K^i_{\theta\theta\theta})\!\left(\frac{z^i_\theta}{I^i_{\theta\theta}} - \frac{z^*_\theta}{I^*}\right)^2$$

$$= Q_5 + Q_6 \quad \text{(say)} \;.$$

Since

$$\mathrm{Cov}(Q_i, Q_j) = 0 \quad (i \neq j;\ i,j = 0,\ 1,\ 4,\ 5,\ 6) \ ,$$

it follows that

$$(2.5) \qquad V(\bar{Q}) = V(Q_0) + V(Q_1) + V(Q_4) + V(Q_5) + V(Q_6) \ .$$

(v) <u>The weighted estimator by $\hat{I}_{\theta\theta}^{i*} = \hat{I}_{\theta\theta}^{i}(\hat{\bar{\theta}}, \hat{\xi}_i)$ $(i = 1,\ldots,m)$.</u>

We consider the estimator

$$\hat{\bar{\theta}}^* = \frac{\displaystyle\sum_{i=1}^{m} \hat{I}_{\theta\theta}^{i*}\, \hat{\theta}_i}{\displaystyle\sum_{i=1}^{m} \hat{I}_{\theta\theta}^{i*}} \ .$$

Then we have

$$\sqrt{n}(\hat{\bar{\theta}}^* - \theta) = \frac{1}{I^*}\, Z_\theta^* + \frac{1}{\sqrt{n}}\, Q_0 + \frac{1}{\sqrt{n}}\, Q_1 + \frac{1}{\sqrt{n}}\, Q_4 + \frac{1}{\sqrt{n}}\, Q_7 + o_p\!\left(\frac{1}{\sqrt{n}}\right)$$

$$= \frac{1}{I^*}\, Z_\theta^* + \frac{1}{\sqrt{n}}\, \bar{Q}^* + o_p\!\left(\frac{1}{\sqrt{n}}\right) \quad (\text{say}) \ ,$$

where

$$Q_7 = \frac{1}{I^*}\left(\sum_{i=1}^{m} \frac{1}{I_{\theta\theta}^{i}}\, Z_{\theta\theta}^{i} Z_\theta^{i} - \frac{1}{I^*}\, Z_{\theta\theta}^* Z_\theta^* - \frac{1}{I^*} \sum_{i=1}^{m} \frac{J_{\theta\theta\theta}^{i}}{I_{\theta\theta}^{i}}\, Z_\theta^{i} Z_\theta^* + \frac{J^*}{I^*}\, Z_\theta^{*2} \right)$$

$$\qquad - \frac{1}{2I^{*2}} \sum_{i=1}^{m} \frac{3J_{\theta\theta\theta}^{i} + K_{\theta\theta\theta}^{i}}{I_{\theta\theta}^{i2}}\, Z_\theta^{i2} + \frac{1}{I^{*3}} \sum_{i=1}^{m} \frac{3J_{\theta\theta\theta}^{i} + K_{\theta\theta\theta}^{i}}{I_{\theta\theta}^{i}}\, Z_\theta^{i} Z_\theta^*$$

$$\qquad - \frac{3J^* + K^*}{2I^{*4}}\, Z_\theta^{*2} \ .$$

Since

$$Q_7 = Q_5 - \frac{1}{2I^{*2}} \sum_{i=1}^{m} (3J_{\theta\theta\theta}^{i} + K_{\theta\theta\theta}^{i})\left(\frac{1}{I_{\theta\theta}^{i}}\, Z_\theta^{i} - \frac{1}{I^*}\, Z_\theta^* \right)^2$$

$$= Q_5 - Q_2$$

and $\mathrm{Cov}(Q_i, Q_j) = 0$ $(i \neq j;\ i,j = 0,1,2,4,5)$, it follows that

$$(2.6) \qquad V(\bar{Q}^*) = V(Q_0) + V(Q_1) + V(Q_2) + V(Q_4) + V(Q_5) \ .$$

(vi) <u>The asymptotically best estimator based on $\hat{\theta}_i$ and $\hat{\xi}_i$ (i = 1,</u>
<u>...,m)</u>.

The estimators $\hat{\bar{\theta}}$ and $\hat{\bar{\theta}}^*$ have the property of being functions of $\hat{\theta}_i$ and $\hat{\xi}_i$ (i = 1,...,m) alone. Now let us construct an estimator

$$\hat{\theta}^{**} = f(\hat{\theta}_1,\ldots,\hat{\theta}_m,\ \hat{\xi}_1,\ldots,\hat{\xi}_m)$$

which is asymptotically best among the estimators which depend only on $\hat{\theta}_i$ and $\hat{\xi}_i$ (i = 1,...,m). We assume that f is twice continuously partially differentiable in $\hat{\theta}_i$ and $\hat{\xi}_i$ (i = 1,...,m). Assuming the condition of consistency

$$f(\theta,\ldots,\theta,\ \xi_1,\ldots,\xi_m) \equiv \theta \ .$$

Denoting

$$a_i = \left.\frac{\partial f}{\partial \hat{\theta}_i}\right|_{\hat{\theta}_1=\theta,\ldots,\hat{\theta}_m=\theta,\ \hat{\xi}_1=\xi_1,\ldots,\hat{\xi}_m=\xi_m} \ ;$$

$$b_k = \left.\frac{\partial f}{\partial \hat{\xi}_k}\right|_{\hat{\theta}_1=\theta,\ldots,\hat{\theta}_m=\theta,\ \hat{\xi}_1=\xi_1,\ldots,\hat{\xi}_m=\xi_m} \ ;$$

$$c_{ij} = \left.\frac{\partial^2 f}{\partial \hat{\theta}_i \partial \hat{\theta}_j}\right|_{\hat{\theta}_1=\theta,\ldots,\hat{\theta}_m=\theta,\ \hat{\xi}_1=\xi_1,\ldots,\hat{\xi}_m=\xi_m} \ ;$$

$$d_{ik} = \left.\frac{\partial^2 f}{\partial \hat{\theta}_i \partial \hat{\xi}_k}\right|_{\hat{\theta}_1=\theta,\ldots,\hat{\theta}_m=\theta,\ \hat{\xi}_1=\xi_1,\ldots,\hat{\xi}_m=\xi_m} \ ;$$

$$e_{\ell k} = \left.\frac{\partial^2 f}{\partial \hat{\xi}_\ell \partial \hat{\xi}_k}\right|_{\hat{\theta}_1=\theta,\ldots,\hat{\theta}_m=\theta,\ \hat{\xi}_1=\xi_1,\ldots,\hat{\xi}_m=\xi_m} \ ,$$

for $i,\ j,\ k,\ \ell = 1,\ \ldots,\ m$, we have from the consistency condition that

$$\sum_{i=1}^{m} a_i = 1; \quad b_k = 0 \quad (k = 1,\ldots,m)$$

and also

$$\sum_{j=1}^{m} c_{ij} = \frac{\partial a_i}{\partial \theta} \; ; \quad \sum_{i=1}^{m} \sum_{j=1}^{m} c_{ij} = 0 \; ;$$

$$\sum_{i=1}^{m} d_{ik} = 0 \; ; \quad d_{ik} = \frac{\partial a_i}{\partial \xi_k} \; ;$$

$$e_{\ell k} = 0$$

for $i, k, \ell = 1, \ldots, m$.

Taylor's expansion of $\hat{\theta}^{**}$ yields that

$$\hat{\theta}^{**} = \theta + \sum_{i=1}^{m} a_i(\hat{\theta}_i - \theta) + \frac{1}{2} \sum_{i=1}^{m} \sum_{j=1}^{m} c_{ij}(\hat{\theta}_i - \theta)(\hat{\theta}_j - \theta)$$

$$+ \sum_{i=1}^{m} \sum_{k=1}^{m} d_{ik}(\hat{\theta}_i - \theta)(\hat{\xi}_k - \xi_k) + o_p\left(\frac{1}{n}\right) \; .$$

The asymptotic variance of $\sqrt{n}(\hat{\theta}^{**} - \theta)$ is given by

$$n \sum_{i=1}^{n} a_i^2 V(\hat{\theta}_i) = \sum_{i=1}^{m} \frac{a_i^2}{I_{\theta\theta}^i}$$

of which minimum value under the condition $\sum_{i=1}^{m} a_i = 1$ is attained when

$$a_i = \frac{I_{\theta\theta}^i}{\sum_{i=1}^{m} I_{\theta\theta}^i} = \frac{I_{\theta\theta}^i}{I^*} \quad (i = 1, \ldots, m) \; .$$

Then we have

$$d_{ik} - \frac{\partial a_i}{\partial \xi_k} = \delta_{ik} \frac{2J_{\theta\xi\theta}^i + K_{\theta\theta\xi}^i}{I^*} - \frac{I_{\theta\theta}^i(2J_{\theta\xi\theta}^k + K_{\theta\theta\xi}^k)}{I^{*2}}$$

for $i, k = 1, \ldots, m$, where δ_{ik} denotes Kronecker's delta.

It follows from these conditions that

$$\sqrt{n}(\hat{\theta}^{**} - \theta) = \frac{1}{I^*} \sum_{i=1}^{m} I_{\theta\theta}^i \sqrt{n}(\hat{\theta}_i - \theta) + \frac{1}{2\sqrt{n}} \sum_{i=1}^{m} \sum_{j=1}^{m} c_{ij} n(\hat{\theta}_i - \theta)(\hat{\theta}_j - \theta)$$

$$+ \frac{1}{\sqrt{n}} \sum_{i=1}^{m} \sum_{k=1}^{m} d_{ik} n(\hat{\theta}_i - \theta)(\hat{\xi}_k - \xi_k) + o_p\left(\frac{1}{\sqrt{n}}\right) ,$$

and substituting the stochastic expansion of $\hat{\theta}_i$ we get

$$
\sqrt{n}(\hat{\theta}^{**}-\theta) = \frac{1}{I^*}\,z_\theta^* + \frac{1}{\sqrt{n}\,I^*}\sum_{i=1}^{m}\frac{z_\theta^i}{I_{\theta\theta}^i}\left(z_{\theta\theta}^i - \frac{J_{\theta\theta\theta}^i}{I_{\theta\theta}^i}\,z_\theta^i\right) + \frac{1}{\sqrt{n}\,I^*}\sum_{i=1}^{m}\frac{z_\xi^i}{I_{\xi\xi}^i}\,z_{\theta\xi}^i
$$

$$
\quad - \frac{1}{2\sqrt{n}\,I^*}\sum_{i=1}^{m}\frac{J_{\theta\xi\xi}^i}{I_{\xi\xi}^i}\,z_\xi^{i2} + \frac{1}{\sqrt{n}\,I^*}\sum_{i=1}^{m}\frac{J_{\theta\xi\theta}^i+K_{\theta\theta\xi}^i}{I_{\theta\theta}^i I_{\xi\xi}^i}\,z_\theta^i z_\xi^i
$$

$$
\quad - \frac{1}{\sqrt{n}\,I^{*2}}\sum_{i=1}^{m}\frac{2J_{\theta\xi\theta}^i+K_{\theta\theta\xi}^i}{I_{\xi\xi}^i}\,z_\xi^i z_\theta^*
$$

$$
\quad + \frac{1}{2\sqrt{n}\,I^*}\sum_{i=1}^{m}\sum_{j=1}^{m}\left(\frac{c_{ij}}{I_{\theta\theta}^i I_{\theta\theta}^j} - \delta_{ij}\,\frac{J_{\theta\theta\theta}^i+K_{\theta\theta\theta}^i}{I^* I_{\theta\theta}^{i2}}\right)z_\theta^i z_\theta^j
$$

$$
\quad + o_p\!\left(\frac{1}{\sqrt{n}}\right)
$$

$$
= \frac{1}{I^*}\,z_\theta^* + \frac{1}{\sqrt{n}}\,Q_0^* + \frac{1}{\sqrt{n}}\,Q_1^* + o_p\!\left(\frac{1}{\sqrt{n}}\right) \quad \text{(say)}\;,
$$

where

$$
Q_1^* = \frac{1}{2I^*}\sum_{i=1}^{m}\sum_{j=1}^{m}\left(\frac{c_{ij}}{I_{\theta\theta}^i I_{\theta\theta}^j} - \delta_{ij}\,\frac{J_{\theta\theta\theta}^i+K_{\theta\theta\theta}^i}{I^* I_{\theta\theta}^{i2}}\right)z_\theta^i z_\theta^j\;.
$$

Since we have

$$
\mathrm{Cov}(Q_0^*,Q_1^*) = 0\;,
$$

we should minimize $V(Q_1^*)$ in order to minimize the asymptotic deficiency of $\hat{\theta}^{**}$ relative to $\hat{\theta}^*$ under the condition

$$
(2.7) \qquad \sum_{j=1}^{m} c_{ij} = \frac{\partial a_i}{\partial\theta} = \frac{2J_{\theta\theta\theta}^i+K_{\theta\theta\theta}^i}{I^*} - \frac{I_{\theta\theta}^i(2J^*+K^*)}{I^{*2}}
$$

for $i = 1, \ldots, m$.

It is expressed by

$$
V(Q_1^*) = \frac{1}{2I^{*2}}\sum_{i=1}^{m}\sum_{j=1}^{m}\left(c_{ij} - \delta_{ij}\,\frac{J_{\theta\theta\theta}^i+K_{\theta\theta\theta}^i}{I^*}\right)\frac{1}{I_{\theta\theta}^i I_{\theta\theta}^j}\;,
$$

and noting that $c_{ij} = c_{ji}$ we get by applying Lagrange's method of constrained minimization

$$c_{ij} = \delta_{ij}\, \frac{J^i_{\theta\theta\theta}+K^i_{\theta\theta\theta}}{I^*} + I^i_{\theta\theta}I^j_{\theta\theta}(\lambda_i+\lambda_j) \quad (i,j = 1,\dots,m) \;,$$

where λ_i $(i = 1,\dots,m)$ are Lagrangean multipliers. From (2.7) we have

$$0 = \sum_{i=1}^{m}\sum_{j=1}^{m} c_{ij} = \frac{J^*+K^*}{I^*} + 2I^*\sum_{i=1}^{m}\lambda_i\, I^i_{\theta\theta} \;.$$

Hence

$$\sum_{i=1}^{m}\lambda_i\, I^i_{\theta\theta} = -\,\frac{J^*+K^*}{2I^{*2}} \;.$$

Again from (2.7) it follows that

$$\sum_{j=1}^{m} c_{ij} = \frac{J^i_{\theta\theta\theta}+K^i_{\theta\theta\theta}}{I^*} + \lambda_i I^i_{\theta\theta}I^* + \left(\sum_{j=1}^{m}\lambda_j I^j_{\theta\theta}\right)I^i_{\theta\theta}$$

$$= \frac{2J^i_{\theta\theta\theta}+K^i_{\theta\theta\theta}}{I^*} - \frac{I^i_{\theta\theta}(2J^*+K^*)}{I^{*2}} \;;$$

$$\lambda_i I^i_{\theta\theta} = \frac{J^i_{\theta\theta\theta}}{I^*} - \frac{I^i_{\theta\theta}(3J^*+K^*)}{2I^{*2}}$$

for $i = 1,\dots,m$.

Consequently the minimum of $V(Q^*_1)$ is attained when

$$c_{ij} = \frac{I^i_{\theta\theta}J^j_{\theta\theta\theta}+I^j_{\theta\theta}J^i_{\theta\theta\theta}}{I^*} - \frac{I^i_{\theta\theta}I^j_{\theta\theta}(3J^*+K^*)}{I^{*2}} + \delta_{ij}\,\frac{J^i_{\theta\theta\theta}+K^i_{\theta\theta\theta}}{I^*}$$

and

$$Q^*_1 = \frac{1}{I^{*2}}\, z^*_\theta \sum_{i=1}^{m} \frac{J^i_{\theta\theta\theta}}{I^i_{\theta\theta}}\, z^i_\theta - \frac{3J^*+K^*}{2I^{*3}}\, z^{*2}_\theta \;.$$

Hence the asymptotically best estimator $\hat{\theta}^{**}$ is expanded, after some rearrangements of terms into the form

$$\sqrt{n}(\hat{\theta}^{**}-\theta) = \frac{1}{I^*}\, z^*_\theta + \frac{1}{\sqrt{n}}\, Q_0 + \frac{1}{\sqrt{n}}\, Q_1 + \frac{1}{\sqrt{n}}\, Q_4 + \frac{1}{\sqrt{n}}\, Q_5 + o_p\!\left(\frac{1}{\sqrt{n}}\right) \;,$$

and the asymptotic deficiency of $\hat{\theta}^{**}$ relative to $\hat{\theta}^{*}$ is given by $I\{V(Q_4) + V(Q_5)\}$. Comparing the above with the stochastic expansion of $\hat{\bar{\theta}}$, we get

$$\hat{\theta}^{**} - \hat{\bar{\theta}} = -\frac{1}{n} Q_6 + o_p\left(\frac{1}{n}\right)$$

$$= -\frac{1}{2I^{*}} \sum_{i=1}^{m} (J_{\theta\theta\theta}^{i}+K_{\theta\theta\theta}^{i})(\hat{\theta}_i-\hat{\bar{\theta}})^2 + o_p\left(\frac{1}{n}\right) .$$

Hence an estimator which is asymptotically equivalent to $\hat{\theta}^{**}$ up to the order n^{-1} is given

$$\hat{\theta}^{***} = \hat{\bar{\theta}} - \frac{1}{2\hat{I}^{*}} \sum_{i=1}^{m} (\hat{J}_{\theta\theta\theta}^{i}+\hat{K}_{\theta\theta\theta}^{i})(\hat{\theta}_i-\hat{\bar{\theta}})^2 ,$$

where $\hat{I}^{*}$, $\hat{J}_{\theta\theta\theta}^{i}$ and $\hat{K}_{\theta\theta\theta}^{i}$ are estimators of I^{*}, $J_{\theta\theta\theta}^{i}$ and $K_{\theta\theta\theta}^{i}$, respectively, obtained substituting $\hat{\bar{\theta}}$ and $\hat{\xi}_i$ for θ and ξ_i ($i = 1, \ldots, m$). The adjusting term

$$\frac{1}{2\hat{I}^{*}} \sum_{i=1}^{m} (\hat{J}_{\theta\theta\theta}^{i}+\hat{K}_{\theta\theta\theta}^{i})(\hat{\theta}_i-\hat{\bar{\theta}})^2$$

may be considered as the estimated non-linear effect in the best combination of $\hat{\theta}_i$'s.

From the above we have established the following theorem.

THEOREM 2.1. Assume that (A.1) $\sim$ (A.5) hold. The stochastic expansions of the estimators $\hat{\theta}_0$, $\hat{\theta}^{*}$, $\tilde{\theta}$, $\hat{\bar{\theta}}$, $\hat{\bar{\theta}}^{*}$ and $\hat{\theta}^{**}$ are given by

$$\sqrt{n}(\hat{\theta}_0-\theta) = \frac{1}{I^{*}} Z_{\theta}^{*} + \frac{1}{\sqrt{n}} Q_0 + o_p\left(\frac{1}{\sqrt{n}}\right) ;$$

$$\sqrt{n}(\hat{\theta}^{*}-\theta) = \frac{1}{I^{*}} Z_{\theta}^{*} + \frac{1}{\sqrt{n}} Q_0 + \frac{1}{\sqrt{n}} Q_1 + o_p\left(\frac{1}{\sqrt{n}}\right) ;$$

$$\sqrt{n}(\tilde{\theta}-\theta) = \frac{1}{I^{*}} Z_{\theta}^{*} + \frac{1}{\sqrt{n}} Q_0 + \frac{1}{\sqrt{n}} Q_1 + \frac{1}{\sqrt{n}} Q_2 + o_p\left(\frac{1}{\sqrt{n}}\right) ;$$

$$\sqrt{n}(\hat{\bar{\theta}}-\theta) = \frac{1}{I^{*}} Z_{\theta}^{*} + \frac{1}{\sqrt{n}} Q_0 + \frac{1}{\sqrt{n}} Q_1 + \frac{1}{\sqrt{n}} Q_4 + \frac{1}{\sqrt{n}} Q_5 + \frac{1}{\sqrt{n}} Q_6 + o_p\left(\frac{1}{\sqrt{n}}\right);$$

$$\sqrt{n}(\hat{\bar{\theta}}^{*}-\theta) = \frac{1}{I^{*}} Z_{\theta}^{*} + \frac{1}{\sqrt{n}} Q_0 + \frac{1}{\sqrt{n}} Q_1 - \frac{1}{\sqrt{n}} Q_2 + \frac{1}{\sqrt{n}} Q_4 + \frac{1}{\sqrt{n}} Q_5 + o_p\left(\frac{1}{\sqrt{n}}\right) ;$$

$$\sqrt{n}(\hat{\theta}^{**}-\theta) = \frac{1}{I^*} Z_\theta^* + \frac{1}{\sqrt{n}} Q_0 + \frac{1}{\sqrt{n}} Q_1 + \frac{1}{\sqrt{n}} Q_4 + \frac{1}{\sqrt{n}} Q_5 + o_p\!\left(\frac{1}{\sqrt{n}}\right) ;$$

respectively, where Q_0, Q_1, Q_2, Q_4, Q_5 and Q_6 are given in the above and $\mathrm{Cov}(Q_i, Q_j) = 0$ for $i \neq j$ $(i,j = 0,1,2,4,5,6)$ except for $i = 2$ and $j = 6$. The asymptotic deficiencies of $\tilde{\theta}$, $\hat{\tilde{\theta}}$, $\hat{\tilde{\theta}}^*$ and $\hat{\theta}^{**}$ relative to $\hat{\theta}^*$ are given in the table below.

Estimator T_n	Asymptotic deficiency of T_n relative to $\hat{\theta}_n^*$
$\tilde{\theta}$	$I^* V(Q_2)$
$\hat{\tilde{\theta}}$	$I^* \{V(Q_4) + V(Q_5) + V(Q_6)\}$
$\hat{\tilde{\theta}}^*$	$I^* \{V(Q_2) + V(Q_4) + V(Q_5)\}$
$\hat{\theta}^{**}$	$I^* \{V(Q_4) + V(Q_5)\}$

From Theorem 2.1 it is seen that $\hat{\theta}^*$ is asymptotically better than $\tilde{\theta}$, $\hat{\tilde{\theta}}$, $\hat{\tilde{\theta}}^*$ and $\hat{\theta}^{**}$ up to order $n^{-\frac{1}{2}}$.

Using the estimator $\tilde{\theta}$ we may construct an estimator $\hat{\tilde{\theta}}^*$ asymptotically equivalent to $\hat{\theta}^*$ as follows:

$$\hat{\tilde{\theta}}^* = \tilde{\theta} - \frac{1}{2 \sum\limits_{i=1}^{m} \mathcal{I}_{\theta\theta}^i} \sum_{i=1}^{m} (3\hat{J}_{\theta\theta\theta}^i + \hat{K}_{\theta\theta\theta}^i)(\hat{\theta}_i - \tilde{\theta})^2 ,$$

where $\hat{J}_{\theta\theta\theta}^i = J_{\theta\theta\theta}^i(\hat{\theta}_i)$ and $\hat{K}_{\theta\theta\theta}^i = K_{\theta\theta\theta}^i(\hat{\theta}_i)$ (see Akahira and Takeuchi, 1981b).

We can formulate still another estimator, i.e., the linear estimator $\tilde{\theta}^*$ given by

$$\tilde{\theta}^* = \frac{\sum\limits_{i=1}^{m} \mathcal{I}_{\theta\theta}^{i*} \hat{\theta}_i}{\sum\limits_{i=1}^{m} \mathcal{I}_{\theta\theta}^{i*}} ,$$

where for each $i = 1, \ldots, m$ the weight $\mathcal{I}_{\theta\theta}^{i*}$ is defined by

$$\mathcal{I}_{\theta\theta}^{i*} = - \frac{1}{n} \frac{\partial^2}{\partial \theta^2} L_i(\tilde{\theta}, \hat{\xi}_i)$$

AKAHIRA-TAKEUCHI $\qquad$ 33

$(\tilde{\theta}$ may be replaced by $\hat{\theta}$, and $\hat{\xi}_i)$. Then we get

$$\sqrt{n}(\tilde{\theta}^* - \theta) = \frac{1}{I^*} Z_\theta^* + \frac{1}{\sqrt{n}} Q_0 + \frac{1}{\sqrt{n}} Q_1 - \frac{1}{\sqrt{n}} Q_2 + o_p\left(\frac{1}{\sqrt{n}}\right) .$$

Therefore the asymptotic deficiency of $\tilde{\theta}^*$ relative to $\hat{\theta}^*$ is equal to that of $\tilde{\theta}$. Moreover, we have from the above,

$$\frac{\tilde{\theta}^* + \tilde{\theta}}{2} = \frac{1}{2} \sum_{i=1}^m \left(\frac{\mathcal{I}_{\theta\theta}^i}{\sum_{i=1}^m \mathcal{I}_{\theta\theta}^i} + \frac{\mathcal{I}_{\theta\theta}^{i*}}{\sum_{i=1}^m \mathcal{I}_{\theta\theta}^{i*}} \right) \hat{\theta}_i = \hat{\theta}^* + o_p(\tfrac{1}{n}) .$$

Thus the estimator $(\tilde{\theta}^* + \tilde{\theta})/2$ is asymptotically equivalent to $\hat{\theta}^*$ up to the third order, i.e., order n^{-1} , that is, the asymptotic deficiency of $(\tilde{\theta}^* + \tilde{\theta})/2$ relative to $\hat{\theta}^*$ is zero, which can also be shown by the more direct way as follows.

Let $\hat{\theta}^*$ be the MLE. Then we have

$$0 = \frac{1}{n} \sum_1^m \frac{\partial}{\partial\theta} L_i(\hat{\theta}^*, \hat{\xi}_i^*)$$

$$= \frac{1}{n} \sum_1^m \frac{\partial^2}{\partial\theta^2} L_i(\hat{\theta}_i, \hat{\xi}_i)(\hat{\theta}^* - \hat{\theta}_i) + \frac{1}{2n} \sum_1^m \frac{\partial^3}{\partial\theta^3} L_i(\hat{\theta}_i, \hat{\xi}_i)(\hat{\theta}^* - \hat{\theta}_i)^2 + o_p(\tfrac{1}{n}) .$$

Similarly we have

$$0 = \frac{1}{n} \sum_1^m \frac{\partial}{\partial\theta} L_i(\hat{\theta}_i, \hat{\xi}_i)$$

$$= \frac{1}{n} \sum_1^m \frac{\partial^2}{\partial\theta^2} L_i(\hat{\theta}^*, \hat{\xi}^*)(\hat{\theta}_i - \hat{\theta}^*) + \frac{1}{2n} \sum_1^m \frac{\partial^3}{\partial\theta^3} L_i(\hat{\theta}^*, \hat{\xi}^*)(\hat{\theta}_i - \hat{\theta}^*)^2 + o_p(\tfrac{1}{n}) .$$

And noting that

$$\frac{1}{n} \sum_1^m \frac{\partial^3}{\partial\theta^3} L_i(\hat{\theta}_i, \hat{\xi}_i) = \frac{1}{n} \sum_1^m \frac{\partial^3}{\partial\theta^3} L_i(\hat{\theta}^*, \hat{\xi}_i^*) + o_p(1) ;$$

$$\frac{1}{n} \sum_1^m \frac{\partial^2}{\partial\theta^2} L_i(\hat{\theta}^*, \hat{\xi}_i) = \frac{1}{n} \sum_1^m \frac{\partial^2}{\partial\theta^2} L_i(\hat{\theta}^*, \hat{\xi}_i^*) + o_p(1) ,$$

we have

$$\frac{1}{n} \sum_1^m \left\{ \frac{\partial^2}{\partial\theta^2} L_i(\hat{\theta}_i, \hat{\xi}_i) + \frac{\partial^2}{\partial\theta^2} L_i(\hat{\theta}^*, \hat{\xi}_i^*) \right\} (\hat{\theta}^* - \hat{\theta}_i) = o_p(\tfrac{1}{n}) .$$

Thus we get

$$\hat{\theta}^* = \frac{\sum_{i}^{m} (\mathcal{J}^i_{\theta\theta} + \mathcal{J}^{i*}_{\theta\theta})\hat{\theta}_i}{\sum_{i}^{m} (\mathcal{J}^i_{\theta\theta} + \mathcal{J}^{i*}_{\theta\theta})} + o_p\left(\frac{1}{n}\right)$$

$$= \frac{1}{2}(\tilde{\theta} + \tilde{\theta}^*) + o_p\left(\frac{1}{n}\right) .$$

Hence we have established the following theorem·

THEOREM 2.2. Under the assumptions (A.1) ~ (A.5), the estimator $\hat{\tilde{\theta}}^*$ is asymptotically equivalent to the MLE $\hat{\theta}^*$ up to the order $n^{-\frac{1}{2}}$ and its asymptotic deficiency relative to $\hat{\theta}^*$ is zero, and the estimator $(\tilde{\theta}^* + \tilde{\theta})/2$ is asymptotically equivalent to $\hat{\theta}^*$ up to the order n^{-1} and its asymptotic deficiency relative to $\hat{\theta}^*$ is zero.

3. Discussion.

$V(Q_i)$ $(i = 0,1,\ldots,6)$ can be interpreted in the following way. $V(Q_0)$ consists of two parts, i.e.,

$$\frac{1}{I^{*4}} (M^* I^* - J^{*2}) \quad \text{and} \quad \frac{(J^* + K^*)^2}{2I^{*4}}$$

the first being the "loss of information" of the MLE and the second due to the condition of (higher order asymptotic) unbiasedness (Amari, 1980). These two factors are unavoidable even when we know the true values of the nuisance parameter. Then for the MLE $\hat{\theta}^*$ we have three additional terms

$$\frac{1}{I^{*2}} \sum_{i=1}^{m} \frac{1}{I^i_{\xi\xi}} \left(M^i_{\theta\xi\theta\xi} - \frac{J^{i2}_{\theta\xi\theta}}{I^i_{\theta\theta}} - \frac{J^{i2}_{\theta\xi\xi}}{I^i_{\xi\xi}} \right) ,$$

$$\frac{1}{2I^{*2}} \sum_{i=1}^{m} \frac{J^{i2}_{\theta\xi\xi}}{I^{i2}_{\xi\xi}}$$

and

$$\frac{1}{I^{*2}} \sum_{i=1}^{m} \left(\frac{1}{I^i_{\theta\theta}} - \frac{1}{I^*} \right) \frac{J^{i2}_{\theta\xi\theta}}{I^i_{\xi\xi}}$$

of which the first is the "loss of information" due to the presence of

AKAHIRA-TAKEUCHI   35

unknown nuisance parameters, the second is the unbiasedness effect and the third represents the loss due to pooling.

$V(Q_2)$ can be considered as representing the effect of the non-linearity of the optimum combination of the MLE's to get the MLE based on the whole sample.

Comparing $\hat{\theta}^*$ with $\tilde{\theta}$, it is shown that the information contained in the statistics $\mathcal{I}^i_{\theta\theta}$ $(i = 1,\ldots,m)$ are lost when we use $\hat{I}^i_{\theta\theta}$ $(i = 1,\ldots,m)$ in its stead, and $V(Q_5)$ represents the loss with respect to θ and $V(Q_4)$ with respect to ξ_i $(i = 1,\ldots,m)$.

Comparison of $\hat{\bar{\theta}}$ and $\hat{\theta}^*$ does not lead to any uniform conclusion, which may seem strange since in $\hat{\theta}^*$ a better estimator $\hat{\bar{\theta}}$ of θ is used instead of $\hat{\theta}_i$ $(i = 1,\ldots,m)$, but it can be interpreted that $\hat{\bar{\theta}}$ could be better since $\hat{I}^i_{\theta\theta}(\hat{\theta}_i,\hat{\xi}_i)$ may be closer to $\mathcal{I}^i_{\theta\theta}$ than $\hat{I}^i_{\theta\theta}(\hat{\bar{\theta}},\hat{\xi}_i)$.

Comparing $\hat{\theta}^{**}$ with $\hat{\theta}^*$, it is shown that the loss of information caused by restricting the estimators to be functions of $\hat{\theta}_i$ and $\hat{\xi}_i$ $(i = 1,\ldots,m)$ alone is expressed by $V(Q_4) + V(Q_5)$. Of the two terms Q_4 and Q_5 the latter is the weighted mean of the asymptotically ancillary statistics

$$Z^i_{\theta\theta} - \frac{J^i_{\theta\theta\theta}}{I^i_{\theta\theta}} Z^i_\theta \qquad (i = 1,\ldots,m)$$

and is easy to interpret, but the former is a weighted sample covariance between $(Z^i_\theta/I^i_{\theta\theta}) - (Z^*_\theta/I^*)$ and $Z^i_\xi/I^i_{\xi\xi}$, but its meaning is harder to grasp.

In case of the symmetric location-scale problem in which the density function $g(x,\theta,\xi_i)$ is expressed as

$$\frac{1}{\xi_i} f\left(\frac{x-\theta}{\xi_i}\right)$$

for each $i = 1, \ldots, m$ and where f is an even function, we have for each $i = 1, \ldots, m$,

$$I^i_{\theta\theta} = \frac{A}{\xi^2_i} \; ; \quad I^i_{\xi\xi} = \frac{B-1}{\xi^2_i} \; ; \quad I^i_{\theta\xi} = E\left[\{\frac{\partial}{\partial\theta} g(X,\theta,\xi)\}\{\frac{\partial}{\partial\xi} g(X,\theta,\xi)\}\right] = 0$$

where A and B are constants determined as

$$A = \int \{\phi'(x)\}^2 f(x)dx \; ; \quad B = \int x^2 \{\phi'(x)\}^2 f(x)dx$$

with $\phi(x) = \log f(x)$. Further we obtain for each $i = 1, \ldots, m$,

$$J^i_{\theta\theta\theta} = 0 \; ; \quad K^i_{\theta\theta\theta} = 0 \; ;$$

$$J^i_{\theta\theta\xi} = \frac{1}{2\xi_i{}^3} (3A+C) \; ; \quad J^i_{\theta\xi\theta} = \frac{1}{2\xi_i{}^3} (-A+C) \; ; \quad K^i_{\theta\theta\xi} = \frac{1}{\xi_i{}^3} (-A-C) \; ,$$

where

$$C = \int x\{\phi'(x)\}^3 f(x)dx \; .$$

And for each $i = 1, \ldots, m$,

$$J^i_{\theta\xi\xi} = 0 \; ; \quad J^i_{\xi\xi\theta} = 0 \; .$$

It is further calculated that for each $i = 1, \ldots, m$,

$$M^i_{\theta\theta\theta\theta} = \frac{1}{\xi_i{}^4} (D-A^2) \; ; \quad M^i_{\theta\xi\theta\xi} = \frac{1}{\xi_i{}^4} (F-C) \; ,$$

where

$$D = \int \{\phi''(x)\}^2 f(x)dx \quad \text{and} \quad F = \int x^2\{\phi''(x)\}^2 f(x)dx \; .$$

Therefore, in the estimation of the common location parameter of different populations of the same shape but with different scale parameters, we have $Q_2 \equiv 0$ and $Q_6 \equiv 0$, thus we have $\tilde{\theta} \equiv \hat{\theta}^*$ and $\hat{\tilde{\theta}} = \hat{\theta}^{**}$. Especially, when the underlying distribution is normal, we have for each $i = 1, \ldots, m$,

$$J^i_{\theta\theta\xi} = \frac{1}{2\xi^3} \left\{3\int x^2 f(x)dx - \int x^4 f(x)dx\right\} = 0$$

and $J^i_{\theta\xi\theta} + K^i_{\theta\theta\xi} = 0$, which implies that $Q_4 \equiv 0$; and also $Z^i_{\theta\theta} = 0$, hence $Q_1 \equiv Q_5 \equiv 0$. Then we see that $\hat{\theta}^*$, $\tilde{\theta}$, $\hat{\tilde{\theta}}$, $\hat{\theta}^*$ and $\hat{\theta}^{**}$ are all asymptotically equivalent, and the most commonly used estimator

AKAHIRA-TAKEUCHI   37

$$\hat{\bar{\theta}} = \frac{\displaystyle\sum_{i=1}^{m} \hat{\theta}_i / \hat{\xi}_i^2}{\displaystyle\sum_{i=1}^{m} 1/\hat{\xi}_i^2}$$

with $\hat{\theta}_i = \bar{X}_i = \frac{1}{n} \sum_{j=1}^{n} X_{ij}$ and $\hat{\xi}_i^2 = \frac{1}{n-1} \sum_{j=1}^{n} (X_{ij} - \bar{X}_i)^2$ ($i = 1,\ldots,m$) is, although not identical with the MLE, asymptotically equivalent to it up to the order n^{-1}.

Acknowledgements.

This paper was written while the first author was at the Limburgs Universitair Centrum in Belgium as a guest professor. The visit was supported by a grant of the Belgian National Scientific Foundation (N.F.W.O.). He is grateful to Professor H. Callaert of the LUC for inviting him and for his interest in this and related work.

References

Akahira, M.: On asymptotic deficiency of estimators. Austral. J. Statist. 23, 67-72 (1981).

Akahira, M. and Takeuchi, K.: Asymptotic Efficiency of Statistical Estimators: Concepts and Higher Order Asymptotic Efficiency, Lecture Notes in Statistics 7, New York-Heidelberg-Berlin, Springer (1981a).

Akahira, M. and Takeuchi, K.: On asymptotic deficiency of estimators in pooled samples. Technical Report of the Limburgs Universitair Centrum, Belgium (1981b).

Amari, S.: Theory of information space: A differential-geometrical foundation of statistics. RAAG Report, 106 (1980).

Fisher, R. A.: Theory of statistical estimation. Proc. Camb. Phi.. Soc. 22, 700-725 (1925).

Ghosh, J. K., Sinha, B. K. and Wieand, H. S.: Second order efficiency of the mle with respect to any bounded bowl-shaped loss function. Ann. Statist. 8, 506-521 (1980).

Hodges, J. L. and Lehmann, E. L.: Deficiency. Ann. Math. Statist. 41, 783-801 (1970).

Pfanzagl, J. and Wefelmeyer, W.: A third order optimum property of the
    maximum likelihood estimator. J. Multivariate Anal. $\underline{8}$, 1-29 (1978).
Takeuchi, K. and Akahira, M.: Third order asymptotic efficiency of max-
    imum likelihood estimator for multiparameter exponential case.
    Rep. Univ. Electro-Comm. $\underline{28}$, 271-293 (1978).

Masafumi Akahira   Kei Takeuchi
Department of Mathematics   Faculty of Economics
University of Electro-Communications   University of Tokyo
Chofu, Tokyo 182   Hongo, Bunkyo-ku
Japan   Tokyo 113
                                        Japan

Ann. Inst. Statist. Math.
37 (1985), Part A, 17–26

ESTIMATION OF A COMMON PARAMETER FOR POOLED SAMPLES FROM THE UNIFORM DISTRIBUTIONS

Masafumi Akahira and Kei Takeuchi

(Received Sept. 6, 1983; revised Mar. 12, 1984)

Summary

The problem to estimate a common parameter for the pooled sample from the uniform distributions is discussed in the presence of nuisance parameters. The maximum likelihood estimator (MLE) and others are compared and it is shown that the MLE based on the pooled sample is not (asymptotically) efficient.

1. Introduction

In regular cases the asymptotic deficiencies of asymptotically efficient estimators were calculated in pooled sample from the same distribution (Akahira [5]) and in the presence of nuisance parameters (Akahira and Takeuchi [7]). In non-regular cases the asymptotic optimality of estimators was discussed in Akahira [4], Akahira and Takeuchi [6], Ibragimov and Has'minskii [10], Jurečková [11] and others, and also recently a Monte Carlo study on the estimator considered in Akahira [1]–[3] has been done by Antoch [9].

In this paper we consider the problem to estimate an unknown real-valued parameter θ based on m samples of size n from the uniform distributions on the interval $(\theta-\xi_i, \theta+\xi_i)$ $(i=1,\cdots,m)$ with different nuisance parameters which is treated as a typical example in non-regular cases. In some cases the MLE and other estimators will be compared and it will be shown that the MLE based on the pooled sample is not better for both a sample of a fixed size and a large sample. Related results can be found in Akai [8].

2. Results

Suppose that it is required to estimate an unknown real-valued

Key words and phrases: Maximum likelihood estimator, weighted estimator, uniform distributions.

parameter θ based on m samples of size n whose values are X_{ij} ($i=1$, $\cdots$, m; $j=1,\cdots$, n) from the uniform distributions on the interval ($\theta-\xi_i$, $\theta+\xi_i$) with different nuisance parameters ξ_i ($i=1,\cdots$, m). For each i let $X_{i(1)}<X_{i(2)}<\cdots<X_{i(n)}$ be order statistics from X_{i1}, $X_{i2},\cdots$, X_{in}. We consider some cases.

Case I. $\xi_i=\xi$ ($i=1,\cdots$, m) *are unknown.*

The MLE $\hat{\theta}_{ML}$ of θ based on the pooled sample $\{X_{ij}\}$ is given by

$$\hat{\theta}_{ML}=\frac{1}{2}(\min_{1\leq i\leq m} X_{i(1)}+\max_{1\leq i\leq m} X_{i(n)}) .$$

An estimator $\hat{\theta}_n$ of θ based on the sample of size n is called to be (asymptotically) median unbiased if

$$\Pr \{\hat{\theta}_n\leq\theta\} = \Pr \{\hat{\theta}_n\geq\theta\} = \frac{1}{2}$$

$\left(\lim_{n\to\infty}\left| \Pr \{\hat{\theta}_n\leq\theta\} -\frac{1}{2}\right| =\lim_{n\to\infty}\left| \Pr \{\hat{\theta}_n\geq\theta\} -\frac{1}{2}\right| =0\right.$ uniformly in some neigh-

borhood of $\theta\bigg)$ (e.g. see Akahira and Takeuchi [6]). Then it is shown that $\hat{\theta}_{ML}$ is one-sided asymptotically efficient in the sense that for any asymptotically median unbiased estimator $\hat{\theta}_n$

$$\lim_{n\to\infty} [\Pr \{n(\hat{\theta}_{ML}-\theta)\leq t\} -\Pr \{n(\hat{\theta}_n-\theta)\leq t\}]\geq 0 \qquad \text{for all } t>0 ;$$

or

$$\overline{\lim_{n\to\infty}} [\Pr \{n(\hat{\theta}_{ML}-\theta)\leq t\} -\Pr \{n(\hat{\theta}_n-\theta)\leq t\}]\leq 0 \qquad \text{for all } t<0 ,$$

since in this case $\hat{\theta}_{ML}$ is actually the MLE from the sample of size mn (see Akahira and Takeuchi [6]).

Case II. ξ_i ($i=1,\cdots$, m) *are known.*

The MLE $\hat{\theta}^*_{ML}$ of θ based on the pooled sample $\{X_{ij}\}$ is given by

$$\hat{\theta}^*_{ML}=\frac{1}{2}\{\max_{1\leq i\leq m} (X_{i(n)}-\xi_i)+\min_{1\leq i\leq m} (X_{i(1)}+\xi_i)\} .$$

Then it can be shown in a similar way as in [4] and [6] that $\hat{\theta}^*_{ML}$ is two-sided asymptotically efficient in the sense that for any asymptotically median unbiased estimator $\hat{\theta}_n$

$$\lim_{n\to\infty} [\Pr \{n|\hat{\theta}^*_{ML}-\theta|\leq t\} -\Pr \{n|\hat{\theta}_n-\theta|\leq t\}]\geq 0 \qquad \text{for all } t>0 .$$

(For the definition see Akahira and Takeuchi [6], page 72 and Akahira [4]).

Case III. ξ_i $(i=1,\cdots,m)$ *are unknown.*

For each i the MLE's $\hat{\theta}_i$ and $\hat{\xi}_i$ of θ and ξ_i based on the sample $X_{i1},\cdots,X_{in}$ is given by

$$\hat{\theta}_i=\frac{1}{2}(X_{i(1)}+X_{i(n)})\ ;\qquad \hat{\xi}_i=\frac{1}{2}(X_{i(n)}-X_{i(1)})\ ,$$

respectively.   Let $\hat{\theta}_{ML}$ be the MLE of θ based on the pooled sample $\{X_{ij}\}$.   Then it will be shown that $\hat{\theta}_{ML}$ is not two-sided asymptotically efficient.   We define

$$\hat{\theta}_n^*=\frac{1}{2}\{\max_{1\leq i\leq m}(\hat{\theta}_i+\hat{\xi}_i-\xi_i^0)+\min_{1\leq i\leq m}(\hat{\theta}_i-\hat{\xi}_i+\xi_i^0)\}\ .$$

Then it can be shown from Case II that $\hat{\theta}_n^*$ is asymptotically locally best estimator of θ at $\xi_i=\xi_i^0$ $(i=1,\cdots,m)$ in the sense that for any asymptotically median unbiased estimator $\hat{\theta}_n$

$$\lim_{n\to\infty}[P_{\theta,\xi_1^0,\cdots,\xi_m^0}\{n|\hat{\theta}_n^*-\theta|\leq t\}-P_{\theta,\xi_1^0,\cdots,\xi_m^0}\{n|\hat{\theta}_n-\theta|\leq t\}]\geq 0$$

for all $t>0$.

First we shall obtain the MLE $\hat{\theta}_{ML}$ based on the pooled sample $\{X_{ij}\}$. Let $f_i(x,\theta)$ be a density function of the uniform distribution on the interval $(\theta-\xi_i,\theta+\xi_i)$.   Since the likelihood function $L(\theta\ ;\ \xi_1,\cdots,\xi_m)$ is given by

$$L(\theta\ ;\ \xi_1,\cdots,\xi_m)=\prod_{i=1}^{m}\prod_{j=1}^{n}f_i(x_j,\theta)$$

$$=\begin{cases} \dfrac{1}{2^n}\left(\dfrac{1}{\prod\limits_{i=1}^{m}\xi_i}\right)^n & \text{for } x_{i(n)}-\xi_i\leq\theta\leq x_{i(1)}+\xi_i \\ & \qquad\qquad (i=1,\cdots,m)\ ; \\[2ex] 0 & \text{otherwise}\ . \end{cases}$$

In order to obtain the MLE $\hat{\theta}_{ML}$ it is enough to find θ minimizing $\prod\limits_{i=1}^{m}\xi_i$ under the condition $\xi_i\geq\max\{\theta-x_{i(1)},x_{i(n)}-\theta\}$ for all i.   Let $\hat{\theta}^*$ be some estimator of θ based on the pooled sample $\{X_{ij}\}$.   For each i we put $\xi_i^*=\max\{\hat{\theta}^*-x_{i(1)},x_{i(n)}-\hat{\theta}^*\}$.   Then we have for each i

$$\xi_i^*=\max\{\hat{\theta}^*-\hat{\theta}_i+\hat{\xi}_i,\hat{\theta}_i+\hat{\xi}_i-\hat{\theta}^*\}=\hat{\xi}_i+|\hat{\theta}_i-\hat{\theta}^*|\ .$$

Hence the MLE $\hat{\theta}_{ML}$ is given by $\hat{\theta}^*$ minimizing

$$\prod_{i=1}^{m}(\hat{\xi}_i+|\hat{\theta}_i-\hat{\theta}^*|)\ ,$$

that is, such an estimator $\hat{\theta}^*$ is given by either of the estimators $\hat{\theta}_i$ $(i=1,\cdots,m)$.   Since

$$\prod_{i=1}^{m}(\hat{\xi}_i+|\hat{\theta}_i-\hat{\theta}^*|)=\prod_{i=1}^{m}\hat{\xi}_i\prod_{i=1}^{m}\left(1+\frac{|\hat{\theta}_i-\hat{\theta}^*|}{\hat{\xi}_i}\right),$$

for sufficiently large n it is asymptotically equivalent to

$$\prod_{i=1}^{m}\hat{\xi}_i\left(1+\sum_{i=1}^{m}\frac{|\hat{\theta}_i-\hat{\theta}^*|}{\hat{\xi}_i}\right).$$

Hence it is seen that for sufficiently large n the MLE $\hat{\theta}_{ML}$ is asymptotically equivalent to a weighted median by the weights $1/\hat{\xi}_i$ ($i=1,\cdots,m$).

Next we shall discuss the comparison among $\hat{\theta}_{ML}$, the weighted estimator and other estimators. We consider the case when $m=2$. Then for each $i=1,2$, $X_{i1},\cdots,X_{in}$ are independently, identically and uniformly distributed random variables on $(\theta-\xi_i,\theta+\xi_i)$. Then for each $i=1,2$, the joint density function $f_n(x,y;\theta,\xi_i)$ of $\hat{\theta}_i$ and $\hat{\xi}_i$ is given by

$$(2.1)\qquad f_n(x,y;\theta,\xi_i)=\begin{cases}\dfrac{n(n-1)}{2\xi_i^{n}}y^{n-2} & \text{for } 0\leq y\leq\xi_i \text{ and}\\[2mm] & \quad\theta-\xi_i+y\leq x\leq\theta+\xi_i-y\,;\\[4mm] 0 & \text{otherwise}.\end{cases}$$

For each $i=1,2$ the density function $f_n(x;\theta,\xi_i)$ of $\hat{\theta}_i$ is given by

$$(2.2)\qquad f_n(x;\theta,\xi_i)=\begin{cases}\dfrac{n}{2\xi_i^{n}}(\xi_i-|x-\theta|)^{n-1} & \text{for } \theta-\xi_i<x<\theta+\xi_i\,;\\[4mm] 0 & \text{otherwise}.\end{cases}$$

Also for each $i=1,2$ the conditional density function $f_n(x|y;\theta,\xi_i)$ of $\hat{\theta}_i$ given $\hat{\xi}_i$ is given by

$$(2.3)\qquad f_n(x|y;\theta,\xi_i)=\begin{cases}\dfrac{1}{2(\xi_i-y)} & \text{for } \theta-\xi_i+y\leq x\leq\theta+\xi_i-y\,;\\[4mm] 0 & \text{otherwise},\end{cases}$$

that is, the conditional distribution of $\hat{\theta}_i$ given $\hat{\xi}_i$ is uniform distribution on the interval $(\theta-(\xi_i-\hat{\xi}_i),\theta+(\xi_i-\hat{\xi}_i))$.

For two (asymptotically) median unbiased estimators $\hat{\theta}^1$ and $\hat{\theta}^2$ of θ, $\hat{\theta}^1$ is called to be (asymptotically) better than $\hat{\theta}^2$ if

$$\Pr\{n|\hat{\theta}^1-\theta|\leq t|\hat{\xi}_1,\hat{\xi}_2\}\geq\Pr\{n|\hat{\theta}^2-\theta|\leq t|\hat{\xi}_1,\hat{\xi}_2\}\ \text{a.e.}\qquad\text{for all } t>0$$

$(\lim_{n\to\infty}[\Pr\{n|\hat{\theta}^1-\theta|\leq t\}-\Pr\{n|\hat{\theta}^2-\theta|\leq t\}]\geq0$ for all $t>0)$, and then for simplicity we denote it symbolically by $\hat{\theta}^1\underset{\text{as.}}{\succ}\hat{\theta}^2$ $(\hat{\theta}^1\succ\hat{\theta}^2)$, where $\Pr\{A|\hat{\xi}_1,\hat{\xi}_2\}$ denotes the conditional probability of A given $\hat{\xi}_1$ and $\hat{\xi}_2$. From (2.3) we see that the conditional density of $\hat{\theta}_1-\theta$ and $\hat{\theta}_2-\theta$ given $\hat{\xi}_1$ and $\hat{\xi}_2$

ESTIMATION OF A COMMON PARAMETER FOR POOLED SAMPLES   21

is given by

$$f_n(x_1, x_2 \mid \hat{\xi}_1, \hat{\xi}_2) = \begin{cases} \dfrac{1}{4\tau_1\tau_2} & \text{for } |x_1|<\tau_1 \text{ and } |x_2|<\tau_2 \,; \\ \\ 0 & \text{otherwise} , \end{cases}$$

where $\tau_i = \xi_i - \hat{\xi}_i$ $(i=1, 2)$. If $c_1\tau_1 > c_2\tau_2$, then the conditional density function $f_n(y \mid \hat{\xi}_1, \hat{\xi}_2)$ of $\hat{\theta}_0 - \theta$ with $\hat{\theta}_0 = c_1\hat{\theta}_1 + c_2\hat{\theta}_2$ given $\hat{\xi}_1$ and $\hat{\xi}_2$ is given by

$$(2.4) \qquad f_n(y \mid \hat{\xi}_1, \hat{\xi}_2) = \begin{cases} \dfrac{1}{4c_1c_2\tau_1\tau_2}(c_1\tau_1+c_2\tau_2+y) & \text{for } y<-c_1\tau_1+c_2\tau_2 \,; \\ \\ \dfrac{1}{2c_1\tau_1} & \text{for } |y|<c_1\tau_1-c_2\tau_2 \,; \\ \\ \dfrac{1}{4c_1c_2\tau_1\tau_2}(c_1\tau_1+c_2\tau_2-y) & \text{for } y>c_1\tau_1-c_2\tau_2 \,. \end{cases}$$

If $c_1\tau_1 < c_2\tau_2$, then the conditional density function is given by

$$(2.5) \qquad f_n(y \mid \hat{\xi}_1, \hat{\xi}_2) = \begin{cases} \dfrac{1}{4c_1c_2\tau_1\tau_2}(c_1\tau_1+c_2\tau_2-y) & \text{for } y>-c_1\tau_1+c_2\tau_2 \,; \\ \\ \dfrac{1}{2c_2\tau_2} & \text{for } |y|<-c_1\tau_1+c_2\tau_2 \,; \\ \\ \dfrac{1}{4c_1c_2\tau_1\tau_2}(c_1\tau_1+c_2\tau_2+y) & \text{for } y<c_1\tau_1-c_2\tau_2 \,. \end{cases}$$

If

$$c_i = \hat{\xi}_j/(\hat{\xi}_1+\hat{\xi}_2) = c_i' \text{ (say)} \qquad (i \neq j; \ i, j=1, 2) ,$$

then

$$(2.6) \qquad \frac{1}{c_1'\tau_1} = \frac{\hat{\xi}_1+\hat{\xi}_2}{\hat{\xi}_2(\xi_1-\hat{\xi}_1)} \,; \qquad \frac{1}{c_2'\tau_2} = \frac{\hat{\xi}_1+\hat{\xi}_2}{\hat{\xi}_1(\xi_2-\hat{\xi}_2)} \,.$$

If

$$c_i = \hat{\xi}_j^2/(\hat{\xi}_1^2+\hat{\xi}_2^2) = c_i'' \text{ (say)} \qquad (i \neq j; \ i, j=1, 2) ,$$

then

$$(2.7) \qquad \frac{1}{c_1''\tau_1} = \frac{\hat{\xi}_1^2+\hat{\xi}_2^2}{\hat{\xi}_2^2(\xi_1-\hat{\xi}_1)} \,; \qquad \frac{1}{c_2''\tau_2} = \frac{\hat{\xi}_1^2+\hat{\xi}_2^2}{\hat{\xi}_1^2(\xi_2-\hat{\xi}_2)} \,.$$

Hence we have the following:

$$(2.8) \qquad \hat{\xi}_2 \gtreqless \hat{\xi}_1 \text{ if and only if } \frac{1}{c_1'\tau_1} = \frac{\hat{\xi}_1+\hat{\xi}_2}{\hat{\xi}_2(\xi_1-\hat{\xi}_1)} \gtreqless \frac{\hat{\xi}_1^2+\hat{\xi}_2^2}{\hat{\xi}_2^2(\xi_1-\hat{\xi}_1)} = \frac{1}{c_1''\tau_1} \,;$$

(2.9) $\hat{\xi}_2 \gtreqless \hat{\xi}_1$ if and only if $\dfrac{1}{c_2'\tau_2} = \dfrac{\hat{\xi}_1+\hat{\xi}_2}{\hat{\xi}_1(\xi_2-\hat{\xi}_2)} \lesseqgtr \dfrac{\hat{\xi}_1^2+\hat{\xi}_2^2}{\hat{\xi}_1^2(\xi_2-\hat{\xi}_2)} = \dfrac{1}{c_2''\tau_2}$.

On the other hand as is seen from the above discussion on the MLE, the MLE $\hat{\theta}_{ML}$ based on the pooled sample $\{X_{ij}\}$ is given by

$$\hat{\theta}_{ML} = \begin{cases} \hat{\theta}_1 & \text{if } \hat{\xi}_1 < \hat{\xi}_2 \,; \\ \hat{\theta}_2 & \text{if } \hat{\xi}_1 > \hat{\xi}_2 \,. \end{cases}$$

The conditional density of the MLE $\hat{\theta}_{ML}$ given $\hat{\xi}_i$ is given by (2.3). Now we consider two cases, i.e., $\xi_1=\xi_2$ and $\xi_1 \neq \xi_2$. Note that the first Case III_1 was also mentioned in the Case I with another optimum criterion.

Case III_1. $\xi_1=\xi_2=\xi$.
In the case when $c_i=c_i'=\hat{\xi}_j/(\hat{\xi}_1+\hat{\xi}_2)$ $(i \neq j;\ i,j=1,2)$,

$\hat{\xi}_1 \lesseqgtr \hat{\xi}_2$ if and only if $\hat{\xi}_1(\xi-\hat{\xi}_2) \lesseqgtr \hat{\xi}_2(\xi-\hat{\xi}_1)$,   i.e., $c_2\tau_2 \lesseqgtr c_1\tau_1$.

Hence we have

(2.10) $\hat{\xi}_1 \lesseqgtr \hat{\xi}_2$ if and only if $c_1'\tau_1+c_2'\tau_2 < \begin{cases} \tau_1 \\ \tau_2 \end{cases}$.

We also obtain

$$c_1''\tau_1+c_2''\tau_2-(c_1'\tau_1+c_2'\tau_2)=\dfrac{\hat{\xi}_1\hat{\xi}_2(\hat{\xi}_1-\hat{\xi}_2)^2}{(\hat{\xi}_1+\hat{\xi}_2)(\hat{\xi}_1^2+\hat{\xi}_2^2)} \geq 0 \ .$$

From (2.4) to (2.10) we have established the following theorem.

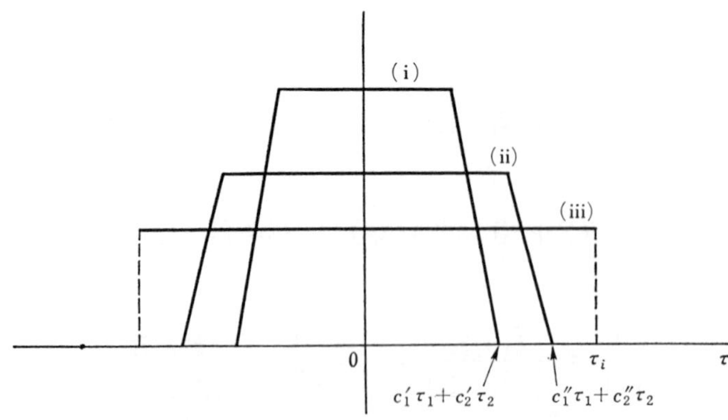

Fig. 2.1. Comparison of the conditional densities of $c_1\hat{\theta}_1+c_2\hat{\theta}_2-\theta$ given $\hat{\xi}_1$ and $\hat{\xi}_2$ with (i) $c_i=c_i'=\hat{\xi}_j/(\hat{\xi}_1+\hat{\xi}_2)$ $(i \neq j;\ i,j=1,2)$ and (ii) $c_i=c_i''=\hat{\xi}_j^2/(\hat{\xi}_1^2+\hat{\xi}_2^2)$ $(i \neq j;\ i,j=1,2)$ and of (iii) the MLE $\hat{\theta}_{ML}$ given $\hat{\xi}_i$. They are given by (2.5) and (2.3), respectively.

THEOREM 2.1. *If* $\hat{\xi}_1=\hat{\xi}_2$ *and* $\hat{\hat{\xi}}_1<\hat{\hat{\xi}}_2$, *then*

$$\frac{\hat{\hat{\xi}}_2\hat{\theta}_1+\hat{\hat{\xi}}_1\hat{\theta}_2}{\hat{\hat{\xi}}_1+\hat{\hat{\xi}}_2} \succ \frac{\hat{\hat{\xi}}_2^2\hat{\theta}_1+\hat{\hat{\xi}}_1^2\hat{\theta}_2}{\hat{\hat{\xi}}_1^2+\hat{\hat{\xi}}_2^2} \succ \hat{\theta}_{ML}=\hat{\theta}_1 \;.$$

If $\hat{\xi}_1=\hat{\xi}_2$ *and* $\hat{\hat{\xi}}_1>\hat{\hat{\xi}}_2$, *then*

$$\frac{\hat{\hat{\xi}}_2\hat{\theta}_1+\hat{\hat{\xi}}_1\hat{\theta}_2}{\hat{\hat{\xi}}_1+\hat{\hat{\xi}}_2} \succ \frac{\hat{\hat{\xi}}_2^2\hat{\theta}_1+\hat{\hat{\xi}}_1^2\hat{\theta}_2}{\hat{\hat{\xi}}_1^2+\hat{\hat{\xi}}_2^2} \succ \hat{\theta}_{ML}=\hat{\theta}_2 \;.$$

We assume that $c_1=c_2=1/2$ and $\xi_1=\xi_2=\xi$. By (2.2) it follows that for each $i=1,2$ the asymptotic density of $n(\hat{\theta}_i-\theta)$ is given by

$$(2.11) \qquad\qquad f_i(x)=\frac{1}{2\xi}e^{-|x|/\xi}\;.$$

Then the characteristic function $\phi(t)$ of the asymptotic density function of $n[\{(\hat{\theta}_1+\hat{\theta}_2)/2\}-\theta]$ is given by

$$\phi(t)=\frac{1}{(1+\xi^2t^2/4)^2}\;.$$

We may represent $\phi(t)$ as follows:

$$(2.12) \qquad\qquad \phi(t)=\frac{1-\xi^2t^2/4}{2(1+\xi^2t^2/4)^2}+\frac{1}{2(1+\xi^2t^2/4)}\;.$$

Since

$$\int_{-\infty}^{\infty}\frac{1}{\xi}\exp\left(-\frac{2}{\xi}|y|\right)\exp\,(ity)dy=\frac{1}{2}\int_{-\infty}^{\infty}\exp\,(-|x|)\exp\left(i\frac{\xi t}{2}x\right)dx$$

$$=\frac{1}{1+\xi^2t^2/4}\;,$$

$$\int_{-\infty}^{\infty}\frac{2}{\xi^2}|y|\exp\left(-\frac{2}{\xi}|y|\right)\exp\,(ity)dy=\frac{1}{2}\int_{-\infty}^{\infty}|x|\exp\,(-|x|)\exp\left(i\frac{\xi t}{2}x\right)dx$$

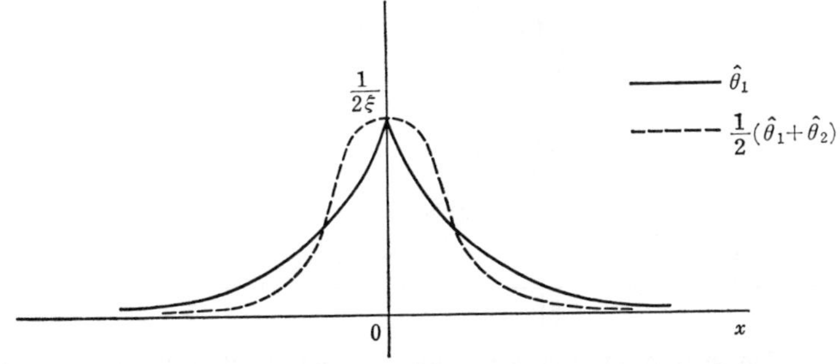

Fig. 2.2. Comparison of the asymptotic densities of $n(\hat{\theta}_1-\theta)$ and $n[\{(\hat{\theta}_1+\hat{\theta}_2)/2\}-\theta]$ given by (2.11) and (2.13), respectively.

24   MASAFUMI AKAHIRA AND KEI TAKEUCHI

$$= \frac{1-\xi^2 t^2/4}{(1+\xi^2 t^2/4)^2} \; ,$$

it follows from (2.12) that the asymptotic density of $n[\{(\hat{\theta}_1+\hat{\theta}_2)/2\} - \theta]$ is given by

$$(2.13) \qquad f(x) = \frac{1}{2\xi}\left(1+2\frac{|x|}{\xi}\right) \exp\left(-\frac{2}{\xi}|x|\right) \; .$$

From (2.11) and (2.13) we have established the following theorem.

THEOREM 2.2. *If* $\xi_1=\xi_2=\xi$, *then*

$$\frac{1}{2}(\hat{\theta}_1+\hat{\theta}_2) \underset{\text{as.}}{\succ} \hat{\theta}_{ML} = \begin{cases} \hat{\theta}_1 & \text{if } \hat{\xi}_1 < \hat{\xi}_2 \; ; \\ \hat{\theta}_2 & \text{if } \hat{\xi}_1 > \hat{\xi}_2 \; . \end{cases}$$

Case III$_2$. $\xi_1 \neq \xi_2$.

We consider only the case when $\xi_1 < \xi_2$. We assume that $c_1\xi_1 \neq c_2\xi_2$. By (2.2) it follows that for each $i=1,2$ the asymptotic density of $n(\hat{\theta}_i -\theta)$ is given by

$$(2.14) \qquad f_i(x) = \frac{1}{2\xi_i} \exp\left(-|x|/\xi_i\right) \; ,$$

and also its characteristic function $\phi_i(t)$ is given by

$$\phi_i(t) = \frac{1}{1+\xi_i^2 t^2} \; .$$

Hence the characteristic function $\phi^*(t)$ of the asymptotic density of $n(\hat{\theta}_0 - \theta) = n(c_1\hat{\theta}_1 + c_2\hat{\theta}_2 - \theta)$ is given by

$$\phi^*(t) = \frac{1}{(1+c_1^2\xi_1^2 t^2)(1+c_2^2\xi_2^2 t^2)} \; .$$

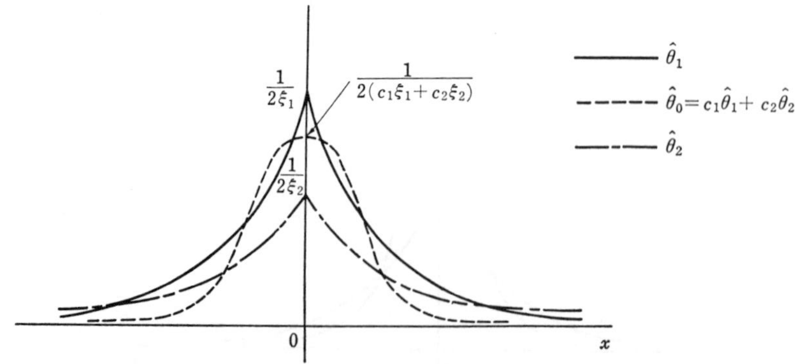

Fig. 2.3. Comparison of the asymptotic densities of $n(\hat{\theta}_1-\theta)$, $n(\hat{\theta}_0-\theta) = n(c_1\hat{\theta}_1+c_2\hat{\theta}_2-\theta)$ and $n(\hat{\theta}_2-\theta)$ given by (2.14), (2.15) and (2.14), respectively, when $\xi_1 < \xi_2$.

ESTIMATION OF A COMMON PARAMETER FOR POOLED SAMPLES 25

Since

$$\phi^*(t) = \frac{1}{(c_1^2\xi_1^2 - c_2^2\xi_2^2)}\left(\frac{c_1^2\xi_1^2}{1+c_1^2\xi_1^2 t^2} - \frac{c_2^2\xi_2^2}{1+c_2^2\xi_2^2 t^2}\right),$$

it follows that the asymptotic density $f(x)$ of $n(\hat\theta_0 - \theta)$ is given by

$$(2.15) \qquad f(x) = \frac{1}{2(c_1^2\xi_1^2 - c_2^2\xi_2^2)}\left\{c_1\xi_1 \exp\left(-\frac{|x|}{c_1\xi_1}\right) - c_2\xi_2 \exp\left(-\frac{|x|}{c_2\xi_2}\right)\right\}.$$

There exists a positive number t_0 such that

$$\int_0^{t_0} f_1(x)dx = \int_0^{t_0} f(x)dx,$$

where $f_1(x)$ and $f(x)$ are given by (2.14) and (2.15), respectively. Hence we have established the following theorem.

THEOREM 2.3. *Suppose that $\xi_1 < \xi_2$ and $c_1\xi_1 \neq c_2\xi_2$. Then*

$$\hat\theta_{ML} = \hat\theta_1 \underset{as.}{\succ} \hat\theta_0$$

in some neighborhood of θ in the sense that

$$\lim_{n\to\infty}[\mathrm{Pr}\,\{n|\hat\theta_{ML} - \theta| \leq t\} - \mathrm{Pr}\,\{n|\hat\theta_0 - \theta| \leq t\}] \geq 0 \qquad for\ all\ t \leq t_0.$$

Further

$$\hat\theta_0 \underset{as.}{\succ} \hat\theta_{ML} = \hat\theta_1$$

in far away from θ in the sense that

$$\lim_{n\to\infty}[\mathrm{Pr}\,\{n|\hat\theta_0 - \theta| \leq t\} - \mathrm{Pr}\,\{n|\hat\theta_{ML} - \theta| \leq t\}] \geq 0 \qquad for\ all\ t > t_0.$$

Remark. From (2.14) and (2.15) it is easily seen that

$$\hat\theta_0 \underset{as.}{\succ} \hat\theta_2 \qquad and \qquad \hat\theta_{ML} = \hat\theta_1 \underset{as.}{\succ} \hat\theta_2.$$

The case $\xi_1 > \xi_2$ may be treated quite similarly.

UNIVERSITY OF ELECTRO-COMMUNICATIONS
UNIVERSITY OF TOKYO

REFERENCES

[1] Akahira, M. (1975). Asymptotic theory for estimation of location in non-regular cases, I: Order of convergence of consistent estimators, *Rep. Statist. Appl. Res., JUSE*, **22**, 8–26.

[2] Akahira, M. (1975). Asymptotic theory for estimation of location in non-regular cases, II: Bounds of asymptotic distributions of consistent estimators, *Rep. Statist. Appl. Res., JUSE*, **22**, 99–115.

[3] Akahira, M. (1976). A remark on asymptotic sufficiency of statistics in non-regular cases, *Rep. Univ. Electro-Comm.*, **27**, 125–128.

[4] Akahira, M. (1982). Asymptotic optimality of estimators in non-regular cases, *Ann. Inst. Statist. Math.*, A, **34**, 69–82.

[5] Akahira, M. (1982). Asymptotic deficiencies of estimators for pooled samples from the same distribution, Probability Theory and Mathematical Statistics, *Lecture Notes in Mathematics*, **1021**, 6–14, Springer, Berlin.

[6] Akahira, M. and Takeuchi, K. (1981). Asymptotic Efficiency of Statistical Estimators: Concepts and Higher Order Asymptotic Efficiency, *Lecture Notes in Statistics* 7, Springer, New York.

[7] Akahira, M. and Takeuchi, K. (1982). On asymptotic deficiency of estimators in pooled samples in the presence of nuisance parameters, *Statistics and Decisions*, **1**, 17–38, Akademische Verlagsgesellschaft.

[8] Akai, T. (1982). A combined estimator of a common parameter, *Keio Science and Technology Reports*, **35**, 93–104.

[9] Antoch, J. (1984). Behaviour of estimators of location in non-regular cases: A Monte Carlo study, Asymptotic Statistics 2, *the Proceedings of the 3rd Prague Symposium on Asymptotic Statistics*, 185–195, North-Holland, Amsterdam.

[10] Ibragimov, I. A. and Has'minskii, R. Z. (1981). *Statistical Estimation: Asymptotic Theory*, Springer, New York.

[11] Jurečková, J. (1981). Tail-behaviour of location estimators in non-regular cases, *Commentationes Mathematicae Universitatis Carolinae*, **22**, 365–375.

Rep. Stat. Appl. Res., JUSE
Vol. 32, No. 3, Sept., 1985
p.p. 17–22

A-Section

A Note on the Minimum Variance Unbiased Estimation When the Fisher Information is Infinity

Masafumi AKAHIRA* and Kei TAKEUCHI**

Abstract

In the case when the Fisher information is infinity, it is shown that the locally minimum variance of unbiased estimators is equal to zero. Some examples are also given.

1. Introduction

In non-regular cases the minimum variance unbiased estimation has been studied by Chapman and Robbins [6], Kiefer [7], Polfeldt [8], Víncze [10], Akahira [1], Akahira and Takeuchi [2], [3], [4], [9], Akahira, Puri and Takeuchi [5] and others.

In the previous paper by Akahira and Takeuchi [2] they introduced the concept of the one-directional distribution and discuss the locally minimum variance unbiased estimation for its family. In this paper, from another point of view we shall treat the case when the Fisher information is infinity and show that the locally minimum variance of unbiased estimators is equal to zero. Further we shall give two examples.

2. Results

Let (X, Y) be a pair of random variables defined on some product space $\mathcal{x} \times \mathcal{y}$ of $\mathcal{x}$ and $\mathcal{y}$ with a joint probability density function (j.p.d.f.) $f_\theta(x, y)$ with respect to a σ-finite measure $\mu_{x, y}$, where θ is a real-valued parameter. Then it follows that for almost all (x, y) $[\mu_{x, y}]$

$$f_\theta(x, y) = f_\theta(x \mid y) f_\theta(y),$$

Received September 11, 1985

* Department of Mathematics, University of Electro-Communications.
1-5-1, Chofugaoka, Chofu-shi, Tokyo 182, Japan.
(The author is visiting Queen's University in Canada from April to July 1985.)
** Faculty of Economics, University of Tokyo.
7-3-1, Hongo, Bunkyo-ku, Tokyo 113, Japan.
(Key words) Locally minimum variance, Unbiased estimator, Fisher information.

M. Akahira and K. Takeuchi

where $f_\theta(x|y)$ and $f_\theta(y)$ denote a conditional p.d.f. of x given y and a marginal p.d.f. of y, respectively. We assume that for almost all (x, y) $[\mu_{x, y}]$, $f_\theta(x,y)$, $f_\theta(x|y)$ and $f_\theta(y)$ are continuously differentiable in θ.

Since

$$\log f_\theta(x, y) = \log f_\theta(x|y) + \log f_\theta(y) \quad \text{a.e. } \mu_{x, y},$$

it follows that

$$I^{X, Y}(\theta) = E_\theta^Y[I^{X|Y}(\theta)] + I^Y(\theta)$$

where $I^{X,Y}(\theta)$, $I^{X|Y}(\theta)$ and $I^Y(\theta)$ denote the amounts of Fisher information with respect to $f_\theta(x,y)$, $f_\theta(x|y)$ and $f_\theta(y)$, respectively, e.g.,

$$I^{X,Y}(\theta) = \iint_{X \times Y} \left\{ \frac{\partial \log f_\theta(x,y)}{\partial \theta} \right\}^2 f_\theta(x,y) d\mu_{x,y},$$

and $E_\theta^Y(.)$ is the expectation w. r. t. $f_\theta(y)$.

We denote $J_\theta(Y) = I^{X|Y}(\theta)$.

Here we assume the following conditions (A.1) to (A.3).

(A.1) $0 < J_\theta(y) < \infty$ for almost all y $[\mu_y]$ and all θ, and $E_{\theta_0}^Y[J_{\theta_0}(Y)] = \infty$ and $I^{X,Y}(\theta_0)$ $= \infty$ for some θ_0.

(A.2) There exists a sequence $\{B_n\}$ of measurable sets of the domain of Y such that

$$b_n \equiv E_\theta^Y[J_{\theta_0}(Y) \chi_{B_n}(Y)] < \infty$$

for all θ, b_n is independent of θ and $b_n \to \infty$ as $n \to \infty$, where $\chi_{B_n}(Y)$ denotes the indicator function of the set B_n.

(A.3) For almost all y $[\mu_y]$, there exists an estimator $\hat{\theta}_y(X)$ of θ such that

$$E_\theta^{X|y}[\hat{\theta}_y(X)] = \theta \quad \text{for all } \theta \ ;$$

$$V_\theta^{X|y}(\hat{\theta}_y(X)) = \frac{1}{J_\theta(y)} \quad \text{for all } \theta,$$

where $E_\theta^{X|y}(.)$ and $V_\theta^{X|y}(.)$ denote the conditional expectation and the conditional variance of X given y .

We shall show that the locally minimum variance of unbiased estimators of θ is equal to zero.

Theorem: Assume that the conditions (A.1) to (A.3) hold. Then

$$\inf_{\hat{\theta}(X,Y) \,:\, \text{unbiased}} V_{\theta_0}(\hat{\theta}(X,Y)) = 0.$$

Proof: Let an estimator $\hat{\theta}^* = \hat{\theta}^*(X,Y)$ be

$$\hat{\theta}^* = \frac{1}{b_n} J_{\theta_0}(Y)\chi_{B_n}(Y) \{\hat{\theta}_y(X) - \theta_0\} + \theta_0.$$

Minimum Variance Unbiased Estimation

By the assumptions (A.2) and (A.3) we have

$$E_\theta^{X,Y}(\hat{\theta}^*) = \frac{1}{b_n} E_\theta^Y [J_{\theta_0}(Y) \chi_{B_n}(Y) \{ E_\theta^{X|Y}(\hat{\theta}_Y(X)) - \theta_0 \}] + \theta_0 = \theta$$

for all θ, hence $\hat{\theta}^*$ is an unbiased estimator of θ, where $E_\theta^{X,Y}(.)$ denotes the expectation w.r.t. $f_\theta(x,y)$. We also obtain by (A.2) and (A.3)

$$E_{\theta_0}^{X,Y}(\hat{\theta}^{*2}) = \frac{1}{b_n^2} E_{\theta_0}^{X,Y}[J_{\theta_0}^2(Y)\chi_{B_n}(Y)\{\hat{\theta}_Y(X) - \theta_0\}^2]$$

$$+ \frac{2\theta_0}{b_n} E_{\theta_0}^{X,Y}[J_{\theta_0}(Y)\chi_{B_n}(Y)\{\hat{\theta}_Y(X) - \theta_0\}] + \theta_0^2$$

$$= \frac{1}{b_n^2} E_{\theta_0}^Y[J_{\theta_0}^2(Y)\chi_{B_n}(Y) E_{\theta_0}^{X|Y}(\hat{\theta}_Y(X) - \theta_0)^2]$$

$$+ \frac{2\theta_0}{b_n} E_{\theta_0}^Y[J_{\theta_0}(Y)\chi_{B_n}(Y)\{E_{\theta_0}^{X|Y}(\hat{\theta}_Y(X)) - \theta_0\}] + \theta_0^2$$

$$= \frac{1}{b_n^2} E_{\theta_0}^Y[J_{\theta_0}(Y)\chi_{B_n}(Y)] + \theta_0^2$$

$$= \frac{1}{b_n} + \theta_0^2.$$

Then it follows that the variance of $\hat{\theta}$ at θ_0 is given by

$$V_{\theta_0}^{X,Y}(\hat{\theta}^*) = \frac{1}{b_n},$$

which tends to zero as $n \to \infty$ because of (A.2). Thus we complete the proof.

Next we shall show some examples.

Example 1. Let $X_i = \theta Y_i + U_i$ $(i = 1, 2, \ldots, n)$, where (Y_i, U_i) $(i = 1, \ldots, n)$ are independently and identically distributed (i.i.d.) random vectors and for each i, Y_i and U_i are mutually independent. Let $f(u)$ be a known density function of U_i. Assume that

$$I^U = \int \frac{\{f'(u)\}^2}{f(u)} \, du < \infty$$

Putting $Y = (Y_1, \ldots, Y_n)$ we have

$$I^Y(\theta) = 0$$

since the distribution of Y is independent of θ, and

$$E^Y(J_\theta(Y)) = nE(Y_1^2)I^U.$$

M. Akahira and K. Takeuchi

Assume that $E(Y_1^2) = \infty$ and $f(u)$ is a density of the standard normal distribution.

For any $K > 0$ and each $i = 1, \ldots, n$ we define

$$Y_{i,K}^* = \begin{cases} Y_i & \text{for } |Y_i| \leq K \; ; \\ 0 & \text{for } |Y_i| > K \,. \end{cases}$$

Since $E(Y_1^2) = \infty$, it is easily seen that $E(Y_{1,K}^{*2})$ tends to infinity as $K \to \infty$. We also define an estimator

$$\hat{\theta}_K = \frac{\displaystyle\sum_{i=1}^n Y_{i,K}^* X_i}{nE(Y_{1,K}^{*2})}$$

It is clear that $\hat{\theta}_K$ is an unbiased estimator of θ . Since the variance of $\hat{\theta}_K$ at $\theta = 0$ is given by

$$V_0(\hat{\theta}_K) = E_0(\hat{\theta}_K^2) = \frac{E(\displaystyle\sum_{i=1}^n Y_{i,K}^{*2})}{n^2 \{ E(Y_{1,K}^{*2}) \}^2} = \frac{1}{nE(Y_{1,K}^{*2})} \quad ,$$

it follows that

$$\lim_{K \to \infty} V_0(\hat{\theta}_K) = 0,$$

hence

$$\inf_{\hat{\theta}\,:\,\text{unbiased}} V_0(\hat{\theta}) = 0.$$

Example 2. Let Y be an unobservable random variable according to the chi-square distribution with one degree of freedom. Assume that $X_1, \ldots, X_n$ are independently, identically, normally and conditionally distributed random variables with mean θ and variance Y given Y , where $n \geqq 5$. Then we have

$$I^{X,Y}(\theta) = nE_\theta^{X_1,Y} \left[\left(\frac{X_1 - \theta}{Y} \right)^2 \right] = nE_\theta^Y \left[\frac{1}{Y^2} E^{X_1 \,|\, Y} \left[(X_1 - \theta)^2 \right] \right] = nE_\theta^Y \left(\frac{1}{Y} \right) = \infty .$$

For any $K > 0$ we define an estimator

$$\hat{\theta}_K = \begin{cases} \dfrac{c_K \bar{X}}{\displaystyle\sum_{i=1}^n (X_i - \bar{X})^2} & \text{for } \displaystyle\sum_{i=1}^n (X_i - \bar{X})^2 \geqq \frac{1}{K} \; ; \\[6pt] 0 & \text{for } \displaystyle\sum_{i=1}^n (X_i - \bar{X})^2 < \frac{1}{K} \, , \end{cases}$$

Minimum Variance Unbiased Estimation

where c_K is a constant such that $\hat{\theta}_K$ is an unbiased estimator of θ.

Then we shall show that

$$\lim_{K \to \infty} V_0(\hat{\theta}_K) = 0.$$

First we easily see that $\sum_{i=1}^{n}(X_i - \bar{X})^2$ is expressed as a product of two independent random variables Y and Z, where Z has the chi-square distribution with $n-1$ degrees of freedom. Let $f_1(y)$ and $f_{n-1}(z)$ be the density functions of Y and Z, respectively. By the unbiasedness condition of $\hat{\theta}_K$ we obtain

$$\iint_{yz \geq \frac{1}{K}} \frac{1}{yz} f_1(y)f_{n-1}(z)dydz = \frac{1}{c_K} \tag{2.1}$$

We also have the variance of $\hat{\theta}_K$ at $\theta = 0$

$$\iint_{yz \geq \frac{1}{K}} \frac{c_K^2}{nyz^2} f_1(y)f_{n-1}(z)\,dydz = v_K \text{ (say)}. \tag{2.2}$$

Putting

$$F(Kz) = \int_{\frac{1}{Kz}}^{\infty} \frac{1}{y} f_1(y)\,dy,$$

we have

$$F(Kz) = c' \int_{\frac{1}{Kz}}^{\infty} y^{-\frac{3}{2}} e^{-\frac{y}{2}}\,dy = c\sqrt{Kz} \quad, \tag{2.3}$$

where c and c' are constants.

By (2.1) and (2.3) we obtain

$$1 = c_K \int_{0}^{\infty} \frac{F(Kz)}{z} f_{n-1}(z)\,dz = A\sqrt{K}c_K \,,$$

which implies $c_K = O(K^{-1/2})$, where A is some constant.

By (2.2) and (2.3) we also have

$$v_K = c_K^2 \int_{0}^{\infty} \frac{F(Kz)}{nz^2} f_{n-1}(z)dz = B\sqrt{K}\,c_K^2 = O(K^{-1/2}),$$

M. Akahira and K. Takeuchi

where B is some constant.

Then we obtain

$$\lim_{K\to\infty} V_0(\hat{\theta}_K) = \lim_{K\to\infty} v_K = 0.$$

Hence we have

$$\inf_{\hat{\theta}\,:\,\text{unbiased}} V_0(\hat{\theta}) = 0.$$

Acknowledgements

The authors wish to thank the referees for their comments. The paper was written while the first author was at Queen's University in Canada as a visiting professor. The visit was supported by a grant of the Natural Sciences and Engineering Research Council of Canada. He is grateful to Professor Colin R. Blyth for inviting him.

REFERENCES

[1] Akahira, M. (1984): "On the Bhattacharyya inequality in non-regular case," *Sûrikaisekikenkyûsho Kôkyûroku (Proc. Symps., Res., Inst. Math. Sci., Kyoto Univ.)* 538, 65-80.

[2] Akahira, M. and Takeuchi, K.: "The lower bound of the variance of unbiased estimators for one-directional distributions" (Submitted for publication).

[3] Akahira, M. and Takeuchi, K.: "Locally minimum variance unbiased estimator in a discontinuous density function" to appear in *the Metrika*.

[4] Akahira, M. and Takeuchi, K. (1985): *Non-Regular Statistical Estimation*, Monograph.

[5] Akahira, M., Puri, M.L. and Takeuchi, K.: "Bhattacharyya bound of variances of unbiased estimators in non-regular cases" to appear in *the Annals of the Institute of Statistical Mathematics*.

[6] Chapman, D.G. and Robbins, H. (1951): "Minimum variance estimation without regularity assumptions", *Ann. Math. Statist.*, 22, 581-586.

[7] Kiefer, J. (1952): "On minimum variance in non-regular estimation", *Ann. Math. Statist.*, 23, 627-630.

[8] Polfeldt, T. (1970): "The order of the minimum variance in a non-regular case", *Ann. Math. Statist.*, 41, 667-672.

[9] Takeuchi, K. and Akahira, M.: "A note on minimum variance" to appear in *the Metrika*.

[10] Víncze, I. (1979): "On the Cramér-fréchet-Rao inequality in the non-regular case", In: *Contributions to Statistics. The Jaroslav Hájek Memorial Volume*, Academia, Prague, 253-263.

Ann. Inst. Statist. Math.
38 (1986), Part A, 35–44

BHATTACHARYYA BOUND OF VARIANCES OF UNBIASED ESTIMATORS IN NONREGULAR CASES

Masafumi Akahira, Madan L. Puri* and Kei Takeuchi

(Received May 19, 1984; revised Jan. 16, 1985)

Summary

Bhattacharyya bound is generalized to nonregular cases when the support of the density depends on the parameter, while it is differentiable several times with respect to the parameter within the support. Some example is discussed, where it is shown that the bound is sharp.

1. Introduction

It is well known that the Cramér-Rao and the Bhattacharyya bounds are most important and very useful for the variances of unbiased estimators. They are, however, not applicable to the non-regular cases when the support of the distribution is dependent on the parameter. Same is true about more general and simpler bounds by Hammersley [6], Chapman and Robbins [2], Kiefer [8], Fraser and Guttman [5], Fend [4] and Chatterji [3], among others. (For an exposition of some of this work along with extensions in different directions, see Polfeldt [10], [11] and the recent papers of Víncze [15], Khatri [7] and Móri [9], among others). In his paper, Polfeldt [10] discussed the lower bound of the variances of the unbiased estimators when the class of probability measures is one-sided, that is, when P_{θ_1} is absolutely continuous with respect to P_{θ_2} (symbolically, $P_{\theta_1} \ll P_{\theta_2}$) when $\theta_1 < \theta_2$ or $\theta_1 > \theta_2$. In this note, our main interest is to obtain the Bhattacharyya bound when for any θ_1, θ_2, with $\theta_1 \neq \theta_2$, neither $P_{\theta_1} \ll P_{\theta_2}$ nor $P_{\theta_2} \ll P_{\theta_1}$.

2. Results

Let $\mathcal{X}$ be an abstract sample space with x as its generic point, $\mathcal{A}$ a σ-field of subsets of $\mathcal{X}$, and let Θ be a parameter space assumed to be an open set in the real line. Let $\mathcal{P} = \{P_\theta : \theta \in \Theta\}$ be a class of prob-

---
* Research supported by NSF Grant MCS-8301409.

Key words and phrases: Cramér-Rao bound, Bhattacharyya bound, unbiased estimator.

ability measures on $(\mathcal{X}, \mathcal{A})$. We assume that for each $\theta \in \Theta$, $P_\theta(\cdot)$ is absolutely continuous with respect to a σ-finite measure μ. We denote $dP_\theta/d\mu$ by $f(x, \theta)$. For each $\theta \in \Theta$, we denote by $A(\theta)$ the set of points in $\mathcal{X}$ for which $f(x, \theta) > 0$.

We shall consider the Bhattacharyya bound of variances of unbiased estimators at some specified point θ_0 in Θ. We make the following assumptions:

(A.1)
$$\mu((\cap_{c>0} (\cup_{|h|<c} A(\theta_0+h))) \triangle A(\theta_0)) = 0 ,$$

where $E \triangle F$ denotes the symmetric difference of two sets E and F.

(A.2)   For every $\theta_0 \in \Theta$ there exists a positive number ε and a positive function $\rho(x)$ such that for every $x \in A(\theta)$ and every $\theta \in (\theta_0 - \varepsilon, \theta_0 + \varepsilon)$, $\rho(x) > f(x, \theta)$, and for every $\theta \in (\theta_0 - \varepsilon, \theta_0 + \varepsilon)$,

$$\int_{A(\theta)} |\gamma(x)| f(x, \theta) d\mu < \infty \quad \text{implies} \quad \int_{\substack{\cup A(\theta) \\ \theta \in (\theta_0-\varepsilon, \theta_0+\varepsilon)}} |\gamma(x)| \rho(x) d\mu < \infty .$$

(A.3)   For some positive integer k

$$\varlimsup_{h \to 0} \sup_{x \in \cup_{j=1}^{i} A(\theta_0+jh) - A(\theta_0)} \frac{\left| \sum_{j=1}^{i} (-1)^j \binom{i}{j} f(x, \theta_0+jh) \right|}{|h|^i \rho(x)} < \infty ,$$
$$i = 1, \cdots, k .$$

First we prove the following lemma.

LEMMA.   *Assume that* (A.1), (A.2) *and* (A.3) *hold. If* $\phi(x)$ *is any measurable function for which* $\int_{\mathcal{X}} |\phi(x)| \rho(x) d\mu < \infty$, *then*

$$\lim_{h \to 0} \frac{1}{h^i} \int_{\cup_{j=1}^{i} A(\theta_0+jh) - A(\theta_0)} \sum_{j=1}^{i} (-1)^j \binom{i}{j} \phi(x) f(x, \theta_0+jh) d\mu = 0 .$$

PROOF.   By (A.2) and (A.3) it follows that for every $i = 1, \cdots, k$ and every $\theta_0 \in \Theta$, there exist positive numbers ε and K_i such that

$$\frac{1}{|h|^i} \left| \sum_{j=1}^{i} (-1)^j \binom{i}{j} f(x, \theta_0+jh) \right| < K_i \rho(x)$$
$$\text{for } |ih| < \varepsilon \text{ and } x \in A(\theta_0)^c ,$$

where $A(\theta_0)^c$ denotes the complement of the set $A(\theta_0)$. Also, it follows from (A.1) that for every $j = 1, \cdots, i$, there exist a sequence $\{\varepsilon_{jn}\}$ of positive numbers converging to zero as $n \to \infty$ and a monotone non-increasing sequence $\{S_{jn}\}$ of measurable sets such that $|jh| < \varepsilon_{jn}$ implies

BHATTACHARYYA BOUND OF VARIANCES OF UNBIASED ESTIMATORS   37

$A(\theta_0+jh)-A(\theta_0)\subset S_{jn}$ and $\mu\left(\bigcap\limits_{n=1}^{\infty} S_{jn}\right)=0$. If for each $j=1,\cdots,i$, $|jh|<\varepsilon_{jn}$, then

$$\left|\frac{1}{h^i}\int\limits_{\bigcup\limits_{j=1}^{i} A(\theta_0+jh)-A(\theta_0)}\sum_{j=1}^{i}(-1)^j\binom{i}{j}\phi(x)f(x,\theta_0+jh)d\mu\right|$$

$$\leq \int\limits_{\bigcup\limits_{j=1}^{i} A(\theta_0+jh)-A(\theta_0)} K_i|\phi(x)|\rho(x)d\mu \leq \int\limits_{\bigcup\limits_{j=1}^{i} S_{jn}} K_i|\phi(x)|\rho(x)d\mu$$

$$\leq \sum_{j=1}^{i}\int\limits_{S_{jn}} K_i|\phi(x)|\rho(x)d\mu ,$$

which tends to zero as $n\to\infty$. The proof follows.

Remark. The assumption (A.2) together with the condition in the above Lemma is satisfied with $p(x)=\sum\limits_{i=1}^{\infty} c_i f(x,\theta_i)$ when the following holds: For each θ_0 and $\varepsilon>0$, there exist countable points $\theta_1,\theta_2,\cdots$ and positive constants $c_1,c_2,\cdots$ such that $\bigcup\limits_{i=1}^{\infty} A(\theta_i)\supset A(\theta)$ for all $\theta\in(\theta_0-\varepsilon,\theta_0+\varepsilon)$ and that $\sum\limits_{i=1}^{\infty} c_i<\infty$ and $\sum\limits_{i=1}^{\infty} c_i f(x,\theta_i)>f(x,\theta)$ for all $\theta\in(\theta_0-\varepsilon,\theta_0+\varepsilon)$ and almost all $x[\mu]$.

We assume the following:

(A.4)   For each $x\in A(\theta_0)$, $f(x,\theta)$ is k-times continuously differentiable in θ at $\theta=\theta_0$.

(A.5)   For each $i=1,\cdots,k$,

$$\varlimsup_{h\to 0}\sup_{x\in A(\theta_0)}\frac{\left|\frac{\partial^i}{\partial\theta^i}f(x,\theta_0+h)\right|}{\rho(x)}<\infty , \qquad \text{where } \rho(x) \text{ is defined in (A.2)}.$$

We now prove our main theorem on the Bhattacharyya bound of the variances of unbiased estimators.

THEOREM 2.1.   *Assume that (A.1) to (A.5) hold. Let $g(\theta)$ be an estimable function which is k-times differentiable over Θ. Let $\hat{g}(x)$ be an unbiased estimator of $g(\theta)$ satisfying*

$$\int\limits_{A(\theta_0)}|\hat{g}(x)|\rho(x)d\mu<\infty .$$

Further, let Λ be a $k\times k$ non-negative definite matrix whose elements are

$$\lambda_{ij}=\int_{A(\theta_0)}\frac{1}{f(x,\theta_0)}\left\{\frac{\partial^i f(x,\theta_0)}{\partial\theta^i}\frac{\partial^j f(x,\theta_0)}{\partial\theta^j}\right\}d\mu\;,\qquad i,j=1,\cdots,k\;.$$

Assume that λ_{ii}, $i=1,\cdots,k$ are finite. If Λ is nonsingular at θ_0, then

$$(2.1)\qquad \operatorname*{Var}_{\theta_0}(\hat g)\geqq(g^{(1)}(\theta_0),\cdots,g^{(k)}(\theta_0))\Lambda^{-1}(g^{(1)}(\theta_0),\cdots,g^{(k)}(\theta_0))'\;,$$

where $g^{(i)}(\theta)$ is the i-th order derivative of $g(\theta)$.

PROOF.   Denote

$$g_0(\theta)=\int_{A(\theta_0)}\hat g(x)f(x,\theta)d\mu\quad\text{and}\quad \Delta_h f(x,\theta)=f(x,\theta+h)-f(x,\theta)\;.$$

Then, by (A.5), we have

$$(2.2)\qquad \left[\frac{\partial^i}{\partial\theta^i}g_0(\theta)\right]_{\theta=\theta_0}=\lim_{h\to0}\frac{1}{h^i}\Delta_h^i g_0(\theta_0)=\lim_{h\to0}\frac{1}{h^i}\int_{A(\theta_0)}\hat g(x)\Delta_h^i f(x,\theta_0)d\mu$$

$$=\lim_{h\to0}\int_{A(\theta_0)}\hat g(x)\frac{1}{h^i}\Delta_h^i f(x,\theta_0)d\mu$$

$$=\lim_{h\to0}\int_{A(\theta_0)}\hat g(x)\left\{\frac{\partial^i}{\partial\theta^i}f(x,\theta_0+\xi h)\right\}d\mu$$

$$=\int_{A(\theta_0)}\hat g(x)\left[\frac{\partial^i}{\partial\theta^i}f(x,\theta)\right]_{\theta=\theta_0}d\mu\;,$$

where $\Delta_h^i g(\theta)=\Delta_h^{i-1}(\Delta_h g(\theta))$, $i=1,\cdots,k$ and $0<\xi<1$. Since $g(\theta)=\int_{A(\theta)}\hat g(x)$
$\cdot f(x,\theta)d\mu$, we obtain for each $i=1,\cdots,k$

$$(2.3)\qquad \Delta_h^i(g(\theta_0)-g_0(\theta_0))$$

$$=\sum_{j=1}^i(-1)^j\binom{i}{j}\{g(\theta_0+jh)-g_0(\theta_0+jh)\}$$

$$=\sum_{j=1}^i(-1)^j\binom{i}{j}\left\{\int_{A(\theta_0+jh)}\hat g(x)f(x,\theta_0+jh)d\mu\right.$$

$$\left.-\int_{A(\theta_0)}\hat g(x)f(x,\theta_0+jh)d\mu\right\}$$

$$=\sum_{j=1}^i(-1)^j\binom{i}{j}\left\{\left(\int_{A(\theta_0+jh)}-\int_{A(\theta_0)\cap A(\theta_0+jh)}-\int_{A(\theta_0)-A(\theta_0+jh)}\right)\right.$$

$$\left.\cdot\hat g(x)f(x,\theta_0+jh)d\mu\right\}$$

$$=\sum_{j=1}^i(-1)^j\binom{i}{j}\left\{\left(\int_{A(\theta_0+jh)}-\int_{A(\theta_0)\cap A(\theta_0+jh)}\right)\hat g(x)f(x,\theta_0+jh)d\mu\right\}$$

$$=\sum_{j=1}^i(-1)^j\binom{i}{j}\int_{A(\theta_0+jh)-A(\theta_0)}\hat g(x)f(x,\theta_0+jh)d\mu$$

BHATTACHARYYA BOUND OF VARIANCES OF UNBIASED ESTIMATORS 39

$$= \sum_{j=1}^{i} (-1)^j \binom{i}{j} \int_{\overset{i}{\underset{k=1}{\bigcup}} A(\theta_0+kh)-A(\theta_0)} \hat{g}(x)f(x,\,\theta_0+jh)d\mu$$

$$= \int_{\overset{i}{\underset{k=1}{\bigcup}} A(\theta_0+kh)-A(\theta_0)} \sum_{j=1}^{i} (-1)^j \binom{i}{j} \hat{g}(x)f(x,\,\theta_0+jh)d\mu \,.$$

By (2.3) and Lemma, we have for each $i=1,\cdots,k$,

$$(2.4) \qquad \left[\frac{\partial^i}{\partial\theta^i}g(\theta)\right]_{\theta=\theta_0} = \lim_{h\to0}\frac{1}{h^i}\varDelta_h^i g(\theta_0)$$

$$= \lim_{h\to0}\frac{1}{h^i}\varDelta_h^i(g(\theta_0)-g_0(\theta_0))+\lim_{h\to0}\frac{1}{h^i}\varDelta_h^i g(\theta_0)$$

$$= \lim_{h\to0}\frac{1}{h^i}\varDelta_h^i g(\theta_0) = \left[\frac{\partial^i}{\partial\theta^i}g_0(\theta)\right]_{\theta=\theta_0}.$$

From (2.2) and (2.4) we obtain for each $i=1,\cdots,k$,

$$(2.5) \qquad \left[\frac{\partial^i}{\partial\theta^i}g(\theta)\right]_{\theta=\theta_0} = \int_{A(\theta_0)} \hat{g}(x)\left[\frac{\partial^i}{\partial\theta^i}f(x,\,\theta)\right]_{\theta=\theta_0} d\mu \,.$$

Proceeding now as in the regular case (see e.g. Zacks [16], page 190), one can show that the Bhattacharyya bound of the variances of the unbiased estimators of $g(\theta)$ is given by (2.1).

3. Example

We consider the location parameter case, i.e., $f(x,\,\theta)=f(x-\theta)$, and unbiased estimators of θ.

We assume that for any $p\geq1$, the density function $f(x)$ is given by

$$f(x)=\begin{cases} c(1-x^2)^{p-1} & \text{if } |x|<1 \\ 0 & \text{if } |x|\geq1 \end{cases}$$

where $c=1/B(1/2,\,p)$ with $B(\alpha,\,\beta)=\int_0^1 x^{\alpha-1}(1-x)^{\beta-1}dx$ $(\alpha>0,\,\beta>0)$.

Case (ⅰ): Let $p=1$. Then the distribution is uniform, and it is easy to check that $\min_{\hat{\theta}:\,\text{unbiased}} \text{Var}_{\theta_0}(\theta)=0$ for any specific value θ_0. (See Takeuchi [14]).

Case (ⅱ): Let $p=2$. In this case, it is easy to check that the Fisher information $\int_{-1}^{1} [f'(x)/f(x)]^2 f(x)dx=\infty$.

For any $\varepsilon>0$, we define an estimator $\hat{\theta}_\varepsilon$ which satisfies

$$\hat{\theta}_\varepsilon(x) = \begin{cases} -c_\varepsilon f'(x)/f(x) & \text{if } |x| \leq 1-\varepsilon , \\ 0 & \text{if } 1-\varepsilon < |x| \leq 1 , \end{cases}$$

where c_ε is a constant determined from the equations

(3.1) $$\int_{-1}^{1} \hat{\theta}_\varepsilon(x) f(x) dx = 0 \quad \text{and} \quad \int_{-1}^{1} \hat{\theta}_\varepsilon(x) f'(x) dx = -1 .$$

We shall determine $\hat{\theta}_\varepsilon(x)$ for x outside the interval $[-1, 1]$ from the unbiasedness condition

(3.2) $$\int_{-1+\theta}^{1+\theta} \hat{\theta}_\varepsilon(x) f(x-\theta) dx = \theta .$$

First consider the case $0 < \theta \leq 1$, and define

(3.3) $$g(\theta) = \int_{-1+\theta}^{1} \hat{\theta}_\varepsilon(x) f(x-\theta) dx .$$

Since $\hat{\theta}_\varepsilon(x)$ and $f'(x)$ are bounded, $g(\theta)$ is differentiable and $g'(\theta) = -\int_{-1+\theta}^{1} \hat{\theta}_\varepsilon(x) f'(x-\theta) dx.$

If we assume that $\hat{\theta}_\varepsilon(x)$ is bounded for $1 < x \leq 2$, the right hand side of (3.3) is also differentiable, and we have by (3.2) and (3.3)

(3.4) $$1 - g'(\theta) = -\int_{1}^{1+\theta} \hat{\theta}_\varepsilon(x) f'(x-\theta) dx .$$

Differentiating (3.4) again, and noting that $\lim_{x \to 1-0} f'(x) = -3/2$, we have

(3.5) $$g''(\theta) = -\frac{3}{2} \hat{\theta}_\varepsilon(1+\theta) - \int_{1}^{1+\theta} \hat{\theta}_\varepsilon(x) f''(x-\theta) dx .$$

If $\hat{\theta}_\varepsilon$ satisfies (3.5), then it also satisfies (3.4) since $\lim_{\theta \to 0} g'(\theta) = -1$; it also satisfies (3.3) since $\lim_{\theta \to 0} g(\theta) = 0$ by (3.1).

Since the integral equation (3.5) is of Volterra's second type, it follows that the solution $\hat{\theta}_\varepsilon(x)$ exists and is bounded. Repeating the same process, we have the solution $\hat{\theta}_\varepsilon(x)$ for all $x > 1$. Similarly, we can construct $\hat{\theta}_\varepsilon(x)$ for $x < -1$. On the other hand,

(3.6) $$\mathrm{Var}_{\theta_0}(\hat{\theta}_\varepsilon) = c_\varepsilon^2 \int_{-1+\varepsilon}^{1-\varepsilon} [f'(x)/f(x)]^2 f(x) dx ,$$

while from (3.1), we have

(3.7) $$c_\varepsilon \int_{-1+\varepsilon}^{1-\varepsilon} f'(x) dx = 0 \quad \text{and} \quad c_\varepsilon \int_{-1+\varepsilon}^{1-\varepsilon} [f'(x)/f(x)]^2 f(x) dx = 1 .$$

Hence $\mathrm{Var}_{\theta_0}(\hat{\theta}_\varepsilon) = c_\varepsilon.$

BHATTACHARYYA BOUND OF VARIANCES OF UNBIASED ESTIMATORS   41

Now, since $\lim\limits_{\varepsilon\to 0}\int_{-1+\varepsilon}^{1-\varepsilon}[f'(x)/f(x)]^2 f(x)dx=\infty$, we have from (3.7), $\lim\limits_{\varepsilon\to 0} c_\varepsilon=0$. Consequently, $\inf\limits_{\hat{\theta}:\,\text{unbiased}}\mathrm{Var}_{\theta_0}(\hat{\theta})=0$ for any specified value θ_0.

Before proceeding on to the next cases, we note the following:

If $k<p/2$, then λ_{ii} $(i=1,\cdots,k)$ given in Theorem 2.1 are finite, since $\int_{-1}^{1}(1-x^2)^{p-2k-1}dx<\infty$.

Also,

$$(3.8)\qquad \lambda_{ij}=\int_{\theta-1}^{\theta+1}\frac{1}{f(x-\theta)}\frac{\partial^i f(x-\theta)}{\partial\theta^i}\frac{\partial^j f(x-\theta)}{\partial\theta^j}dx$$

$$=\int_{-1}^{1}(-1)^{i+j}\frac{1}{f(x)}f^{(i)}(x)f^{(j)}(x)dx;\qquad i,j=1,\cdots,k,$$

We also obtain for $|x|<1$,

$$f^{(1)}(x)=-2c(p-1)x(1-x^2)^{p-2};$$

$$(3.9)\qquad f^{(2)}(x)=-2c(p-1)\{(1-x^2)^{p-2}-2(p-2)x^2(1-x^2)^{p-3}\};$$

$$f^{(3)}(x)=4c(p-1)(p-2)\{3x(1-x^2)^{p-3}-2(p-3)x^3(1-x^2)^{p-4}\};\cdots.$$

If $i+j$ is an odd number, if follows by (3.8) and (3.9) that $\lambda_{ij}=0$ since $f^{(i)}(x)f^{(j)}(x)$ is an odd function.

From (3.8) and (3.9), we have

$$\lambda_{11}=4c(p-1)^2 B\left(\frac{3}{2},\,p-2\right);$$

$$\lambda_{13}=8c(p-1)^2(p-2)\left\{2(p-3)B\left(\frac{5}{2},\,p-4\right)-3B\left(\frac{3}{2},\,p-3\right)\right\};$$

$$\lambda_{22}=4c(p-1)^2\left\{B\left(\frac{1}{2},\,p-2\right)-4(p-2)B\left(\frac{3}{2},\,p-3\right)\right.$$

$$\left.+4(p-2)^2 B\left(\frac{5}{2},\,p-4\right)\right\};$$

$$\lambda_{33}=16c(p-1)^2(p-2)^2\left\{9B\left(\frac{3}{2},\,p-4\right)-12(p-3)B\left(\frac{5}{2},\,p-5\right)\right.$$

$$\left.+4(p-3)^2 B\left(\frac{7}{2},\,p-6\right)\right\};\cdots.$$

Case (iii): Let $p=3,4$. Then, we have, for any unbiased estimator $\hat{\theta}(X)$ of θ,

$$(3.10)\qquad \int_{-1}^{1}\hat{\theta}(x)f(x)dx=0$$

$$(3.11) \qquad \int_{-1}^{1} \hat{\theta}(x) f'(x) dx = -1$$

$$(3.12) \qquad \int_{-1}^{1} \hat{\theta}(x) f^{(k)}(x) dx = 0 , \qquad k = 2, \cdots, p-1 .$$

Noting that $\int_{-1}^{1} \left[\left\{ \sum_{k=1}^{p-1} c_k f^{(k)}(x) \right\}^2 \Big/ f(x) \right] dx < \infty$ implies $c_2 = \cdots = c_{p-1} = 0$, we have by Takeuchi and Akahira [13], that the infimum of $\mathrm{Var}_0(\hat{\theta}) = \int_{-1}^{1} \hat{\theta}^2(x) f(x) dx$ under (3.10), (3.11) and (3.12) is given by $\inf_{\hat{\theta}: (3.10)\sim(3.12)} \mathrm{Var}_0(\hat{\theta}) = 1/\lambda_{11}$, where $\lambda_{11} = \int_{-\infty}^{\infty} [f'(x)/f(x)]^2 f(x) dx = (p-1)(2p-1)/(p-2)$ and for any $\varepsilon > 0$ there exists $\hat{\theta}_\varepsilon(x)$ in $(-1, 1)$ satisfying (3.10), (3.11) and (3.12), and $\int_{-1}^{1} \hat{\theta}_\varepsilon^2(x) f(x) dx < (1/\lambda_{11}) + \varepsilon$. We can extend $\hat{\theta}_\varepsilon(x)$ for x outside $(-1, 1)$ from the unbiasedness condition $\int_{-1+\theta}^{1+\theta} \hat{\theta}_\varepsilon(x) f(x-\theta) dx = \theta$. First, we consider the case when $0 \leq \theta < 1$, and define $g(\theta) = \int_{-1+\theta}^{1} \hat{\theta}_\varepsilon(x) f(x-\theta) dx$. In a similar way to the case (ii), we have for $k = 2, \cdots, p-1$,

$$(3.13) \qquad (-1)^{k+1} g^{(k+1)}(\theta) = B_k \hat{\theta}_\varepsilon(1+\theta) - \int_{1}^{1+\theta} \hat{\theta}_\varepsilon(x) f^{(k+1)}(x-\theta) dx ,$$

$$B_k = \lim_{x \to 1-0} f^{(k)}(x) .$$

Since the integral equation (3.13) is again of Volterra's second type, it follows that the solution $\hat{\theta}_\varepsilon(x)$ exists. Repeating the process, we can construct an unbiased estimator $\hat{\theta}_\varepsilon(x)$ for all values of x. Then, it follows that $\inf_{\hat{\theta}: \text{unbiased}} \mathrm{Var}_{\theta_0}(\hat{\theta}) = 1/\lambda_{11}$ for any specified value θ_0.

Case (iv): Let $p = 5, 6$. Note that

$$\int_{-1}^{1} [f'(x) f''(x)/f(x)] dx = 0 ,$$

$$\int_{-1}^{1} [f''(x)/f(x)]^2 f(x) dx = \frac{(p-1)(2p-1)(2p-3)}{(p-2)(p-3)(p-4)} (2p^2 - 7p + 8)$$
$$= \lambda_{22} \qquad (\text{say}) ,$$

and

$$\int_{-1}^{1} \left\{ \sum_{k=1}^{p-1} c_k f^{(k)}(x)/f(x) \right\}^2 f(x) dx < \infty$$

imply $c_3 = \cdots = c_{p-1} = 0$. Then we see that

$$\mathrm{Var}_{\theta_0}(\hat{\theta}) \geq (1, 0) \begin{pmatrix} \lambda_{11} & 0 \\ 0 & \lambda_{22} \end{pmatrix}^{-1} \begin{pmatrix} 1 \\ 0 \end{pmatrix} = \frac{1}{\lambda_{11}}$$

for any specified θ_0, where λ_{11} is defined above (3.8). Here again, as in the previous case $\inf_{\hat{\theta}:\,\text{unbiased}} \mathrm{Var}_{\theta_0}(\hat{\theta}) = 1/\lambda_{11}$ for any specific θ_0.

Case (v): Let $p=7$. In this case, we see that $k=3$. Using Theorem 2.1, (3.8) and (3.9), we obtain

$$\mathrm{Var}_{\theta_0}(\hat{\theta}) \geq \frac{1}{|\varLambda|}\begin{vmatrix}\lambda_{22} & 0 \\ 0 & \lambda_{33}\end{vmatrix} = \left[\lambda_{11}\left(1 - \frac{\lambda_{13}^2}{\lambda_{11}\lambda_{33}}\right)\right]^{-1}$$

where

$$\varLambda = \begin{pmatrix}\lambda_{11} & 0 & \lambda_{13} \\ 0 & \lambda_{22} & 0 \\ \lambda_{13} & 0 & \lambda_{33}\end{pmatrix}$$

with

$$\lambda_{11} = 144c\,B\left(\frac{3}{2},\,5\right)\ ;$$

$$\lambda_{13} = 1440c\left\{8B\left(\frac{5}{2},\,3\right) - 3B\left(\frac{3}{2},\,4\right)\right\}\ ;$$

$$\lambda_{22} = 144c\left\{B\left(\frac{1}{2},\,5\right) - 20B\left(\frac{3}{2},\,4\right) + 100B\left(\frac{5}{2},\,3\right)\right\}\ ;$$

$$\lambda_{33} = 14400c\left\{9B\left(\frac{3}{2},\,3\right) - 48B\left(\frac{5}{2},\,2\right) + 64B\left(\frac{7}{2},\,1\right)\right\}\ .$$

We also obtain

$$\frac{\lambda_{13}^2}{\lambda_{11}\lambda_{33}} = \frac{\{8B(5/2,\,3) - 3B(3/2,\,4)\}^2}{B(3/2,\,5)\{9B(3/2,\,3) - 48B(5/2,\,2) + 64B(7/2,\,1)\}}\ .$$

Here again $\inf_{\hat{\theta}:\,\text{unbiased}} \mathrm{Var}_{\theta_0}(\hat{\theta}) = [\lambda_{11}(1 - (\lambda_{13}^2/\lambda_{11}\lambda_{13}))]^{-1}$ for any specific θ_0, i.e. we have a sharp bound.

Case (vi): For $p \geq 8$ we can continue in a similar manner by choosing $k = [(p-1)/2]$, where $[s]$ denotes the largest integer less than or equal to s.

The above discussion establishes that here the bound is sharp but generally it is not attainable.

Acknowledgement

The authors are indebted to Professor Michael Perlman for the very careful and critical examination of the first version of the paper.

44   MASAFUMI AKAHIRA, MADAN L. PURI AND KEI TAKEUCHI

His constructive comments and suggestions for improvements are gratefully acknowledged.

They are also grateful to the editor and the referee for their helpful comments.

UNIVERSITY OF ELECTRO-COMMUNICATIONS
INDIANA UNIVERSITY
UNIVERSITY OF TOKYO

REFERENCES

[1] Bhattacharyya, A. (1946). On some analogues of the amount of information and their use in statistical estimation, *Sankhyā*, **8**, 1–32.

[2] Chapman, D. G. and Robbins, H. (1951). Minimum variance estimation without regularity assumptions, *Ann. Math. Statist.*, **22**, 581–586.

[3] Chatterji, S. D. (1982). A remark on the Cramér-Rao inequality, in *Statistics and Probability: Essays in Honor of C. R. Rao*, North Holland Publishing Company, 193–196.

[4] Fend, A. V. (1959). On the attainment of the Cramér-Rao and Bhattacharyya bounds for the variance of an estimate, *Ann Math. Statist.*, **30**, 381–388.

[5] Fraser, D. A. S. and Guttman, I. (1951). Bhattacharyya bounds without regularity assumptions, *Ann. Math. Statist.*, **23**, 629–632.

[6] Hammersley, J. M. (1950). On estimating restricted parameters, *J. Roy. Statist. Soc.*, B, **12**, 192–240.

[7] Khatri, C. G. (1980). Unified treatment of Cramér-Rao bound for the non-regular density functions, *J. Statist. Plann. Inf.*, **4**, 75–79.

[8] Kiefer, J. (1952). On minimum variance in non-regular estimation, *Ann. Math. Statist.*, **23**, 627–630.

[9] Móri, T. F. (1983). Note on the Cramér-Rao inequality in the non-regular case: The family of uniform distributions, *J. Statist. Plann. Inf.*, **7**, 353–358.

[10] Polfeldt, T. (1970). Asymptotic results in non-regular estimation, *Skand. Akt. Tidskr. Suppl.*, 1-2.

[11] Polfeldt, T. (1970). The order of the minimum variance in a non-regular case, *Ann. Math. Statist.*, **41**, 667–672.

[12] Rao, C. R. (1973). *Linear Statistical Inference and Its Applications*, Wiley, New York.

[13] Takeuchi, K. and Akahira, M. (1986). A note on minimum variance, in press in *Metrika*.

[14] Takeuchi, K. (1961). On the fallacy of a theory of Gunner Blom, *Rep. Stat. Appl. Res.*, *JUSE*, **9**, 34–35.

[15] Víncze, I. (1979). On the Cramér-Fréchet-Rao inequality in the non-regular case, in *Contributions to Statistics*, The Jaroslav Hájek Memorial Volume, Academia, Prague, 253–262.

[16] Zacks, S. (1971). *Theory of Statistical Inference*, John Wiley.

Metrika, Volume 33, 1986, page 85–91

A Note on Minimum Variance

By K. Takeuchi[1] and M. Akahira[2]

Summary: Minimizing $\int \{\hat{\theta}(x)\}^2 f(x) d\mu$ is discussed under the unbiasedness condition: $\int \hat{\theta}(x) f_i(x) d\mu = c_i$ $(i = 1, \ldots, p)$ and the condition (A): $f_i(x)$ $(i = 1, \ldots, p)$ are linearly independent,

$\int \{f_i(x)\}^2/f(x) d\mu < \infty$ $(i = 1, \ldots, k; k \leqslant p)$, and $\int \{\sum_{i=1}^{p} a_i f_i(x)\}^2/f(x) d\mu < \infty$ implies $a_{k+1} = \ldots = a_p = 0$.

1 Introduction

Let $(\chi, \mathcal{B})$ be a sample space. We consider a family $\mathcal{P} = \{P_\theta : \theta \in \Theta\}$ of probability measures on $\mathcal{B}$, where the index set Θ is called a parameter space.

In minimum variance unbiased estimation theory, the locally best unbiased estimator γ^* of $\gamma = g(\theta)$ at $\theta = \theta_0$ is obtained by minimizing

$$\int_{\chi} \{\hat{\gamma}(x)\}^2 dP_{\theta_0}(x)$$

under the condition that

$$\int_{\chi} \hat{\gamma}(x) dP_\theta(x) = g(\theta) \quad \text{for all } \theta \in \Theta. \tag{1.1}$$

When the parameter space Θ is a finite set $\{\theta_0, \theta_1, \ldots, \theta_p\}$, the condition (1.1) is reduced to

$$\int_{\chi} \hat{\gamma}(x) dP_{\theta_i}(x) = g(\theta_i) = c_i \text{ (say)} \quad (i = 0, 1, \ldots, p)$$

And it is easily derived that the minimizing solution $\hat{\gamma}^*$ is obtained as

$$\hat{\gamma}^*(x) = \sum_{i=0}^{p} \lambda_i f_i(x),$$

[1] Kei Takeuchi, Faculty of Economics, University of Tokyo, Hongo, Bunkyo-ku, Tokyo 113, Japan.
[2] Masafumi Akahira, Statistical Laboratory, Department of Mathematics, University of Electro-Communications, Chofu, Tokyo 182, Japan.

0026–1335/86/020085–091 $2.50 © 1986 Physica-Verlag, Vienna

86   K. Takeuchi and M. Akahira

where

$$f_i(x) = \frac{dP_{\theta_i}}{dP_{\theta_0}}(x),$$

provided that

For each $i = 1, \ldots, p$, P_{θ_i} is absolutely continuous with respect to P_{θ_0}. $\qquad$ (C.1)

$$\int_X \left(\frac{dP_{\theta_i}}{dP_{\theta_0}}\right)^2 dP_{\theta_0}(x) < \infty \quad (i = 1, \ldots, p). \qquad (C.2)$$

$f_i(x)$ $(i = 1, \ldots, p)$ are linearly independent. $\qquad$ (C.3)

The purpose of this note is to investigate the situation where for some i the first and/or the second condition does not hold. Remark that the third condition is obviously necessary. First we consider the case when we assume the first condition but not the second. To deal with the case we need a more subtle formulation of the condition in place of (C.2) which is given in the next section.

2 Results

Let $f(x)$ and $f_i(x)$ $(i = 1, \ldots, p)$ be measurable functions. Then we consider to minimize

$$\int_X \{\hat{\theta}(x)\}^2 f(x) d\mu$$

under the unbiasedness conditions:

$$\int_X \hat{\theta}(x) f_i(x) d\mu = c_i \quad (i = 1, \ldots, p)$$

and the condition (A): $f_i(x)$ $(i = 1, \ldots, p)$ are linearly independent,

$$\int_X \frac{\{f_i(x)\}^2}{f(x)} d\mu < \infty \quad (i = 1, \ldots, k; k \leqslant p),$$

and

$$\int_X \frac{\{\sum_{i=1}^{p} a_i f_i(x)\}^2}{f(x)} d\mu < \infty$$

implies

$$a_{k+1} = \ldots = a_p = 0.$$

Note that in the above f and f_i $(i = 1, \ldots, p)$ may not necessarily be density functions. In fact, in some applications we may take the derivatives of the density with respect to the parameter θ as f and obtain some generalization of Bhattacharyya inequality. Also generally f is equal to one of f_i's but it is not essential for the subsequent discussion.

Let Λ_{11} be a $k \times k$ non-negative definite matrix whose elements are

$$\lambda_{ij} = \int_X \frac{f_i(x)f_j(x)}{f(x)} \, d\mu \quad (i, j = 1, \ldots, k).$$

Theorem: Under the condition (A),

$$\inf_{\hat{\theta}: \text{unbiased}} \int_X \{\hat{\theta}(x)\}^2 f(x) d\mu = c'_{(k)} \Lambda_{11}^{-1} c_{(k)}$$

holds, where $c_{(k)} = (c_1, \ldots, c_k)'$ is given in the above.

The equality is not usually attained.

Proof: We put

$$A_n = \left\{ x : \max_{1 \leq i \leq p} \frac{|f_i(x)|}{f(x)} < n \right\} \quad (n = 1, 2, \ldots).$$

We define the class C_n of estimators such that for any $\hat{\theta}_n \in C_n$, $\hat{\theta}_n(x) = 0$ for $x \notin A_n$.

We also put

$$\lambda_{ij}^{(n)} = \int_{A_n} \frac{f_i(x)f_j(x)}{f(x)} \, d\mu \quad (i, j = 1, \ldots, p)$$

and denote the matrix $(\lambda_{ij}^{(n)})$ by Λ_n.

Then we have

$$\min_{\hat{\theta}_n \in C_n: \text{unbiased}} \text{Var}_\theta (\hat{\theta}_n) = c' \Lambda_n^{-1} c,$$

where $c = (c_1, \ldots, c_p)'$ is given in the condition (A).

On the other hand we may make a partition of the matrix Λ_n into blocks as

$$\Lambda_n = \begin{pmatrix} \Lambda_{11}^{(n)} & \Lambda_{12}^{(n)} \\ \Lambda_{21}^{(n)} & \Lambda_{22}^{(n)} \end{pmatrix},$$

where $\Lambda_{11}^{(n)}$ and $\Lambda_{22}^{(n)}$ are $k \times k$ and $(p-k) \times (p-k)$ matrixes, respectively.

88   K. Takeuchi and M. Akahira

Let $\underset{\sim}{c} = (\underset{\sim}{c}_1, \underset{\sim}{c}_2)'$ with $\underset{\sim}{c}_1 \in R^k$ and $\underset{\sim}{c}_2 \in R^{p-k}$.
Then the condition (A) implies that

$$\lim_{n \to \infty} \underset{\sim}{c}' \Lambda_n \underset{\sim}{c} = \infty \tag{2.1}$$

for every vector $\underset{\sim}{c} = (\underset{\sim}{c}_1, \underset{\sim}{c}_2)'$ such that $\underset{\sim}{c}_2 \neq 0$.

To prove the theorem it is necessary and sufficient to show that

$$\Lambda_n^{-1} \to \begin{pmatrix} \Lambda_{11}^{-1} & 0 \\ 0 & 0 \end{pmatrix} \quad \text{as } n \to \infty$$

and since the matrix Λ_n is non-negative definite it is enough to show that

$$(\Lambda_{22}^{(n)} - \Lambda_{21}^{(n)} \Lambda_{11}^{(n)}{}^{-1} \Lambda_{12}^{(n)})^{-1} \to 0 \quad \text{as } n \to \infty,$$

that is, the minimum characteristic roots of the matrix $\widetilde{\Lambda}_n = \Lambda_{22}^{(n)} - \Lambda_{21}^{(n)} \Lambda_{11}^{(n)}{}^{-1} \Lambda_{12}^{(n)}$ diverges to infinity as n tends to be large.

Assume otherwise, then we have a sequence $\{\underset{\sim}{c}_n^*\}$ of vectors such that $\underset{\sim}{c}_n^* = (\underset{\sim}{c}_{1n}^*, \underset{\sim}{c}_{2n}^*)'$ with $\|\underset{\sim}{c}_{2n}^*\| = 1$ and

$$\underset{\sim}{c}_n^{*\prime} \Lambda_n \underset{\sim}{c}_n^* = \underset{\sim}{c}_{2n}^{*\prime} \widetilde{\Lambda}_n \underset{\sim}{c}_{2n}^* = \rho_n \text{ (say)}$$

and $\rho_n \uparrow \rho^*$ as $n \to \infty$.

Then there exists a subsequence $\{\underset{\sim}{c}_{n_j}^*\}$ of $\{\underset{\sim}{c}_n^*\}$ such that $\underset{\sim}{c}_{n_j}^* = (\underset{\sim}{c}_{1n_j}^*, \underset{\sim}{c}_{2n_j}^*)'$ with $\|\underset{\sim}{c}_{2n_j}^*\| = 1$ and

$$\underset{\sim}{c}_{n_j}^* \to \underset{\sim}{c}^* \quad \text{as } j \to \infty,$$

where $\underset{\sim}{c}^* = (\underset{\sim}{c}_1^*, \underset{\sim}{c}_2^*)'$ with $\|\underset{\sim}{c}_2^*\| = 1$.
We can show that

$$\underset{\sim}{c}^{*\prime} \Lambda_n \underset{\sim}{c}^* \to \rho^* \quad \text{as } n \to \infty,$$

which contradicts (2.1).
Assume otherwise, then there exist $\epsilon > 0$ and n_0 such that

$$\underset{\sim}{c}^{*\prime} \Lambda_{n_0} \underset{\sim}{c}^* > \rho^* + \epsilon.$$

Since Λ_n is monotone increasing in n in the sense that $\underset{\sim}{c}' \Lambda_n \underset{\sim}{c}$ is so for any vector $\underset{\sim}{c}$, it follows that for $n \geq n_0$

$$\underset{\sim}{c}_n^{*\prime} \Lambda_{n_0} \underset{\sim}{c}_n^* < c_n^{*\prime} \Lambda_n c_n^* \leq \rho^*.$$

Hence

$$\rho^* + \epsilon \leqslant \underset{\sim}{c}^{*\prime} \Lambda_{n_0} \underset{\sim}{c}^* \leqslant \rho^*,$$

which is a contradiction.

Thus we complete the proof.

Next we shall discuss the case when the first condition (C.1) does not hold true. We consider a family $P = \{P_\theta : \theta \in \Theta\}$ of probability measures on (X, B), where $\Theta = \{\theta_1, ..., \theta_p\}$. We assume that for each $i = 1, ..., p, P_{\theta_i}$ is absolutely continuous with respect to a σ-finite measure μ. For each $i = 1, ..., p$ we denote $dP_{\theta_i}/d\mu$ by $f_i(x)$. Let $f(x)$ be one of $f_1(x), ..., f_p(x)$. We denote by S the support of $f(x)$.

Then we have the following problem: Minimize

$$\int_S \{\hat\theta(x)\}^2 f(x) d\mu$$

under the unbiasedness condition

$$\int_X \hat\theta(x) f_i(x) d\mu = (\int_S + \int_{S^c}) \hat\theta(x) f_i(x) d\mu = c_i \quad (i = 1, ..., p),$$

where $f_i(x)$ $(i = 1, ..., p)$ are linearly independent.

Assume that we have a condition similar to (A) above:

$$\int_S \frac{\{f_i(x)\}^2}{f(x)} d\mu < \infty \quad (i = 1, ..., k),$$

and

$$\int_S \frac{\{\sum_{i=1}^p a_i f_i(x)\}^2}{f(x)} d\mu < \infty$$

implies

$$a_{k+1} = ... = a_p = 0,$$

and further we assume that $f_i(x)$ $(i = 1, ..., p)$ are linearly independent within S.

Let Λ_{11}^* be a $k \times k$ non-negative matrix whose elements are

$$\lambda_{ij} = \int_S \frac{f_i(x) f_j(x)}{f(x)} d\mu \quad (i, j = 1, ..., k).$$

For any estimator $\hat\theta(x)$ we define a vector $\underset{\sim}{d} = (d_1, ..., d_p)'$ by

$$d_i = \int_{S^c} \hat\theta(x) f_i(x) d\mu \quad (i = 1, ..., p).$$

90   K. Takeuchi and M. Akahira

Suppose that $\underset{\sim}{d}$ is given, then the unbiasedness condition is reduced to

$$\int_S \hat{\theta}(x) f_i(x) d\mu = c_i - d_i \quad (i = 1, \ldots, p)$$

and under this condition we have from above theorem that

$$\inf_{\substack{\hat{\theta}:\,\text{unbiased with} \\ \text{a fixed } \underset{\sim}{d}}} \int_S \{\hat{\theta}(x)\}^2 f(x) d\mu = (\underset{\sim}{c}_1 - \underset{\sim}{d}_1)' \Lambda_{11}^{*-1} (\underset{\sim}{c}_1 - \underset{\sim}{d}_1),$$

where $\underset{\sim}{c} = (c_1, \ldots, c_p)' = (\underset{\sim}{c}_1, \underset{\sim}{c}_2)'$ and $d = (\underset{\sim}{d}_1, \underset{\sim}{d}_2)'$ with $\underset{\sim}{c}_1, \underset{\sim}{d}_1 \in R^k$ and $\underset{\sim}{c}_2, \underset{\sim}{d}_2 \in R^{p-k}$.

It is obvious that the set of possible vectors $\underset{\sim}{d}_1$ is a linear space which is denoted by $\mathcal{D}_1$. Then we have

$$\inf_{\hat{\theta}:\,\text{unbiased}} \int_S \{\hat{\theta}(x)\}^2 f(x) d\mu = \min_{\underset{\sim}{d}_1 \in \mathcal{D}_1} (\underset{\sim}{c}_1 - \underset{\sim}{d}_1)' \Lambda_{11}^{*-1} (\underset{\sim}{c}_1 - \underset{\sim}{d}_1).$$

If the dimension of $\mathcal{D}_1$ is l, we have a basis $\underset{\sim}{d}_1^*, \ldots, \underset{\sim}{d}_l^*$ of $\mathcal{D}_1$ and any $\underset{\sim}{d}_1 \in \mathcal{D}_1$ can be expressed as

$$\underset{\sim}{d}_1 = b_1 \underset{\sim}{d}_1^* + \ldots + b_l \underset{\sim}{d}_l^* = D^* \underset{\sim}{b},$$

where $D^* = (\underset{\sim}{d}_1^{*\prime}, \ldots, \underset{\sim}{d}_l^{*\prime})$ is a $k \times l$ matrix and $\underset{\sim}{b} = (b_1, \ldots, b_l)'$.

Then it is straightforward to show that

$$\min_{\underset{\sim}{d}_1 \in \mathcal{D}_1} (\underset{\sim}{c}_1 - \underset{\sim}{d}_1)' \Lambda_{11}^{*-1} (\underset{\sim}{c}_1 - \underset{\sim}{d}_1)$$

$$= \min_b (\underset{\sim}{c}_1 - D^* \underset{\sim}{b})' \Lambda_{11}^{*-1} (\underset{\sim}{c}_1 - D^* \underset{\sim}{b})$$

$$= \underset{\sim}{c}_1' \Lambda_{11}^{*-1} D^{*\prime} (D^{*\prime} \Lambda_{11}^{*-1} D^*)^{-1} D^* \Lambda_{11}^{*-1} \underset{\sim}{c}_1$$

and the minimum is attained by

$$\underset{\sim}{b}^* = (D^{*\prime} \Lambda_{11}^{*-1} D^*)^{-1} D^* \Lambda_{11}^{*-1} \underset{\sim}{c}_1.$$

If, moreover, $f_i\ (i = 1, \ldots, p)$ are not linearly independent in S, the vector $\underset{\sim}{c} - \underset{\sim}{d}$ is restricted to a linear space E and

$$\inf_{\substack{\hat{\theta}:\,\text{unbiased with } \underset{\sim}{d}; \\ c - d \in E}} \int_S \{\hat{\theta}(x)\}^2 f(x) d\mu = (\underset{\sim}{c}_0 - \underset{\sim}{d}_0)' \Lambda_0^{*-1} (\underset{\sim}{c}_0 - \underset{\sim}{d}_0),$$

where Λ_0^* is the maximum rank principal minor of Λ_{11}^* and $\underset{\sim}{c}_0$ and $\underset{\sim}{d}_0$ are corresponding subvectors of $\underset{\sim}{c}$ and $\underset{\sim}{d}$. Then $\underset{\sim}{d}_0$ is on a hyperplane H of some dimensionality and

$$\inf_{\hat{\theta}:\,\text{unbiased}} \int_S \{\hat{\theta}(x)\}^2 f(x) d\mu = \min_{\underset{\sim}{d}_0 \in H} (\underset{\sim}{c}_0 - \underset{\sim}{d}_0)' \Lambda_0^{*-1} (\underset{\sim}{c}_0 - \underset{\sim}{d}_0).$$

A Note on Minimum Variance 91

Note that even when the parameter space Θ is infinite, minimizing the variance with a finite number of conditions is often applied to obtain a lower bound of the variances of unbiased estimators, as is the case of Cramér-Rao and Bhattacharyya. (See Hammersley; Chapman/Robbins; Kiefer; Fraser/Guttman; Fend; Sen/Ghosh; Chatterji; Akahira/ Puri/Takeuchi and also Isii with relation to mathematical programming.)

References

Akahira M, Puri ML, Takeuchi K (1986) Bhattacharyya bound of variances of unbiased estimators in non-regular cases. To appear in the Ann Inst Statist Math 38

Chapman DG, Robbins H (1951) Minimum variance estimation without regularity assumptions. Ann Math Statist 22:581–586

Chatterji SD (1982) A remark on the Cramér-Rao inequality. In: Kallianpur, Krishnaiah, Ghosh (eds) Statistics and probability: Essays in honor of C R Rao. North Holland, New York, pp 193–196

Fend AV (1959) Bounds for the variance of an estimate. Ann Math Statist 30:381–388

Fraser DAS, Guttman I (1952) Bhattacharyya bounds without regularity assumptions. Ann Math Statist 23:629–632

Hammersley JM (1950) On estimating restricted parameters. J Roy Statist Soc (B) 12:192–240

Isii K (1964) Inequalities of the types of Chebyshev and Cramér-Rao and mathematical programming. Ann Inst Statist Math 16:277–293

Kiefer J (1952) On minimum variance in non-regular estimation. Ann Math Statist 23:627–629

Sen PK, Ghosh BK (1976) Comparison of some bounds in estimation theory. Ann Statist 4: 755–765; Correction. Ann Statist 5:593

Received August 11, 1983

Metrika, Volume 33, 1986, page 217–222

A Note on Optimum Spacing of Observations
from a Continuous Time Simple Markov Process

By K. Takeuchi[1] and M. Akahira[2]

Summary: Assume that $X(\tau)$ is a continuous time simple Markov process with a parameter θ. The problem is to choose observation points $\tau_0 < \tau_1 < \ldots < \tau_T$ which provide with the maximum possible information on θ. Suppose that the observation points are equally spaced, that is, for $t = 1, \ldots, T, \tau_t - \tau_{t-1} = s$ is constant. Then the optimum value for s is obtained.

1 Introduction

In various practical situations especially in engineering applications of time series analysis, it often happens that the process itself is continuous in time, but measurements can be made only at discrete time points. Then the problem of choosing sample points arises, and its most natural formulations would be "what is the best set of sample points in continuous time points which would give the largest possible amount of information with respect to some unknown parameter given the size of the sample?" In this paper we deal with the problem in nearly the simplest possible case, that is, the underlying process is simple Markov and the measurement points are equally spaced, and calculate the optimum spacing. It is conjectured that the solution given is also optimum even when the sample points can be arbitrarily spaced, but the authors have been unable to prove it yet. The model can be also generalized in various ways, which will be discussed in subsequent papers.

As far as the authors are aware, there seems to be little literature which has dealt with this problem, and we mention Taga (1966) who dealt with a similar problem in connection with the least squares prediction.

2 Optimum Spacing of Observations

Suppose that $X(\tau)$ is a stationary Gaussian process with a continuous time parameter τ, and also suppose that $E(X(\tau)) = 0$, $\mathrm{Var}\,(X(\tau)) = \sigma^2$ and $\mathrm{Cov}\,(X(\tau_1), X(\tau_2)) = \theta^{|\tau_2 - \tau_1|}\sigma^2$, that is, $X(\tau)$ is a simple Markov process with a parameter θ, where $0 < \theta < 1$. We as-

---

[1] Kei Takeuchi, Faculty of Economics, University of Tokyo, Hongo, Bunkyo-ku, Tokyo 113, Japan.
[2] Masafumi Akahira, Department of Mathematics, University of Electro-Communications, Chofu, Tokyo 182, Japan.

0026–1335/86/03040217–222 $2.50 © 1986 Physica-Verlag, Vienna

sume that θ and σ^2 are unknown, and it is required to estimate θ based on $T+1$ measurements on $X(\tau)$, which are denoted as $X(\tau_0), X(\tau_1), \ldots, X(\tau_T)$.

The problem is to choose such observation points $\tau_0 < \tau_1 < \ldots < \tau_T$ that provide with the maximum possible information on θ.

For simplicity's sake we assume that for $t = 1, \ldots, T, \tau_t - \tau_{t-1} = s$ is constant, that is, the observation points are equally spaced, and we want to get the optimum value for s. In order to simplify the notation we denote $X_t = X(\tau_t)$ $(t = 0, 1, \ldots, T)$ and also put $\rho = \theta^s$.

Then we have the following simple Markov process

$$X_t = \rho X_{t-1} + U_t \quad (t = 1, 2, \ldots),$$

where $\{U_t\}$ is a sequence of independently, identically and normally distributed random variables with mean 0 and variance $\sigma^2(1 - \rho^2)$, and X_0 is a normal random variable with mean 0 and variance σ^2 and for each t, X_0 is independent of U_t.

The log-likelihood function of ρ and σ^2 based on the sample $(X_0, X_1, \ldots, X_T)$ is given by

$$\log L(\rho, \sigma^2) = -\frac{1}{2(1-\rho^2)\sigma^2} \left\{ x_0^2 + x_T^2 + (1 + \rho^2) \sum_{t=1}^{T-1} x_t^2 - 2\rho \sum_{t=1}^{T} x_t x_{t-1} \right\}$$

$$-\frac{T}{2} \log(1 - \rho^2) - \frac{T+1}{2} \log \sigma^2 - \frac{T+1}{2} \log 2\pi.$$

The first and second order partial derivatives with respect to ρ and σ^2 are also given by

$$\frac{\partial}{\partial \rho} \log L(\rho, \sigma^2) = -\frac{\rho}{(1-\rho^2)^2 \sigma^2} \left\{ x_0^2 + X_T^2 + (1 + \rho^2) \sum_{t=1}^{T-1} x_t^2 - 2\rho \sum_{t=1}^{T} x_t x_{t-1} \right\}$$

$$-\frac{1}{(1-\rho^2)\sigma^2} \left(\rho \sum_{t=1}^{T-1} x_t^2 - \sum_{t=1}^{T} x_t x_{t-1} \right) + \frac{T\rho}{1-\rho^2};$$

$$\frac{\partial}{\partial \sigma^2} \log L(\rho, \sigma^2)$$

$$= \frac{1}{2(1-\rho^2)\sigma^4} \left\{ x_0^2 + x_T^2 + (1 + \rho^2) \sum_{t=1}^{T-1} x_t^2 - 2\rho \sum_{t=1}^{T} x_t x_{t-1} \right\} - \frac{T+1}{2\sigma^2};$$

$$-\frac{\partial^2}{\partial \rho^2} \log L(\rho, \sigma^2) = \frac{1 + 3\rho^2}{(1-\rho^2)^3 \sigma^2} \left\{ x_0^2 + x_T^2 + (1 + \rho^2) \sum_{t=1}^{T-1} x_t^2 - 2\rho \sum_{t=1}^{T} x_t x_{t-1} \right\}$$

$$+ \frac{4\rho}{(1-\rho^2)^2 \sigma^2} \left(\rho \sum_{t=1}^{T-1} x_t^2 - \sum_{t=1}^{T} x_t x_{t-1} \right)$$

$$+ \frac{1}{(1-\rho^2)\sigma^2} \sum_{t=1}^{T-1} x_t^2 - \frac{T(1+\rho^2)}{(1-\rho^2)^2}; \tag{1}$$

$$-\frac{\partial^2}{\partial(\sigma^2)^2}\log L(\rho,\sigma^2)$$

$$=\frac{1}{(1-\rho^2)\sigma^6}\left\{x_0^2+x_T^2+(1+\rho^2)\sum_{t=1}^{T-1}x_t^2-2\rho\sum_{t=1}^{T}x_tx_{t-1}\right\}-\frac{T+1}{2\sigma^4};\qquad(2)$$

$$-\frac{\partial^2}{\partial\rho\partial\sigma^2}\log L(\rho,\sigma^2)=\frac{1}{\sigma^2}\left\{\frac{\partial}{\partial\rho}\log L(\rho,\sigma^2)+\frac{T\rho}{1-\rho^2}\right\}.\qquad(3)$$

It follows by (1), (2) and (3) that the information amounts on ρ and σ^2 are obtained by

$$I_{\rho\rho}=-E\left[\frac{\partial^2}{\partial\rho^2}\log L(\rho,\sigma^2)\right]=\frac{T(1+\rho^2)}{(1-\rho^2)^2};$$

$$I_{\sigma^2\sigma^2}=-E\left[\frac{\partial^2}{\partial(\sigma^2)^2}\log L(\rho,\sigma^2)\right]=\frac{T+1}{2\sigma^4};$$

$$I_{\rho\sigma^2}=-E\left[\frac{\partial^2}{\partial\rho\partial\sigma^2}\log L(\rho,\sigma^2)\right]=\frac{T\rho}{(1-\rho^2)\sigma^2}.$$

Hence the inverse of the information matrix is given by

$$\cdot\begin{pmatrix}I_{\rho\rho}&I_{\rho\sigma^2}\\[1em]I_{\rho\sigma^2}&I_{\sigma^2\sigma^2}\end{pmatrix}^{-1}=\begin{pmatrix}\dfrac{T(1+\rho^2)}{(1-\rho^2)^2}&\dfrac{T\rho}{(1-\rho^2)\sigma^2}\\[1em]\dfrac{T\rho}{(1-\rho^2)\sigma^2}&\dfrac{T+1}{2\sigma^4}\end{pmatrix}^{-1}$$

$$=\begin{pmatrix}J(\rho)^{-1}&*\\[0.5em]*&**\end{pmatrix},$$

where

$$J(\rho)=\frac{T}{1-\rho^2}\left\{1+\frac{2\rho^2}{(T+1)(1-\rho^2)}\right\}.$$

We can regard $J(\rho)$ as the amount of information obtainable from the sample with respect to the parameter ρ, and when T is large, for any best asymptotically normal estimator $\hat{\rho}$ (which may be the maximum likelihood estimator, the least squares estimator, etc.). $\sqrt{T+1}\,(\hat{\rho}-\rho)$ is asymptotically normal with mean 0 and variance $(T+1)J(\rho)^{-1}$. And for the initial parameter θ with which we are concerned, the in-

220   K. Takeuchi and M. Akahira

formation is given by

$$I(\theta) = \left(\frac{\partial \rho}{\partial \theta}\right)^2 J(\theta^s) = \frac{s^2 \theta^{2s-2}}{1 - \theta^{2s}}\left(T + \frac{2T}{T+1} \cdot \frac{\theta^{2s}}{1 - \theta^{2s}}\right) \tag{4}$$

since $\rho = \theta^s$.

Now we shall obtain such value of s which maximizes $I(\theta)$.

Since

$$\theta^2 (\log \theta)^2 I(\theta) = \frac{(\log \rho)^2 \rho^2}{1 - \rho^2}\left(T + \frac{2T}{T+1} \cdot \frac{\rho^2}{1 - \rho^2}\right),$$

Putting $\xi = (1 - \rho^2)/\rho^2$ we have

$$4\theta^2 (\log \theta)^2 I(\theta) = \frac{T}{\xi}\{\log (1 + \xi)\}^2 + O(1) = I_\xi + O(1) \quad \text{(say)}.$$

In order to get s maximizing $I(\theta)$ given by (4), it is enough to take the solution ξ_0 of $dI_\xi/d\xi = 0$, i.e., $\log (1 + \xi) = 2\xi/(1 + \xi)$ which is about 3.926. It is also noted that for $\xi = \xi_0$, $\rho \doteq 0.451$. Hence s maximizing $I(\theta)$ is given by $s = -\{\log (1 + \xi_0)\}/(2 \log \theta) \doteq -0.797/\log \theta$.

Even when T is small, the amount of the information provided by the sample may have some value, and it is interesting to calculate the value of β which maximizes

$$\frac{4}{T} \theta^2 (\log \theta)^2 I(\theta) = \frac{\beta^2 e^\beta}{1 - e^\beta}\left\{1 + \frac{2e^\beta}{(T + 1)(1 - e^\beta)}\right\}, \tag{5}$$

where $\beta = 2s \log \theta$.

Then the optimum solution is given by the negative root of the equation

$$\frac{2}{\beta} - \frac{1}{1 - ce^\beta} + \frac{2}{1 - e^\beta} = 0$$

which is independent of θ, where $c = (T - 1)/(T + 1)$.

Table 1 provides us with the values of β which maximize (5) for some values of T.

Note that for $T = 1$, i.e. the case of two observations, the optimum is given by $s = 0$. For comparison if $T = 3$ we have $c = 1/2$, hence $\beta = -1.151$, while for $T \to \infty$, i.e. $c = 1$, we have $\beta = -1.594$ which says that for small T, the optimum spacing is smaller than for large, but is at least 73% of the asymptotically optimum spacing if $T \geqslant 3$.

The optimum value of s depends on the unknown parameter θ, hence we can not apply the result to the practical case but the autocorrelation of the successive measurements for the optimum spacing is independent of θ, and we may get some rough idea about the optimum spacing since in general cases we may have some prior information about the possible speed of decreasing of Cov $(X(t), X(t + s))$.

Table 1

T	2	3	4	5	6	7	8
β	-0.926	-1.151	-1.261	-1.327	-1.371	-1.402	-1.426

T	9	10	20	50	100	∞
β	-1.444	-1.459	-1.526	-1.566	-1.580	-1.594

Otherwise we may resort to a two-stage procedure by first observing T_1 values based on an initial guess and then observing $T_2 (= T - T_1)$ values corresponding to the estimated optimum spacing based on the first sample.

Now let us consider the problem of testing the hypothesis $\theta = \theta_0$ against the alternative $\theta \neq \theta_0$. Then asymptotically, the locally most powerful test and the locally most powerful unbiased test are equivalent to the one-sided and the two-sided tests respectively of the maximum likelihood estimator (e.g. Rao 1965).

When the sample spacing is given, the hypothesis is transformed into $H : \rho = \rho_0 = \theta_0^s$. Then if T is large, the tests based on the maximum likelihood estimator $\hat{\rho}$ are given by the procedure:

One-sided case;   Reject H iff $\sqrt{T+1}\,(\hat{\rho} - \rho_0) > u_\alpha / \sqrt{J(\rho_0)}$

$$(\text{or } \sqrt{T+1}\,(\hat{\rho} - \rho_0) < u_\alpha / \sqrt{J(\rho_0)});$$

Two-sided case;   Reject H iff $\sqrt{T+1}\,|\hat{\rho} - \rho_0| > u_{\alpha/2} / \sqrt{J(\rho_0)},$

where u_α denotes the α-upper quantile of the standard normal distribution. Under the contiguous alternative $K : \rho = \rho_0 + r/\sqrt{T+1}$ the asymptotic powers of the above tests are represented as monotone increasing functions of $r\sqrt{J(\rho_0)}$, where r is a real number. Transforming back to the original parameter, the asymptotic power for the alternative $\theta = \theta_0 + t/\sqrt{T+1}$ is given as a function of $t\,|d\rho/d\theta\,|\sqrt{J(\rho_0)} = t\sqrt{I(\theta_0)}$. Hence the power is maximized when $I(\theta_0)$ is maximized. Therefore the optimum spacing is also given by $s \doteq -0.797/\log\theta_0$.

Remark: In the above discussion we excluded the possibility of θ being equal to zero. It may be of some importance, however, in practice to test the hypothesis $\theta = 0$ against $\theta > 0$. Then it can be shown that the local power is larger when s is smaller and the maximum of the local power can not be attained, which can be also seen from the fact that $s \to 0$ as $\theta \to 0$.

Acknowledgements: The authors are indebted to the referees for their helpful comments which resulted in improvements at various points of the paper. They also wish to thank Mr. H. Osawa of the University of Electro-Communications for making the numerical calculation of Table 1.

222   K. Takeuchi and M. Akahira: A Note on Optimum Spacing of Observations

References

Rao CR (1965) Linear statistical inference and its applications. Wiley, New York
Taga Y (1966) Optimum time sampling in stochastic processes (in Japanese). Proc Inst Statist Math 14:59–61

Received February 8, 1984
(Revised version July 10, 1984)

ON THE BOUND OF THE ASYMPTOTIC DISTRIBUTION OF ESTIMATORS WHEN THE MAXIMUM ORDER OF CONSISTENCY DEPENDS ON THE PARAMETER *

Masafumi Akahira[*] and Kei Takeuchi[**]

Abstract

In this papar, the case when the order of consistency depends on the parameter is discussed, and in the simple unstable process the asymptotic means and variances of the log-likelihood ratio test statistic are obtained under the null and the alternative hypotheses. Further its asymptotic distribution is also discussed.

(*) Department of Mathematics, University of Electro-Communications, Chofu, Tokyo 182, Japan.

The author is on leave and visiting Queen's University from April to July 1985.

(**) Faculty of Economics, University of Tokyo, Hongo Bunkyo-ku, Tokyo 113, Japan.

AMS Subject Classification (1980). 62F11, 62M10.

Key words and phrases: Order of consistency, Asymptotically median unbiased estimator, Autoregressive process, Log-likelihood ratio, Asymptotic distribution.

---

*This paper is retyped with the correction of typographical errors.

1. INTRODUCTION

In the regular case it is known that the order of consistency is equal to $\sqrt{n}$, but in the non-regular case it is not always so, e.g. $n^{1/\alpha}$ $(0 < \alpha < 2)$, $\sqrt{n \log n}$ etc., which are independent of the unknown parameter (e.g. see Akahira, 1975a; Akahira and Takeuchi, 1981; Vostrikova, 1984). However, the order of consistency may depend on it in the case of the unstable process. Here a discussion on that will be done.

In the autoregressive process $\{X_t\}$ with $X_t = \theta X_{t-1} + U_t$ $(t = 1, 2, \ldots)$, where $\{U_t\}$ is a sequence of independently and identically distributed random variables and $X_0 = 0$, it is known that the asymptotic distribution of the least squares estimator of θ is normal for $|\theta| < 1$, Cauchy for $|\theta| > 1$ (e.g. White, 1958; Anderson, 1959) and some one for $|\theta| = 1$ (Rao, 1978). Further, in the case when $|\theta| < 1$ the asymptotic efficiency of estimators was studied by Akahira (1976) and Kabaila (1983) for the ARMA process, and their higher order asymptotic efficiency has been discussed by Akahira (1975b, 1979, 1982, 1984) and Taniguchi (1983) for the ARMA process. In this paper in the first order AR process with $|\theta| \geq 1$ the asymptotic means and variances of the log-likelihood ratio test statistic are obtained under the null and the alternative hypothesis.

Further its asymptotic distribution is also discussed.

2.   RESULTS

Let $(\mathcal{X}, \mathcal{B})$ be a sample space and Θ be a parameter space, which is assumed to be an open set in an Euclidean 1-space $\mathbf{R}^1$. We shall denote by $(\mathcal{X}^{(T)}, \mathcal{B}^{(T)})$ the T-fold direct product of $(\mathcal{X}, \mathcal{B})$. For each $T = 1, 2, \ldots$, the points of $\mathcal{X}^{(T)}$ will be denoted by $\tilde{x}_T = (x_1, \ldots, x_T)$. We consider a sequence of classes of probability measures $\{P_{\theta,T} : \theta \in \Theta\}$ $(T = 1, 2, \ldots)$ each defined on $(\mathcal{X}^{(T)}, \mathcal{B}^{(T)})$ such that for each $T = 1, 2, \ldots$ and each $\theta \in \Theta$ the following holds:

$$P_{\theta,T}\left(B^{(T)}\right) = P_{\theta,T+1}\left(B^{(T)} \times \mathcal{X}\right)$$

for all $B^{(T)} \in \mathcal{B}^{(T)}$.

An estimator of θ is defined to be a sequence $\{\hat{\theta}_T\}$ of $B^{(T)}$-measurable functions. For simplicity we may denote an estimator $\hat{\theta}_T$ instead of $\{\hat{\theta}_T\}$. For an increasing sequence $\{c_T\}$ (c_T tending to infinity) an estimator $\hat{\theta}_T$ is called consistent with order $\{c_T\}$ (or c_T-consistent for short) if for every $\varepsilon > 0$ and every $\vartheta \in \Theta$ there exists a positive number L such that

$$\lim_{L \to \infty} \varlimsup_{T \to \infty} \sup_{\theta : |\theta - \vartheta| < \delta} P_{\theta,T}\left\{c_T |\hat{\theta}_T - \theta| \geq L\right\} = 0$$

(Akahira, 1975a).

It is known that the order of convergence of consistent estimators is equal to $\sqrt{n}$ is the regular case, but not always so, e.g. $n^{1/\alpha}$ $(0 < \alpha < 2)$, $\sqrt{n \log n}$ etc., in the non-regular case when the support of the density depends on the parameter θ. In both cases the order of consistency is independent of θ.

If c_T can not be decided independently of θ, we may change $c_T(\theta)$ instead of c_T in the above definition. However, in such a definition we shall not be able to determine uniquely the value of $c_T(\theta)$ at θ_0 (Takeuchi, 1974). Then a similar phenomenon to "superefficiency" happens. Indeed, if $\hat{\theta}_T$ has an order c_T of consistency independent of θ, then for a specified value θ_0 we define an estimator $\hat{\theta}_T^*$ as

$$\hat{\theta}_T^* = \begin{cases} d_T^{-1}\left(\hat{\theta}_T - \theta_0\right) + \theta_0 & \text{for } |\hat{\theta}_T - \theta_0| \leq c_T^{-1/2}; \\ \hat{\theta}_T & \text{for } |\hat{\theta}_T - \theta_0| > c_T^{-1/2}, \end{cases}$$

where for each $T = 1, 2, \ldots$, d_T is a constant with $d_T > 1$. We define $c_T^*(\theta)$ as follows:

$$c_T^*(\theta_0) = c_T d_T$$

and for any $\theta \neq \theta_0$

$$c_T^*(\theta) = \begin{cases} c_T & \text{for } |\theta - \theta_0| \leq c_T^{-1/2} \\ c_T \left\{ 1 - \frac{c_T^{-1/2}}{|\theta - \theta_0|} \right\} & \text{for } |\theta - \theta_0| > c_T^{-1/2} \end{cases}$$

<u>Case (i) $\theta = \theta_0$.</u> Since $|\hat{\theta}_T - \theta_0| < c_T^{-1/2}$ implies $c_T d_T |\hat{\theta}_T^* - \theta_0| = c_T |\hat{\theta}_T - \theta_0|$, it follows that

$$P_{\theta_0,T} \left\{ c_T^*(\theta_0)|\hat{\theta}_T^* - \theta_0| > L \right\}$$

$$\leq P_{\theta_0,T} \left\{ c_T |\hat{\theta}_T - \theta_0| > L \right\} + P_{\theta_0,T} \left\{ |\hat{\theta}_T - \theta_0| > c_T^{-1/2} \right\}.$$

<u>Case (ii) $\theta \neq \theta_0$.</u> In the case when $|\theta - \theta_0| \leq c_T^{-1/2}$, $|\hat{\theta}_T - \theta_0| < c_T^{-1/2}$ implies $c_T^*(\theta)|\hat{\theta}_T^* - \theta| < c_T^*(\theta)|\hat{\theta}_T - \theta_0| < 1$, and $|\hat{\theta}_T - \theta_0| \geq c_T^{-1/2}$ implies $c_T^*(\theta)|\hat{\theta}_T^* - \theta| = c_T^*(\theta)|\hat{\theta}_T - \theta| < c_T|\hat{\theta}_T - \theta|$. Then we have for $L > 1$

$$P_{\theta,T} \left\{ c_T^*(\theta)|\hat{\theta}_T^* - \theta| > L \right\} < P_{\theta,T} \left\{ c_T|\hat{\theta}_T - \theta| > L \right\}.$$

In the case when $|\theta - \theta_0| > c_T^{-1/2}$, $|\hat{\theta}_T - \theta_0| \leq c_T^{-1/2}$ implies $c_T^*(\theta)|\hat{\theta}_T^* - \theta| < c_T^*(\theta)|\hat{\theta}_T - \theta_0| < c_T|\hat{\theta}_T - \theta|$, and $|\hat{\theta}_T - \theta_0| > c_T^{-1/2}$ implies

$$c_T^*(\theta)|\hat{\theta}_T^* - \theta| = c_T \left\{ 1 - \frac{c_T^{-1/2}}{|\theta - \theta_0|} \right\} |\hat{\theta}_T - \theta_0| < c_T|\hat{\theta}_T - \theta|.$$

Then we have for $L > 1$

$$P_{\theta,T}\left\{c_T^*(\theta)|\hat{\theta}_T^* - \theta| > L\right\} < P_{\theta,T}\left\{c_T|\hat{\theta}_T - \theta| > L\right\}.$$

Hence in both cases (i) and (ii) we obtain for $L > 1$

$$P_{\theta,T}\left\{c_T^*(\theta)|\hat{\theta}_T^* - \theta| > L\right\}$$
$$< P_{\theta,T}\left\{c_T(\theta)|\hat{\theta}_T - \theta| > L\right\} + P_{\theta_0,T}\left\{c_T|\hat{\theta}_T - \theta_0| > c_T^{1/2}\right\}.$$

Letting $n \to \infty$ and $L \to \infty$, we see that the right-hand side of the above inequality tends to zero locally uniformly. Since $\{d_T\}$ can be a sequence tending to infinity as $T \to \infty$ and for any $\theta \neq \theta_0$, $c_T^*(\theta)/c_T \to 1$ as $T \to \infty$, it is possible to make the order of convergence arbitrarily large at $\theta = \theta_0$. Hence we can not decide uniquely the value $c_T(\theta)$ at the point.

A c_T-consistent estimator $\hat{\theta}_T$ is defined to be asymptotically median unbiased (AMU) if for any $\vartheta \in \Theta$ there exists a positive number δ such that

$$\lim_{T\to\infty}\sup_{\theta:|\theta-\vartheta|<\delta}\left|P_{\theta,T}\left\{\hat{\theta}_T \leq \theta\right\} - \frac{1}{2}\right| = 0;$$

$$\lim_{T\to\infty}\sup_{\theta:|\theta-\vartheta|<\delta}\left|P_{\theta,T}\left\{\hat{\theta}_T \geq \theta\right\} - \frac{1}{2}\right| = 0.$$

Let $\hat{\theta}_T$ be an AMU estimator satisfying the following:

(2.0)

$$\lim_{T\to\infty}\left[P_{\theta,T}\left\{c_T\left(\hat{\theta}_T-\theta\right)\leq t\right\}-\frac{1}{2}\right]>0 \quad\text{for some } t>0;$$

$$\overline{\lim_{T\to\infty}}\left[P_{\theta,T}\left\{c_T\left(\hat{\theta}_T-\theta\right)\leq t\right\}-\frac{1}{2}\right]<0 \quad\text{for some } t<0.$$

If for any sequence $\{c'_T\}$ of order of consistency with (2.0), $\varliminf_{T\to\infty} c'_T/c_T < \infty$, then $\{c_T\}$ is called maximum order of consistency.

There are most cases when the maximum order is uniquely determined, but there may not exist an estimator which satisfies (2.0) and uniformly attain the bound of

$$P_{\theta,T}\left\{c_T\left(\hat{\theta}_T-\theta\right)\leq t\right\}$$

in the class of AMU estimators with (2.0). It is noted that the condition (2.0) means non-constant of the concentration probability of the estimator in some interval involving the origin.

We consider a simple autoregressive (AR) process $\{X_t\}$ which is defined by $X_t = \theta X_{t-1} + U_t$ $(t = 1, 2, \ldots)$, where $\{U_t\}$ is a sequence of independently, identically and normally distributed random variables with mean 0 and variance 1 and $X_0 = 0$.

In the case when $|\theta| < 1$, the bound of the asymptotic distribution of the all AMU estimators is obtained up to the $n^{-1/2}$ and it is shown that a (modified) least squares estimator is (second order) asymptotically efficient (e.g., see Akahira, 1975b, 1976, 1979, 1982, 1984).

In the subsequent discussion we shall treat the case when $|\theta| \geq 1$. Then it is known that the order $\{c_T(\theta)\}$ of consistency is given by

$$c_T(\theta) = \begin{cases} |\theta|^T & \text{for } |\theta| > 1; \\ T & \text{fot } |\theta| = 1, \end{cases}$$

(e.g. see Anderson, 1959; Rao, 1978).
Letting $|\theta_0| \geq 1$, we deal with the problem of testing the hypothesis $H : \theta = \theta_0 + (u/c_T(\theta_0))$ against the alternative $K : \theta = \theta_0$. Putting $\theta_1 = \theta_0 + (u/c_T(\theta_0))$ we consider the log-likelihood ratio L_T given by

$$(2.1) \quad L_T = (\theta_0 - \theta_1) \left(\sum_{t=2}^{T} x_t x_{t-1} - \frac{\theta_0 + \theta_1}{2} \sum_{t=2}^{T} x_{t-1}^2 \right).$$

Then we shall obtain the asymptotic mean and variance of L_T under $H : \theta = \theta_1$, and $K : \theta = \theta_0$.

For each $k = 0, 1, \ldots, t$ we have

(2.2)
$$E_\theta \left(X_t X_{t-k} \right) = \theta E_\theta \left(X_{t-1} X_{t-k} \right) = \cdots = \theta^k V_\theta \left(X_{t-k} \right) = \theta^k \sigma_{t-k}^2 \ (\text{say}).$$

Since for each $t = 1, 2, \ldots,$

$$V_\theta \left(X_t \right) = \theta^2 V_\theta \left(X_{t-1} \right) + V \left(U_t \right),$$

it follows that

(2.3)
$$\sigma_t^2 = \theta^2 \sigma_{t-1}^2 + 1, \quad (t = 1, 2, \ldots),$$

where $\sigma_0^2 = 0$. First we obtain

(2.4)
$$V_\theta \left(L_T \right) = \left(\theta_0 - \theta_1 \right)^2 \left\{ V_\theta \left(\sum_{t=1}^{T} X_t X_{t-1} \right) \right.$$
$$- \left(\theta_0 + \theta_1 \right) \mathrm{Cov}_\theta \left(\sum_{t=2}^{T} X_t X_{t-1}, \sum_{t=2}^{T} X_{t-1}^2 \right)$$
$$\left. + \frac{\left(\theta_0 + \theta_1 \right)^2}{4} V_\theta \left(\sum_{t=2}^{T} X_{t-1}^2 \right) \right\}.$$

By (2.2) and (2.3) we have

(2.5)
$$V_\theta \left(\sum_{t=2}^{T} X_t X_{t-1} \right)$$
$$= \sum_{t=2}^{T} V_\theta \left(X_t X_{t-1} \right) + 2 \sum \sum_{t<t'} \mathrm{Cov}_\theta \left(X_t X_{t-1}, X_{t'} X_{t'-1} \right)$$
$$= \sum_{t=2}^{T} \left(\sigma_t^2 \sigma_{t-1}^2 + \theta^2 \sigma_{t-1}^4 \right) + 4 \sum \sum_{t<t'} \sigma_t^2 \sigma_{t-1}^2 \theta^{2(t'-t)}.$$

We also obtain by (2.2) and (2.3)

$$(2.6) \quad V_\theta \left(\sum_{t=2}^{T} X_{t-1}^2 \right) = 2 \sum_{t=2}^{T} \sigma_{t-1}^4 + 4 \sum\sum_{t<t'} \sigma_{t-1}^4 \theta^{2(t'-t)};$$

$$(2.7) \quad \mathrm{Cov}_\theta \left(\sum_{t=2}^{T} X_t X_{t-1}, \sum_{t=2}^{T} X_{t-1}^2 \right)$$

$$= 2 \left\{ \sum\sum_{t<t'} \sigma_t^2 \sigma_{t-1}^2 \theta^{2(t'-t)-1} + \sum\sum_{t\le t'} \sigma_{t-1}^4 \theta^{2(t'-t)+1} + \sum_{t=2}^{T} \sigma_{t-1}^4 \theta \right\}.$$

<u>Case (I): $\theta_0 > 1$.</u> Since by (2.2)

$$E_\theta \left(\sum_{t=1}^{T} X_t X_{t-1} \right) = \sum_{t=2}^{T} E_\theta \left(X_t X_{t-1} \right) = \theta \sum_{t=2}^{T} \sigma_{t-1}^2 = \theta \sum_{t=1}^{T-1} \sigma_t^2,$$

$$E_\theta \left(\sum_{t=1}^{T} X_{t-1}^2 \right) = \sum_{t=2}^{T-1} \sigma_t^2,$$

it follows from (2.1) that

$$(2.8) \quad E_\theta \left(L_T \right) = (\theta_0 - \theta_1) \left(\theta \sum_{t=1}^{T-1} \sigma_t^2 - \frac{\theta_0 + \theta_1}{2} \sum_{t=1}^{T-1} \sigma_t^2 \right)$$

$$= (\theta_0 - \theta_1) \left(\theta - \frac{\theta_0 + \theta_1}{2} \right) \sum_{t=1}^{T-1} \sigma_t^2.$$

Since by (2.3)

$$\sigma_t^2 = 1 + \theta^2 + \cdots + \theta^{2(t-1)} = \frac{\theta^{2t} - 1}{\theta^2 - 1},$$

it follows that

$$(2.9) \quad \sum_{t=1}^{T-1} \sigma_t^2 = \frac{\theta^{2(T-1)} - 1}{(\theta^2 - 1)^2} - \frac{T-1}{\theta^2 - 1}.$$

By (2.8) and (2.9) we have

$$(2.10) \qquad E_{\theta_0}(L_T) = \frac{u^2}{2\theta_0^2 \left(\theta_0^2 - 1\right)^2} + o(1);$$

$$(2.11) \qquad E_{\theta_1}(L_T) = -\frac{u^2}{2\theta_0^2 \left(\theta_0^2 - 1\right)^2} + o(1).$$

Since $\theta_0 - \theta_1 = -u\theta_0^{-T}$ and $\theta_0 + \theta_1 = 2\theta_0 + u\theta_0^{-T} = 2\theta_0 + o(1)$, it follows from (2.3) to (2.7) and (2.9) that

$$(2.12) \quad V_{\theta_0}(L_T) = u^2 \theta_0^{-2T} \sum_{t=2}^{T} \left(\sigma_t^2 \sigma_{t-1}^2 - \theta_0^2 \sigma_{t-1}^4\right) + o(1)$$

$$= u^2 \theta_0^{-2T} \sum_{t=2}^{T} \sigma_{t-1}^2 + o(1)$$

$$= \frac{u^2}{\theta_0^2 \left(\theta_0^2 - 1\right)^2} + o(1).$$

Similarly we have

$$(2.13) \qquad V_{\theta_1}(L_T) = \frac{u^2}{\theta_0^2 \left(\theta_0^2 - 1\right)^2} + o(1).$$

In the case when $\theta_0 < -1$, we have similar asymptotic means and variances of L_T under H and K to those of this case.

Case (II): $\theta_0 = 1$. Since by (2.2) and (2.3)

$$\sum_{t=2}^{T} E_1\left(X_t X_{t-1}\right) = \sum_{t=1}^{T-1} E_1\left(X_t^2\right) = \sum_{t=1}^{T-1} t = \frac{(T-1)T}{2},$$

it follows from (2.1) that

$$(2.14) \qquad E_1\left(L_T\right) = \frac{u^2}{4} + o(1).$$

In a similar way as the case (I), we obtain under $H:$ $\theta = \theta_1 = 1 + (u/T)$

(2.15)

$$
\begin{aligned}
E_{\theta_1}\left(L_T\right) &= -\frac{u^2}{2T^2}\left\{\frac{1 - \theta_1^{2(T-1)}}{\left(1 - \theta_1^{-2}\right)\left(1 - \theta_1^2\right)} - \frac{(T-1)\theta_1^{-2}}{1 - \theta_1^{-2}}\right\} \\
&= -\frac{u^2}{2T^2}\left\{\frac{\theta_1^{2T} - \theta_1^2}{\left(\theta_1^2 - 1\right)^2} - \frac{T-1}{\theta_1^2 - 1}\right\}.
\end{aligned}
$$

Since

$$\frac{1}{\theta_1^2 - 1} = \frac{T}{2u + \frac{u^2}{T}},$$

it follows from (2.15) that

(2.16)

$$
\begin{aligned}
E_{\theta_1}\left(L_T\right) &= -\frac{u^2}{2T^2}\left[\left\{\left(1 + \frac{u}{T}\right)^{2T} - \left(1 + \frac{u}{T}\right)^2\right\} \frac{T^2}{\left(2u + \frac{u^2}{T}\right)^2} \right. \\
&\qquad\qquad \left. - \frac{T(T-1)}{2u + \frac{u^2}{T}}\right] \\
&= -\frac{1}{8}\left(e^{2u} - 1\right) + \frac{u}{4} + o(1).
\end{aligned}
$$

We also have the asymptotic variances as

(2.17)
$$V_1\left(L_T\right) = \frac{u^2}{T^2}\sum_{t=2}^{T}\sigma_{t-1}^2 = \frac{u^2}{T^2}\sum_{t=1}^{T-1}t = \frac{u^2}{2} + o(1);$$

(2.18)
$$V_{\theta_1}\left(L_T\right) = \frac{u^2}{T^2}\sum_{t=2}^{T}\sigma_{t-1}^2 = \frac{u^2}{T^2}\left\{\frac{\theta_1^{2(T-1)}-1}{\left(\theta_1^2-1\right)^2} - \frac{T-1}{\theta_1^2-1}\right\}$$

$$= \frac{u^2}{T^2}\left[\left\{\left(1+\frac{u}{T}\right)^{2(T-1)}-1\right\}\cdot\frac{T^2}{\left(2u+\frac{u^2}{T}\right)^2} - \frac{(T-1)T}{2u+\frac{u^2}{T}}\right]$$

$$= u^2\left(\frac{e^{2u}-1}{4u^2} - \frac{1}{2u}\right) + o(1)$$

$$= \frac{1}{4}\left(e^{2u}-1\right) - \frac{u}{2} + o(1).$$

In the case when $\theta_0 = -1$, we have similar asymptotic means and variances of L_T under H and K to those of this case.

Hence we established the following theorem.

<u>Theorem</u>. In the AR process the asymptotic means and variances of the log-likelihood ratio L_T under $H : \theta = \theta_1$ and $K : \theta = \theta_0$ for $|\theta_0| \geq 1$ are given by the following:

	$|\theta_0| > 1$	$|\theta_0| = 1$
$E_{\theta_0}(L_T)$	$\dfrac{u^2}{2\theta_0^2(\theta_0^2-1)^2} + o(1)$	$\dfrac{u^2}{4} + o(1)$
$E_{\theta_1}(L_T)$	$-\dfrac{u^2}{2\theta_0^2(\theta_0^2-1)^2} + o(1)$	$-\frac{1}{8}(e^{2u} - 1) + \frac{u}{4} + o(1)$
$V_{\theta_0}(L_T)$	$\dfrac{u^2}{\theta_0^2(\theta_0^2-1)^2} + o(1)$	$\dfrac{u^2}{2} + o(1)$
$V_{\theta_1}(L_T)$	$\dfrac{u^2}{\theta_0^2(\theta_0^2-1)^2} + o(1)$	$\frac{1}{4}(e^{2u} - 1) - \frac{u}{2} + o(1)$

Table 2.1

<u>Remark.</u> By the Taylor expansion it is shown that for $|\theta_0| = 1$

$$E_{\theta_1}(L_T) = -\frac{1}{8}(e^{2u} - 1) + \frac{u}{2} + o(1) = -\frac{u^2}{4} + \cdots$$

$$V_{\theta_1}(L_T) = \frac{u^2}{2} + \cdots ,$$

in which the leading terms are equal to those by $E_{\theta_0}(L_T)$ and $V_{\theta_0}(L_T)$, respectively.

Next we shall consider the asymptotic distribution of L_T under H and K. Using the discussion by Basawa and Brockwell (1984, page 165), we see that L_T converges in distribution to

$$\{a_{\theta_i}(u)Y + b_{\theta_i}(u)Z\}\, Y \qquad (i = 0, 1)$$

under H and K, respectively, where for each $i = 0, 1$, $a_{\theta_i}(u)$ and $b_{\theta_i}(u)$ are constants, and where Y and Z are mutually, independently and normally distributed random variables with mean 0 and variance 1. By the fact we may obtain the bound of the asymptotic distribution of the all AMU estimators, but there might not exist an AMU estimator whose asymptotic distribution uniformly attains it.

ACKNOWLEDGEMENTS

The paper was written while the first author was at Queen's University in Canada as a visiting professor. The visit was supported by a grant of the Natural Sciences and Engineering Council of Canada. He is grateful to Professor Colin R. Blyth for inviting him.

REFERENCES

Akahira, M. (1975a). Asymptotic theory for estimation of location in non-regular cases, I: Order of convergence of consistent estimators. *Rep. Stat. Appl. Res., JUSE*, **22**, 8–26.

Akahira, M. (1975b). A note on the second order asymptotic efficiency of estimators in an autoregressive process. *Rep. Univ. Electro-Comm.*, **26**, 143–149.

Akahira, M. (1976). On the asymptotic efficiency of estimators in an autoregressive process. *Ann. Inst. Statist. Math.*, **28**, 35–48.

Akahira, M. (1979). On the second order asymptotic optimality of estimators in an autoregressive process. *Rep. Univ. Electro-Comm.*, **29**, 213–218.

Akahira, M. (1982). Second order asymptotic optimality of estimators in an autoregressive process with unknown mean. *Selecta Statistica Canadiana VI*, 19–36.

Akahira, M. (1984). Asymptotic deficiency of the estimator of a parameter of an autoregressive process with the missing observation. *Rep. Stat. Appl. Res., JUSE*, **31**, 1–13.

Akahira, M. and Takeuchi, K. (1981). *Asymptotic Efficiency of Statistical Estimators: Concepts and Higher Order Asymptotic Efficiency.* Lecture Notes in Statistics 7, Springer-Verlag, New York.

Anderson, T. W. (1959). On asymptotic distributions of estimates of stochastic difference equations. *Ann. Math. Statist.*, **30**, 676–687.

Basawa, I. V. and Brockwell, P. J. (1984). Asymptotic conditional inference for regular nonergodic models with an application to autoregressive processes. *Ann. Statist.*, **12**, 161–171.

Kabaila, P. (1983). On the asymptotic efficiency of the estimators of the parameters of an ARMA process. *Journal of Time Series Analysis*, **4**, 37–47.

Rao, M. M. (1978). Asymptotic distribution of an estimator of the boundary parameter of an unstable process. *Ann. Statist.*, **6**, 185–190.

Takeuchi, K. (1974). Tôkeiteki suitei no Zenkinriron (Asymptotic Theory of Statistical Estimation). (In Japanese) Kyôiku-Shuppan, Tokyo.

Taniguchi, M. (1983). On the second order asymptotic efficiency of estimators of Gaussian ARMA processes. *Ann. Statist.*, **11**, 157–169.

Vostrikova, L. Ju. (1984). On criteria for c_n-consistency of estimators. *Stochastics* **11**, 265–290.

White, J. S. (1958). The limiting distribution of the serial correlation coefficient in the explosive case. *Ann. Math. Statist.*, **29**, 1188–1197.

Reçu en Mai 1985

Masafumi Akahira [1] and Kei Takeuchi [2]

ON THE DEFINITION OF ASYMPTOTIC EXPECTATION

ABSTRACT

In this paper a definition of asymptotic expectation is given and its fundamental properties are discussed.

1. INTRODUCTION

In the asymptotic theory of estimation, the concept of asymptotic expectation is widely used (e.g., Akahira and Takeuchi, 1981; Ibragimov and Has'minskii, 1981; Lehmann, 1982), and it is usually remarked that it can be different from the asymptotic value of expectation. It is, however, not sufficiently accurately defined in the literature, especially when the asymptotic distribution does not exist.

In the paper we shall give a definition of the asymptotic expectation and show its properties, e.g., its linearity and a Markov type inequality. We shall also obtain the necessary and sufficient conditions for the convergences in probability and distribution.

2. RESULTS

Let $\{X_n\}$ be a sequence of non-negative random variables. For any sequence $\{A_n\}$ of positive numbers we define an A_n-censored sequence as

$$X_n^*(A_n) = \min\{X_n, A_n\}, \qquad n = 1, 2, \ldots.$$

For $\{X_n\}$ we denote by $\mathcal{A}(X_n)$ a set of all the sequences $\{A_n\}$ of positive

---

[1] Department of Mathematics, University of Electro-Communications, Chofu, Tokyo 182, Japan

[2] Faculty of Economics, University of Tokyo, Hongo, Bunkyo-ku, Tokyo 113, Japan.

199

I. B. MacNeill and G. J. Umphrey (eds.), Foundations of Statistical Inference, 199–208.

200 M. AKAHIRA AND K. TAKEUCHI

numbers satisfying

$$\Pr\{X_n > A_n\} \to 0 \quad \text{as} \quad n \to \infty.$$

We define the upper and the lower asymptotic expectations of $\{X_n\}$ as

$$\overline{As}.E(X_n) = \inf_{\{A_n\} \in \mathcal{A}(X_n)} \limsup_{n \to \infty} E(X_n^*(A_n))$$

and

$$\underline{As}.E(X_n) = \inf_{\{A_n\} \in \mathcal{A}(X_n)} \liminf_{n \to \infty} E(X_n^*(A_n)),$$

respectively. Note that they can be infinity.

Definition 2.1. If $\overline{As}.E(X_n) = \underline{As}.E(X_n) < \infty$, then we call it the asymptotic expectation of $\{X_n\}$ and denote it by $As.E(X_n)$.

The following theorem establishes that the above definition is reduced to the usual concept when X_n has an asymptotic distribution.

Theorem 2.1. If X_n has a proper asymptotic distribution F, that is

$$\Pr\{X_n \leq x\} \to F(x) \quad \text{as} \quad n \to \infty$$

for every continuity point x of F, then

$$As.E(X_n) = \int_0^\infty x \, dF(x),$$

provided that the right-hand side is finite.

Proof. Denote by $F_n(x)$ the distribution of X_n. We put

$$\mu = \int_0^\infty x \, dF(x).$$

Since for any fixed positive number A

$$|E(X_n^*(A)) - E(X^*(A))|$$

$$\leq \left| \int_0^A x \, dF_n(x) - \int_0^A x \, dF(x) \right| + A|F_n(A) - F(A)|,$$

we have

$$|E(X_n^*(A)) - E(X^*(A))| \to 0 \quad \text{as} \quad n \to \infty. \tag{2.1}$$

Since μ is finite, it follows that

$$|E(X^*(A)) - \mu| \to 0 \quad \text{as} \quad A \to \infty. \tag{2.2}$$

Then it follows from (2.1) and (2.2) that for any sequence $\{\varepsilon_m\}$ of positive numbers there exists a sequence $\{A_m\}$ of positive numbers such that

$$|E(X^*(A_m)) - \mu| < \varepsilon_m \qquad (m = 1, 2, \ldots). \tag{2.3}$$

For the sequence $\{A_m\}$ there is a monotone increasing sequence $\{N(m)\}$ of positive integers such that for $n \geq N(m)$

$$|E(X_n^*(A_m)) - E(X^*(A_m))| < \varepsilon_m. \tag{2.4}$$

For n satisfying $N(m) \leq n < N(m+1)$ we define $m = N^{-1}(n)$. Denoting $A_{N^{-1}(n)}$ and $\varepsilon_{N^{-1}(n)}$ by A_n and ε_n, respectively, we have by (2.3) and (2.4)

$$|E(X_n^*(A_n)) - \mu| < 2\varepsilon_n$$

for sufficiently large n.

Letting $\varepsilon_n \to 0$ as $n \to \infty$, we obtain

$$|E(X_n^*(A_n)) - \mu| \to 0 \quad \text{as} \quad n \to \infty.$$

This completes the proof.

In order to obtain some properties on the asymptotic expectation we have to prove some lemmas.

Lemma 2.1. If $\overline{As}.E(X_n) < \infty$, then for any sequence $\{A_n\}$ of positive numbers tending to infinity as $n \to \infty$, we have that $\{A_n\} \in \mathcal{A}(X_n)$.

Proof. Assume that there exists a sequence $\{A_n'\}$ of positive numbers such that $\lim_{n\to\infty} A_n' = \infty$ and $\{A_n'\} \notin \mathcal{A}(X_n)$. For some $\varepsilon > 0$ there exists a subsequence $\{A_{n_j}'\}$ of $\{A_n'\}$ such that

$$\Pr\{X_{n_j} > A_{n_j}'\} > \varepsilon.$$

Since for any $\{A_n\} \in \mathcal{A}(X_n)$, $\Pr\{X_n > A_n\} \to 0$ as $n \to \infty$, it follows that for sufficiently large j

$$\Pr\{X_{n_j} > A_{n_j}\} < \varepsilon,$$

202   M. AKAHIRA AND K. TAKEUCHI

hence $A_{n_j} > A'_{n_j}$. And we have for sufficiently large j

$$E(X^*_{n_j}(A_{n_j})) \geq E(X^*_{n_j}(A'_{n_j}))$$
$$\geq A'_{n_j}\Pr\{X_{n_j} \geq A'_{n_j}\}$$
$$> A'_{n_j}\varepsilon,$$

hence the last term tends to infinity as $j \to \infty$. Hence we have

$$\limsup_{n\to\infty} E(X^*_n(A_n)) = \infty,$$

which contradicts the condition of the lemma. This completes the proof.

We denote by $\mathcal{A}_\infty(X_n)$ a subset of $\mathcal{A}(X_n)$ whose element $\{A_n\}$ satisfies $\lim_{n\to\infty} A_n = \infty$.

Lemma 2.2. The following hold:

$$\overline{As}.E(X_n) = \inf_{\{A_n\}\in\mathcal{A}_\infty(X_n)} \limsup_{n\to\infty} E(X^*_n(A_n));$$
$$\underline{As}.E(X_n) = \inf_{\{A_n\}\in\mathcal{A}_\infty(X_n)} \liminf_{n\to\infty} E(X^*_n(A_n)).$$

Proof. If the left-hand side of either of the two equalities above is infinity, it is obvious that the right-hand side is also infinity, hence we may assume that the right-hand side is finite.

For any $\{A_n\} \in \mathcal{A}(X_n)$ we put $p_n = \Pr\{X_n > A_n\}$. Since $p_n \to 0$ as $n \to \infty$, putting $A'_n = \max\{A_n, p_n^{-1/2}\}$ we have $A'_n \to \infty$ as $n \to \infty$, hence $\{A'_n\} \in \mathcal{A}_\infty(X_n)$. Since $A'_n \geq A_n$, we obtain

$$E(X^*_n(A'_n)) \geq E(X^*_n(A_n)). \tag{2.5}$$

Since

$$E(X^*_n(A'_n)) - E(X^*_n(A_n))$$
$$\leq (A'_n - A_n)\Pr\{X_n > A_n\}$$
$$\leq p_n^{1/2},$$

it follows from (2.5) that

$$\lim_{n\to\infty}\{E(X^*_n(A'_n)) - E(X^*_n(A_n))\} = 0.$$

Hence for any $\{A_n\} \in \mathcal{A}(X_n)$ there exists a sequence $\{A'_n\} \in \mathcal{A}_\infty(X_n)$ such that

$$\lim_{n\to\infty}\{E(X^*_n(A'_n)) - E(X^*_n(A_n))\} = 0.$$

This fact leads to the conclusion of the lemma.

Lemma 2.3. If $\overline{As}.E(X_n) < \infty$, then there exists a sequence $\{A_n^*\} \in \mathcal{A}_\infty(X_n)$ such that

$$\limsup_{n\to\infty} E(X_n^*(A_n^*)) = \overline{As}.E(X_n);$$
$$\liminf_{n\to\infty} E(X_n^*(A_n^*)) = \underline{As}.E(X_n).$$

Proof. For a sequence $\{\varepsilon_m\}$ of positive numbers, such that $\varepsilon_m \downarrow 0$, there exists a sequence $\{A_{n,m}\}$ of positive numbers such that

$$\limsup_{n\to\infty} E(X_n^*(A_{n,m})) \leq \overline{As}.E(X_n) + \varepsilon_m.$$

Without loss of generality we assume that, for each n, $A_{n,m}$ is monotone decreasing in m. We put $p_{n,m} = \Pr\{X_n > A_{n,m}\}$. Then for each n, $p_{n,m}$ is monotone increasing in m and for each m, $\lim_{n\to\infty} p_{n,m} = 0$. For any m there exists $n(m)$ such that $p_{n,m} < \varepsilon_m$ for $n > n(m)$. For any n let $m = m^{-1}(n)$ be a maximum value satisfying $n(m) \geq n$. Since $m^{-1}(n) \to \infty$ as $n \to \infty$, putting $A_n^{**} = A_{n,m^{-1}(n)}$ we have

$$\Pr\{X_n > A_n^{**}\} < \varepsilon_{m^{-1}(n)},$$

of which the right-hand side tends to 0 as $n \to \infty$. For any fixed n_0 we obtain

$$\limsup_{n\to\infty} E(X_n^*(A_n^{**})) \leq \limsup_{n\to\infty} E(X_n^*(A_{n,m^{-1}(n_0)}))$$
$$\leq \overline{As}.E(X_n) + \varepsilon_{m^{-1}(n_0)}.$$

Since $\varepsilon_{m^{-1}(n_0)} \to 0$ as $n_0 \to \infty$, we have

$$\limsup_{n\to\infty} E(X_n^*(A_n^{**})) \leq \overline{As}.E(X_n). \tag{2.6}$$

On the other hand it follows from the definition of $\overline{As}.E(X_n)$ that the inverse inequality of (2.6) holds. Hence it is seen that the equality in (2.6) holds.

In a way similar to the above, it is shown that there exists a sequence $\{A_n^{***}\} \in \mathcal{A}_\infty(X_n)$ such that

$$\liminf_{n\to\infty} E(X_n^*(A_n^{***})) = \underline{As}.E(X_n).$$

We put $A_n^* = \min\{A_n^{**}, A_n^{***}\}$. Then we have

$$\Pr\{X_n > A_n^*\} \to 0 \quad \text{as} \quad n \to \infty$$

and

$$\limsup_{n\to\infty} E(X_n^*(A_n^*)) \le \limsup_{n\to\infty} E(X_n^*(A_n^{**})) = \overline{As}.E(X_n); \qquad (2.7)$$

$$\liminf_{n\to\infty} E(X_n^*(A_n^*)) \le \liminf_{n\to\infty} E(X_n^*(A_n^{***})) = \underline{As}.E(X_n). \qquad (2.8)$$

Returning to the definitions of $\overline{As}.E(X_n)$ and $\underline{As}.E(X_n)$ we see that the inverse inequalities of (2.7) and (2.8) hold. This completes the proof.

Lemma 2.4. Assume that $\overline{As}.E(X_n) < \infty$. Let $\{A_n^*\}$ be a sequence satisfying the condition of Lemma 2.3. Let $\{B_n^*\}$ be any sequence such that $B_n^* \to \infty$ as $n \to \infty$ and $B_n^* \le A_n^*$ for all $n \ge n_0$ with some n_0. Then $\{B_n^*\}$ satisfies the condition of Lemma 2.3.

Proof. By Lemmas 2.1 and 2.2 we have $\{B_n^*\} \in \mathcal{A}(X_n)$ and

$$\limsup_{n\to\infty} E(X_n^*(B_n^*)) \le \limsup_{n\to\infty} E(X_n^*(A_n^*)); \qquad (2.9)$$

$$\liminf_{n\to\infty} E(X_n^*(B_n^*)) \le \liminf_{n\to\infty} E(X_n^*(A_n^*)). \qquad (2.10)$$

From the definitions of $\overline{As}.E(X_n)$ and $\underline{As}.E(X_n)$ we see that the inverse inequalities of (2.9) and (2.10) hold. This completes the proof.

Lemma 2.5. Assume that $\overline{As}.E(X_n) < \infty$ and $\overline{As}.E(Y_n) < \infty$. Then

(i) $\overline{As}.E(X_n) + \overline{As}.E(Y_n) \ge \overline{As}.E(X_n + Y_n);$

(ii) $\underline{As}.E(X_n) + \underline{As}.E(Y_n) \le \underline{As}.E(X_n + Y_n);$

(iii) $\underline{As}.E(X_n + Y_n) \le \overline{As}.E(X_n) + \underline{As}.E(Y_n) \le \overline{As}.E(X_n + Y_n).$

Further if $X_n \ge Y_n$, then

$$\overline{As}.E(X_n - Y_n) \le \overline{As}.E(X_n) - \underline{As}.E(Y_n);$$

$$\underline{As}.E(X_n - Y_n) \ge \underline{As}.E(X_n) - \overline{As}.E(Y_n).$$

Proof. (i) By Lemma 2.3 we take $\{A_n\} \in \mathcal{A}_\infty(X_n)$ and $\{B_n\} \in \mathcal{A}_\infty(Y_n)$ such that

$$\overline{As}.E(X_n) = \limsup_{n\to\infty} E(X_n^*(A_n));$$

$$\underline{As}.E(Y_n) = \limsup_{n\to\infty} E(Y_n^*(B_n)).$$

Let $c_n = \min\{A_n, B_n\}$. Since

$$E((X_n + Y_n)^*(c_n)) \le E(X_n^*(A_n)) + E(Y_n^*(B_n)),$$

it follows from Lemmas 2.3 and 2.4 that

$$\limsup_{n\to\infty} E((X_n + Y_n)^*(c_n)) \leq \overline{As}.E(X_n) + \overline{As}.E(Y_n).$$

By Lemma 2.2 we obtain

$$\overline{As}.E(X_n + Y_n) \leq \overline{As}.E(X_n) + \overline{As}.E(Y_n).$$

(ii) By Lemma 2.3 we take a sequence $\{c_n\} \in \mathcal{A}_\infty(X_n + Y_n)$ such that

$$\overline{As}.E(X_n + Y_n) = \limsup_{n\to\infty} E((X_n + Y_n)^*(c_n)).$$

Since

$$E\left(X_n^*\left(\frac{c_n}{2}\right)\right) + E\left(Y_n^*\left(\frac{c_n}{2}\right)\right) \leq E((X_n + Y_n)^*(c_n)),$$

it follows from Lemmas 2.3 and 2.4 that

$$\liminf_{n\to\infty} E\left(X_n^*\left(\frac{c_n}{2}\right)\right) + \liminf_{n\to\infty} E\left(Y_n^*\left(\frac{c_n}{2}\right)\right) \leq \underline{As}.E(X_n + Y_n).$$

By Lemma 2.2 we have

$$\underline{As}.E(X_n) + \underline{As}.E(Y_n) \leq \underline{As}.E(X_n + Y_n).$$

(iii) By Lemma 2.3 we take $\{c_n\} \in \mathcal{A}_\infty(X_n + Y_n)$ such that

$$\overline{As}.E(X_n + Y_n) = \limsup_{n\to\infty} E((X_n + Y_n)^*(c_n)).$$

Since

$$E((X_n + Y_n)^*(c_n)) \geq E\left(X_n^*\left(\frac{c_n}{2}\right)\right) + E\left(Y_n^*\left(\frac{c_n}{2}\right)\right),$$

it follows that

$$\limsup_{n\to\infty} E((X_n + Y_n)^*(c_n))$$
$$\geq \limsup_{n\to\infty} E\left(X_n^*\left(\frac{c_n}{2}\right)\right) + \liminf_{n\to\infty} E\left(Y_n^*\left(\frac{c_n}{2}\right)\right).$$

By Lemmas 2.3 and 2.4 we have

$$\overline{As}.E(X_n + Y_n) \geq \overline{As}.E(X_n) + \underline{As}.E(Y_n).$$

Similarly we obtain

$$\underline{As}.E(X_n + Y_n) \leq \underline{As}.E(X_n) + \overline{As}.E(Y_n).$$

206   M. AKAHIRA AND K. TAKEUCHI

The proof of the rest follows directly from (iii). This completes the proof.

From the above lemmas we have the following theorems.

Theorem 2.2. Assume that $As.E(X_n)$ and $As.E(Y_n)$ exist. Then

$$As.E(X_n + Y_n) = As.E(X_n) + As.E(Y_n).$$

Further if $X_n \geq Y_n$, then

$$As.E(X_n - Y_n) = As.E(X_n) - As.E(Y_n).$$

The proof of the theorem is directly derived from Lemma 2.5.

Theorem 2.3. (Markov type inequality.) If $\overline{As}.E(X_n) < \infty$, then for any $c > 0$

$$\limsup_{n \to \infty} \Pr\{X_n \geq c\} \leq \frac{\overline{As}.E(X_n)}{c};$$

$$\liminf_{n \to \infty} \Pr\{X_n \geq c\} \leq \frac{\underline{As}.E(X_n)}{c}.$$

Proof. There exists a sequence $\{A_n\}$ such that

$$\overline{As}.E(X_n) = \limsup_{n \to \infty} E(X_n^*(A_n))$$

and $A_n \to \infty$ as $n \to \infty$. Then we can take n_0 such that for $n \geq n_0$, $A_n > c$ and

$$\Pr\{X_n \geq c\} = \Pr\{X_n^*(A_n) \geq c\} \leq \frac{E(X_n^*(A_n))}{c}.$$

By Lemma 2.3 we have

$$\limsup_{n \to \infty} \Pr\{X_n \geq c\} \leq \frac{\overline{As}.E(X_n)}{c}$$

The other inequality is obtained similarly. This completes the proof.

Theorem 2.4. X_n converges in probability to zero if and only if $As.E(X_n) = 0$.

Proof. The proof of necessity is clear. If X_n does not converge in probability to zero, then there exists a positive number c such that $\lim_{n \to \infty} \Pr\{X_n > c\} > 0$. It follows from Theorem 2.3 that this contradicts the condition. Thus the proof is completed.

Theorem 2.5. Assume that random variables X_n and X have the distribution functions F_n and F, respectively, and the moment generating function (m.g.f.) of X,

$$g(\theta) = \int e^{\theta x} dF(x),$$

exists for all θ in some open interval I including the origin. Then X_n converges in distribution to X as $n \to \infty$, i.e., $X_n \xrightarrow{D} X$ if and only if $As.E(e^{\theta X_n}) = g(\theta)$ for all $\theta \in I$.

Proof. If $X_n \xrightarrow{D} X$, then $e^{\theta X_n} \xrightarrow{D} e^{\theta X}$, hence from Theorem 2.1

$$As.E(e^{\theta X_n}) = g(\theta) \quad \text{for all} \quad \theta \in I.$$

If $As.E(e^{\theta X_n}) = g(\theta)$ for all $\theta \in I$, then it follows from Theorem 2.3 that the sequence $\{F_n\}$ is tight, hence for any $\varepsilon > 0$ there exist $n_0 > 0$ and $K > 0$ such that for $n \geq n_0$

$$\Pr\{X_n > K\} < \varepsilon.$$

Then there exists a subsequence $\{n_i\}$ of $\{n\}$ such that F_{n_i} converges in distribution to some distribution F_0 as $i \to \infty$ and

$$As.E(X_{n_i}) = \lim_{i \to \infty} \int e^{\theta x} dF_{n_i}(x) = g(\theta) \quad \text{for all} \quad \theta \in I.$$

If a subsequence $\{F_{m_i}\}$ of $\{F_n\}$ does not converge to F_0, then there is some positive number δ such that a set $A_\delta = \{n : \lambda(F_n, F_0) > \delta\}$ is infinite, where $\lambda(F, G)$ denotes the Levy distance between two one-dimensional distributions. We also have a subsequence $\{F_{m'_i}\}$ of $\{F_{m_i}\}$ such that $m'_i \in A_\delta$ ($i = 1, 2, \ldots$) and $F_{m'_i}$ converges in distribution to some distribution F'_0 as $i \to \infty$. By the first part of the proof we have $F'_0 = F_0$, which contradicts the assumption. This completes the proof.

Further, if Y_n converges in probability to zero as $n \to \infty$ and a function $f(x, y)$ is continuous in y, then it may be shown that $As.E(f(X_n, Y_n)) = As.E(f(X_n, 0))$.

For general real random variables X_n we have $X_n = X_n^+ - X_n^-$, where $X_n^+ = \max(X_n, 0)$ and $X_n^- = \max(-X_n, 0)$. Then we define the upper and the lower asymptotic expectations of $\{X_n\}$ as

$$\overline{As}.E(X_n) = \overline{As}.E(X_n^+) - \underline{As}.E(X_n^-);$$
$$\underline{As}.E(X_n) = \underline{As}.E(X_n^+) - \overline{As}.E(X_n^-),$$

respectively, and if $-\infty < \overline{As}.E(X_n) = \underline{As}.E(X_n) < \infty$ we call it the asymptotic expectation of $\{X_n\}$ and denote it by $As.E(X_n)$.

M. AKAHIRA AND K. TAKEUCHI

In a way similar to the case of the non-negative random variable it may be shown that

$$\overline{As}.E(-X_n) = -\underline{As}.E(X_n);$$
$$\underline{As}.E(-X_n) = -\overline{As}.E(X_n);$$
$$\overline{As}.E(X_n + Y_n) \leq \overline{As}.E(X_n) + \overline{As}.E(Y_n);$$
$$\underline{As}.E(X_n + Y_n) \geq \underline{As}.E(X_n) + \underline{As}.E(Y_n),$$

provided that $\overline{As}.E(|X_n|) < \infty$ and $\overline{As}.E(|Y_n|) < \infty$. Hence

$$As.E(X_n + Y_n) = As.E(X_n) + As.E(Y_n),$$

provided that $As.E(X_n)$ and $As.E(Y_n)$ exist.

ACKNOWLEDGMENTS

The paper was written while the first author was at Queen's University in Canada as a visiting Professor. The visit was supported by a grant of the Natural Sciences and Engineering Research Council of Canada. He is grateful to Professor Colin R. Blyth for inviting him.

REFERENCES

Akahira, M., and K. Takeuchi (1981). *Asymptotic Efficiency of Statistical Estimators: Concepts and Higher Order Asymptotic Efficiency*. Lecture Notes in Statistics 7. New York: Springer.

Ibragimov, I. A., and R. Z. Has'minskii (1981). *Statistical Estimation: Asymptotic Theory*. New York: Springer.

Lehmann, E. L. (1982). *Theory of Point Estimation*. New York: Wiley and Sons.

Metrika, Volume 34, 1987, page 1–15

Locally Minimum Variance Unbiased Estimator in a Discontinuous Density Function

By M. Akahira[1] and K. Takeuchi[2]

Abstract: The exact forms of the locally minimum variance unbiased estimators and their variances are given in the case of a discontinuous density function.

Key words and phrases: Locally minimum variance unbiased estimator, Non-regular cases, Discontinuous density function.

1 Introduction

In the theory of minimum variance unbiased estimation in regular cases, it is usually assumed that the density function is continuous with respect to the unknown parameter θ to be estimated.

In non-regular cases the minimum variance unbiased estimation was studied by Chapman and Robbins (1951), Kiefer (1952), Polfeldt (1970), Vincze (1979), Takeuchi and Akahira (1986), Akahira, Puri and Takeuchi (1986) and others. Though the Chapman-Robbins-Kiefer type discussion does not require the continuity of the density function in θ, their bounds are not usually sharp and the precise form of the locally minimum variance unbiased (LMVU) estimator is not given.

In this paper we shall give an example where the density function is discontinuous in θ, and obtain the exact forms of the LMVU estimators of the parameter and their variances, and illustrate some aspects of the LMVU estimators in such a situation.

2 Results

Suppose that the density function $f(x, \theta)$ (with respect to a σ-finite measure μ) is not continuous in θ, while the support $A(\theta) = \{x \mid f(x, \theta) > 0\}$ can be defined to be independent of θ. Note that this condition does not affect the existence of an LMVU

---

[1] M. Akahira, Statistical Laboratory, Department of Mathematics, University of Electro-Communications, Chofu, Tokyo 182, Japan.
[2] K. Takeuchi, Faculty of Economics, University of Tokyo, Hongo, Bunkyo-ku, Tokyo 113, Japan.

2   M. Akahira and K. Takeuchi

estimator, because it can be shown that an LMVU estimator at $\theta = \theta_0$ exists if

$$\int_{A(\theta_0)} \frac{\{f(x, \theta)\}^2}{f(x, \theta_0)} d\mu(x) < \infty$$

(e.g. see Barankin 1949 and Stein 1950). But the LMVU estimator sometimes behaves very strangely.

Let $X_1, ..., X_n$ be independently and identically distributed random variables according to the following density function w.r.t. the Lebesgue measure

$$f(x, \theta) = \begin{cases} p & \text{for } 0 \leqslant x \leqslant \theta \text{ and } \theta + 1 \leqslant x \leqslant 2; \\ q & \text{for } \theta < x < \theta + 1; \\ 0 & \text{otherwise,} \end{cases} \tag{2.1}$$

where the parameter has the range $0 \leqslant \theta \leqslant 1$, and p and q with $0 < p < q$ and $p + q = 1$ are fixed constants.

Since the support of $f(x, \theta)$ is the closed interval $[0, 2]$, in the subsequent discussion it is enough to consider it as the domain of x.

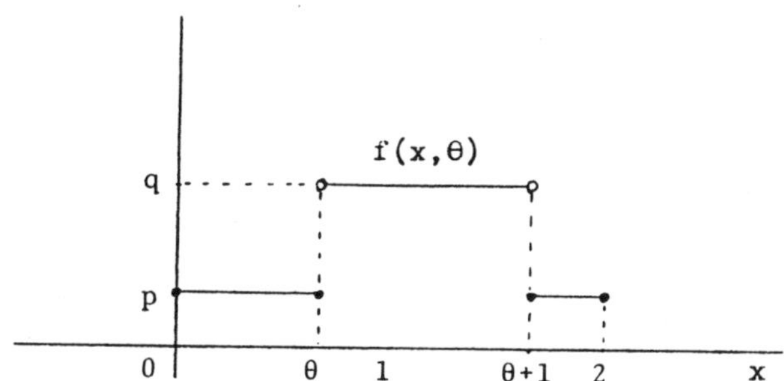

Fig. 2.1 The density function $f(x, \theta)$ given by (2.1)

Then the LMVU estimator $\hat{\theta}_n^* = \hat{\theta}_n^*(x_1, ..., x_n)$ at $\theta = \theta_0$ has the form

$$\hat{\theta}_n^*(x_1, ..., x_n) = \int_0^1 \prod_{i=1}^n \frac{f(x_i, \eta)}{f(x_i, \theta_0)} dG_n(\eta), \tag{2.2}$$

where $G_n(\eta)$ is some signed measure over the closed interval $[0, 1]$ (e.g. see Stein 1950).

First we shall construct the LMVU estimator at $\theta = \theta_0$ given by the form (2.2) in the case when $n = 1$, i.e., $x_1 = x$. We put

$$K_{\theta_0}(\theta, \eta) = \int_0^2 \pi_{\theta_0, \eta}(x) f(x, \theta) dx \tag{2.3}$$

with

$$\pi_{\theta_0, \eta}(x) = \frac{f(x, \eta)}{f(x, \theta_0)}.$$

In order to obtain the LMVU estimator $\hat{\theta}_1^*(x)$ at $\theta = \theta_0$ it is enough to take some signed measure G over $[0, 1]$ such that

$$E_\theta[\hat{\theta}_1^*(X)] = \int_0^2 \left\{ \int_0^1 \frac{f(x, \eta)}{f(x, \theta_0)}\, dG(\eta) \right\} f(x, \theta)\, dx$$

$$= \int_0^1 K_{\theta_0}(\theta, \eta)\, dG(\eta)$$

$$= \theta$$

for all $\theta \in [0, 1]$.

For $\eta > \theta_0$ we have

$$\pi_{\theta_0, \eta}(x) = \begin{cases} \dfrac{p}{q} & \text{for } \theta_0 < x \leq \eta; \\[2ex] \dfrac{q}{p} & \text{for } \theta_0 + 1 \leq x < \eta + 1; \\[2ex] 1 & \text{otherwise.} \end{cases} \qquad (2.4)$$

For $\eta < \theta_0$ we also obtain

$$\pi_{\theta_0, \eta}(x) = \begin{cases} \dfrac{q}{p} & \text{for } \eta < x \leq \theta_0; \\[2ex] \dfrac{p}{q} & \text{for } \eta + 1 \leq x < \theta_0 + 1; \\[2ex] 1 & \text{otherwise.} \end{cases} \qquad (2.5)$$

Note that for $\eta = \theta_0$, $\pi_{\theta_0, \eta}(x) = 1$.

In order to calculate $K_{\theta_0}(\theta, \eta)$ we consider two cases $\theta > \theta_0$ and $\theta < \theta_0$. Note that for $\theta = \theta_0$, $K_{\theta_0}(\theta_0, \eta) = 0$.

Case 1: $\theta > \theta_0$. If $\eta < \theta_0$, then we have by (2.5)

$$K_{\theta_0}(\theta, \eta) = \int_0^2 \pi_{\theta_0, \eta}(x) f(x, \theta)\, dx$$

$$= \int_0^\eta p\, dx + \int_\eta^{\theta_0} q\, dx + \int_{\theta_0}^\theta p\, dx + \int_\theta^{\eta+1} q\, dx + \int_{\eta+1}^{\theta_0+1} p\, dx + \int_{\theta_0+1}^{\theta+1} q\, dx + \int_{\theta+1}^2 p\, dx$$

$$= 1.$$

4   M. Akahira and K. Takeuchi

If $\theta_0 < \eta < \theta$, then we obtain by (2.4)

$$K_{\theta_0}(\theta, \eta) = \int_0^2 \pi_{\theta_0, \eta}(x) f(x, \theta) dx$$

$$= \int_0^{\theta_0} p\, dx + \int_{\theta_0}^{\eta} \frac{p^2}{q} dx + \int_{\eta}^{\theta} p\, dx + \int_{\theta}^{\theta_0+1} q\, dx + \int_{\theta_0+1}^{\eta+1} \frac{q^2}{p} dx + \int_{\eta+1}^{\theta+1} q\, dx + \int_{\theta+1}^{2} p\, dx$$

$$= 1 + c(\eta - \theta_0),$$

where

$$c = \frac{1}{pq} - 4 \quad (> 0).$$

If $\theta \leq \eta$, then we have by (2.4)

$$K_{\theta_0}(\theta, \eta) = \int_0^2 \pi_{\theta_0, \eta}(x) f(x, \theta) dx$$

$$= \int_0^{\theta_0} p\, dx + \int_{\theta_0}^{\theta} \frac{p^2}{q} dx + \int_{\theta}^{\eta} p\, dx + \int_{\eta}^{\theta_0+1} q\, dx + \int_{\theta_0+1}^{\theta+1} \frac{q^2}{p} dx + \int_{\theta+1}^{\eta+1} q\, dx + \int_{\eta+1}^{2} p\, dx$$

$$= 1 + c(\theta - \theta_0).$$

Hence we obtain for $\theta > \theta_0$

$$K_{\theta_0}(\theta, \eta) = \begin{cases} 1 & \text{for } \eta < \theta_0; \\ 1 + c(\eta - \theta_0) & \text{for } \theta_0 < \eta \leq \theta; \\ 1 + c(\theta - \theta_0) & \text{for } \theta < \eta, \end{cases} \tag{2.6}$$

where

$$c = \frac{1}{pq} - 4.$$

Case 2: $\theta < \theta_0$. In a similar way to the Case 1 we have by (2.4) and (2.5)

$$K_{\theta_0}(\theta, \eta) = \begin{cases} 1 & \text{for } \theta_0 < \eta; \\ 1 + c(\theta_0 - \eta) & \text{for } \theta \leq \eta < \theta_0; \\ 1 + c(\theta_0 - \theta) & \text{for } \eta < \theta. \end{cases} \tag{2.7}$$

We take a signed measure G^* over $[0, 1]$ satisfying $G^*(\{0\}) = -1/c$, $G^*(\{1\}) = 1/c$, $G^*(\{\theta_0\}) = \theta_0$ and $G^*((0, \theta_0) \cup (\theta_0, 1)) = 0$.

Locally Minimum Variance Unbiased Estimator in a Discontinuous Density Function 5

Since by (2.6) and (2.7)

$$K_{\theta_0}(\theta, 0) = \begin{cases} 1 & \text{for } \theta > \theta_0; \\ 1 + c(\theta_0 - \theta) & \text{for } \theta < \theta_0; \end{cases}$$

and

$$K_{\theta_0}(\theta, 1) = \begin{cases} 1 + c(\theta - \theta_0) & \text{for } \theta > \theta_0; \\ 1 & \text{for } \theta < \theta_0, \end{cases}$$

we obtain

$$\begin{aligned} E_\theta[\hat{\theta}_1^*(X)] &= \int_0^2 \left\{ \int_0^1 \pi_{\theta_0,\eta}(x)dG^*(\eta) \right\} f(x, \theta)dx \\ &= \int_0^1 K_{\theta_0}(\theta, \eta)dG^*(\eta) \\ &= \frac{1}{c}\{K_{\theta_0}(\theta, 1) - K_{\theta_0}(\theta, 0)\} + K_{\theta_0}(\theta, \theta_0)\theta_0 \\ &= \theta \end{aligned}$$

for all $\theta \in [0, 1]$, which shows the unbiasedness.

Hence the LMVU estimator $\hat{\theta}_1^*(x)$ at $\theta = \theta_0$ is given by

$$\begin{aligned} \hat{\theta}_1^*(x) &= \int_0^1 \pi_{\theta_0,\eta}(x)dG^*(\eta) \\ &= \frac{1}{c}\{\pi_{\theta_0,1}(x) - \pi_{\theta_0,0}(x)\} + \theta_0, \end{aligned}$$

where

$$\pi_{\theta_0,1}(x) - \pi_{\theta_0,0}(x) = \begin{cases} 1 - \dfrac{q}{p} & \text{for } 0 \leq x \leq \theta_0; \\[2ex] \dfrac{p}{q} - 1 & \text{for } \theta_0 < x < 1; \\[2ex] 1 - \dfrac{p}{q} & \text{for } 1 < x < \theta_0 + 1; \\[2ex] \dfrac{q}{p} - 1 & \text{for } \theta_0 + 1 \leq x \leq 2. \end{cases}$$

6   M. Akahira and K. Takeuchi

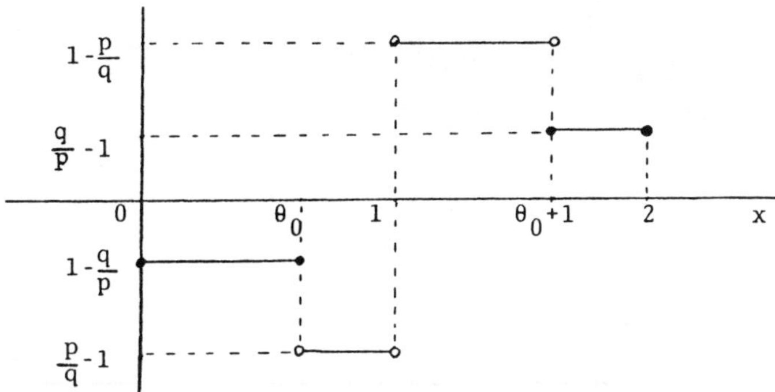

Fig. 2.2. The values of $\pi_{\theta_0,1}(x) - \pi_{\theta_0,0}(x)$ for all $x \in [0, 2]$

Note that $\hat{\theta}_1^*(x)$ can take values outside of the interval $[0, 1]$ which is the range of the parameter θ.

Since for $\theta > \theta_0$,

$$E_\theta[\pi_{\theta_0,1}(X) - \pi_{\theta_0,0}(X)] = c(\theta - \theta_0),$$

$$E_\theta[\{\pi_{\theta_0,1}(X) - \pi_{\theta_0,0}(X)\}^2] = c\left\{1 + \frac{(p-q)^2}{pq}(\theta - \theta_0)\right\},$$

we obtain for $\theta > \theta_0$

$$V_\theta(\hat{\theta}_1^*(X)) = \frac{1}{c} + (\theta - \theta_0)\{1 - (\theta - \theta_0)\},$$

where V_θ designates the variance at θ.

In a similar way to the case when $\theta > \theta_0$, we have for $\theta < \theta_0$

$$V_\theta(\hat{\theta}_1^*(X)) = \frac{1}{c} + (\theta_0 - \theta)\{1 - (\theta_0 - \theta)\}.$$

Hence the variance of the LMVU estimator $\hat{\theta}_1^*(x)$ is given by

$$V_\theta(\hat{\theta}_1^*(X)) = \frac{1}{c} + |\theta - \theta_0|(1 - |\theta - \theta_0|). \tag{2.8}$$

Note that

$$V_\theta(\hat{\theta}_1^*(X)) \geq \frac{1}{c} = V_{\theta_0}(\hat{\theta}_1^*(X))$$

for all $\theta \in [0, 1]$.

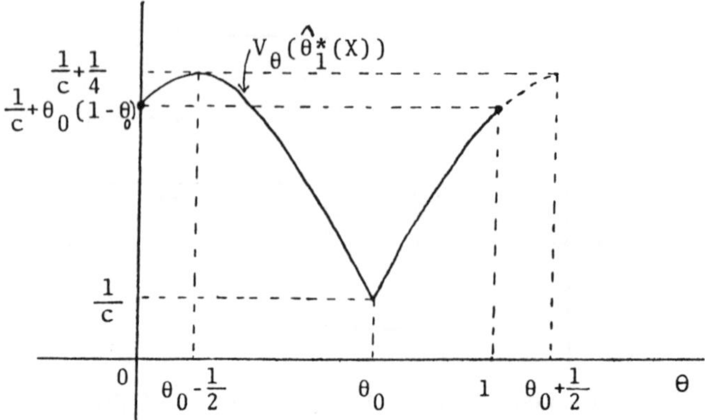

Fig. 2.3. The variance of the LMVU estimator $\hat{\theta}_1^*(x)$ at $\theta = \theta_0$ given by (2.8) when $\theta_0 > 1/2$

Next we shall construct the LMVU estimator at $\theta = \theta_0$ given by the form (2.2) in the case when $n \geqslant 1$.

From (2.3) we have

$$\int_0^2 \cdots \int_0^2 \left\{ \prod_{i=1}^n \frac{f(x_i, \eta)}{f(x_i, \theta_0)} \right\} \prod_{i=1}^n f(x_i, \theta) dx_1 \ldots dx_n$$

$$= \{K_{\theta_0}(\theta, \eta)\}^n = K_{\theta_0}^n(\theta, \eta) \quad \text{(say)} \tag{2.9}$$

By (2.6) and (2.7) we obtain for $\theta > \theta_0$

$$K_{\theta_0}^n(\theta, \eta) = \begin{cases} 1 & \text{for } \eta < \theta_0 \\ \{1 + c(\eta - \theta_0)\}^n & \text{for } \theta_0 < \eta \leqslant \theta; \\ \{1 + c(\theta - \theta_0)\}^n & \text{for } \theta < \eta, \end{cases} \tag{2.10}$$

and for $\theta < \theta_0$

$$K_{\theta_0}^n(\theta, \eta) = \begin{cases} 1 & \text{for } \theta_0 < \eta; \\ \{1 + c(\theta_0 - \eta)\}^n & \text{for } \theta \leqslant \eta < \theta_0; \\ \{1 + c(\theta_0 - \theta)\}^n & \text{for } \eta < \theta, \end{cases} \tag{2.11}$$

where

$$c = \frac{1}{pq} - 4.$$

Note that for $\theta = \theta_0$, $K_{\theta_0}^n(\theta_0, \eta) = 1$.

8   M. Akahira and K. Takeuchi

We take a signed measure G_n^* over $[0, 1]$ satisfying

$$\frac{dG_n^*}{d\eta} = \left(1 - \frac{1}{n}\right) \frac{1}{(1 + c|\eta - \theta_0|)^n} \operatorname{sgn}(\eta - \theta_0) = g_n(\eta) \quad \text{(say)}$$

$$\text{for } 0 < \eta < \theta_0 , \ \theta_0 < \eta < 1;$$

$$G_n^*(\{0\}) = -\frac{1}{nc(1 + c\theta_0)^{n-1}}; \tag{2.12}$$

$$G_n^*(\{1\}) = \frac{1}{nc\{1 + c(1 - \theta_0)\}^{n-1}};$$

$$G_n^*(\{\theta_0\}) = \theta_0.$$

Note that G_1^* is consistent with the G^* given in above case $n = 1$. Then the estimator

$$\hat{\theta}_n^* = \hat{\theta}_n^*(x_1, \ldots, x_n) = \int_0^1 \prod_{i=1}^n \frac{f(x_i, \eta)}{f(x_i, \theta_0)} dG_n^*(\eta) \tag{2.13}$$

is unbiased.

Indeed, we have by (2.9), (2.10) and (2.11) and for $\theta > \theta_0$

$$E_\theta(\hat{\theta}_n^*) = \int_0^2 \cdots \int_0^2 \left\{ \int_0^1 \prod_{i=1}^n \frac{f(x_i, \eta)}{f(x_i, \theta_0)} dG_n^*(\eta) \right\} \prod_{i=1}^n f(x_i, \theta) dx_1 \ldots dx_n$$

$$= \int_0^1 K_{\theta_0}^n(\theta, \eta) dG_n^*(\eta)$$

$$= G_n^*(\{0\}) K_{\theta_0}^n(\theta, 0) + G_n^*(\{1\}) K_{\theta_0}^n(\theta, 1) + G_n^*(\{\theta_0\}) K_{\theta_0}^n(\theta, \theta_0)$$

$$\quad + \int_0^{\theta_0} g_n(\eta) d\eta + \int_{\theta_0}^\theta \{1 + c(\eta - \theta_0)\}^n g_n(\eta) d\eta + \{1 + c(\theta - \theta_0)\}^n \int_\theta^1 g_n(\eta) d\eta$$

$$= -\frac{1}{nc(1 + c\theta_0)^{n-1}} + \frac{\{1 + c(\theta - \theta_0)\}^n}{nc\{1 + c(1 - \theta_0)\}^{n-1}} - \left(1 - \frac{1}{n}\right) \int_0^{\theta_0} \frac{1}{\{1 + c(\theta_0 - \eta)\}^n} d\eta$$

$$\quad + \left(1 - \frac{1}{n}\right) \int_{\theta_0}^\theta d\eta + \left(1 - \frac{1}{n}\right) \{1 + c(\theta - \theta_0)\}^n \int_\theta^1 \frac{1}{\{1 + c(\eta - \theta_0)\}^n} d\eta + \theta_0$$

$$= \theta.$$

Similary we obtain for $\theta < \theta_0$

$$E_\theta(\hat{\theta}_n^*) = \theta$$

and for $\theta = \theta_0$

$$E_{\theta_0}(\hat{\theta}_n^*) = \theta_0.$$

Hence the estimator $\hat{\theta}_n^*$ is the LMVU estimator.

Next we shall construct the estimator. Putting

$$\pi_{\theta_0,\eta}^n(\underline{x}) = \prod_{i=1}^n \frac{f(x_i,\eta)}{f(x_i,\theta_0)} = \prod_{i=1}^n \pi_{\theta_0,\eta}(x_i)$$

with $\underline{x} = (x_1,\ldots,x_n)$, we have from (2.13)

$$\hat{\theta}_n^*(\underline{x}) = \int_0^1 \pi_{\theta_0,\eta}^n(x)dG_n^*(\eta)$$

$$= \pi_{\theta_0,0}^n(\underline{x})G_n^*(\{0\}) + \pi_{\theta_0,1}^n(\underline{x})G_n^*(\{1\}) + \theta_0$$

$$+ \int_0^{\theta_0} \pi_{\theta_0,\eta}^n(\underline{x})g_n(\eta)d\eta + \int_{\theta_0}^1 \pi_{\theta_0,\eta}^n(\underline{x})g_n(\eta)d\eta. \tag{2.14}$$

For $\eta = 0, 1$ we have

$$\pi_{\theta_0,0}^n(\underline{x}) = \prod_{i=1}^n \pi_{\theta_0,0}(x_i) = \left(\frac{q}{p}\right)^a\left(\frac{p}{q}\right)^b = \left(\frac{q}{p}\right)^{a-b}$$

and

$$\pi_{\theta_0,1}^n(\underline{x}) = \prod_{i=1}^n \pi_{\theta_0,1}(x_i) = \left(\frac{p}{q}\right)^{a'}\left(\frac{q}{p}\right)^{b'} = \left(\frac{p}{q}\right)^{a'-b'}$$

with

$$a = \#\{x_i \mid 0 < x_i < \theta_0\}, \quad b = \#\{x_i \mid 1 < x_i < \theta_0 + 1\},$$

$$a' = \#\{x_i \mid \theta_0 < x_i < 1\} \quad \text{and} \quad b' = \#\{x_i \mid \theta_0 + 1 < x_i < 2\},$$

where $\#\{\ \}$ designates the number of the elements of the set $\{\ \}$.

Note that $0 \leqslant a + b \leqslant n$ and $0 \leqslant a' + b' \leqslant n$. If $0 < \eta < \theta_0$, then

$$\pi_{\theta_0,\eta}(\underline{x}) = \left(\frac{q}{p}\right)^s\left(\frac{p}{q}\right)^t = \left(\frac{q}{p}\right)^{s-t},$$

where $s = \#\{x_i \mid \eta < x_i < \theta_0\}$ and $t = \#\{x_i \mid \eta + 1 < x_i < \theta_0 + 1\}$.

10   M. Akahira and K. Takeuchi

If $\theta_0 < \eta < 1$, then

$$\pi_{\theta_0,\eta}(\underset{\sim}{x}) = \left(\frac{p}{q}\right)^{s'}\left(\frac{q}{p}\right)^{t'} = \left(\frac{p}{q}\right)^{s'-t'},$$

where $s' = \#\{x_i \mid \theta_0 < x_i < \eta\}$ and $t' = \#\{x_i \mid \theta_0 + 1 < x_i < \eta + 1\}$.

Note that $0 \leqslant s + t \leqslant a + b$ and $0 \leqslant s' + t' \leqslant a' + b'$.

Hence we have from (2.12), (2.13) and (2.14)

$$\hat{\theta}_n^*(\underset{\sim}{x}) = \left(\frac{q}{p}\right)^{a-b} G_n^*(\{0\}) + \left(\frac{p}{q}\right)^{a'-b'} G_n^*(\{1\}) + \theta_0$$

$$+ \int_0^{\theta_0}\left(\frac{q}{p}\right)^{s-t} g_n(\eta)d\eta + \int_{\theta_0}^1\left(\frac{p}{q}\right)^{s'-t'} g_n(\eta)d\eta$$

$$= -\frac{\left(\dfrac{q}{p}\right)^{a-b}}{nc(1+c\theta_0)^{n-1}} + \frac{\left(\dfrac{p}{q}\right)^{a'-b'}}{nc\{1+c(1-\theta_0)\}^{n-1}} + \theta_0$$

$$+ \int_0^{\theta_0}\left(\frac{q}{p}\right)^{s-t} g_n(\eta)d\eta + \int_{\theta_0}^1\left(\frac{p}{q}\right)^{s'-t'} g_n(\eta)d\eta, \tag{2.15}$$

where

$$g_n(\eta) = \left(1-\frac{1}{n}\right)\frac{1}{(1+c\mid\eta-\theta_0\mid)^n} \operatorname{sgn}(\eta-\theta_0) \quad (0 < \eta < \theta_0, \theta_0 < \eta < 1).$$

For $n \geqslant 2$, it may be difficult to calculate directly the variance of the LMVU estimator $\hat{\theta}_n^*$ given by (2.15), but it may be possible to do it by (2.13).

Putting

$$K_{\theta_0}(\theta, \eta, \eta') = \int_0^2 \pi_{\theta,\eta,\eta'}(x)f(x,\theta)dx$$

with

$$\pi_{\theta_0,\eta,\eta'}(x) = \pi_{\theta_0,\eta}(x)\pi_{\theta_0,\eta'}(x),$$

we have

$$E_\theta(\hat{\theta}_n^{*2}) = \int_0^2 \cdots \int_0^2 \left\{ \int_0^1 \int_0^1 \prod_{i=1}^n \frac{f(x_i, \eta)}{f(x_i, \theta_0)} \right.$$

$$\left. \cdot \prod_{i=1}^n \frac{f(x_i, \eta')}{f(x_i, \theta_0)} dG_n^*(\eta) dG_n^*(\eta') \right\} \cdot \prod_{i=1}^n f(x_i, \theta) \prod_{i=1}^n dx_i$$

$$= \int_0^1 \int_0^1 \left\{ \int_0^2 \cdots \int_0^2 \prod_{i=1}^n \frac{f(x_i, \eta) f(x_i, \eta')}{(f(x_i, \theta_0))^2} f(x_i, \theta) \prod_{i=1}^n dx_i \right\} dG_n^*(\eta) dG_n^*(\eta')$$

$$= \int_0^1 \int_0^1 \left\{ \int_0^2 \frac{f(x, \eta) f(x, \eta')}{(f(x, \theta_0))^2} f(x_i, \theta) dx \right\}^n dG_n^*(\eta) dG_n^*(\eta')$$

$$= \int_0^1 \int_0^1 K_{\theta_0}^n(\theta, \eta, \eta') dG_n^*(\eta) dG_n^*(\eta') \tag{2.16}$$

Here we shall obtain the variance of the estimator $\hat{\theta}_n^*$ at $\theta = \theta_0$. By (2.4) and (2.5) we have for $\eta' < \eta \leqslant \theta_0$

$$\pi_{\theta_0, \eta, \eta'}(x) = \begin{cases} 1 & \text{for } 0 \leqslant x \leqslant \eta'; \\[2mm] \dfrac{q}{p} & \text{for } \eta' < x \leqslant \eta; \\[2mm] \left(\dfrac{q}{p}\right)^2 & \text{for } \eta < x \leqslant \theta_0; \\[2mm] 1 & \text{for } \theta_0 < x < \eta' + 1; \\[2mm] \dfrac{p}{q} & \text{for } \eta' + 1 \leqslant x < \eta + 1; \\[2mm] \left(\dfrac{p}{q}\right)^2 & \text{for } \eta + 1 \leqslant x < \theta_0 + 1; \\[2mm] 1 & \text{for } \theta_0 + 1 \leqslant x \leqslant 2. \end{cases} \tag{2.17}$$

Since $\pi_{\theta_0, \eta, \eta'}(x) = \pi_{\theta_0, \eta', \eta}(x)$, it is easy from (2.17) to obtain $\pi_{\theta_0, \eta, \eta'}(x)$ for $\eta < \eta' \leqslant \theta_0$.

From (2.17) we have

$$K_{\theta_0}(\theta_0, \eta, \eta') = \begin{cases} 1 + c(\theta_0 - \eta) & \text{if } \eta' < \eta \leqslant \theta_0; \\ 1 + c(\theta_0 - \eta') & \text{if } \eta < \eta' \leqslant \theta_0. \end{cases} \tag{2.18}$$

12   M. Akahira and K. Takeuchi

By (2.4) and (2.5) we have for $\eta \leqslant \theta_0 \leqslant \eta'$

$$\pi_{\theta_0,\eta,\eta'}(x) = \begin{cases} 1 & \text{for } 0 \leqslant x \leqslant \eta; \\[2mm] \dfrac{q}{p} & \text{for } \eta < x \leqslant \theta_0; \\[2mm] \dfrac{p}{q} & \text{for } \theta_0 < x \leqslant \eta'; \\[2mm] 1 & \text{for } \eta' < x < \eta + 1; \\[2mm] \dfrac{p}{q} & \text{for } \eta + 1 \leqslant x < \theta_0 + 1; \\[2mm] \dfrac{q}{p} & \text{for } \theta_0 + 1 \leqslant x < \eta' + 1; \\[2mm] 1 & \text{for } \eta' + 1 \leqslant x \leqslant 2. \end{cases}$$

Then we obtain for $\eta \leqslant \theta_0 \leqslant \eta'$

$$K_{\theta_0}(\theta_0, \eta, \eta') = 1. \tag{2.19}$$

From (2.18) and (2.19) we have

$$K_{\theta_0}(\theta_0, \eta, \eta') = \begin{cases} 1 + c(\theta_0 - \eta) & \text{if } \eta' < \eta \leqslant \theta_0; \\ 1 + c(\theta_0 - \eta') & \text{if } \eta < \eta' \leqslant \theta_0; \\ 1 & \text{if } \eta \leqslant \theta_0 \leqslant \eta'. \end{cases} \tag{2.20}$$

In a similar way to the above we have

$$K_{\theta_0}(\theta_0, \eta, \eta') = \begin{cases} 1 & \text{if } \eta' \leqslant \theta_0 \leqslant \eta; \\ 1 + c(\eta' - \theta_0) & \text{if } \theta_0 \leqslant \eta' < \eta; \\ 1 + c(\eta - \theta_0) & \text{if } \theta_0 \leqslant \eta < \eta'. \end{cases} \tag{2.21}$$

In order to calculate $E_{\theta_0}(\hat{\theta}_n^{*2})$ we obtain from (2.16)

$$E_{\theta_0}(\hat{\theta}_n^{*2}) = G_n^*(\{0\}) \int_0^1 K_{\theta_0}^n(\theta_0, 0, \eta') dG_n^*(\eta')$$

$$+ G_n^*(\{1\}) \int_0^1 K_{\theta_0}^n(\theta_0, 1, \eta') dG_n^*(\eta')$$

$$+ G_n^*(\{\theta_0\}) \int_0^1 K_{\theta_0}^n(\theta_0, \theta_0, \eta') dG_n^*(\eta')$$

$$+ \int_0^1 \int_0^{\theta_0} K_{\theta_0}^n(\theta_0, \eta, \eta') dG_n^*(\eta) dG_n^*(\eta')$$

$$+ \int_0^1 \int_{\theta_0}^1 K_{\theta_0}^n(\theta_0, \eta, \eta') dG_n^*(\eta) dG_n^*(\eta'). \tag{2.22}$$

By (2.20) and (2.21) we have

$$K_{\theta_0}(\theta_0, 0, \eta') = \begin{cases} 1 + c(\theta_0 - \eta') & \text{for } \eta' \leqslant \theta_0; \\ 1 & \text{for } \theta_0 \leqslant \eta', \end{cases} \tag{2.23}$$

$$K_{\theta_0}(\theta_0, 1, \eta') = \begin{cases} 1 & \text{for } \eta' \leqslant \theta_0; \\ 1 + c(\eta' - \theta_0) & \text{for } \theta_0 \leqslant \eta'. \end{cases} \tag{2.24}$$

It is also clear that

$$K_{\theta_0}(\theta_0, \theta_0, \eta') = 1 \quad \text{for all } \eta' \in [0, 1]. \tag{2.25}$$

By (2.12), (2.23), (2.24) and (2.25) we have

$$G_n^*(\{0\}) \int_0^1 K_{\theta_0}^n(\theta_0, 0, \eta') dG_n^*(\eta') = 0; \tag{2.26}$$

$$G_n^*(\{1\}) \int_0^1 K_{\theta_0}^n(\theta_0, 1, \eta') dG_n^*(\eta') = \frac{1}{cn\{1 + c(1 - \theta_0)\}^{n-1}} \tag{2.27}$$

$$G_n^*(\{\theta_0\}) \int_0^1 K_{\theta_0}^n(\theta_0, \theta_0, \eta') dG_n^*(\eta') = \theta_0^2. \tag{2.28}$$

By (2.12) and (2.25) we also have for $0 < \eta < \theta_0$

$$\int_0^1 K_{\theta_0}^n(\theta_0, \eta, \eta') dG_n^*(\eta')$$

$$= K_{\theta_0}^n(\theta_0, \eta, 0) G_n^*(\{0\}) + K_{\theta_0}^n(\theta_0, \eta, 1) G_n^*(\{1\}) + K_{\theta_0}^n(\theta_0, \eta, \theta_0) G_n^*(\{\theta_0\})$$

$$+ \left(\int_0^\eta + \int_\eta^{\theta_0} + \int_{\theta_0}^1 \right) K_{\theta_0}(\theta_0, \eta, \eta') dG_n^*(\eta')$$

$$= \eta.$$

14   M. Akahira and K. Takeuchi

Then we have for $n \geqslant 2$

$$\int_0^1 \int_0^{\theta_0} K_{\theta_0}^n(\theta_0, \eta, \eta')dG_n^*(\eta)dG_n^*(\eta')$$

$$= \int_0^{\theta_0} \int_0^1 K_{\theta_0}^n(\theta_0, \eta, \eta')dG_n^*(\eta')dG_n^*(\eta)$$

$$= \int_0^{\theta} \eta \, dG_n^*(\eta)$$

$$= \begin{cases} -\dfrac{\theta_0}{2c} + \dfrac{1}{2c^2}\log(1 + c\theta_0) & \text{for } n = 2; \\[2ex] -\dfrac{\theta_0}{cn} + \dfrac{1}{c^2 n(n-2)} - \dfrac{1}{c^2 n(n-2)(1+c\theta_0)^{n-2}} & \text{for } n > 2. \end{cases} \tag{2.29}$$

In a similar way to the above we obtain by (2.12) and (2.21)

$$\int_0^1 \int_{\theta_0}^1 K_{\theta_0}^n(\theta_0, \eta, \eta')dG_n^*(\eta)dG_n^*(\eta')$$

$$= \begin{cases} \dfrac{1}{2c}\left[\theta_0 - \dfrac{1}{1 + c(1-\theta_0)} + \dfrac{1}{c}\log\{1 + c(1-\theta_0)\}\right] & \text{for } n = 2; \\[2ex] \dfrac{1}{cn}\left[\theta_0 - \dfrac{1}{\{1 + c(1-\theta_0)\}^{n-1}} - \dfrac{1}{c^2(n-2)}\left\{\dfrac{1}{(1 + c(1-\theta_0))^{n-2}} - 1\right\}\right] \end{cases} \tag{2.30}$$

$$\text{for } n > 2.$$

From (2.22), (2.26) $\sim$ (2.30) we have

$$E_{\theta_0}(\hat{\theta}_n^{*2}) = \begin{cases} \theta_0^2 + \dfrac{1}{2c^2}\log\{(1 + c\theta_0)(1 + c(1-\theta_0))\} & \text{for } n = 2; \\[2ex] \theta_0^2 + \dfrac{1}{c^2 n(n-2)}\left\{2 - \dfrac{1}{(1 + c\theta_0)^{n-2}} - \dfrac{1}{(1 + c(1-\theta_0))^{n-2}}\right\} \end{cases}$$

$$\text{for } n > 2.$$

Hence for $n \geqslant 2$ the variance of the LMVU estimator $\hat{\theta}_n^*$ is given by

$$V_{\theta_0}(\hat{\theta}_n^*) = \begin{cases} \dfrac{1}{2c^2}\log\{(1 + c\theta_0)(1 + c(1-\theta_0))\} & \text{for } n = 2; \\[2ex] \dfrac{1}{c^2 n(n-2)}\left\{2 - \dfrac{1}{(1 + c\theta_0)^{n-2}} - \dfrac{1}{(1 + c(1-\theta_0))^{n-2}}\right\} & \text{for } n > 2. \end{cases}$$

Locally Minimum Variance Unbiased Estimator in a Discontinuous Density Function   15

Note that for sufficiently large n

$$V_{\theta_0}(\hat{\theta}_n^*) = \frac{2}{c^2 n^2} + o\left(\frac{1}{n^2}\right),$$

i.e., the variance of the LMVU estimator $\hat{\theta}_n^*$ is the order of n^{-2}. The calculation of the variance $V_\theta(\hat{\theta}_n^*)$ at any θ is more complicate than that of $V_{\theta_0}(\hat{\theta}_n^*)$.

References

Akahira M, Puri ML, Takeuchi K (1986) Bhattacharyya bound of variances of unbiased estimators in non-regular cases. Ann Inst Statist Math 38A:35–44

Barankin EW (1949) Locally best unbiased estimates. Ann Math Statist 20:477–501

Chapman DG, Robbins H (1951) Minimum variance estimation without regularity assumptions. Ann Math Statist 22:581–586

Kiefer J (1952) On minimum variance in non-regular estimation. Ann Math Statist 23:627–630

Polfeldt T (1970) The order of the minimum variance in a non-regular case. Ann Math Statist 41:667–672

Stein C (1950) Unbiased estimates of minimum variance. Ann Math Statist 21:406–415

Takeuchi K, Akahira M (1986) A note on minimum variance. Metrika 33:85–91

Vincze I (1979) On the Cramér-Fréchet-Rao inequality in the non-regular case. In: Contributions to statistics. The Jaroslav Hájek Memorial Volume. Academia, Prague, 253–263

Received June 12, 1984

Ann. Inst. Statist. Math.
39 (1987), Part A, 593–610

THE LOWER BOUND FOR THE VARIANCE OF UNBIASED ESTIMATORS FOR ONE-DIRECTIONAL FAMILY OF DISTRIBUTIONS

MASAFUMI AKAHIRA AND KEI TAKEUCHI

(Received Sept. 13, 1985; revised Feb. 15, 1986)

Summary

In this paper we introduce the concept of one-directionality which includes both cases of location (and scale) parameter and selection parameter and also other cases, and establish some theorems for sharp lower bounds and for the existence of zero variance unbiased estimator for this class of non-regular distributions.

1. Introduction

For the lower bound for the variance of unbiased estimators, the most famous is the so-called Cramér-Rao bound. But the Cramér-Rao bound and its Bhattacharyya extension assume a set of regularity conditions. Chapman and Robbins [2], Kiefer [4] and Fraser and Guttman [3] obtained bounds with much less stringent assumptions, but they still require the independence of the support of the parameter θ or almost equivalently that the distribution with $\theta \neq \theta_0$ is absolutely continuous with respect to that with $\theta = \theta_0$ when θ_0 is the specified parameter value at which the variance is evaluated. Recently in the non-regular cases the Cramér-Rao bound has been discussed by Víncze [8], Móri [5] and others.

In the previous paper, Akahira, Puri and Takeuchi [1] get the Bhattacharyya type bound for the variance of unbiased estimators in non-regular cases.

In this paper we introduce the concept of one-directionality which includes both cases of location (and scale) parameter and selection parameter and also other cases, and show that the bound for the variance of unbiased estimators is sharp in the sense that the actual infimum of the variance of unbiased estimators is equal to the bound for a specified θ_0, for this class of non-regular distributions. We also estab-

Key words and phrases: Cramér-Rao bound, Bhattacharyya bound, unbiased estimator, one-directional family of distributions, sharp lower bound.

lish that for a wide class of the non-regular distributions the infimum of the variance of unbiased estimators can be zero when the sample size is not smaller than 2.

A simple but rather general case of the class of distributions of which none dominates another is that it is characterized by a real parameter θ and the distribution shifts monotonically as θ changes. For one dimensional random variables the case can be visualized by the following example.

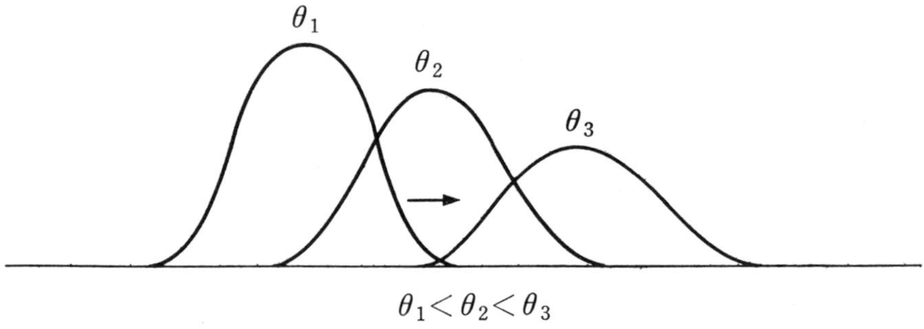

Mathematical definition for such cases in a rather general set-up is given in Section 2 and is termed as one-directional family of distributions.

2. Definition of the one-directional family of distributions

We assume that we are given a model consisting of a sample space (χ, β) and a family $\mathcal{P} = \{P_\theta : \theta \in \Theta\}$ of probability measures, where a parameter space Θ is an open subset in a Euclidean 1-space R^1.

Throughout the subsequent discussion we shall assume the following:

(A.2.1) For each $\theta \in \Theta$, P_θ is absolutely continuous with respect to a σ-finite measure μ and the corresponding density w.r.t. μ is $f(x, \theta)$.

Let $A(\theta)$ be a support of $f(x, \theta)$, that is, $A(\theta) = \{x : f(x, \theta) > 0\}$. The determination of $A(\theta)$ is not unique so far as any null set may be added to it, but in the sequel we take one and fixed determination of $A(\theta)$ for every $\theta \in \Theta$ which satisfies the following:

(A.2.2) For any disjoint points θ_1 and θ_2 in Θ, neither $A(\theta_1) \supset A(\theta_2)$ nor $A(\theta_1) \subset A(\theta_2)$.

(A.2.3) For $\theta_1 < \theta_2 < \theta_3$,

$$A(\theta_1) \cap A(\theta_3) \subset A(\theta_1) \cap A(\theta_2) \, ,$$

$$A(\theta_1) \cap A(\theta_3) \subset A(\theta_2) \cap A(\theta_3) \, .$$

(A.2.4) If θ_n tends to θ as $n \to \infty$, then

$$\mu\left(\left(\bigcup_{n=1}^{\infty}\bigcap_{i\geq n}A(\theta_i)\right)\varDelta A(\theta)\right)=\mu\left(\left(\bigcap_{n=1}^{\infty}\bigcup_{i\geq n}A(\theta_i)\right)\varDelta A(\theta)\right)=0\,,$$

where $E\varDelta F$ denotes the symmetric difference of two sets E and F.

(A.2.5)   For any two points θ_1 and θ_2 in Θ with $\theta_1<\theta_2$, there exists a finite number of ξ_i $(i=1,\cdots,k)$ such that $\theta_1=\xi_0<\xi_1<\cdots<\xi_k=\theta_2$ and $\mu(A(\xi_i)\cap A(\xi_{i-1}))>0$ $(i=1,\cdots,k)$.

Then $\mathscr{P}$ is called to be a one-directional family of distributions if the conditions (A.2.1) to (A.2.5) hold.

3. The lower bound for the variance of unbiased estimators when X is a real random variable

Now suppose that X is a real random variable with a density function $f(x,\theta)$ whose support is an open interval $(a(\theta),b(\theta))$, then the condition of one-directionality means that $a(\theta)$ and $b(\theta)$ are both monotone and continuous functions. Without loss of generality we assume that $a(\theta)$ and $b(\theta)$ are monotone increasing functions. Therefore we can formulate the following problem. Let $\chi=\Theta=R^1$. Let X be a real random variable with a density function $f(x,\theta)$ (with respect to the Lebesgue measure μ) satisfying the following conditions (A.3.1) to (A.3.7):

(A.3.1) $$f(x,\theta)>0\qquad\text{for}\qquad a(\theta)<x<b(\theta)\,,$$

$$f(x,\theta)=0\qquad\text{for}\qquad x\leq a(\theta),\ x\geq b(\theta)\,,$$

where $f(x,\theta)$ is continuous in x and θ for which $a(\theta)<x<b(\theta)$, and $(p+1)$-times continuously differentiable in θ for a.a. $x\,[\mu]$ for some non-negative integer p, both of functions $a(\theta)$ and $b(\theta)$ are p-times continuously differentiable and $a'(\theta)>0$, $b'(\theta)>0$ for all θ.

(A.3.2) $$\lim_{x\to a(\theta)+0}f(x,\theta)=\lim_{x\to b(\theta)-0}f(x,\theta)=0\,,$$

and for some positive integer p

$$\lim_{x\to a(\theta)+0}\frac{\partial^i}{\partial\theta^i}f(x,\theta)=\lim_{x\to b(\theta)-0}\frac{\partial^i}{\partial\theta^i}f(x,\theta)=0\qquad(i=1,\cdots,p-1)\,,$$

$$\lim_{x\to a(\theta)+0}\frac{\partial^p}{\partial\theta^p}f(x,\theta)=A_p(\theta)\,,\qquad\lim_{x\to b(\theta)-0}\frac{\partial^p}{\partial\theta^p}f(x,\theta)=B_p(\theta)\,,$$

where $A_p(\theta)$ and $B_p(\theta)$ are non-zero, finite and continuous in θ.

(A.3.3) $(\partial^i/\partial\theta^i)f(x,\theta)$ $(i=1,\cdots,p)$ are linearly independent.

(A.3.4)   For some $\theta_0\in\Theta$

$$0<\int_{a(\theta_0)}^{b(\theta_0)}\frac{\left\{\dfrac{\partial^i}{\partial\theta^i}f(x,\theta_0)\right\}^2}{f(x,\theta_0)}d\mu(x)$$

is finite for each $i=1,\cdots,k$, and

$$\int_{a(\theta_0)}^{b(\theta_0)} \frac{\left\{\sum_{i=k+1}^{p} c_i \frac{\partial^i}{\partial \theta^i} f(x,\theta_0)\right\}^2}{f(x,\theta_0)} d\mu$$

is infinite for each $i=k+1,\cdots,p$ unless $c_{k+1}=\cdots=c_p=0$.

(A.3.5)   For $\theta_0 \in \Theta$ there exists a positive number ε and a positive-valued measurable function $\rho(x)$ such that for every $x \in A(\theta)$ and every $\theta \in (\theta_0-\varepsilon, \theta_0+\varepsilon)$, $\rho(x) > f(x,\theta)$, and for every $\theta \in (\theta_0-\varepsilon, \theta_0+\varepsilon)$, $\int_{A(\theta)} |r(x)| \times f(x,\theta)d\mu < \infty$ implies $\int_{\underset{\theta \in (\theta_0-\varepsilon,\theta_0+\varepsilon)}{\cup} A(\theta)} |r(x)| \rho(x)d\mu < \infty$.

(A.3.6)   For each $i=1,\cdots,p+1$,

$$\varlimsup_{h \to 0} \sup_{x \in \underset{j=1}{\overset{i}{\cup}} A(\theta_0+jh)-A(\theta_0)} \frac{\left|\sum_{j=1}^{i} (-1)^j \binom{i}{j} f(x,\theta_0+jh)\right|}{|h|^i \rho(x)} < \infty .$$

(A.3.7)   For each $i=1,\cdots,p+1$,

$$\varlimsup_{h \to 0} \sup_{x \in A(\theta_0)} \frac{\left|\frac{\partial^i}{\partial \theta^i} f(x,\theta_0+h)\right|}{\rho(x)} < \infty .$$

Note that the conditions (A.3.5), (A.3.6) and (A.3.7) are assumed to obtain the Bhattacharyya bound for the variance of unbiased estimators (see Akahira et al. [1]).

First we consider the special case when $p=0$, then we have to modify slightly the condition (A.3.2) as follows:

(A.3.2)' $\displaystyle\lim_{x \to a(\theta)+0} f(x,\theta)=A_0(\theta)>0 , \qquad \lim_{x \to b(\theta)-0} f(x,\theta)=B_0(\theta)>0 .$

In the following theorem we shall have a lower bound.

THEOREM 3.1.   *Let $g(\theta)$ be continuously differentiable over Θ. Let $\hat{g}(x)$ be an unbiased estimator of $g(\theta)$. If for $p=0$ and a fixed θ_0, the conditions (A.3.1), (A.3.2)', (A.3.5), (A.3.6) and (A.3.7) hold, then*

$$\min_{\hat{g}:\text{ unbiased}} V_{\theta_0}(\hat{g})=0 ,$$

where $V_{\theta_0}(\hat{g})$ denotes the variance of $\hat{g}$ at $\theta=\theta_0$.

PROOF.   We first define

$$\hat{g}(x)=g(\theta_0) \qquad \text{for} \qquad a(\theta_0)<x<b(\theta_0) .$$

Since $a(\theta_0)<a(\theta)<b(\theta_0)<b(\theta)$ for $\theta>\theta_0$, it follows from the unbiasedness condition that

THE LOWER BOUND FOR THE VARIANCE OF UNBIASED ESTIMATORS   597

$$(3.1) \quad g(\theta) = \int_{a(\theta)}^{b(\theta)} \hat{g}(x) f(x,\theta) d\mu = \int_{a(\theta)}^{b(\theta_0)} \hat{g}(x) f(x,\theta) d\mu + \int_{b(\theta_0)}^{b(\theta)} \hat{g}(x) f(x,\theta) d\mu .$$

Putting

$$h(\theta) = \int_{a(\theta)}^{b(\theta_0)} \hat{g}(x) f(x,\theta) d\mu ,$$

we have by (3.1)

$$(3.2) \qquad \int_{b(\theta_0)}^{b(\theta)} \hat{g}(x) f(x,\theta) d\mu = g(\theta) - h(\theta) .$$

Differentiating both sides of (3.2) with respect to θ, we obtain

$$(3.3) \qquad b'(\theta) B_0(\theta) \hat{g}(b(\theta)) + \int_{b(\theta_0)}^{b(\theta)} \hat{g}(x) \left\{ \frac{\partial}{\partial \theta} f(x,\theta) \right\} d\mu = g'(\theta) - h'(\theta) .$$

Differentiation under the integral sign is admitted because of (A.3.7) with $p=0$. If $\hat{g}(x)$ satisfies (3.3), then it also satisfies (3.1) since $g(\theta_0) = h(\theta_0)$. Since by (A.3.1) and (A.3.2)', $b'(\theta) B_0(\theta) > 0$, it follows that the integral equation (3.3) is of Volterra's second type, hence the solution $\hat{g}(x)$ exists for $b(\theta) > x \geq b(\theta_0)$. Similarly we can construct $\hat{g}(x)$ for $x < a(\theta_0)$. Repeating the same process we can define $\hat{g}(x)$ for all x. Hence we have

$$\min_{\hat{g}:\,\text{unbiased}} V_{\theta_0}(\hat{g}) = 0 .$$

Thus we complete the proof.

The following useful lemma is a special case of the result by Takeuchi and Akahira [7].

LEMMA 3.1. *Let $g(\theta)$ be p-times differentiable over Θ. Suppose that the conditions (A.3.3) and (A.3.4) hold and G is the class of the all estimators $\hat{g}(x)$ of $g(\theta)$ for which*

$$\int_{a(\theta_0)}^{b(\theta_0)} \hat{g}(x) f(x,\theta_0) d\mu = g(\theta_0) ,$$

$$\int_{a(\theta_0)}^{b(\theta_0)} \hat{g}(x) \left\{ \frac{\partial^i}{\partial \theta^i} f(x,\theta_0) \right\} d\mu = g^{(i)}(\theta_0) \qquad (i=1,\cdots,p) ,$$

where $g^{(i)}(\theta)$ is the i-th order derivative of $g(\theta)$ with respect to θ. Then

$$\inf_{\hat{g} \in G} V_{\theta_0}(\hat{g}) = (g^{(1)}(\theta_0), \cdots, g^{(k)}(\theta_0)) \Lambda^{-1} (g^{(1)}(\theta_0), \cdots, g^{(k)}(\theta_0))'$$

where Λ is a $k \times k$ matrix whose elements are

$$\lambda_{ij} = \int_{a(\theta_0)}^{b(\theta_0)} \frac{\dfrac{\partial^i}{\partial \theta^i} f(x,\theta_0) \dfrac{\partial^j}{\partial \theta^j} f(x,\theta_0)}{f(x,\theta_0)} d\mu \qquad (i,j=1,\cdots,k) .$$

The proof is omitted. In the following theorem we shall get a sharp lower bound.

THEOREM 3.2. *Let $g(\theta)$ be $(p+1)$-times differentiable over Θ. Let $\hat{g}(x)$ be an unbiased estimator of $g(\theta)$. If for $p \geq 1$ and a fixed θ_0, the conditions (A.3.1) to (A.3.7) hold, then*

$$\inf_{\hat{g}:\,\text{unbiased}} V_{\theta_0}(\hat{g}) = v_k(\theta_0) \,,$$

that is, the bound $v_k(\theta_0)$ is sharp, where

$$v_k(\theta_0) = (g^{(1)}(\theta_0), \cdots, g^{(k)}(\theta_0)) \Lambda^{-1} (g^{(1)}(\theta_0), \cdots, g^{(k)}(\theta_0))' \,,$$

with a $k \times k$ matrix Λ given in Lemma 3.1.

PROOF. From the unbiasedness condition of $\hat{g}(x)$, (A.3.2) and (A.3.7) we have

$$(3.4) \qquad \int_{a(\theta_0)}^{b(\theta_0)} \hat{g}(x) f(x, \theta_0) d\mu = g(\theta_0) \,,$$

$$(3.5) \qquad \int_{a(\theta_0)}^{b(\theta_0)} \hat{g}(x) \left\{ \frac{\partial^i}{\partial \theta^i} f(x, \theta_0) \right\} d\mu = g^{(i)}(\theta_0) \qquad (i = 1, \cdots, p) \,.$$

By Lemma 3.1 it follows that the sharp lower bound of $V_{\theta_0}(\hat{g}) = \int_{a(\theta_0)}^{b(\theta_0)}$ $\{\hat{g}(x) - g(\theta_0)\}^2 f(x, \theta_0) d\mu(x)$ under (3.4) and (3.5) is given by $(g^{(1)}(\theta_0), \cdots, g^{(k)}(\theta_0)) \Lambda^{-1} (g^{(1)}(\theta_0), \cdots, g^{(k)}(\theta_0))'$, i.e.,

$$(3.6) \qquad \inf_{\hat{g}:\,(3.4),\,(3.5)} V_{\theta_0}(\hat{g}) = (g^{(1)}(\theta_0), \cdots, g^{(k)}(\theta_0)) \Lambda^{-1} (g^{(1)}(\theta_0), \cdots, g^{(k)}(\theta_0))'$$
$$= v_k(\theta_0) \ (\text{say}) \,.$$

Note that the right-hand side of (3.6) is the Bhattacharyya bound for the variance of unbiased estimators at $\theta = \theta_0$ (see Akahira et al. [1]). From (3.6) it follows that for any $\varepsilon > 0$ there exists $\hat{g}_\varepsilon(x)$ in the interval $(a(\theta_0), b(\theta_0))$ satisfying (3.4), (3.5) and

$$V_{\theta_0}(\hat{g}_\varepsilon) \leq v_k(\theta_0) + \varepsilon \,.$$

We can extend $\hat{g}_\varepsilon(x)$ for x outside $(a(\theta_0), b(\theta_0))$ from the unbiasedness condition

$$(3.7) \qquad \int_{a(\theta)}^{b(\theta)} \hat{g}_\varepsilon(x) f(x, \theta) d\mu(x) = g(\theta) \qquad \text{for all} \qquad \theta \in \Theta \,.$$

For $\theta > \theta_0$, i.e., $b(\theta) \geq b(\theta_0)$ we put

$$h(\theta) = \int_{a(\theta)}^{b(\theta_0)} \hat{g}_\varepsilon(x) f(x, \theta) d\mu(x) \,.$$

THE LOWER BOUND FOR THE VARIANCE OF UNBIASED ESTIMATORS   599

By (3.7) we obtain

$$(3.8) \qquad \int_{b(\theta_0)}^{b(\theta)} \hat{g}_\varepsilon(x) f(x, \theta) d\mu(x) = g(\theta) - h(\theta) .$$

Differentiating $(p+1)$-times both sides of (3.8) with respect to θ, recursively, we have by (A.3.1), (A.3.2), (A.3.5), (A.3.6) and (A.3.7)

$$\int_{b(\theta_0)}^{b(\theta)} \hat{g}_\varepsilon(x) \left\{ \frac{\partial}{\partial \theta} f(x, \theta) \right\} d\mu(x) = g^{(1)}(\theta) - h^{(1)}(\theta) ,$$

$$\int_{b(\theta_0)}^{b(\theta)} \hat{g}_\varepsilon(x) \left\{ \frac{\partial^p}{\partial \theta^p} f(x, \theta) \right\} d\mu(x) = g^{(p)}(\theta) - h^{(p)}(\theta) ,$$

$$(3.9) \quad B_p(\theta) b^{(1)}(\theta) \hat{g}_\varepsilon(b(\theta)) + \int_{b(\theta_0)}^{b(\theta)} \hat{g}_\varepsilon(x) \left\{ \frac{\partial^{p+1}}{\partial \theta^{p+1}} f(x, \theta) \right\} d\mu = g^{(p+1)}(\theta) - h^{(p+1)}(\theta) .$$

Differentiation under the integral sign is admitted because of (A.3.7). If $\hat{g}_\varepsilon(x)$ satisfies (3.9), then it also satisfies (3.8) since $g^{(i)}(\theta_0) = h^{(i)}(\theta_0)$ $(i=1, \cdots, p)$ and $g(\theta_0) = h(\theta_0)$. Note that $h^{(p+1)}(\theta)$ is determined by the values of $\hat{g}_\varepsilon(x)$ for $a(\theta) < x < b(\theta_0)$, where it is already given. Since the integral equation (3.9) is of Volterra's second type, it follows that the solution $\hat{g}_\varepsilon(x)$ exists for $b(\theta) > x \geq b(\theta_0)$. Similarly we can construct $\hat{g}_\varepsilon(x)$ for $x < a(\theta_0)$. Repeating the same process we can define $\hat{g}_\varepsilon(x)$ for all x. Hence we have

$$\inf_{\hat{g}: \text{ unbiased}} V_{\theta_0}(\hat{g}) = v_k(\theta_0)$$

i.e., $v_k(\theta_0)$ is a sharp bound. Thus we have completed the proof.

We shall give one example corresponding to each of the situations where the conditions (A.3.1), (A.3.2)/(A.3.2)' to (A.3.7) are assumed.

Example 3.1. Let X be a real random variable with a density function $f(x, \theta)$ (with respect to the Lebesgue measure μ) satisfying each case.

(i) Location parameter case. The density function is of the form $f(x-\theta)$ and satisfies the following:

$$f(x) > 0 \qquad \text{for} \qquad a < x < b ,$$

$$f(x) = 0 \qquad \text{for} \qquad x \leq a, \ x \geq b ,$$

and $\lim_{x \to a+0} f(x) > 0$, $\lim_{x \to b-0} f(x) > 0$ and $f(x)$ is continuously differentiable in the open interval (a, b).

(ii) The case on estimation of $g(\theta) = \theta$ with a density function

$$f(x-\theta)=\begin{cases} c(1-(x-\theta)^2)^{q-1} & \text{for} \quad |x-\theta|<1 , \\ 0 & \text{for} \quad |x-\theta|\geq 1 , \end{cases}$$

is discussed in Akahira et al. [1], where $q>1$ and c is some constant.

(iii) Scale parameter case. The density function of the form $f(x/\theta)/\theta$ satisfies the following:

$$f(x)>0 \quad \text{for} \quad 0<a<x<b ,$$

$$f(x)=0 \quad \text{otherwise} ,$$

and satisfies the same condition in (i).

(iv) Selection parameter case (e.g., Morimoto and Sibuya [6]). Consider a family of density functions whose supporting intervals depend on a selection parameter θ and are of the form $(\theta, b(\theta))$, where $-\infty<\theta<b(\theta)<\infty$ and $b(\theta)$ is a nondecreasing function of θ and almost everywhere differentiable. Such a family of density functions is specified by

$$f(x, \theta)=\begin{cases} \dfrac{p(x)}{F(\theta)} & \text{for} \quad \theta\leq x\leq b(\theta) , \\ \\ 0 & \text{otherwise} , \end{cases}$$

where $p(x)>0$ a.e. and $F(\theta)=\int_\theta^{b(\theta)} p(x)d\mu(x)$. Note that the cases (i), (iii) and (iv) correspond to the case $p=0$, and (ii) corresponds to the case $p=q-1$, where q is an integer.

4. The lower bound for the variance of unbiased estimators for a sample of size n of real-valued observations

Now suppose that we have a sample of size n, $(X_1,\cdots, X_n)$ of which X_i's are independently and identically distributed according to the distribution characterized in the previous section. Then we can define statistics

$$Y=\max_{1\leq i\leq n} X_i+\min_{1\leq i\leq n} X_i , \qquad Z=\max_{1\leq i\leq n} X_i-\min_{1\leq i\leq n} X_i$$

and we may concentrate our attention on the estimators depending only on Y and Z. Since they are not sufficient statistics, we may lose some information by doing so. More generally, for a sample of size n, $(X_1, \cdots, X_n)$ from a population in a one-directional family of densities with a support $A(\theta)$, we can define two statistics

$$\bar{\theta}=\sup \{\theta \,|\, X_i \in A(\theta) \ (i=1,\cdots, n)\} ,$$

$$\underline{\theta}=\inf \{\theta \,|\, X_i \in A(\theta) \ (i=1,\cdots, n)\}$$

and also define

$$Y=\frac{1}{2}(\bar{\theta}+\underline{\theta})\,,\qquad Z=\frac{1}{2}(\bar{\theta}-\underline{\theta})\,.$$

There are various ways of defining the pair of statistics Y and Z, but disregarding their construction we assume that there exists a pair (Y, Z) which satisfies the following.

Let Y and Z be real-valued statistics based on a sample $(X_1,\cdots, X_n)$ of size n for $n\geq2$. We assume that (Y, Z) has a joint probability density function $f_\theta(y, z)$ (with respect to the Lebesgue measure $\mu_{y,z}$) satisfying

$$f_\theta(y, z)=f_\theta(y\,|\,z)h_\theta(z)\,,\qquad \text{a.e.},$$

where $f_\theta(y\,|\,z)$ is a conditional density function of y given z with respect to the Lebesgue measure μ_y and $h_\theta(z)$ is a density function of z with respect to the Lebesgue measure μ_z. Note that if Z is ancillary, $h_\theta(z)$ is independent of θ. We assume the following condition:

(A.4.1)   For almost all $z\ [\mu_z]$

$$f_\theta(y\,|\,z)>0\qquad \text{for}\qquad a_z(\theta)<y<b_z(\theta)\,,$$

$$f_\theta(y\,|\,z)=0\qquad \text{for}\qquad y\leq a_z(\theta),\ y\geq b_z(\theta)\,,$$

where $a_z(\theta)$ and $b_z(\theta)$ are strictly monotone increasing functions of θ for almost all $z\ [\mu_z]$ which depend on z, and

$$h_\theta(z)>0\qquad \text{for}\qquad c<z<d\,,$$

$$h_\theta(z)=0\qquad \text{for}\qquad z\leq c,\ z\geq d\,,$$

where c and d are constants independent of θ. We also assume that for almost all $z\,[\mu_z]$, $f_\theta(y\,|\,z)$ instead of $f(x, \theta)$ satisfies the conditions (A.3.2) to (A.3.7), and we call the corresponding conditions (A.4.2) to (A.4.7).

Let $\hat{g}(y, z)$ be any unbiased estimator of $g(\theta)$. We define

$$(4.1)\qquad \phi_z(\theta)=\int_{a_z(\theta)}^{b_z(\theta)}\hat{g}(y, z)f_\theta(y\,|\,z)d\mu_y\qquad \text{for}\qquad \text{a.a.}\ z\,[\mu_z]\,.$$

Further we assume the following condition:

$$(A.4.8)\qquad \phi_z(\theta_0)=g(\theta_0)\qquad \text{for}\qquad \text{a.a.}\ z\,[\mu_z]\,,$$

$$\phi_z^{(i)}(\theta_0)=0\qquad \text{for}\qquad \text{a.a.}\ z\,[\mu_z]\ (i=1,\cdots, k)\,,$$

where $\phi_z^{(i)}(\theta)$ is the k-th order derivative of $\phi_z(\theta)$ with respect to θ.

In the following theorem we shall show that the sharp bound is equal to zero.

THEOREM 4.1. *Let $g(\theta)$ be $(p+1)$-times differentiable over Θ. Let $\hat{g}(X_1,\cdots,X_n)$ be an unbiased estimator of $g(\theta)$. If $n\geq2$ and for a fixed θ_0, the conditions (A.4.1) to (A.4.8) hold, then,*

$$\inf_{\hat{g}\,:\,\text{unbiased}} V_{\theta_0}(\hat{g})=0 \ .$$

PROOF. From the unbiasedness condition we obtain

$$(4.2)\qquad \int_c^d \phi_z(\theta)h_\theta(z)d\mu_z=g(\theta)\qquad \text{for all}\qquad \theta\in\Theta \ .$$

First we assume that $\phi_z(\theta)$ is given, then under (A.4.1) to (A.4.7) we have by Lemma 3.1

$$(4.3)\qquad \inf_{\hat{g}\,:\,(4.1)} \int_{a_z(\theta_0)}^{b_z(\theta_0)} \{\hat{g}(y,\,z)-\phi_z(\theta_0)\}^2 f_{\theta_0}(y\,|\,z)d\mu_y$$
$$=(\phi_z^{(1)}(\theta_0),\cdots,\phi_z^{(k)}(\theta_0))\,\varLambda^{-1}(\phi_z^{(1)}(\theta_0),\cdots,\phi_z^{(k)}(\theta_0))'$$
$$=v_k(\theta_0\,|\,z)\qquad \text{(say)}\ ,$$

where $\phi_z^{(i)}(\theta)$ is the i-th order derivative of $\phi_z(\theta)$ with respect to θ and $\varLambda$ is a $k\times k$ matrix whose elements are

$$\lambda_{ij}=\int_{a_z(\theta_0)}^{b_z(\theta_0)} \frac{1}{f_{\theta_0}(y\,|\,z)} \left\{\frac{\partial^i}{\partial\theta^i}f_{\theta_0}(y\,|\,z)\right\} \left\{\frac{\partial^j}{\partial\theta^j}f_{\theta_0}(y\,|\,z)\right\}d\mu_y \ .$$

From (4.3) it follows that for any $\varepsilon>0$ there exists $\hat{g}_\varepsilon(y,\,z)$ such that

$$(4.4)\qquad \int_{a_z(\theta)}^{b_z(\theta)} \hat{g}_\varepsilon(y,\,z)f_\theta(y\,|\,z)d\mu_y=\phi_z(\theta)\qquad \text{for all}\qquad \theta\in\Theta\ ,$$

$$(4.5)\qquad \int_{a_z(\theta_0)}^{b_z(\theta_0)} \{\hat{g}(y,\,z)-\phi_z(\theta_0)\}^2 f_{\theta_0}(y\,|\,z)d\mu_y<v_k(\theta_0\,|\,z)+\varepsilon \ .$$

Since by (4.2)

$$\int_c^d\int_{a_z(\theta)}^{b_z(\theta)} \hat{g}_\varepsilon(y,\,z)f_\theta(y\,|\,z)h_\theta(z)d\mu_y\,d\mu_z=g(\theta)$$

for all $\theta\in\Theta$, it follows from (4.5) that

$$(4.6)\qquad \int_c^d\int_{a_z(\theta_0)}^{b_z(\theta_0)} \{\hat{g}_\varepsilon(y,\,z)-\phi_z(\theta_0)\}^2 f_{\theta_0}(y\,|\,z)h_{\theta_0}(z)d\mu_y\,d\mu_z$$
$$<\int_c^d v_k(\theta_0\,|\,z)h_{\theta_0}(z)d\mu_z+\varepsilon \ .$$

By the condition (A.4.8)

$$v_k(\theta_0\,|\,z)=0\qquad \text{for}\qquad \text{a.a.}\ \ z\,[\mu_z] \ .$$

From (4.6) we obtain

$$(4.7) \qquad \int_c^d \int_{a_z(\theta_0)}^{b_z(\theta_0)} \{\hat{g}_\varepsilon(y, z) - \theta_0\}^2 f_{\theta_0}(y \mid z) h_{\theta_0}(z) d\mu_y d\mu_z < \varepsilon .$$

Putting $\hat{g}_\varepsilon(x_1, \cdots, x_n) = \hat{g}_\varepsilon(y, z)$, we have by (4.4) and (4.7)

$$E_\theta(\hat{g}_\varepsilon) = g(\theta) \qquad \text{for all} \qquad \theta \in \Theta ,$$

$$V_{\theta_0}(\hat{g}_\varepsilon) < \varepsilon .$$

Letting $\varepsilon \to 0$ we obtain

$$\inf_{\hat{g}:\text{ unbiased}} V_{\theta_0}(\hat{g}) = 0 .$$

Thus we complete the proof.

Now we shall find a function $\phi_z(\theta)$ satisfying the condition (A.4.8) when $\Theta = R^1$, $g(\theta) = \theta$ and μ_z is a Lebesgue measure. Without loss of generality, we put $\theta_0 = 0$. We define

$$(4.8) \qquad \phi_z(\theta) = \frac{M \operatorname{sgn} \theta |\theta|^{k+2} (z-c)^k}{h_\theta(z)\{(d-z)^{k+2} + |\theta|^{k+2}(z-c)^{k+2}\}} ,$$

where M is a constant and k is a positive integer. Then we have

$$\int_c^d \phi_z(\theta) h_\theta(z) dz = M \int_c^d \frac{\operatorname{sgn} \theta |\theta|^{k+2} (z-c)^k}{(d-z)^{k+2} + |\theta|^{k+2}(z-c)^{k+2}} dz$$

$$= \frac{M}{d-c} \operatorname{sgn} \theta |\theta|^{k+2} \int_0^\infty \frac{u^k}{1 + |\theta|^{k+2} u^{k+2}} du$$

$$\left(\text{after transformation} \quad u = \frac{z-c}{d-z} \right)$$

$$= \frac{M}{d-c} \operatorname{sgn} \theta |\theta| \int_0^\infty \frac{v^k}{1 + v^{k+2}} dv = \frac{MK}{d-c} \theta ,$$

where $K = \int_0^\infty \frac{v^k}{1 + v^{k+2}} dv$ is a constant. If we put $M = (d-c)/K$, then

$$\int_c^d \phi_z(\theta) h_\theta(z) dz = \theta .$$

And it is easily seen that

$$\phi_z(0) = 0 \qquad \text{for} \qquad \text{a.a.} \quad z ,$$

$$\phi_z^{(i)}(0) = 0 \qquad \text{for} \qquad \text{a.a.} \quad z \ (i = 1, \cdots, k) .$$

Thus it is shown that $\phi_z(\theta)$ given by (4.8) satisfies the condition (A.4.8).

We consider the estimation on the location parameter θ. Let X_1 and X_2 be independently and identically distributed with a density function $f(x, \theta)$ of the form $f(x - \theta)$ which satisfies the following:

$$f(x)>0 \quad \text{for} \quad a<x<b,$$

$$f(x)=0 \quad \text{for} \quad x\leq a,\ x\geq b,$$

and $\lim_{x\to a+0} f(x)>0$, $\lim_{x\to b-0} f(x)>0$ and $f(x)$ is continuously differentiable in the open interval (a, b). We define

$$Y=\frac{1}{2}(X_1+X_2)\,, \qquad Z=\frac{1}{2}(X_1-X_2)\,.$$

Then if the conditions (A.4.1) to (A.4.8) are assumed, we have

$$\inf_{\hat{\theta}:\text{unbiased}} V_{\theta_0}(\hat{\theta})=0\,.$$

Note that Z is an ancillary statistic, but that (Y, Z) is not sufficient unless $f(x)$ is constant for $a<x<b$.

5. A second type approach to obtain the lower bound for the variance of unbiased estimators

Suppose that (X, Y) is a pair of real random variables according to a joint density function $f(x, y, \theta)$ (with respect to the Lebesgue measure μ) which has the product set $(0, a(\theta))\times(0, b(\theta))$ of two open intervals as its support $A(\theta)$, where $a(\theta)$ is a monotone increasing function and $b(\theta)$ is a monotone decreasing function. We assume the condition

$$(A.5.1) \qquad \inf_{(x,y)\in A(\theta)} f(x, y, \theta)>0\,.$$

Let the marginal density functions of X and Y be $f_1(x, \theta)$ and $f_2(y, \theta)$ with respect to the Lebesgue measures μ_x and μ_y, respectively. We further make the following assumption:

(A.5.2) The density functions $f_1(x, \theta)$ and $f_2(y, \theta)$ are continuously differentiable in θ and satisfy the conditions (A.3.5), (A.3.6) and (A.3.7) for $k=1$ when $f_1(x, \theta)$ and $f_2(y, \theta)$ are substituted instead of $f(x, \theta)$ in them.

In the following theorem we shall show that the sharp bound is equal to zero.

THEOREM 5.1. *Let* $\Theta=R^1$. *Suppose that X and Y are random variables with a joint density function $f(x, y, \theta)$ (with respect to a σ-finite measure μ) satisfying* (A.5.1) *and* (A.5.2) *for a fixed* θ_0. *Let* $g(\theta)$ *be continuously differentiable over* Θ. *Let* $\hat{g}(x, y)$ *be an unbiased estimator of* $g(\theta)$. *Then*

$$\inf_{\hat{g}:\text{unbiased}} V_{\theta_0}(\hat{g})=0\,.$$

THE LOWER BOUND FOR THE VARIANCE OF UNBIASED ESTIMATORS   605

PROOF.   We first define

$$\hat{g}(x, y) = g(\theta_0) \qquad \text{for} \qquad (x, y) \in A(\theta_0) .$$

In order to extend $\hat{g}(x, y)$ for (x, y) outside $A(\theta_0)$ using the unbiasedness condition, we consider unbiased estimators $\hat{g}_1(x)$ and $\hat{g}_2(y)$ of $g(\theta)$ with respect to $f_1(x, \theta)$ and $f_2(y, \theta)$, respectively, such that $\hat{g}_1(x) = g(\theta_0)$ for $0 < x < a(\theta_0)$ and $\hat{g}_2(y) = g(\theta_0)$ for $0 < y < b(\theta_0)$. For $\theta > \theta_0$, i.e., $a(\theta) \geqq a(\theta_0)$ we put

$$h_1(\theta) = g(\theta_0) \int_0^{a(\theta_0)} f_1(x, \theta) d\mu_x$$

and also for $\theta \leqq \theta_0$, i.e., $b(\theta) \geqq b(\theta_0)$

$$h_2(\theta) = g(\theta_0) \int_0^{b(\theta_0)} f_2(y, \theta) d\mu_y .$$

Since $\hat{g}_1(X)$ and $\hat{g}_2(Y)$ are unbiased estimator of $g(\theta)$, it follows that

$$(5.1) \qquad \begin{aligned} &\int_{a(\theta_0)}^{a(\theta)} \hat{g}_1(x) f_1(x, \theta) d\mu_x = g(\theta) - h_1(\theta) \qquad \text{for all} \qquad \theta \geqq \theta_0 , \\[2mm] &\int_{b(\theta_0)}^{b(\theta)} \hat{g}_2(y) f_2(y, \theta) d\mu_y = g(\theta) - h_2(\theta) \qquad \text{for all} \qquad \theta \leqq \theta_0 . \end{aligned}$$

Since the supports of the density functions $f_1(x, \theta)$ and $f_2(y, \theta)$ are open intervals $(0, a(\theta))$ and $(0, b(\theta))$, respectively, it follows from (A.5.1) that

$$(5.2) \qquad \begin{aligned} &0 < \lim_{x \to a(\theta)-0} f_1(x, \theta) = \alpha_1(\theta) \qquad \text{(say)} , \\[2mm] &0 < \lim_{y \to b(\theta)-0} f_2(y, \theta) . \end{aligned}$$

Differentiating both sides of (5.1) we have by (A.5.2)

$$(5.3) \qquad \hat{g}_1(a(\theta)) \alpha_1(\theta) + \int_{a(\theta_0)}^{a(\theta)} \hat{g}_1(x) \left\{ \frac{\partial}{\partial \theta} f_1(x, \theta) \right\} d\mu_x = g'(\theta) - h_1'(\theta) ,$$

for all $\theta \geqq \theta_0$. Since the equation (5.3) is of Volterra's second type, it follows that the solution $\hat{g}_1(x)$ exists for all $x \geqq a(\theta_0)$. If $\hat{g}_1(x)$ satisfies (5.3), then it also satisfies (5.1) since $g(\theta_0) = h_1(\theta_0)$. Similarly we can construct the unbiased estimator $\hat{g}_2(y)$ for all $y \geqq b(\theta_0)$. We define an estimator

$$(5.4) \qquad \hat{g}(x, y) = \begin{cases} g(\theta_0) & \text{for} \quad 0 < x < a(\theta_0),\ 0 < y < b(\theta_0) , \\[2mm] \hat{g}_1(x) & \text{for} \quad a(\theta_0) \leqq x,\ 0 < y < b(\theta_0) , \\[2mm] \hat{g}_2(y) & \text{for} \quad 0 < x < a(\theta_0),\ b(\theta_0) \leqq y . \end{cases}$$

Then $\hat{g}(X, Y)$ is an unbiased estimator of $g(\theta)$ with variance 0 at $\theta = \theta_0$.

Indeed, we have from (5.1) and (5.4)

$$E_\theta[\hat{g}(X, Y)] = \int_0^{b(\theta)} \int_0^{a(\theta_0)} g(\theta_0) f(x, y, \theta) d\mu + \int_0^{b(\theta)} \int_{a(\theta_0)}^{a(\theta)} \hat{g}_1(x) f(x, y, \theta) d\mu$$

$$= g(\theta_0) \int_0^{a(\theta_0)} f_1(x, \theta) d\mu_x + \int_{a(\theta_0)}^{a(\theta)} \hat{g}_1(x) f_1(x, \theta) d\mu_x ,$$

for all $\theta \geq \theta_0$. Similarly we have that $E_\theta[\hat{g}(X, Y)] = g(\theta)$ for all $\theta \leq \theta_0$. Hence we see that $\hat{g}(X, Y)$ is an unbiased estimator of $g(\theta)$. We also have, for all $\theta \geq \theta_0$

$$V_\theta(\hat{g}(X, Y)) = g^2(\theta_0) \int_0^{a(\theta_0)} f_1(x, \theta) d\mu_x - \int_{a(\theta_0)}^{a(\theta)} \hat{g}_1^2(x) f_1(x, \theta) d\mu_x - g^2(\theta) .$$

When $\theta = \theta_0$, we obtain

$$V_{\theta_0}(\hat{g}(X, Y)) = 0 .$$

Thus we complete the proof.

We can give the following example.

Example 5.1. Let $X_1, \cdots, X_n$ and $Y_1, \cdots, Y_n$ be independently and identically distributed random variables with the uniform distributions $U(0, \theta)$ and $U(0, 1/\theta)$, respectively. Put

$$T_1 = \max_{1 \leq i \leq n} X_i , \qquad T_2 = \max_{1 \leq i \leq n} Y_i .$$

Then the unbiased estimator $\hat{\theta}(T_1, T_2)$ of θ with variance 0 at $\theta = 1$ is given by

$$(5.5) \qquad \hat{\theta}(t_1, t_2) = \begin{cases} \hat{\theta}_1(t_1) & \text{for} & 1 \leq t_1, \ 0 < t_2 < 1 , \\ \hat{\theta}_2(t_2) & \text{for} & 1 \leq t_2, \ 0 < t_1 < 1 , \\ 1 & \text{for} & t_1 < 1, \ t_2 < 1 , \end{cases}$$

where

$$\hat{\theta}_1(t_1) = \left(1 + \frac{1}{n}\right) t_1 \qquad \text{for} \qquad 1 \leq t_1 ,$$

$$\hat{\theta}_2(t_2) = \left(1 - \frac{1}{n}\right) t_2 \qquad \text{for} \qquad 1 \leq t_2 .$$

Indeed, we can easily see that the estimator $\hat{\theta}(T_1, T_2)$ is unbiased. We also have, for $0 < \theta \leq 1$

$$(5.6) \qquad V_\theta(\hat{\theta}(T_1, T_2)) = \left\{1 - \frac{(n-1)^2}{n(n-2)}\right\} (\theta^n - \theta^2) .$$

We also have, for $\theta \geq 1$

$$(5.7) \qquad V_\theta(\hat\theta(T_1, T_2)) = \left\{1 - \frac{(n+1)^2}{n(n-2)}\right\}(\theta^{-n} - \theta^2) .$$

From (5.6) and (5.7), we obtain

$$V_1(\hat\theta(T_1, T_2)) = 0 .$$

Thus we see that the unbiased estimator $\hat\theta(T_1, T_2)$ given by (5.5) with variance 0 at $\theta = 1$.

As a further case of this situation, let X and Y be independent real random variables according to density functions $(1/\theta)f(x/\theta)$ and $\theta f(\theta y)$ (with respect to the Lebesgue measure μ) with a positive valued parameter θ, respectively, which satisfy the following:

$$(A.5.3) \qquad f(x) > 0 \qquad \text{for} \qquad 0 < x < 1 ,$$

$$f(x) = 0 \qquad \text{otherwise} ,$$

and $f(x)$ is $(p+1)$-times continuously differentiable in the open interval $(0, 1)$ and for each $i = 0, 1, \cdots, p$

$$0 < \lim_{x \to 0+0}\left|\frac{f^{(i)}(x)}{x^{p-i}}\right| < \infty , \qquad 0 < \lim_{x \to 1-0}\left|\frac{f^{(i)}(x)}{(1-x)^{p-i}}\right| < \infty .$$

By Theorem 4.1 we have the following:

THEOREM 5.2. *Let $g(\theta)$ be an estimable function which is $(p+1)$-times differentiable over R^1. Let $\hat g(X, Y)$ be an unbiased estimator of $g(\theta)$. If the conditions (A.5.3) and (A.4.8) on $p_\theta(x|t)$ hold, then*

$$\inf_{\hat g:\ \text{unbiased}} V_{\theta_0}(\hat g) = 0 ,$$

where $p_\theta(x|t)$ denotes the conditional density function of X given $XY = t$.

PROOF. Letting $T = XY$, we have the conditional density function $p_\theta(x|t)$ of X given $T = t$:

$$(5.8) \qquad p_\theta(x|t) = \begin{cases} \dfrac{\dfrac{1}{x}f\left(\dfrac{x}{\theta}\right)f\left(\dfrac{\theta t}{x}\right)}{\displaystyle\int_{\theta t}^{\theta}\frac{1}{x}f\left(\frac{x}{\theta}\right)f\left(\frac{\theta t}{x}\right)dx} & \text{for} \qquad \theta t < x < \theta , \\[20pt] 0 & \text{otherwise} , \end{cases}$$

for almost all $t\,[\mu]$. Since for $\theta = 1$ and almost all $t\,[\mu]$

$$p_1(x|t) = \begin{cases} \dfrac{c_t}{x}f(x)f\left(\dfrac{t}{x}\right) & \text{for} \qquad t < x < 1 , \\[16pt] 0 & \text{otherwise} , \end{cases}$$

                MASAFUMI AKAHIRA AND KEI TAKEUCHI

we obtain from (5.8)

$$p_\theta(x\,|\,t)=\frac{1}{\theta}\,p_1\!\left(\frac{x}{\theta}\,\Big|\,t\right)\qquad\text{for all}\qquad\theta,\ \text{a.a.}\ \ t\ [\mu]$$

where

$$c_t=\left(\int_t^1\frac{1}{x}\,f(x)f\!\left(\frac{t}{x}\right)dx\right)^{-1}.$$

Putting

$$g_t(x)=\begin{cases}\dfrac{c_t}{x}\,f\!\left(\dfrac{t}{x}\right)&\text{for}\qquad t<x<1\,,\\[2ex]0&\text{otherwise}\,,\end{cases}$$

for almost all $t\ [\mu]$, we have

$$(5.9)\qquad\qquad p_1(x\,|\,t)=f(x)g_t(x)\,,\qquad\text{a.a.}\ \ t\ [\mu]\,,$$

hence the same conditions on $p_1(x\,|\,t)$ as (A.4.1) and (A.4.2) hold. Indeed, we have by (A.5.3) and (5.9)

$$(5.10)\qquad\begin{aligned}&\lim_{x\to t+0} p_1(x\,|\,t)=f(t+0)g_t(t+0)=\frac{c_t}{t}\,f(t+0)f(1+0)=0\,,\\[1ex]&\lim_{x\to 1-0} p_1(x\,|\,t)=f(1-0)g_t(1-0)=0\end{aligned}$$

for almost all $t\ [\mu]$. By the Leibniz's formula, we obtain

$$(5.11)\qquad\begin{aligned}\frac{\partial^i p_1(x\,|\,t)}{\partial x^i}&=\sum_{j=0}^{i}\binom{i}{j}f^{(j)}(x)g_t^{(i-j)}(x)\\[1ex]&=\sum_{j=0}^{i}\binom{i}{j}f^{(i-j)}(x)g_t^{(j)}(x)\,,\qquad\text{a.a.}\ \ t\ [\mu]\,.\end{aligned}$$

By (A.5.3) we have for $i=1,\cdots,p-1$ and a.a. $t\ [\mu]$

$$(5.12)\qquad\lim_{x\to 1-0}\frac{\partial^i p_1(x\,|\,t)}{\partial x^i}=0\,,\qquad\lim_{x\to 1-0}\frac{\partial^p p_1(x\,|\,t)}{\partial x^p}=c_t\neq 0$$

where c_t is finite. Since

$$g^{(i)}(x)=c_t\sum_{j=0}^{i}(-1)^j\,j!\,\frac{1}{x^{j+1}}\,\frac{\partial^{j-i}}{\partial x^{j-i}}f\!\left(\frac{t}{x}\right)\,,$$

it follows by (A.5.3) and (5.11) that for almost all $t\ [\mu]$,

$$(5.13)\qquad\begin{aligned}&\lim_{x\to t+0}\frac{\partial^i p_1(x\,|\,t)}{\partial x^i}=0\qquad(i=1,\cdots,p-1)\,,\\[1ex]&\lim_{x\to t+0}\frac{\partial^p p_1(x\,|\,t)}{\partial x^p}=D_t\neq 0\,,\end{aligned}$$

THE LOWER BOUND FOR THE VARIANCE OF UNBIASED ESTIMATORS   609

where D_t is finite. It is seen by (5.9), (5.10), (5.12) and (5.13) that the same condition on $p_1(x|t)$ as (A.4.2) holds.

For $i=1,\cdots,[p/2]$

$$0<\int_t^1 \frac{\left\{\frac{\partial^i}{\partial x^i}p_1(x|t)\right\}^2}{p_1(x|t)}dx<\infty\,, \qquad \text{a.a.}\quad t\ [\mu]$$

and

$$\int_t^1 \frac{\left\{\sum_{i=[p/2]+1}^{p}c_i\frac{\partial^i}{\partial x^i}p_1(x|t)\right\}^2}{p_1(x|t)}dx\,, \qquad \text{a.a.}\quad t\ [\mu]$$

is infinite unless $c_{[p/2]+1}=\cdots=c_p$, where $[s]$ denotes the largest integer less than or equal to s, since when $x\to 0+0$ or $x\to t-0$ the numerator of the integrand approaches to a polynomial in x or $t-x$ of the degree $p-i^*$ if $c_{i^*}\neq 0$ and $c_{i^*+1}=\cdots=c_p=0$ and the denominator tends to that of the degree p. Hence the same condition on $p_1(x|t)$ as (A.4.4) holds for $k=[p/2]$. And also from (A.5.1) it is seen that when $x\to 0+0$ or $x\to t-0$, $\{(\partial^i/\partial x^i)p_1(x|t)\}/p_1(x|t)$ $(i=0,1,\cdots,k)$ approaches to polynomials of different degrees, hence they are linearly independent. Putting

$$\rho_t(x)=\sup_{x':|x'-x|<c}p_1(x'|t)$$

for appropriate $c>0$, the same conditions on $p_1(x|t)$ as (A.4.5) to (A.4.7) hold. By Theorem 4.1 we obtain the conclusion of Theorem 5.2.

When $g(\theta)=\theta$, Example 5.1 can be also an example of Theorem 5.2.

Acknowledgements

The authors wish to thank the associate editor and the referee for useful comments.

UNIVERSITY OF ELECTRO-COMMUNICATIONS, TOKYO*
UNIVERSITY OF TOKYO

REFERENCES

[1] Akahira, M., Puri, M. L. and Takeuchi, K. (1984). Bhattacharyya bound of variances of unbiased estimators in non-regular cases, *Ann. Inst. Statist. Math.*, **38**, 35–44.

[2] Chapman, D. G. and Robbins, H. (1951). Minimum variance estimation without regularity assumptions, *Ann. Math. Statist.*, **22**, 581–586.

[3] Fraser, D. A. S. and Guttman, I. (1952). Bhattacharyya bounds without regularity assumptions, *Ann. Math. Statist.*, **23**, 629–632.

---

* New at University of Tsukuba.

610   MASAFUMI AKAHIRA AND KEI TAKEUCHI

[4] Kiefer, J. (1952). On minimum variance in non-regular estimation, *Ann. Math. Statist.*, **23**, 627–629.

[5] Móri, T. F. (1983). Note on the Cramér-Rao inequality in the non-regular cases: The family of uniform distributions, *J. Statist. Plann. Inference*, **7**, 353–358.

[6] Morimoto, H. and Sibuya, M. (1967). Sufficient statistics and unbiased estimation of restricted selection parameter, *Sankhyā*, A27, 15–40.

[7] Takeuchi, K. and Akahira, M. (1983). A note on minimum variance, *Metrika*, **33**, 85–91.

[8] Víncze, I. (1979). On the Cramér-Fréchet-Rao inequality in the non-regular case, In *Contributions to Statistics*, the Jaroslav Hájek Memorial Volume, Academia, Prague, 253–262.

STATISTICAL THEORY AND DATA ANALYSIS II
K. Matusita (Editor)
© Elsevier Science Publishers B.V. (North-Holland), 1988

191

SECOND ORDER ASYMPTOTIC EFFICIENCY IN TERMS OF ASYMPTOTIC VARIANCES OF THE SEQUENTIAL MAXIMUM LIKELIHOOD ESTIMATION PROCEDURES

Kei TAKEUCHI and Masafumi AKAHIRA

Under suitable regularity conditions, the Bhattacharyya type bound for asymptotic variances of estimation procedures is obtained. It is also shown that the modified maximum likelihood estimation procedure attains the bound if the stopping rule is properly determined.

1. INTRODUCTION

Sequential estimation procedures can be defined in two stages (a) definition of a stopping rule, and (b) definition of the estimation procedure once the stopping rule is determined.

In this paper, we restrict ourselves to the class of estimation procedures related with a sequence of sequential sampling procedures where the sample size tends stochastically to infinity and is asymptotically constant in the sense that its coefficient of variation tends to zero. Then we can show that the asymptotic variance of the estimator must satisfy the Bhattacharyya type bound, and for one parameter case, the modified maximum likelihood (ML) estimation procedure attains the bound, if the stopping rule is properly defined.

The above results can be interpreted to mean that the asymptotic loss (deficiency) of the ML estimation procedure can be reduced to zero when we apply an appropriate sequential estimation procedure. This fact can be established also for the asymptotic distribution of the sequential ML estimation procedure which will be dealt with in a subsequent paper by the same authors.

2. BHATTACHARYYA BOUND AND MAXIMUM LIKELIHOOD ESTIMATION PROCEDURE

Let $X_1, X_2, \cdots, X_n, \cdots$ be a sequence of independent and identically distributed random variables with a density function $f(x, \theta)$ with respect to a σ-finite measure μ, where θ is a real-valued parameter. Suppose that the sample size n is determined according to some sequential rule. Actually, we consider a sequence of sequential estimation procedures $\{\Pi_\alpha : \alpha = 1,2, \cdots \}$ such that $E_{\theta,\alpha}(n) = v_\alpha(\theta)$ becomes large uniformly in θ as $\alpha \to \infty$, where for each α we define a stopping rule and estimators based on it and consider the asymptotic distribution of $\sqrt{v_\alpha}(\hat{\theta}_\alpha - \theta)$ as $\alpha \to \infty$.

For simplicity, we denote $v_\alpha(\theta)$ by v. In order to consider the second order asymptotic efficiency, we assume the following conditions.

(A.1) $E_\theta(n) = v + o(1)$, $\quad V_\theta(n)/v = O(1)$, $\quad E_\theta(n^k/v^k) = O(1)$ $(k = 2,3,4)$,
$\quad \{(\partial/\partial\theta)v_\alpha(\theta)\}/v_\alpha(\theta) = O(1)$ and $\{(\partial^2/\partial\theta^2)v_\alpha(\theta)\}/v_\alpha(\theta) = O(1)$.

(A.2) $\{x : f(x,\theta) > 0\}$ does not depend on θ.

(A.3) For almost all $x[\mu]$, $f(x, \theta)$ is four times continuously differentiable in θ. In the Taylor expansion

$$\log \frac{f(x,\theta+h)}{f(x,\theta)} = \sum_{i=1}^{4} \frac{h^i}{i!} \, \ell^{(i)}(\theta,x) + h^4 R(x,h),$$

$R(x,h)$ is uniformly bounded by a function $G(x)$ which has moments up to the fourth order, where $\ell^{(i)}(\theta,x) = (\partial^i/\partial\theta^i)\ell(\theta,x)$ $(i=1,2,3,4)$ and $\ell(\theta,x) = \log f(x,\theta)$.

(A.4) For each θ,
$$0 < I(\theta) = E_\theta[\{\ell^{(1)}(\theta,X)\}^2] = -E_\theta[\ell^{(2)}(\theta,X)] < \infty$$
and $I(\theta)$ is twice continuously differentiable in θ.

(A.5) There exist
$$J(\theta) = E_\theta[\ell^{(1)}(\theta,X)\ell^{(2)}(\theta,X)], \quad K(\theta) = E_\theta[\{\ell^{(1)}(\theta,X)\}^3],$$
$$L(\theta) = E_\theta[\ell^{(1)}(\theta,X)\ell^{(3)}(\theta,X)],$$
$$M(\theta) = E_\theta[\{\ell^{(2)}(\theta,X)\}^2] - I^2(\theta),$$
$$N(\theta) = E_\theta[\{\ell^{(1)}(\theta,X)\}^2\ell^{(2)}(\theta,X)] + I^2(\theta),$$
$$H(\theta) = E_\theta[\{\ell^{(1)}(\theta,X)\}^4] - 3I^2(\theta),$$

and both $J(\theta)$ and $K(\theta)$ are differentiable in θ, and $E_\theta[\ell^{(3)}(\theta,X)] = -3J(\theta) - K(\theta)$ and $E_\theta[\ell^{(4)}(\theta,X)] = 4L(\theta) + 3M(\theta) + 6N(\theta) + H(\theta)$.

We put

$$Z_{1,v} = \frac{1}{\sqrt{v}} \sum_{i=1}^{n} \ell^{(1)}(\theta,X_i), \quad Z_{2,v} = \frac{1}{\sqrt{v}} \sum_{i=1}^{n} \{\ell^{(2)}(\theta,X_i) + I(\theta)\},$$

and

$$Z_{3,v} = \frac{1}{\sqrt{v}} \sum_{i=1}^{n} \{\ell^{(3)}(\theta,X_i) - 3J(\theta) - K(\theta)\}.$$

The following lemma is very useful for calculations of cumulants.

LEMMA 2.1.   Suppose that Y_θ is a function of $X_1, \cdots, X_n$ and θ and is differentiable in θ. Then

$$E_\theta(Z_{1,v}Y_\theta) = \frac{1}{\sqrt{v}} \frac{d}{d\theta} E_\theta(Y_\theta) - \frac{1}{\sqrt{v}} E_\theta\left(\frac{\partial Y_\theta}{\partial\theta}\right)$$

and

$$E_\theta(Z_{1,v}^2 Y_\theta) = \frac{1}{\sqrt{v}} \frac{d}{d\theta} E_\theta(Z_{1,v}Y_\theta) - \frac{1}{\sqrt{v}} E_\theta\left(Y_\theta \frac{\partial Z_{1,v}}{\partial\theta}\right) - \frac{1}{\sqrt{v}} E_\theta\left(Z_{1,v} \frac{\partial Y_\theta}{\partial\theta}\right),$$

provided that differentiation under the integral signs of $E_\theta(Y_\theta)$ and $E_\theta(Z_{1,v}Y_\theta)$ is allowed.

The proof is omitted since the lemma is similar to Lemmas 5.1.1 and 5.1.2 in Akahira and Takeuchi [2](see also Lemmas 2.1.1 and 2.1.2 in Akahira [1]).

In the following theorem, we have the Bhattacharyya type bound for asymptotic variances.

THEOREM 2.1.   Assume that the conditions (A.1) to (A.4) hold. Then for any asymptotically unbiased estimator $\hat{\theta}_n$ of θ, i.e., $E_\theta(\hat{\theta}_n) = \theta + o(1/v)$, it holds

$$V_\theta(\sqrt{v}\,(\hat\theta_n-\theta))\geqq \frac{1}{I(\theta)}+\frac{1}{2v(\theta)I^2(\theta)}\left\{\frac{J(\theta)+K(\theta)}{I(\theta)}+\frac{2v'(\theta)}{v(\theta)}\right\}^2+o(\frac{1}{v(\theta)}),$$

where $E_\theta(\cdot)$ and $V_\theta(\cdot)$ designate the asymptotic mean and the asymptotic variance, respectively, and $v'(\theta)=(\partial/\partial\theta)v(\theta)$.

REMARK.   In the above theorem, for each $k=0,1,2,\cdots$ we mean by the term asymptotic mean of Z_v is equal to μ_v up to the order $v-k$ if $Z_v=\tilde Z_v+o_p(v-k)$ and $E(\tilde Z_v)=\mu_v$, and note that the above lemma also holds the asymptotic mean.

PROOF.   Since $E_\theta(\hat\theta_n)=\theta+o(1/v)$, $\sqrt{v}E_\theta(\hat\theta_n Z_{1,v})=1$ and

$$\sum_{i=1}^{n}\frac{\partial^2}{\partial\theta^2}\log f(X_i,\theta)=\sqrt{v}\,Z_{2,v}-nI,$$

it follows that

$$E_\theta[\sqrt{v}\,(\hat\theta_n-\theta)Z_{1,v}]=o(1),$$

$$E_\theta[\sqrt{v}\,(\hat\theta_n-\theta)(\sqrt{v}\,Z_{1,v}^2-\frac{n}{\sqrt{v}}I+Z_{2,v})]=o(1),$$

where v, v' and I denote $v(\theta)$, $v'=v'(\theta)=(d/d\theta)v(\theta)$ and $I(\theta)$, respectively. Similarly, J and K below stand for $J(\theta)$ and $K(\theta)$. Then we define the extended information matrix B as is given by

$$B=\begin{pmatrix} E_\theta(Z_{1,v}^2) & E_\theta[Z_{1,v}(\sqrt{v}\,Z_{1,v}^2+Z_{2,v}-\frac{n}{\sqrt{v}}I)] \\[2mm] E_\theta[Z_{1,v}(\sqrt{v}\,Z_{1,v}^2+Z_{2,v}-\frac{n}{\sqrt{v}}I)] & E_\theta[\sqrt{v}\,Z_{1,v}^2+Z_{2,v}-\frac{n}{\sqrt{v}}I)^2] \end{pmatrix}$$

Since by Lemma 2.1 and Wald's identity [3]

$$E_\theta(Z_{1,v}^2)=I(\theta)$$

$$E_\theta[Z_{1,v}(\sqrt{v}\,Z_{1,v}^2+Z_{2,v}-\frac{n}{\sqrt{v}}I)]=J(\theta)+K(\theta)+2I\frac{v'}{v}+o(1),$$

$$E_\theta[(\sqrt{v}\,Z_{1,v}^2+Z_{2,v}-\frac{n}{\sqrt{v}}I)^2]$$

$$=E_\theta[vZ_{1,v}^4-2InZ_{1,v}^2+\frac{n^2}{v}I^2+2\sqrt{v}\,Z_{1,v}^2 Z_{2,v}-\frac{2In}{\sqrt{v}}Z_{2,v}+Z_{1,v}^2]$$
$$=3vI2-2vI2+vI2+o(1)$$
$$=2vI2+o(1),$$

it follows that

$$B^{-1}=\begin{pmatrix} B^{11} & * \\ * & ** \end{pmatrix},$$

where $B^{11}=\left\{I-\frac{(J+K+2I\frac{v'}{v})^2}{2vI^2}\right\}^{-1}+o(1)$

194   *K. Takeuchi and M. Akahira*

$$= \frac{1}{I} + \frac{1}{2vI^2}(\frac{J+K}{I} + \frac{2v'}{v})^2 + o\,(1)\,.$$

Hence we have $V_\theta\,(\sqrt{v}(\hat{\theta}_n - \theta)) \geqq B^{11}$. This completes the proof.

In order to show that the modified ML estimation procedure attains the Bhattacharyya type bound given in the above theorem, we need the following lemma.

  LEMMA 2.2. Assume that the condition (A.1) holds.
If

$$\sqrt{\bar{v}}\,(\hat{\theta}_n - \theta) = \frac{1}{I(\theta)}Z_{1,v} + \frac{1}{\sqrt{\bar{v}}}Q + o_p(\frac{1}{\sqrt{\bar{v}}})\,,$$

then

$$V_\theta(\sqrt{\bar{v}}\,(\hat{\theta}_n - \theta)) = \frac{1}{I(\theta)} + \frac{1}{v}V_\theta(Q) + o\,(\frac{1}{v})\,,$$

where $Q = O_p(1)$.

  PROOF. Putting $T = \sqrt{v}\,(\hat{\theta}_n - \theta)$, we have

$$V_\theta(\sqrt{\bar{v}}\,(\hat{\theta}_n - \theta)) = E_\theta(T^2) = E_\theta[(T - \frac{Z_{1,v}}{I})^2 + \frac{2}{I}Z_{1,v}T - \frac{1}{I^2}Z^2_{1,v}]$$

$$= \frac{1}{v}E_\theta(Q^2) + \frac{2}{I(\theta)}E_\theta(Z_{1,v}T) - \frac{1}{I^2(\theta)}E_\theta(Z^2_{1,v})$$

Since by Lemma 2.1

$$E_\theta(Z_{1,v}T) = \frac{1}{\sqrt{\bar{v}}}\frac{\partial}{\partial\theta}E_\theta(T) - \frac{1}{\sqrt{\bar{v}}}E_\theta(\frac{\partial T}{\partial\theta})$$

$$= -\frac{1}{\sqrt{\bar{v}}}E_\theta[\frac{v'}{2\sqrt{\bar{v}}}(\hat{\theta}_n - \theta) + \sqrt{\bar{v}}]$$

$$= 1 + o(1),$$

the result follows.

  Denote by $\hat{\theta}_{ML}$ the maximum likelihood estimator of θ based on the sample $(X_1, \cdots, X_n)$. Let θ^*_{ML} be the ML estimator modified so that $E_\theta(\theta^*_{ML}) = \theta + o(v^{-1})$, and generally it automatically ensures that $E_\theta(\theta^*_{ML}) = \theta + o(v^{-3/2})$.

  THEOREM 2.2. If the stopping rule is so determined that sampling is stopped at n and satisfies

$$-\sum_{i=1}^{n} \ell^{(2)}(\hat{\theta}^*_{ML}, X_i) = v(\hat{\theta}^*_{ML})I(\hat{\theta}^*_{ML}) + c(\hat{\theta}^*_{ML}) + \varepsilon \tag{2.1}$$

with

$$c(\theta) = \frac{J(\theta)v'(\theta)}{I(\theta)v(\theta)} + \frac{L(\theta)}{I(\theta)} - \frac{v''(\theta)}{2v(\theta)} - \frac{1}{2I(\theta)}\{2L(\theta) + M(\theta) + N(\theta)\}$$

and some random variable ε with $E_\theta(\varepsilon) = o(1)$, then the asymptotic variance of $\hat{\theta}^*_{ML}$ is given by

$$V_\theta(\sqrt{\bar{v}}(\hat{\theta}^*_{ML} - \theta)) = \frac{1}{I(\theta)} + \frac{1}{2v(\theta)I^2(\theta)}\left\{\frac{J(\theta) + K(\theta)}{I(\theta)} + \frac{2v'(\theta)}{v(\theta)}\right\}^2 + o(\frac{1}{v(\theta)})\,,$$

which is equal to the Bhattacharyya type bound, that is, the modified maximum likelihood estimation procedure has the second order asymptotic efficiency property in terms of asymptotic variance.

PROOF. Since by (2.1)

$$-\sum_{i=1}^{n} \ell^{(2)}(\hat{\theta}_{ML}, X_i) = v(\hat{\theta}_{ML})I(\hat{\theta}_{ML}) + c(\hat{\theta}_{ML}),$$

it follows by the Taylor expansion that

$$-\sum_{i=1}^{n} \ell^{(2)}(\theta, X_i) - \left\{\sum_{i=1}^{n} \ell^{(3)}(\theta, X_i)\right\}(\hat{\theta}_{ML}-\theta) - \frac{1}{2}\left\{\sum_{i=1}^{n} \ell^{(4)}(\theta, X_i)\right\}(\hat{\theta}_{ML}-\theta)^2$$

$$= vI + (v'I + vI')(\hat{\theta}_{ML}-\theta) + \frac{1}{2}(v''I + 2v'I' + vI'')(\hat{\theta}_{ML}-\theta)^2 + c(\theta) + o_p(1), \tag{2.2}$$

where I' and v'' are defined by $I' = I'(\theta) = (d/d\theta)I(\theta)$ and $v'' = v''(\theta) = (\partial^2/\partial\theta^2)v(\theta)$. Since $I' = 2J + K$, we obtain from (2.2)

$$nI - \sqrt{v}\, Z_{2,v} + \{\frac{n}{\sqrt{v}}(3J+K) + Z_{3,v}\}\sqrt{v}\,(\hat{\theta}_{ML}-\theta)$$

$$+ \frac{n}{2v}(4L + 3M + 6N + H)v(\hat{\theta}_{ML}-\theta)^2 \qquad .$$

$$= vI + \frac{1}{\sqrt{v}}\{v'I + v(2J+K)\}\sqrt{v}\,(\hat{\theta}_{ML}-\theta)$$

$$+ \frac{1}{2v}(v''I + 2v'I' + vI'')v(\hat{\theta}_{ML}-\theta)^2 + c(\theta) + o_p(1),$$

which implies

$$(n-v)I = \sqrt{v}\,(Z_{2,v} - \frac{J}{I}Z_{1,v}) - \frac{3J+K}{I}\cdot\frac{n-v}{\sqrt{v}}Z_{1,v} + \frac{v'}{\sqrt{v}}Z_{1,v} - \frac{1}{I}Z_{1,v}Z_{3,v} - \frac{4L+3M+6N+H}{2I^2v}\cdot nZ_{1,v}^2$$

$$+ \frac{1}{2I^2v}\{v''I + 2v'(2J+K) + v(2L+2M+5N+H)\}Z_{1,v}^2 + c(\theta) + o_p(1), \tag{2.3}$$

since $\sqrt{v}(\hat{\theta}_{ML}-\theta) = (Z_{1,v}/I) + O_p(1/\sqrt{v})$, $J' = (d/d\theta)J(\theta) = L(\theta) + M(\theta) + N(\theta) = L + M + N$ and $K' = (d/d\theta)K(\theta) = H(\theta) + 3N(\theta) = H + 3N$.
In order to determine n so as to satisfy $E_\theta(n) = v + o(1)$, we have from (2.3)

$$c(\theta) = \frac{Jv'}{Iv} + \frac{L}{I} - \frac{v''}{2v} + \frac{1}{2I}(2L + M + N),$$

since $\quad L = L(\theta) = E_\theta[Z_{1,v}Z_{3,v}] + o(1), E_\theta(nZ_{1,v}) = v'/\sqrt{v}$ and $E_\theta(nZ_{1,v}^2) = vI + o(v)$.

We also obtain from (2.3)

$$\frac{n-v}{Iv}Z_{1,v} = \frac{Z_{1,v}}{I^2\sqrt{v}}(Z_{2,v} - \frac{J}{I}Z_{1,v}) + \frac{v'}{I^2v\sqrt{v}}Z_{1,v}^2 + o_p(\frac{1}{\sqrt{v}}). \tag{2.4}$$

On the other hand we have

$$0 = \frac{1}{\sqrt{v}}\sum_{i=1}^{n} \ell^{(1)}(\hat{\theta}_{ML}, X_i)$$

$$= \frac{1}{\sqrt{v}}\sum_{i=1}^{n} \ell^{(1)}(\theta, X_i) + \frac{1}{v}\left\{\sum_{i=1}^{n} \ell^{(2)}(\theta, X_i)\right\}\sqrt{v}\,(\hat{\theta}_{ML}-\theta)$$

196   *K. Takeuchi and M. Akahira*

$$+ \frac{1}{\nu^{3/2}}\left\{\sum_{i=1}^{n} \ell^{(3)}(\theta, X_i)\right\} \nu(\hat{\theta}_{ML} - \theta)^2 + o_p\left(\frac{1}{\sqrt{\nu}}\right)$$

$$= Z_{1,\nu} + \left(\frac{1}{\sqrt{\nu}} Z_{2,\nu} - \frac{nI}{\nu}\right)\sqrt{\nu}\,(\hat{\theta}_{ML} - \theta) - \frac{1}{2\sqrt{\nu}}(3J + K)\nu(\hat{\theta}_{ML} - \theta)^2 + o_p\left(\frac{1}{\sqrt{\nu}}\right)$$

$$= Z_{1,\nu} + \left\{\frac{1}{\sqrt{\nu}} Z_{2,\nu} - I - \frac{I(n - \nu)}{\nu}\right\}\sqrt{\nu}\,(\hat{\theta}_{ML} - \theta) - \frac{1}{2\sqrt{\nu}}(3J + K)\nu(\hat{\theta}_{ML} - \theta)^2 + o_p\left(\frac{1}{\sqrt{\nu}}\right)$$

which implies

$$\sqrt{\nu}\,(\hat{\theta}_{ML} - \theta) = \frac{1}{I} Z_{1,\nu} - \frac{n - \nu}{I\nu} Z_{1,\nu} + \frac{1}{I^2\sqrt{\nu}} Z_{1,\nu} Z_{2,\nu}$$

$$- \frac{3J + K}{2I^3\sqrt{\nu}} Z_{1,\nu}^2 + o_p\left(\frac{1}{\sqrt{\nu}}\right). \tag{2.5}$$

From (2.2), (2.4) and (2.5) we obtain

$$\sqrt{\nu}\,(\hat{\theta}_{ML} - \theta) = \frac{1}{I} Z_{1,\nu} - \frac{J + K}{2I^3\sqrt{\nu}} Z_{1,\nu}^2 - \frac{\nu'}{I^2\nu\sqrt{\nu}} Z_{1,\nu}^2 + o_p\left(\frac{1}{\sqrt{\nu}}\right),$$

hence, by Lemma 2.2,

$$V_\theta(\sqrt{\nu}\,(\hat{\theta}_{ML} - \theta)) = \frac{1}{I} + \frac{1}{2\nu I^2}\left(\frac{J + K}{I} + \frac{2\nu'}{\nu}\right)^2 + o\left(\frac{1}{\nu}\right).$$

Since the asymptotic variances of $\hat{\theta}_{ML}$ and $\hat{\theta}^*_{ML}$ are equal up to the order of $\nu-1$, the conclusion of the theorem follows.

REFERENCES

[1] Akahira, M. (1986). <u>The Structure of Asymptotic Deficiency of Estimators.</u> Queen's Papers in Pure and Applied Mathematics No.75, Queen's University Press, Kingston, Ontario, Canada.

[2] Akahira, M. and Takeuchi, K. (1981). <u>Asymptotic Efficiency of Statistical Estimators : Concepts and Higher Order Asymptotic Efficiency.</u> Lecture Notes in Statistics 7, Springer-Verlag, New York.

[3] Wald, A. (1973). <u>Sequential Analysis.</u> Dover Publications, Inc., New York.

University of Tokyo
University of Tsukuba

SECOND AND THIRD ORDER ASYMPTOTIC COMPLETENESS
OF THE CLASS OF ESTIMATORS*

Masafumi Akahira, Fumiko Hirakawa, and Kei Takeuchi

1. Introduction

The concept of higher order asymptotic efficiency of estimator $\hat{\theta}_n^*$ of a parameter θ in an open subset Θ of $\mathbb{R}^p$ is usually defined by the property that for any other estimator $\hat{\theta}_n$ satisfying some condition we have

$$\lim_{n \to \infty} n^{(k-1)/2} \left[P_{\theta,n} \left\{ \sqrt{n} \left(\hat{\theta}_n^* - \theta \right) \in C \right\} - P_{\theta,n} \left\{ \sqrt{n} \left(\hat{\theta}_n - \theta \right) \in C \right\} \right] \geq 0$$

for all $\theta \in \Theta$ and any convex set C containing the origin. And usually the condition imposed is k-th order asymptotically median unbiased and the like, and $\hat{\theta}_n^*$ is usually "modified" maximum likelihood estimator (MLE) (see [4], [6], [7] and [8]).

However, the condition of the asymptotic median unbiasedness is rather arbitrary, and a more meaningful property will be that of higher order "asymptotic completeness" which is defined as follows: $\hat{\theta}_n^*$ is called k-th order asymptotically complete if for any estimator $\hat{\theta}_n$ within some class of estimators we can construct

$$\hat{\theta}_n^{**} = \hat{\theta}_n^* + h_n \left(\hat{\theta}_n^* \right),$$

$\{h_n\}$ being a sequence of functions depending on $\hat{\theta}_n$ so that we have

$$\lim_{n \to \infty} n^{(k-1)/2} \left[P_{\theta,n} \left\{ \sqrt{n} \left(\hat{\theta}_n^{**} - \theta \right) \in C \right\} - P_{\theta,n} \left\{ \sqrt{n} \left(\hat{\theta}_n - \theta \right) \in C \right\} \right] \geq 0$$

for all $\theta \in \Theta$ and any convex set C containing the origin. Actually, the definition was given in the monograph [4] by Akahira and Takeuchi. But there the discussion was limited to the class of asymptotically median unbiased estimators in its application. The concept is also discussed in [10], but again the class of estimators was limited to that of regular functions of sufficient statistics.

The purpose of this paper is to show that, for any class of estimators which admit Edgeworth expansions but are not necessarily asymptotically median unbiased, we get the second order asymptotic completeness of the MLE, and the third order asymptotic completeness of the MLE $\hat{\theta}_{ML}$ together with the second order derivative of the log-likelihood function evaluated at the MLE which is denoted by $Z_2 \left(\hat{\theta}_{ML} \right)$ later.

---

*This paper is retyped with the correction of typographical errors.

12

It follows from this higher order asymptotic completeness property that, in any "decision theoretic" set-up, where we consider a sequence of loss functions of the type $L_n(u) \sim L^*(\sqrt{n}u)$ for a sufficiently large n, for any Bayes decision rule δ_n with smooth prior we can construct a $\delta_n^* = \hat{\theta}_{ML} + h_n\left(\hat{\theta}_{ML}, Z_2\left(\hat{\theta}_{ML}\right)\right)$, $\{h_n\}$ being a sequence of functions depending on $\hat{\theta}_n$, such that

$$\lim_{n \to \infty} n\left(E\left[L_n(\delta_n - \theta)\right] - E\left[L_n(\delta_n^* - \theta)\right]\right) \geq 0$$

for all $\theta \in \Theta$.

The concept of asymptotic completeness is related to that of asymptotic sufficiency. A sequence $\{T_n\}$ of statistics is called to be higher order asymptotically sufficient if for any sequence $\{T_n'\}$ we can construct $T_n'' = g_n(T_n', U_n)$, $\{g_n\}$ being a sequence of functions depending on T_n, where U_n is a random variable independent of θ, such that the asymptotic distribution of T_n'' coincides with that of T_n' up to the order $n^{-(k-1)/2}$. (The concept of higher order asymptotic sufficiency may be given in other ways but basically it amounts to the above.)

Asymptotic sufficiency, however, has generally been derived from the statement that we have

$$T_n' - T_n'' = o_p\left(n^{-k/2}\right),$$

which is actually much stronger than the equivalence of asymptotic distributions up to the order $n^{-(k-1)/2}$, where $X_n = o_p(Y_n)$ means that X_n/Y_n converges to zero in probability as $n \to \infty$. For example in Takeuchi [9], Bickel, Götze and van Zwet [5], it was discussed that for any Bayes decision procedure δ_n it is possible to get $\delta_n^* = g_n\left(\hat{\theta}_{ML}\right)$ such that $\delta_n^* - \delta_n = o_p\left(n^{-3/2}\right)$ only if the loss function is symmetric. However, when the loss is <u>not</u> symmetric, it was shown in [9] that even with Z_2 we can not construct $\delta_n^* = g_n\left(\hat{\theta}_{ML}, Z_2\left(\hat{\theta}_{ML}\right)\right)$ such that $\delta_n^* - \delta_n = o_p\left(n^{-3/2}\right)$. But, in order that δ_n^* and δ_n have asymptotically equivalent distributions up to the same order, hence have asymptotically the same risk, it is not necessary that δ_n^* and δ_n are stochastically equivalent up to the same order. As a simple illustration of this fact, let us assume that

$$\delta_n = \hat{\theta}_n + \frac{1}{n}\varepsilon_n,$$

where ε_n has the property that $E(\varepsilon_n|\hat{\theta}_n) = o(1)$. Then

$$\delta_n - \hat{\theta}_n = \frac{1}{n}O_p(\varepsilon_n) = O_p\left(\frac{1}{n}\right),$$

13

but

$$
\begin{aligned}
E\left[\exp(it\delta_n)\right] &= E\left[\exp\left\{it\left(\hat{\theta}_n + \frac{1}{n}\varepsilon_n\right)\right\}\right] \\
&= E\left[\exp\left(it\hat{\theta}_n\right)\right] + \frac{it}{n}E\left[\varepsilon_n\exp\left(it\hat{\theta}_n\right)\right] + o\left(\frac{1}{n}\right) \\
&= E\left[\exp\left(it\hat{\theta}_n\right)\right] + o\left(\frac{1}{n}\right),
\end{aligned}
$$

which means that the distributions of $\hat{\theta}_n$ and δ_n are asymptotically equivalent up to the order n^{-1}. Actually, from what is derived in the paper, it is shown that, under a sequence of decision problems with a proper set of regularity conditions, for any Bayes decision rule δ_n with some smooth prior there exists a $\delta_n^*\left(\hat{\theta}_{ML}, Z_2\left(\hat{\theta}_{ML}\right)\right)$ such that

$$
P_{\theta,n}\left\{\delta_n^* \in A\right\} = P_{\theta,n}\left\{\delta_n \in A\right\} + o\left(\frac{1}{n}\right)
$$

for all measurable set A in the decision space.

The main purpose of this paper is to establish that $\hat{\theta}_{ML}$ is generally second order asymptotically complete, and $\hat{\theta}_{ML}$ together with $Z_2\left(\hat{\theta}_{ML}\right)$ is third order asymptotically complete. Although this result is given in the framework of "point" estimation theory, it is also applicable to other problems of inference including testing hypothesis and interval estimation, which will be discussed in subsequent papers.

2. Preliminaries

Let $\mathcal{X}$ be an abstract space, elements of which are denoted by x, and let $\mathcal{B}$ be a σ-field of subsets of $\mathcal{X}$. Let Θ be a parameter space, which is assumed to be an open subset of the Euclidean p-space $\mathbb{R}^p$ with the usual norm denoted by $\|\cdot\|$. We shall denote by $(\mathcal{X}^{(n)}, \mathcal{B}^{(n)})$ the n-fold direct product of $(\mathcal{X}, \mathcal{B})$. We consider a sequence of classes of probability measures $\{P_{\theta,n} : \theta \in \Theta\}$ $(n = 1, 2, \ldots)$, each defined on $(\mathcal{X}^{(n)}, \mathcal{B}^{(n)})$, such that for each n and each $\theta \in \Theta$ the following holds:

$$
P_{\theta,n}\left(B^{(n)}\right) = P_{\theta,n+1}\left(B^{(n)} \times \mathcal{X}\right)
$$

for all $B^{(n)} \in \mathcal{B}^{(n)}$.

An estimator of θ is defined to be a sequence $\{\hat{\theta}_n\}$ where $\hat{\theta}_n$ is a $\mathcal{B}^{(n)}$-measurable function from $\mathcal{X}^{(n)}$ into Θ $(n = 1, 2, \ldots)$. For simplicity we denote an estimator as $\hat{\theta}_n$ instead of $\{\hat{\theta}_n\}$. For an increasing sequence of positive numbers $\{c_n\}$ (c_n tending to infinity) an estimator $\hat{\theta}_n$ is called to be c_n-consistent if for every η of Θ, there exists a sufficiently small positive number δ such that

14

$$\lim_{L\to\infty}\overline{\lim_{n\to\infty}}\sup_{\theta:||\theta-\eta||<\delta} P_{\theta,n}\left\{c_n\left\|\hat{\theta}_n-\theta\right\|\geq L\right\}=0 \quad\text{(Akahira [1])}.$$

3. One parameter case

Suppose that Θ is an open subset of $\mathbb{R}^1$. Let $X_1, X_2, \ldots, X_n, \ldots$ be a sequence of independently and identically distributed (i.i.d.) real random variables with a density function $f(x,\theta)$ with respect to a σ-finite measure μ, where $\theta\in\Theta$. In the subsequent discussion it is enough to consider only the case $c_n=\sqrt{n}$. For each $k=1,2,\ldots$, $0<\alpha(\theta)<1$ and a continuously differentiable function $a(\theta)$ of θ, a c_n-consistent estimator $\hat{\theta}_n$ is called k-th order asymptotically (α,a)-biased (as. (α,a)-biased) estimator if, for any $\eta\in\Theta$, there exists a positive number δ such that

$$\lim_{n\to\infty}\sup_{\theta:|\theta-\eta|<\delta} n^{(k-1)/2}\left|P_{\theta,n}\left\{\sqrt{nI(\theta)}\left(\hat{\theta}_n-\theta\right)\leq a(\theta)\right\}-\alpha(\theta)\right|=0,$$

$$\lim_{n\to\infty}\sup_{\theta:|\theta-\eta|<\delta} n^{(k-1)/2}\left|P_{\theta,n}\left\{\sqrt{nI(\theta)}\left(\hat{\theta}_n-\theta\right)\geq a(\theta)\right\}-1+\alpha(\theta)\right|=0,$$

where $I(\theta)$ denotes the Fisher information of f where $\alpha(\theta)$ and $a(\theta)$ are determined from the outset. Now we assume that $\alpha(\theta)$ is a constant α. Alternatively, we may fix $a(\theta)$ to be constant letting $\alpha(\theta)$ depend on θ, and the subsequent discussions apply exactly in a similar way.

For each $k=1,2,\ldots$ we denote by $B_k(\alpha,a)$ the class of k-th order as. (α,a)-biased estimators for which the distribution of $\sqrt{nI(\theta)}\left(\hat{\theta}_n-\theta\right)-a(\theta)$ admits the Edgeworth expansion up to the k-th order. If an estimator $\hat{\theta}_n$ belongs to the class $B_k(\alpha,a)$, then we call it a $B_k(\alpha,a)$-estimator.

For each $k=1,2,\ldots$, a $B_k(\alpha,a)$-estimator $\hat{\theta}_n^*$ is called k-th order asymptotically efficient in the class $B_k(\alpha,a)$ if for any $B_k(\alpha,a)$-estimator $\hat{\theta}_n$

$$P_{\theta,n}\left\{-t_1<\sqrt{nI(\theta)}\left(\hat{\theta}_n^*-\theta\right)-a(\theta)<t_2\right\}\geq P_{\theta,n}\left\{-t_1<\sqrt{nI(\theta)}\left(\hat{\theta}_n-\theta\right)-a(\theta)<t_2\right\}+o\left(n^{-(k-1)/2}\right)$$

for all $t_1, t_2>0$ and $\theta\in\Theta$.

We assume the following conditions.

(A.1) $\{x:f(x,\theta)>0\}$ does not depend on θ.

(A.2)$_k$ For almost all $x[\mu]$, $f(x,\theta)$ is k times continuously differentiable in θ. In the Taylor expansion

$$\log\frac{f(x,\theta+h)}{f(x,\theta)}=\sum_{i=1}^{k}\frac{h^i}{i!}\ell^{(i)}(\theta,x)+h^k R(x,h)$$

$R(x,h)$ is uniformly bounded by a function $G(x)$ which has moments up to the k-th order, where

$$\ell^{(i)}(\theta, x) = (\partial^i / \partial\theta^i)\ell(\theta, x) \quad (i = 1, \ldots, k)$$

with $\ell(\theta, x) = \log f(x, \theta)$.

(A.3) For each $\theta \in \Theta$

$$0 < I(\theta) = E_\theta \left[\left\{ \ell^{(1)}(\theta, X) \right\}^2 \right] = -E_\theta \left[\ell^{(2)}(\theta, X) \right] < \infty,$$

and $I(\theta)$ is differentiable in θ.

(A.4) There exist

$$J(\theta) = E_\theta \left[\ell^{(1)}(\theta, X)\ell^{(2)}(\theta, X) \right],$$

$$K(\theta) = E_\theta \left[\left\{ \ell^{(1)}(\theta, X) \right\}^3 \right],$$

and both $J(\theta)$ and $K(\theta)$ are differentiable in θ, and

$$E_\theta \left[\ell^{(3)}(\theta, X) \right] = -3J(\theta) - K(\theta).$$

In the following theorem we obtain the asymptotic distribution of a $B_2(\alpha, a)$-estimator up to the second, i.e., the order $n^{-1/2}$.

<u>Theorem 3.1.</u> Assume that the conditions (A.1), (A.2)$_3$, (A.3), and (A.4) hold and that the asymptotic cumulants of a biased best asymptotically normal (BAN) estimator $\hat{\theta}_n$ are given as follows:

For $T_n = \sqrt{n}\left(\hat{\theta}_n - \theta \right)$

$$E_\theta(T_n) = C_0(\theta) + \frac{C_1(\theta)}{\sqrt{n}} + o\left(\frac{1}{\sqrt{n}} \right);$$

$$V_\theta(T_n) = \frac{1}{I(\theta)} + \frac{C_2(\theta)}{\sqrt{n}} + o\left(\frac{1}{\sqrt{n}} \right);$$

$$\kappa_{3,\theta}(T_n) = \frac{C_3(\theta)}{\sqrt{n}} + o\left(\frac{1}{\sqrt{n}} \right);$$

where $C_i(\theta)$ ($i = 0, 1, 2, 3$) are differentiable functions of θ. Then the second order asymptotic distribution of the second order as. (α, a)-biased BAN estimator $\hat{\theta}_n^*$, of the form $\hat{\theta}_n^* = \hat{\theta}_n - n^{-1/2}u_1\left(\hat{\theta}_n \right) - n^{-1}u_2\left(\hat{\theta}_n \right)$, is given by

$$(3.1) \qquad P_{\theta,n}\left\{ \sqrt{nI(\theta)}\left(\hat{\theta}_n^* - \theta \right) \leq t + a(\theta) \right\}$$

$$= \Phi(t + \nu_\alpha) + \frac{1}{\sqrt{n}}\phi(t + \nu_\alpha)\left\{ -\frac{1}{6}I^{3/2}(\theta)C_3(\theta)t^2 \right.$$

$$- \frac{1}{3}I^{3/2}(\theta)C_3(\theta)\nu_\alpha t + t\left(C_0'(\theta) - \frac{I(\theta)C_2(\theta)}{2} - \frac{a'(\theta)}{\sqrt{I(\theta)}} \right.$$

$$\left. \left. + \frac{2J(\theta) + K(\theta)}{2I^{3/2}(\theta)}(a(\theta) - \nu_\alpha) \right) \right\} + o\left(\frac{1}{\sqrt{n}} \right),$$

where $\Phi(u) = \int_{-\infty}^u \phi(x)dx$ with $\phi(x) = (1/\sqrt{2\pi})e^{-x^2/2}$ and $\Phi(\nu_\alpha) = \alpha$.

16

Remark: In the above theorem, $u_1(\theta)$ and $u_2(\theta)$ are given as follows:

$$u_1(\theta) = C_0(\theta) - \frac{a(\theta) - \nu_\alpha}{I(\theta)};$$

$$u_2(\theta) = C_1(\theta) - C_0(\theta)u_1'(\theta) + \frac{1}{2}\sqrt{I(\theta)}\left\{C_2(\theta) - \frac{2u_1'(\theta)}{I(\theta)}\right\}[a(\theta)$$
$$- \sqrt{I(\theta)}\{C_0(\theta) - u_1(\theta)\}] + \frac{1}{6}I(\theta)C_3(\theta)[\{a(\theta)$$
$$- \sqrt{I(\theta)}(C_0(\theta) - u_1(\theta))\}^2 - 1].$$

In the following theorem we shall obtain the bound for the asymptotic distributions of $B_2(\alpha, a)$-estimators up to the order $n^{-1/2}$.

Theorem 3.2. Assume that the conditions (A.1), (A.2)$_3$, (A.3), and (A.4) hold. Then the bound $F^*(t, \theta)$ for the asymptotic distributions of $\sqrt{nI(\theta)}\left(\hat{\theta}_n - \theta\right)$ for the all $B_2(\alpha, a)$-estimators $\hat{\theta}_n$ is given by

$$F^*(t, \theta) = \Phi(t + \nu_\alpha) + \frac{1}{\sqrt{n}}\phi(t + \nu_\alpha)\left[\frac{\{3J(\theta) + 2K(\theta)\}t^2}{6I^{3/2}(\theta)}\right.$$
$$\left. - \frac{t}{\sqrt{I(\theta)}}\left\{a'(\theta) - \frac{a(\theta)(2J(\theta) + K(\theta))}{2I(\theta)} - \frac{\nu_\alpha K(\theta)}{6I(\theta)}\right\}\right]$$
$$+ o\left(\frac{1}{\sqrt{n}}\right)$$

in the sense that for any $B_2(\alpha, a)$-estimator $\hat{\theta}_n$

$$P_{\theta,n}\left\{\sqrt{nI(\theta)}\left(\hat{\theta}_n - \theta\right) \leq t + a(\theta)\right\} \leq F^*(t, \theta)$$

for all $t > 0$ and all $\theta \in \Theta$,

$$P_{\theta,n}\left\{\sqrt{nI(\theta)}\left(\hat{\theta}_n - \theta\right) \leq t + a(\theta)\right\} \geq F^*(t, \theta)$$

for all $t < 0$ and all $\theta \in \Theta$.

From Theorems 3.1 and 3.2 we have the following:

Theorem 3.3. Under the conditions (A.1), (A.2)$_3$, (A.3), and (A.4) it holds that for the biased BAN estimator $\hat{\theta}_n^*$

$$C_3(\theta) = -\frac{3J(\theta) + 2K(\theta)}{I^3(\theta)}, \quad C_0'(\theta) \leq \frac{I(\theta)C_2(\theta)}{2} \quad \text{for all } \theta \in \Theta,$$

and further if $C_0'(\theta) \equiv I(\theta)C_2(\theta)/2$, the second order as. (α, a)-biased BAN estimator $\hat{\theta}_n^*$ is second order asymptotically efficient in the class $B_2(\alpha, a)$.

Sketches of the proofs of Theorems 3.1, 3.2 and 3.3 are given below, since their detailed ones are done in Akahira [2].

Sketch of the proof of Theorem 3.1.

Let $\hat{\theta}_n^* = \hat{\theta}_n - \dfrac{u_1\left(\hat{\theta}_n\right)}{\sqrt{n}} - \dfrac{u_2\left(\hat{\theta}_n\right)}{n},$

17

where $u_1(\theta)$ is continuously differentiable in θ and $u_2(\theta)$ is a function of θ. Then we have the following asymptotic cumulants of $T_n^* = \sqrt{nI(\theta)}\left(\hat{\theta}_n^* - \theta\right)$:

$$E_\theta(T_n^*) = \sqrt{I}(C_0 - u_1) + \sqrt{I/n}(C_1 - C_0 u_1' - u_2) + o\left(\frac{1}{\sqrt{n}}\right); \tag{3.2}$$

$$V_\theta(T_n^*) = 1 + \frac{I}{\sqrt{n}}\left(C_2 - \frac{2u_1'}{I}\right) + o\left(\frac{1}{\sqrt{n}}\right); \tag{3.3}$$

$$\kappa_{3,\theta}(T_n^*) = \frac{1}{\sqrt{n}}I^{3/2}C_3 + o\left(\frac{1}{\sqrt{n}}\right), \tag{3.4}$$

where $I, C_0, u_1, u_1', u_2, C_2$ and C_3 denote $I(\theta)$, $C_0(\theta)$, $u_1(\theta)$, $u_1'(\theta) = du_1(\theta)/d\theta$, $u_2(\theta)$, $C_2(\theta)$ and $C_3(\theta)$.

From (3.2) to (3.4) it follows that the Edgeworth expansion of the distribution of $\sqrt{nI(\theta)}\left(\hat{\theta}_n^* - \theta\right)$ is given by

$$P_{\theta,n}\left\{\sqrt{nI}\left(\hat{\theta}_n^* - \theta\right) \le t + \sqrt{I}(C_0 - u_1)\right\} \tag{3.5}$$

$$= \Phi(t) - \frac{1}{\sqrt{n}}\phi(t)\left\{\sqrt{I}(C_1 - C_0 u_1' - u_2) + \frac{I}{2}\left(C_2 - \frac{2}{I}u_1'\right)t + \frac{1}{6}I^{3/2}C_3(t^2 - 1)\right\}$$

$$+ o\left(\frac{1}{\sqrt{n}}\right).$$

By the second order as. (α, a)-biased condition we have the following u_1 and u_2:

$$u_1 = C_0 - \frac{a - \nu_\alpha}{I}; \tag{3.6}$$

$$u_2 = C_1 - C_0 u_1' + \frac{\sqrt{I}}{2}\left(C_2 - \frac{2u_1'}{I}\right)\{a - \sqrt{I}(C_0 - u_1)\} \tag{3.7}$$

$$+ \frac{IC_3}{6}[\{a - \sqrt{I}(C_0 - u_1)\}^2 - 1],$$

where $\Phi(\nu_\alpha) = \alpha$.

Since $I'(\theta) = 2J(\theta) + K(\theta)$, it follows from (3.5), (3.6) and (3.7) that (3.1) holds. This completes the proof.

<u>Sketch of the proof of Theorem 3.2.</u> Let θ_0 be arbitrary but fixed in Θ. Consider the problem of testing the hypothesis H : $\theta = \theta_1$ against alternative A : $\theta = \theta_0$, where $\theta_1 = \theta_0 + O(1/\sqrt{n})$. In order to obtain the upper bound of

$$P_{\theta_0,n}\left\{\sqrt{nI(\theta_0)}\left(\hat{\theta}_n - \theta_0\right) \le t + a(\theta)\right\}$$

for each $t > 0$ and all $B_2(\alpha, a)$-estimators $\hat{\theta}_n$, that is, under the second order as. (α, a)-biased condition

$$P_{\theta_1,n}\left\{\sqrt{nI(\theta_1)}\left(\hat{\theta}_n - \theta_1\right) \le a(\theta_1)\right\} = \alpha + o\left(\frac{1}{\sqrt{n}}\right),$$

18

we first take

$$\theta_1 = \theta_0 + \frac{t}{\sqrt{nI}}\left[1 - \frac{1}{\sqrt{nI}}\left\{a' - \frac{a(2J+K)}{2I}\right\}\right],$$

where a, a', I, J and K denote $a(\theta_0)$, $a'(\theta_0)$, $I(\theta_0)$, $J(\theta_0)$ and $K(\theta_0)$.

Setting

$$T_n = \sum_{i=1}^{n} \log \frac{f(x_i, \theta_0)}{f(x_i, \theta_1)},$$

by the Edgeworth expansion of the distribution of T_n we take the following a_n so that the test with a rejection region $\{T_n \geq a_n\}$ has the level $\alpha + o(1/\sqrt{n})$:

$$a_n = -\frac{t^2}{2} - t\nu_\alpha + \frac{t}{\sqrt{nI}}\left\{a' - \frac{a(2J+k)}{2I}\right\}(t + \nu_\alpha)$$
$$- \frac{t}{6\sqrt{n}I^{3/2}}\left\{(3J+2K)t^2 + 3(J+K)\nu_\alpha t + K(\nu_\alpha^2 - 1)\right\} + o\left(\frac{1}{\sqrt{n}}\right).$$

In a similar way to the above we have the asymptotic power of the test with the rejection region $\{T_n \geq a_n\}$ as follows:

$$(3.8) \qquad P_{\theta_0, n}\{T_n \geq a_n\}$$
$$= \Phi(t + \nu_\alpha) + \frac{1}{\sqrt{n}}\phi(t + \nu_\alpha)\left[\frac{(3J+2K)t^2}{6I^{3/2}} - \frac{t^2}{\sqrt{I}}\left\{a' - \frac{a(2J+K)}{2I} - \frac{\nu_\alpha K}{6I}\right\}\right]$$
$$+ o\left(\frac{1}{\sqrt{n}}\right)$$
$$= F^*(t, \theta_0) \quad \text{(say)}.$$

By the fundamental lemma of Neyman and Pearson it is seen that the asymptotic power series of the most powerful test of the level $\alpha + o(1/\sqrt{n})$ is given by (3.8). Since θ_0 is arbitrary, we have the desired result for all $t > 0$. In a similar way to the case $t > 0$, we also obtain the conclusion for all $t < 0$.

Sketch of the proof of Theorem 3.3. By Theorems 3.1 and 3.2 we have

$$(3.9) \qquad F^*(t, \theta) - P_{\theta, n}\left\{\sqrt{nI(\theta)}\left(\hat{\theta}_n^* - \theta\right) \leq t + a(\theta)\right\}$$
$$= \frac{1}{\sqrt{n}}\phi(t + \nu_\alpha)\left\{\frac{3J(\theta) + 2K(\theta) + I^3 C_3(\theta)}{6I^{3/2}(\theta)}(t^2 + 2\nu_\alpha t) - \left(C_0' - \frac{IC_2}{2}\right)t\right\}$$
$$+ o\left(\frac{1}{\sqrt{n}}\right).$$

If one assumes that $3J(\theta_0) + 2K(\theta_0) + I^3(\theta_0)C_3(\theta_0) \neq 0$ for some $\theta_0 \in \Theta$, it leads to a contradiction to the bound $F^*(t, \theta)$. Hence it follows that

19

$$(3.10) \qquad C_3(\theta) = -\frac{3J(\theta) + 2K(\theta)}{I^3(\theta)} \quad \text{for all } \theta \in \Theta.$$

Since $F^*(t, \theta)$ is the bound for the asymptotic distributions of $\sqrt{nI(\theta)}\left(\hat{\theta}_n - \theta\right)$ for all $B_2(\alpha, a)$-estimators $\hat{\theta}_n$, it follows from (3.9) and (3.10) that

$$C_0'(\theta) \leq I(\theta)C_2(\theta)/2 \quad \text{for all } \theta \in \Theta.$$

If $C_0'(\theta) \equiv I(\theta)C_2(\theta)/2$, then it is seen that the asymptotic distribution of $\sqrt{nI(\theta)}\left(\hat{\theta}_n^* - \theta\right)$ attains the bound $F^*(t, \theta)$ for all t and all $\theta \in \Theta$. Hence the desired result follows.

Let $\hat{\theta}_{ML}$ be the maximum likelihood estimator (MLE) of θ. Then the second order asymptotic efficiency of the MLE is given as follows:

Theorem 3.4. Assume that the conditions (A.1), (A.2)$_3$, (A.3), and (A.4) hold. Then the estimator $\hat{\theta}_{ML}^*$ modified from the MLE to be in $B_2(\alpha, a)$ is second order asymptotically efficient in the class $B_2(\alpha, a)$.

The proof follows from Theorems 3.1 and 3.3 since $C_0(\theta) = C_2(\theta) \equiv 0$ and $C_3(\theta) \equiv -\{3J(\theta) + 2K(\theta)\}/I^3(\theta)$ in the asymptotic cumulants of the MLE. (See Akahira and Takeuchi [4], page 90).

Theorem 3.5. Assume that the conditions (A.1), (A.2)$_3$, (A.3), and (A.4) hold. Suppose that $\hat{\theta}_n$ is any BAN estimator of which the Edgeworth expansion up to the order $n^{-1/2}$ is valid and the coefficients $C_i(\theta)$ ($i = 0, 1, 2, 3$) in its asymptotic cumulants as in Theorem 3.1 are continuously differentiable in θ. Then the MLE is second order asymptotically complete in the sense that there exists a modified MLE $\hat{\theta}_{ML}^*$, of the form

$$\hat{\theta}_{ML}^* = \hat{\theta}_{ML} + \frac{1}{\sqrt{n}}g_1\left(\hat{\theta}_{ML}\right) + \frac{1}{n}g_2\left(\hat{\theta}_{ML}\right),$$

such that

$$P_{\theta,n}\left\{-t_1 < \sqrt{nI(\theta)}\left(\hat{\theta}_{ML}^* - \theta\right) - a(\theta) < t_2\right\}$$
$$\geq P_{\theta,n}\left\{-t_1 < \sqrt{nI(\theta)}\left(\hat{\theta}_n - \theta\right) - a(\theta) < t_2\right\} + o\left(\frac{1}{\sqrt{n}}\right)$$

for all $t_1, t_2 > 0$ and all $\theta \in \Theta$.

Sketch of the proof. For fixed α, define $a_n(\theta)$ by

$$P_{\theta,n}\left\{\sqrt{nI(\theta)}\left(\hat{\theta}_n - \theta\right) \leq a_n(\theta)\right\} = \alpha + o\left(\frac{1}{\sqrt{n}}\right),$$
$$P_{\theta,n}\left\{\sqrt{nI(\theta)}\left(\hat{\theta}_n - \theta\right) \geq a_n(\theta)\right\} = 1 - \alpha + o\left(\frac{1}{\sqrt{n}}\right),$$

locally uniformly in θ, then $a_n(\theta)$ can be expanded as $a_n(\theta) =$

$a(\theta) + n^{-1/2}b(\theta) + o\left(n^{-1/2}\right)$, where $a(\theta)$ and $b(\theta)$ are continuously differentiable. Consider the class $B_2(\alpha, a_n)$ with $a_n(\theta)$ thus defined. Analogously to the proof of Theorem 3.4 we can show that within the class there exists a modified MLE $\hat{\theta}^*_{ML}$, of the form

$$\hat{\theta}^*_{ML} = \hat{\theta}_{ML} + \frac{1}{\sqrt{n}}g_1\left(\hat{\theta}_{ML}\right) + \frac{1}{n}g_2\left(\hat{\theta}_{ML}\right),$$

which is second order asymptotically efficient in the class $B_2(\alpha, a)$. This implies the desired result.

For the third order asymptotic completeness we further assume the following:

(A.5) There exist $E_\theta\left[\ell^{(1)}(\theta, X)\ell^{(3)}(\theta, X)\right]$, $E_\theta\left[\{\ell^{(2)}(\theta, X)\}^2\right]$,
$E_\theta\left[\{\ell^{(1)}(\theta, X)\}^2\ell^{(2)}(\theta, X)\right]$ and $E_\theta\left[\{\ell^{(1)}(\theta, X)\}^4\right]$.

<u>Theorem 3.6.</u> Assume that the conditions (A.1), (A.2)$_4$, (A.3), (A.4), and (A.5) hold. Suppose that $\hat{\theta}_n$ is second order asymptotically efficient in $B_2(\alpha, a)$, of which the Edgeworth expansion up to the order n^{-1} is valid. Then the pair of statistics $\left(\hat{\theta}_{ML}, Z_2\left(\hat{\theta}_{ML}\right)\right)$ is third order asymptotically complete in the sense that there exists a modified MLE $\hat{\theta}^*_{ML}$, of the form

$$\hat{\theta}^*_{ML} = \hat{\theta}_{ML} + \frac{1}{\sqrt{n}}g_1\left(\hat{\theta}_{ML}\right) + \frac{1}{n}g_2\left(\hat{\theta}_{ML}, Z_2\left(\hat{\theta}_{ML}\right)\right) + \frac{1}{n\sqrt{n}}g_3\left(\hat{\theta}_{ML}, Z_2\left(\hat{\theta}_{ML}\right)\right),$$

such that

$$P_{\theta,n}\left\{-t_1 < \sqrt{nI(\theta)}\left(\hat{\theta}^*_{ML} - \theta\right) - a(\theta) < t_2\right\}$$

$$\geq P_{\theta,n}\left\{-t_1 < \sqrt{nI(\theta)}\left(\hat{\theta}_n - \theta\right) - a(\theta) < t_2\right\} + o\left(\frac{1}{n}\right)$$

for all $t_1, t_2 > 0$ and $\theta \in \Theta$, where $Z_2(\theta) = \sum_{i=1}^n \ell^{(2)}(\theta, X_i)/\sqrt{n} + \sqrt{n}I(\theta)$.

<u>Sketch of the proof.</u> The asymptotic cumulants κ_i $(i = 1, 2, 3, 4)$ of $T_n = \sqrt{n}\left(\hat{\theta}_n - \theta\right)$ have the following form:

$$\kappa_1 = E_\theta(T_n) = \mu_{10}(\theta) + \frac{\mu_{11}(\theta)}{\sqrt{n}} + \frac{\mu_{12}(\theta)}{n} + o\left(\frac{1}{n}\right),$$

$$\kappa_2 = V_\theta(T_n) = \frac{1}{I(\theta)} + \frac{\mu_{21}(\theta)}{\sqrt{n}} + \frac{\mu_{22}(\theta)}{n} + o\left(\frac{1}{n}\right),$$

$$\kappa_3 = \kappa_{3,\theta}(T_n) = E_\theta\left[\{T_n - E_\theta(T_n)\}^3\right]$$

$$= \frac{\mu_{31}(\theta)}{\sqrt{n}} + \frac{\mu_{32}(\theta)}{n} + o\left(\frac{1}{n}\right),$$

21

$$\kappa_4 = \kappa_{4,\theta}(T_n) = E_\theta\left[\{T_n - E_\theta(T_n)\}^4\right] - 3\{V_\theta(T_n)\}^2$$

$$= \frac{\mu_{42}(\theta)}{n} + o\left(\frac{1}{n}\right).$$

Then it is seen that the coefficients $\mu_{10}(\theta)$, $\mu_{11}(\theta)$, $\mu_{21}(\theta)$, $\mu_{31}(\theta)$ and $\mu_{42}(\theta)$ in the above are determined from the second order asymptotic efficiency of $\hat{\theta}_n$, but $\mu_{22}(\theta)$ and $\mu_{32}(\theta)$ are not so (e.g. see Akahira and Takeuchi, [4]). And it follows that, in the Edgeworth expansion of the distribution $P_{\theta,n}\left\{\sqrt{nI(\theta)}\left(\hat{\theta}_n - \theta\right) - a(\theta) \leq t\right\}$, the only undecided term has the form $3\mu_{32}(\theta) + \{\mu_{22}(\theta)/\sqrt{I(\theta)}\}w(t)$, where $w(t)$ is some linear function of t. It follows by the discretized likelihood method in Akahira and Takeuchi [3], [4] that for any $\hat{\theta}_n \in B_2(\alpha, a)$ with $\mu_{32}(\theta)$ and $\mu_{22}(\theta)$ in its asymptotic cumulants, any fixed $\theta_0 \in \Theta$ and each real t, there exists a second order as. (α, a)-biased estimator $\hat{\theta}^t$, with $\mu_{32}^t(\cdot)$ and $\mu_{22}^t(\cdot)$ in its asymptotic cumulants, such that

$$(3.11) \qquad 3\mu_{32}^t(\theta_0) + \frac{\mu_{22}^t(\theta_0)}{I(\theta_0)}w(t) \leq 3\mu_{32}(\theta_0) + \frac{\mu_{22}(\theta_0)}{I(\theta_0)}w(t).$$

For any given $\hat{\theta}_n$, let $\mu_{22}^0(\theta)$ be the corresponding $\mu_{22}(\theta)$, then it is shown that there exists $t_0 = t_0(\theta_0)$ such that $\mu_{22}^0(\theta_0) = \mu_{22}^{t_0}(\theta_0)$. Since, by (3.8), for any $\mu_{32}^0(\cdot)$

$$3\mu_{32}^{t_0}(\theta_0) + \frac{\mu_{22}^{t_0}(\theta_0)}{\sqrt{I(\theta_0)}}w(t_0) \leq 3\mu_{32}^0(\theta_0) + \frac{\mu_{22}^0(\theta_0)}{\sqrt{I(\theta_0)}}w(t_0),$$

it follows that

$$(3.12) \qquad \mu_{32}^{t_0}(\theta_0) \leq \mu_{32}^0(\theta_0).$$

From (3.11) and (3.12) we have

$$3\mu_{32}^{t_0}(\theta_0) + \frac{\mu_{22}^0(\theta_0)}{\sqrt{I(\theta_0)}}w(t) \leq 3\mu_{32}^0(\theta_0) + \frac{\mu_{22}^0(\theta_0)}{\sqrt{I(\theta_0)}}w(t)$$

for all real t.

Hence, for any second order asymptotically efficient estimator $\hat{\theta}_n$ in $B_2(\alpha, a)$

$$(3.13) \qquad P_{\theta_0,n}\left\{-t_1 < \sqrt{n}\left(\hat{\theta}^{t_0} - \theta_0\right) - a(\theta_0) < t_2\right\}$$

$$\geq P_{\theta_0,n}\left\{-t_1 < \sqrt{n}\left(\hat{\theta}_n - \theta_0\right) - a(\theta_0) < t_2\right\} + o\left(\frac{1}{n}\right)$$

for all $t_1, t_2 > 0$.

22

Now define

$$\hat{\theta}_{ML}^0 = \hat{\theta}_{ML} + \frac{1}{\sqrt{n}}h_1\left(\hat{\theta}_{ML}\right) + \frac{1}{n}h_2\left(\hat{\theta}_{ML}\right) + \frac{1}{n\sqrt{n}}h_3\left(\hat{\theta}_{ML}\right)$$

such that $E_\theta\left(\hat{\theta}_{ML}^0\right)$ is asymptotically equal to $E_\theta\left(\hat{\theta}_n\right)$ up to the order n^{-1}, then $\hat{\theta}_{ML}^0$ and $\hat{\theta}_n$ have the same values for $\mu_{10}(\theta)$, $\mu_{11}(\theta)$, $\mu_{21}(\theta)$, $\mu_{31}(\theta)$ and $\mu_{42}(\theta)$, but the coefficient $\hat{\theta}_{32}^*(\theta)$ of the order n^{-1} in the third order cumulant for $\hat{\theta}_{ML}^0$ is identically equal to zero.

By the discretized likelihood method in Akahira and Takeuchi [3], [4] it follows that

$$\hat{\theta}^{t_0} = \hat{\theta}_{ML}^0 + \frac{t_0}{2nI(\theta_0)}\left\{Z_2(\theta_0) - \frac{J(\theta_0)}{I(\theta_0)}Z_1(\theta_0)\right\} + o_p\left(\frac{1}{n\sqrt{n}}\right).$$

Now define

$$\hat{\theta}^{\hat{t}_0} = \hat{\theta}_{ML}^0 + \frac{\hat{t}_0}{2nI\left(\hat{\theta}_{ML}\right)}Z_2\left(\hat{\theta}_{ML}\right),$$

where $\hat{t}_0 = t_0\left(\hat{\theta}_{ML}\right)$.

Then it can be shown that

$$(3.14) \qquad \hat{\theta}^{\hat{t}_0} = \hat{\theta}^{t_0} + \frac{1}{n\sqrt{n}}W + o_p\left(\frac{1}{n\sqrt{n}}\right),$$

where W is some stochastic quantity of magnitude of order 1 and $E_{\theta_0}(W) = o(1)$. Since $\hat{\theta}^{t_0}$ can be stochastically expanded as

$$\sqrt{n}\left(\hat{\theta}^{t_0} - \theta_0\right) = \frac{1}{I(\theta_0)}Z_1(\theta_0) + \frac{1}{\sqrt{n}}Q + \frac{1}{n}R + o_p\left(\frac{1}{n}\right),$$

it follows from (3.14) that

$$\sqrt{n}\left(\hat{\theta}^{\hat{t}_0} - \theta_0\right) = \frac{1}{I(\theta_0)}Z_1(\theta_0) + \frac{1}{\sqrt{n}}Q + \frac{1}{n}(R + W) + o_p\left(\frac{1}{n}\right),$$

where both Q and R are certain stochastic quantities of magnitude of order 1. Then it is shown in Akahira and Takeuchi [4] that the asymptotic distributions of the above two estimators $\hat{\theta}^{t_0}$ and $\hat{\theta}^{\hat{t}_0}$ coincide up to the order n^{-1} if $E_{\theta_0}(W) = 0$. Hence it follows from (3.13) that for any second order asymptotically efficient estimator $\hat{\theta}_n$ in $B_2(\alpha, a)$

$$P_{\theta_0,n}\left\{-t_1 < \sqrt{n}\left(\hat{\theta}^{\hat{t}_0} - \theta_0\right) - a(\theta_0) < t_2\right\}$$

$$\geq P_{\theta_0,n}\left\{-t_1 < \sqrt{n}\left(\hat{\theta}_n - \theta_0\right) - a(\theta_0) < t_2\right\} + o\left(\frac{1}{n}\right)$$

for all $t_1, t_2 > 0$.

23

Since $\hat{\theta}^{i_0}$ is independent of θ_0 and θ_0 is arbitrary, we have the conclusion of the theorem.

4. Multiparameter case

Assume that Θ is an open subset of $\mathbb{R}^p$. Let $X_1, X_2, \ldots, X_n, \ldots$ be a sequence of i.i.d. real random variables with a density function $f(x, \theta)$ with respect to a σ-finite measure μ, where $\theta \in \Theta$. Denote by $\hat{\theta}_n = \left(\hat{\theta}_{n1}, \ldots, \hat{\theta}_{np}\right)'$ an estimator of $\theta = (\theta_1, \ldots, \theta_p)' \in \Theta$. We assume that the distribution of $\sqrt{n}\left(\hat{\theta}_n - \theta\right)$ admits the Edgeworth expansion up to the order n^{-1}, and $\hat{\theta}_n$ has the stochastic expansion

$$\sqrt{n}\left(\hat{\theta}_{n\alpha} - \theta_\alpha\right) = U_\alpha + \frac{1}{2\sqrt{n}}Q_\alpha + \frac{1}{n}W_\alpha + o_p\left(\frac{1}{n}\right) \quad (\alpha = 1, \ldots, p),$$

with $U_\alpha = \sum_{\beta=1}^p I^{\alpha\beta}(\theta) \sum_{i=1}^n (\partial/\partial\theta_\beta) \log f(X_i, \theta)/\sqrt{n}$, $Q_\alpha = O_p(1)$ and $W_\alpha = O_p(1)$ $(\alpha = 1, \ldots, p)$, whose asymptotic cumulants are the following:

for $Y_\alpha = \sqrt{n}\left(\hat{\theta}_{n\alpha} - \theta_\alpha\right)$ $(\alpha = 1, \ldots, p)$

$$E_\theta(Y_\alpha) = \mu_{0\alpha}(\theta) + \frac{\mu_{1\alpha}(\theta)}{\sqrt{n}} + \frac{\mu_{2\alpha}(\theta)}{n} + o\left(\frac{1}{n}\right);$$

$$\mathrm{Cov}(Y_\alpha, Y_\beta) = I^{\alpha\beta}(\theta) + \frac{C^{(1)}_{\alpha\beta}(\theta)}{\sqrt{n}} + \frac{C^{(2)}_{\alpha\beta}(\theta)}{n} + o\left(\frac{1}{n}\right);$$

$$\kappa_{3,\theta}(Y_\alpha, Y_\beta, Y_\gamma) = \frac{\kappa^{(1)}_{\alpha\beta\gamma}(\theta)}{\sqrt{n}} + \frac{\kappa^{(2)}_{\alpha\beta\gamma}(\theta)}{n} + o\left(\frac{1}{n}\right);$$

$$\kappa_{4,\theta}(Y_\alpha, Y_\beta, Y_\gamma, Y_\delta) = \frac{\kappa_{\alpha\beta\gamma\delta}(\theta)}{n} + o\left(\frac{1}{n}\right)$$

for $\alpha, \beta, \gamma, \delta = 1, \ldots, p$, where $I^{\alpha\beta}(\theta)$ is an (α, β)-element of the inverse matrix of the Fisher information matrix.

For each $k = 1, 2, \ldots$, an estimator $\hat{\theta}_n^*$ is called k-th order asymptotically efficient in the class A_k of the estimators $\hat{\theta}_n$ with the same bias

$$E_\theta\left(\hat{\theta}_{n\alpha}\right) = \theta_\alpha + \sum_{i=0}^{k-1} \mu_{i\alpha}(\theta)n^{-(i+1)/2} + o\left(n^{-k/2}\right) \quad (\alpha = 1, \ldots, p)$$

up to the order $n^{-k/2}$ if for any estimator $\hat{\theta}_n \in A_k$

$$P_{\theta,n}\left\{\sqrt{n}\left(\hat{\theta}_n^* - \theta\right) - \mu_0(\theta) \in C\right\} \geq P_{\theta,n}\left\{\sqrt{n}\left(\hat{\theta}_n - \theta\right) - \mu_0(\theta) \in C\right\} + o\left(n^{-(k-1)/2}\right)$$

for any convex set C of $\mathbb{R}^p$ containing the origin and all $\theta \in \Theta$.

We shall use the following notations: for $\alpha, \beta, \gamma, \delta = 1, \ldots, p$

24

$$J_{\alpha\beta\cdot\gamma} = E_\theta\left[\left\{\frac{\partial^2}{\partial\theta_\alpha\partial\theta_\beta}\log f(X,\theta)\right\}\left\{\frac{\partial}{\partial\theta_\gamma}\log f(X,\theta)\right\}\right];$$

$$K_{\alpha\beta\gamma} = E_\theta\left[\left\{\frac{\partial}{\partial\theta_\alpha}\log f(X,\theta)\right\}\left\{\frac{\partial}{\partial\theta_\beta}\log f(X,\theta)\right\}\left\{\frac{\partial}{\partial\theta_\gamma}\log f(X,\theta)\right\}\right];$$

$$J^{\alpha\beta\cdot\gamma} = \sum_{\alpha'=1}^{p}\sum_{\beta'=1}^{p}\sum_{\gamma'=1}^{p} I^{\alpha\alpha'} I^{\beta\beta'} I^{\gamma\gamma'} J_{\alpha'\beta'\cdot\gamma'};$$

$$K^{\alpha\beta\gamma} = \sum_{\alpha'=1}^{p}\sum_{\beta'=1}^{p}\sum_{\gamma'=1}^{p} I^{\alpha\alpha'} I^{\beta\beta'} I^{\gamma\gamma'} K_{\alpha'\beta'\gamma'};$$

$$M_{\alpha\beta\cdot\gamma\delta} = E_\theta\left[\left\{\frac{\partial^2}{\partial\theta_\alpha\partial\theta_\beta}\log f(X,\theta)\right\}\left\{\frac{\partial^2}{\partial\theta_\gamma\partial\theta_\delta}\log f(X,\theta)\right\}\right] - I_{\alpha\beta}I_{\gamma\delta};$$

$$M^{\alpha\beta\cdot\gamma\delta} = \sum_{\alpha'}\sum_{\beta'}\sum_{\gamma'}\sum_{\delta'} I^{\alpha\alpha'} I^{\beta\beta'} I^{\gamma\gamma'} I^{\delta\delta'} M_{\alpha'\beta'\cdot\gamma'\delta'};$$

$$J^{\alpha\beta}_{\cdot\gamma} = \sum_{\alpha'}\sum_{\beta'} I^{\alpha\alpha'} I^{\beta\beta'} J_{\alpha'\beta'\cdot\gamma}; \quad J^{\alpha\cdot\gamma}_{\beta} = \sum_{\alpha'}\sum_{\gamma'} I^{\alpha\alpha'} I^{\gamma\gamma'} J_{\alpha'\beta\cdot\gamma'};$$

$$K^{\alpha\beta}_{\gamma} = \sum_{\alpha'}\sum_{\beta'} I^{\alpha\alpha'} I^{\beta\beta'} K_{\alpha'\beta'\gamma}.$$

<u>**Theorem 4.1.**</u> Assume that $\hat{\theta}_n$ is any asymptotically efficient estimator in the class of the estimators with the same bias $\mu_{0\alpha}(\theta)$ ($\alpha = 1,\ldots,p$) up to the order $n^{-1/2}$ in the above, which admits the Edgeworth expansion of its distribution up to the order $n^{-1/2}$. Then the MLE $\hat{\theta}_{ML}$ is second order asymptotically complete in the sense that there exists a modified MLE $\hat{\theta}^*_{ML}$, with the same asymptotic bias as $\hat{\theta}_n$ up to the order n^{-1}, of the form

$$\hat{\theta}^*_{ML} = \hat{\theta}_{ML} + \frac{1}{\sqrt{n}}\underset{\sim}{g_1}\left(\hat{\theta}_{ML}\right) + \frac{1}{n}\underset{\sim}{g_2}\left(\hat{\theta}_{ML}\right),$$

such that

$$P_{\theta,n}\left\{\sqrt{n}\left(\hat{\theta}^*_{ML} - \theta\right) - \mu_0(\theta) \in C\right\} \geq P_{\theta,n}\left\{\sqrt{n}\left(\hat{\theta}_n - \theta\right) - \mu_0(\theta) \in C\right\} + o\left(\frac{1}{\sqrt{n}}\right)$$

for any convex set C of $\mathbb{R}^p$ containing the origin and all $\theta \in \Theta$, where $\mu_0(\theta)$ $= (\mu_{01}(\theta),\ldots,\mu_{0p}(\theta))'$.

The proof is omitted since it is done in a similar way to that of Theorem 3.4.

<u>**Theorem 4.2.**</u> Assume that $\hat{\theta}_n$ is second order asymptotically efficient in the class of the estimators with the same bias up to the order n^{-1} in the above, which admits the Edgeworth expansion of its distribution up to the order n^{-1}. Then the pair of statistics $\left(\hat{\theta}_{ML}, \underset{\sim}{Z_2}\left(\hat{\theta}_{ML}\right)\right)$ is third order asymptotically complete in the sense that there exists a modified MLE $\hat{\theta}^*_{ML}$, with the same asymptotic bias as $\hat{\theta}_n$ up to the order $n^{-3/2}$, of the form

25

$$\hat{\theta}^*_{ML} = \hat{\theta}_{ML} + \frac{1}{\sqrt{n}} \underset{\sim}{g}_1\left(\hat{\theta}_{ML}\right) + \frac{1}{n} \underset{\sim}{g}_2\left(\hat{\theta}_{ML}, \underset{\sim}{Z}_2\left(\hat{\theta}_{ML}\right)\right) + \frac{1}{n\sqrt{n}} \underset{\sim}{g}_3\left(\hat{\theta}_{ML}, \underset{\sim}{Z}_2\left(\hat{\theta}_{ML}\right)\right)$$

such that

$$P_{\theta,n}\left\{\sqrt{n}\left(\hat{\theta}^*_{ML} - \theta\right) - \mu_0(\theta) \in C\right\} \geq P_{\theta,n}\left\{\sqrt{n}\left(\hat{\theta}_n - \theta\right) - \mu_0(\theta) \in C\right\} + o\left(\frac{1}{n}\right)$$

for any convex set C of $\mathbb{R}^p$ containing the origin and all $\theta \in \Theta$, where

$$\underset{\sim}{Z}_2(\theta) = \frac{1}{\sqrt{n}} \sum_{i=1}^{n} \frac{\partial^2}{\partial\theta\partial\theta'} \log f(X_i, \theta) + \sqrt{n}\underset{\sim}{I}(\theta)$$

with the Fisher information matrix $\underset{\sim}{I}(\theta)$.

$\underline{\text{Sketch of the proof.}}$ First it is noted that the coefficients $\mu_{0\alpha}(\theta)$, $\mu_{1\alpha}(\theta)$, $C^{(1)}_{\alpha\beta}(\theta)$, $\kappa^{(1)}_{\alpha\beta\gamma}(\theta)$ and $\kappa_{\alpha\beta\gamma\delta}(\theta)$ $(\alpha, \beta, \gamma, \delta = 1, \ldots, p)$ in the above asymptotic cumulants of $\hat{\theta}_n$ are determined from the second order asymptotic efficiency of $\hat{\theta}_n$. In order to obtain $\hat{\theta}^*_{ML}$ with more concentration probability than $\hat{\theta}_n$ up to the order n^{-1} by the Edgeworth expansion of the distribution of $\hat{\theta}_n$, it is enough to find an estimator $\hat{\theta}_n = \hat{\theta}^*_n$ which minimizes $C^{(2)}_{\alpha\beta}(\theta) = \text{Cov}_\theta(Q_\alpha, Q_\beta)$, given $\kappa^{(2)}_{\alpha\beta\gamma}(\theta)$. Putting $V_{\alpha\beta} = \sum_{\alpha=1}^{p} I^{\alpha\beta}(\partial/\partial\theta_\gamma)U_\beta$ with $I^{\alpha\beta} = I^{\alpha\beta}(\theta)$ $(\alpha, \beta = 1, \ldots, p)$, we have for $\alpha, \beta, \gamma = 1, \ldots, p$

$$(4.1) \qquad \kappa^{(2)}_{\alpha\beta\gamma}(\theta) = E_\theta\left(Q_\alpha V_{\beta\gamma}\right) + E_\theta\left(Q_\beta V_{\gamma\alpha}\right) + E_\theta\left(Q_\gamma V_{\alpha\beta}\right).$$

We also have for $\alpha, \beta, \gamma, \delta = 1, \ldots, p$,

$$(4.2) \qquad E_\theta\left(U_\beta U_\gamma Q_\alpha\right) = 2\left(\mu_{1\alpha}I^{\beta\gamma} - K^{\alpha\beta\gamma} - J^{\beta\gamma\cdot\alpha}\right) + o(1);$$

$$(4.3) \qquad E_\theta\left(U_\beta V_{\gamma\delta} Q_\alpha\right) = 2\left\{-\mu_{1\alpha}\left(K^{\gamma\delta\beta} + J^{\gamma\beta\cdot\delta}\right) + \sum_{\delta'=1}^{p}\left(K^{\beta\alpha}_{\delta'} + J^{\beta\cdot\alpha}_{\delta'}\right)\right.$$
$$\left. \cdot\left(K^{\gamma\delta\cdot\delta'} + J^{\gamma\delta'\cdot\delta}\right) - \sum_{\delta'=1}^{p} J^{\beta\alpha}_{\cdot\delta'} J^{\gamma\delta\cdot\delta'} + M^{\beta\alpha\cdot\gamma\delta}\right\}$$
$$+ o(1),$$

where for each $\alpha = 1, \ldots, p$, $\mu_{1\alpha}$ denotes $\mu_{1\alpha}(\theta)$.

From the second order asymptotic efficiency of $\hat{\theta}_n$ we obtain

$$(4.4) \qquad E_\theta\left(U_\alpha Q_\alpha\right) = o(1) \quad (\alpha = 1, \ldots, p).$$

Then it follows by the Lagrange method that, under the conditions (4.1) to (4.4) and $E_\theta\left(Q_\alpha\right) = 2\mu_{1\alpha} + o(1)$, $Q_\alpha = Q^*_\alpha$ minimizing $\text{Cov}_\theta\left(Q_\alpha, Q_\beta\right)$ have the following: for $\alpha = 1, \ldots, p$

26

$$(4.5) \qquad Q_\alpha^* = 2\mu_{1\alpha} + \sum_{\beta=1}^{p} \sum_{\gamma=1}^{p} \eta_\alpha^{\beta\gamma}(\theta) \left\{ V_{\beta\gamma} - \sum_{\beta=1}^{p} \sum_{\gamma=1}^{p} \mu_{\beta\gamma}^{\alpha}(\theta) U_\alpha \right\}$$

$$+ \sum_{\beta=1}^{p} \sum_{\gamma=1}^{p} \lambda_\alpha^{\beta\gamma}(\theta) \left(U_\beta U_\gamma - I^{\beta\gamma} \right) + \sum_{\beta=1}^{p} \sum_{\gamma=1}^{p} \sum_{\delta=1}^{p} \lambda_\alpha^{\beta\gamma\delta}(\theta)$$

$$\cdot \left(U_\beta V_{\gamma\delta} + K^{\alpha\beta\gamma} + J^{\alpha\gamma\cdot\beta} \right),$$

where $\eta_\alpha^{\beta\gamma}(\theta)$, $\mu_{\beta\gamma}^{\alpha}(\theta)$, $\lambda_\alpha^{\beta\gamma}(\theta)$ and $\lambda_\alpha^{\beta\gamma\delta}(\theta)$ $(\alpha, \beta, \gamma, \delta = 1, \ldots, p)$ are Lagrangian multipliers determined so that the conditions are satisfied. Hence it is seen the desired estimator $\hat{\theta}_n^*$ has Q_α^* in its stochastic expansion.

Substituting $\hat{\theta}_{ML}$ for θ in (4.5) and making it to have the same asymptotic bias as $\hat{\theta}_n$ up to the order $n^{-3/2}$, we obtain the desired result.

References

[1] M. Akahira: Asymptotic theory for estimation of location in non-regular cases, I: Order of convergence of consistent estimators, *Rep. Stat. Appl. Res., JUSE* **22** (1975), 8–26.

[2] M. Akahira: *The structure of asymptotic deficiency of estimators*, to appear in Queen's papers in Pure and Applied Mathematics.

[3] M. Akahira and K. Takeuchi: Discretized likelihood methods—Asymptotic properties of discretized likelihood estimators (DLE's), *Ann. Inst. Statist. Math.*, **31** (1979), 39–56.

[4] M. Akahira and K. Takeuchi: *Asymptotic Efficiency of Statistical Estimators: Concepts and Higher Order Asymptotic Efficiency*, Lecture Notes in Statistics 7, Springer-Verlag, New York-Heidelberg-Berlin (1981).

[5] P. J. Bickel, F. Götze and W. R. van Zwet: A simple analysis of third-order efficiency of estimates, Proceedings of the Berkeley Conference in Honor of Jerzy Neyman and Jack Kiefer, Vol. II (L. M. LeCam and R. A. Olshen eds.), Wadsworth (1985), 749–768.

[6] J. K. Ghosh, B. K. Sinha and H. S. Wieand: Second order efficiency of the mle with respect to any bounded bowl-shaped loss function, *Ann. Statist.*, **8** (1980), 506–521.

[7] J. Pfanzagl and W. Wefelmeyer: A third order optimum property of the maximum likelihood estimator, *J. Multivariate Anal.*, **8** (1978), 1–29.

[8] J. Pfanzagl and W. Wefelmeyer: *Asymptotic Expansions for General Statistical Models*, Lecture Notes in Statistics 31, Springer-Verlag, Berlin-Heidelberg-New York-Tokyo (1985).

27

[9] K. Takeuchi: Higher order asymptotic efficiency of estimators in decision procedures, *Proc. III Purdue Symp. on Decision Theory and Related Topics*, Vol. II, (J. Berger and S. Gupta eds.), Academic Press, New York (1982), 351–361.

[10] K. Takeuchi and K. Morimune: Third-order efficiency of the extended maximum likelihood estimators in a simultaneous equation system, *Econometrica* **53** (1985), 177–200.

M. Akahira
Department of Mathematics
University of Electro-
Communications
Chofu, Tokyo 182
Japan

F. Hirakawa
Department of Mathematics
Science University of Tokyo
Yamazaki, Noda-shi
Chiba 278, Japan

K. Takeuchi
Faculty of Economics
University of Tokyo
Hongo, Bunkyo-ku
Tokyo 113, Japan

Ann. Inst. Statist. Math.
Vol. 41, No. 4, 725–752 (1989)

HIGHER ORDER ASYMPTOTICS IN ESTIMATION FOR TWO-SIDED WEIBULL TYPE DISTRIBUTIONS

MASAFUMI AKAHIRA[1] AND KEI TAKEUCHI[2]

[1]*Institute of Mathematics, University of Tsukuba, Ibaraki 305, Japan*
[2]*Research Center for Advanced Science and Technology, University of Tokyo,
Komaba, Meguro-ku, Tokyo 156, Japan*

(Received May 18, 1988; revised January 25, 1989)

Abstract.   We consider the estimation problem of a location parameter θ on a sample of size n from a two-sided Weibull type density $f(x - \theta) = C(\alpha) \exp\left(-|x - \theta|^{\alpha}\right)$ for $-\infty < x < \infty$, $-\infty < \theta < \infty$ and $1 < \alpha < 3/2$, where $C(\alpha) = \alpha/\{2\Gamma(1/\alpha)\}$. Then the bound for the distribution of asymptotically median unbiased estimators is obtained up to the 2α-th order, i.e., the order $n^{-(2\alpha-1)/2}$. The asymptotic distribution of a maximum likelihood estimator (MLE) is also calculated up to the 2α-th order. It is shown that the MLE is not 2α-th order asymptotically efficient. The amount of the loss of asymptotic information of the MLE is given.

Key words and phrases: 2α-th order asymptotically median unbiased estimator, 2α-th order asymptotic distribution, 2α-th order asymptotic efficiency, Edgeworth expansion, maximum likelihood estimator.

1.   Introduction

Higher order asymptotics have been studied by Pfanzagl and Wefelmeyer (1978, 1985), Ghosh *et al.* (1980) and Akahira and Takeuchi (1981), Akahira (1986), Akahira *et al.* (1988), among others, under suitable regularity conditions.

In non-regular cases when the regularity conditions do not necessarily hold, higher order asymptotics were discussed by Akahira and Takeuchi (1981), Akahira (1987, 1988*a*, 1988*b*), Pfanzagl and Wefelmeyer (1985), Sugiura and Naing (1987) and others.

In this paper we consider the estimation problem of a location parameter θ on a sample of size n from a two-sided Weibull type density $f(x - \theta) = C(\alpha) \exp\left(-|x - \theta|^{\alpha}\right)$ for $-\infty < x < \infty$, $-\infty < \theta < \infty$ and $1 < \alpha < 3/2$, where $C(\alpha) = \alpha/\{2\Gamma(1/\alpha)\}$. It is noted that there is a Fisher information amount and a first order derivative of $f(x)$ at $x = 0$, but there is no second order one of $f(x)$ at $x = 0$. It is also seen in Akahira (1975)

726   MASAFUMI AKAHIRA AND KEI TAKEUCHI

that the order of consistency is equal to $n^{-1/2}$ in this situation. Then we shall obtain the bound for the distribution of asymptotically median unbiased estimators of θ up to the 2α-th order, i.e., the order $n^{-(2\alpha-1)/2}$. We shall also get the asymptotic distribution of the maximum likelihood estimator (MLE) of θ up to the 2α-th order and see that the MLE is not generally 2α-th order asymptotically efficient. Further, we shall obtain the amount of the loss of asymptotic information of the MLE.

2. The 2α-th order asymptotic bound for the distribution of second order AMU estimators

Let $X_1,\ldots,X_n,\ldots$ be a sequence of independent and identically distributed (i.i.d.) random variables with a two-sided Weibull type density $f(x-\theta) = C(\alpha) \exp\{-|x-\theta|^\alpha\}$ for $-\infty < x < \infty$ where θ is a real-valued parameter, $1 < \alpha < 3/2$ and $C(\alpha) = \alpha/\{2\Gamma(1/\alpha)\}$ with a Gamma function $\Gamma(u)$, i.e., $\Gamma(u) = \int_0^\infty x^{u-1}e^{-x}dx$ $(u > 0)$.

We denote by $P_{\theta,n}$ the n-fold products of probability measure P_θ with the above density $f(x-\theta)$. An estimator $\hat{\theta}_n$ of θ based on $X_1,\ldots,X_n$ is called a 2α-th order asymptotically median unbiased (AMU) estimator if for any $\eta \in R^1$, there exists a positive number δ such that

$$\lim_{n\to\infty} \sup_{\theta:|\theta-\eta|<\delta} n^{(2\alpha-1)/2} \left| P_{\theta,n}\{\hat{\theta}_n \leq \theta\} - \frac{1}{2} \right| = 0,$$

$$\lim_{n\to\infty} \sup_{\theta:|\theta-\eta|<\delta} n^{(2\alpha-1)/2} \left| P_{\theta,n}\{\hat{\theta}_n \geq \theta\} - \frac{1}{2} \right| = 0.$$

We denote by $A_{2\alpha}$ the class of all best asymptotically normal and 2α-th order AMU estimators. For a $\hat{\theta}_n$ 2α-th order AMU, $G_0(t,\theta) + n^{-(2\alpha-1)/2}G_1(t,\theta)$ is defined to be the 2α-th order asymptotic distribution of $\sqrt{n}(\hat{\theta}_n - \theta)$ (or $\hat{\theta}_n$ for short) if for any $t \in R^1$ and each $\theta \in R^1$

$$\lim_{n\to\infty} n^{(2\alpha-1)/2} | P_{\theta,n}\{\sqrt{n}(\hat{\theta}_n - \theta) \leq t\} - G_0(t,\theta) - n^{-(2\alpha-1)/2}G_1(t,\theta)| = 0.$$

In order to obtain the bound for the distribution of 2α-th order AMU estimators of θ, for arbitrary but fixed θ_0, we consider the problem of testing hypothesis $H: \theta = \theta_0 + tn^{-1/2}$ $(t > 0)$ against the alternative $K: \theta = \theta_0$. Then the log-likelihood ratio test statistic Z_n is given by

$$Z_n = \sum_{i=1}^n \log\{f(X_i - \theta_0)/f(X_i - \theta_0 - tn^{-1/2})\}$$

$$= -\sum_{i=1}^n (|X_i - \theta_0|^\alpha - |X_i - \theta_0 - tn^{-1/2}|^\alpha).$$

In order to obtain the asymptotic cumulants of Z_n, we need the following lemma.

LEMMA 2.1. *If $h_\Delta(x) = (x + \Delta)^\alpha - x^\alpha$ for $\Delta > 0$, then*

$$\int_0^\infty h_\Delta^2(x)e^{-x^\alpha}dx = \alpha\Gamma\left(2 - \frac{1}{\alpha}\right)\Delta^2 + \alpha(\alpha - 1)\Gamma\left(2 - \frac{2}{\alpha}\right)\Delta^3$$

$$- \frac{1 + \gamma}{2\alpha + 1}\Delta^{2\alpha + 1} + o(\Delta^{2\alpha + 1}),$$

$$\int_0^\infty h_\Delta^3(x)e^{-x^\alpha}dx = \alpha^2\Gamma\left(3 - \frac{2}{\alpha}\right)\Delta^3 + O(\Delta^4),$$

where

$$\gamma = \frac{\alpha(\alpha - 1)\Gamma(\alpha - 1)\Gamma(3 - 2\alpha)}{2(2\alpha - 1)\Gamma(2 - \alpha)}.$$

PROOF. First we have

$$(2.1) \qquad \int_0^\infty h_\Delta^2(x)e^{-x^\alpha}dx = \int_0^\infty (x + \Delta)^{2\alpha}e^{-x^\alpha}dx - 2\int_0^\infty (x + \Delta)^\alpha x^\alpha e^{-x^\alpha}dx$$

$$+ \int_0^\infty x^{2\alpha}e^{-x^\alpha}dx.$$

Since for $\beta > 0$

$$(2.2) \qquad \int_0^\infty x^{\beta - 1}e^{-x^\alpha}dx = \frac{1}{\alpha}\Gamma\left(\frac{\beta}{\alpha}\right),$$

it follows that

$$(2.3) \qquad \int_0^\infty (x + \Delta)^{2\alpha}e^{-x^\alpha}dx$$

$$= -\frac{\Delta^{2\alpha + 1}}{2\alpha + 1} + \frac{\alpha}{2\alpha + 1}\int_0^\infty (x + \Delta)^{2\alpha + 1}x^{\alpha - 1}e^{-x^\alpha}dx$$

$$= -\frac{\Delta^{2\alpha + 1}}{2\alpha + 1} + \frac{\alpha}{2\alpha + 1}$$

$$\cdot \int_0^\infty \left\{ x^{3\alpha} + (2\alpha + 1)x^{3\alpha - 1}\Delta + \alpha(2\alpha + 1)x^{3\alpha - 2}\Delta^2 \right.$$

$$\left. + \frac{1}{3}\alpha(2\alpha - 1)(2\alpha + 1)x^{3\alpha - 3}\Delta^3 \right\} e^{-x^\alpha}dx + O(\Delta^4)$$

$$= \frac{1}{2\alpha + 1} \Gamma\left(3 + \frac{1}{\alpha}\right) + 2\Delta + \alpha\Gamma\left(3 - \frac{1}{\alpha}\right)\Delta^2$$

$$+ \frac{1}{3}\alpha(2\alpha - 1)\Gamma\left(3 - \frac{2}{\alpha}\right)\Delta^3 - \frac{1}{2\alpha + 1}\Delta^{2\alpha+1} + O(\Delta^4) \ .$$

From (2.2), we obtain

$$\int_0^\infty (x + \Delta)^\alpha x^\alpha e^{-x^\alpha} dx$$

$$= -\frac{\alpha}{\alpha + 1}\int_0^\infty (x + \Delta)^{\alpha+1}(x^{\alpha-1} - x^{2\alpha-1})e^{-x^\alpha}dx$$

$$= \frac{\alpha}{\alpha + 1}\int_0^\infty \left\{ x^{\alpha+1} + (\alpha + 1)x^\alpha\Delta + \frac{1}{2}\alpha(\alpha + 1)x^{\alpha-1}\Delta^2 \right.$$

$$\left. + \frac{1}{6}\alpha(\alpha - 1)(\alpha + 1)x^{\alpha-2}\Delta^3 \right\} (x^{2\alpha-1} - x^{\alpha-1})e^{-x^\alpha}dx$$

$$- \frac{\alpha}{\alpha + 1}\int_0^\infty R(\Delta)x^{\alpha-1}e^{-x^\alpha}dx + O(\Delta^4) \ ,$$

where

$$R(\Delta) = (x + \Delta)^{\alpha+1} - x^{\alpha+1} - (\alpha + 1)\Delta x^\alpha - \frac{1}{2}\alpha(\alpha + 1)\Delta^2 x^{\alpha-1}$$

$$- \frac{1}{6}\alpha(\alpha - 1)(\alpha + 1)\Delta^3 x^{\alpha-2} \ .$$

Then the remainder term $R(\Delta)$ of the Taylor expansion is represented by

$$R(\Delta) = K_\alpha \int_0^\Delta (\Delta - t)^3(x + t)^{\alpha-3}dt \ ,$$

where $0 \le t \le \Delta$ and $K_\alpha = \alpha(\alpha + 1)(\alpha - 1)(\alpha - 2)/6$. Since $1 - x^\alpha < e^{-x^\alpha} < 1$, it follows that

$$\int_0^\infty (x + t)^{\alpha-3}x^{\alpha-1}(1 - e^{-x^\alpha})dx \le \int_0^\infty x^{2\alpha-4}(1 - e^{-x^\alpha})dx$$

$$= \int_0^1 x^{2\alpha-4}(1 - e^{-x^\alpha})dx + \int_1^\infty x^{2\alpha-4}(1 - e^{-x^\alpha})dx$$

$$\le \int_0^1 x^{3\alpha-4}dx + \int_1^\infty x^{2\alpha-4}dx$$

$$= \frac{\alpha}{3(\alpha - 1)(3 - 2\alpha)} \, .$$

Since $\int_0^\infty (x + t)^{\alpha - 3} x^{\alpha - 1} dx = t^{2\alpha - 3} B(\alpha, 3 - 2\alpha)$, we have

$$\int_0^\infty R(\Delta) x^{\alpha - 1} e^{-x^\alpha} dx = \int_0^\infty R(\Delta) x^{\alpha - 1} dx - \int_0^\infty R(\Delta) x^{\alpha - 1} (1 - e^{-x^\alpha}) dx$$

$$= K_\alpha \int_0^\Delta (\Delta - t)^3 \left\{ \int_0^\infty (x + t)^{\alpha - 3} x^{\alpha - 1} dx \right\} dt + o(\Delta^{2\alpha + 1})$$

$$= K_\alpha B(\alpha, 3 - 2\alpha) \int_0^\Delta t^{2\alpha - 3} (\Delta - t)^3 dt + o(\Delta^{2\alpha + 1})$$

$$= K_\alpha B(\alpha, 3 - 2\alpha) B(2\alpha - 2, 4) \Delta^{2\alpha + 1} + o(\Delta^{2\alpha + 1})$$

$$= - \frac{(\alpha + 1)(\alpha - 1)\Gamma(\alpha - 1)\Gamma(3 - 2\alpha)}{4(2\alpha + 1)(2\alpha - 1)\Gamma(2 - \alpha)} \Delta^{2\alpha + 1} + o(\Delta^{2\alpha + 1}) \, .$$

Hence, we obtain

$$(2.4) \qquad \int_0^\infty (x + \Delta)^\alpha x^\alpha e^{-x^\alpha} dx$$

$$= \frac{1}{\alpha + 1} \left\{ \Gamma\left(3 + \frac{1}{\alpha}\right) - \Gamma\left(2 + \frac{1}{\alpha}\right) \right\}$$

$$+ \Delta + \frac{\alpha}{2} \left\{ \Gamma\left(3 - \frac{1}{\alpha}\right) - \Gamma\left(2 - \frac{1}{\alpha}\right) \right\} \Delta^2$$

$$+ \frac{1}{6} \alpha(\alpha - 1) \left\{ \Gamma\left(3 - \frac{2}{\alpha}\right) - \Gamma\left(2 - \frac{2}{\alpha}\right) \right\} \Delta^3$$

$$+ \frac{\alpha(\alpha - 1)\Gamma(\alpha - 1)\Gamma(3 - 2\alpha)}{4(2\alpha + 1)(2\alpha - 1)\Gamma(2 - \alpha)} \Delta^{2\alpha + 1} + o(\Delta^{2\alpha + 1})$$

$$= \frac{1}{\alpha} \Gamma\left(2 + \frac{1}{\alpha}\right) + \Delta + \left(\frac{\alpha - 1}{2}\right) \Gamma\left(2 - \frac{1}{\alpha}\right) \Delta^2$$

$$+ \frac{1}{6} (\alpha - 1)(\alpha - 2)\Gamma\left(2 - \frac{2}{\alpha}\right) \Delta^3$$

$$+ \frac{\alpha(\alpha - 1)\Gamma(\alpha - 1)\Gamma(3 - 2\alpha)}{4(2\alpha + 1)(2\alpha - 1)\Gamma(2 - \alpha)} \Delta^{2\alpha + 1} + o(\Delta^{2\alpha + 1}) \, ,$$

and by (2.2), $\int_0^\infty x^{2\alpha} e^{-x^\alpha} dx = \Gamma(2 + 1/\alpha)/\alpha$. From (2.1) to (2.4), we have

730   MASAFUMI AKAHIRA AND KEI TAKEUCHI

$$\int_0^\infty h_\Delta^2(x)e^{-x^\alpha}dx = \alpha\Gamma\left(2 - \frac{1}{\alpha}\right)\Delta^2 + \alpha(\alpha - 1)\Gamma\left(2 - \frac{2}{\alpha}\right)\Delta^3$$

$$- \frac{1 + \gamma}{2\alpha + 1}\Delta^{2\alpha+1} + o(\Delta^{2\alpha+1}),$$

where $\gamma = \alpha(\alpha - 1)\Gamma(\alpha - 1)\Gamma(3 - 2\alpha)/2(2\alpha - 1)\Gamma(2 - \alpha)$. We also obtain

$$\int_0^\infty h_\Delta^3(x)e^{-x^\alpha}dx = \int_0^\infty \{(x + \Delta)^\alpha - x^\alpha\}^3 e^{-x^\alpha}dx$$

$$= \alpha^3\Delta^3\int_0^\infty x^{3\alpha-3}e^{-x^\alpha}dx + O(\Delta^4)$$

$$= \alpha^2\Gamma\left(3 - \frac{2}{\alpha}\right)\Delta^3 + O(\Delta^4).$$

Thus we complete the proof.

In the following lemma we obtain the asymptotic mean, variance and third-order cumulant of Z_n, under H and K.

LEMMA 2.2.   *The asymptotic mean, variance and third-order cumulant of Z_n are given as follows: Under K: $\theta = \theta_0$,*

$$E_{\theta_0}(Z_n) = \frac{I}{2}t^2 - \frac{k}{2}t^{2\alpha+1}n^{-(2\alpha-1)/2} + o(n^{-(2\alpha-1)/2}),$$

$$V_{\theta_0}(Z_n) = It^2 - kt^{2\alpha+1}n^{-(2\alpha-1)/2} + o(n^{-(2\alpha-1)/2}),$$

$$K_{3,\theta_0}(Z_n) = o(n^{-(2\alpha-1)/2}),$$

and under H: $\theta = \theta_0 + tn^{-1/2}$

$$E_{\theta_0 + tn^{-1/2}}(Z_n) = -\frac{I}{2}t^2 + \frac{k}{2}t^{2\alpha+1}n^{-(2\alpha-1)/2} + o(n^{-(2\alpha-1)/2}),$$

$$V_{\theta_0 + tn^{-1/2}}(Z_n) = It^2 - kt^{2\alpha+1}n^{-(2\alpha-1)/2} + o(n^{-(2\alpha-1)/2}),$$

$$K_{3,\theta_0 + tn^{-1/2}}(Z_n) = o(n^{-(2\alpha-1)/2}),$$

where

$$I = E_\theta\left[\left\{\frac{\partial}{\partial\theta}\log f(X - \theta)\right\}^2\right]$$

$$= -E_\theta\left[\frac{\partial^2}{\partial\theta^2}\log f(X - \theta)\right] = \alpha(\alpha - 1)\Gamma(1 - (1/\alpha))/\Gamma(1/\alpha)$$

and

$$k = \alpha \left\{ B(\alpha + 1, \alpha + 1) + \frac{\gamma}{2\alpha + 1} \right\} \Big/ \Gamma(1/\alpha)$$

with

$$B(u, v) = \int_0^1 x^{u-1}(1 - x)^{v-1} dx \quad (u, v > 0) .$$

PROOF. Without loss of generality, we assume that $\theta_0 = 0$. Putting $\Psi_\Delta(x) = |x|^\alpha - |x - \Delta|^\alpha$ with $\Delta > 0$, we have $Z_n = - \sum_{i=1}^{n} \Psi_\Delta(X_i)$ where $\Delta = tn^{-1/2}$. Since

$$(2.5) \qquad \Psi_\Delta(x) = \begin{cases} x^\alpha - (x - \Delta)^\alpha & \text{for} \quad x \geq \Delta , \\ x^\alpha - (\Delta - x)^\alpha & \text{for} \quad 0 \leq x < \Delta , \\ (-x)^\alpha - (\Delta - x)^\alpha & \text{for} \quad x < 0 , \end{cases}$$

it follows that

$$(2.6) \qquad E_0[\Psi_\Delta(X)] = C(\alpha)\left[-\int_0^\infty \{(x + \Delta)^\alpha - x^\alpha\}e^{-x^\alpha}dx \right.$$

$$+ \int_0^\Delta \{x^\alpha - (\Delta - x)^\alpha\}e^{-x^\alpha}dx$$

$$\left. + \int_\Delta^\infty \{x^\alpha - (x - \Delta)^\alpha\}e^{-x^\alpha}dx \right]$$

$$= C(\alpha)(I_1 + I_2 + I_3) \quad \text{(say)} .$$

Putting $h_\Delta(x) = (x + \Delta)^\alpha - x^\alpha$, we have from Lemma 2.1

$$(2.7) \quad I_1 + I_3 = -\int_0^\infty \{(x + \Delta)^\alpha - x^\alpha\}e^{-x^\alpha}dx + \int_\Delta^\infty \{x^\alpha - (x - \Delta)^\alpha\}e^{-x^\alpha}dx$$

$$= \int_0^\infty \{(x + \Delta)^\alpha - x^\alpha\}\{e^{-(x+\Delta)^\alpha} - e^{-x^\alpha}\}dx$$

$$= \int_0^\infty h_\Delta(x)\{e^{-h_\Delta(x)} - 1\}e^{-x^\alpha}dx$$

$$= -\int_0^\infty h_\Delta^2(x)e^{-x^\alpha}dx + \frac{1}{2}\int_0^\infty h_\Delta^3(x)e^{-x^\alpha}dx + O(\Delta^4)$$

$$= -\alpha\Gamma\left(2 - \frac{1}{\alpha}\right)\Delta^2 + \frac{1 + \gamma}{2\alpha + 1}\Delta^{2\alpha+1} + o(\Delta^{2\alpha+1}) .$$

We also obtain

$$(2.8) \quad I_2 = \int_0^{\Delta} \{x^{\alpha} - (\Delta - x)^{\alpha}\} e^{-x^{\alpha}} dx$$

$$= \int_0^{\Delta} \{x^{\alpha} - (\Delta - x)^{\alpha}\} \left\{ 1 - x^{\alpha} + \frac{1}{2} x^{2\alpha} + O(x^{3\alpha}) \right\} dx$$

$$= -\frac{\Delta^{2\alpha+1}}{2\alpha+1} + \Delta^{2\alpha+1} \int_0^{\Delta} \left(\frac{x}{\Delta}\right)^{\alpha} \left(1 - \frac{x}{\Delta}\right)^{\alpha} \frac{1}{\Delta} dx + O(\Delta^{3\alpha+1})$$

$$= \left\{ B(\alpha + 1, \alpha + 1) - \frac{1}{2\alpha+1} \right\} \Delta^{2\alpha+1} + O(\Delta^{3\alpha+1}),$$

where $B(u, v)$ denotes the Beta function. From (2.7) and (2.8), we have

$$I_1 + I_2 + I_3 = -\alpha \Gamma\left(2 - \frac{1}{\alpha}\right) \Delta^2$$

$$+ \left\{ B(\alpha + 1, \alpha + 1) + \frac{\gamma}{2\alpha+1} \right\} \Delta^{2\alpha+1} + o(\Delta^{2\alpha+1}).$$

Since $C(\alpha) = \alpha/\{2\Gamma(1/\alpha)\}$, it follows from (2.6), (2.7) and (2.8) that

$$(2.9) \quad E_0[\Psi_{\Delta}(x)] = C(\alpha)(I_1 + I_2 + I_3)$$

$$= -\frac{\alpha(\alpha - 1)\Gamma\left(1 - \dfrac{1}{\alpha}\right)}{2\Gamma\left(\dfrac{1}{\alpha}\right)} \Delta^2$$

$$+ \frac{\alpha\{B(\alpha + 1, \alpha + 1) + (\gamma/(2\alpha + 1))\}}{2\Gamma\left(\dfrac{1}{\alpha}\right)} \Delta^{2\alpha+1} + o(\Delta^{2\alpha+1}).$$

From (2.5), we obtain

$$(2.10) \quad E_0[\Psi_{\Delta}^2(X)] = \int_{-\infty}^{0} \{(-x)^{\alpha} - (\Delta - x)^{\alpha}\}^2 f(x) dx$$

$$+ \int_0^{\Delta} \{x^{\alpha} - (\Delta - x)^{\alpha}\}^2 f(x) dx$$

$$+ \int_{\Delta}^{\infty} \{x^{\alpha} - (x - \Delta)^{\alpha}\}^2 f(x) dx$$

$$= \int_0^{\infty} \{x^{\alpha} - (x + \Delta)^{\alpha}\}^2 f(x) dx$$

$$+\int_0^\Delta \{x^\alpha - (\Delta - x)^\alpha\}^2 f(x)\,dx$$

$$+\int_\Delta^\infty \{x^\alpha - (x - \Delta)^\alpha\}^2 f(x)\,dx$$

$$= C(\alpha)(I_1' + I_2' + I_3') \quad \text{(say)}\,.$$

Since $h_\Delta(x) = (x + \Delta)^\alpha - x^\alpha$, it follows from Lemma 2.1 that

$$(2.11) \quad I_1' + I_3' = \int_0^\infty \{(x + \Delta)^\alpha - x^\alpha\}^2 e^{-x^\alpha}\,dx + \int_\Delta^\infty \{x^\alpha - (x - \Delta)^\alpha\}^2 e^{-x^\alpha}\,dx$$

$$= \int_0^\infty \{(x + \Delta)^\alpha - x^\alpha\}^2 \{e^{-x^\alpha} + e^{-(x+\Delta)^\alpha}\}\,dx$$

$$= \int_0^\infty h_\Delta^2(x)\{e^{-h_\Delta(x)} + 1\}e^{-x^\alpha}\,dx$$

$$= 2\int_0^\infty h_\Delta^2(x)e^{-x^\alpha}\,dx - \int_0^\infty h_\Delta^3(x)e^{-x^\alpha}\,dx + O(\Delta^4)$$

$$= 2\alpha\Gamma\left(2 - \frac{1}{\alpha}\right)\Delta^2 - \frac{2(1 + \gamma)}{2\alpha + 1}\Delta^{2\alpha+1} + o(\Delta^{2\alpha+1})\,.$$

Since

$$I_2' = \int_0^\Delta \{x^\alpha - (\Delta - x)^\alpha\}^2 e^{-x^\alpha}\,dx$$

$$= \int_0^\Delta \{x^\alpha - (\Delta - x)^\alpha\}^2 (1 - x^\alpha)\,dx + O(\Delta^{3\alpha+1})$$

$$= \int_0^\Delta \{x^{2\alpha} - 2x^\alpha(\Delta - x)^\alpha + (\Delta - x)^{2\alpha}\}\,dx + O(\Delta^{3\alpha+1})$$

$$= \frac{2\Delta^{2\alpha+1}}{2\alpha + 1} - 2\Delta^{2\alpha+1}\int_0^\Delta \left(\frac{x}{\Delta}\right)^\alpha \left(1 - \frac{x}{\Delta}\right)^\alpha \frac{1}{\Delta}\,dx + O(\Delta^{3\alpha+1})$$

$$= \frac{2\Delta^{2\alpha+1}}{2\alpha + 1} - 2\Delta^{2\alpha+1}B(\alpha + 1, \alpha + 1) + O(\Delta^{3\alpha+1})\,,$$

we obtain from (2.10) and (2.11)

$$(2.12) \quad E_0[\Psi_\Delta^2(X)] = C(\alpha)(I_1' + I_2' + I_3')$$

$$= \frac{\alpha(\alpha - 1)\Gamma\left(1 - \dfrac{1}{\alpha}\right)}{\Gamma\left(\dfrac{1}{\alpha}\right)}\Delta^2$$

$$- \frac{\alpha\{B(\alpha + 1, \alpha + 1) + (\gamma/(2\alpha + 1))\}}{\Gamma\left(\frac{1}{\alpha}\right)} \Delta^{2\alpha+1} + o(\Delta^{2\alpha+1}) \,.$$

From (2.5), we have

$$(2.13) \quad E_0[\Psi_\Delta^3(X)] = \int_{-\infty}^{0} \{(-x)^\alpha - (\Delta - x)^\alpha\}^3 f(x)\,dx$$

$$+ \int_0^\Delta \{x^\alpha - (\Delta - x)^\alpha\}^3 f(x)\,dx$$

$$+ \int_\Delta^\infty \{x^\alpha - (x - \Delta)^\alpha\}^3 f(x)\,dx$$

$$= \int_0^\infty \{x^\alpha - (x + \Delta)^\alpha\}^3 f(x)\,dx$$

$$+ \int_0^\Delta \{x^\alpha - (\Delta - x)^\alpha\}^3 f(x)\,dx$$

$$+ \int_\Delta^\infty \{x^\alpha - (x - \Delta)^\alpha\}^3 f(x)\,dx$$

$$= C(\alpha)(I_1'' + I_2'' + I_3'') \quad \text{(say)}\,.$$

Since $h_\Delta(x) = (x + \Delta)^\alpha - x^\alpha$, it follows that

$$(2.14) \quad I_1'' + I_3'' = -\int_0^\infty \{(x + \Delta)^\alpha - x^\alpha\}^3 e^{-x^\alpha}\,dx + \int_\Delta^\infty \{x^\alpha - (x - \Delta)^\alpha\}^3 e^{-x^\alpha}\,dx$$

$$= \int_0^\infty \{(x + \Delta)^\alpha - x^\alpha\}^3 \{e^{-(x+\Delta)^\alpha} - e^{-x^\alpha}\}\,dx$$

$$= \int_0^\infty h_\Delta^3(x)\{e^{-h_\Delta(x)} - 1\}e^{-x^\alpha}\,dx\,.$$

Since

$$\int_0^\infty h_\Delta^4(x) e^{-x^\alpha}\,dx = \int_0^\infty \{(x + \Delta)^\alpha - x^\alpha\}^4 e^{-x^\alpha}\,dx$$

$$= \alpha^4 \Delta^4 \int_0^\infty x^{4\alpha-4} e^{-x^\alpha}\,dx + o(\Delta^4) = O(\Delta^4)\,,$$

it follows from (2.14) that

$$(2.15) \quad I_1'' + I_3'' = O(\Delta^4)\,.$$

We also have

$$(2.16) \qquad I_2'' = \int_0^{\Delta} \{x^{\alpha} - (\Delta - x)^{\alpha}\}^3 e^{-x^{\alpha}} dx = O(\Delta^{3\alpha+1}) \ .$$

From (2.13), (2.15) and (2.16), we obtain

$$(2.17) \qquad E_0[\Psi_{\Delta}^3(X)] = O(\Delta^4) \ .$$

Putting $\Delta = tn^{-1/2}$ with $t > 0$, we obtain from (2.9), (2.12) and (2.17)

$$E_0[Z_n] = -nE_0[\Psi_{tn^{-1/2}}(X)]$$

$$= \frac{\alpha(\alpha - 1)\Gamma\left(1 - \dfrac{1}{\alpha}\right)}{2\Gamma\left(\dfrac{1}{\alpha}\right)} t^2$$

$$- \frac{\alpha\{B(\alpha + 1, \alpha + 1) + (\gamma/(2\alpha + 1))\}}{2\Gamma\left(\dfrac{1}{\alpha}\right)} t^{2\alpha+1} n^{-(2\alpha-1)/2}$$

$$+ o(n^{-(2\alpha-1)/2})$$

$$= \frac{I}{2} t^2 - \frac{k}{2} t^{2\alpha+1} n^{-(2\alpha-1)/2} + o(n^{-(2\alpha-1)/2}) \ ,$$

$$V_0(Z_n) = nV_0(\Psi_{tn^{-1/2}}(X))$$

$$= \frac{\alpha(\alpha - 1)\Gamma\left(1 - \dfrac{1}{\alpha}\right)}{\Gamma\left(\dfrac{1}{\alpha}\right)} t^2$$

$$- \frac{\alpha\{B(\alpha + 1, \alpha + 1) + (\gamma/(2\alpha + 1))\}}{\Gamma\left(\dfrac{1}{\alpha}\right)} t^{2\alpha+1} n^{-(2\alpha-1)/2}$$

$$+ o(n^{-(2\alpha-1)/2})$$

$$= It^2 - kt^{2\alpha+1} n^{-(2\alpha-1)/2} + o(n^{-(2\alpha-1)/2}) \ ,$$

$$K_{3,0}(Z_n) = -nK_{3,0}(\Psi_{tn^{-1/2}}(X))$$

$$= -nE_0[\{\Psi_{tn^{-1/2}}(X) - E_0[\Psi_{tn^{-1/2}}(X)]\}^3]$$

$$= o(n^{-(2\alpha-1)/2}) \ ,$$

where

MASAFUMI AKAHIRA AND KEI TAKEUCHI

$$I = E_\theta\left[\left\{\frac{\partial}{\partial\theta}\log f(X-\theta)\right\}^2\right] = -E_\theta\left[\frac{\partial^2}{\partial\theta^2}\log f(X-\theta)\right]$$

$$= \alpha(\alpha-1)\Gamma(1-(1/\alpha))/\Gamma(1/\alpha)$$

and

$$k = \alpha\left\{B(\alpha+1,\alpha+1)+\frac{\gamma}{2\alpha+1}\right\}\bigg/\Gamma(1/\alpha).$$

In a similar way to the case under K: $\theta = 0$, we can obtain the asymptotic mean, variance and third-order cumulants under H: $\theta = tn^{-1/2}$. Thus we complete the proof.

In order to get the bound for the 2α-th order asymptotic distribution of 2α-th order AMU estimators, we need the following.

LEMMA 2.3. *Assume that the asymptotic mean, variance and third order cumulant of Z_n, under the distributions $P_{\theta,n}(\theta = \theta_0, \theta_0 + tn^{-1/2})$, are given by the following form.*

$$E_\theta(Z_n) = \mu(t,\theta) + n^{-(2\alpha-1)/2}C_1(t,\theta) + o(n^{-(2\alpha-1)/2}),$$

$$V_\theta(Z_n) = v^2(t,\theta) + n^{-(2\alpha-1)/2}C_2(t,\theta) + o(n^{-(2\alpha-1)/2}),$$

$$K_{3,\theta}(Z_n) = o(n^{-(2\alpha-1)/2}).$$

Then

$$P_{\theta,n}\{Z_n \le \alpha_0\} = \frac{1}{2} + o(n^{-(2\alpha-1)/2}),$$

if and only if

$$\alpha_0 = \mu(t,\theta) + C_1(t,\theta)n^{-(2\alpha-1)/2} + o(n^{-(2\alpha-1)/2}).$$

The proof is essentially given in Akahira and Takeuchi ((1981), pp. 132, 133).

In the following theorem we obtain the 2α-th asymptotic bound for the distribution of 2α-th order AMU estimators of θ.

THEOREM 2.1. *The bound for the 2α-th order asymptotic distribution of 2α-th order AMU estimators of θ is given by*

$$\Phi(t) - C_0|t|^{2\alpha}\phi(t)n^{-(2\alpha-1)/2}\,\mathrm{sgn}\,t + o(n^{-(2\alpha-1)/2}),$$

that is, for any $\hat{\theta}_n \in A_{2\alpha}$

$$P_{\theta,n}\{\sqrt{In}\,(\hat{\theta}_n - \theta) \leq t\} \leq \Phi(t) - C_0 t^{2\alpha}\phi(t)n^{-(2\alpha-1)/2} + o(n^{-(2\alpha-1)/2})$$

for all $\quad t > 0$,

$$P_{\theta,n}\{\sqrt{In}\,(\hat{\theta}_n - \theta) \leq t\} \geq \Phi(t) + C_0 |t|^{2\alpha}\phi(t)n^{-(2\alpha-1)/2} + o(n^{-(2\alpha-1)/2})$$

for all $\quad t < 0$,

where

$$C_0 = \frac{\alpha\{B(\alpha + 1, \alpha + 1) + (\gamma/(2\alpha + 1))\}}{2I^{\alpha+(1/2)}\Gamma(1/\alpha)}$$

and $\Phi(t)$ and $\phi(t)$ denote the standard normal distribution function and its density function, respectively.

PROOF. Without loss of generality, we assume that $\theta_0 = 0$. We consider the case when $t > 0$. In order to choose α_0 such that

$$(2.18) \qquad P_{tn^{-1/2},n}\{Z_n \leq \alpha_0\} = \frac{1}{2} + o(n^{-(2\alpha-1)/2}) ,$$

we have by Lemmas 2.2 and 2.3

$$\alpha_0 = -\frac{I}{2}\,t^2 + \frac{k}{2}\,t^{2\alpha+1}n^{-(2\alpha-1)/2} + o(n^{-(2\alpha-1)/2}) .$$

Since

$$(2.19) \qquad P_{0,n}\{Z_n \geq \alpha_0\} = P_{0,n}\{-(Z_n - It^2 - \alpha_0) \leq It^2\} ,$$

putting $W_n = -(Z_n - It^2 - \alpha_0)$, we have from Lemma 2.2

$$(2.20) \qquad E_0(W_n) = kt^{2\alpha+1}n^{-(2\alpha-1)/2} + o(n^{-(2\alpha-1)/2}) ,$$

$$(2.21) \qquad V_0(W_n) = It^2 - kt^{2\alpha+1}n^{-(2\alpha-1)/2} + o(n^{-(2\alpha-1)/2}) ,$$

$$(2.22) \qquad K_{3,0}(W_n) = o(n^{-(2\alpha-1)/2}) .$$

We obtain by (2.19) to (2.22) and the Edgeworth expansion

$$(2.23) \quad P_{0,n}\{Z_n \geq \alpha_0\} = P_{0,n}\{W_n \leq It^2\}$$

$$= \Phi(\sqrt{I}\,t) - \frac{k}{2\sqrt{I}}\,t^{2\alpha}\phi(\sqrt{I}\,t)n^{-(2\alpha-1)/2} + o(n^{-(2\alpha-1)/2}) .$$

Here, from (2.18), the definition of Z_n and the fundamental lemma of Neyman-Pearson it is noted that a test with the rejection region $\{Z_n \geq \alpha_0\}$ is the most powerful test of level $1/2 + o(n^{-(2\alpha-1)/2})$.

Let $\hat{\theta}_n$ be any 2α-th order AMU estimator. Putting $A_{\hat{\theta}_n} = \{\sqrt{n}\,\hat{\theta}_n \leq t\}$, we have

$$P_{tn^{-1},n}(A_{\hat{\theta}_n}) = P_{tn^{-1/2},n}\{\hat{\theta}_n \leq tn^{-1/2}\} = \frac{1}{2} + o(n^{-(2\alpha-1)/2})\,.$$

Then it is seen that $\chi_{A_{\hat{\theta}_n}}$ of the indicator of $A_{\hat{\theta}_n}$ is a test of level $1/2 + o(n^{-(2\alpha-1)/2})$. From (2.23), we obtain for any $\hat{\theta}_n \in A_{2\alpha}$

$$P_{0,n}\{\sqrt{n}\,\hat{\theta}_n \leq t\} \leq P_{0,n}\{W_n \leq It^2\}$$

$$= \Phi(\sqrt{I}\,t) - \frac{k}{2\sqrt{I}}\,t^{2\alpha}\phi(\sqrt{I}\,t)n^{-(2\alpha-1)/2} + o(n^{-(2\alpha-1)/2})\,,$$

that is,

$$(2.24) \qquad P_{0,n}\{\sqrt{In}\,\hat{\theta}_n \leq t\} \leq \Phi(t) - C_0 t^{2\alpha}\phi(t)n^{-(2\alpha-1)/2} + o(n^{-(2\alpha-1)/2})$$

for all $t > 0$, where

$$C_0 = \frac{k}{2I^{\alpha+(1/2)}} = \frac{\alpha\{B(\alpha+1, \alpha+1) + (\gamma/(2\alpha+1))\}}{2I^{\alpha+(1/2)}\Gamma(1/\alpha)}\,.$$

Hence we see that the bound for the 2α-th order distribution of 2α-th order AMU estimators for all $t > 0$ is given by (2.24). In a similar way to the case $t > 0$, we can obtain the 2α-th order bound for all $t < 0$. Thus we complete the proof.

Remark 2.1.   The result of Theorem 2.1 holds for $2/3 < \alpha < 1$, where the information amount I must be expressed as $\alpha^2\Gamma(2 - (1/\alpha))/\Gamma(1/\alpha)$. The proof is omitted since it is essentially similar to the above.

3.   The 2α-th order asymptotic distribution of the maximum likelihood estimator

In this section we obtain the 2α-th order asymptotic distribution of the maximum likelihood estimator (MLE) and compare it with the 2α-th order asymptotic bound obtained in the previous section.

We denote by θ_0 and $\hat{\theta}_{ML}$ the true parameter and the MLE, respectively. It is seen that for real t, $\hat{\theta}_{ML} < \theta_0 + tn^{-1/2}$ if and only if

$$(\partial/\partial\theta) \sum_{i=1}^{n} \log f(X_i - \theta_0 - tn^{-1/2}) < 0 \, .$$

Without loss of generality, we assume that $\theta_0 = 0$. Hence we see that for each real t

$$(3.1) \quad \hat{\theta}_{\text{ML}} < tn^{-1/2} \quad \text{if and only if} \quad \frac{1}{\sqrt{n}} \sum_{i=1}^{n} (d/dx) \log f(X_i - tn^{-1/2}) > 0 \, .$$

Since

$$\frac{d}{dx} \log f(x) = - \alpha |x|^{\alpha - 1} \operatorname{sgn} x \, ,$$

we put

$$(3.2) \qquad U_n = - \frac{1}{\sqrt{n}} \sum_{i=1}^{n} (d/dx) \log f(X_i - tn^{-1/2})$$

$$= \frac{\alpha}{\sqrt{n}} \sum_{i=1}^{n} |X_i - tn^{-1/2}|^{\alpha - 1} \operatorname{sgn} (X_i - tn^{-1/2}) \, .$$

In order to obtain the asymptotic cumulants of U_n, we need the following lemma.

LEMMA 3.1. *If $h_\Delta(x) = (x + \Delta)^\alpha - x^\alpha$ for $\Delta > 0$, then*

$$\int_0^\infty x^{\alpha - 1} e^{-x^\alpha} h_\Delta(x) dx = \Gamma\left(2 - \frac{1}{\alpha}\right) \Delta + \frac{\alpha - 1}{2} \Gamma\left(2 - \frac{2}{\alpha}\right) \Delta^2$$

$$- \frac{\gamma}{2\alpha} \Delta^{2\alpha} + o(\Delta^{2\alpha}) \, ,$$

$$(3.3) \quad \int_0^\infty x^{2\alpha - 2} e^{-x^\alpha} h_\Delta(x) dx = \Gamma\left(3 - \frac{2}{\alpha}\right) \Delta + \frac{1}{2}(\alpha - 1)\Gamma\left(3 - \frac{3}{\alpha}\right) \Delta^2$$

$$+ o(\Delta^2) \, ,$$

$$(3.4) \quad \int_0^\infty x^{3\alpha - 3} e^{-x^\alpha} h_\Delta(x) dx = \left(3 - \frac{3}{\alpha}\right) \Gamma\left(3 - \frac{3}{\alpha}\right) \Delta + O(\Delta^2) \, ,$$

$$\int_0^\infty x^{\alpha - 1} e^{-x^\alpha} h_\Delta^2(x) dx = \alpha\Gamma\left(3 - \frac{2}{\alpha}\right) \Delta^2 + O(\Delta^3) \, ,$$

$$(3.5) \quad \int_0^\infty x^{2\alpha - 2} e^{-x^\alpha} h_\Delta^2(x) dx = \alpha\Gamma\left(4 - \frac{3}{\alpha}\right) \Delta^2 + O(\Delta^3) \, ,$$

740   MASAFUMI AKAHIRA AND KEI TAKEUCHI

$$(3.6) \qquad \int_0^\infty x^{3\alpha-3} e^{-x^\alpha} h_\Delta^2(x)\,dx = O(\Delta^2)\,,$$

$$\int_0^\infty x^{\alpha-1} e^{-x^\alpha} h_\Delta^3(x)\,dx = O(\Delta^3)\,,$$

$$(3.7) \qquad \int_0^\infty x^{2\alpha-2} e^{-x^\alpha} h_\Delta^3(x)\,dx = O(\Delta^3)\,.$$

PROOF. From (2.2), we have

$$\int_0^\infty x^{\alpha-1} e^{-x^\alpha} h_\Delta(x)\,dx$$

$$= \int_0^\infty (x+\Delta)^\alpha x^{\alpha-1} e^{-x^\alpha}\,dx - \int_0^\infty x^{2\alpha-1} e^{-x^\alpha}\,dx$$

$$= -\frac{1}{\alpha+1} \int_0^\infty (x+\Delta)^{\alpha+1} \{(\alpha-1)x^{\alpha-2} - \alpha x^{2\alpha-2}\} e^{-x^\alpha}\,dx - \frac{1}{\alpha}$$

$$= -\frac{1}{\alpha+1} \int_0^\infty \left\{ x^{\alpha+1} + (\alpha+1)\Delta x^\alpha + \frac{\alpha(\alpha+1)}{2} \Delta^2 x^{\alpha-1} \right\}$$

$$\cdot \{(\alpha-1)x^{\alpha-2} - \alpha x^{2\alpha-2}\} e^{-x^\alpha}\,dx$$

$$-\frac{\alpha-1}{\alpha+1} \int_0^\infty R^*(\Delta) x^{\alpha-2} e^{-x^\alpha}\,dx - \frac{1}{\alpha} + O(\Delta^3)\,,$$

where

$$R^*(\Delta) = (x+\Delta)^{\alpha+1} - x^{\alpha+1} - (\alpha+1)\Delta x^\alpha - \frac{1}{2}\alpha(\alpha+1)\Delta^2 x^{\alpha-1}\,.$$

In a similar way to (2.4), we obtain

$$\int_0^\infty R^*(\Delta) x^{\alpha-2} e^{-x^\alpha}\,dx$$

$$= \frac{1}{2}\alpha(\alpha+1)(\alpha-1) \int_0^\Delta (\Delta-t)^2 \left(\int_0^\infty (x+t)^{\alpha-2} x^{\alpha-2} e^{-x^\alpha}\,dx \right) dt$$

$$= \frac{1}{2}\alpha(\alpha+1)(\alpha-1) B(2\alpha-2,3) B(\alpha-1,3-2\alpha)\Delta^{2\alpha} + o(\Delta^{2\alpha})$$

$$= \frac{(\alpha+1)\Gamma(\alpha-1)\Gamma(3-2\alpha)}{4(2\alpha-1)\Gamma(2-\alpha)}\Delta^{2\alpha} + o(\Delta^{2\alpha})\,.$$

Hence, we have

$$\int_0^\infty x^{\alpha-1} e^{-x^\alpha} h_\Delta(x)$$

$$= -\frac{1}{\alpha+1} \left\{ (\alpha-1) \int_0^\infty x^{2\alpha-1} e^{-x^\alpha} dx - \alpha \int_0^\infty x^{3\alpha-1} e^{-x^\alpha} dx \right.$$

$$+ (\alpha-1)(\alpha+1)\Delta \int_0^\infty x^{2\alpha-2} e^{-x^\alpha} dx$$

$$- \alpha(\alpha+1)\Delta \int_0^\infty x^{3\alpha-2} e^{-x^\alpha} dx$$

$$+ \frac{1}{2}\alpha(\alpha-1)(\alpha+1)\Delta^2 \int_0^\infty x^{2\alpha-3} e^{-x^\alpha} dx$$

$$\left. - \frac{1}{2}\alpha^2(\alpha+1)\Delta^2 \int_0^\infty x^{3\alpha-3} e^{-x^\alpha} dx \right\}$$

$$- \frac{1}{\alpha} - \frac{(\alpha-1)\Gamma(\alpha-1)\Gamma(3-2\alpha)}{4(2\alpha-1)\Gamma(2-\alpha)} \Delta^{2\alpha} + o(\Delta^{2\alpha})$$

$$= -\frac{1}{\alpha+1} \left\{ \frac{\alpha-1}{\alpha} - 2 + \frac{1}{\alpha}(\alpha-1)(\alpha+1)\Gamma\left(2-\frac{1}{\alpha}\right)\Delta \right.$$

$$- (\alpha+1)\Gamma\left(3-\frac{1}{\alpha}\right)\Delta$$

$$+ \frac{1}{2}(\alpha-1)(\alpha+1)\Gamma\left(2-\frac{2}{\alpha}\right)\Delta^2$$

$$\left. - \frac{1}{2}\alpha(\alpha+1)\Gamma\left(3-\frac{2}{\alpha}\right)\Delta^2 \right\}$$

$$- \frac{1}{\alpha} - \frac{(\alpha-1)\Gamma(\alpha-1)\Gamma(3-2\alpha)}{4(2\alpha-1)\Gamma(2-\alpha)} \Delta^{2\alpha} + O(\Delta^3)$$

$$= \Gamma\left(2-\frac{1}{\alpha}\right)\Delta + \frac{\alpha-1}{2}\Gamma\left(2-\frac{2}{\alpha}\right)\Delta^2 - \frac{\gamma}{2\alpha}\Delta^{2\alpha} + o(\Delta^{2\alpha}).$$

In a similar way to the above, we can obtain (3.3) and (3.4). We also have

$$\int_0^\infty x^{\alpha-1} e^{-x^\alpha} h_\Delta^2(x) dx$$

$$= \int_0^\infty (x+\Delta)^{2\alpha} x^{\alpha-1} e^{-x^\alpha} dx - 2 \int_0^\infty (x+\Delta)^\alpha x^{2\alpha-1} e^{-x^\alpha} dx$$

$$+ \int_0^\infty x^{3\alpha-1} e^{-x^\alpha} dx$$

742   MASAFUMI AKAHIRA AND KEI TAKEUCHI

$$= \int_0^\infty \{x^{2\alpha} + 2\alpha x^{2\alpha-1}\Delta + \alpha(2\alpha - 1)x^{2\alpha-2}\Delta^2\} x^{\alpha-1}e^{-x^\alpha}dx$$

$$- 2\int_0^\infty \left\{ x^\alpha + \alpha x^{\alpha-1}\Delta + \frac{1}{2}\alpha(\alpha-1)x^{\alpha-2}\Delta^2 \right\} x^{2\alpha-1}e^{-x^\alpha}dx$$

$$+ \frac{2}{\alpha} + O(\Delta^3)$$

$$= \alpha\Gamma\left(3 - \frac{2}{\alpha}\right)\Delta^2 + O(\Delta^3),$$

and similarly get (3.5) and (3.6). From (2.2), we obtain

$$\int_0^\infty x^{\alpha-1}e^{-x^\alpha}h_\Delta^3(x)dx = \int_0^\infty x^{\alpha-1}e^{-x^\alpha}(\alpha x^{\alpha-1}\Delta)^3\,dx + o(\Delta^3)$$

$$= \alpha^2\Gamma\left(4 - \frac{3}{\alpha}\right)\Delta^3 + o(\Delta^3)$$

$$= O(\Delta^3),$$

and similarly have (3.7). Thus we complete the proof.

In the following lemma, we obtain the asymptotic mean, variance and third order cumulant of U_n.

LEMMA 3.2.   *The asymptotic mean, variance and third order cumulant of U_n are given for $t > 0$ as follows:*

$$E_0(U_n) = -It + \frac{k'}{2}t^{2\alpha}n^{-(2\alpha-1)/2} + o(n^{-(2\alpha-1)/2}),$$

$$V_0(U_n) = I + o(n^{-(2\alpha-1)/2}),$$

$$K_{3,0}(U_n) = o(n^{-(2\alpha-1)/2}),$$

where I is given in Lemma 2.2 and

$$k' = \frac{\alpha^2\{B(\alpha, \alpha + 1) + (\gamma/\alpha)\}}{\Gamma(1/\alpha)}.$$

PROOF.   Putting $\Delta = tn^{-1/2}$, we obtain

(3.8) $E_0[|X - \Delta|^{\alpha-1}\,\mathrm{sgn}\,(X - \Delta)]$

$$= C(\alpha)\left\{ \int_\Delta^\infty (x - \Delta)^{\alpha-1}e^{-x^\alpha}dx - \int_0^\Delta (\Delta - x)^{\alpha-1}e^{-x^\alpha}dx \right.$$

$$-\int_{-\infty}^{0}(\Delta - x)^{\alpha-1}e^{-|x|^\alpha}dx\Bigg\}$$

$$= C(\alpha)\Bigg\{\int_{0}^{\infty}x^{\alpha-1}e^{-(x+\Delta)^\alpha}dx - \int_{0}^{\infty}(x+\Delta)^{\alpha-1}e^{-x^\alpha}dx$$

$$-\int_{0}^{\Delta}(\Delta - x)^{\alpha-1}e^{-x^\alpha}dx\Bigg\}$$

$$= C(\alpha)\Bigg[\int_{0}^{\infty}x^{\alpha-1}\{e^{-(x+\Delta)^\alpha} - e^{-x^\alpha}\}dx - \int_{0}^{\infty}\{(x+\Delta)^{\alpha-1} - x^{\alpha-1}\}e^{-x^\alpha}dx$$

$$-\int_{0}^{\Delta}(\Delta - x)^{\alpha-1}e^{-x^\alpha}dx\Bigg]$$

$$= C(\alpha)(J_1 + J_2 + J_3) \quad \text{(say)}.$$

Putting $h_\Delta(x) = (x+\Delta)^\alpha - x^\alpha$, we have from Lemma 3.1

$$(3.9) \quad J_1 = \int_{0}^{\infty}x^{\alpha-1}\{e^{-(x+\Delta)^\alpha} - e^{-x^\alpha}\}dx$$

$$= \int_{0}^{\infty}x^{\alpha-1}e^{-x^\alpha}\{e^{-h_\Delta(x)} - 1\}dx$$

$$= -\int_{0}^{\infty}x^{\alpha-1}e^{-x^\alpha}h_\Delta(x)dx + \frac{1}{2}\int_{0}^{\infty}x^{\alpha-1}e^{-x^\alpha}h_\Delta^2(x)dx + O(\Delta^3)$$

$$= -\Gamma\left(2 - \frac{1}{\alpha}\right)\Delta - \frac{\alpha-1}{2}\Gamma\left(2 - \frac{2}{\alpha}\right)\Delta^2 + \frac{\alpha}{2}\Gamma\left(3 - \frac{2}{\alpha}\right)\Delta^2$$

$$+ \frac{\gamma}{2\alpha}\Delta^{2\alpha} + o(\Delta^{2\alpha})$$

$$= -\left(1 - \frac{1}{\alpha}\right)\Gamma\left(1 - \frac{1}{\alpha}\right)\Delta + \frac{\alpha-1}{2}\Gamma\left(2 - \frac{2}{\alpha}\right)\Delta^2$$

$$+ \frac{\gamma}{2\alpha}\Delta^{2\alpha} + o(\Delta^{2\alpha}).$$

From (2.2), we obtain

$$(3.10) \qquad -J_2 = \int_{0}^{\infty}(x+\Delta)^{\alpha-1}e^{-x^\alpha}dx - \int_{0}^{\infty}x^{\alpha-1}e^{-x^\alpha}dx$$

$$= -\frac{\Delta^\alpha}{\alpha} + \int_{0}^{\infty}(x+\Delta)^\alpha x^{\alpha-1}e^{-x^\alpha}dx - \frac{1}{\alpha}.$$

Since by a similar way to (2.4)

$$\int_0^\infty (x + \varDelta)^\alpha x^{\alpha-1} e^{-x^\alpha} dx$$

$$= \frac{1}{\alpha} + \Gamma\left(2 - \frac{1}{\alpha}\right)\varDelta + \frac{\alpha-1}{2}\Gamma\left(2 - \frac{2}{\alpha}\right)\varDelta^2 - \frac{\gamma}{2\alpha}\varDelta^{2\alpha} + o(\varDelta^{2\alpha}),$$

it follows from (3.10) that

$$(3.11) \qquad J_2 = \frac{\varDelta^\alpha}{\alpha} - \Gamma\left(2 - \frac{1}{\alpha}\right)\varDelta - \frac{\alpha}{2}\left(1 - \frac{1}{\alpha}\right)\Gamma\left(2 - \frac{2}{\alpha}\right)\varDelta^2$$

$$+ \frac{\gamma}{2\alpha}\varDelta^{2\alpha} + o(\varDelta^{2\alpha}).$$

We also have

$$(3.12) \qquad -J_3 = \int_0^\varDelta (\varDelta - x)^{\alpha-1} e^{-x^\alpha} dx$$

$$= \int_0^\varDelta (\varDelta - x)^{\alpha-1}(1 - x^\alpha) dx + O(\varDelta^{3\alpha})$$

$$= \frac{\varDelta^\alpha}{\alpha} - \varDelta^{2\alpha}\int_0^1 \left(1 - \frac{x}{\varDelta}\right)^{\alpha-1}\left(\frac{x}{\varDelta}\right)^\alpha \frac{1}{\varDelta}\, dx + O(\varDelta^{3\alpha})$$

$$= \frac{\varDelta^\alpha}{\alpha} - B(\alpha, \alpha + 1)\varDelta^{2\alpha} + O(\varDelta^{3\alpha}).$$

From (3.8), (3.9), (3.11) and (3.12), we obtain

$$(3.13) \qquad E_0[\,|X - \varDelta|^{\alpha-1}\, \mathrm{sgn}\,(X - \varDelta)]$$

$$= C(\alpha)\left\{ -\frac{2(\alpha-1)}{\alpha}\Gamma\left(\frac{\alpha-1}{\alpha}\right)\varDelta \right.$$

$$+ \left. \left(B(\alpha, \alpha + 1) + \frac{\gamma}{\alpha}\right)\varDelta^{2\alpha}\right\} + o(\varDelta^{2\alpha})$$

$$= -(\alpha - 1)\frac{\Gamma(1 - (1/\alpha))}{\Gamma(1/\alpha)}\varDelta$$

$$+ \frac{\alpha\{B(\alpha, \alpha + 1) + (\gamma/\alpha)\}}{2\Gamma(1/\alpha)}\varDelta^{2\alpha} + o(\varDelta^{2\alpha}).$$

Next, we have

$$(3.14) \quad E_0[\,|X - \varDelta|^{2\alpha-2}]$$

$$= C(\alpha) \left\{ \int_{\Delta}^{\infty} (x - \Delta)^{2\alpha-2} e^{-x^{\alpha}} dx + \int_{-\infty}^{\Delta} (\Delta - x)^{2\alpha-2} e^{-|x|^{\alpha}} dx \right\}$$

$$= C(\alpha) \left\{ \int_{0}^{\infty} x^{2\alpha-2} e^{-(x+\Delta)^{\alpha}} dx + \int_{0}^{\infty} (x + \Delta)^{2\alpha-2} e^{-x^{\alpha}} dx \right.$$

$$\left. + \int_{0}^{\Delta} (\Delta - x)^{2\alpha-2} e^{-x^{\alpha}} dx \right\}$$

$$= C(\alpha) \left[2\int_{0}^{\infty} x^{2\alpha-2} e^{-x^{\alpha}} dx + \int_{0}^{\infty} \{e^{-(x+\Delta)^{\alpha}} - e^{-x^{\alpha}}\} x^{2\alpha-2} dx \right.$$

$$+ \int_{0}^{\infty} \{(x + \Delta)^{2\alpha-2} - x^{2\alpha-2}\} e^{-x^{\alpha}} dx$$

$$\left. + \int_{0}^{\Delta} (\Delta - x)^{2\alpha-2} e^{-x^{\alpha}} dx \right]$$

$$= C(\alpha)[\, J_1' + J_2' + J_3' + J_4'\,] \quad \text{(say)}.$$

From (2.2), we obtain

$$(3.15) \quad J_1' = 2\int_{0}^{\infty} x^{2\alpha-2} e^{-x^{\alpha}} dx = \frac{2}{\alpha} \Gamma\left(2 - \frac{1}{\alpha}\right) = \frac{2}{\alpha}\left(1 - \frac{1}{\alpha}\right)\Gamma\left(1 - \frac{1}{\alpha}\right).$$

Since

$$J_2' = \int_{0}^{\infty} \{e^{-(x+\Delta)^{\alpha}} - e^{-x^{\alpha}}\} x^{2\alpha-2} dx$$

$$= \int_{0}^{\infty} x^{2\alpha-2} e^{-x^{\alpha}} \{e^{-h_{\Delta}(x)} - 1\} dx$$

$$= -\int_{0}^{\infty} x^{2\alpha-2} e^{-x^{\alpha}} h_{\Delta}(x) dx + \frac{1}{2}\int_{0}^{\infty} x^{2\alpha-2} e^{-x^{\alpha}} h_{\Delta}^{2}(x) dx$$

$$- \frac{1}{6}\int_{0}^{\infty} x^{2\alpha-2} e^{-x^{\alpha}} h_{\Delta}^{3}(x) dx + O(\Delta^{3}),$$

it follows from Lemma 3.1 that

$$(3.16) \quad J_2' = -\Gamma\left(3 - \frac{2}{\alpha}\right)\Delta + (\alpha - 1)\Gamma\left(3 - \frac{3}{\alpha}\right)\Delta^{2} + O(\Delta^{3}).$$

From (2.2), we have

$$(3.17) \quad J_3' = \int_{0}^{\infty} \{(x + \Delta)^{2\alpha-2} - x^{2\alpha-2}\} e^{-x^{\alpha}} dx$$

$$= -\frac{1}{2\alpha - 1}\Delta^{2\alpha-1} + \frac{\alpha}{2\alpha - 1}\int_0^\infty (x + \Delta)^{2\alpha-1}x^{\alpha-1}e^{-x^\alpha}dx$$

$$-\frac{1}{\alpha}\Gamma\left(\frac{2\alpha-1}{\alpha}\right)$$

$$= -\frac{1}{2\alpha - 1}\Delta^{2\alpha-1} - \frac{1}{\alpha}\Gamma\left(\frac{2\alpha-1}{\alpha}\right)$$

$$+ \frac{\alpha}{2\alpha - 1}\left\{\int_0^\infty x^{3\alpha-2}e^{-x^\alpha}dx + (2\alpha - 1)\Delta\int_0^\infty x^{3\alpha-3}e^{-x^\alpha}dx\right.$$

$$\left. + (\alpha - 1)(2\alpha - 1)\Delta^2\int_0^\infty x^{3\alpha-4}e^{-x^\alpha}dx\right\} + o(\Delta^2)$$

$$= -\frac{1}{2\alpha - 1}\Delta^{2\alpha-1} - \frac{1}{\alpha}\Gamma\left(2 - \frac{1}{\alpha}\right) + \frac{1}{2\alpha - 1}\Gamma\left(3 - \frac{1}{\alpha}\right)$$

$$+ \Gamma\left(3 - \frac{2}{\alpha}\right)\Delta + (\alpha - 1)\Gamma\left(3 - \frac{3}{\alpha}\right)\Delta^2 + O(\Delta^3)$$

$$= -\frac{1}{2\alpha - 1}\Delta^{2\alpha-1} + \Gamma\left(3 - \frac{2}{\alpha}\right)\Delta + (\alpha - 1)\Gamma\left(3 - \frac{3}{\alpha}\right)\Delta^2$$

$$+ o(\Delta^2)\,.$$

We also obtain

$$(3.18)\quad J_4' = \int_0^\Delta (\Delta - x)^{2\alpha-2}e^{-x^\alpha}dx$$

$$= \int_0^\Delta (\Delta - x)^{2\alpha-2}dx - \int_0^\Delta x^\alpha(\Delta - x)^{2\alpha-2}dx + O(\Delta^{4\alpha-1})$$

$$= \frac{\Delta^{2\alpha-1}}{2\alpha - 1} - \Delta^{3\alpha-1}\int_0^\Delta \left(\frac{x}{\Delta}\right)^\alpha\left(1 - \frac{x}{\Delta}\right)^{2\alpha-2}\frac{1}{\Delta}dx + O(\Delta^{4\alpha-1})$$

$$= \frac{\Delta^{2\alpha-1}}{2\alpha - 1} - B(\alpha + 1, 2\alpha - 1)\Delta^{3\alpha-1} + O(\Delta^{4\alpha-1})\,.$$

From (3.14) to (3.18), we have

$$E_0[|X - \Delta|^{2\alpha-2}] = \frac{\alpha - 1}{\alpha\Gamma(1/\alpha)}\Gamma\left(1 - \frac{1}{\alpha}\right)$$

$$+ \frac{\alpha(\alpha - 1)}{\Gamma(1/\alpha)}\Gamma\left(3 - \frac{3}{\alpha}\right)\Delta^2 + o(\Delta^2)\,,$$

hence, by (3.13)

$$(3.19) \quad V_0(|X - \Delta|^{\alpha-1} \operatorname{sgn}(X - \Delta))$$

$$= \frac{\alpha - 1}{\alpha \Gamma(1/\alpha)} \Gamma\left(1 - \frac{1}{\alpha}\right)$$

$$+ (\alpha - 1)\left[\frac{\alpha}{\Gamma(1/\alpha)} \Gamma\left(3 - \frac{3}{\alpha}\right)\right.$$

$$\left. - (\alpha - 1)\left\{\frac{\Gamma(1 - (1/\alpha))}{\Gamma(1/\alpha)}\right\}^2\right]\Delta^2 + o(\Delta^2).$$

Third, we have

$$(3.20) \quad E_0[|X - \Delta|^{3\alpha-3} \operatorname{sgn}(X - \Delta)]$$

$$= C(\alpha)\left\{\int_\Delta^\infty (x - \Delta)^{3\alpha-3} e^{-x^\alpha} dx - \int_0^\Delta (\Delta - x)^{3\alpha-3} e^{-x^\alpha} dx\right.$$

$$\left. - \int_{-\infty}^0 (\Delta - x)^{3\alpha-3} e^{-|x|^\alpha} dx\right\}$$

$$= C(\alpha)\left[\int_0^\infty x^{3\alpha-3}\{e^{-(x+\Delta)^\alpha} - e^{-x^\alpha}\} dx\right.$$

$$- \int_0^\infty \{(x + \Delta)^{3\alpha-3} - x^{3\alpha-3}\} e^{-x^\alpha} dx$$

$$\left. - \int_0^\Delta (\Delta - x)^{3\alpha-3} e^{-x^\alpha} dx\right]$$

$$= C(\alpha)(J_1'' + J_2'' + J_3'') \quad \text{(say)}.$$

Since

$$J_1'' = \int_0^\infty x^{3\alpha-3}\{e^{-(x+\Delta)^\alpha} - e^{-x^\alpha}\} dx = \int_0^\infty x^{3\alpha-3} e^{-x^\alpha}\{e^{-h_\Delta(x)} - 1\} dx$$

$$= -\int_0^\infty x^{3\alpha-3} e^{-x^\alpha} h_\Delta(x) dx + \frac{1}{2}\int_0^\infty x^{3\alpha-3} e^{-x^\alpha} h_\Delta^2(x) dx + o(\Delta^2),$$

it follows by Lemma 3.1 that

$$(3.21) \quad J_1'' = -3\left(1 - \frac{1}{\alpha}\right)\Gamma\left(3 - \frac{3}{\alpha}\right)\Delta + O(\Delta^2).$$

From (2.2), we have

MASAFUMI AKAHIRA AND KEI TAKEUCHI

$$(3.22) \qquad -J_2'' = \int_0^\infty \{(x+\varDelta)^{3\alpha-3} - x^{3\alpha-3}\} e^{-x^\alpha} dx$$

$$= 3(\alpha-1)\varDelta \int_0^\infty x^{3\alpha-4} e^{-x^\alpha} dx + o(\varDelta)$$

$$= 3\left(1 - \frac{1}{\alpha}\right) \Gamma\left(3 - \frac{3}{\alpha}\right) \varDelta + o(\varDelta),$$

and also

$$(3.23) \quad -J_3'' = \int_0^\varDelta (\varDelta - x)^{3\alpha-3} e^{-x^\alpha} dx$$

$$= \frac{\varDelta^{3\alpha-2}}{3\alpha-2} - \varDelta^{4\alpha-2} \int_0^\varDelta \left(\frac{x}{\varDelta}\right)^\alpha \left(1 - \frac{x}{\varDelta}\right)^{3\alpha-3} \frac{1}{\varDelta} dx + O(\varDelta^{5\alpha-2})$$

$$= \frac{\varDelta^{3\alpha-2}}{3\alpha-2} + O(\varDelta^{4\alpha-2}).$$

From (3.20) to (3.23), we obtain

$$E_0[|X-\varDelta|^{3\alpha-3} \operatorname{sgn}(X-\varDelta)] = -\frac{3(\alpha-1)}{\Gamma(1/\alpha)} \Gamma\left(3 - \frac{3}{\alpha}\right) \varDelta + o(\varDelta),$$

hence

$$(3.24) \quad K_{3,0}(|X-\varDelta|^{\alpha-1} \operatorname{sgn}(X-\varDelta))$$

$$= E_0[\{|X-\varDelta|^{\alpha-1} \operatorname{sgn}(X-\varDelta) - E_0[|X-\varDelta|^{\alpha-1}\operatorname{sgn}(X-\varDelta)]\}^3]$$

$$= O(\varDelta).$$

Putting $\varDelta = tn^{-1/2}$ with $t > 0$, we have from (3.2), (3.13), (3.19) and (3.24)

$$E_0(U_n) = \alpha\sqrt{n}\, E_0[|X - tn^{-1/2}|^{\alpha-1} \operatorname{sgn}(X - tn^{-1/2})]$$

$$= -\alpha(\alpha-1)\frac{\Gamma(1-(1/\alpha))}{\Gamma(1/\alpha)} t$$

$$+ \frac{\alpha^2\{B(\alpha, \alpha+1) + (\gamma/\alpha)\}}{2\Gamma(1/\alpha)} t^{2\alpha} n^{-(2\alpha-1)/2}$$

$$+ o(n^{-(2\alpha-1)/2})$$

$$= -It + \frac{k'}{2} t^{2\alpha} n^{-(2\alpha-1)/2} + o(n^{-(2\alpha-1)/2}),$$

$$V_0(U_n) = \alpha^2 V_0(|X - tn^{-1/2}|^{\alpha-1} \operatorname{sgn}(X - tn^{-1/2}))$$

$$= \alpha(\alpha - 1)\frac{\Gamma(1 - (1/\alpha))}{\Gamma(1/\alpha)} + o(n^{-(2\alpha-1)/2})$$

$$= I + o(n^{-(2\alpha-1)/2}),$$

$$K_{3,0}(U_n) = \frac{\alpha^3}{\sqrt{n}} K_{3,0}(|X - tn^{-1/2}|^{\alpha-1} \operatorname{sgn}(X - tn^{-1/2}))$$

$$= o(n^{-(2\alpha-1)/2}),$$

where $I = \alpha(\alpha - 1)\Gamma(1 - (1/\alpha))/\Gamma(1/\alpha)$ and $k' = \alpha^2\{B(\alpha, \alpha + 1) + (\gamma/\alpha)\}/\Gamma(1/\alpha)$. This completes the proof.

In the following theorem, we obtain the 2α-th order asymptotic distribution of the MLE of θ.

THEOREM 3.1. *The 2α-th order asymptotic distribution of the MLE $\hat{\theta}_{\mathrm{ML}}$ of θ is given by*

$$(3.25) \qquad P_{\theta,n}\{\sqrt{In}\,(\hat{\theta}_{\mathrm{ML}} - \theta) \le t\}$$

$$= \Phi(t) - C_1|t|^{2\alpha}\phi(t)n^{-(2\alpha-1)/2}\operatorname{sgn} t + (n^{-(2\alpha-1)/2}),$$

where $C_1 = (2\alpha + 1)C_0$ with $C_0 = \alpha\{B(\alpha + 1, \alpha + 1) + (\gamma/(2\alpha + 1))\}/\{2I^{\alpha+(1/2)}\Gamma(1/\alpha)\}$, and also the MLE is not 2α-th order asymptotically efficient in the sense that its 2α-th order asymptotic distribution does not uniformly attain the bound given in Theorem 2.1.

PROOF. Since the density $f(x) = C(\alpha)e^{-|x|^\alpha}$ is symmetric about the origin, we see that the MLE of θ is a 2α-th order AMU. We consider the case when $t > 0$. Using the Edgeworth expansion, we have by (3.1), (3.2) and Lemma 3.2

$$P_{\theta,n}\{\sqrt{n}\,(\hat{\theta}_{\mathrm{ML}} - \theta) \le t\}$$

$$= P_{\theta,n}\{U_n \le 0\}$$

$$= \Phi(\sqrt{I}\,t) - \frac{k'}{2\sqrt{I}}\,t^{2\alpha}\phi(\sqrt{I}\,t)n^{-(2\alpha-1)/2} + o(n^{-(2\alpha-1)/2}),$$

that is,

$$(3.26) \qquad P_{\theta,n}\{\sqrt{In}\,(\hat{\theta}_{\mathrm{ML}} - \theta) \le t\}$$

$$= \Phi(t) - \frac{k'}{2I^{\alpha+(1/2)}}\,t^{2\alpha}\phi(t)n^{-(2\alpha-1)/2} + o(n^{-(2\alpha-1)/2})$$

$$= \Phi(t) - C_1 t^{2\alpha} \phi(t) n^{-(2\alpha-1)/2} + o(n^{-(2\alpha-1)/2}) \, ,$$

where $C_1 = k'/\{2I^{\alpha+(1/2)}\}$.

In a similar way to the case $t > 0$, we obtain for $t < 0$

$$(3.27) \qquad P_{\theta,n}\{\sqrt{In}\,(\hat{\theta}_{\mathrm{ML}} - \theta) \le t\}$$

$$= \Phi(t) + C_1 |t|^{2\alpha} \phi(t) n^{-(2\alpha-1)/2} + o(n^{-(2\alpha-1)/2}) \, .$$

Hence (3.26) and (3.27) imply (3.25). Since $k' = \alpha^2\{B(\alpha, \alpha + 1) + (\gamma/\alpha)\}/\Gamma(1/\alpha)$ and $C_0 = \alpha\{B(\alpha + 1, \alpha + 1) + (\gamma/(2\alpha + 1))\}/\{2I^{\alpha+(1/2)}\Gamma(1/\alpha)\}$, it is seen that $C_1 = (2\alpha + 1)C_0$. Since $C_1 > C_0$ for $1 < \alpha < 3/2$, it follows from Theorem 2.1 and (3.25) that the MLE is not 2α-th order asymptotically efficient in the sense that its 2α-th order asymptotic distribution does not uniformly attain the bound given in Theorem 2.1. This completes the proof.

Remark 3.1.   In the double exponential distribution case, that is, the case when $\alpha = 1$, it is shown in Akahira and Takeuchi (1981) that the bound for the second order asymptotic distribution of the second order AMU estimators of θ is given by

$$\Phi(t) - \frac{t^2}{6} \phi(t) n^{-1/2} \operatorname{sgn} t + o(n^{-1/2}) \, ,$$

and the second order distribution of the MLE of θ, i.e., the median of $X_1,\ldots, X_n$, is given by

$$(3.28) \qquad \Phi(t) - \frac{t^2}{2} \phi(t) n^{-1/2} \operatorname{sgn} t + o(n^{-1/2}) \, .$$

The results coincide with the case when $\alpha = 1$ is substituted in the formulae of Theorems 2.1 and 3.1, but note that the proofs of these theorems do not include the case for $\alpha = 1$ since it does not automatically hold that $\Gamma(\alpha) = (\alpha - 1)\Gamma(\alpha - 1)$ for $\alpha = 1$.

4. The amount of the loss of asymptotic information of the maximum likelihood estimator

In the section we obtain the amount of the loss of asymptotic information of the MLE $\hat{\theta}_{\mathrm{ML}}$ using its second order asymptotic distribution (3.25). Differentiating the right-hand side of (3.25) w.r.t. t, we have the second order asymptotic density $g(t)$ of $\sqrt{In}\,(\hat{\theta}_{\mathrm{ML}} - \theta)$ as follows:

$$(4.1) \quad g(t) = \phi(t)\{1 - C_1(2\alpha|t|^{2\alpha-1} - |t|^{2\alpha+1})n^{-(2\alpha-1)/2}\} + o(n^{-(2\alpha-1)/2})$$

$$\text{for } -\infty < t < \infty.$$

In general, we obtain for $\alpha > 0$

$$(4.2) \qquad \int_{-\infty}^{\infty} |t|^{\alpha} \phi(t)\, dt = \frac{2}{\sqrt{2\pi}} \int_{0}^{\infty} t^{\alpha} e^{-t^2/2}\, dt$$

$$= \frac{2^{\alpha/2}}{\sqrt{\pi}} \int_{0}^{\infty} u^{\{(\alpha+1)/2\}-1} e^{-u}\, du$$

$$(\text{after transformation } u = t^2/2)$$

$$= \frac{2^{\alpha/2}}{\sqrt{\pi}} \Gamma\left(\frac{\alpha+1}{2}\right).$$

Since, for sufficiently large n,

$$\frac{d}{dt} \log g(t) = -t - C_1\{2\alpha(2\alpha-1)|t|^{2\alpha-2} - (2\alpha+1)|t|^{2\alpha}\}$$

$$\cdot n^{-(2\alpha-1)/2} \operatorname{sgn} t + o(n^{-(2\alpha-1)/2}),$$

it follows from (4.1) and (4.2) that the asymptotic information amount I_{ML} of the MLE is given by

$$I_{\mathrm{ML}} = nI \int_{-\infty}^{\infty} \left\{ \frac{d}{dt} \log g(t) \right\}^2 g(t)\, dt$$

$$= nI \int_{-\infty}^{\infty} \phi(t)[t^2 + C_1\{4\alpha(2\alpha-1)|t|^{2\alpha-1}$$

$$- 2(3\alpha+1)|t|^{2\alpha+1} + |t|^{2\alpha+3}\} n^{-(2\alpha-1)/2}]\, dt$$

$$+ o(n^{(3/2)-\alpha})$$

$$= nI \left\{ 1 - \frac{2^{\alpha+(3/2)}}{\sqrt{\pi}} C_1 \Gamma(\alpha+1) n^{-(2\alpha-1)/2} \right\} + o(n^{(3/2)-\alpha}).$$

Hence, the amount of the loss L of asymptotic information of the MLE is given by

$$L = nI - I_{\mathrm{ML}} = \frac{2^{\alpha+(3/2)}}{\sqrt{\pi}} C_1 I \Gamma(\alpha+1) n^{(3/2)-\alpha} + o(n^{(3/2)-\alpha})$$

$$= \frac{2^{\alpha+(1/2)}\alpha(2\alpha+1)\Gamma(\alpha+1)\{B(\alpha+1,\alpha+1) + (\gamma/(2\alpha+1))\}}{\sqrt{\pi}\, I^{\alpha-(1/2)}\Gamma(1/\alpha)} n^{(3/2)-\alpha}$$

$$+ o(n^{(3/2)-\alpha}).$$

In a similar way to the above, it follows from (3.28) that, in the double-exponential distribution case, namely, when $\alpha = 1$, the amount of the loss of asymptotic information of the MLE is given by $2\sqrt{2n/\pi} + o(\sqrt{n})$, since $I = 1$.

Acknowledgement

The authors wish to thank the referee for pointing out two errors in the original version.

References

Akahira, M. (1975). Asymptotic theory for estimation of location in non-regular cases, II: Bounds of asymptotic distributions of consistent estimators, *Rep. Statist. Appl. Res. JUSE*, **22**, 99–115.

Akahira, M. (1986). *The Structure of Asymptotic Deficiency of Estimators*, Queen's Papers in Pure and Appl. Math., No. 75, Queen's University Press, Kingston, Ontario, Canada.

Akahira, M. (1987). Second order asymptotic comparison of estimators of a common parameter in the double exponential case, *Ann. Inst. Statist. Math.*, **39**, 25–36.

Akahira, M. (1988a). Second order asymptotic properties of the generalized Bayes estimators for a family of non-regular distributions, *Statistical Theory and Data Analysis II*, Proceedings of the Second Pacific Area Statistical Conference, (ed. K. Matusita), 87–100, North-Holland, Amsterdam-New York.

Akahira, M. (1988b). Second order asymptotic optimality of estimators for a density with finite cusps, *Ann. Inst. Statist. Math.*, **40**, 311–328.

Akahira, M. and Takeuchi, K. (1981). *Asymptotic Efficiency of Statistical Estimators: Concepts and Higher Order Asymptotic Efficiency*, Lecture Notes in Statistics 7, Springer, New York.

Akahira, M., Hirakawa, F. and Takeuchi, K. (1988). Second and third order asymptotic completeness of the class of estimators, *Probability Theory and Mathematical Statistics*, Proceedings of the Fifth Japan–USSR Symposium on Probability Theory, (eds. S. Watanabe and Yu. V. Prokhorov), Lecture Notes in Mathematics 1299, 11–27, Springer, Berlin.

Ghosh, J. K., Sinha, B. K. and Wieand, H. S. (1980). Second order efficiency of the mle with respect to any bounded bowl-shaped loss function, *Ann. Statist.*, **8**, 506–521.

Pfanzagl, J. and Wefelmeyer, W. (1978). A third-order optimum property of maximum likelihood estimator, *J. Multivariate Anal.*, **8**, 1–29.

Pfanzagl, J. and Wefelmeyer, W. (1985). *Asymptotic Expansions for General Statistical Models*, Lecture Notes in Statistics 31, Springer, Berlin.

Sugiura, N. and Naing, M. T. (1987). Improved estimators for the location of double exponential distribution, Contributed Papers of 46 Session of ISI, Tokyo, 427–428.

SEQUENTIAL ANALYSIS, 8(4), 333-359 (1989)

THIRD ORDER ASYMPTOTIC EFFICIENCY OF THE SEQUENTIAL MAXIMUM LIKELIHOOD ESTIMATION PROCEDURE

Masafumi Akahira

Kei Takeuchi

Institute of Mathematics
University of Tsukuba
Ibaraki   305
Japan

Research Center for Advanced
Science and Technology
University of Tokyo
4-6-1   Komaba,   Meguro-ku
Tokyo   156,   Japan

Key words and Phrases :   *Sequential estimation procedure ; stopping rule ; third order asymptotic efficiency ; maximum likelihood estimation procedure ; Wald identity ; asymptotically median unbiased estimator ; Edgeworth expansion.*

ABSTRACT

Under suitable regularity conditions, the third order asymptotic bounds for distributions of regular estimators are obtained. It is shown that the modified maximum likelihood estimation procedure combined with appropriate stopping rule is uniformly third order asymptotically efficient in the sense that its asymptotic distribution attains the bound uniformly in stopping rules up to the third order.

1. INTRODUCTION

We consider a class of sequences of sequential estimation procedures $\{\Pi_\alpha : \alpha = 1,2, \cdots\}$, where for each α, Π_α denotes a sequential estimation procedure, that is, an estimating method combined with a stopping rule. We consider such a sequence that the expected sample size tends to infinity as $\alpha \to \infty$, also that the sample size n is almost constant in the sense that $V_\alpha(n)/\{E_\alpha(n)\}^2 \to 0$ as $\alpha \to \infty$ or more precisely $V_\alpha(n)/E_\alpha(n) = O(1)$ as $\alpha \to \infty$. In the previous paper by the same authors (1987), the Bhattacharyya type bound for asymptotic variances of estimation procedures is obtained, up to the second order, and it is also shown that the modified maximum likelihood estimation procedure attains the bound if the stopping rule is properly determined.

In this paper it is shown that the third order asymptotic bounds for distributions of regular estimators which are completely similar to the fixed sample case, can be obtained, and it is also shown that for a choice of appropriate stopping rule, we may attain the third order bound uniformly which is generally impossible for the fixed sample case.

2.   Wald's identity and moments

Suppose that $X_1, \cdots, X_n$ are independently and identically distributed (i. i. d.) random variables according to a distribution with a density $f(x, \theta)$ with respect to a σ-finite measure μ, where θ is a real-valued parameter.

We assume an usual set of regularity conditions on $f(x, \theta)$. And now we assume that a sequential sampling rule is given, with which we continue to observe $X_1, X_2, \cdots, X_n$ until $n = N_0$, and make an inference (estimation, test, etc.) based on $X_1, \cdots, X_{N_0}$. Whether or not we stop sampling at n is decided based on $X_1, \cdots, X_n$.

In the subsequent discussions we consider asymptotic case where we actually consider a sequence of sequential inference rules $\{\Pi_a\}(a = 1, 2, \cdots)$ in which expected sample size $v_a(\theta) = E_{\theta,a}(n)$ approaches to infinity as $a \to \infty$, and consider limiting properties of inference taken.

In what follows we actually consider asymptotically almost fixed sample size rule, that is, we assume that $n/v_a(\theta)$ converges in probability to 1 uniformly in θ as $a \to \infty$. More precisely we assume the following

(A.1) For any fixed point θ_0, $v_a(\theta)/v_a(\theta_0) = O(1)$, $E_{\theta,a}(n) = v_a(\theta) + O(1/v_a(\theta))$,

$V_{\theta,a}(n)/v_a(\theta) = O(1)$, $E_{\theta,a}(n^k)/\{v_a(\theta)\}^k = O(1)$ $(k = 2,3,4)$,

and

$\{(\partial^k/\partial\theta^k)v_a(\theta)\}/v_a(\theta) = O(1)$ $(k = 1,2)$, uniformly in a neighborhood of θ_0,

where $E_{\theta,a}(\cdot)$ and $V_{\theta,a}(\cdot)$ designate expectation and variance, respectively.

As the basic tool of the analysis we shall make frequent use of the following fundamental lemma of A.Wald (1959). Suppose that $X_1, \cdots, X_n$ are i.i.d. random variables.

<u>Proposition 2.1.</u>   Let $Z_n = g(X_1) + \cdots + g(X_n)$ and $\phi(t) = E[\exp\{itg(X_1)\}]$. Then we have

$$E[e^{itZ_n}/\{\phi(t)\}^n] = 1.$$

This proposition can be easily generalized to the following :

<u>Proposition 2.2.</u>   Let $Z_n{}^j = g_j(X_1) + \cdots + g_j(X_n)$ $(j = 1, \cdots, k)$, and $\phi(t_1, \cdots, t_k)$ $= E[\exp\{it_1g_1(X_1) + \cdots + it_kg_k(X_1)\}]$. Then we have

$$E\left[\frac{exp(it_1 Z_n^1 + \cdots + it_k Z_n^k)}{\{\phi(t_1, \cdots, t_k)\}^n}\right] = 1. \tag{2.1}$$

From this we have the following lemma.

$\underline{\text{Lemma 2.1.}}$ Let $Z_n{}^j$ $(j=1,\cdots,k)$ be defined as above, and denote $\mu_j = E[g_j(X_1)]$ $(j=1,\cdots,k)$ and

$$\mu(j_1,\cdots,j_{k'}) = E[\{g_{j_1}(X_1)-\mu_{j_1}\}\cdots\{g_{j_k}(X_1)-\mu_{j_k}\}], \text{ where } \{j_1,\cdots,j_k\}\subset\{1,\cdots,k\}.$$

Then we have

$$E[Z_n{}^j] = E(n)\mu_j \quad (j=1,\cdots,k);$$

$$E[(Z_n{}^{j_1}-\mu_{j_1})(Z_n{}^{j_2}-\mu_{j_2})] = E(n)\mu(j_1,j_2);$$

$$E[(Z_n{}^{j_1}-\mu_{j_1})(Z_n{}^{j_2}-\mu_{j_2})(Z_n{}^{j_3}-\mu_{j_3})] = E(n)\mu(j_1,j_2,j_3) + E(nZ_n{}^{j_1})\mu(j_2,j_3)$$
$$+ E(nZ_n{}^{j_2})\mu(j_1,j_3) + E(nZ_n{}^{j_3})\mu(j_1,j_2);$$

$$E[(Z_n{}^{j_1}-\mu_{j_1})(Z_n{}^{j_2}-\mu_{j_2})(Z_n{}^{j_3}-\mu_{j_3})(Z_n{}^{j_4}-\mu_{j_4})]$$
$$= E(n)\mu(j_1,j_2,j_3,j_4) - E(n^2)\{\mu(j_1,j_2)\mu(j_3,j_4) + \mu(j_1,j_3)\mu(j_2,j_4)$$
$$+ \mu(j_1,j_4)\mu(j_2,j_3)\} + E(nZ_n{}^{j_1}Z_n{}^{j_2})\mu(j_3,j_4) + E(nZ_n{}^{j_2}Z_n{}^{j_3})\mu(j_1,j_4)$$
$$+ E(nZ_n{}^{j_3}Z_n{}^{j_4})\mu(j_1,j_2) + E(nZ_n{}^{j_1}Z_n{}^{j_3})\mu(j_2,j_4)$$
$$+ E(nZ_n{}^{j_2}Z_n{}^{j_4})\mu(j_1,j_3) + E(nZ_n{}^{j_1}Z_n{}^{j_4})\mu(j_2,j_3)$$
$$+ E(nZ_n{}^{j_1})\mu(j_2,j_3,j_4) + E(nZ_n{}^{j_2})\mu(j_1,j_3,j_4)$$
$$+ E(nZ_n{}^{j_3})\mu(j_1,j_2,j_4) + E(nZ_n{}^{j_4})\mu(j_1,j_2,j_3),$$

where $\{j_1,\cdots,j_k\}\subset\{1,\cdots,k\}$, provided that the differentiation under the integral sign of the left-hand side of (2.1) is allowed.

$\underline{\text{Proof.}}$ We shall give the proof of the last equality since the previous ones are far easier to prove. For simplicity we shall write 1,2,3,4 instead of j_1, j_2, j_3, j_4, and also assume $\mu_{j_\alpha}=0$ $(\alpha=1,2,3,4)$ without loss of generality. Putting $t_j=it_j$ $(j=1,2,3,4)$ in Proposition 2.2 we have

$$1 = E\left[\frac{exp(t_1 Z_n^1 + t_2 Z_n^2 + t_3 Z_n^3 + t_4 Z_n^4)}{\{\phi(t_1,t_2,t_3,t_4)\}^n}\right]$$
$$= E[\exp\{t_1 Z_n{}^1 + t_2 Z_n{}^2 + t_3 Z_n{}^3 + t_4 Z_n{}^4 - n\psi(t_1,t_2,t_3,t_4)\}]$$
$$= E(e^\Phi) \quad \text{(say)}, \tag{2.2}$$

where $\psi(t_1,t_2,t_3,t_4) = \log\phi(t_1,t_2,t_3,t_4)$. Differentiating both sides of (2.2) we obtain

$$\frac{\partial^4}{\partial t_1 \partial t_2 \partial t_3 \partial t_4} E(e^\Phi) = 0$$

Since differentiation and expectation can be interchanged , we have

$$0 = E\left[\frac{\partial^4}{\partial t_1 \partial t_2 \partial t_3 \partial t_4} e^\Phi\right]$$

$$= E\left[\frac{\partial^3}{\partial t_1 \partial t_2 \partial t_3}\left(\frac{\partial\Phi}{\partial t_4} e^\Phi\right)\right]$$

336

$$= E\left[\frac{\partial^2}{\partial t_1 \partial t_2} \left\{ \left(\frac{\partial^2 \phi}{\partial t_3 \partial t_4} + \frac{\partial \phi}{\partial t_3} \frac{\partial \phi}{\partial t_4} \right) e^\phi \right\} \right]$$

$$= E\left[\frac{\partial}{\partial t_1} \left\{ \left(\frac{\partial^3 \phi}{\partial t_2 \partial t_3 \partial t_4} + \frac{\partial^2 \phi}{\partial t_3 \partial t_4} \cdot \frac{\partial \phi}{\partial t_2} + \frac{\partial^2 \phi}{\partial t_2 \partial t_4} \cdot \frac{\partial \phi}{\partial t_3} + \frac{\partial \phi}{\partial t_4} \cdot \frac{\partial^2 \phi}{\partial t_2 \partial t_3} + \frac{\partial \phi}{\partial t_4} \cdot \frac{\partial \phi}{\partial t_3} \cdot \frac{\partial \phi}{\partial t_2} \right) e^\phi \right\} \right]$$

$$= E\left[\frac{\partial}{\partial t_1} \left\{ \left(\phi_{234} + \phi_{34}\phi_2 + \phi_{24}\phi_3 + \phi_4\phi_{23} + \phi_4\phi_3\phi_2 \right) e^\phi \right\} \right]$$

$$= E\Bigg[\bigg(\phi_{1234} + \phi_{134}\phi_2 + \phi_{34}\phi_{12} + \phi_{124}\phi_3 + \phi_{24}\phi_{13} + \phi_{14}\phi_{23} + \phi_4\phi_{123}$$

$$+ \phi_{14}\phi_3\phi_2 + \phi_4\phi_{13}\phi_2 + \phi_4\phi_3\phi_{12} + \phi_1\phi_{234} + \phi_1\phi_2\phi_{34}$$

$$+ \phi_1\phi_3\phi_{24} + \phi_1\phi_4\phi_{23} + \phi_1\phi_2\phi_3\phi_4 \bigg) e^\phi \Bigg],$$

where $\phi_\alpha = \partial\phi/\partial t_\alpha$, $\phi_{\alpha\beta} = \partial^2\phi/\partial t_\alpha \partial t_\beta$, $\phi_{\alpha\beta\gamma} = \partial^3\phi/\partial t_\alpha \partial t_\beta \partial t_\gamma$ and $\phi_{\alpha\beta\gamma\delta} = \partial^4\phi/\partial t_\alpha \partial t_\beta \partial t_\gamma \partial t_\delta$ $(\alpha,\beta,\gamma,\delta = 1,\cdots,4)$. Then we obtain

$$0 = E[-n\mu(1,2,3,4) - n\mu(1,3,4)Z_n{}^2 + n^2\mu(3,4)\mu(1,2) - n\mu(1,2,4)Z_n{}^3$$
$$+ n^2\mu(2,4)\mu(1,3) + n^2\mu(1,4)\mu(2,3) - n\mu(1,2,3)Z_n{}^4 - n\mu(1,4)Z_n{}^2 Z_n{}^3$$
$$- n\mu(1,3)Z_n{}^2 Z_n{}^4 - n\mu(1,2)Z_n{}^3 Z_n{}^4 - n\mu(2,3,4)Z_n{}^1 - n\mu(3,4)Z_n{}^1 Z_n{}^2$$
$$- n\mu(2,4)Z_n{}^1 Z_n{}^3 - n\mu(2,3)Z_n{}^1 Z_n{}^4 + Z_n{}^1 Z_n{}^2 Z_n{}^3 Z_n{}^4],$$

hence

$$E[Z_n{}^1 Z_n{}^2 Z_n{}^3 Z_n{}^4] = E(n)\mu(1,2,3,4) - E(n^2)\{\mu(1,2)\mu(3,4) + \mu(1,3)\mu(2,4)$$
$$+ \mu(1,4)\mu(2,3)\} + E(nZ_n{}^1 Z_n{}^2)\mu(3,4) + E(nZ_n{}^2 Z_n{}^3)\mu(1,4)$$
$$+ E(nZ_n{}^3 Z_n{}^4)\mu(1,2) + E(nZ_n{}^1 Z_n{}^3)\mu(2,4) + E(nZ_n{}^2 Z_n{}^4)\mu(1,3)$$
$$+ E(nZ_n{}^1 Z_n{}^4)\mu(2,3) + E(nZ_n{}^1)\mu(2,3,4) + E(nZ_n{}^2)\mu(1,3,4)$$
$$+ E(nZ_n{}^3)\mu(1,2,4) + E(nZ_n{}^4)\mu(1,2,3).$$

This completes the proof.

<u>Remark 2.1.</u>   From Lemma 2.1 it follows that the fourth order cumulant of $Z_n{}^{j_1}$, $Z_n{}^{j_2}$, $Z_n{}^{j_3}$ and $Z_n{}^{j_4}$ is given by

$$\kappa(Z_n{}^{j_1}, Z_n{}^{j_2}, Z_n{}^{j_3}, Z_n{}^{j_4})$$
$$= E(n)\mu(j_1, j_2, j_3, j_4) - V(n)\{\mu(j_1, j_2)\mu(j_3, j_4) + \mu(j_1 j_3)\mu(j_2 j_4)$$
$$+ \mu(j_1 j_4)\mu(j_2 j_3)\} + \mathrm{Cov}(n, Z_n{}^{j_1} Z_n{}^{j_2})\mu(j_3 j_4)$$
$$+ \mathrm{Cov}(n, Z_n{}^{j_2} Z_n{}^{j_3})\mu(j_1 j_4) + \mathrm{Cov}(n, Z_n{}^{j_3} Z_n{}^{j_4})\mu(j_1 j_2)$$
$$+ \mathrm{Cov}(n, Z_n{}^{j_1} Z_n{}^{j_3})\mu(j_2 j_4) + \mathrm{Cov}(n, Z_n{}^{j_2} Z_n{}^{j_4})\mu(j_1 j_3)$$
$$+ \mathrm{Cov}(n, Z_n{}^{j_1} Z_n{}^{j_4})\mu(j_2 j_3) + E(nZ_n{}^{j_1})\mu(j_2 j_3 j_4)$$

$$+ \mathrm{E}(nZ_n j_2)\mu(j_1 j_3 j_4) + \mathrm{E}(nZ_n j_3)\mu(j_1 j_2 j_4)$$
$$+ \mathrm{E}(nZ_n j_4)\mu(j_1 j_2 j_3).$$

3. Asymptotic cumulants of sequential estimation procedures

In the same situation as in the previous section, we shall obtain asymptotic cumulants of sequential estimation procedures. For simplicity we denote $v_a(\theta)$ by v. Let $\textcircled{H}$ be a parameter space which is assumed to be an open subset of Euclidean 1-space R^1. In order to obtain the asymptotic cumulants of sequential estimation procedures, we assume the following conditions.

(A.2) $\{x:f(x,\theta)>0\}$ does not depend on θ, and $f(x,\theta_1)/f(x,\theta_2)$ is not equal to constant for all disjoint points θ_1, θ_2.

(A.3) For almost all $x[\mu]$, $f(x,\theta)$ is four times continuously differentiable in θ. In the Taylor expansion

$$\log \frac{f(x,\theta+h)}{f(x,\theta)} = \sum_{i=1}^{4} \frac{h^i}{i!} \ell^{(i)}(\theta,x) + h^4 R(x,h)$$

$R(x,h)$ is uniformly bounded by a function $G(x)$ which has moments up to the fourth, where

$\ell^{(i)}(\theta,x) = (\partial^i/\partial\theta^i)\ell(\theta,x)(i=1,2,3,4)$ with $\ell(\theta,x) = \log f(x,\theta)$

(A.4)   For each θ, $\mathrm{E}_\theta[\ell^{(1)}(\theta,X)] = 0$,

$$0 < I(\theta) = \mathrm{E}_\theta[\{\ell^{(1)}(\theta,X)\}^2] = -\mathrm{E}_\theta[\ell^{(2)}(\theta,X)] < \infty$$

and $I(\theta)$ is twice continuously differentiable in θ.

(A.5)   There exist

$J(\theta) = \mathrm{E}_\theta[\ell^{(1)}(\theta,X)\ell^{(2)}(\theta,X)]$, $K(\theta) = \mathrm{E}_\theta[\{\ell^{(1)}(\theta,X)\}^3]$,

$L(\theta) = \mathrm{E}_\theta[\ell^{(1)}(\theta,X)\ell^{(3)}(\theta,X)]$, $M(\theta) = \mathrm{E}_\theta[\{\ell^{(2)}(\theta,X)\}^2] - I^2(\theta)$,

$N(\theta) = \mathrm{E}_\theta[\{\ell^{(1)}(\theta,X)\}^2\ell^{(2)}(\theta,X)] + I^2(\theta)$,

$H(\theta) = \mathrm{E}_\theta[\{\ell^{(1)}(\theta,X)\}^4] - 3I^2(\theta)$,

and both of $J(\theta)$ and $K(\theta)$ are differentiable in θ, and $\mathrm{E}_\theta[\ell^{(3)}(\theta,X)] = -3J(\theta) - K(\theta)$ and $\mathrm{E}_\theta[\ell^{(4)}(\theta,X)]=-4L(\theta)-3M(\theta)-6N(\theta)-H(\theta)$.

We put

$$Z_{1,v} = \frac{1}{\sqrt{v}} \sum_{i=1}^{n} \ell^{(1)}(\theta,X_i), \quad Z_{2,v} = \frac{1}{\sqrt{v}} \sum_{i=1}^{n} \{\ell^{(2)}(\theta,X_i) + I(\theta)\},$$

and

$$Z_{3,v} = \frac{1}{\sqrt{v}} \sum_{i=1}^{n} \{\ell^{(3)}(\theta,X_i) - 3J(\theta) - K(\theta)\}.$$

The following lemma is very useful for calculations of asymptotic cumulants.

AKAHIRA AND TAKEUCHI

<u>Lemma 3,1.</u>   Suppose that Y_θ is a function of $X_1, \cdots, X_n$ and θ and is differentiable in θ. Then

$$E_\theta(Z_{1,v} Y_\theta) = \frac{1}{\sqrt{v}} \frac{d}{d\theta} E_\theta(Y_\theta) - \frac{1}{\sqrt{v}} E_\theta\left(\frac{\partial Y_\theta}{\partial \theta} \right)$$

and

$$E_\theta(Z^2_{1,v} Y_\theta) = \frac{1}{\sqrt{v}} \frac{d}{d\theta} E_\theta(Z_{1,v} Y_\theta) - \frac{1}{\sqrt{v}} E_\theta\left(Y_\theta \frac{\partial Z_{1,v}}{\partial \theta} \right) - \frac{1}{\sqrt{v}} E_\theta\left(Z_{1,v} \frac{\partial Y_\theta}{\partial \theta} \right),$$

provided that the differentiations under the integral signs of $E_\theta(Y_\theta)$ and $E_\theta(Z_{1,v} Y_\theta)$ are allowed, respectively.

The proof is omitted since the lemma is similar to Lemmas 5.1.1. and 5.1.2 in Akahira and Takeuchi (1981) (see also Lemmas 2.1.1 and 2.1.2 in Akahira, 1986).

An estimator $\hat\theta_a = \hat\theta_a (X_1, \cdots, X_n)$ of θ is called to be $\sqrt{v}$-consistent if for any $\varepsilon > 0$ and any η of $Ⓗ$, there exist a sufficiently small positive number δ and a sufficiently large positive number L satisfying the following :

$$\overline{\lim_{a\to\infty}} \sup_{\theta : |\theta - \eta| < \delta} P_\theta\{\sqrt{v} |\hat\theta_a - \theta| \geq L\} < \varepsilon.$$

For each $k = 1, 2, \cdots$, a $\sqrt{v}$-consistent estimator $\hat\theta_a$ is called k-th order asymptotically median unbiased (or k-th order AMU) estimator if for any $\eta \in Ⓗ$ there exists a positive number δ such that

$$\lim_{a\to\infty} \sup_{\theta : |\theta - \eta| < \delta} v^{(k-1)/2} |P_\theta\{\hat\theta_a \leq \theta\} - \frac{1}{2}| = 0,$$

$$\lim_{a\to\infty} \sup_{\theta : |\theta - \eta| < \delta} v^{(k-1)/2} |P_\theta\{\hat\theta_a \geq \theta\} - \frac{1}{2}| = 0.$$

Let C be the class of the all bias-adjusted BAN estimators $\hat\theta_a$ which are third order AMU and asymptotically expanded as

$$\sqrt{v} (\hat\theta_a - \theta) = \frac{Z_{1,v}(\theta)}{I(\theta)} + \frac{1}{\sqrt{v}} Q + \frac{1}{v} R + o_p(\frac{1}{v}),$$

where $Q = O_p(1)$, $\partial Q/\partial \theta = O_p(1)$, $R = O_p(1)$, $\partial R/\partial \theta = O_p(1)$, and the distribution of $\sqrt{v}(\hat\theta_a - \theta)$ admits the Edgeworth expansion up to the order $v - 1$. If an estimator $\hat\theta_a$ belongs to the class C, then we call it C-estimator. However, in the subsequent discussion, the term R is not explicitly needed, hence we simply write

$$\sqrt{v} (\hat\theta_a - \theta) = \frac{Z_{1,v}(\theta)}{I(\theta)} + \frac{1}{\sqrt{v}} Q + o_p(\frac{1}{\sqrt{v}}), \tag{3.1}$$

although the above stochastic expansion is necessary in order to validate the proofs, since they actually depends on two facts that (i) the asymptotic distribution of $\sqrt{v}(\hat\theta_a - \theta)$ is equivalent to that of

$$\frac{Z_{1,v}(\theta)}{I(\theta)} + \frac{1}{\sqrt{v}}Q + \frac{1}{v}R$$

up to the order $v-1$, and that (ii) the asymptotic moments of R and their derivatives are of $0(1)$.

Theorem 3.1. Assume that the conditions (A.1) to (A.5) hold. Let $\hat{\theta}_a$ be an C-estimator with (3.1). Then its asymptotic cumulants κ_i ($i=1,2,3,4$) up to the fourth and the order $o(v-1)$ are given as follows : For $T_a = \sqrt{v}(\hat{\theta}_a - \theta)$

$$\kappa_1 = E_\theta(T_a) = \frac{\mu_1(\theta)}{\sqrt{v}} + \frac{\mu_2(\theta)}{v} + o(\frac{1}{v}),$$

$$\kappa_2 = V_\theta(T_a) = \frac{1}{I(\theta)} + \frac{2\mu_1'(\theta)}{vI(\theta)} - \frac{2v\mu_1(\theta)}{v^2 I(\theta)} + \frac{\tau(\theta)}{v} + o(\frac{1}{v}),$$

$$\kappa_3 = \kappa_{3,\theta}(T_a) = E_\theta[\{T_a - E_\theta(T_a)\}^3] = \frac{\beta_3(\theta)}{\sqrt{v}} + \frac{r_3(\theta)}{v} + o(\frac{1}{v}),$$

$$\kappa_4 = \kappa_{4,\theta}(T_a) = E_\theta[\{T_a - E_\theta(T_a)\}^4] - 3\{V_\theta(T_a)\}^2 = \frac{\beta_4(\theta)}{v} + \frac{r_4(\theta)}{v} + o(\frac{1}{v})$$

where

$$\mu_1(\theta) = \frac{1}{6}I(\theta)\beta_3(\theta) \quad with \quad \beta_3(\theta) = -\frac{3J(\theta) + 2K(\theta)}{I^3(\theta)} - \frac{3v'}{vI^2(\theta)},$$

$$\mu_2(\theta) = \frac{1}{6}I(\theta)r_3(\theta) \quad with \quad r_3(\theta) = \frac{3}{2I(\theta)}E_\theta[Z_{1,v}(Q - \mu_1(\theta))^2],$$

$$\beta_4(\theta) = \frac{12(2J(\theta) + K(\theta))(J(\theta) + K(\theta))}{I^5(\theta)} - \frac{3H(\theta) + 4L(\theta) + 12N(\theta)}{I^4(\theta)} - \frac{3}{I^4(\theta)}\left(M(\theta) - \frac{J^2(\theta)}{I(\theta)}\right)$$

$$+ \frac{5v'}{3I^4(\theta)v}(30J(\theta) + 19K(\theta)) + \frac{27v'^2}{I^3(\theta)v^2} - \frac{4v''}{I^3(\theta)v},$$

$$r_4(\theta) = \frac{3}{I^4(\theta)}E_\theta(W^2) \quad with \quad W = Z_{2,v} - (\frac{J}{I} - \frac{v'}{v})Z_{1,v} - \frac{(n-v)I}{\sqrt{v}},$$

$\tau(\theta) = V_\theta(Q)$, $v' = (\partial/\partial\theta)v_a(\theta)$, and $v'' = (\partial^2/\partial\theta^2)v_a(\theta)$.

Remark 3.1. If $v = v_a(\theta)$ is independent of θ, then it is seen that the cumulants in Theorem 3.1 coincide with those in non-sequential case given in Theorem 2.1.1 in Akahira (1986), since $v' = v'' = 0$ and $E_\theta(W2) = M(\theta) - \{J2(\theta)/I(\theta)\}$.

Remark 3.2.   Note that $\mu_1(\theta)$, $\beta_3(\theta)$ and $\beta_4(\theta)$ in Theorem 3.1 are independent of the specific estimator $\hat{\theta}_a$.

Theorem 3.2. Assume that the conditions (A.1) to (A.5) hold. Let $\hat{\theta}_a$ be an C-estimator with (3.1). Then the Edgeworth expansion of the distribution of $\hat{\theta}_a$ up to the order v^{-1} is given by

$$P_\theta\{\sqrt{\overline{I(\theta)v}}\,(\hat{\theta}_a-\theta)\leqq t\}$$

$$=\Phi(t)-\frac{I(\theta)\sqrt{I(\theta)}\,\beta_3(\theta)}{6\sqrt{\overline{v}}}\,t^2\phi(t)-\frac{I^2(\theta)\{\beta_4(\theta)+\gamma_4(\theta)\}}{24v}(t^3-3t)\phi(t)$$

$$-\frac{I^3(\theta)\beta_3^2(\theta)}{72v}(t^5-10t^3+15t)\phi(t)-\frac{I(\theta)\{\tau(\theta)+\mu_1^2(\theta)\}}{2v}t\phi(t)$$

$$-\frac{I(\theta)\sqrt{I(\theta)}\,\gamma_3(\theta)}{6v}\,t^2\phi(t)-\frac{I^2(\theta)\,\beta_3(\theta)\mu_1(\theta)}{6v}t(t^2-3)\phi(t)+o\left(\frac{1}{v}\right),\tag{3.2}$$

where $\mu_1(\theta)$, $\mu_2(\theta)$, $\beta_3(\theta)$, $\beta_4(\theta)$, $\gamma_3(\theta)$, $\gamma_4(\theta)$ and $\tau(\theta)$ are given in Theorem 3.1, and

$$\Phi(t)=\int_{-\infty}^{t}\phi(x)dx \text{ with } \phi(x)=(1/\sqrt{\overline{2\pi}})\,e^{-x^2/2}.$$

Theorem 3.3 Assume that the conditions (A.1) to (A.5) hold. Let $\hat{\theta}_a$ be any C-estimator with (3.1). Let $E_\theta(WQ^*)=rE_\theta(W^2)/I+o(1)$ with $W=Z_{2,v}-\{(J/I)-(v'/v)\}Z_{1,v}-\{(n-v)I/\sqrt{\overline{v}}\,\}$.   Then the asymptotic bound for the distributions of $\hat{\theta}_a$ and a fixed stopping rule up to the order v^{-1} is given as follows:

$$P_\theta\{\sqrt{\overline{I(\theta)v}}\,(\hat{\theta}_a-\theta)\leqq t\}$$

$$\begin{array}{c}\leqq\\(\geqq)\end{array}\ \Phi(t)-\frac{I(\theta)\sqrt{I(\theta)}\,\beta_3(\theta)}{6\sqrt{\overline{v}}}\,t^2\phi(t)-\frac{I^2(\theta)\,\beta_4(\theta)}{24v}(t^3-3t)\phi(t)$$

$$-\frac{I^3(\theta)\beta_3^2(\theta)}{72v}(t^5-10t^3+15t)\phi(t)-\frac{I(\theta)\mu_1^2(\theta)}{2v}t\phi(t)$$

$$-\frac{I^2(\theta)\,\beta_3(\theta)\mu_1(\theta)}{6v}t(t^2-3)\phi(t)-\frac{E_\theta(W^2)}{2vI^2(\theta)}t\phi(t)\{(\frac{t}{2}+r\sqrt{\overline{I(\theta)}})^2+\frac{1}{4}\}$$

$$-\frac{1}{4vI^3(\theta)}t\phi(t)\{J(\theta)+K(\theta)+\frac{2v'}{v}I(\theta)\}^2+o\left(\frac{1}{v}\right)\tag{3.3}$$

for all $t>0$ ($t<0$), where $\mu_1(\theta)$, $\beta_3(\theta)$ and $\beta_4(\theta)$ are given in Theorem 3.1, and $Q^*=Q-\mu_1(\theta)$.

Corollary 3.1. Assume that the conditions (A.1) to (A.5) hold. Let $\hat{\theta}_a$ be any C-estimator with (3.1). Then the asymptotic bound for the distributions of $\hat{\theta}_a$ and any stopping rule satisfying (A.1) up to the order $v-1$ is given by

$$P_\theta\{\sqrt{\overline{I(\theta)v}}\,(\hat{\theta}_a-\theta)\leq t\}$$

$$\underset{(\geq)}{\leq}\ \Phi(t)-\frac{I(\theta)\sqrt{\overline{I(\theta)}}\,\beta_3(\theta)}{6\sqrt{v}}t^2\phi(t)-\frac{I^2(\theta)\beta_4(\theta)}{24v}(t^3-3t)\phi(t)$$

$$-\frac{I^3(\theta)\beta_3^2(\theta)}{72v}(t^5-10t^3+15t)\phi(t)-\frac{I(\theta)\mu_1^2(\theta)}{2v}t\phi(t)$$

$$-\frac{I^2(\theta)\beta_3(\theta)\mu_1(\theta)}{6v}t(t^2-3)\phi(t)-\frac{1}{4vI^3(\theta)}t\phi(t)\{J(\theta)+K(\theta)+\frac{2v'}{v}I(\theta)\}^2+o(\frac{1}{v}) \qquad (3.4)$$

for all $t>0 (t<0)$.

Denote by $\hat{\theta}_{ML}$ the maximum likelihood(ML) estimator of θ based on the sample $(X_1,\cdots,X_n)$. Let $\hat{\theta}_{ML}^*$ be the estimator modified from the ML estimator to be third order AMU. Then it follows from a similar way to Akahira and Takeuchi (1981) that under the condition (A.1)

$$\sqrt{v}(\hat{\theta}_{ML}-\theta)=\frac{Z_{1,v}}{I}-(\frac{n-v}{v})\frac{Z_{1,v}}{I}+\frac{1}{I^2\sqrt{v}}(Z_{1,v}Z_{2,v}-\frac{3J+K}{2I}Z_{1,v}^2)+o_p(\frac{1}{\sqrt{v}}). \qquad (3.5)$$

Theorem 3.4. Assume that the conditions (A.1) to (A.5) hold. If the stopping rule is so determined that the observation is stopped at n satisfying

$$-\sum_{i=1}^{n}\ell^{(2)}(\hat{\theta}_{ML}^*,X_i)=v_a(\hat{\theta}_{ML}^*)I(\hat{\theta}_{ML}^*)+c(\hat{\theta}_{ML}^*)+\varepsilon \qquad (3.6)$$

with

$$c(\theta)=\frac{J(\theta)v'(\theta)}{I(\theta)v(\theta)}+\frac{L(\theta)}{I(\theta)}-\frac{v''(\theta)}{2v(\theta)}-\frac{1}{2I(\theta)}\{2L(\theta)+M(\theta)+N(\theta)\} \qquad (3.7)$$

and some random variable ε with $E_\theta(\varepsilon)=o(1)$, and the stopping rule of (3.6) satisfies (A.1), then the modified maximum likelihood estimation procedure combined with the stopping rule is uniformly third order asymptotically efficient in the sense that its asymptotic distribution attains the bound (3.4), up to the order $v-1$, uniformly in t and stopping rule.

It has been well known that for fixed sample estimations $E_\theta(W^2)$ $=E_\theta[\{Z_{2,v}-(J/I)Z_{1,v}\}^2]$ represents the minimum loss of information for any best asymptotically normal estimator (Fisher, 1925; Rao, 1961) or the curvature of the model (Efron, 1975 ; Amari, 1985). The above result establishes that by the choice of appropriate stopping rule the loss of information can be reduced to zero (!), or the model can be made flat (!), and it is also to be noted that the right-hand side of (3.6) can be interpreted as the "realized" or ex post amount of information, and the stopping rule concerned is that we continue to sample until we get the "realized" amount of information is (nearly) equal to the expected amount of information corresponding to the predetermined expected sample size. In practical cases it may be troublesome but not really quite difficult to determine ε precisely, and also to check the condition (A.1), but we may rather stop sampling as soon as the right-hand side of (3.6) exceeds $v(\hat\theta^*_{ML})I(\hat\theta^*_{ML})+c(\hat\theta^*_{ML})$. Then the expected sample size will not be exactly equal to 1, but the difference $E_\theta(n)-v(\theta)$ will be of small magnitude of $o(1)$.

<u>Proof of Theorem 3.1.</u>   Putting $U=T_a-E_\theta(T_a)$, we have

$$U=\sqrt{v}(\hat\theta_a-E_\theta(\hat\theta_a))$$

$$=\frac{Z_{1,v}}{I}+\frac{1}{\sqrt{v}}(Q-\mu_1)+o(\frac{1}{\sqrt{v}}), \tag{3.8}$$

where $Z_{1,v}$, I and μ_1 denote $Z_{1,v}(\theta)$, $I(\theta)$ and $\mu_1(\theta)$, respectively.
Since

$$U^2=(U-\frac{Z_{1,v}}{I})^2+\frac{2}{I}Z_{1,v}U-\frac{1}{I^2}Z_{1,v}^2 ,$$

it follows that

$$E_\theta(U^2)=\frac{1}{v}V_\theta(Q^*)+\frac{2}{I}E_\theta(Z_{1,v}U)-\frac{1}{I}+o(\frac{1}{v}), \tag{3.9}$$

where $Q^*=Q-\mu_1$.
Since

$$\frac{\partial U}{\partial\theta}=\frac{\partial}{\partial\theta}(\sqrt{v}(\hat\theta_n-E_\theta(\hat\theta_n)))$$

$$=\frac{v'}{2v}U-\sqrt{v}\frac{\partial}{\partial\theta}E_\theta(\hat\theta_n)$$

$$=\frac{v'}{2v}U-\sqrt{v}\left\{\frac{\partial}{\partial\theta}E_\theta(\hat\theta_n-\theta)+1\right\}$$

$$=\frac{v'}{2v}U-\sqrt{v}\left\{\frac{\partial}{\partial\theta}E_\theta\left(\frac{Z_{1,v}}{\sqrt{v}I}+\frac{1}{v}Q+o_p(\frac{1}{v})\right)+1\right\}$$

$$= \frac{\nu'}{2\nu}U - \sqrt{\bar{\nu}}\left\{\frac{\partial}{\partial\theta}\left(\frac{1}{\nu}E_\theta(Q)\right) + 1 + o\left(\frac{1}{\nu}\right)\right\}$$

$$= \frac{\nu'}{2\nu}U - \sqrt{\bar{\nu}}\left(-\frac{\nu'}{\nu^2}\mu_1 + \frac{1}{\nu}\mu'_1\right) - \sqrt{\bar{\nu}} + o\left(\frac{1}{\sqrt{\bar{\nu}}}\right)$$

$$= \frac{\nu'}{2\nu}U + \frac{\nu'}{\nu\sqrt{\bar{\nu}}}\mu_1 - \frac{1}{\sqrt{\bar{\nu}}}\mu'_1 - \sqrt{\bar{\nu}} + o\left(\frac{1}{\sqrt{\bar{\nu}}}\right), \tag{3.10}$$

it follows by Lemma 3.1 that

$$E_\theta(Z_{1,\nu}U) = \frac{1}{\sqrt{\bar{\nu}}}\frac{\partial}{\partial\theta}E_\theta(U) - \frac{1}{\sqrt{\bar{\nu}}}E_\theta\left(\frac{\partial U}{\partial\theta}\right)$$

$$= -\frac{1}{\sqrt{\bar{\nu}}}\left\{E_\theta\left(\frac{\nu'}{2\nu}U\right) + \frac{\nu'}{\nu\sqrt{\bar{\nu}}}\mu_1 - \frac{1}{\sqrt{\bar{\nu}}}\mu'_1 - \sqrt{\bar{\nu}}\right\} + o\left(\frac{1}{\nu}\right)$$

$$= 1 + \frac{1}{\nu}\mu'_1 - \frac{\nu'}{\nu^2}\mu_1 + o\left(\frac{1}{\nu}\right), \tag{3.11}$$

where ν' and μ_1' denote $\nu'(\theta) = (\partial/\partial\theta)\nu_a(\theta)$ and $\mu_1' = (\partial/\partial\theta)\mu_1(\theta)$, respectively.
From (3.8), (3.9) and (3.11) we have

$$\kappa_2 = V_\theta(T_a) = E_\theta(U^2)$$

$$= \frac{1}{I} + \frac{2}{\nu I}\mu'_1 - \frac{2\nu'}{\nu^2 I}\mu_1 + \frac{1}{\nu}V_\theta(Q^*) + o\left(\frac{1}{\nu}\right)$$

$$= \frac{1}{I} + \frac{2}{\nu I}\mu'_1 - \frac{2\nu'}{\nu^2 I}\mu_1 + \frac{1}{\nu}\tau(\theta) + o\left(\frac{1}{\nu}\right) \text{ (say).} \tag{3.12}$$

Since

$$U^3 = (U - \frac{Z_{1,\nu}}{I})^3 + \frac{3}{2}(U - \frac{Z_{1,\nu}}{I})^2 Z_{1,\nu} + \frac{3}{2}Z_{1,\nu}U^2 - \frac{1}{2I^3}Z_{1,\nu}^3,$$

it follows that

$$E_\theta(U^3) = \frac{3}{2\nu I}E_\theta(Z_{1,\nu}Q^{*2}) + \frac{3}{2I}E_\theta(Z_{1,\nu}U^2) - \frac{1}{2I^3}E_\theta(Z_{1,\nu}^3) + o\left(\frac{1}{\nu}\right). \tag{3.13}$$

Since

$$E_\theta(U^2) = \frac{1}{I} + O\left(\frac{1}{\nu}\right) \text{ and } I'(\theta) = \frac{dI(\theta)}{d\theta} = 2J(\theta) + K(\theta),$$

it follows from (3.10) and Lemma 3.1 that

$$E_\theta(Z_{1,\nu}U^2) = \frac{1}{\sqrt{\bar{\nu}}}\frac{\partial}{\partial\theta}E_\theta(U^2) - \frac{2}{\sqrt{\bar{\nu}}}E_\theta\left(U\frac{\partial U}{\partial\theta}\right) = -\frac{2J+K}{\sqrt{\bar{\nu}}I^2} - \frac{\nu'}{\nu^{3/2}I} + o\left(\frac{1}{\nu}\right) \tag{3.14}$$

where J and K denote $J(\theta)$ and $K(\theta)$, respectively.
Since

$$E_\theta(Z_{1,\nu}^3) = \frac{K}{\sqrt{\bar{\nu}}} + \frac{3\nu'}{\nu^{3/2}}I + o\left(\frac{1}{\nu}\right),$$

we have from (3.13) and (3.14)

$$\kappa_3 = \kappa_{3,\theta}(T_a) = E_\theta(U^3)$$

$$= \frac{3}{2\nu I} E_\theta(Z_{1,\nu} Q^{*2}) + \frac{3}{2I}\left(-\frac{2J+K}{\sqrt{\nu}I^2} - \frac{\nu'}{\nu^{3/2}I}\right) - \frac{1}{2I^3}\left(\frac{K}{\sqrt{\nu}} + \frac{3\nu'}{\nu^{3/2}}I\right) + o\left(\frac{1}{\nu}\right)$$

$$= -\frac{3J+2K}{\sqrt{\nu}I^3} - \frac{3\nu'}{\nu^{3/2}I^2} + \frac{3}{2\nu I} E_\theta(Z_{1,\nu} Q^{*2}) + o\left(\frac{1}{\nu}\right).$$

$$= \frac{1}{\sqrt{\nu}}\beta_3 + \frac{1}{\nu}\gamma_3(\theta) + o\left(\frac{1}{\nu}\right), \tag{3.15}$$

where $\quad \beta_3 = \beta_3(\theta) = -\sqrt{\nu}\{(3J+2K)/(\sqrt{\nu}I3)\} - \{3\nu'/(\nu3/2I2)\}$
and $\quad \gamma_3(\theta) = 3E_\theta(Z_{1,\nu} Q^{*2})/(2\ I).$

Since

$$U^4 = \left(U - \frac{Z_{1,\nu}}{I}\right)^4 + \frac{4}{3I} Z_{1,\nu}\left(U - \frac{Z_{1,\nu}}{I}\right)^3 - \frac{2}{I^2} Z_{1,\nu}^2 U^2 + \frac{8}{3I} Z_{1,\nu} U^3 + \frac{1}{3I^4} Z_{1,\nu}^4$$

it follows that

$$E_\theta(U^4) = -\frac{2}{I^2} E_\theta(Z_{1,\nu}^2 U^2) + \frac{8}{3I} E_\theta(Z_{1,\nu} U^3) + \frac{1}{3I^4} E_\theta(Z_{1,\nu}^4) \tag{3.16}$$

From (3.10), (3.15) and Lemma 3.1 we obtain

$$E_\theta(Z_{1,\nu} U^3) = \frac{1}{\sqrt{\nu}}\frac{\partial}{\partial\theta} E_\theta(U^3) - \frac{3}{\sqrt{\nu}} E_\theta\left(U^2\frac{\partial U}{\partial\theta}\right) = \frac{1}{\sqrt{\nu}}\frac{\partial}{\partial\theta}\left(-\frac{3J+2K}{\sqrt{\nu}I^3} - \frac{3\nu'}{\nu^{3/2}I^2}\right)$$

$$- \frac{3}{\sqrt{\nu}} E_\theta\left[U^2\left\{\frac{\nu'}{2\nu}U - \sqrt{\nu}\left(1 + \frac{1}{\nu}\mu_1' - \frac{\nu'}{\nu^2}\mu_1\right)\right\}\right] + o\left(\frac{1}{\nu}\right).$$

$$= \frac{1}{\sqrt{\nu}}\frac{\partial}{\partial\theta}\left(-\frac{3J+2K}{\sqrt{\nu}I^3} - \frac{3\nu'}{\nu^{3/2}I^2}\right)$$

$$- \frac{3\nu'}{2\nu^{3/2}} E_\theta(U^3) + 3\left(1 + \frac{1}{\nu}\mu_1' - \frac{\nu'}{\nu^2}\mu_1\right)E_\theta(U^2) + o\left(\frac{1}{\nu}\right). \tag{3.17}$$

Since

$$\frac{\partial Z_{1,\nu}}{\partial\theta} = Z_{2,\nu} - \frac{\nu'}{2\nu}Z_{1,\nu} - \frac{nI}{\sqrt{\nu}}, \tag{3.18}$$

it follows from (3.10), (3.14) and Lemma 3.1 that

$$E_\theta(Z_{1,\nu}^2 U^2) = \frac{1}{\sqrt{\nu}}\frac{d}{d\theta} E_\theta(Z_{1,\nu} U^2) - \frac{1}{\sqrt{\nu}} E_\theta\left(\frac{\partial Z_{1,\nu}}{\partial\theta}U^2\right) - \frac{2}{\sqrt{\nu}} E_\theta\left(Z_{1,\nu} U\frac{\partial U}{\partial\theta}\right)$$

$$= -\frac{1}{\sqrt{\nu}}\frac{d}{d\theta}\left(\frac{2J+K}{\sqrt{\nu}I^2} + \frac{\nu'}{\nu^{3/2}I}\right) - \frac{1}{\sqrt{\nu}} E_\theta\left[\left(Z_{2,\nu} - \frac{\nu'}{2\nu}Z_{1,\nu} - \frac{nI}{\sqrt{\nu}}\right)U^2\right]$$

THIRD ORDER ASYMPTOTIC EFFICIENCY 345

$$-\frac{2}{\sqrt{\bar{\nu}}}E_{\theta}\left[Z_{1,\nu}U\left(\frac{\nu'}{2\nu}U+\frac{\nu'}{\nu^{3/2}}\mu_1-\frac{1}{\sqrt{\bar{\nu}}}\mu_1'-\sqrt{\bar{\nu}}\right)\right]+o\left(\frac{1}{\nu}\right). \tag{3.19}$$

Putting

$$W=Z_{2,\nu}-(\frac{J}{I}-\frac{\nu'}{\nu})Z_{1,\nu}-\frac{(n-\nu)I}{\sqrt{\bar{\nu}}}. \tag{3.20}$$

we have by Lemma 3.1

$$E_{\theta}(WU^2)=E_{\theta}\left[W(U-\frac{Z_{1,\nu}}{I})^2\right]+\frac{2}{I}E_{\theta}(Z_{1,\nu}UW)-\frac{1}{I^2}E_{\theta}(Z_{1,\nu}^2 W)$$

$$=\frac{2}{I}\cdot\frac{1}{\sqrt{\bar{\nu}}}\frac{\partial}{\partial\theta}E_{\theta}(UW)-\frac{2}{I}\cdot\frac{1}{\sqrt{\bar{\nu}}}E_{\theta}(\frac{\partial U}{\partial\theta}W)-\frac{2}{I}\cdot\frac{1}{\sqrt{\bar{\nu}}}E_{\theta}(U\frac{\partial W}{\partial\theta})$$

$$-\frac{1}{I^2}E_{\theta}(Z_{1,\nu}^2 W)+o\left(\frac{1}{\sqrt{\bar{\nu}}}\right). \tag{3.21}$$

Since by Lemma 3.1

$$E_{\theta}(nZ_{1,\nu})=\frac{1}{\sqrt{\bar{\nu}}}\frac{\partial}{\partial\theta}E_{\theta}(n)=\frac{\nu'}{\sqrt{\bar{\nu}}}, \tag{3.22}$$

we obtain from (3.11)

$$E_{\theta}(UW)=E_{\theta}(Z_{2,\nu}U)-\frac{J}{I}E_{\theta}(Z_{1,\nu}U)-E_{\theta}(\frac{n-\nu}{\sqrt{\bar{\nu}}}IU)+\frac{\nu'}{\nu}E_{\theta}(Z_{1,\nu}U)$$

$$=E_{\theta}(\frac{1}{I}Z_{1,\nu}Z_{2,\nu})-\frac{J}{I}-E_{\theta}(\frac{n-\nu}{\sqrt{\bar{\nu}}}Z_{1,\nu})+\frac{\nu'}{\nu I}E_{\theta}(Z_{1,\nu}^2)+o(1)=o(1). \tag{3.23}$$

From (3.8) we have

$$\frac{\partial U}{\partial\theta}=\frac{d}{d\theta}\{\sqrt{\bar{\nu}}(\hat{\theta}_n-E_{\theta}(\hat{\theta}_n))\}$$

$$=\frac{\nu'}{2\nu}U-\sqrt{\bar{\nu}}\frac{d}{d\theta}E_{\theta}(\hat{\theta}_n-\theta)-\sqrt{\bar{\nu}}$$

$$=\frac{\nu'}{2\nu}U-\sqrt{\bar{\nu}}\frac{d}{d\theta}\{\frac{1}{\nu}E_{\theta}(Q)\}-\sqrt{\bar{\nu}}+o\left(\frac{1}{\sqrt{\bar{\nu}}}\right)$$

$$=\frac{\nu'}{2\nu}U-\sqrt{\bar{\nu}}+o\left(\frac{1}{\sqrt{\bar{\nu}}}\right). \tag{3.24}$$

Then by (3.20) and (3.22)

$$E_{\theta}(\frac{\partial U}{\partial\theta}W)=E_{\theta}\left[(\frac{\nu'}{2\nu}U-\sqrt{\bar{\nu}})\{Z_{2,\nu}-(\frac{J}{I}-\frac{\nu'}{\nu})Z_{1,\nu}-\frac{(n-\nu)I}{\sqrt{\bar{\nu}}}\}\right]+o(1)$$

$$=E_{\theta}\left[\frac{\nu'}{2\nu}\{Z_{2,\nu}U-(\frac{J}{I}-\frac{\nu'}{\nu})Z_{1,\nu}U-\frac{(n-\nu)I}{\sqrt{\bar{\nu}}}U\}\right]+o(1)$$

$$=\frac{\nu'}{2\nu}E\left[\frac{1}{I}Z_{1,\nu}Z_{2,\nu}-\frac{1}{I}(\frac{J}{I}-\frac{\nu'}{\nu})Z_{1,\nu}^2-\frac{(n-\nu)}{\sqrt{\bar{\nu}}}Z_{1,\nu}\right]+o(1)$$

$$=o(1). \tag{3.25}$$

Since $I'(\theta) = 2J(\theta) + K(\theta)$ and $J'(\theta) = L(\theta) + M(\theta) + N(\theta)$, it follows that

$$\frac{\partial Z_{2,\nu}}{\partial \theta} = -\frac{\nu'}{2\nu} Z_{2,\nu} + Z_{3,\nu} - \frac{nJ}{\sqrt{\nu}} + o_p(1)$$

$$\frac{\partial}{\partial \theta}\left((\frac{J}{I} - \frac{\nu'}{\nu})Z_{1,\nu}\right) = \{\frac{1}{I}(L+M+N) - \frac{J(2J+K)}{I^2} + \frac{\nu'^2}{\nu^2} - \frac{\nu''}{\nu}\}Z_{1,\nu} + (\frac{J}{I} - \frac{\nu'}{\nu})(Z_{2,\nu} - \frac{\nu'}{2\nu}Z_{1,\nu} - \frac{nI}{\sqrt{\nu}}) + o_p(1),$$

and

$$\frac{\partial}{\partial \theta}\left(\frac{(n-\nu)I}{\sqrt{\nu}}\right) = -\frac{\nu'I}{\sqrt{\nu}} + \frac{(n-\nu)(2J+K)}{\sqrt{\nu}} - \frac{(n-\nu)\nu'I}{2\nu^{3/2}} + o_p(1)$$

Hence

$$\frac{\partial W}{\partial \theta} = \frac{\partial Z_{2,\nu}}{\partial \theta} - \frac{\partial}{\partial \theta}((\frac{J}{I} - \frac{\nu'}{\nu})Z_{1,\nu}) - \frac{\partial}{\partial \theta}\left(\frac{(n-\nu)I}{\sqrt{\nu}}\right)$$

$$= -\frac{\nu'}{2\nu} Z_{2,\nu} + Z_{3,\nu} - \frac{nJ}{\sqrt{\nu}}$$

$$-\frac{1}{I}(L+M+N) Z_{1,\nu} - (\frac{J}{I} - \frac{\nu'}{\nu})(Z_{2,\nu} - \frac{\nu'}{2\nu}Z_{1,\nu} - \frac{nI}{\sqrt{\nu}}) + \frac{J(2J+K)}{I^2} Z_{1,\nu}$$

$$+ (\frac{\nu''}{\nu} - \frac{\nu'^2}{\nu^2})Z_{1,\nu} + \frac{\nu'I}{\sqrt{\nu}} - \frac{(n-\nu)(2J+K)}{\sqrt{\nu}} + \frac{(n-\nu)\nu'I}{2\nu^{3/2}} + o_p(1),$$

which implies by (3.22)

$$E_\theta[U\frac{\partial W}{\partial \theta}] = \frac{1}{I} E_\theta\Bigg[-\frac{\nu'}{2\nu} Z_{1,\nu}Z_{2,\nu} + Z_{1,\nu}Z_{3,\nu} - \frac{nJ}{\sqrt{\nu}} Z_{1,\nu}$$

$$-\frac{1}{I}(L+M+N) Z_{1,\nu}^2 - (\frac{J}{I} - \frac{\nu'}{\nu})(Z_{1,\nu}Z_{2,\nu} - \frac{\nu'}{2\nu}Z_{1,\nu}^2 - \frac{nZ_{1,\nu}}{\sqrt{\nu}}I)$$

$$+ \frac{J(2J+K)}{I^2} Z_{1,\nu}^2 + (\frac{\nu''}{\nu} - \frac{\nu'^2}{\nu^2})Z_{1,\nu}^2 - \frac{(n-\nu)(2J+K)}{\sqrt{\nu}} Z_{1,\nu} + \frac{(n-\nu)\nu'I}{2\nu^{3/2}} Z_{1,\nu}\Bigg]$$

$$= \frac{1}{I}\Bigg\{ -\frac{\nu'}{2\nu}J + L - \frac{J}{\sqrt{\nu}}\cdot\frac{\nu'}{\sqrt{\nu}} - (L+M+N)$$

$$-(\frac{J}{I} - \frac{\nu'}{\nu})(J - \frac{\nu'}{2\nu}I - \frac{I}{\sqrt{\nu}}\cdot\frac{\nu'}{\sqrt{\nu}}) + \frac{J(2J+K)}{I} + (\frac{\nu''}{\nu} - \frac{\nu'^2}{\nu^2})I$$

$$-\frac{(2J+K)\nu'}{\nu} + \frac{I\nu'^2}{2\nu^2}\Bigg\}$$

$$= \frac{1}{I}\Bigg\{\frac{J(J+K)}{I} - (M+N) - (J+K)\frac{\nu'}{\nu} + \frac{I\nu''}{\nu} - \frac{2I\nu'^2}{\nu^2}\Bigg\}. \tag{3.26}$$

We also have from (3.20) and (3.24)

$$E_\theta\left(\frac{\partial U}{\partial\theta}W\right)=E_\theta\left[\left(\frac{v'}{2v}U-\sqrt{\bar{v}}\right)\left\{Z_{2,v}-\left(\frac{J}{I}-\frac{v'}{v}\right)Z_{1,v}-\frac{(n-v)I}{\sqrt{\bar{v}}}\right\}\right]+o(1)$$

$$=E_\theta\left[\frac{v'}{2v}\left\{Z_{2,v}U-\left(\frac{J}{I}-\frac{v'}{v}\right)Z_{1,v}U-\frac{(n-v)I}{\sqrt{\bar{v}}}U\right\}\right]+o(1)$$

$$=\frac{v'}{2v}E\left[\frac{Z_{1,v}Z_{2,v}}{I}-\frac{J}{I^2}Z_{1,v}^2+\frac{vZ_{1,v}^2}{vI}-\frac{n-v}{\sqrt{\bar{v}}}Z_{1,v}\right]+o(1)$$

$$=o(1). \qquad (3.27)$$

By Lemma 3.1, (3.23) and (3.26) we obtain

$$E_\theta(Z_{1,v}^2W)=\frac{1}{\sqrt{\bar{v}}}\frac{d}{d\theta}E_\theta(Z_{1,v}W)-\frac{1}{\sqrt{\bar{v}}}E_\theta\left(W\frac{\partial Z_{1,v}}{\partial\theta}\right)-\frac{1}{\sqrt{\bar{v}}}E_\theta\left(Z_{1,v}\frac{\partial W}{\partial\theta}\right)$$

$$=\frac{1}{\sqrt{\bar{v}}}\frac{d}{d\theta}\left(-\frac{v'}{v}I\right)-\frac{1}{\sqrt{\bar{v}}}E_\theta\left[W\left\{W+\left(\frac{J}{I}-\frac{v'}{2v}\right)Z_{1,v}-\sqrt{\bar{v}}I\right\}\right]$$

$$-\frac{1}{\sqrt{\bar{v}}}\left\{\frac{J(J+K)}{I}-(M+N)-(J+K)\frac{v'}{v}+\frac{Iv''}{v}-\frac{2Iv'^2}{v^2}\right\}+o\left(\frac{1}{\sqrt{\bar{v}}}\right)$$

$$=-\frac{1}{\sqrt{\bar{v}}}\left(\frac{v'I}{v}\right)'-\frac{1}{\sqrt{\bar{v}}}E_\theta(W^2)-\frac{1}{\sqrt{\bar{v}}}\left(\frac{J}{I}-\frac{v'}{2v}\right)E_\theta(Z_{1,v}W)$$

$$-\frac{1}{\sqrt{\bar{v}}}\left\{\frac{J(J+K)}{I}-(M+N)-(J+K)\frac{v'}{v}+\frac{Iv''}{v}-\frac{2Iv'^2}{v^2}\right\}+o\left(\frac{1}{\sqrt{\bar{v}}}\right)$$

$$=-\frac{1}{\sqrt{\bar{v}}}\left\{\frac{v''}{v}I+\frac{v'}{v}(J+K)-\frac{v'^2}{v^2}I\right\}-\frac{1}{\sqrt{\bar{v}}}E_\theta(W^2)$$

$$-\frac{1}{\sqrt{\bar{v}}}\left\{\frac{J(J+K)}{I}-(M+N)-(J+K)\frac{v'}{v}+\frac{Iv''}{v}-\frac{2Iv'^2}{v^2}\right\}+o\left(\frac{1}{\sqrt{\bar{v}}}\right)$$

$$=-\frac{1}{\sqrt{\bar{v}}}\left\{\frac{J(J+K)}{I}-(M+N)+\frac{Jv'}{v}-\frac{3Iv'^2}{v^2}+2I\frac{v''}{v}\right\}-\frac{1}{\sqrt{\bar{v}}}E_\theta(W^2)+o\left(\frac{1}{\sqrt{\bar{v}}}\right). \qquad (3.28)$$

From (3.21), (3.23), (3.25), (3.26), (3.27) and (3.28) we have

$$E_\theta(WU^2)=-\frac{2}{I^2\sqrt{\bar{v}}}\left\{\frac{J(J+K)}{I}-(M+N)-(J+K)\frac{v'}{v}+\frac{Iv''}{v}-\frac{2Iv'^2}{v^2}\right\}$$

$$+\frac{1}{I^2\sqrt{\bar{v}}}\left\{\frac{J(J+K)}{I}-(M+N)+\frac{Jv'}{v}-\frac{3Iv'^2}{2v^2}+2I\frac{v''}{v}\right\}+\frac{1}{I^2\sqrt{\bar{v}}}E_\theta(W^2)+o\left(\frac{1}{\sqrt{\bar{v}}}\right)$$

$$=-\frac{v'^2}{Iv^2\sqrt{\bar{v}}}-\frac{1}{I^2\sqrt{\bar{v}}}\left\{\frac{J(J+K)}{I}-(M+N)-(3J+K)\frac{v'}{v}\right\}+\frac{1}{I^2\sqrt{\bar{v}}}E_\theta(W^2)+o\left(\frac{1}{\sqrt{\bar{v}}}\right). \qquad (3.29)$$

From (3.20) we obtain

$$E_\theta\left[\left(Z_{2,\nu} - \frac{\nu'}{2\nu}Z_{1,\nu} - \frac{nI}{\sqrt{\nu}}\right)U^2\right]$$

$$= E_\theta\left[\left\{W + (\frac{J}{I} - \frac{3\nu'}{2\nu})Z_{1,\nu} - \sqrt{\nu}\,I\right\}U^2\right]$$

$$= E_\theta(WU^2) + (\frac{J}{I} - \frac{3\nu'}{2\nu})E_\theta(Z_{1,\nu}U^2) - \sqrt{\nu}\,IE(U^2). \tag{3.30}$$

Hence we have from (3.19) and (3.30)

$$E_\theta(Z_{1,\nu}^2 U^2) = -\frac{1}{\sqrt{\nu}}\frac{d}{d\theta}\left(\frac{2J+K}{\sqrt{\nu}\,I^2} + \frac{\nu'}{\nu^{3/2}I}\right)$$

$$-\frac{1}{\sqrt{\nu}}E_\theta(WU^2) - \frac{1}{\sqrt{\nu}}(\frac{J}{I} - \frac{3\nu'}{2\nu})E_\theta(Z_{1,\nu}U^2) + IE(U^2)$$

$$-\frac{2}{\sqrt{\nu}}\left\{\frac{\nu'}{2\nu}E_\theta(Z_{1,\nu}U^2) + (\frac{\nu'}{\nu^{3/2}}\mu_1 - \frac{1}{\sqrt{\nu}}\mu_1' - \sqrt{\nu})E_\theta(Z_{1,\nu}U)\right\} + o(\frac{1}{\nu})$$

$$= -\frac{1}{\sqrt{\nu}}\frac{d}{d\theta}\left(\frac{2J+K}{\sqrt{\nu}\,I^2} + \frac{\nu'}{\nu^{3/2}I}\right)$$

$$-\frac{1}{\sqrt{\nu}}E_\theta(WU^2) - \frac{1}{\sqrt{\nu}}(\frac{J}{I} - \frac{3\nu'}{2\nu})E_\theta(Z_{1,\nu}U^2) + IE_\theta(U^2)$$

$$-2(\frac{\nu'}{\nu^2}\mu_1 - \frac{1}{\nu}\mu_1' - 1)E_\theta(Z_{1,\nu}U) + o(\frac{1}{\nu}). \tag{3.31}$$

Since by Lemma 2.1 and (3.22)

$$E_\theta(Z_{1,\nu}^4) = \frac{H}{\nu} - \frac{3I^2}{\nu^2}E_\theta(n^2) + \frac{6I}{\nu}E_\theta(nZ_{1,\nu}^2) + \frac{3K}{\nu^{3/2}}E_\theta(nZ_{1,\nu}) + o(\frac{1}{\nu})$$

$$= \frac{H}{\nu} + \frac{3K\nu'}{\nu^2} - \frac{3I^2}{\nu^2}E_\theta(n^2) + \frac{6I}{\nu}E_\theta(nZ_{1,\nu}^2) + o(\frac{1}{\nu}), \tag{3.32}$$

it follows from (3.16), (3.17), (3.31) and (3.32) that

$$E_\theta(U^4) = -\frac{2}{I^2}\left\{-\frac{1}{\sqrt{\nu}}\frac{d}{d\theta}\left(\frac{2J+K}{\sqrt{\nu}\,I^2} + \frac{\nu'}{\nu^{3/2}I}\right) - \frac{1}{\sqrt{\nu}}E_\theta(WU^2)\right.$$

$$-\frac{1}{\sqrt{\nu}}(\frac{J}{I} - \frac{3\nu'}{2\nu})E_\theta(Z_{1,\nu}U^2) + IE_\theta(U^2)$$

$$\left.-2(\frac{\nu'}{\nu^2}\mu_1 - \frac{1}{\nu}\mu_1' - 1)E_\theta(Z_{1,\nu}U)\right\}$$

$$+ \frac{8}{3I}\left\{ \frac{1}{\sqrt{\bar v}}\frac{d}{d\theta}\left(-\frac{3J+2K}{\sqrt{\bar v}\,I^3} - \frac{3v'}{v^{3/2}I^2}\right) - \frac{3v'}{2v^{3/2}}E_\theta(U^3)\right.$$

$$\left. + 3(1+\frac{1}{v}\mu_1' - \frac{v'}{v^2}\mu_1)E_\theta(U^2)\right\}$$

$$+ \frac{1}{3I^4}\left\{ \frac{H}{v} + \frac{3Kv'}{v^2} - \frac{3I^2}{v^2}E_\theta(n^2) + \frac{6I}{v}E_\theta(nZ_{1,v}^2)\right\} + o(\tfrac{1}{v}). \tag{3.33}$$

Since by Lemma 3.1, (3.18) and (3.22)

$$E_\theta(nZ_{1,v}^2) = \frac{1}{\sqrt{\bar v}}\frac{d}{d\theta}E_\theta(nZ_{1,v}) - \frac{1}{\sqrt{\bar v}}E_\theta(n\frac{\partial Z_{1,v}}{\partial\theta})$$

$$= \frac{1}{\sqrt{\bar v}}\frac{d}{d\theta}(\frac{v'}{\sqrt{\bar v}}) - \left\{E_\theta(nZ_{2,v}) - \frac{v^2}{2v^{3/2}} - \frac{I}{\sqrt{\bar v}}E_\theta(n^2)\right\} + o(1)$$

$$= \frac{1}{\sqrt{\bar v}}\frac{d}{d\theta}(\frac{v'}{\sqrt{\bar v}}) + \frac{v^2}{2v^2} + \frac{I}{\sqrt{\bar v}}E_\theta(nZ_{2,v}) + o(1),$$

it follows from (3.23) that

$$\frac{1}{I^2\bar v}\left\{ \frac{2}{I}E_\theta(nZ_{1,v}^2) - \frac{1}{v}E_\theta(n^2)\right\}$$

$$= \frac{1}{I^2\bar v}\left\{ \frac{2}{I\sqrt{\bar v}}\frac{d}{d\theta}(\frac{v'}{\sqrt{\bar v}}) + \frac{v^2}{Iv^2} + \frac{2}{v}E_\theta(n^2) - \frac{2}{I\sqrt{\bar v}}E_\theta(nZ_{2,v}) - \frac{1}{v}E_\theta(n^2)\right\}$$

$$= \frac{1}{I^3\bar v}\left\{ \frac{2}{\sqrt{\bar v}}\frac{d}{d\theta}(\frac{v'}{\sqrt{\bar v}}) + \frac{v^2}{v^2}\right\} + \frac{1}{I^2\bar v}\left\{ \frac{1}{v}E_\theta(n^2) - \frac{2}{I\sqrt{\bar v}}E_\theta(nZ_{2,v})\right\}$$

$$= \frac{1}{I^3\bar v}\left\{ \frac{2}{\sqrt{\bar v}}\frac{d}{d\theta}(\frac{v'}{\sqrt{\bar v}}) + \frac{v^2}{v^2}\right\} + \frac{1}{I^2\bar v}\left\{ E_\theta\left[\left(\frac{Z_{2,v}}{I} - \frac{n-v}{\sqrt{\bar v}}\right)^2\right] - \frac{1}{I^2}E_\theta(Z_{2,v}^2) + v\right\}$$

$$= \frac{1}{I^3\bar v}\left\{ \frac{2}{\sqrt{\bar v}}\frac{d}{d\theta}(\frac{v'}{\sqrt{\bar v}}) + \frac{v^2}{v^2}\right\} + \frac{1}{I^4\bar v}E_\theta\left[\left(W+(\frac{J}{I} - \frac{v'}{v})Z_{1,v}\right)^2\right] - \frac{M}{I^4\bar v} + \frac{1}{I^2} + o(\tfrac{1}{v})$$

$$= \frac{1}{I^3\bar v}\left\{ \frac{2}{\sqrt{\bar v}}\frac{d}{d\theta}(\frac{v'}{\sqrt{\bar v}}) + \frac{v^2}{v^2}\right\} + \frac{1}{I^4\bar v}E_\theta\left(W^2 + 2(\frac{J}{I} - \frac{v'}{v})Z_{1,v}W + (\frac{J}{I} - \frac{v'}{v})^2 Z_{1,v}^2\right) - \frac{M}{I^4\bar v} + \frac{1}{I^2} + o(\tfrac{1}{v})$$

$$= \frac{1}{I^2} + \frac{1}{I^3\bar v}\left\{ \frac{2}{\sqrt{\bar v}}\frac{d}{d\theta}(\frac{v'}{\sqrt{\bar v}}) + \frac{2v^2}{v^2} - \frac{2Jv'}{Iv} + \frac{J^2}{I^2} - \frac{M}{I}\right\} + \frac{1}{I^4\bar v}E_\theta(W^2) + o(\tfrac{1}{v}). \tag{3.34}$$

Since $J'(\theta)=L(\theta)+M(\theta)+N(\theta)$ and $K'(\theta)=H(\theta)+3N(\theta)$, we have from (3.33) and (3.34)

$$
\begin{aligned}
E_\theta(U^4)=&\frac{6}{I}E_\theta(U^2)+\frac{2}{I^2}\Big\{2\Big(-1+\frac{v'}{v^2}\mu_1-\frac{1}{v}\mu_1'\Big)E_\theta(Z_{1,v}U)\\
&+\frac{1}{\sqrt{v}}\frac{d}{d\theta}\Big(\frac{2J+K}{\sqrt{v}I^2}+\frac{v'}{v^{3/2}I}\Big)+\frac{1}{\sqrt{v}}E_\theta(WU^2)+\frac{1}{\sqrt{v}}\Big(\frac{J}{I}-\frac{3v'}{2v}\Big)E_\theta(Z_{1,v}U^2)\Big\}\\[4pt]
&+\frac{8}{3I}\Big\{\frac{1}{\sqrt{v}}\frac{d}{d\theta}\Big(-\frac{3J+2K}{\sqrt{v}I^3}-\frac{3v'}{v^{3/2}I^2}\Big)-\frac{3v'}{2v^{3/2}}E_\theta(U^3)\\
&+3\Big(\frac{1}{v}\mu_1'-\frac{v'}{v^2}\mu_1\Big)E_\theta(U^2)\Big\}+\frac{1}{3I^4}\Big(\frac{H}{v}+\frac{3Kv'}{v^2}\Big)\\[4pt]
&+\frac{1}{I^2}+\frac{1}{I^3v}\Big\{\frac{2}{\sqrt{v}}\frac{d}{d\theta}\Big(\frac{v'}{\sqrt{v}}\Big)+\frac{2v^2}{v^2}-\frac{2Jv'}{Iv}+\frac{J^2}{I^2}-\frac{M}{I}\Big\}+\frac{1}{I^4v}E_\theta(W^2)+o\Big(\frac{1}{v}\Big)\\[8pt]
=&\frac{6}{I}E_\theta(U^2)+\frac{4}{I^2}\Big(-1+\frac{v'}{v^2}\mu_1-\frac{1}{v}\mu_1'\Big)\Big(1+\frac{1}{v}\mu_1'-\frac{v'}{v^2}\mu_1\Big)\\[4pt]
&+\frac{2}{I^2\sqrt{v}}\frac{d}{d\theta}\Big(\frac{2J+K}{\sqrt{v}I^2}+\frac{v'}{v^{3/2}I}\Big)+\frac{2}{I^2\sqrt{v}}\Big[\frac{v'}{Iv^2\sqrt{v}}-\frac{1}{I^2\sqrt{v}}\Big\{\frac{J(J+K)}{I}-M-N\\
&-(3J+K)\frac{v'}{v}\Big\}+\frac{1}{I^2\sqrt{v}}E_\theta(W^2)\Big]+\frac{2}{I^2\sqrt{v}}\Big(\frac{J}{I}-\frac{3v'}{2v}\Big)\Big(-\frac{2J+K}{\sqrt{v}I^2}-\frac{v'}{v^{3/2}I}\Big)\\[4pt]
&+\frac{8}{3I}\Big\{\frac{1}{\sqrt{v}}\frac{d}{d\theta}\Big(-\frac{3J+2K}{\sqrt{v}I^3}-\frac{3v'}{v^{3/2}I^2}\Big)-\frac{3v'}{2v^{3/2}}\Big(-\frac{3J+2K}{\sqrt{v}I^3}-\frac{3v'}{v^{3/2}I^2}\Big)\\[4pt]
&+3\Big(\frac{1}{v}\mu_1'-\frac{v'}{v^2}\mu_1\Big)\frac{1}{I}\Big\}+\frac{1}{3I^4}\Big(\frac{H}{v}+\frac{3Kv'}{v^2}\Big)\\[4pt]
&+\frac{1}{I^2}+\frac{1}{I^3v}\Big\{\frac{2}{\sqrt{v}}\frac{d}{d\theta}\Big(\frac{v'}{\sqrt{v}}\Big)+\frac{2v^2}{v^2}-\frac{2Jv'}{Iv}+\frac{J^2}{I^2}-\frac{M}{I}\Big\}+\frac{1}{I^4v}E_\theta(W^2)+o\Big(\frac{1}{v}\Big)\\[8pt]
=&\frac{6}{I}E_\theta(U^2)-\frac{3}{I^2}+\frac{8}{I^2}\Big(\frac{v'}{v^2}\mu_1-\frac{1}{v}\mu_1'\Big)+\frac{2}{I^2\sqrt{v}}\frac{d}{d\theta}\Big(\frac{2J+K}{\sqrt{v}I^2}+\frac{v'}{v^{3/2}I}\Big)+\frac{2v^2}{I^3v^3}\\[4pt]
&-\frac{2}{I^4v}\Big\{\frac{J(J+K)}{I}-M-N-(3J+2K)\frac{v'}{v}\Big\}+\frac{2}{I^2\sqrt{v}}\Big(\frac{J}{I}-\frac{3v'}{2v}\Big)\Big(-\frac{2J+K}{\sqrt{v}I^2}-\frac{v'}{v^{3/2}I}\Big)\\[4pt]
&+\frac{8}{3I}\Big\{\frac{1}{\sqrt{v}}\frac{d}{d\theta}\Big(-\frac{3J+2K}{\sqrt{v}I^3}-\frac{3v'}{v^{3/2}I^2}\Big)-\frac{3v'}{2v^{3/2}}\Big(-\frac{3J+2K}{\sqrt{v}I^3}-\frac{3v'}{v^{3/2}I^2}\Big)
\end{aligned}
$$

$$+3(\frac{1}{v}\mu_1' - \frac{v'}{v^2}\mu_1)\frac{1}{I}\} + \frac{1}{3I^4}(\frac{H}{v} + \frac{3Kv'}{v^2})$$

$$+ \frac{1}{I^3v}\left\{\frac{2}{\sqrt{v}}\frac{d}{d\theta}(\frac{v'}{\sqrt{v}}) + \frac{2v'^2}{v^2} - \frac{2Jv'}{Iv} + \frac{J^2}{I^2} - \frac{M}{I}\right\} + \frac{3}{I^4v}E_\theta(W^2) + o(\frac{1}{v})$$

$$= \frac{6}{I}E_\theta(U^2) - \frac{3}{I^2} + \frac{2(2J'+K')}{I^4v} - \frac{4(2J+K)^2}{I^5v} - \frac{J(5J+4K)}{I^5v}$$

$$+ \frac{M}{I^4v} + \frac{2N}{I^4v} + \frac{H}{3I^4v} - \frac{8(3J'+2K')}{3I^4v} + \frac{8(3J+2K)(2J+K)}{I^5v}$$

$$+ \frac{v'}{3I^4v^2}(150J+95K) + \frac{27v'^2}{I^3v^3} - \frac{4v''}{I^3v^2} + \frac{3}{I^4v}E_\theta(W^2) + o(\frac{1}{v})$$

$$= \frac{6}{I}E_\theta(U^2) - \frac{3}{I^2} + \frac{12(2J+K)(J+K)}{I^5v} - \frac{3H+4L+12N}{I^4v}$$

$$- \frac{3}{I^4v}(M - \frac{J^2}{I}) + \frac{5v'}{3I^4v^2}(30J+19K) + \frac{27v'^2}{I^3v^3} - \frac{4v''}{I^3v^2}$$

$$+ \frac{3}{I^4v}E_\theta(W^2) + o(\frac{1}{v})$$

$$= \frac{6}{I}E_\theta(U^2) - \frac{3}{I^2} + \frac{\beta_4}{v} + \frac{3}{I^4v}E_\theta(W^2) + o(\frac{1}{v}) \qquad \text{(say)}. \tag{3.35}$$

Since by (3.12)

$$\{V_\theta(T_a)\}^2 = \{E_\theta(U^2)\}^2 = \frac{2}{I}E_\theta(U^2) - \frac{1}{I^2} + o(\frac{1}{v}),$$

it follows from (3.35) that

$$\kappa_4 = E_\theta[\{T_a - E_\theta(T_a)\}^4] - 3\{V_\theta(T_a)\}^2$$

$$= E_\theta(U^4) - \frac{6}{I}E_\theta(U^2) + \frac{3}{I^2}$$

$$= \frac{\beta_4}{v} + \frac{3}{I^4v}E_\theta(W^2)$$

$$= \frac{\beta_4}{v} + \frac{1}{v}r_4(\theta) \qquad \text{(say)}. \tag{3.36}$$

We put

$$E_\theta(T_a) = \frac{\mu_1(\theta)}{\sqrt{\nu}} + \frac{\mu_2(\theta)}{\nu} + o\left(\frac{1}{\nu}\right).$$

It follows from (3.12), (3.15) and (3.36) that the Edgeworth expansion of the distribution of $\hat{\theta}_a$ up to the order $\nu-1$ is given by

$$P_\theta\{\sqrt{\nu I(\theta)}\,(\hat{\theta}_a - \theta) \leqq t\}$$

$$= \Phi(t) - \frac{\sqrt{I}\,\mu_1}{\sqrt{\nu}}\phi(t) - \frac{\sqrt{I}\,\mu_2}{\nu}\phi(t) - \frac{I\sqrt{I}\,\beta_3}{6\sqrt{\nu}}(t^2-1)\phi(t) - I\left(\frac{\mu_1'}{\nu} - \frac{\nu'\mu_1}{\nu^2}\right)t\phi(t)$$

$$- \frac{I^2}{24\nu}(\beta_4 + \gamma_4)(t^3 - 3t)\phi(t) - \frac{I^3\beta_3^2}{72\nu}(t^5 - 10t^3 + 15t)\phi(t)$$

$$- \frac{It + I\mu_1^2}{2\nu}t\phi(t) - \frac{I\sqrt{I}\,\gamma_3}{6\nu}(t^2-1)\phi(t) - \frac{I^2\beta_3\mu_1}{6\nu}t(t^2-3)\phi(t) + o\left(\frac{1}{\nu}\right) \qquad (3.37)$$

where $\quad \Phi(t) = \displaystyle\int_{-\infty}^{t}\phi(t)dx \; with \; \phi(t) = (1/\sqrt{2\pi})e^{-x^2/2}.$

(See also Kendall and Stuart, 1969). By the third order AMU condition of $\hat{\theta}_a$ we have from (3.37)

$$P_\theta\{\hat{\theta}_a \leqq \theta\} = \frac{1}{2} - \frac{\sqrt{I}\,\mu_1}{\sqrt{\nu}}\phi(0) - \frac{\sqrt{I}\,\mu_2}{\nu}\phi(0) + \frac{I\sqrt{I}\,\beta_3}{6\sqrt{\nu}}\phi(0) + \frac{I\sqrt{I}\,\gamma_3}{6\nu}\phi(0) + o\left(\frac{1}{\nu}\right)$$

$$= \frac{1}{2} + o\left(\frac{1}{\nu}\right),$$

hence

$$\mu_1 = \frac{I\beta_3}{6} \quad and \quad \mu_2 = \frac{I\gamma_3}{6}. \qquad (3.38)$$

Therefore we obtain

$$E_\theta(T_a) = \frac{I(\theta)\beta_3(\theta)}{6\sqrt{\nu}} + \frac{I(\theta)\gamma_3(\theta)}{6\nu} + o\left(\frac{1}{\nu}\right).$$

Thus we complete the proof.

<u>Proof of Theorem 3.2.</u>   It follows from (3.37) with (3.38).

<u>Proof of Theorem 3.3.</u>   From (3.2) we have

$$P_\theta\{\sqrt{I(\theta)\nu}\,(\hat{\theta}_a - \theta) \leqq t\}$$

$$= \Phi(t) - \frac{I^{3/2}\beta_3}{6\sqrt{\nu}} t^2 \phi(t) - \frac{I^2\beta_4}{24\nu}(t^3 - 3t)\phi(t) - \frac{I^3\beta_3^2}{72\nu}(t^5 - 10t^3 + 15t)\phi(t)$$

$$- \frac{I\mu_1^2}{2\nu} t\phi(t) - \frac{I^2\beta_3\mu_1}{6\nu} t(t^2 - 3)\phi(t) - \frac{1}{8\nu I^2}E_\theta(W^2)(t^3 - 3t)\phi(t)$$

$$- \frac{I}{2\nu}V_\theta(Q^*)t\phi(t) - \frac{\sqrt{I}}{4\nu}E_\theta(Z_{1,\nu}Q^{*2})t^2\phi(t) + o(\frac{1}{\nu}), \tag{3.39}$$

where $Q^* = Q - \mu_1$, and I, μ_1, μ_2, β_3 and β_4 denote $I(\theta)$, $\mu_1(\theta)$, $\mu_2(\theta)$, $\beta_3(\theta)$ and $\beta_4(\theta)$.

By Lemma 3.1 we obtain

$$E_\theta(Z_{1,\nu}Q^{*2}) = \frac{1}{\sqrt{\nu}}\frac{\partial}{\partial\theta}E_\theta(Q^{*2}) - \frac{2}{\sqrt{\nu}}E_\theta(\frac{\partial Q^*}{\partial\theta}Q^*)$$

$$= -\frac{2}{\sqrt{\nu}}E_\theta\left[\left\{\frac{\partial}{\partial\theta}\left(\sqrt{\nu}(T_a - \frac{Z_{1,\nu}}{I})\right)\right\}Q^*\right] + o(1)$$

$$= -\frac{2}{\sqrt{\nu}}E_\theta\left[\frac{\nu'}{2\sqrt{\nu}}(T_a - \frac{Z_{1,\nu}}{I})Q^* + \sqrt{\nu}\left\{\frac{\partial}{\partial\theta}(T_a - \frac{Z_{1,\nu}}{I})\right\}Q^*\right] + o(1)$$

$$= -\frac{\nu'}{\nu}E_\theta\left[(T_a - \frac{Z_{1,\nu}}{I})Q^*\right]$$

$$-2E_\theta\left[Q^*\left\{\frac{\nu'}{2\nu}T_a - \sqrt{\nu} - \frac{\mu_1'}{\nu} + \frac{1}{I^2}\left((2J+K)Z_{1,\nu} - I\frac{\partial Z_{1,\nu}}{\partial\theta}\right)\right\}\right] + o(1)$$

$$= \frac{2}{I}E_\theta(Q^*\frac{\partial Z_{1,\nu}}{\partial\theta}) + o(1)$$

$$= \frac{2}{I}E_\theta\left[Q^*(Z_{2,\nu} - \frac{\nu'}{2\nu}Z_{1,\nu} - \frac{nI}{\sqrt{\nu}})\right] + o(1)$$

$$= \frac{2}{I}E_\theta\left[Q^*\left\{W + (\frac{J}{I} - \frac{3\nu'}{2\nu})Z_{1,\nu} - \sqrt{\nu}I\right\}\right] + o(1)$$

$$= \frac{2}{I}E_\theta(WQ^*) + o(1), \tag{3.40}$$

where

$$W = Z_{2,\nu} - (\frac{J}{I} - \frac{\nu'}{\nu})Z_{1,\nu} - \frac{(n-\nu)I}{\sqrt{\nu}}.$$

We also have from Lemma 3.1

$$E_\theta(Z_{1,\nu} W Q^*) = \frac{1}{\sqrt{\nu}} \frac{\partial}{\partial\theta} E_\theta(WQ^*) - \frac{1}{\sqrt{\nu}} E_\theta(\frac{\partial W}{\partial\theta} Q^*) - \frac{1}{\sqrt{\nu}} E_\theta(W \frac{\partial Q^*}{\partial\theta})$$

$$= -\frac{1}{\sqrt{\nu}} E_\theta(\frac{\partial W}{\partial\theta} Q^*) - \frac{1}{\sqrt{\nu}} E_\theta(W \frac{\partial Q^*}{\partial\theta}) + o(1). \tag{3.41}$$

Since

$$\frac{\partial W}{\partial\theta} = -\frac{\nu'}{2\nu} Z_{2,\nu} + Z_{3,\nu} - \frac{nJ}{\sqrt{\nu}} - \frac{1}{I}(L+M+N)Z_{1,\nu}$$

$$-(\frac{J}{I} - \frac{\nu'}{\nu})(Z_{2,\nu} - \frac{\nu'}{2\nu}Z_{1,\nu} - \frac{nI}{\sqrt{\nu}}) + \frac{J(2J+K)}{I^2}Z_{1,\nu} + (\frac{\nu''}{\nu} - \frac{\nu'^2}{\nu^2})Z_{1,\nu} + \frac{\nu'I}{\sqrt{\nu}}$$

$$-\frac{(n-\nu)(2J+K)}{\sqrt{\nu}} + \frac{(n-\nu)\nu'I}{2\nu^{3/2}} + o_p(1) \ ,$$

it follows Lemma 3.1 that

$$\frac{1}{\sqrt{\nu}} E_\theta(\frac{\partial W}{\partial\theta} Q^*) = o(1). \tag{3.42}$$

Since

$$\frac{1}{\sqrt{\nu}} \frac{\partial Q^*}{\partial\theta} = \frac{1}{\sqrt{\nu}} \frac{\partial}{\partial\theta}\left(\sqrt{\nu}(T_a - \frac{Z_{1,\nu}}{I})\right)$$

$$= \frac{\nu'}{2\nu}(T_a - \frac{Z_{1,\nu}}{I}) + \frac{\partial T_a}{\partial\theta} + \frac{1}{I^2}\left\{(2J+K)Z_{1,\nu} - I\frac{\partial Z_{1,\nu}}{\partial\theta}\right\}$$

$$= \frac{\nu'}{2\nu}\frac{1}{\sqrt{\nu}}Q^* + \frac{\nu'}{2\nu}T_a - \sqrt{\nu} - \frac{\mu_1'}{\sqrt{\nu}} + \frac{2J+K}{I^2}Z_{1,\nu}$$

$$-\frac{1}{I}(Z_{2,\nu} - \frac{\nu'}{2\nu}Z_{1,\nu} - \frac{nI}{\sqrt{\nu}}) + \frac{\nu'}{2\nu}\left\{\frac{\mu_1}{\sqrt{\nu}} + \frac{\mu_2}{\nu} + o(\frac{1}{\nu})\right\}$$

$$= -\frac{1}{I}\left\{W + (\frac{J}{I} - \frac{3\nu'}{2\nu})Z_{1,\nu} - \sqrt{\nu}I\right\} + \frac{\nu'}{2\nu}T_a + (\frac{2J+K}{I^2} + \frac{\nu'}{2I\nu})Z_{1,\nu} - \nu + o_p(1)$$

$$= -\frac{1}{I}W + \frac{\nu'}{2\nu}T_a + \left(\frac{J+K}{I^2} + \frac{2\nu'}{I\nu}\right)Z_{1,\nu} + o_p(1)$$

it follows from (3.23), (3.41) and (3.42) that

$$E_\theta(Z_{1,\nu} W Q^*) = -\frac{1}{\sqrt{\nu}} E_\theta(W \frac{\partial Q^*}{\partial\theta}) + o(1)$$

$$= \frac{1}{I} E_\theta(W^2) - \frac{v'}{2v} E_\theta(WT_a) - \left(\frac{J+K}{I^2} + \frac{2v'}{Iv}\right) E_\theta(WZ_{1,v}) + o(1)$$

$$= \frac{1}{I} E_\theta(W^2) + o(1).$$

By Lemma 3.1 we have

$$E_\theta(Z_{1,v}^2 Q^*) = \frac{1}{\sqrt{v}} \frac{\partial}{\partial\theta} E_\theta(Z_{1,v} Q^*) - \frac{1}{\sqrt{v}} E_\theta\left(\frac{\partial Z_{1,v}}{\partial\theta} Q^*\right) - \frac{1}{\sqrt{v}} E_\theta\left(Z_{1,v} \frac{\partial Q^*}{\partial\theta}\right)$$

$$= -\frac{1}{\sqrt{v}} E_\theta\left[Q^*\left\{W + (\frac{J}{I} - \frac{3v'}{2v})Z_{1,v} - \sqrt{v}I\right\}\right]$$

$$- \frac{1}{\sqrt{v}} E_\theta\left[Z_{1,v}\left\{\frac{\partial}{\partial\theta}\left(\sqrt{v}(T_a - \frac{Z_{1,v}}{I})\right)\right\}\right] + o(1)$$

$$= -\frac{1}{\sqrt{v}} E_\theta(WQ^*) - \frac{1}{\sqrt{v}} E_\theta\left[\frac{v'}{2\sqrt{v}}(T_a - \frac{Z_{1,v}}{I})Z_{1,v}\right.$$

$$\left. + \sqrt{v}\left\{\frac{\partial}{\partial\theta}(T_a - \frac{Z_{1,v}}{I})\right\}Z_{1,v}\right] + o(1)$$

$$= -\frac{v'}{2v} E_\theta\left[(T_a - \frac{Z_{1,v}}{I})Z_{1,v}\right]$$

$$- E_\theta\left[\left\{\frac{v'}{2v}T_a - \sqrt{v} - \frac{\mu_1'}{v} + \frac{1}{I^2}\left((2J+K)Z_{1,v} - I\frac{\partial Z_{1,v}}{\partial\theta}\right)\right\}Z_{1,v}\right] + o(1)$$

$$= -\frac{v'}{2v} E_\theta(T_a Z_{1,v}) - \frac{2J+K}{I^2} E_\theta(Z_{1,v}^2) + \frac{1}{I} E_\theta\left(\frac{\partial Z_{1,v}}{\partial\theta} Z_{1,v}\right) + o(1)$$

$$= -\frac{v'}{2v} - \frac{2J+K}{I} + \frac{1}{I} E_\theta\left[Z_{1,v}\left\{W + (\frac{J}{I} - \frac{v'}{v})Z_{1,v} - \sqrt{v}I\right\}\right] + o(1)$$

$$= -\frac{v'}{2v} - \frac{2J+K}{I} + \frac{J}{I} - \frac{3v'}{2v} + o(1)$$

$$= -\frac{J+K}{I} - \frac{2v'}{v} + o(1).$$

If we take Q_0^* as Q^* such that $V_\theta(Q^*)$ is minimized under the conditions

$$E_\theta(WQ^*) = \frac{r}{I} E_\theta(W^2) + o(1), \tag{3.43}$$

$$E_\theta(Z_{1,v} WQ^*) = \frac{1}{I} E_\theta(W^2) + o(1), \tag{3.44}$$

$$E_\theta(Z_{1,v}^2 Q^*) = -\frac{J+K}{I} - \frac{2v'}{v} + o(1) \tag{3.45}$$

for real r, then Q_0^* must have the form
$$Q_0^* = A_0 + A_1 W + A_2 W Z_{1,v} + A_3 Z_1^2, v ,$$

where A_i ($i=0,1,2,3$) are constants. Since $\mathrm{Cov}_\theta(W, Z_{1,v} W) = o(1)$, $\mathrm{Cov}_\theta(W, Z_1^2, v) = o(1)$, and $\mathrm{Cov}_\theta(Z_{1,v} W, Z_1^2, v) = o(1)$, it follows that

$$V_\theta(Q_0^*) = \frac{1}{I^2}(r^2 + \frac{1}{I}) E_\theta(W^2) + \frac{1}{2I^4}(J+K+\frac{2v'}{v}I)^2 + o(1). \tag{3.46}$$

Hence we have from (3.39), (3.40), (3.43), (3.44), (3.45) and (3.46)

$$P_\theta\{\sqrt{I(\theta)v}\,(\hat\theta_a - \theta) \leq t\}$$

$$\begin{array}{c}\leq \\ (\geq)\end{array} \Phi(t) - \frac{I^{3/2}\beta_3}{6\sqrt{v}} t^2\phi(t) - \frac{I^2\beta_4}{24v}(t^3 - 3t)\phi(t) - \frac{I^3\beta_3^2}{72v}(t^5 - 10t^3 + 15t)\phi(t)$$

$$- \frac{I\mu_1^2}{2v} t\phi(t) - \frac{I^2\beta_3\mu_1}{6v} t(t^2 - 3)\phi(t) - \frac{t\phi(t)}{2v}\left\{ \frac{1}{4I^2}E_\theta(W^2)(t^2 - 3) \right.$$

$$\left. + IV_\theta(Q_0^*) + \frac{\sqrt{I}}{2}E_\theta(Z_{1,v}Q_0^{*2})t \right\} + o(\frac{1}{v})$$

$$= \cdots - \frac{t\phi(t)}{2v}\left\{ \frac{1}{4I^2}E_\theta(W^2)(t^2-3) + \frac{1}{I}(r^2+\frac{1}{I})E_\theta(W^2) + \frac{1}{2I^3}(J+K+\frac{2v'}{v}I)^2 \right.$$

$$\left. + \frac{r}{I\sqrt{I}}E_\theta(W^2)t \right\} + o(\frac{1}{v})$$

$$= \cdots - \frac{E_\theta(W^2)}{2vI^2} t\phi(t)\left\{ \frac{t^2-3}{4} + (Ir^2+1) + \sqrt{I}\,rt \right\} - \frac{1}{4vI^3}t\phi(t)(J+K+\frac{2v'}{v}I)^2 + o(\frac{1}{v})$$

$$= \cdots - \frac{E_\theta(W^2)}{2vI^2} t\phi(t)\left\{ (\frac{t}{2}+\sqrt{I}\,r)^2 + \frac{1}{4} \right\} - \frac{1}{4vI^3}t\phi(t)(J+K+\frac{2v'}{v}I)^2 + o(\frac{1}{v})$$

for all $t > 0$ ($t < 0$), where "$\cdots$" denotes first six terms of the above right-hand side. This completes the proof.

<u>Proof of Corollary 3.1.</u>   It is straightforward from Theorem 3.3.

<u>Proof of Theorem 3.4.</u>   From (3.5) we have

$$\sqrt{v}\,(\hat\theta_{ML} - \theta) = \frac{Z_{1,v}}{I} + \frac{Z_{1,v}}{I^2\sqrt{v}}(Z_{2,v} - \frac{n-v}{\sqrt{v}}I) - \frac{3J+K}{2I^3\sqrt{v}}Z_{1,v}^2 + o_p(\frac{1}{\sqrt{v}})$$

$$= \frac{Z_{1,v}}{I} + \frac{Z_{1,v}}{I^2\sqrt{v}}\{W + (\frac{J}{I} - \frac{v'}{v})Z_{1,v}\} - \frac{3J+K}{2I^3\sqrt{v}}Z^2_{1,v} + o_p(\frac{1}{\sqrt{v}})$$

$$= \frac{Z_{1,v}}{I} + \frac{1}{I^2\sqrt{v}}Z_{1,v}W - \frac{J+K}{2I^3\sqrt{v}}Z^2_{1,v} - \frac{v'}{I^2 v\sqrt{v}}Z^2_{1,v} + o_p(\frac{1}{\sqrt{v}}). \tag{3.47}$$

In a similar way to Theorem 2.2 in Takeuchi and Akahira (1988), we obtain by Taylor expansion of both sides of (3.6) about θ

$$(n-v)I = \sqrt{v}(Z_{2,v} - \frac{J}{I}Z_{1,v}) - \frac{3J+K}{I} \cdot \frac{n-v}{\sqrt{v}}Z_{1,v} + \frac{v'}{\sqrt{v}}Z_{1,v}$$

$$- \frac{1}{I}Z_{1,v}Z_{3,v} - \frac{4L+3M+6N+H}{2I^2 v} \cdot nZ^2_{1,v}$$

$$+ \frac{1}{2I^2 v}\{v''I + 2v'(2J+K) + v(2L+2M+5N+H)\}Z^2_{1,v} + c(\theta) + o(1), \tag{3.48}$$

where $v = v_a(\theta)$, $v' = v'_a(\theta) = (\partial/\partial\theta)v_a(\theta)$, $v'' = v''_a(\theta) = (\partial 2/\partial\theta 2)v_a(\theta)$, $Z_{1,v} = Z_{1,v}(\theta)$, $Z_{2,v} = Z_{2,v}(\theta)$, $Z_{3,v} = Z_{3,v}(\theta)$, $I = I(\theta)$, $J = J(\theta)$, $K = K(\theta)$, $L = L(\theta)$, $M = M(\theta)$, $N = N(\theta)$ and $H = H(\theta)$. In order to determine n satisfying $E_\theta(n) = v + o(1)$, we have from (3.48)

$$c(\theta) = \frac{Jv'}{Iv} + \frac{L}{I} - \frac{v''}{2v} + \frac{1}{2I}(2L + M + N).$$

From (3.48) we also obtain

$$Z_{2,v} - (\frac{J}{I} - \frac{v'}{v})Z_{1,v} - \frac{n-v}{\sqrt{v}}I = o_p(1),$$

that is,

$$W = o_p(1), \tag{3.49}$$

hence by (3.47)

$$\sqrt{v}(\hat{\theta}_{ML} - \theta) = \frac{Z_{1,v}}{I} - \frac{J+K}{2I^3\sqrt{v}}Z^2_{1,v} - \frac{v'}{I^2 v\sqrt{v}}Z^2_{1,v} + o_p(\frac{1}{\sqrt{v}}).$$

Since, by Takeuchi and Akahira (1988),

$$V_\theta(\sqrt{v}(\hat{\theta}_{ML} - \theta)) = \frac{1}{I} + \frac{1}{2vI^2}(\frac{J+K}{I} + \frac{2v'}{v})^2 + o(\frac{1}{v}),$$

the estimator $\hat{\theta}^*_{ML}$ modified from the ML estimator $\hat{\theta}_{ML}$ to be third order AMU has the same asymptotic variance as that of $\hat{\theta}_{ML}$. Then it follows from Akahira (1986) and Akahira and Takeuchi (1981) that the modified ML estimator $\hat{\theta}^*_{ML}$ has the

following asymptotic distribution up to the order v^{-1}:

$$P_\theta\{\sqrt{\overline{I(\theta)v(\theta)}}\,(\hat{\theta}^*_{ML}-\theta)\leqq t\}$$

$$=\Phi(t)-\frac{I(\theta)\sqrt{I(\theta)}\,\beta_3(\theta)}{6\sqrt{v}}t^2\phi(t)-\frac{I^2(\theta)\beta_4(\theta)}{24v}(t^3-3t)\phi(t)$$

$$-\frac{I^3(\theta)\beta_3^2(\theta)}{72v}(t^5-10t^3+15t)\phi(t)-\frac{I(\theta)\mu_1^2(\theta)}{2v}t\phi(t)$$

$$-\frac{I^2(\theta)\beta_3(\theta)\mu_1(\theta)}{6v}t(t^2-3)\phi(t)-\frac{1}{4vI^3(\theta)}t\phi(t)\{J(\theta)+K(\theta)+\frac{2v'}{v}I(\theta)\}^2+o\left(\frac{1}{v}\right),$$

where $\mu_1(\theta)$ $\beta_3(\theta)$ and $\beta_4(\theta)$ are given in Theorem 3.1.

Since, by (3.49), $E_\theta(W^2)=o(1)$, it is seen from (3.4) that the asymptotic distribution of $\hat{\theta}^*_{ML}$ coincides with the bound up to the order v^{-1} uniformly in t and stopping rules. This completes the proof.

References

Akahira, M. (1986).   The Structure of Asymptotic Deficiency of Estimators. Queen's Papers in Pure and Applied Mathematics No.75, Queen's University Press, Kingston, Ontario, Canada.

Akahira, M. and Takeuchi, K. (1981).   Asymptotic Efficiency of Statistical Estimators : Concepts and Higher Order Asymptotic Efficiency. Lecture Notes in Statistics 7, Springer-Verlag, NewYork.

Amari, S. (1985).   Differential-Geometrical Methods in Statistics. Lecture Notes in Statistics 28, Springer-Verlag, Berlin.

Efron, B. (1975).   Defining the curvature of a statistical problem (with application to second order efficiency). Ann. Statist., 3, 1189-1242.

Fisher, R. A. (1925).   Theory of statistical estimation. Proc. Comb. Phil. Soc., 22, 700-725.

Kendall, M. G. and Stuart, A. (1969).   The Advanced Theory of Statistics, Vol.1, 3rd ed., Griffin, London.

Rao, C. R. (1961).   Asymptotic efficiency and limiting information. Proc. Fourth Berkeley Symp. Math. Statist. Prob., 1, 531-545.

Takeuchi, K. and Akahira, M. (1988).   Second order asymptotic efficiency in terms of asymptotic variances of the sequential maximum likelihood estimation procedures. In : 2nd Pacific Area Statistical Conference, Statistical Theory and Data Analysis II, North-Holland, Amsterdam, 191-196.

Wald, A. (1959).   Sequential Analysis. John Wiley, New York.

Received December 1987; Revised June 1989
Recommended by J.K. Ghosh

Pub. Inst. Stat. Univ. Paris
XXXV, fasc. 1, 1990, 3 à 9

First order asymptotic efficiency in semiparametric models implies infinite asymptotic deficiency *

Masafumi Akahira[(*)] and Kei Takeuchi[(**)]

Abstract

For semiparametric models, it is shown that, under fairly regularity conditions, the asymptotic deficiency of the maximum likelihood estimator or any regular best asymptotically normal estimator is infinity.

1. Introduction

For various types of semiparametric models, it has been shown that uniformly asymptotically efficient estimators may be obtained under some regularity conditions (e.g. see Hájek, 1962 and Takeuchi, 1971). More precisely let us assume that $X_1, \ldots, X_n$ are independent

---

(*) Institute of Mathematics, University of Tsukuba, Ibaraki 305, Japan

(**) Research Center for Advanced Science and Technology, University of Tokyo, Komaba, Meguro-ku, Tokyo 156, Japan

Key words and phrases: Semiparametric models, asymptotic deficiency, maximum likelihood estimator.

AMS Subject Classification (1980): 62F12, 62F10

---

*This paper is retyped with the correction of typographical errors.

and identically distributed random variables according to a density function $f(x, \theta)$ with respect to a σ-finite measure μ, where θ is a real-valued parameter, and, given θ, f belongs to some class F which can not be identified by a finite number of parameters. It is naturally required that $f_1(x, \theta_1) \neq f_2(x, \theta_2)$ for any disjoint θ_1 and θ_2 and f_1, $f_2 \in F$. Then we can construct estimators $\hat{\theta}_n$ such that $\hat{\theta}_n$ is asymptotically efficient for the model with known f, that is, in regular cases, $\sqrt{n}\left(\hat{\theta}_n - \theta\right)$ is asymptotically normal $N\left(0, I_f^{-1}\right)$, where

$$I_f = \int_{-\infty}^{\infty} \left\{ \frac{\partial}{\partial \theta} \log f(x, \theta) \right\}^2 f(x, \theta) d\mu(x)$$

whatever true $f \in F$ be.

The pertinent question here is that when we can also achieve uniform second order asymptotic efficiency ? or does the distribution of $\sqrt{n}\left(\hat{\theta}_n - \theta\right)$ attain the second order asymptotic bound ? In the subsequent discussion we will show that, in a sense, loss of third order asymptotic efficiency, or asymptotic deficiency, is generally infinity under fairly general conditions for semiparametric models, which although does not directly answer the above question, may shed some light on the possible higher order asymptotic efficiency of estimators under semiparametric models.

2. Infinite deficiency in semiparametric models

Suppose that $X_1, \ldots, X_n$ are independently and identically distributed (i.i.d.) real random variables with a density function $f(x, \theta, \xi)$ with respect to a σ-finite measure μ, where θ is a real-valued parameter to be estimated and ξ is a real-valued nuisance parameter. We assume the following conditions (A.1) to (A.5).

(A.1) The set $\{x : f(x, \theta, \xi) > 0\}$ does not depend on θ.

(A.2) For almost all $x[\mu]$, $f(x,\theta,\xi)$ is three times continuously differentiable in θ and ξ.

(A.3) For each θ and each ξ

$$0 < I_{00}(\theta,\xi) = E\left[\{\ell_0(\theta,\xi,X)\}^2\right] = -E\left[\ell_{00}(\theta,\xi,X)\right] < \infty,$$

$$0 < I_{11}(\theta,\xi) = E\left[\{\ell_1(\theta,\xi,X)\}^2\right] = -E\left[\ell_{11}(\theta,\xi,X)\right] < \infty,$$

where $\ell_0(\theta,\xi,x) = (\partial/\partial\theta)\ell(\theta,\xi,x)$, $\ell_{00}(\theta,\xi,x) = (\partial^2/\partial\theta^2)\ell(\theta,\xi,x)$, $\ell_1(\theta,\xi,x) = (\partial/\partial\xi)\ell(\theta,\xi,x)$ and $\ell_{11}(\theta,\xi,x) = (\partial^2/\partial\xi^2)\ell(\theta,\xi,x)$ with $\ell(\theta,\xi,x) = \log f(x,\theta,\xi)$.

(A.4) The parameters are defined to be "orthogonal" in the sense that

$$E\left[\ell_{01}(\theta,\xi,X)\right] = 0,$$

where

$$\ell_{01}(\theta,\xi,x) = (\partial^2/\partial\theta\partial\xi)\ell(\theta,\xi,x).$$

Note that the condition (A.4) is not necessarily restricted, because otherwise we can redefine the parameter $\xi' = g(\theta,\xi)$ so that we have the above orthogonality.

(A.5) There exist

$$J_{000} = E\left[\ell_{00}(\theta,\xi,X)\ell_0(\theta,\xi,X)\right], \quad J_{001} = E\left[\ell_{00}(\theta,\xi,X)\ell_1(\theta,\xi,X)\right],$$

$$J_{010} = E\left[\ell_{01}(\theta,\xi,X)\ell_0(\theta,\xi,X)\right], \quad J_{011} = E\left[\ell_{01}(\theta,\xi,X)\ell_1(\theta,\xi,X)\right],$$

$$J_{110} = E\left[\ell_{11}(\theta,\xi,X)\ell_0(\theta,\xi,X)\right], \quad K_{000} = E\left[\{\ell_0(\theta,\xi,X)\}^3\right],$$

$$K_{001} = E\left[\{\ell_0(\theta,\xi,X)\}^2\ell_1(\theta,\xi,X)\right],$$

$$M_{0000} = E\left[\{\ell_{00}(\theta,\xi,X)\}^2\right] - I_{00}^2, \quad M_{0101} = E\left[\{\ell_{01}(\theta,\xi,X)\}^2\right],$$

and the following holds.

$$E\left[\ell_{000}(\theta,\xi,X)\right] = -3J_{000} - K_{000}, \quad E\left[\ell_{001}(\theta,\xi,X)\right] = -J_{010},$$

$$E\left[\ell_{011}(\theta,\xi,X)\right] = -J_{011},$$

where $\ell_{000}(\theta,\xi,x) = (\partial^3/\partial\theta^3)\ell(\theta,\xi,x)$, $\ell_{001}(\theta,\xi,x) = (\partial^3/\partial\theta^2\partial\xi)\ell(\theta,\xi,x)$ and $\ell_{011}(\theta,\xi,x) = (\partial^3/\partial\theta\partial\xi^2)\ell(\theta,\xi,x)$.

From the condition (A.5) it is noted that $K_{001} = J_{010} - J_{001}$. We put

$$Z_0 = \frac{1}{\sqrt{n}} \sum_{i=1}^{n} \ell_0(\theta, \xi, X_i), \quad Z_1 = \frac{1}{\sqrt{n}} \sum_{i=1}^{n} \ell_1(\theta, \xi, X_i),$$

$$Z_{00} = \frac{1}{\sqrt{n}} \sum_{i=1}^{n} \left\{ \ell_{00}(\theta, \xi, X_i) + I_{00} \right\}, \quad Z_{01} = \frac{1}{\sqrt{n}} \sum_{i=1}^{n} \ell_{01}(\theta, \xi, X_i).$$

Let $\hat{\theta}^*$ and $\hat{\xi}^*$ be the maximum likelihood estimators (MLEs) of θ and ξ, respectively. Then we have the following.

<u>Theorem 2.1.</u> Assume that the conditions (A.1) to (A.5) hold. Then the MLE $\hat{\theta}^*$ of θ has the following stochastic expansion.

$$\sqrt{n}\left(\hat{\theta}^* - \theta\right) = \frac{1}{I_{00}} Z_0 + \frac{1}{\sqrt{n}I_{00}^2} Z_{00} Z_0 - \frac{3J_{000} + K_{000}}{2\sqrt{n}I_{00}^3} Z_0^2 + \frac{1}{\sqrt{n}I_{00}I_{11}} Z_{01} Z_1$$
$$- \frac{J_{010}}{\sqrt{n}I_{00}^2 I_{11}} Z_0 Z_1 - \frac{J_{011}}{2\sqrt{n}I_{00}I_{11}^2} Z_1^2 + o_p\left(\frac{1}{\sqrt{n}}\right).$$

The proof is given in the paper by Akahira and Takeuchi (1982) and also in section 4.2 of the monograph by Akahira (1986). We put

$$Q_0 = \frac{1}{I_{00}^2} Z_{00} Z_0 - \frac{3J_{000} + K_{000}}{2I_{00}^3} Z_0^2,$$

$$Q_1 = \frac{1}{I_{00}I_{11}} Z_1 \left(Z_{01} - \frac{J_{010}}{I_{00}} Z_0 - \frac{J_{011}}{I_{11}} Z_1 \right) + \frac{J_{011}}{2I_{00}I_{11}^2} Z_1^2.$$

Then we have the following:

<u>Theorem 2.2.</u> Assume that the conditions (A.1) to (A.5) hold. Then the asymptotic deficiency of the MLE $\hat{\theta}^*$ is given by

$$d = I_{00} \left\{ V(Q_0) + V(Q_1) \right\}$$
$$= \frac{1}{I_{00}^3} \left(M_{0000} I_{00} - J_{000}^2 \right) + \frac{(J_{000} + K_{000})^2}{2I_{00}^3} + \frac{1}{I_{00}I_{11}} \left(M_{0101} - \frac{J_{010}^2}{I_{00}} - \frac{J_{011}^2}{I_{11}} \right)$$
$$+ \frac{J_{011}^2}{2I_{00}I_{11}^2},$$

where $V(\cdot)$ designates the asymptotic variance.

The proof is given in the paper by Akahira and Takeuchi (1982) and also in section 4.2 of the monograph by Akahira (1986). Now, in order to consider a semiparametric model, we assume the condition (A.6).

(A.6) For each θ, each ξ and almost all $x[\mu]$, $f(x,\theta,\xi) = f(x-\theta,\xi)$, and $f(x,\xi) = f(-x,\xi)$.

Then we have the following corollary.

Corollary 2.1. Assume that the conditions (A.1) to (A.6) hold. Then the asymptoic deficiency of the MLE $\hat{\theta}^*$ is given by

(2.1)
$$d = \frac{1}{I_{00}^3}\left(M_{0000}I_{00} - J_{000}^2\right) + \frac{(J_{000} + K_{000})^2}{2I_{00}^3} + \frac{1}{I_{00}^2 I_{11}}\left(M_{0101}I_{00} - J_{010}^2\right).$$

Since, by the condition (A.6), $J_{011} = 0$, the proof is easily derived from Theorem 2.2. It is also noted that the inequalities $M_{0000}I_{00} - J_{000}^2 \geq 0$ and $M_{0101}I_{00} - J_{010}^2 \geq 0$ hold from the Schwarz inequality.

Next we consider a semiparametric model of the type:

(2.2) $f(x-\theta,\xi) = C_\xi\left\{f_0(x-\theta)\right\}^{1-\xi}\left\{g(x-\theta)\right\}^\xi$ for $0 \leq \xi \leq 1$,

where C_ξ is some constant, depending on ξ, with $C_0 = C_1 = 1$, and $f_0(x)$ and $g(x)$ are certain density functions. Then we assume the condition (A.7).

(A.7) $\displaystyle\int_{-\infty}^{\infty} \frac{\{f_0'(x)\}^2}{f_0(x)} d\mu(x) = 1.$

It is easily seen that the condition (A.7) implies $I_{00} = 1$. From (2.2) and (A.7) it follows that, by the Schwarz inequality,

$$M_{0101} - J_{010}^2 = \int_{-\infty}^{\infty}\left\{\frac{g'(x)}{g(x)}\right\}^2 f_0(x)d\mu - \left\{\int_{-\infty}^{\infty}\frac{g'(x)f_0'(x)}{g(x)f_0(x)}f_0(x)d\mu\right\}^2$$
$$= \rho\left(f_0, g\right) \text{ (say)}$$

is nonnegative. Then we have the following.

Theorem 2.3. Assume that the conditions (A.1) to (A.7) in the semiparametric model (2.2) hold. Let $g(x)$ be of the form $K_\alpha g_0^\alpha(x)$ with some constant K_α for $\alpha > 0$ and some function $g_0(x)$. If $\rho(f_0, g_0) > 0$, then the asymptotic deficiency of the MLE $\hat{\theta}^*$ tends to infinity as $\alpha \to \infty$.

Proof. From the assumptions we have

$$\lim_{\alpha \to \infty} \left(M_{0101} - J_{010}^2 \right) = \lim_{\alpha \to \infty} \alpha^2 \rho(f_0, g_0) = \infty,$$

hence, by (2.1), it follows that the asymptotic deficiency of the MLE $\hat{\theta}^*$ tends to infinity as $\alpha \to \infty$. This completes the proof.

Denote by F a semiparametric model consisting of density functions $f(x)$ with $f(x) = f(-x)$. Then we have the following.

Corollary 2.2. Assume that for any $f_0 \in F$ there exists $g_0 \in F$ such that $\rho(f_0, g_0) > 0$, and that for any ξ with $0 < \xi < 1$ and any $\alpha > 0$, $C(\xi, \alpha) f_0^{1-\xi} g_0^{\alpha\xi} \in F$, which is denoted by $f_\alpha(x, \xi)$, where $C(\xi, \alpha)$ is some constant. If the conditions (A.1) to (A.7) on $f_\alpha(x - \theta, \xi)$ for every $\alpha > 0$ hold in the case when θ is a location parameter, then the asymptotic deficiency of the MLE $\hat{\theta}^*$ of θ tends to infinity as $\alpha \to \infty$.

The proof is straightforward from Theorem 2.3.

Example. Let $f_0(x)$ be a non-normal density and $g_0(x)$ be a normal density with mean 0 and variance 1. Then we consider a mixture f_σ of densities f_0 and g_0 as

$$f_\sigma(x, \xi) = C(\xi, \sigma) \left\{ f_0(x) \right\}^{1-\xi} \left\{ g_0(x) \right\}^{\xi/\sigma^2} \quad \text{for } 0 \le \xi \le 1,$$

where $C(\xi, \sigma)$ is some constant with $C(0, \sigma) = 1$. If the assumptions of Corollary 2.2 are satisfied for $1/\sigma^2$ instead of α in the case when θ is a location parameter, i.e., $f_\sigma(x - \theta, \xi)$ and $g_0(x - \theta)$, then the asymptotic deficiency of the MLE of θ tends to infinity as $\sigma \to 0$.

References

Akahira, M. (1986). *The Structure of Asymptotic Deficiency of Estimators.* Queen's Papers in Pure and Applied Mathematics No.75, Queen's University Press, Kingston, Ontario, Canada

Akahira, M. and Takeuchi, K. (1982). On asymptotic deficiency of estimators in pooled samples in the presence of nuisance parameters. *Statistics & Decisions* **1**, 17–38.

Hájek, J. (1962). Asymptotically most powerful rank order tests. *Ann. Math. Statist.*, **33**, 1124–1147.

Takeuchi, K. (1971). A uniformly asymptotically efficient estimator of a location parameter. *Jour. Amer. Statist. Assoc.*, **66**, 292–301.

Reçu en Juin 1988

Austral. J. Statist., **32**(3), 1990, 281–291

LOSS OF INFORMATION ASSOCIATED WITH THE ORDER STATISTICS AND RELATED ESTIMATORS IN THE DOUBLE EXPONENTIAL DISTRIBUTION CASE

MASAFUMI AKAHIRA[1] AND KEI TAKEUCHI[2]

University of Tsukuba and University of Tokyo

Summary

Fisher (1934) derived the loss of information of the maximum likelihood estimator (MLE) of the location parameter in the case of the double exponential distribution. Takeuchi & Akahira (1976) showed that the MLE is not second order asymptotically efficient. This paper extends these results by obtaining the (asymptotic) losses of information of order statistics and related estimators, and by comparing them via their asymptotic distributions up to the second order.

Key words: Loss of information; double exponential distribution; second order asymptotic efficiency; maximum likelihood estimator; median.

1. Introduction

Fisher (1934), starting from his fundamental paper (1922), discussed estimators of the location parameter of a double exponential (two-sided exponential) distribution as a typical example of non-regular estimation. He showed that the maximum likelihood estimator (MLE), which is equal to the sample median in this case, has asymptotic loss of information of order $\sqrt{n}$, as compared to constant order in regular cases. In Takeuchi & Akahira (1976) we showed that the MLE does not have second order asymptotic efficiency, unlike regular cases. This paper extends these results by showing that it is possible to construct an estimator which is asymptotically

Received May 1989; revised September 1989.

[1]Institute of Mathematics, University of Tsukuba, Ibaraki 305, Japan.

[2]Research Center for Advanced Science and Technology, University of Tokyo, 4-6-1 Komaba, Meguro-ku, Tokyo 156, Japan.

Acknowledgements. The authors thank the associate editor and the referee for valuable comments, in particular for pointing out the form of proof used after Lemma 2.2.

    MASAFUMI AKAHIRA AND KEI TAKEUCHI

better than the MLE both in terms of asymptotic variance and asymptotic loss of information in the second order, but that it is impossible to have an estimator which is uniformly better than the MLE in the second order expansion of the distribution. We thus conclude that *there is no second order asymptotically efficient estimator.*

2. The Loss of Information

Supose that $X_1, \ldots, X_n$ are independently and identically distributed (i.i.d.) random variables with the double exponential density function.

$$f(x, \theta) = f_0(x - \theta) = \tfrac{1}{2}\exp(-|x - \theta|).$$

First we compute the amount of information contained in the order statistics around the median, or equivalently the loss of information by discarding all other order statistics. We use the following lemma.

Lemma 2.1. (Fisher, 1925; Rao 1961) *The loss of information associated with any statistic T, which is a function of a sample of size n obtained from the population with the density function $f(x, \theta)$, is given by*

$$\mathrm{E}_\theta\left[\mathrm{V}_\theta\left(\sum_{i=1}^{n} \frac{\partial}{\partial \theta} \log f(X_i, \theta) \mid T\right)\right],$$

where $\mathrm{E}_\theta(\cdot)$ denotes unconditional expectation and $\mathrm{V}_\theta(\cdot \mid T)$ the conditional variance given T.

Now let T_k be the set of the central $2k+1$ order statistics $X_{(s-k+1)} \leq \cdots \leq X_{(s+k+1)}$ obtained from a sample of size $n = 2s + 1$ from the double exponential distribution. Then we have the following lemma.

Lemma 2.2. *For each $k = 0, 1, \ldots, s - 1$ the loss of information L_k associated with T_k is given by*

$$L_k = \mathrm{E}_\theta\left[\mathrm{V}_\theta\left(\sum_{i=1}^{2s+1} \mathrm{sgn}(X_i - \theta) \mid T_k\right)\right] \tag{2.1}$$

$$= \frac{2(2s+1)!}{(s-k-1)!(s+k)!} \int_{\frac{1}{2}}^{1} (2u^{s-k-1} - u^{s-k-2})(1 - u)^{s+k}\, du. \tag{2.2}$$

Proof. Equation (2.1) follows directly from Lemma 2.1 and the assumed form of the density function. Given T_k, the set of the first $s-k$ order statistics $X_{(1)}, \ldots, X_{(s-k)}$ is a random ordered sample from the distribution with

LOSS OF INFORMATION IN THE DOUBLE EXPONENTIAL DISTRIBUTION 283

a density $f_0(x)/F_0(x_{(s-k+1)})$ on $x \le x_{(s-k+1)}$, where $F_0(x)$ is the double exponential distribution function of the X_i. Similarly, the last $s - k$ are from $f_0(x)/[1 - F_0(x_{(s+k+1)})]$ on $x \ge x_{(s+k+1)}$. Define $u = F_0(X_{(s-k+1)})$, $v = 1 - F_0(X_{(s+k+1)})$, $w_i = F_0(X_{(i)})$, $z_i = 1 - F_0(X_{(i)})$. Then, ignoring the ordering of the first and last $(s-k)$ X s, $w_1, \ldots, w_{s-k}$ and $z_{s+k+2}, \ldots, z_{2s+1}$ are random samples from uniform distributions on $[0, u]$ and $[0, v]$ respectively. Given T_k, each $\mathrm{sgn}(X_i - \theta)$ is determined for $s - k + 1 \le i \le s + k + 1$, so we have

$$
\mathrm{V}\left\{ \sum_{i=1}^{2s+1} \mathrm{sgn}(X_i - \theta) \mid T_k \right\} = (s - k)(\mathrm{V}_1 + \mathrm{V}_2),
$$

where $\mathrm{V}_1 = 1 - \left\{ \int_0^u \mathrm{sgn}(x - \theta) \, dw/u \right\}^2$, which corresponds to the conditional variance of X for $X \le X_{(s-k+1)}$, V_2 corresponds to that of X for $X \ge X_{(s+k+1)}$, and

$$
w = F_0(x - \theta) = \begin{cases} \frac{1}{2} e^{-|x-\theta|} & \text{for } x \le \theta, \\ 1 - \frac{1}{2} e^{-|x-\theta|} & \text{for } x > \theta. \end{cases}
$$

Then we get

$$
\mathrm{V}_1 = \begin{cases} 0 & \text{for } X_{(s-k+1)} \le \theta, \\ (2u - 1)/u^2 & \text{for } X_{(s-k+1)} > \theta, \end{cases}
$$

$$
\mathrm{V}_2 = \begin{cases} 0 & \text{for } X_{(s+k+1)} \ge \theta, \\ (2v - 1)/v^2 & \text{for } X_{(s+k+1)} < \theta, \end{cases}
$$

and hence

$$
L_k = (s - k)\{\mathrm{E}(\mathrm{V}_1) + \mathrm{E}(\mathrm{V}_2)\}.
$$

Symmetry now implies that

$$
\mathrm{E}(\mathrm{V}_1) = \mathrm{E}(\mathrm{V}_2) = \frac{(2s + 1)!}{(s - k)!(s + k)!} \int_{\frac{1}{2}}^1 \frac{(2u - 1)}{u^2} u^{s-k}(1 - u)^{s+k} \, du,
$$

and from this result equation (2.2) follows as required.

This result can also be expressed in the following form.

Theorem 2.1. *For each $k = 0, 1, \ldots, s - 2$, the loss of information L_k associated with T_k is given by*

$$
\frac{L_k}{2(2s + 1)} = \binom{2s}{s}(\tfrac{1}{2})^{2s} - \frac{k + 1}{s - k - 1} + \sum_{j=0}^k \frac{2(k - j + 1)}{s - k - 1}\binom{2s}{s - j}(\tfrac{1}{2})^{2s}.
$$

$$
(2.3)
$$

Proof. The essence of the proof lies in recognizing the integrals in (2.2) as incomplete beta functions, which in turn are tail probabilities of binomial random variables. Specifically, using $Y(n)$ to denote a symmetric binomial random variable with parameter n, i.e., $\Pr\{Y(n) = p\} = \binom{n}{p}(\frac{1}{2})^n$, we have

$$\int_{\frac{1}{2}}^{1} \frac{n!}{(p-1)!(n-p)!} x^{p-1}(1-x)^{n-p}\, dx = \Pr\{Y(n) \le p - 1\}$$

for integers $p = 1, \ldots, n$. Then (2.2) gives

$$\frac{L_k}{2(2s+1)} = 2\Pr\{Y(2s) \le s - k - 1\} - \frac{2s\Pr\{Y(2s-1) \le s - k - 2\}}{s - k - 1}.$$

Using symmetry twice, this equals

$$-\Pr\{|Y(2s) - s| \le k\} + \frac{s\Pr\{s - k - 1 \le Y(2s-1) \le s + k\} - k - 1}{s - k - 1}$$

$$= \Pr\{Y(2s) = s\} - \frac{k+1}{s - k - 1}$$

$$- 2\sum_{j=0}^{k} \left[\Pr\{Y(2s) = s - j\} - \frac{s\Pr\{Y(2s-1) = s - 1 - j\}}{s - k - 1} \right]. \quad (2.4)$$

Since

$$s\Pr\{Y(2s-1) = i - 1\} = i\Pr\{Y(2s) = i\},$$

the sum in (2.4) equals

$$2\sum_{j=0}^{k} \left(1 - \frac{s-j}{s-k-1}\right) \Pr\{Y(2s) = s - j\}.$$

Substitution in (2.4) completes the proof of (2.3).

Remark 2.1. For the case $k = 0$, Theorem 2.1 yields

$$\frac{L_0}{2(2s+1)} = \frac{1}{s-1}\left[(s+1)\binom{2s}{s}(\tfrac{1}{2})^{2s} - 1\right].$$

This expression is consistent with a result in Fisher (1934, p.300).

3. The Asymptotic Loss of Information

Evaluating the three terms in Theorem 2.1 for fixed k and large s shows that asymptotically the loss of information is

$$\frac{4\sqrt{s}}{\sqrt{\pi}}[1 + o(1)] - 4(k+1) + O\left(\frac{k^2}{\sqrt{s}}\right). \quad (3.1)$$

LOSS OF INFORMATION IN THE DOUBLE EXPONENTIAL DISTRIBUTION 285

But in order to improve this quantity we must increase k with s, and (3.1) suggests that we take $k = r\sqrt{s} + o(\sqrt{s})$ with a non-negative constant r.

From Theorem 2.1 and the Cramér–Rao inequality it is follows that for any unbiased estimator $\hat{\theta}$ which is a function of T_k only, we have for fixed k

$$V_\theta(\hat{\theta}) \geq \frac{1}{I_{\hat{\theta}}} \geq \frac{1}{I_{T_k}} = \frac{1}{n - L_k} = \frac{1}{n}\left\{1 + \frac{1}{n}L_k + o\left(\frac{1}{n}\right)\right\}, \qquad (3.2)$$

where $I_{\hat{\theta}}$ and I_{T_k} denote the Fisher information at $\hat{\theta}$ and T_k respectively.

Theorem 3.1. *For $k = r\sqrt{s} + o(\sqrt{s}) = \frac{1}{2}\rho\sqrt{n} + o(\sqrt{n})$ for $\rho = r\sqrt{2} \geq 0$ and $n = 2s + 1$, the loss of information, L_k, associated with T_k is given by*

$$L_k = 4\big[\phi(\rho) - \rho\{1 - \Phi(\rho)\}\big]\sqrt{n} + o(\sqrt{n}), \qquad (3.3)$$

where $\Phi(x) = \int_{-\infty}^{x} \phi(u)\,du$ with $\phi(u) = e^{-u^2/2}/\sqrt{2\pi}$.

Proof. Using the normal approximation to the binomial distribution we have

$$\Pr\{Y(2s) = s\} = \frac{1}{\sqrt{(\pi s)}}\left(1 - \frac{1}{8s} + O(s^{-2})\right). \qquad (3.4)$$

The summation term in (2.3) equals

$$2\sum_{j=0}^{k} \frac{k - j + 1}{s - k - 1}\Pr\{Y(2s) = s + j\}$$

$$= \frac{2(k + 1)}{s - k - 1}\sum_{j=0}^{k}\Pr\{Y(2s) = s + j\} - \frac{2}{s - k - 1}\sum_{j=0}^{k} j\Pr\{Y(2s) = s + j\}$$

$$= \frac{2(r\sqrt{s} + 1)}{s - r\sqrt{s} - 1}\int_0^\rho \phi(u)\,du - \frac{\sqrt{(2s)}}{s - r\sqrt{s} - 1}\int_0^\rho u\phi(u)\,du + o\left(\frac{1}{\sqrt{s}}\right)$$

after approximation by Riemann integrals,

$$= \frac{2(r\sqrt{s} + 1)}{s - r\sqrt{s} - 1}\int_0^\rho \phi(u)\,du - \frac{1 - e^{-r^2}}{\sqrt{(2\pi)}} \cdot \frac{\sqrt{(2s)}}{s - r\sqrt{s} - 1} + o\left(\frac{1}{\sqrt{s}}\right). \qquad (3.5)$$

By (2.3), (3.4) and (3.5) we obtain

$$L_k = 2(2s + 1)\left\{-\frac{r}{\sqrt{s}} + \frac{2r}{\sqrt{s}}\int_0^\rho \phi(u)\,du + \frac{1}{\sqrt{(\pi s)}}e^{-r^2}\right\} + o(\sqrt{s})$$

$$= 4\big[\phi(\rho) - \rho\{1 - \Phi(\rho)\}\big]\sqrt{n} + o(\sqrt{n}),$$

which completes the proof.

We now consider the explicit form of estimators depending on T_k. As the simplest one, we put $\hat{\theta}_k = (X_{(s-k+1)} + X_{(s+k+1)})/2$. The proof of the next result is in Section 4.

Theorem 3.2. *For $k = r\sqrt{s} + o(\sqrt{s}) = \frac{1}{2}\rho\sqrt{n} + o(\sqrt{n})$ where $\rho = r\sqrt{2} \geq 0$ and $n = 2s + 1$, the estimator $\hat{\theta}_k$ has the stochastic expansion*

$$\sqrt{n}(\hat{\theta}_k - \theta) = Z + \frac{(Z - \rho)|Z - \rho| + (Z + \rho)|Z + \rho|}{4\sqrt{n}} + o_p\left(n^{-\frac{1}{2}}\right)$$

and asymptotic variance

$$V_\theta(\sqrt{n}\hat{\theta}_k) = 1 + c_\rho n^{-\frac{1}{2}}(1 + o(1)), \tag{3.6}$$

where $Z = \sqrt{n}\{F_0(X_{(s-k+1)} - \theta) + F_0(X_{(s+k+1)} - \theta) - 1\}$ with $F_0(\cdot)$ the distribution function of X_i, equivalently, $Z = \sqrt{n}(U_{(s-k+1)} + U_{(s+k+1)} - 1)$ with $U_{(1)}, \ldots, U_{(n)}$ the order statistics from the uniform distribution on the interval $(0, 1)$, and

$$c_\rho = 2\left[2\phi(\rho) + \rho\{2\Phi(\rho) - \tfrac{3}{2}\}\right]. \tag{3.7}$$

Remark 3.1. Since $\phi(x) \geq x\{1 - \Phi(x)\}$ for $x \geq 0$, it follows for any $\rho \geq 0$ that $c_\rho \geq \rho$. Define ρ_0 by $\Phi(\rho_0) = 3/4$, so that $\rho_0 \approx 0.67$. Then c_ρ has a minimum value at $\rho = \rho_0$. Hence $\hat{\theta}_k$ has minimum variance at $\rho = \rho_0$.

Remark 3.2. In particular, for $k = 0$ i.e., $\rho = 0$, we obtain from Theorem 3.2,

$$V_\theta(\sqrt{n}\,\hat{\theta}_0) = 1 + \frac{2\sqrt{2}}{\sqrt{(\pi n)}} + o\left(\frac{1}{\sqrt{n}}\right). \tag{3.8}$$

Note that $\hat{\theta}_0$ is the median, i.e., the MLE in this case.

Remark 3.3. Since

$$V_\theta(\hat{\theta}_k) = n^{-1}\left\{1 + c_\rho n^{-\frac{1}{2}}[1 + o(1)]\right\},$$

it follows from (3.2) that $c_\rho \geq L_k/\sqrt{n}$. In this case we have from (3.3) and Theorem 3.2,

$$c_\rho - \frac{1}{\sqrt{n}}L_k = \rho.$$

Corollary 3.1. *For $k = r\sqrt{s} + o(\sqrt{s}) = \frac{1}{2}\rho\sqrt{n} + o(\sqrt{n})$ where $\rho = r\sqrt{2} \geq 0$ and $n = 2s + 1$, the bound of the loss of information L_k' associated with the estimator $\hat{\theta}_k$ is given by $L_k' \leq c_{\rho_0}\sqrt{n} + o(\sqrt{n})$, where ρ_0 is defined in Remark 3.1.*

Proof. From (3.2) we have $V_\theta(\hat{\theta}_k) \geq (n - L_k')^{-1}$, which implies that

$$L_k' \leq n - \frac{1}{V_\theta(\hat{\theta}_k)} = n - n\left(1 + \frac{c_\rho}{\sqrt{n}}\right)^{-1} + o(\sqrt{n}) = c_\rho\sqrt{n} + o(\sqrt{n}).$$

LOSS OF INFORMATION IN THE DOUBLE EXPONENTIAL DISTRIBUTION 287

Therefore, by putting $\rho = \rho_0$ we have

$$L'_k \leq c_{\rho_0}\sqrt{n} + o(\sqrt{n}).$$

Remark 3.4. Since $\rho_0 \approx 0.67$, it is seen that

$$L'_k \leq c_{\rho_0}\sqrt{n} + o(\sqrt{n}) = 1.27\sqrt{n} + o(\sqrt{n}). \tag{3.9}$$

On the other hand, it follows from (3.1) that the loss L'_0 of the MLE $\hat{\theta}_0$, i.e., the median, is given by

$$L'_0 = L_0 = \frac{2\sqrt{2}}{\sqrt{\pi}}\sqrt{n} + o(\sqrt{n}) = 1.60\sqrt{n} + o(\sqrt{n}). \tag{3.10}$$

From (3.9) and (3.10) we see that the estimator $\hat{\theta}_k$ is asymptotically better than the MLE $\hat{\theta}_0$ in the sense of $L'_k \leq L'_0 + o(\sqrt{n})$.

In Akahira & Takeuchi (1981, p.97) it is shown that the asymptotic distribution of $\hat{\theta}_0$ up to the order $n^{-1/2}$ is given by

$$\begin{aligned}
\Pr\left\{\sqrt{n}(\hat{\theta}_0 - \theta) \leq t\right\} &= \Phi(t) - \tfrac{1}{2}n^{-\frac{1}{2}}t^2\phi(t)\operatorname{sgn} t + o\left(n^{-\frac{1}{2}}\right) \\
&= F_n(t) \quad \text{(say)}.
\end{aligned} \tag{3.11}$$

Then the asymptotic density $f_n(t)$ of $\hat{\theta}_0$ is given by

$$f_n(t) = F'_n(t) = \phi(t) - \tfrac{1}{2}n^{-\frac{1}{2}}(2t - t^3)\phi(t)\operatorname{sgn} t + o(n^{-\frac{1}{2}}).$$

Hence the asymptotic variance of $\hat{\theta}_0$ is given by

$$V_\theta(\sqrt{n}\hat{\theta}_0) = \int_{-\infty}^{\infty} t^2 f_n(t)\, dt = 1 + \frac{2\sqrt{2}}{\sqrt{(\pi n)}}\left(1 + o(1)\right),$$

which is consistent with (3.8), and identically so when $n = 2s + 1$.

Next we obtain the asymptotic distribution of $\hat{\theta}_k$ up to the order $n^{-1/2}$. The proof is given in Section 4.

Theorem 3.3. *For $k = r\sqrt{s} + o(\sqrt{s}) = \frac{1}{2}\rho\sqrt{n} + o(\sqrt{n})$ where $\rho = r\sqrt{2} \geq 0$ and $n = 2s + 1$, the asymptotic distribution of $\hat{\theta}_k$ up to the order $n^{-\frac{1}{2}}$ is given by*

$$\Pr\left\{\sqrt{n}(\hat{\theta}_k - \theta) \leq t\right\}$$
$$= \Phi(t) - \frac{\phi(t)}{2\sqrt{n}}\left[-\rho t + \tfrac{1}{2}\{(t - \rho)|t - \rho| + (t + \rho)|t + \rho|\}\right] + o(n^{-\frac{1}{2}}).$$

Remark 3.5. Letting $r = 0$, we see that, for $k = 0$, the asymptotic distribution of the estimator $\hat{\theta}_0$ is consistent with that given by (3.11).

From (3.11) and Theorem 3.3 we have the following (the proof is straightforward).

MASAFUMI AKAHIRA AND KEI TAKEUCHI

Corollary 3.2. *For estimators $\hat{\theta}_k$ and $\hat{\theta}_0$,*

$$\Pr\left\{\sqrt{n}(\hat{\theta}_0 - \theta) \le t\right\} \le \Pr\left\{\sqrt{n}(\hat{\theta}_k - \theta) \le t\right\} + o(1/\sqrt{n}) \quad \text{for } t \ge \rho,$$

$$\Pr\left\{\sqrt{n}(\hat{\theta}_0 - \theta) \le t\right\} \ge \Pr\left\{\sqrt{n}(\hat{\theta}_k - \theta) \le t\right\} + o(1/\sqrt{n}) \quad \text{for } t \le -\rho,$$

$$\Pr\left\{\sqrt{n}(\hat{\theta}_0 - \theta) \le t\right\} \ge \Pr\left\{\sqrt{n}(\hat{\theta}_k - \theta) \le t\right\} + o(1/\sqrt{n}) \quad \text{for } 0 \le t < \rho,$$

$$\Pr\left\{\sqrt{n}(\hat{\theta}_0 - \theta) \le t\right\} \le \Pr\left\{\sqrt{n}(\hat{\theta}_k - \theta) \le t\right\} + o(1/\sqrt{n}) \quad \text{for } -\rho < t \le 0.$$

Remark 3.6. From Corollary 3.2 we see that, for $0 < t < \rho$, $\hat{\theta}_0$ is asymptotically better than $\hat{\theta}_k$ in the sense of the concentration probability. That is,

$$\Pr\left\{\sqrt{n}|\hat{\theta}_0 - \theta| \le t\right\} \ge \Pr\left\{\sqrt{n}|\hat{\theta}_k - \theta| \le t\right\} + o(1/\sqrt{n}) \quad \text{for } 0 < t < \rho,$$

and, for $t \ge \rho$, $\hat{\theta}_k$ is asymptotically better than $\hat{\theta}_0$ in the same sense as the above, i.e.,

$$\Pr\left\{\sqrt{n}|\hat{\theta}_0 - \theta| \le t\right\} \le \Pr\left\{\sqrt{n}|\hat{\theta}_k - \theta| \le t\right\} + o(1/\sqrt{n}) \quad \text{for } t \ge \rho.$$

Recently, Sugiura & Naing (1989) extended the above results to the other estimators using a weighted linear combination of the sample median and pairs of order statistics.

4. Proofs

In this section we give the proofs of Theorems 3.2 and 3.3, based on the following lemmas.

Lemma 4.1. *Let $Y_1, \ldots, Y_n$ be i.i.d. random variables with an exponential density function*

$$f(y, \theta) = \begin{cases} e^{-y/\theta}/\theta & \text{for } y > 0, \\ 0 & \text{for } y \le 0, \end{cases} \tag{4.1}$$

and let $g(Y_1, \ldots, Y_n)$ be homogeneous in $Y_1, \ldots, Y_n$ of degree 0 so that $g(\kappa Y_1, \ldots, \kappa Y_n) \equiv g(Y_1, \ldots, Y_n)$ for all non-zero κ. Then $g(Y_1, \ldots, Y_n)$ and $\sum_{i=1}^{n} Y_i$ are independent.

Proof. It is clear that $\sum_{i=1}^{n} Y_i$ is a complete sufficient statistic for θ. Since the distribution of $g(Y_1, \ldots, Y_n)$ is independent of θ, then, by the theorem of Basu (1955), it is independent of $\sum_{i=1}^{n} Y_i$.

LOSS OF INFORMATION IN THE DOUBLE EXPONENTIAL DISTRIBUTION 289

Lemma 4.2. *Under the same conditions as Lemma 4.1,*

$$\mathrm{E}\big[g(Y_1,\ldots,Y_n)\big] = \frac{\mathrm{E}\big[g(Y_1,\ldots,Y_n)(\sum_{i=1}^{n} Y_i)^{\alpha}\big]}{\mathrm{E}\big[(\sum_{i=1}^{n} Y_i)^{\alpha}\big]}$$

for all real numbers α.

The proof is straightforward from Lemma 4.1.

Proof of Theorem 3.2. Without loss of generality, we assume that $\theta = 0$. Let $F_0(x)$ be a distribution function of X_i $(i = 1,\ldots,n)$. Then we have $X_{(i)} = F_0^{-1}(U_{(i)})$ $(i = 1,\ldots,n)$, where the $U_{(i)}$ s are order statistics from the uniform distribution on the interval $(0,1)$. We also obtain

$$F_0^{-1}(u) = -\big\{\mathrm{sgn}\big(u - \tfrac{1}{2}\big)\big\}\log\big(1 - |2u - 1|\big).$$

We put $U = U_{(s-k+1)}$ and $V = U_{(s+k+1)}$. Then we have

$$\begin{aligned}
n^{\frac{1}{2}}\hat{\theta}_k &= \tfrac{1}{2}n^{\frac{1}{2}}\big\{F^{-1}(U) + F^{-1}(V)\big\} \\
&= n^{\frac{1}{2}}(U + V - 1) + n^{-\frac{1}{2}}\big\{n^{\frac{1}{2}}(U - \tfrac{1}{2})|n^{\frac{1}{2}}(U - \tfrac{1}{2})|\big\} \\
&\quad + n^{-\frac{1}{2}}\big\{n^{\frac{1}{2}}(V - \tfrac{1}{2})|n^{\frac{1}{2}}(V - \tfrac{1}{2})|\big\} + o_p\big(n^{-\frac{1}{2}}\big).
\end{aligned}$$

Putting $X = n^{\frac{1}{2}}(U - \tfrac{1}{2})$, $Y = n^{\frac{1}{2}}(V - \tfrac{1}{2})$, we obtain

$$n^{\frac{1}{2}}\hat{\theta}_k = X + Y + n^{-\frac{1}{2}}(X|X| + Y|Y|) + o_p\big(n^{-\frac{1}{2}}\big). \tag{4.2}$$

If $Y_1,\ldots,Y_{2s+1}$ are i.i.d. random variables with the exponential density function as at (4.1) with $\theta = 1$, then it is known that

$$U_{(j)} = \sum_{i=1}^{j} Y_i \Big/ \sum_{i=1}^{2s+1} Y_i$$

for each $j = 1,\ldots,2s + 1$. Using this fact we have, by Lemma 4.2,

$$\begin{aligned}
\mathrm{E}[(U + V - 1)^2] &= \mathrm{E}\left[\left(\frac{\sum_{i=1}^{s-k+1} Y_i - \sum_{i=s+k+2}^{2s+1} Y_i}{\sum_{i=1}^{2s+1} Y_i}\right)^2\right] \\
&= \frac{\mathrm{E}\big[(\sum_{i=1}^{s-k+1} Y_i - \sum_{i=s+k+1}^{2s+1} Y_i)^2\big]}{\mathrm{E}\big[(\sum_{i=1}^{2s+1} Y_i)^2\big]} \\
&= \frac{s - k + 1}{(s + 1)(2s + 1)}.
\end{aligned}$$

290　　MASAFUMI AKAHIRA AND KEI TAKEUCHI

Hence we obtain

$$V(X+Y) = E[(X+Y)^2] = E\big[\{\sqrt{n}(U+V-1)\}^2\big]$$
$$= \frac{s-k+1}{s+1} = 1 - \frac{\rho}{\sqrt{n}} + o(n^{-\frac{1}{2}}). \qquad (4.3)$$

Since

$$E(X-Y) = E[n^{\frac{1}{2}}(U-V)] = -\rho,$$

it follows similarly that

$$V(X-Y) = E[\{\sqrt{n}(U-V)\}^2] - \rho^2 = n^{-\frac{1}{2}}(\rho + o(1)).$$

We put

$$Z = X+Y; \qquad W = s^{1/4}(X-Y+\rho).$$

Since

$$X = \tfrac{1}{2}(X+Y+X-Y) = \tfrac{1}{2}(Z-\rho+s^{-1/4}W);$$
$$Y = \tfrac{1}{2}\{(X+Y)-(X-Y)\} = \tfrac{1}{2}(Z+\rho-s^{-1/4}W),$$

we have from (4.2)

$$n^{\frac{1}{2}}\hat{\theta}_k = Z + \tfrac{1}{4}n^{-\frac{1}{2}}\big[(Z-\rho)|Z-\rho| + (Z+\rho)|Z+\rho|\big] + o_p(n^{-\frac{1}{2}}). \qquad (4.4)$$

Hence

$$V(\sqrt{n}\hat{\theta}_k) = E(Z^2) + \tfrac{1}{2}n^{-\frac{1}{2}}\big\{E\big[Z(Z-\rho)|Z-\rho|\big]$$
$$+ E\big[Z(Z+\rho)|Z+\rho|\big]\big\} + o(n^{-\frac{1}{2}}). \qquad (4.5)$$

For any constant c we have

$$E\big[Z(Z-c)|Z-c|\big] = E\big[|Z-c|^3\big] + cE\big[(Z-c)|Z-c|\big]. \qquad (4.6)$$

We obtain

$$\int_{-\infty}^{\infty} |x-c|^3 \phi(x)\,dx = 2(c^2+c)\phi(c) + (c^3+3c)\{2\Phi(c)-1\}, \qquad (4.7)$$
$$\int_{-\infty}^{\infty} (x-c)|x-c|\phi(x)\,dx = -2c\phi(c) + (c^2+1)\{1-2\Phi(c)\}. \qquad (4.8)$$

LOSS OF INFORMATION IN THE DOUBLE EXPONENTIAL DISTRIBUTION 291

Since Z is asymptotically normally distributed with mean 0 and variance $1 - r/\sqrt{s} = 1 - \rho/\sqrt{n}$, we have from (4.6)–(4.8) that

$$E[Z(Z - c)|Z - c|] = 4\phi(c) + 2c\{2\Phi(c) - 1\}\,.$$

Hence we obtain

$$E[Z(Z - \rho)|Z - \rho|] = 4\phi(\rho) + 2\rho\{2\Phi(\rho) - 1\}\,. \tag{4.9}$$

From (4.3), (4.5) and (4.9) we have

$$V(n^{\frac{1}{2}}\hat{\theta}_k) = 1 + 2n^{-\frac{1}{2}}\left[2\phi(\rho) + \rho\{2\Phi(\rho) - \tfrac{3}{2}\}\right] + o(n^{-\frac{1}{2}})\,.$$

The coefficient of $n^{-\frac{1}{2}}$ here yields c_ρ as at (3.7), proving (3.6).

Proof of Theorem 3.3. Assume without loss of generality that $\theta = 0$. From (4.4) we have that $\Pr\{\sqrt{n}\hat{\theta}_k \le t\}$ equals

$$\Pr\left\{Z + \tfrac{1}{4}n^{-\frac{1}{2}}\left[(Z - \rho)|Z - \rho| + (Z + \rho)|Z + \rho|\right] \le t + o_p\left(n^{-\frac{1}{2}}\right)\right\}$$

$$= \Pr\left\{Z \le t - \tfrac{1}{4}n^{-\frac{1}{2}}\left[(t - \rho)|t - \rho| + (t + \rho)|t + \rho|\right] + o_p\left(n^{-\frac{1}{2}}\right)\right\}\,.$$

We see that Z is asymptotically normally distributed with mean 0 and variance $1 + o(1/\sqrt{s}) = 1 + o(1/\sqrt{n})$. Hence we obtain

$$\Pr\{\sqrt{n}\hat{\theta}_k \le t\} = \Pr\left\{Z \le t - \tfrac{1}{2}n^{-\frac{1}{2}}\left\{-\rho t + \tfrac{1}{2}[(t - \rho)|t - \rho|\right.\right.$$

$$\left.\left. + (t + \rho)|t + \rho|]\right\} + o_p\left(n^{-\frac{1}{2}}\right)\right\}$$

$$= \Phi(t) - \tfrac{1}{2}n^{-\frac{1}{2}}\phi(t)\left(\rho t + \operatorname{sgn}(t)[(|t| - \rho)_+]^2\right) + o\left(n^{-\frac{1}{2}}\right)\,.$$

References

AKAHIRA, M. & TAKEUCHI, K. (1981). *Asymptotic Efficiency of Statistical Estimators: Concepts and Higher Order Asymptotic Efficiency.* Lecture Notes in Statistics **7**. New York: Springer-Verlag.

BASU, D. (1955). On statistics independent of a complete sufficient statistic. *Sankhyā* **15**, 377–380.

FISHER, R.A. (1922). On the mathematical foundations of theoretical statistics. *Philos. Trans. Roy. Soc. London Ser. A* **222**, 309–368.

FISHER, R.A. (1925). Theory of statistical estimation. *Proc. Cambridge Philos. Soc.* **22**, 700–725.

FISHER, R.A. (1934). Two new properties of mathematical likelihood. *Proc. Roy. Soc. London Ser. A* **144**, 285–307.

RAO, C.R. (1961). Asymptotic efficiency and limiting information. *Proc. Fourth Berkeley Symp. Math. Statist. Probab.* **1**, 531–545.

SUGIURA, N. & NAING, M.T. (1989). Improved estimators for the location of double exponential distribution. *Comm. Statist. A—Theory Methods* **18**, 541–554.

TAKEUCHI, K. & AKAHIRA, M. (1976). On the second order asymptotic efficiencies of estimators. In *Proc. Third Japan–USSR Symp. Probab. Theory*, eds. G. Maruyama and J.V. Prokhorov, 604–638. Lecture Notes in Mathematics **550**. Berlin: Springer-Verlag.

Ann. Inst. Statist. Math.
Vol. 43, No. 2, 297–310 (1991)

BOOTSTRAP METHOD AND EMPIRICAL PROCESS

Masafumi Akahira[1] and Kei Takeuchi[2]

[1] *Institute of Mathematics, University of Tsukuba, Tsukuba, Ibaraki 305, Japan*
[2] *Research Center for Advanced Science and Technology, University of Tokyo,
4-6-1 Komaba, Meguro-ku, Tokyo 156, Japan*

(Received August 24, 1989; revised February 22, 1990)

Abstract. In this paper we consider the sampling properties of the bootstrap process, that is, the empirical process obtained from a random sample of size n (with replacement) of a fixed sample of size n of a continuous distribution. The cumulants of the bootstrap process are given up to the order n^{-1} and their unbiased estimation is discussed. Furthermore, it is shown that the bootstrap process has an asymptotic minimax property for some class of distributions up to the order $n^{-1/2}$.

Key words and phrases: Bootstrap process, cumulants, unbiased estimators, asymptotic minimax property.

1. Introduction

The bootstrap method may be reviewed from different viewpoints. In this paper, we intend to consider the sampling properties of the bootstrap process, that is, the empirical process derived from the bootstrap sampling, i.e. that obtained from a random sample of size n (with replacement) of a fixed sample of size n of a certain distribution. Let $X_1, \ldots, X_n$ be a sample of size n from a population with the distribution function $F(t)$ and let $X_1^*, \ldots, X_n^*$ be a bootstrap sample of size n, that is, a random sample of size n from $X_1, \ldots, X_n$. Let the empirical distribution functions obtained from $(X_1, \ldots, X_n)$ and $(X_1^*, \ldots, X_n^*)$ be denoted by $F_n(t)$ and $F_n^*(t)$, respectively. It is well known that $\sqrt{n}(F_n(t) - F(t))$ approaches a Gaussian process as $n \to \infty$, and, given $F_n(t)$, $\sqrt{n}(F_n^*(t) - F_n(t))$ conditionally approaches a Gaussian process with the same variance and covariance of $\sqrt{n}(F_n(t) - F(t))$ with $F(t)$ replaced by $F_n(t)$. Hence $\sqrt{n}(F_n^*(t) - F_n(t))$ can be considered to be the consistent estimator of $\sqrt{n}(F_n(t) - F(t))$. Note that $\sqrt{n}(F_n(t) - F(t))$ can not be usually observed since $F(t)$ is unknown, whereas the distribution of $\sqrt{n}(F_n^*(t) - F_n(t))$ can be completely computed from the sample. We further investigate how $\sqrt{n}(F_n^*(t) - F_n(t))$ will differ from $\sqrt{n}(F_n(t) - F(t))$ in higher order terms and we discuss possible improvements on $F_n^*(t)$.

In many problems of statistical inference, the procedures will depend on the distribution of a statistic T_n under an unknown distribution $F(t)$, which in many

          MASAFUMI AKAHIRA AND KEI TAKEUCHI

cases can be discussed in terms of $\sqrt{n}(F_n(t) - F(t))$, at least asymptotically. Hence $\sqrt{n}(F_n^*(t) - F_n(t))$ can be used instead of $\sqrt{n}(F_n(t) - F(t))$ in the derivative of the asymptotic distribution of the statistic. It will be shown that the method is in a sense asymptotically efficient in a nonparametric (or semiparametric) framework. A first order approximation was considered in the work of Efron (1979, 1982) and Beran (1982) considered a second order approximation from a different viewpoint. The purpose of this paper is to compute the cumulants up to the order n^{-1} and to show that the bootstrap process is in a sense, asymptotically, the best estimator of the empirical process up to the order $n^{-1/2}$, whereas in terms of the order n^{-1} there are many complications and although a slight improvement is possible over the usual bootstrap process, no uniformly optimal results seem to be obtainable. The bootstrap method is used to estimate the distribution of some statistic T_n under a general unknown population distribution and it is shown that it is asymptotically best up to the second order in the sense that the estimator of the asymptotic variance as well as that of the asymptotic distribution of T_n can not be uniformly improved if the class of possible population distributions is sufficiently wide.

2. Unbiased estimation of cumulants of the empirical process

In the framework of Section 1, we put $W_n(t) = \sqrt{n}(F_n(t) - F(t))$ and $W_n^*(t) = \sqrt{n}(F_n^*(t) - F_n(t))$. Consequently we have the following.

LEMMA 2.1. *The cumulants of $W_n(t)$ are given, up to the fourth order, as follows*:

$$
\begin{aligned}
&E[W_n(t)] = 0, \\
&\mathrm{Cov}(W_n(t_1), W_n(t_2)) = F(t_1)(1 - F(t_2)) && for \quad t_1 \le t_2, \\
&\kappa_3(W_n(t_1), W_n(t_2), W_n(t_3)) \\
&\qquad = (1/\sqrt{n})F(t_1)(1 - 2F(t_2))(1 - F(t_3)) && for \quad t_1 \le t_2 \le t_3, \\
&\kappa_4(W_n(t_1), W_n(t_2), W_n(t_3), W_n(t_4)) \\
&\qquad = (1/n)F(t_1)(1 - F(t_4))(1 - 4F(t_2) - 2F(t_3) + 6F(t_2)F(t_3)) \\
&\qquad\qquad\qquad\qquad\qquad\qquad for \quad t_1 \le t_2 \le t_3 \le t_4.
\end{aligned}
$$

The proof is given in Section 4, but Lemma 2.1 may be also derived from Lemma 3.1 of Withers (1983). From Lemma 2.1 we have the following.

LEMMA 2.2. *Given $F_n(t)$, the conditional cumulants of $W_n^*(t)$ are given, up to the fourth order, as follows*:

$$
\begin{aligned}
&E[W_n^*(t) \mid F_n(t)] = 0, \\
&\mathrm{Cov}(W_n^*(t_1), W_n^*(t_2) \mid F_n(t_1), F_n(t_2)) = F_n(t_1)(1 - F_n(t_2)) \\
&\qquad\qquad\qquad\qquad\qquad\qquad for \quad t_1 \le t_2, \\
&\kappa_3(W_n^*(t_1), W_n^*(t_2), W_n^*(t_3) \mid F_n(t_1), F_n(t_2), F_n(t_3)) \\
&\qquad = (1/\sqrt{n})F_n(t_1)(1 - 2F_n(t_2))(1 - F_n(t_3)) && for \quad t_1 \le t_2 \le t_3,
\end{aligned}
$$

$$\kappa_4(W_n^*(t_1),\, W_n^*(t_2),\, W_n^*(t_3),\, W_n^*(t_4) \mid F_n(t_1),\, F_n(t_2),\, F_n(t_3),\, F_n(t_4))$$
$$= (1/n)F_n(t_1)(1 - F_n(t_4))(1 - 4F_n(t_2) - 2F_n(t_3) + 6F_n(t_2)F_n(t_3))$$
$$for \quad t_1 \le t_2 \le t_3 \le t_4.$$

The proof is straightforward from Lemma 2.1. On the other hand, we also have the following.

LEMMA 2.3.

$$E[F_n(t_1)(1 - F_n(t_2))] = \{1 - (1/n)\}F(t_1)(1 - F(t_2)) \quad for \quad t_1 \le t_2,$$
$$E[F_n(t_1)(1 - 2F_n(t_2))(1 - F_n(t_3))]$$
$$= \{1 - (1/n)\}\{1 - (2/n)\}F(t_1)(1 - 2F(t_2))(1 - F(t_3))$$
$$for \quad t_1 \le t_2 \le t_3,$$
$$E[F_n(t_1)(1 - F_n(t_4))(1 - 4F_n(t_2) - 2F_n(t_3) + 6F_n(t_2)F_n(t_3))]$$
$$= \{1 - (1/n)\}\{1 - (2/n)\}\{1 - (3/n)\}F(t_1)(1 - F(t_4))$$
$$\cdot (1 - 4F(t_2) - 2F(t_3) + 6F(t_2)F(t_3))$$
$$- (1/n)\{1 - (1/n)\}F(t_1)(1 - F(t_4)) \quad for \quad t_1 \le t_2 \le t_3 \le t_4.$$

The proof is given in Section 4. From Lemmas 2.2 and 2.3 it is seen that, given $F_n(t)$, the conditional cumulants of $W_n^*(t)$ are not unbiased estimators of the corresponding cumulants of $W_n(t)$.

LEMMA 2.4.   *The (unconditional) cumulants of $W_n^*(t)$ are given, up to the fourth order, as follows:*

$$E[W_n^*(t)] = 0,$$
$$\mathrm{Cov}(W_n^*(t_1),\, W_n^*(t_2)) = \{1 - (1/n)\}F(t_1)(1 - F(t_2)) \quad for \quad t_1 \le t_2,$$
$$\kappa_3(W_n^*(t_1),\, W_n^*(t_2),\, W_n^*(t_3))$$
$$= (1/\sqrt{n})\{1 - (1/n)\}\{1 - (2/n)\}F(t_1)(1 - 2F(t_2))(1 - F(t_3))$$
$$for \quad t_1 \le t_2 \le t_3,$$
$$\kappa_4(W_n^*(t_1),\, W_n^*(t_2),\, W_n^*(t_3),\, W_n^*(t_4))$$
$$= (1/n)\{1 - (1/n)\}\{1 - (2/n)\}\{1 - (3/n)\}F(t_1)(1 - F(t_4))$$
$$\cdot (1 - 4F(t_2) - 2F(t_3) + 6F(t_2)F(t_3))$$
$$- (1/n^2)\{1 - (1/n)\}F(t_1)(1 - F(t_4))$$
$$+ (1/n)F(t_1)(1 - F(t_4))(3 - 8F(t_2) - 4F(t_3) + 12F(t_2)F(t_3))$$
$$- (2/n^2)F(t_1)(1 - F(t_4))(3 - 10F(t_2) - 5F(t_3) + 15F(t_2)F(t_3))$$
$$+ (3/n^3)F(t_1)(1 - F(t_4))(1 - 4F(t_2) - 2F(t_3) + 6F(t_2)F(t_3)) + o(1/n^3)$$
$$for \quad t_1 \le t_2 \le t_3 \le t_4.$$

The proof is given in Section 4. From Lemmas 2.1, 2.2 and 2.3 we also have the following.

                    MASAFUMI AKAHIRA AND KEI TAKEUCHI

THEOREM 2.1. *The unbiased estimators of the covariance* $\mathrm{Cov}(W_n(t_1),$ $W_n(t_2))$ *and the third order cumulant* $\kappa_3(W_n(t_1), W_n(t_2), W_n(t_3))$ *are given by*

$$\{n/(n-1)\}\,\mathrm{Cov}(W_n^*(t_1), W_n^*(t_2) \mid F_n(t_1), F_n(t_2))$$
$$= \{1/(n-1)\}F_n(t_1)(1 - F_n(t_2)) \qquad \textit{for} \quad t_1 \leq t_2,$$
$$\{n^2/(n-1)(n-2)\}\kappa_3(W_n^*(t_1), W_n^*(t_2), W_n^*(t_3) \mid F_n(t_1), F_n(t_2), F_n(t_3))$$
$$= \{n\sqrt{n}/(n-1)(n-2)\}F_n(t_1)(1 - 2F_n(t_2))(1 - F_n(t_3)) \textit{ for } \quad t_1 \leq t_2 \leq t_3,$$

respectively.

PROOF. From Lemmas 2.1, 2.2 and 2.3 we have for $t_1 \leq t_2$

$$E[\mathrm{Cov}(W_n^*(t_1), W_n^*(t_2) \mid F_n(t_1), F_n(t_2))] = E[F_n(t_1)(1 - F_n(t_2))]$$
$$= \{1 - (1/n)\}F(t_1)(1 - F(t_2))$$
$$= \{(n-1)/n\}\,\mathrm{Cov}(W_n(t_1), W_n(t_2)),$$

hence

$$\{n/(n-1)\}\,\mathrm{Cov}(W_n^*(t_1), W_n^*(t_2) \mid F_n(t_1), F_n(t_2)) = \{n/(n-1)\}F_n(t_1)(1 - F_n(t_2))$$

is an unbiased estimator of $\mathrm{Cov}(W_n(t_1), W_n(t_2))$. In a similar way we obtain for $t_1 \leq t_2 \leq t_3$

$$E[\kappa_3(W_n^*(t_1), W_n^*(t_2), W_n^*(t_3) \mid F_n(t_1), F_n(t_2), F_n(t_3))]$$
$$= E[(1/\sqrt{n})F_n(t_1)(1 - 2F_n(t_2))(1 - F_n(t_3))]$$
$$= (1/\sqrt{n})\{(n-1)(n-2)/n^2\}F(t_1)(1 - 2F(t_2))(1 - F(t_3))$$
$$= \{(n-1)(n-2)/n^2\}\kappa_3(W_n(t_1), W_n(t_2), W_n(t_3)),$$

hence

$$\{n^2/(n-1)(n-2)\}\kappa_3(W_n^*(t_1), W_n^*(t_2), W_n^*(t_3) \mid F_n(t_1), F_n(t_2), F_n(t_3))$$
$$= \{n\sqrt{n}/(n-1)(n-2)\}F_n(t_1)(1 - 2F_n(t_2))(1 - F_n(t_3))$$

is an unbiased estimator of $\kappa_3(W_n(t_1), W_n(t_2), W_n(t_3))$. Thus we complete the proof.

Remark 2.1. Let $X_1', \ldots, X_{n-1}'$ be a bootstrap sample of size $n-1$, that is, a random sample of size $n-1$ from $X_1, \ldots, X_n$. We put $\tilde{W}_{n-1}^*(t) = \sqrt{n}(F_{n-1}^*(t) - F_n(t))$ with the empirical distribution $F_{n-1}^*(t)$ of $X_1', \ldots, X_{n-1}'$. Then it follows from Lemmas 2.1, 2.2, 2.3 and Theorem 2.1 that

$$\mathrm{Cov}(\tilde{W}_{n-1}^*(t_1), \tilde{W}_{n-1}^*(t_2) \mid F_n(t_1), F_n(t_2)) = \{n/(n-1)\}F_n(t_1)(1 - F_n(t_2))$$

is also an unbiased estimator of $\mathrm{Cov}(W_n(t_1), W_n(t_2))$, but $\mathrm{Cov}(W_n^*(t_1), W_n^*(t_2) \mid F_n(t_1), F_n(t_2))$ is not unbiased for it. Hence it is desirable to use the bootstrap sample of size $n-1$ in place of size n. And also the biases of higher order cumulants become smaller.

3.   Minimax property of the bootstrap estimator

In this section we consider the estimation problem based on the i.i.d. sample $X_1, \ldots, X_n$ on some real parameter θ which can be defined as a functional $\theta = \Psi(F)$ of a continuous distribution F. Then the natural estimator is $\hat{\theta}_n = \Psi(F_n)$, where F_n is the empirical distribution function. We shall show that the bootstrap estimator of the distribution of $\hat{\theta}_n$ has a minimax property for some parametric family of distributions. We assume the following condition.

(A.1)   The functional Ψ is Fréchet differentiable up to the third order, that is, there are functions $\partial\Psi/\partial F$, $\partial^2\Psi/\partial F\partial F$ and $\partial^3\Psi/\partial F\partial F\partial F$ such that

$$(3.1) \quad \Psi(G) - \Psi(F) = \int_{-\infty}^{\infty} (\partial\Psi/\partial F)d(G - F)$$

$$+ (1/2)\int_{-\infty}^{\infty}\int_{-\infty}^{\infty} (\partial^2\Psi/\partial F\partial F)d(G - F)d(G - F)$$

$$+ (1/6)\int_{-\infty}^{\infty}\int_{-\infty}^{\infty}\int_{-\infty}^{\infty} (\partial^3\Psi/\partial F\partial F\partial F)$$

$$\cdot dF(G - F)d(G - F)d(G - F)$$

$$+ o(\|G - F\|^3),$$

where $\|G - F\| = \sup_x |G(x) - F(x)|$.

Putting $W_n(x) = \sqrt{n}(F_n(x) - F(x))$, we have from (3.1)

$$(3.2) \quad \sqrt{n}(\hat{\theta}_n - \theta) = \int_{-\infty}^{\infty} \phi_1(x)dW_n(x)$$

$$+ (1/2\sqrt{n})\int_{-\infty}^{\infty}\int_{-\infty}^{\infty} \phi_2(x, y)dW_n(x)dW_n(y)$$

$$+ (1/6n)\int_{-\infty}^{\infty}\int_{-\infty}^{\infty}\int_{-\infty}^{\infty} \phi_3(x, y, z)dW_n(x)dW_n(y)dW_n(z)$$

$$+ o_p(1/n),$$

where $\phi_1(x) = (\partial\Psi/\partial F)(x)$, $\phi_2(x, y) = (\partial^2\Psi/\partial F\partial F)(x, y)$ and $\phi_3(x, y, z) = (\partial\Psi^3/\partial F\partial F\partial F)(x, y, z)$. We also assume that the following holds.

$$(A.2) \quad \int \phi_1(x)dF(x) = 0, \qquad \int \phi_2(x, t)dF(t) = \int \phi_2(s, y)dF(s) = 0,$$

$$\int \phi_3(x, y, u)dF(u) = \int \phi_3(x, t, z)dF(t) = \int \phi_3(s, y, z)dF(s) = 0,$$

and the functions $\phi_2(x, y)$ and $\phi_3(x, y, z)$ are symmetric in (x, y) and (x, y, z), respectively.

Furthermore, using $T_n = \sqrt{n}(\hat{\theta}_n - \theta)$, we assume the following condition.

$$(A.3) \qquad\qquad\qquad E(T_n^4) < \infty.$$

LEMMA 3.1.   *Assume that the conditions* (A.1), (A.2) *and* (A.3) *hold. Then the asymptotic cumulants of T_n are given as follows.*

          MASAFUMI AKAHIRA AND KEI TAKEUCHI

$$E(T_n) = (1/2\sqrt{n})\left\{\int_{-\infty}^{\infty}\phi_2(x, x)dF(x) - \int_{-\infty}^{\infty}\int_{-\infty}^{\infty}\phi_2(x, y)dF(x)dF(y)\right\}$$
$$+ o(1/n)$$
$$= (1/\sqrt{n})b_1 + o(1/n) \quad (say),$$
$$V(T_n) = \int_{-\infty}^{\infty}\phi_1^2(x)dF(x) - \left\{\int_{-\infty}^{\infty}\phi_1(x)dF(x)\right\}^2$$
$$+ (1/n)\left\{\int_{-\infty}^{\infty}\phi_1(x)\phi_2(x, x)dF(x)\right.$$
$$- 3\int_{-\infty}^{\infty}\int_{-\infty}^{\infty}\phi_1(x)\phi_2(x, y)dF(x)dF(y)$$
$$+ 2\left(\int_{-\infty}^{\infty}\phi_1(x)dF(x)\right)\int_{-\infty}^{\infty}\int_{-\infty}^{\infty}\phi_2(x, y)dF(x)dF(y)\right\}$$
$$+ (1/2n)\left\{\int_{-\infty}^{\infty}\phi_2(x, x)dF(x) - \int_{-\infty}^{\infty}\int_{-\infty}^{\infty}\phi_2(x, y)dF(x)dF(y)\right\}^2$$
$$+ (1/n)\left\{\int_{-\infty}^{\infty}\int_{-\infty}^{\infty}\phi_1(x)\phi_3(x, y, y)dF(x)dF(y)\right.$$
$$- 2\int_{-\infty}^{\infty}\int_{-\infty}^{\infty}\int_{-\infty}^{\infty}\phi_1(x)\phi_3(x, y, z)dF(x)dF(y)dF(z)$$
$$+ \int_{-\infty}^{\infty}\int_{-\infty}^{\infty}\int_{-\infty}^{\infty}\int_{-\infty}^{\infty}\phi_1(x)\phi_3(y, z, u)$$
$$\left. \cdot dF(x)dF(y)dF(z)dF(u)\right\} + o(1/n)$$
$$= v_0 + (1/n)v_1 + o(1/n) \quad (say),$$
$$\kappa_3(T_n) = E[\{T_n - E(T_n)\}^3]$$
$$= (1/\sqrt{n})\left\{\int_{-\infty}^{\infty}\phi_1^3(x)dF(x) - 3\left(\int_{-\infty}^{\infty}\phi_1^2(x)dF(x)\right)\left(\int_{-\infty}^{\infty}\phi_1(x)dF(x)\right)\right.$$
$$+ 2\left(\int_{-\infty}^{\infty}\phi_1(x)dF(x)\right)^3$$
$$+ 3\left(\int_{-\infty}^{\infty}\phi_1^2(x)dF(x)\right)\int_{-\infty}^{\infty}\phi_2(x, x)dF(x)$$
$$- 3\left(\int_{-\infty}^{\infty}\phi_1^2(x)dF(x)\right)\int_{-\infty}^{\infty}\int_{-\infty}^{\infty}\phi_2(x, y)dF(x)dF(y)$$
$$+ 3\left(\int_{-\infty}^{\infty}\phi_1(x)dF(x)\right)^2\int_{-\infty}^{\infty}\int_{-\infty}^{\infty}\phi_2(x, y)dF(x)dF(y)$$
$$\left. + (3/2)\left(\int_{-\infty}^{\infty}\phi_1(x)dF(x)\right)^2\int_{-\infty}^{\infty}\phi_2(x, x)dF(x)\right\}$$
$$+ o(1/n),$$
$$= (1/\sqrt{n})\beta_3 + o(1/n) \quad (say),$$
$$\kappa_4(T_n) = E[\{T_n - E(T_n)\}^4] - 3\{V(T_n)\}^2 = O(1/n).$$

The proof is omitted since Lemma 3.1 is similar to Theorem 3.1 of Withers (1983).

Remark 3.1.   With the condition (A.3) and from the fact that there exists a finite positive constant c such that

$$P\left\{\sup_x \sqrt{n}|F_n(x) - F(x)| > r\right\} < ce^{-2r^2} \qquad \text{(Dvoretzky } et~al.~(1956)),$$

holds for all $r \geq 0$ and all positive integers n, it follows that the above expansion of the remainder term is valid.

For an estimator $\hat{\theta}_n^*$ based on the bootstrap sample $X_1^*, \ldots, X_n^*$ of size n, we put $T_n^* = \sqrt{n}(\hat{\theta}_n^* - \theta)$.

LEMMA 3.2.   *Assume that the conditions* (A.1), (A.2) *and* (A.3) *hold. Then the conditional asymptotic cumulants of* T_n^*, *given the empirical distribution function* F_n, *have the following form.*

$$E[T_n^* \mid F_n] = (1/\sqrt{n})b_1 + (1/n)\xi_1 + o_p(1/n),$$
$$V(T_n^* \mid F_n) = v_0 + (1/\sqrt{n})\xi_2 + (1/n)v_1 + o_p(1/n),$$
$$\kappa_3(T_n^* \mid F_n) = (1/\sqrt{n})\beta_3 + (1/n)\xi_3 + o_p(1/n),$$
$$\kappa_4(T_n^* \mid F_n) = \kappa_4(T_n) + o_p(1/n),$$

where $\xi_1 = O_p(1)$, $\xi_2 = O_p(1)$, $\xi_3 = O_p(1)$, *and* b_1, v_0, v_1 *and* β_3 *are constants given in Lemma* 3.1.

The proof is given in Section 4.

Remark 3.2.   In order to evaluate the bootstrap estimator $\hat{\theta}_n^*$, it is seen from Lemmas 3.1 and 3.2 that the variance of $\xi_2 = \sqrt{n}(V(T_n^* \mid F_n) - V(T_n)) + o_p(1/\sqrt{n})$ plays an important part.

LEMMA 3.3.   *Under the conditions* (A.1), (A.2) *and* (A.3), *the variance of* ξ_2 *is given by*

$$V(\xi_2) = \int_{-\infty}^{\infty} \phi_1^2(x)\{\phi_1(x) - 2m\}^2 dF(x) - \left\{\int_{-\infty}^{\infty} \phi_1^2(x)dF(x) - 2m^2\right\}^2,$$

where $m = \int_{-\infty}^{\infty} \phi_1(x)dF(x)$.

The proof is given in Section 4. Now we consider a parametric family $\mathcal{F} = \{F_\theta : \theta \in \Theta\}$ of distribution functions, where Θ is an open set of R^1 involving the origin. Take F_{θ_0} as the previous distribution function F. We assume that, for each $\theta \in \Theta$, the distribution function F_θ is absolutely continuous with respect to a σ-finite measure μ, and denote $dF_\theta(x)/d\mu(x)$ by $f_\theta(x)$. For each $\theta \in \Theta$, we put

$$v_\theta = \int_{-\infty}^{\infty} \phi_1^2(x)f_\theta(x)d\mu - \left\{\int_{-\infty}^{\infty} \phi_1(x)f_\theta(x)d\mu\right\}^2.$$

Since $\hat{\theta}_n = \Psi(F_n)$ is an asymptotically unbiased estimator of θ, we have by Taylor's expansion of v_θ around $\theta = \theta_0$

$$v_{\hat{\theta}_n} = v_{\theta_0} + [\partial v_\theta/\partial\theta]_{\theta=\theta_0}(\hat{\theta}_n - \theta_0) + o_p(1/\sqrt{n}),$$

hence the variance of $v_{\hat{\theta}_n}$ is given by

$$V_{\theta_0}(v_{\hat{\theta}_n}) = ([\partial v_\theta/\partial\theta]_{\theta=\theta_0})^2 V_{\theta_0}(\hat{\theta}_n) + o(1/n).$$

Assume that the Fisher information amount $I(\theta)$ exists, i.e.

$$0 < I(\theta) = \int_{-\infty}^{\infty} \{\partial \log f_\theta(x)/\partial\theta\}^2 f_\theta(x)d\mu < \infty,$$

then we have by Cramér-Rao's inequality that

$$(3.3) \qquad n V_{\theta_0}(v_{\hat{\theta}_n}) \geq ([\partial v_\theta/\partial\theta]_{\theta=\theta_0})^2 / I(\theta) + o(1),$$

provided that the differentiation under the integral sign is allowed. We further restrict our attention to a family of subclasses $\mathcal{F}_\psi = \{F_\theta : dF_\theta(x)/d\mu = f_\theta(x)$ with the form $\log(f_\theta(x)/f_{\theta_0}(x)) = c(\theta) + \theta\psi(x)$ a.e. $[\mu]$ with $c(0) = 0\}$ of $\mathcal{F}$, where $\psi(x)$ is a function with finite variance at f_{θ_0}. Then we have the following.

THEOREM 3.1.   *Assume that the conditions* (A.1), (A.2) *and* (A.3) *hold. Then the bootstrap estimator* $\hat{\theta}_n^*$ *has a minimax property in the above family, i.e.*

$$\max_{\mathcal{F}_\psi} \min_{\hat{\theta}_n} n V_{\theta_0}(v_{\hat{\theta}_n}) = n V_{\theta_0}(v_{\hat{\theta}_n^*}) + o(1),$$

provided that the differentiation under the integral sign is allowed.

The proof is given in Section 4.

Remark 3.3.   From Theorem 3.1 we see that the maximum of relative efficiency of the bootstrap estimator $\hat{\theta}_n^*$ is equal to $1 + o(1)$, i.e.

$$\max_{\mathcal{F}_\psi} \left[\left\{ \min_{\hat{\theta}_n} n V_{\theta_0}(v_{\hat{\theta}_n}) \right\} \Big/ n V_{\theta_0}(v_{\hat{\theta}_n^*}) \right] = 1 + o(1).$$

It also follows from Theorem 3.1 that in a semiparametric situation where the class of distributions is sufficiently wide to include $\mathcal{F}_\psi$, it is impossible to get an estimator with a smaller asymptotic variance than $v_{\hat{\theta}_n^*}$.

4. Proofs

In this section the proofs of lemmas and theorems are given. In order to prove Lemma 2.1 we have the following.

LEMMA 4.1. *Let Z be a real random variable. Assume that, for each $i = 1, 2, 3, 4$, $Y_i = 1$ for $Z \leq c_i$, $Y_i = 0$ for $Z > c_i$, where $c_1 \leq c_2 \leq c_3 \leq c_4$. Then*

$$\kappa_3(Y_1, Y_2, Y_3) = E[(Y_1 - p_1)(Y_2 - p_2)(Y_3 - p_3)] = p_1(1 - 2p_2)(1 - p_3),$$

$$\kappa_4(Y_1, Y_2, Y_3, Y_4) = E[(Y_1 - p_1)(Y_2 - p_2)(Y_3 - p_3)(Y_4 - p_4)]$$
$$- \operatorname{Cov}(Y_1, Y_2) \operatorname{Cov}(Y_3, Y_4) - \operatorname{Cov}(Y_1, Y_3) \operatorname{Cov}(Y_2, Y_4)$$
$$- \operatorname{Cov}(Y_1, Y_4) \operatorname{Cov}(Y_2, Y_3)$$
$$= p_1(1 - p_4)(1 - 4p_2 - 2p_3 + 6p_2p_3),$$

where for each $i = 1, 2, 3, 4$, $p_i = P\{Z \leq c_i\}$ and $\operatorname{Cov}(\cdot, \cdot)$ denotes the covariance.

PROOF. It is seen that $p_1 \leq p_2 \leq p_3 \leq p_4$. Since $E(Y_i) = p_i$ $(i = 1, 2, 3)$, $E(Y_1Y_2) = E(Y_1) = p_1$, $E(Y_2Y_3) = E(Y_2) = p_2$, $E(Y_1Y_2Y_3) = E(Y_1) = p_1$, it follows that

$$\kappa_3(Y_1, Y_2, Y_3) = E[(Y_1 - p_1)(Y_2 - p_2)(Y_3 - p_3)]$$
$$= E(Y_1Y_2Y_3) - p_1 E(Y_2Y_3) - p_2 E(Y_1Y_3) - p_3 E(Y_1Y_2) + 2p_1p_2p_3$$
$$= p_1(1 - 2p_2)(1 - p_3).$$

In a similar way, we have

$$E[(Y_1 - p_1)(Y_2 - p_2)(Y_3 - p_3)(Y_4 - p_4)]$$
$$= \{p_1(1 - 2p_2)(1 - p_3) + p_1p_2p_3\}(1 - p_4),$$
$$\operatorname{Cov}(Y_i, Y_j) = p_i(1 - p_j) \quad (1 \leq i \leq j \leq 4).$$

Hence we obtain

$$\kappa_4(Y_1, Y_2, Y_3, Y_4) = E[(Y_1 - p_1)(Y_2 - p_2)(Y_3 - p_3)(Y_4 - p_4)]$$
$$- \operatorname{Cov}(Y_1, Y_2) \operatorname{Cov}(Y_3, Y_4) - \operatorname{Cov}(Y_1, Y_3) \operatorname{Cov}(Y_2, Y_4)$$
$$- \operatorname{Cov}(Y_1, Y_4) \operatorname{Cov}(Y_2, Y_3)$$
$$= p_1(1 - p_4)(1 - 4p_2 - 2p_3 + 6p_2p_3).$$

Thus we complete the proof.

PROOF OF LEMMA 2.1. Since $W_n(t) = \sqrt{n}(F_n(t) - F(t))$, it is easily seen that $E[W_n(t)] = 0$ and $\operatorname{Cov}(W_n(t_1), W_n(t_2)) = F(t_1)(1 - F(t_2))$ for $t_1 \leq t_2$. From Lemma 4.1 we have

$$\kappa_3(W_n(t_1), W_n(t_2), W_n(t_3)) = (1/\sqrt{n})F(t_1)(1 - 2F(t_2))(1 - F(t_3))$$
$$\text{for} \quad t_1 \leq t_2 \leq t_3,$$

$$\kappa_4(W_n(t_1), W_n(t_2), W_n(t_3), W_n(t_4))$$
$$= (1/n)F(t_1)(1 - F(t_4))(1 - 4F(t_2) - 2F(t_3) + 6F(t_2)F(t_3))$$
$$\text{for} \quad t_1 \leq t_2 \leq t_3 \leq t_4.$$

This completes the proof.

In order to prove Lemma 2.3 we have the following.

LEMMA 4.2. *Suppose that, for each $i = 1, 2, 3, 4$, Y_i is a real random variable with mean $E(Y_i) = m_i$. Then*

$$E[Y_1(1 - Y_2)] = m_1(1 - m_2) - \sigma_{12},$$
$$E[Y_1(1 - 2Y_2)(1 - Y_3)] = m_1 - m_1 m_3 - 2m_1 m_2 + 2m_1 m_2 m_3$$
$$- \sigma_{13} - 2\sigma_{12} + 2m_1\sigma_{23} + 2m_2\sigma_{13} + 2m_3\sigma_{12} + 2\kappa_{123},$$
$$E[Y_1(1 - Y_4)(1 - 4Y_2 - 2Y_3 + 6Y_2 Y_3)]$$
$$= m_1 - 4(\sigma_{12} + m_1 m_2) - 2(\sigma_{13} + m_1 m_3) - (\sigma_{14} + m_1 m_4)$$
$$+ 6(\kappa_{123} + m_1\sigma_{23} + m_2\sigma_{13} + m_3\sigma_{12} + m_1 m_2 m_3)$$
$$+ 4(\kappa_{124} + m_1\sigma_{24} + m_2\sigma_{14} + m_4\sigma_{12} + m_1 m_2 m_4)$$
$$+ 2(\kappa_{134} + m_1\sigma_{34} + m_3\sigma_{14} + m_4\sigma_{13} + m_1 m_3 m_4)$$
$$- 6(\kappa_{1234} + m_4\kappa_{123} + m_3\kappa_{124} + m_2\kappa_{134} + m_1\kappa_{234}$$
$$+ m_1 m_4\sigma_{23} + m_2 m_4\sigma_{13} + m_3 m_4\sigma_{12} + m_1 m_3\sigma_{24} + m_2 m_3\sigma_{14}$$
$$+ m_1 m_2\sigma_{34} + m_1 m_2 m_3 m_4 + \sigma_{12}\sigma_{34} + \sigma_{13}\sigma_{24} + \sigma_{14}\sigma_{23}),$$

where, for $1 \le i \le j \le k \le r \le 4$, $\sigma_{ij} = \mathrm{Cov}(Y_i, Y_j)$, $\kappa_{ijk} = \kappa_3(Y_i, Y_j, Y_k)$ and $\kappa_{ijkr} = \kappa_4(Y_i, Y_j, Y_k, Y_r)$.

PROOF. The first one is easily derived. Since

$$E(Y_1 Y_2 Y_3) = \kappa_{123} + m_1\sigma_{23} + m_2\sigma_{13} + m_3\sigma_{12} + m_1 m_2 m_3,$$

it follows that

$$E[Y_1(1 - 2Y_2)(1 - Y_3)] = E(Y_1) - E(Y_1 Y_3) - 2E(Y_1 Y_2) + 2E(Y_1 Y_2 Y_3)$$
$$= m_1 - m_1 m_3 - 2m_1 m_2 + 2m_1 m_2 m_3$$
$$- \sigma_{13} - 2\sigma_{12} + 2m_1\sigma_{23} + 2m_2\sigma_{13} + 2m_3\sigma_{12} + 2\kappa_{123}.$$

Since

$$E[Y_1 Y_2 Y_3 Y_4]$$
$$= \kappa_{1234} + \sigma_{12}\sigma_{34} + \sigma_{13}\sigma_{24} + \sigma_{14}\sigma_{23} + m_4\kappa_{123} + m_3\kappa_{124} + m_2\kappa_{134} + m_1\kappa_{234}$$
$$+ m_1 m_4\sigma_{23} + m_2 m_4\sigma_{13} + m_3 m_4\sigma_{12} + m_1 m_3\sigma_{24} + m_2 m_3\sigma_{14} + m_1 m_2\sigma_{34}$$
$$+ m_1 m_2 m_3 m_4,$$

it follows that

$$E[Y_1(1 - Y_4)(1 - 4Y_2 - 2Y_3 + 6Y_2 Y_3)]$$
$$= m_1 - 4E(Y_1 Y_2) - 2E(Y_1 Y_3) - E(Y_1 Y_4) + 6E(Y_1 Y_2 Y_3) + 4E(Y_1 Y_2 Y_4)$$
$$+ 2E(Y_1 Y_3 Y_4) - 6E(Y_1 Y_2 Y_3 Y_4),$$

hence the desired result follows.

PROOF OF LEMMA 2.3. For $i = 1, 2, 3, 4$, we put $Y_i = F_n(t_i)$ and $m_i = F(t_i) = E[F_n(t_i)]$. Then we have $\sigma_{ij} = (1/n)m_i(1 - m_j)$, $\kappa_{ijk} = (1/n\sqrt{n})m_i(1 - 2m_j)(1 - m_k)$, $\kappa_{ijkr} = (1/n^2)m_i(1 - m_r)(1 - 4m_j - 2m_k + 6m_j m_k)$ for $1 \le i \le j \le k \le r \le 4$. From Lemmas 4.1 and 4.2 we have the conclusion of Lemma 2.3.

PROOF OF LEMMA 2.4. From Lemmas 2.2 and 2.3 it follows that

$$E[W_n^*(t)] = E[E[W_n^*(t) \mid F_n(t)]] = 0,$$
$$\begin{aligned}
\mathrm{Cov}(W_n^*(t_1), W_n^*(t_2)) &= E[\mathrm{Cov}(W_n^*(t_1), W_n^*(t_2) \mid F_n(t_1), F_n(t_2))] \\
&= \{1 - (1/n)\}F(t_1)(1 - F(t_2)) \quad \text{for} \quad t_1 \le t_2,
\end{aligned}$$
$$\begin{aligned}
\kappa_3(W_n^*(t_1), &W_n^*(t_2), W_n^*(t_3)) \\
&= E[\kappa_3(W_n^*(t_1), W_n^*(t_2), W_n^*(t_3) \mid F_n(t_1), F_n(t_2), F_n(t_3))] \\
&= (1/\sqrt{n})\{1 - (1/n)\}\{1 - (2/n)\}F(t_1)(1 - 2F(t_2))(1 - F(t_3)) \\
&\qquad\qquad\qquad\qquad\qquad\qquad\qquad\qquad \text{for} \quad t_1 \le t_2 \le t_3.
\end{aligned}$$

In a similar way, we have

$$\begin{aligned}
\kappa_4(&W_n^*(t_1), W_n^*(t_2), W_n^*(t_3), W_n^*(t_4)) \\
&= E[\kappa_4(W_n^*(t_1), W_n^*(t_2), W_n^*(t_3), W_n^*(t_4) \mid F_n(t_1), F_n(t_2), F_n(t_3), F_n(t_4))] \\
&\quad + \mathrm{Cov}(\mathrm{Cov}(W_n^*(t_1), W_n^*(t_2) \mid F_n(t_1), F_n(t_2)), \\
&\qquad\qquad \mathrm{Cov}(W_n^*(t_3), W_n^*(t_4) \mid F_n(t_3), F_n(t_4))) \\
&\quad + \mathrm{Cov}(\mathrm{Cov}(W_n^*(t_1), W_n^*(t_3) \mid F_n(t_1), F_n(t_3)), \\
&\qquad\qquad \mathrm{Cov}(W_n^*(t_2), W_n^*(t_4) \mid F_n(t_2), F_n(t_4))) \\
&\quad + \mathrm{Cov}(\mathrm{Cov}(W_n^*(t_1), W_n^*(t_4) \mid F_n(t_1), F_n(t_4)), \\
&\qquad\qquad \mathrm{Cov}(W_n^*(t_2), W_n^*(t_3) \mid F_n(t_2), F_n(t_3))) \\
&= E[\kappa_4(W_n^*(t_1), W_n^*(t_2), W_n^*(t_3), W_n^*(t_4) \mid F_n(t_1), F_n(t_2), F_n(t_3), F_n(t_4))] \\
&\quad + \gamma_4 \quad (\text{say}).
\end{aligned}$$

Since, by Lemma 2.2,

$$\mathrm{Cov}(W_n^*(t_i), W_n^*(t_j) \mid F_n(t_i), F_n(t_j)) = F_n(t_i)(1 - F_n(t_j)) \quad \text{for} \quad t_i \le t_j,$$

it follows that

$$\begin{aligned}
\mathrm{Cov}(\mathrm{Cov}(&W_n^*(t_1), W_n^*(t_2) \mid F_n(t_1), F_n(t_2)), \\
&\mathrm{Cov}(W_n^*(t_3), W_n^*(t_4) \mid F_n(t_3), F_n(t_4))) \\
&= \mathrm{Cov}(F_n(t_1)(1 - F_n(t_2)), F_n(t_3)(1 - F_n(t_4))) \\
&= E[F_n(t_1)(1 - F_n(t_2))F_n(t_3)(1 - F_n(t_4))] \\
&\quad - E[F_n(t_1)(1 - F_n(t_2))]E[F_n(t_3)(1 - F_n(t_4))].
\end{aligned}$$

       MASAFUMI AKAHIRA AND KEI TAKEUCHI

We put $Y_i = F_n(t_i)$ and $m_i = F(t_i) = E[F_n(t_i)]$ for $i = 1, 2, 3, 4$. Then we have from Lemma 4.2

$$
\begin{aligned}
\gamma_4 &= E[Y_1(1 - Y_2)Y_3(1 - Y_4)] - E[Y_1(1 - Y_2)]E[Y_3(1 - Y_4)] \\
&\quad + E[Y_1(1 - Y_3)Y_2(1 - Y_4)] - E[Y_1(1 - Y_3)]E[Y_2(1 - Y_4)] \\
&\quad + E[Y_1(1 - Y_4)Y_2(1 - Y_3)] - E[Y_1(1 - Y_4)]E[Y_2(1 - Y_3)] \\
&= (1/n)m_1(1 - m_4)(3 - 8m_2 - 4m_3 + 12m_2 m_3) \\
&\quad - (2/n^2)m_1(1 - m_4)(3 - 10m_2 - 5m_3 + 15m_2 m_3) \\
&\quad + (3/n^3)m_1(1 - m_4)(1 - 4m_2 - 2m_3 + 6m_2 m_3) + o(1/n^2).
\end{aligned}
$$

Hence we obtain from Lemmas 2.2 and 2.3

$$
\begin{aligned}
\kappa_4(&W_n^*(t_1), W_n^*(t_2), W_n^*(t_3), W_n^*(t_4)) \\
&= (1/n)\{1 - (1/n)\}\{1 - (2/n)\}\{1 - (3/n)\}F(t_1)(1 - F(t_4)) \\
&\quad \cdot (1 - 4F(t_2) - 2F(t_3) + 6F(t_2)F(t_3)) \\
&\quad - (1/n^2)\{1 - (1/n)\}F(t_1)(1 - F(t_4)) \\
&\quad + (1/n)F(t_1)(1 - F(t_4))(3 - 8F(t_2) - 4F(t_3) + 12F(t_2)F(t_3)) \\
&\quad - (2/n^2)F(t_1)(1 - F(t_4))(3 - 10F(t_2) - 5F(t_3) + 15F(t_2)F(t_3)) \\
&\quad + (3/n^3)F(t_1)(1 - F(t_4))(1 - 4F(t_2) - 2F(t_3) + 6F(t_2)F(t_3)) + o(1/n^3).
\end{aligned}
$$

Thus we complete the proof.

PROOF OF LEMMA 3.2.   From Lemma 3.1, it follows that the conditional cumulants of T_n^*, given $F_n(t)$, have the form of

$$
\begin{aligned}
E(T_n^* \mid F_n) &= (1/\sqrt{n})b_1^* + o_p(1/n) = (1/\sqrt{n})b_1 + (1/n)\xi_1 + o_p(1/n), \\
V(T_n^* \mid F_n) &= v_0^* + (1/n)v_1^* + o_p(1/n) = v_0 + (1/\sqrt{n})\xi_2 + (1/n)v_1 + o_p(1/n), \\
\kappa_3(T_n^* \mid F_n) &= (1/\sqrt{n})\beta_3^* + o_p(1/n) = (1/\sqrt{n})\beta_3 + (1/n)\xi_3 + o_p(1/n), \\
\kappa_4(T_n^* \mid F_n) &= (1/n)\beta_4^* + o_p(1/n) = (1/n)\beta_4 + o_p(1/n).
\end{aligned}
$$

This completes the proof.

PROOF OF LEMMA 3.3.   Since $W_n(x) = \sqrt{n}(F_n(x) - F(x))$, it follows that

$$
\begin{aligned}
\xi_2 &= \int_{-\infty}^{\infty} \phi_1^2(x)dW_n(x) - (1/\sqrt{n})\int_{-\infty}^{\infty}\int_{-\infty}^{\infty} \phi_1(x)\phi_1(y)dW_n(x)dW_n(y) \\
&\quad - \int_{-\infty}^{\infty}\int_{-\infty}^{\infty} \phi_1(x)\phi_1(y)dW_n(x)dF(y) - \int_{-\infty}^{\infty}\int_{-\infty}^{\infty} \phi_1(x)\phi_1(y)dF(x)dW_n(y) \\
&= \int_{-\infty}^{\infty} \phi_1^2(x)dW_n(x) - 2m\int_{-\infty}^{\infty} \phi_1(x)dW_n(x) \\
&\quad - (1/\sqrt{n})\left(\int_{-\infty}^{\infty} \phi_1(x)dW_n(x)\right)^2,
\end{aligned}
$$

where $m = \int_{-\infty}^{\infty} \phi_1(x)dF(x)$. Then we have

$$E(\xi_2) = -(1/\sqrt{n})E\left[\int_{-\infty}^{\infty}\int_{-\infty}^{\infty}\phi_1(x)\phi_1(y)dW_n(x)dW_n(y)\right]$$

$$= -(1/\sqrt{n})\left\{\int_{-\infty}^{\infty}\phi_1^2(x)dF(x) - \left(\int_{-\infty}^{\infty}\phi_1(x)dF(x)\right)^2\right\}$$

$$= O(1/\sqrt{n}).$$

We also obtain

$$E(\xi_2^2) = E\left[\int_{-\infty}^{\infty}\int_{-\infty}^{\infty}\phi_1(x)\phi_1(y)\{\phi_1(x)\phi_1(y) - 4m\phi_1(x) + 4m^2\}dW_n(x)dW_n(y)\right]$$

$$+ o(1/\sqrt{n})$$

$$= \int_{-\infty}^{\infty}\phi_1^2(x)\{\phi_1^2(x) - 4m\phi_1(x) + 4m^2\}dF(x)$$

$$- \int_{-\infty}^{\infty}\int_{-\infty}^{\infty}\phi_1(x)\phi_1(y)\{\phi_1(x)\phi_1(y) - 4m\phi_1(x) + 4m^2\}dF(x)dF(y)$$

$$+ o(1/\sqrt{n})$$

$$= \int_{-\infty}^{\infty}\phi_1^2(x)\{\phi_1(x) - 2m\}^2 dF(x) - \left\{\int_{-\infty}^{\infty}\phi_1^2(x)dF(x) - 2m^2\right\}^2$$

$$+ o(1/\sqrt{n}).$$

Since $V(\xi_2) = E(\xi_2^2) + o(1/\sqrt{n})$, we have the desired result.

PROOF OF THEOREM 3.1.   Since the scaling of θ is arbitrary, without loss of generality, we assume that $\theta_0 = 0$ and $I(0) = 1$. It is enough to obtain $\psi(x)$ which maximizes $[\partial v_\theta/\partial\theta]_{\theta=0}$ under the condition

$$1 = \int_{-\infty}^{\infty}\{[(\partial/\partial\theta)\log f_\theta(x)]_{\theta=0}\}^2 f_0(x)d\mu - \int_{-\infty}^{\infty}\{c'(0) + \psi(x)\}^2 f_0(x)d\mu,$$

that is, to get $\psi(x)$ which minimizes

$$\int_{-\infty}^{\infty}\{c'(0) + \psi(x)\}^2 f_0(x)d\mu$$

under the condition

$$\int_{-\infty}^{\infty}\phi_1(x)\{\phi_1(x) - 2m\}\{c'(0) + \psi(x)\}f_0(x)d\mu = 1.$$

We put $h(x) = \phi_1(x)\{\phi_1(x) - 2m\}$. With the Lagrange multipliers λ_0 and λ_1, we have $\psi(x) + c'(0) = \lambda_0 h(x) + \lambda_1$ and it follows that

$$\lambda_0\int_{-\infty}^{\infty}h^2(x)f_0(x)d\mu + \lambda_1\int_{-\infty}^{\infty}h(x)f_0(x)d\mu = 1,$$

$$\lambda_0\int_{-\infty}^{\infty}h(x)f_0(x)d\mu + \lambda_1 = 0,$$

hence

$$\lambda_0 = 1 \Big/ \left[\int_{-\infty}^{\infty} h^2(x) f_0(x) d\mu - \left\{ \int_{-\infty}^{\infty} h(x) f_0(x) d\mu \right\}^2 \right],$$

$$\lambda_1 = -\left\{ \int_{-\infty}^{\infty} h(x) f_0(x) d\mu \right\} \Big/ \left[\int_{-\infty}^{\infty} h^2(x) f_0(x) d\mu - \left\{ \int_{-\infty}^{\infty} h(x) f_0(x) d\mu \right\}^2 \right].$$

From Lemma 3.3 we have

$$\int_{-\infty}^{\infty} \{c'(0) + \psi(x)\}^2 h(x) d\mu = 1 \Big/ \left[\int_{-\infty}^{\infty} h^2(x) f_0(x) d\mu - \left\{ \int_{-\infty}^{\infty} h(x) f_0(x) d\mu \right\}^2 \right]$$
$$= 1/\{V(\xi_2) + o(1)\},$$

hence, by (3.3),

$$\max_{\mathcal{F}_\psi} \min_{\hat{\theta}_n} n V_{\theta_0}(v_{\hat{\theta}_n}) = V(\xi_2) + o(1) = n V_{\theta_0}(v_{\hat{\theta}_n^*}) + o(1).$$

This completes the proof.

Acknowledgements

The authors wish to thank the referees for useful comments and the Shundoh International Foundation for a grant which enabled the first author to present the results of this paper at the 47th Session of the International Statistical Institute in Paris, 1989.

References

Beran, R. (1982). Estimated sampling distributions: bootstrap and competitors, *Ann. Statist.*, **10**, 212–225.

Dvoretzky, A., Kiefer, J. and Wolfowitz, J. (1956). Asymptotic minimax character of the sample distribution function and of the classical multinomial estimator, *Ann. Math. Statist.*, **27**, 642–669.

Efron, B. (1979). Bootstrap methods: another look at the jackknife, *Ann. Statist.*, **7**, 1–26.

Efron, B. (1982). *The Jackknife, the Bootstrap and Other Resampling Plans*, CBMS Regional Conference Series in Applied Mathematics 38, SIAM, Philadelphia.

Withers, C. S. (1983). Expansions for the distribution and quantiles of regular functional of the empirical distribution with applications to nonparametric confidence intervals, *Ann. Statist.*, **11**, 577–587.

SEQUENTIAL ANALYSIS, 10(1&2), 27-43 (1991)

SECOND ORDER ASYMPTOTIC EFFICIENCY IN TERMS OF THE ASYMPTOTIC VARIANCE OF SEQUENTIAL ESTIMATION PROCEDURES IN THE PRESENCE OF NUISANCE PARAMETERS

Masafumi Akahira

Institute of Mathematics
University of Tsukuba
Ibaraki 305
Japan

Kei Takeuchi

Research Center for Advanced
Science and Tochnology
University of Tokyo
4-6-1 Komaba, Meguro-ku
Tokyo 156, Japan

Key words and phrases : Sequential estimation procedure ; stopping rule; Bhattacharyya type bound ; asymptotically median unbiased estimator ; maximum likelihood estimation procedure ; second order asymptotic efficiency.

Abstract

In the presence of nuisance parameters, the Bhattacharyya type bound for the asymptotic variance of estimation procedures is obtained. It is shown that the modified maximum likelihood (ML) estimation procedures together with any stopping rule does not attain the bound. Further it is shown that the modified ML estimation procedure with the appropriate stopping rule is second order asymptotically efficient in some class of estimation procedures in the sense that it attains the lower bound for the asymptotic variance in the class.

1. Introduction

In the previous papers by Takeuchi and Akahira (1988) and Akahira and Takeuchi (1989), for one parameter case, the second and third order asymptotic efficiencies have been investigated in terms of the asymptotic variance and the asymptotic distribution of sequential estimation procedures, respectively.

In this paper, in the presence of nuisance parameters, the second order asymptotic efficiency is discussed in the class of estimation procedures related to a sequence of sequential sampling procedures where the size of sample tends to stochastically infinity and is asymptotically constant in the sense that its coefficient of variation tends to zero. Then it is shown that the asymptotic variance of asymptotically unbiased estimators satisfies the Bhattacharyya type inequality, and that the modified maximum likelihood estimation procedures together with any stopping rule does not generally attain the bound. The second fact is essentially different from that in one parameter case. Further, if the class of asymptotically unbiased estimators is restricted, the modified maximum likelihood estimation procedure together with the stopping rule is second order asymptotically efficient in the class in the sense that its asymptotic variance attains the lower bound for the asymptotic variance of estimators in the class. The related discussion from the viewpoint of differential geometry is done by Okamoto et al. (1990).

2. Notations and assumptions

Let X_1, $X_2, \cdots$, $X_n, \cdots$, be a sequence of independent and identically distributed random variables with a density function $f(x, \theta, \xi)$ with respect to a σ-finite measure μ, where θ is a real-valued parameter to be estimated and ξ is a real-valued nuisance parameter, Suppose that the size n of sample is determined according to some sequential rule, Actually, we consider a sequence of sequential estimation procedures $\{\Pi_a: a=1, 2.\cdots\}$ such that $E_{\theta, \xi, a}(n) = v_a(\theta, \xi)$ becomes large uniformly in θ and ξ as $a \to \infty$, where for each a we define a stopping rule and estimators based on it and consider the asymptotic distribution of

$$\sqrt{v_a}(\hat{\theta}_a - \theta) \text{ as } a \to \infty.$$

For simplicity, we denote $v_a(\theta, \xi)$ by v. In order to consider the second order asymptotic efficiency we assume the following conditions (A. 1) to (A. 6).

(A. 1) $E_{\theta, \xi}(n) = v + o(1)$, $V_{\theta, \xi}(n)/v = O(1)$, $E_{\theta, v}(n^k)/v^k = O(1)$ $(k=2,3,4)$, and $\{(\partial^k/\partial\theta^k)v\}/v = O(1)$ $(k=1,2)$, uniformly in θ and ξ.

SECOND ORDER ASYMPTOTIC EFFICIENCY 29

(A. 2) The set $\{x : f(x, \theta, \xi) > 0\}$ does not depend on θ and ξ.

(A. 3) For almost all $x[\mu]$, $f(x, \theta, \xi)$ is three times continuously differentiable in θ and ξ .

(A. 4) For each θ and each ξ
$$0 < I_{00}(\theta, \xi) = E[\{l_0(\theta, \xi, X)\}^2] = -E[l_{00}(\theta, \xi, X)] < \infty,$$
$$0 < I_{11}(\theta, \xi) = E[\{l_1(\theta, \xi, X)\}^2] = -E[l_{11}(\theta, \xi, X)] < \infty,$$
where $l_0(\theta, \xi, x) = (\partial/\partial\xi) l(\theta, \xi, x)$, $l_{00}(\theta, \xi, x) = (\partial^2/\partial\theta^2) l(\theta, \xi, x)$
$l_1(\theta, \xi, x) = (\partial/\partial\xi) l(\theta, \xi, x)$ and $l_{11}(\theta, \xi, x) = (\partial^2/\partial\xi^2) l(\theta, \xi, x)$
with $l(\theta, \xi, x) = \log f(x, \theta, \xi)$.

(A. 5) The parameters are defined to be "orthogonal" in the sense that
$$E[l_{01}(\theta, \xi, x)] = 0,$$
where $l_{01}(\theta, \xi, x) = (\partial^2/\partial\theta\partial\xi) l(\theta, \xi, x)$.

It is noted that this assumption is not necessarily restrictive, since otherwise we can redefine the parameter $\xi' = g(\theta, \xi)$ so that we have the above orthogonality.

(A. 6) There exist
$$J_{000} = E[l_{00}(\theta, \xi, X)l_0(\theta, \xi, X)],$$
$$J_{001} = E[l_{00}(\theta, \xi, X)l_1(\theta, \xi, X)],$$
$$J_{010} = E[l_{01}(\theta, \xi, X)l_0(\theta, \xi, X)],$$
$$J_{011} = E[l_{01}(\theta, \xi, X)l_1(\theta, \xi, X)],$$
$$J_{110} = E[l_{11}(\theta, \xi, X)l_0(\theta, \xi, X)],$$
$$J_{111} = E[l_{11}(\theta, \xi, X)l_1(\theta, \xi, X)],$$
$$K_{000} = E[\{l_0(\theta, \xi, X)\}^3],$$
$$K_{001} = E[\{l_0(\theta, \xi, X)\}^2 l_1(\theta, \xi, X)],$$
$$K_{111} = E[\{l_1(\theta, \xi, X)\}^3],$$
$$M_{0101} = E[\{l_{01}(\theta, \xi, X)\}^2],$$
and the following holds.
$$E[l_{000}(\theta, \xi, X)] = -3J_{000} - K_{000},$$
$$E[l_{111}(\theta, \xi, X)] = -3J_{111} - K_{111},$$
$$E[l_{001}(\theta, \xi, X)] = -J_{010},$$
$$E[l_{011}(\theta, \xi, X)] = -J_{011},$$
where $l_{000}(\theta, \xi, x) = (\partial^3/\partial\theta^3) l(\theta, \xi, x)$, $l_{111}(\theta, \xi, x) = (\partial^3/\partial\xi^3) l(\theta, \xi, x)$, $l_{001}(\theta, \xi, x) = (\partial^2/\partial\theta^2\partial\xi) l(\theta, \xi, x)$ and $l_{011}(\theta, \xi, x) = (\partial^3/\partial\theta\partial\xi^2) l(\theta, \xi, x)$.

From the condition (A. 5) it is noted that $K_{001} = -J_{010} - J_{001}$.
We put
$$Z_0 = \frac{1}{\sqrt{v}} \sum_{i=1}^{n} l_0(\theta, \xi, X_i), \qquad Z_1 = \frac{1}{\sqrt{v}} \sum_{i=1}^{n} l_1(\theta, \xi, X_i),$$

$$Z_{00} = \frac{1}{\sqrt{v}} \sum_{i=1}^{n} \{l_{00}(\theta, \xi, X_i) + I_{00}\}, \qquad Z_{11} = \frac{1}{\sqrt{v}} \sum_{i=1}^{n} \{l_{11}(\theta, \xi, X_i) + I_{11}\},$$

$$Z_{01} = \frac{1}{\sqrt{v}} \sum_{i=1}^{n} l_{01}(\theta, \xi, X_i),$$

where I_{00} and I_{11} denote $I_{00}(\theta, \xi)$ and $I_{11}(\theta, \xi)$, respectively.

3. Bhattacharyya type bound for the asymptotic variance of asymptotically unbiased estimators

In order to obtain the Bhattacharyya type bound, we need the following useful lemma for calculations of asymptotic cumulants.

LEMMA 3.1. Suppose $Y_{\theta,\xi}$ is a function of $X_1, \cdots, X_n$, θ and ξ and is differentiable in θ and ξ. Then

$$E(Z_0 Y_{\theta,\xi}) = \frac{1}{\sqrt{v}} \frac{\partial}{\partial\theta} E(Y_{\theta,\xi}) - \frac{1}{\sqrt{v}} E\left(\frac{\partial Y_{\theta,\xi}}{\partial\theta}\right),$$

$$E(Z_1 Y_{\theta,\xi}) = \frac{1}{\sqrt{v}} \frac{\partial}{\partial\xi} E(Y_{\theta,\xi}) - \frac{1}{\sqrt{v}} E\left(\frac{\partial Y_{\theta,\xi}}{\partial\xi}\right),$$

$$E(Z_0^2 Y_{\theta,\xi}) = \frac{1}{\sqrt{v}} \frac{\partial}{\partial\theta} E(Z_0 Y_{\theta,\xi}) - \frac{1}{\sqrt{v}} E\left(Y_{\theta,\xi} \frac{\partial Z_0}{\partial\theta}\right) - \frac{1}{\sqrt{v}} E\left(Z_0 \frac{\partial Y_{\theta,\xi}}{\partial\xi}\right),$$

and

$$E(Z_1^2 Y_{\theta,\xi}) = \frac{1}{\sqrt{v}} \frac{\partial}{\partial\xi} E(Z_1 Y_{\theta,\xi}) - \frac{1}{\sqrt{v}} E\left(Y_{\theta,\xi} \frac{\partial Z_1}{\partial\xi}\right) - \frac{1}{\sqrt{v}} E\left(Z_1 \frac{\partial Y_{\theta,\xi}}{\partial\xi}\right),$$

provided that partial differentiation under the integral signs of $E(Y_{\theta,\xi})$, $E(Z_i Y_{\theta,\xi})$ ($i = 0, 1$) with respect to θ and ξ is allowed.

The proof is omitted since the lemma is similar to Lemma 5.1.1 and 5.1.2 in Akahira and Takeuchi (1981) (see also Lemmas 2.1.1 and 2.1.2 in Akahira, 1986). In the following theorem, we have the Bhattacharyya type bound for the asymptotic variance of asymptotically unbiased estimators of θ.

THEOREM 3.1. Assume that the conditions (A.1) to (A.6) hold. Then for any asymptotically unbiased estimator $\hat{\theta}$ of θ, i, e., $E(\hat{\theta}) = \theta + o(v-1)$,

$$V\left(\sqrt{v}(\hat{\theta} - \theta)\right) \geq \frac{1}{I_{00}} + \frac{1}{2I_{00}^2 v}\left(\frac{J_{000} + K_{000}}{I_{00}} + \frac{2v_0}{v}\right)^2 + \frac{1}{I_{00}I_{11}v}\left(\frac{J_{001}}{I_{00}} - \frac{v_1}{v}\right)^2$$

$$+ \frac{J_{011}^2}{2I_{00}^2 I_{11}^2 v} + o\!\left(\frac{1}{v}\right)$$

$$= \frac{1}{I_{00}} + \frac{1}{v}\beta + o\!\left(\frac{1}{v}\right) \quad (say),$$

where $E(\cdot)$, $V(\cdot)$, v_0 and v_1 designate the asymptotic mean, the asymptotic variance, $(\partial/\partial\theta)\, v\,(\theta, \xi)$ and $(\partial/\partial\xi)\, v\,(\theta, \xi)$, respectively, and $(1/I_{00}) + (\beta/v)$ is called the Bhattacharyya type bound for the asymptotic variance of asymptotically unbiased estimators of θ.

<u>PROOF</u>. Putting

$$L = \prod_{i=1}^{n} f(X_i, \theta, \xi), \quad i, e.$$

$$\log L = \sum_{i=1}^{n} \log f(X_i, \theta, \xi),$$

we have the Bhattacharyya type bound for the asymptotic variance of asymptotically unbiased estimators of θ analogously as in the one parameter case (Takeuchi and Akahira, 1988), which is given by

$$\begin{pmatrix} 1 & 0 & 0 & 0 & 0 \\ 0 & 1 & 0 & 0 & 0 \end{pmatrix} \begin{pmatrix} B_{11} & B_{12} \\ B_{12}^t & B_{22} \end{pmatrix}^{-1} \begin{pmatrix} 1 & 0 \\ 0 & 1 \\ 0 & 0 \\ 0 & 0 \\ 0 & 0 \end{pmatrix},$$

where

$$B_{11} = E\!\left[\frac{1}{vL^2} \begin{pmatrix} L_0^2 & L_0 L_1 \\ L_0 L_1 & L_1^2 \end{pmatrix} \right],$$

$$B_{12} = E\!\left[\frac{1}{vL^2} \begin{pmatrix} L_0 L_{00} & L_0 L_{01} & L_0 L_{11} \\ L_1 L_{11} & L_1 L_{10} & L_1 L_{00} \end{pmatrix} \right],$$

$$B_{22} = E\!\left[\frac{1}{vL^2} \begin{pmatrix} L_{00}^2 & L_{00} L_{01} & L_{00} L_{11} \\ L_{00} L_{01} & L_{01}^2 & L_{01} L_{11} \\ L_{00} L_{11} & L_{01} L_{11} & L_{11}^2 \end{pmatrix} \right]$$

with $L_0 = (\partial/\partial\theta)L$, $L_{00} = (\partial^2/\partial\theta^2)L$, $L_{01} = L_{10} = (\partial^2/\partial\theta\partial\xi)L$, $L_1 = (\partial/\partial\xi)L$, $L_{11} = (\partial^2/\partial\xi^2)L$, and $B_{12}{}^t$ denotes the transposed matrix of B_{12} (e. g. see Zacks, 1971, pages 189, 190).

Then we have

$$(3.1) \qquad B_{11} = \begin{pmatrix} I_{00} & 0 \\ 0 & I_{11} \end{pmatrix} + (o(1)),$$

(3.2)

$$B_{12} = \begin{pmatrix} E\left[Z_0\left(Z_{00} + \sqrt{\bar{v}}\,Z_0^2 - \dfrac{n}{v}I_{00}\right)\right], & E\left[Z_0\left(Z_{01} + \sqrt{\bar{v}}\,Z_0 Z_1\right)\right], & E\left[Z_0\left(Z_{11} + \sqrt{\bar{v}}\,Z_1^2 - \dfrac{n}{v}I_{11}\right)\right] \\ E\left[Z_1\left(Z_{11} + \sqrt{\bar{v}}\,Z_1^2 - \dfrac{n}{v}I_{11}\right)\right], & E\left[Z_1\left(Z_{01} + \sqrt{\bar{v}}\,Z_0 Z_1\right)\right], & E\left[Z_1\left(Z_{00} + \sqrt{\bar{v}}\,Z_0^2 - \dfrac{n}{v}I_{00}\right)\right] \end{pmatrix}$$

$$(3.3) \qquad B_{22} = \begin{pmatrix} 2vI_{00}^2 & 0 & 0 \\ 0 & vI_{00}I_{11} & 0 \\ 0 & 0 & 2vI_{11}^2 \end{pmatrix} + (o(1)).$$

Since by Lemma 3.1

$$E(nZ_0) = \frac{1}{\sqrt{\bar{v}}}\frac{\partial v}{\partial\theta},$$

we have

$$(3.4) \qquad E\left[Z_0\left(Z_{00} + \sqrt{\bar{v}}\,Z_0^2 - \frac{n}{v}I_{00}\right)\right]$$

$$= J_{000} + \sqrt{\bar{v}}\,E(Z_0^3) - \frac{v_0}{v}I_{00},$$

where $v_0 = (\partial/\partial\theta)v$.

Since $(\partial I_{00}/\partial\theta) = 2J_{000} + K_{000}$, it follows from Lemma 3.1 that

$$(3.5) \qquad E(Z_0^3) = \frac{1}{\sqrt{\bar{v}}}\frac{\partial}{\partial\theta}E(Z_0^2) - \frac{2}{\sqrt{\bar{v}}}E\left(Z_0\frac{\partial Z_0}{\partial\theta}\right)$$

$$= \frac{1}{\sqrt{\bar{v}}}\frac{\partial}{\partial\theta}I_{00} - \frac{2}{\sqrt{\bar{v}}}E\left[Z_0\left(-\frac{v_0}{2v}Z_0 + Z_{00} - \frac{n}{\sqrt{\bar{v}}}I_{00}\right)\right]$$

$$= 2J_{000} + K_{000} - E\left[-\frac{v_0}{v}Z_0^2 + 2Z_0Z_{00} - \frac{2}{\sqrt{\bar{v}}}I_{00}nZ_0\right]$$

SECOND ORDER ASYMPTOTIC EFFICIENCY   33

$$= K_{000} + \frac{3v_0}{v} I_{00} + o(1).$$

From (3. 4) and (3. 5) we have

(3. 6) $\qquad E\left[Z_0\left(Z_{00} + \sqrt{v}\, Z_0^2 - \frac{n}{v} I_{00}\right)\right] = J_{000} + K_{000} + \frac{2v_0}{v} I_{00} + o(1).$

Since.by Lemma 3. 1

$$E(Z_0^2 Z_1) = \frac{1}{\sqrt{v}} \frac{\partial}{\partial \theta} E(Z_0 Z_1) - \frac{1}{\sqrt{v}} E\left(Z_1 \frac{\partial Z_0}{\partial \theta}\right) - \frac{1}{\sqrt{v}} E\left(Z_0 \frac{\partial Z_1}{\partial \theta}\right)$$

$$= -\frac{1}{\sqrt{v}} E\left[Z_1\left(-\frac{v_0}{2v} Z_0 + Z_{00} - \frac{n}{\sqrt{v}} I_{00}\right)\right] - \frac{1}{\sqrt{v}} E\left[Z_0\left(-\frac{v_0}{2v} Z_1 + Z_{01}\right)\right]$$

$$= -\frac{J_{001}}{\sqrt{v}} + \frac{I_{00}}{\sqrt{v}} \frac{v_1}{v} - \frac{J_{010}}{\sqrt{v}} + o\left(\frac{1}{\sqrt{v}}\right),$$

we obtain

(3. 7) $\qquad E\left[Z_0\left(Z_{01} + \sqrt{v}\, Z_0 Z_{01}\right)\right]$

$$= E(Z_0 Z_{01}) + \sqrt{v}\, E[Z_0^2 Z_{01}]$$

$$= -J_{001} + I_{00} \frac{v_1}{v} + o(1),$$

where $v_1 = \partial v / \partial \xi$.

Since by Lemma 3. 1

$$E(Z_0 Z_1^2) = \frac{1}{\sqrt{v}} \frac{\partial}{\partial \theta} E(Z_1^2) - \frac{1}{\sqrt{v}} E\left(2Z_1 \frac{\partial Z_1}{\partial \theta}\right)$$

$$= \frac{1}{\sqrt{v}} \frac{\partial I_{11}}{\partial \theta} - \frac{1}{\sqrt{v}} E\left[2Z_1\left(-\frac{v_0}{2v} Z_1 + Z_{01}\right)\right]$$

$$= \frac{1}{\sqrt{v}} (J_{011} - J_{110}) - \frac{1}{\sqrt{v}}\left(-\frac{v_0}{v} I_{11} + 2J_{011}\right) + o\left(\frac{1}{\sqrt{v}}\right)$$

$$= -\frac{1}{\sqrt{\nu}}(J_{011}+J_{110})+I_{11}\frac{\nu_0}{\nu\sqrt{\nu}}+o\left(\frac{1}{\sqrt{\nu}}\right),$$

it follows that

$$(3.8)\qquad E\left[Z_0\left(Z_{11}+\sqrt{\nu}\,Z_1^2-\frac{n}{\nu}I_{11}\right)\right]$$

$$=J_{110}-\sqrt{\nu}\,E(Z_0Z_1^2)-\frac{I_{11}}{\nu}E(nZ_0)=-J_{110}+o(1).$$

In a similar way to the above we have

$$(3.9)\qquad E\left[Z_1\left(Z_{11}+\sqrt{\nu}\,Z_1^2-\frac{n}{\nu}I_{11}\right)\right]=J_{111}+K_{111}+2I_{11}\frac{\nu_1}{\nu}+o(1),$$

$$(3.10)\qquad E\left[Z_1\left(Z_{01}+\sqrt{\nu}\,Z_0Z_1\right)\right]=-J_{110}+I_{11}\frac{\nu_0}{\nu}+o(1),$$

$$(3.11)\qquad E\left[Z_1\left(Z_{00}+\sqrt{\nu}\,Z_0^2-\frac{n}{\nu}I_{00}\right)\right]=-J_{010}+o(1).$$

From (3.6) to (3.11) we obtain

$$B_{12}=\begin{pmatrix} J_{000}+K_{000}+2I_{00}\dfrac{\nu_0}{\nu}, & -J_{001}+I_{00}\dfrac{\nu_1}{\nu}, & -J_{011}\\[2ex] J_{111}+K_{111}+2I_{11}\dfrac{\nu_1}{\nu}, & -J_{110}+I_{11}\dfrac{\nu_0}{\nu}, & -J_{010}\end{pmatrix}+o((1)),$$

hence by (3.1) to (3.3)

$$\left(B_{11}-B_{12}B_{22}^{-1}B_{21}\right)^{-1}$$

$$=\begin{pmatrix} I_{00}-\dfrac{1}{2\nu I_{00}^2}\left(J_{000}+K_{000}+2I_{00}\dfrac{\nu_0}{\nu}\right)^2-\dfrac{1}{\nu I_{00}I_{11}}\left(J_{001}-I_{00}\dfrac{\nu_1}{\nu}\right)^2-\dfrac{J_{011}^2}{2\nu I_{11}^2}+o\left(\dfrac{1}{\nu}\right), & O\left(\dfrac{1}{\nu}\right)\\[3ex] O\left(\dfrac{1}{\nu}\right) & , \; I_{11}+o\left(\dfrac{1}{\nu}\right)\end{pmatrix}^{-1}$$

$$=\begin{pmatrix} B^{11} & *\\ * & **\end{pmatrix},$$

where

$$B^{11}=\frac{1}{I_{00}}+\frac{1}{2\nu I_{00}^2}\left(\frac{J_{000}+K_{000}}{I_{00}}+\frac{2\nu_0}{\nu}\right)^2+\frac{1}{\nu I_{00}I_{11}}\left(\frac{J_{001}}{I_{00}}-\frac{\nu_1}{\nu}\right)^2+\frac{J_{011}^2}{2\nu I_{00}^2I_{11}^2}+o\left(\frac{1}{\nu}\right).$$

Therefore we have

$$V\left(\sqrt{n}\,(\hat{\theta}-\theta)\right)\geqq B^{11}\,.$$

This completes the proof.

4. The second order asymptotic efficiency of maximum likelihood estimation procedure

In order to obtain the asymptotic variance of the modified maximum likelihood (ML) estimator, we need the following lemma.

LEMMA 4. 1. Assume that the condition (A. 1) holds. If for any asymptotically unbiased estimator $\hat{\theta}_n$, i. e., $E(\hat{\theta}_n)=\theta+o(v-1)$,

$$\sqrt{v}\,(\hat{\theta}_n-\theta)=\frac{1}{I_{00}}Z_0+\frac{1}{\sqrt{v}}Q+o_p\!\left(\frac{1}{\sqrt{v}}\right),$$

then

$$V(\sqrt{v}\,(\hat{\theta}_n-\theta))=\frac{1}{I_{00}}+\frac{1}{v}V(Q)+o\!\left(\frac{1}{v}\right),$$

where $Q=O_p(1)$.

The proof is omitted since the lemma is given in Akahira (1986) and Takeuchi and Akahira (1988).

Let $\hat{\theta}^*$ and $\hat{\xi}^*$ be the ML estimators of θ and ξ, respectively. Under suitable regularity conditions it will be shown that $\sqrt{n}\,(\hat{\theta}^*-\theta)$ and $\sqrt{n}\,(\hat{\xi}^*-\xi)$ are of order $O_p(1)$. In order to establish this proposition we first need to prove the consistency of the ML estimators, from which the asymptotic joint normality of $(\sqrt{n}\,(\hat{\theta}^*-\theta),\sqrt{n}\,(\hat{\xi}^*-\xi))$ will follow easily, and for the consistency we need a separate set of regularity conditions similar to those given by Wald (1949). Since this part is outside of our main concern, the detailed discussion is omitted. By Taylor's expansion we have

$$(4.1)\qquad 0=\frac{1}{\sqrt{v}}\sum_{i=1}^{n}l_0(\hat{\theta}^*,\hat{\xi}^*,X_i)$$

$$=\frac{1}{\sqrt{v}}\sum_{i=1}^{n}l_0(\theta,\xi,X_i)+\frac{1}{v}\sum_{i=1}^{n}l_{00}(\theta,\xi,X_i)\sqrt{v}\,(\hat{\theta}^*-\theta)$$

$$+\frac{1}{v}\sum_{i=1}^{n}l_{01}(\theta,\xi,X_i)\sqrt{v}\,(\hat{\xi}^*-\xi)+\frac{1}{2v\sqrt{v}}\sum_{i=1}^{n}l_{000}(\theta,\xi,X_i)v(\hat{\theta}^*-\theta)^2$$

$$+\frac{1}{2v\sqrt{v}}\sum_{i=1}^{n}l_{011}(\theta,\xi,X_i)v(\hat{\xi}^*-\xi)^2$$

$$+\frac{1}{v\sqrt{v}}\sum_{i=1}^{n}l_{001}(\theta,\xi,X_i)\{v(\hat{\theta}^*-\theta)(\hat{\xi}^*-\xi)\}+o_p\!\left(\frac{1}{\sqrt{v}}\right),$$

$$(4.2)\qquad 0=\frac{1}{\sqrt{v}}\sum_{i=1}^{n}l_1(\hat{\theta}^*,\hat{\xi}^*,X_i)$$

$$=\frac{1}{\sqrt{v}}\sum_{i=1}^{n}l_1(\theta,\xi,X_i)+\frac{1}{v}\sum_{i=1}^{n}l_{11}(\theta,\xi,X_i)\sqrt{v}\,(\hat{\xi}^*-\xi)$$

$$+\frac{1}{v}\sum_{i=1}^{n}l_{01}(\theta,\xi,X_i)\sqrt{v}\,(\hat{\theta}^*-\theta)+\frac{1}{2v\sqrt{v}}\sum_{i=1}^{n}l_{111}(\theta,\xi,X_i)v(\hat{\xi}^*-\xi)^2$$

$$+\frac{1}{2v\sqrt{v}}\sum_{i=1}^{n}l_{001}(\theta,\xi,X_i)v(\hat{\theta}^*-\theta)^2$$

$$+\frac{1}{v\sqrt{v}}\sum_{i=1}^{n}l_{011}(\theta,\xi,X_i)\{v(\hat{\theta}^*-\theta)(\hat{\xi}^*-\xi)\}+o_p\!\left(\frac{1}{\sqrt{v}}\right),$$

From (4.1) and (4.2) we obtain

$$0=Z_0+\left(\frac{1}{\sqrt{v}}Z_{00}-\frac{n}{v}I_{00}\right)\sqrt{v}\,(\hat{\theta}^*-\theta)+\frac{1}{\sqrt{v}}Z_{01}\sqrt{v}\,(\hat{\xi}^*-\xi)$$

$$-\frac{1}{2\sqrt{v}}(3J_{000}+K_{000})v(\hat{\theta}^*-\theta)^2-\frac{1}{2\sqrt{v}}J_{011}v(\hat{\xi}^*-\xi)^2$$

$$-\frac{1}{\sqrt{v}}J_{010}\{v(\hat{\theta}^*-\theta)(\hat{\xi}^*-\xi)\}+o_p\!\left(\frac{1}{\sqrt{v}}\right),$$

$$0=Z_1+\left(\frac{1}{\sqrt{v}}Z_{11}-\frac{n}{v}I_{11}\right)\sqrt{v}\,(\hat{\xi}^*-\xi)+\frac{1}{\sqrt{v}}Z_{01}\sqrt{v}\,(\hat{\theta}^*-\theta)$$

$$-\frac{1}{2\sqrt{v}}(3J_{111}+K_{111})v(\hat{\xi}^*-\xi)^2-\frac{1}{2\sqrt{v}}J_{010}v(\hat{\theta}^*-\theta)^2$$

$$- \frac{1}{\sqrt{v}} J_{011}\{v(\hat{\theta}^*-\theta)(\hat{\xi}^*-\xi)\} + o_p\!\left(\frac{1}{\sqrt{v}}\right),$$

hence

$$(4.3) \qquad \sqrt{v}\,(\hat{\theta}^*-\theta) = \frac{1}{I_{00}} Z_0 - \frac{n-v}{v}\sqrt{v}\,(\hat{\theta}^*-\theta) + \frac{1}{I_{00}\sqrt{v}} Z_{00}\sqrt{v}\,(\hat{\theta}^*-\theta)$$

$$+ \frac{1}{I_{00}\sqrt{v}} Z_{01}\sqrt{v}\,(\hat{\xi}^*-\xi) - \frac{1}{2I_{00}\sqrt{v}}(3J_{000}+K_{000})v(\hat{\theta}^*-\theta)^2$$

$$- \frac{1}{2I_{00}\sqrt{v}} J_{011} v(\hat{\xi}^*-\xi)^2 - \frac{1}{I_{00}\sqrt{v}} J_{010}\{v(\hat{\theta}^*-\theta)(\hat{\xi}^*-\xi)\} + o_p\!\left(\frac{1}{\sqrt{v}}\right)$$

$$= \frac{1}{I_{00}} Z_0 - \left(\frac{n-v}{v}\right)\frac{Z_0}{I_{00}} + \frac{1}{I_{00}^2\sqrt{v}} Z_0 Z_{00} + \frac{1}{I_{00}I_{11}\sqrt{v}} Z_1 Z_{01}$$

$$- \frac{1}{2I_{00}^3\sqrt{v}}(3J_{000}+K_{000})Z_0^2 - \frac{1}{2I_{00}I_{11}^2\sqrt{v}} J_{011} Z_1^2$$

$$- \frac{1}{I_{00}^2 I_{11}\sqrt{v}} J_{010} Z_0 Z_1 + o_p\!\left(\frac{1}{\sqrt{v}}\right).$$

Let $\hat{\theta}^{**}$ be the ML estimator of θ modified so that $E(\hat{\theta}^{**})=\theta+o(v^{-1})$, and generally it automatically ensures that $E(\hat{\theta}^{**})=\theta+o(v^{-3/2})$.

From (4.3) we have the following.

THEOREM 4.1. Assume that the conditions (A. 1) to (A. 6) hold. If the stopping rule is so determined that sampling is stopped at n and satisfies

$$(4.4) \qquad -\sum_{i=1}^{n} I_{00}(\hat{\theta}^{**}, \xi, X_i) = v(\hat{\theta}^{**}, \xi) + I_{00}(\hat{\theta}^{**}, \xi) + o_p(\sqrt{v}),$$

$$(4.5) \qquad E[v(\hat{\theta}^{**}, \xi)] = v(\theta, \xi) + o(v(\theta, \xi)),$$

then the asymptotic variance of the modified ML estimator $\hat{\theta}^{**}$ is given by

$$V\!\left(\sqrt{v}\,(\hat{\theta}^{**}-\theta)\right) = \frac{1}{I_{00}} + \frac{1}{2I_{00}^2 v}\left(\frac{J_{000}+K_{000}}{I_{00}} + \frac{2v_0}{v}\right)^2 + \frac{1}{I_{00}I_{11}v}\left(\frac{J_{001}}{I_{00}} - \frac{v_1}{v}\right)^2$$

$$+ \frac{J_{011}^2}{2I_{00}^2 I_{11}^2 v} + \frac{1}{I_{00}^2 I_{11} v}\left(M_{0101} - \frac{J_{010}^2}{I_{00}} - \frac{J_{011}^2}{I_{11}}\right) + \left(\frac{1}{v}\right)$$

$$= \frac{1}{I_{00}} + \frac{1}{v}\beta + \frac{1}{I_{00}^2 I_{11} v}\left(M_{0101} - \frac{J_{010}^2}{I_{00}} - \frac{J_{011}^2}{I_{11}}\right) + \left(\frac{1}{v}\right),$$

and it does not generally attain the Bhattacharyya type bound given in Theorem 3. 1.

REMARK 4. 1. In one parameter case, it is shown by Takeuchi and Akahira (1988) that the modified ML estimation procedure with some stopping rule attains the Bhattacharyya type bound. However, in the presence of nuisance parameters, the fact is not always true , as is stated in Theorem 4. 1. That is an essential difference between two cases. It is noted that the modified ML estimation procedure with the stopping rule (4. 4) attains the bound if and only if

$$l_{01}(\theta, \xi, x) = a_0(\theta, \xi) + a_1(\theta, \xi)l_0(\theta, \xi, x) + a_2(\theta, \xi)l_1(\theta, \xi, x) \ a.\,e.\,[\mu]$$

for all θ and all ξ, where $a_i(\theta, \xi)$ $(i = 0.\,1.\,2)$ are certain functions of θ and ξ, which is equivalent to

$$M_{0101} = (J_{010}^2/I_{00}) + (J_{011}^2/I_{11}).$$

REMARK 4. 2. The equation (4. 4) with (4. 5) is attained if

$$-\sum_{i=1}^{n} l_{00}(\hat{\theta}^{**}, \hat{\xi}^{*}, X_i) = v(\hat{\theta}^{**}, \hat{\xi}^{*})I_{00}(\hat{\theta}^{**}, \hat{\xi}^{*}) + c(\hat{\theta}^{**}, \hat{\xi}^{*}),$$

where $c(\theta, \xi)$ is of constant order and determined to satisfy (4. 5).

Further, in Corollary 4. 1 below, it is shown that the modified ML estimator has the minimum asymptotic variance in some class of asymptotically unbiased estimators.

PROOF of THEOREM 4. 1. Since by (4. 4) and (4. 5)

$$-\sum_{i=1}^{n} l_{00}(\hat{\theta}^{*}, \xi, X_i) = v(\hat{\theta}^{*}, \hat{\xi}^{*})I_{00}(\hat{\theta}^{*}, \hat{\xi}^{*}) + o_p(\sqrt{v}),$$

it follows by the Taylor expansion that

$$(4.\,6) \qquad -\sum_{i=1}^{n} I_{00}(\theta, \xi, X_i) - \left\{\sum_{i=1}^{n} l_{000}(\theta, \xi, X_i)\right\}(\hat{\theta}^{*} - \theta) - \left\{\sum_{i=1}^{n} l_{001}(\theta, \xi, X_i)\right\}(\hat{\xi}^{*} - \xi)$$

SECOND ORDER ASYMPTOTIC EFFICIENCY   39

$$= v(\theta,\xi)I_{00}(\theta,\xi)+I_{00}(\theta,\xi)\{v_0(\hat{\theta}^*-\theta)+v_1(\hat{\xi}^*-\xi)\}$$

$$+ v(\theta,\xi)\left[\left\{\frac{\partial}{\partial\theta}I_{00}(\theta,\xi)\right\}(\hat{\theta}^*-\theta)+\left\{\frac{\partial}{\partial\xi}I_{00}(\theta,\xi)\right\}(\hat{\xi}^*-\xi)\right]+o_p(\sqrt{v}),$$

where v_0 and v_1 denote $(\partial/\partial\theta)v(\theta,\xi)$ and $(\partial/\partial\xi)v(\theta,\xi)$, respectively. From (4. 6) we have

$$nI_{00}-\sqrt{v}\,Z_{00}+\sqrt{v}\,(3J_{000}+K_{000})\frac{Z_0}{I_{00}}+\sqrt{v}\,J_{010}\frac{Z_1}{I_{11}}$$

$$=vI_{00}+\frac{1}{\sqrt{v}}\left(v_0Z_0+\frac{I_{00}}{I_{11}}v_1Z_1\right)+\sqrt{v}\left\{\frac{2J_{000}+K_{000}}{I_{00}}Z_0+\frac{2J_{001}+K_{001}}{I_{11}}Z_1\right\}+o_p(\sqrt{v}),$$

hence

(4. 7)
$$\left(\frac{n-v}{\sqrt{v}}\right)I_{00}=Z_{00}-\frac{J_{000}}{I_{00}}Z_0-\frac{J_{001}}{I_{11}}Z_1+\frac{v_0}{v}Z_0+\frac{v_1I_{00}}{vI_{11}}Z_1+o_p(1).$$

From (4. 3) and (4. 7) we obtain

(4. 8)
$$\sqrt{v}\,(\hat{\theta}^*-\theta)=\frac{Z_0}{I_{00}}-\frac{Z_0}{I_{00}^2\sqrt{v}}\left(Z_{00}-\frac{J_{000}}{I_{00}}Z_0-\frac{J_{001}}{I_{11}}Z_1+\frac{v_0}{v}Z_0+\frac{v_1I_{00}}{vI_{11}}Z_1\right)$$

$$+\frac{1}{I_{00}^2\sqrt{v}}Z_0Z_{00}+\frac{1}{I_{00}I_{11}\sqrt{v}}Z_1Z_{01}-\frac{1}{2I_{00}^3\sqrt{v}}(3J_{000}+K_{000})Z_0^2$$

$$-\frac{1}{2I_{00}I_{11}^2\sqrt{v}}J_{011}Z_1^2-\frac{1}{I_{00}^2I_{11}\sqrt{v}}J_{010}Z_0Z_1+o_p\left(\frac{1}{\sqrt{v}}\right)$$

$$=\frac{Z_0}{I_{00}}-\frac{J_{000}+K_{000}}{2I_{00}^3\sqrt{v}}Z_0^2+\frac{1}{I_{00}I_{11}\sqrt{v}}Z_1\left(Z_{01}-\frac{J_{010}}{I_{00}}Z_0-\frac{J_{011}}{I_{11}}Z_1\right)$$

$$-\frac{J_{001}}{I_{00}^2I_{11}\sqrt{v}}Z_0Z_1+\frac{J_{011}}{2I_{00}I_{11}^2\sqrt{v}}Z_1^2-\frac{v_0}{I_{00}^2v\sqrt{v}}Z_0^2$$

$$-\frac{v_1}{I_{00}I_{11}v\sqrt{v}}Z_0Z_1+o_p\left(\frac{1}{\sqrt{v}}\right)$$

$$= \frac{Z_0}{I_{00}} - \frac{1}{2I_{00}^2\sqrt{v}}\left(\frac{J_{000}+K_{000}}{I_{00}} + \frac{2v_0}{v}\right)Z_0^2$$

$$+ \frac{1}{I_{00}I_{11}\sqrt{v}}Z_1\left(Z_{01} - \frac{J_{001}}{I_{00}}Z_0 - \frac{J_{011}}{I_{11}}Z_1\right)$$

$$+ \frac{1}{I_{00}I_{11}\sqrt{v}}\left(\frac{J_{001}}{I_{00}} - \frac{v_1}{v}\right)Z_0Z_1 + \frac{J_{011}}{2I_{00}I_{11}^2\sqrt{v}}Z_1^2$$

$$+ o_p\left(\frac{1}{\sqrt{v}}\right).$$

Hence it follows from (4. 8) and Lemma 4. 1 that the asymptotic variance of the modified ML estimator $\hat{\theta}^{**}$ of θ is given by

$$V(\sqrt{v}\,(\hat{\theta}^{**}-\theta)) = \frac{1}{I_{00}} + \frac{1}{2I_{00}^2\sqrt{v}}\left(\frac{J_{000}+K_{000}}{I_{00}} + \frac{2v_0}{v}\right)^2 + \frac{1}{I_{00}I_{11}v}\left(\frac{J_{001}}{I_{00}} - \frac{v_1}{v}\right)^2$$

$$+ \frac{J_{011}^2}{2I_{00}^2I_{11}^2v} + \frac{1}{I_{00}^2I_{11}v}\left(M_{0101} - \frac{J_{010}^2}{I_{00}} - \frac{J_{011}^2}{I_{11}}\right) + o\left(\frac{1}{v}\right)$$

$$= \frac{1}{I_{00}} + \frac{1}{v}\beta + \frac{1}{I_{00}^2I_{11}v}\left(M_{0101} - \frac{J_{010}^2}{I_{00}} - \frac{J_{011}^2}{I_{11}}\right) + o\left(\frac{1}{v}\right).$$

It is also seen from Theorem 3.1 that $\hat{\theta}^{**}$ does not generally attain the Bhattacharyya type bound. Thus we complete the proof.

Next we consider a class of asymptotically unbiased estimators of θ. Let C be the subclass of all asymptotically best asymptotically normal and unbiased estimators $\hat{\theta}_n$, with $E(\hat{\theta}_n)=\theta+o(v^{-1})$ of those which are asymptotically expanded as

$$\sqrt{v}\,(\hat{\theta}_n-\theta) = \frac{Z_0}{I_{00}} + \frac{1}{\sqrt{v}}Q_0 + o_p\left(\frac{1}{\sqrt{v}}\right),$$

where $Q_0=O_p(1)$ and $\partial Q_0/\partial\theta=O_p(1)$. Then we have the following theorem.

THEOREM 4. 2. Assume that the conditions (A. 1) to (A. 6) hold. Then, for any estimator $\hat{\theta}_n\in C$,

SECOND ORDER ASYMPTOTIC EFFICIENCY   41

$$(4.9) \qquad V(\sqrt{\bar{v}}\,(\hat{\theta}_n - \theta)) \geqq \frac{1}{I_{00}} + \frac{1}{v}\beta + \frac{1}{vI_{00}^2 I_{11}} E(W_{00}^2) + \frac{1}{vI_{00}^2 I_{11}} E(W_{01}^2) + \left(\frac{1}{v}\right),$$

where β is given in Theorem 3. 1, and

$$W_{00} = Z_{00} - \left(\frac{J_{000}}{I_{00}} - \frac{v_0}{v}\right) Z_0 - \frac{(n-v)I_{00}}{\sqrt{\bar{v}}},$$

$$W_{01} = Z_{01} - \frac{J_{010}}{I_{00}} Z_0 - \frac{J_{011}}{I_{11}} Z_1.$$

REMARK 4. 3. In the lower bound of (4. 9),as is shown above from the formula the ML estimators $E(W_{00}^2)$ can be reduced to zero by the stopping rule, while

$$E(W_{01}^2) = M_{0101} - (J_{010}^2/I_{00}) - (J_{011}^2/I_{11})$$

is independent of the stopping rule.

Sketch of the proof of Theorem 4.2 is given as follows, since the proof is similar to that given in Akahira and Takeuchi (1989). Since

$$E(Z_0^2 Q_0) = -\frac{J_{000} + K_{000}}{I_{00}} - \frac{2v_0}{v} + o(1),$$

$$E(Z_0 Z_1 Q_0) = \frac{J_{001}}{I_{00}} - \frac{v_1}{v} + o(1),$$

$$E(Z_1^2 Q_0) = \frac{J_{011}}{I_{00}} + o(1),$$

$$E(Z_0 W_{00} Q_0) = \frac{1}{I_{00}} E(W_{00}^2) + o(1),$$

$$E(Z_1 W_{01} Q_0) = \frac{1}{I_{00}} E(W_{01}^2) + o(1),$$

it can be shown that under these conditions

$$V(Q_0) \geqq \frac{1}{2I_{00}^2}\left(\frac{J_{000} + K_{000}}{I_{00}} + \frac{2v_0}{v}\right)^2 + \frac{1}{I_{00}I_{11}}\left(\frac{J_{001}}{I_{00}} - \frac{v_1}{v}\right)^2$$

$$+ \frac{J_{011}^2}{2I_{00}^2 I_{11}^2} + \frac{1}{I_{00}^2 I_{11}} E(W_{00}^2) + \frac{1}{I_{00}^2 I_{11}} E(W_{01}^2) + o(1).$$

Then the conclusion of the theorem follows from Lemma 4. 1.

CorollARY 4. 1. Assume that the conditions (A. 1) to (A. 6) hold. Then the modified ML estimation procedure together with the stopping rule (4. 4) is second order asymptotically efficient in the class C in the sense that it attains the lower bound of (4. 9).

Proof. It is seen from (4. 3) that the modified ML estimator belongs to the class C. Since, by the stopping rule (4. 4), $E(W_{00}^2) = o(1)$, the conclusion of the corollary follows from Theorem 4. 1 and Remark 4. 2.

Remark 4. 4. Almost all "regular" type estimators in particular cases belong to the class C, as in the non-sequential cases.

References

Akahira, M. (1986). The Structure of Asymptotic Deficiency of Estimators. Queen's Papers in Pure and Applied Mathematics No. 75, Queen's University Press, Kingston, Ontario, Canada.

Akahira, M. and Takeuchi, K. (1981). Asymptotic Efficiency of Statistical Estimators : Concepts and Higher Order Asymptotic Efficiency. Lecture Notes in Statistics 7, Spriger-Verlag, New York.

Akahira, M. and Takeuchi, K. (1989). Third order asymptotic efficiency of the sequential maximum likelihood estimation procedure. Sequential Analysis 8, 333-359.

Okamoto, I., Amari, S. and Takeuchi, K. (1990). Asymptotic theory of sequential estimation : Differential geometrical approach. To be published in the Annals of Statistics.

Takeuchi, K. and Akahira, M. (1988). Second order asymptotic efficiency in terms of asymptotic variances of the sequential maximum likelihood estimation procedures. In : Statistical Theory and Data Analysis II, Proceedings of the Second Pacific Area Statistical Conference (K. Matusita, ed.), 191-196, North-Holland.

Wald, A. (1949). Note on the consistency of the maximum likelihood estimator. Ann. Math. Statist., 20, 595-601.

Zacks, S. (1971). The Theory of Statistical Inference. John Wiley, New York.

Received November 1989; Revised November 1990

Recommended by J.K. Ghosh

Rep. Stat. Appl. Res., JUSE
Vol. 38, No. 1, 1991
pp. 1-9

A-Section

Asymptotic Efficiency of Estimators for a Location Parameter Family of Densities with the Bounded Support

Masafumi AKAHIRA* and Kei TAKEUCHI**

Abstract

We consider the estimation problem on a location parameter of the density function with a support of a finite interval and contact of the power α-1 at both endpoints, where $1<\alpha<2$. Then the bound for the asymptotic distribution of asymptotically median unbiased estimators of θ based on a random sample from the density $f_0(x-\theta)$ is obtained. It is also shown that the bias-adjusted maximum likelihood estimator is not asymptotically efficient in the sense that its asymptotic distribution does not uniformly attain the bound.

1. Introduction

Let $f_0(x)$ be a density function which vanishes on the exterior of the open interval $(-1, 1)$ and twice continuously differentiable in $(-1, 1)$, and consider a location parameter family of densities $f(x, \theta)$, $\theta \in R^1$, defined by $f(x, \theta) = f_0(x-\theta)$, $x \in R^1$. We assume that $f_0(x) \sim A(1+x)^{\alpha-1}$ as $x \to -1+0$, $f_0(x) \sim B(1-x)^{\alpha-1}$ as $x \to 1-0$. Suppose that $X_1, ..., X_n$ are independently and identically distributed random variables according to the density $f_0(x-\theta)$. In a similar setup to the above, the order of minimum variance was obtained by Polfeldt [4], and it was shown by Akahira [1] and Woodroofe [7] that, for $\alpha=2$, the maximum likelihood estimator has the asymptotic normal distribution. And also it was shown by Akahira [1] that the order of consistency is equal to $n^{1/\alpha}$, $\sqrt{n\log n}$ and $\sqrt{n}$ for $0 < \alpha < 2$, $\alpha=2$ and $\alpha > 2$, respectively, and, for $0 < \alpha \le 1$ and $\alpha \ge 2$, the asymptotic accuracy of estimators in terms of their asymptotic distributions was discussed by Akahira [2]. Related results can be found in Smith [5]. However, in the case when $1 < \alpha < 2$, it seemed to be difficult to discuss the asymptotic efficiency of estimators.

In this paper the bound for the asymptotic distribution of asymptotically median unbiased estimators based on the sample is obtained and it is shown that the bias-adjusted maximum likelihood estimator is not asymptotically efficient in the sense that its asymptotic distribution does not uniformly attain the bound.

Received February 16, 1990
* Institute of Mathematics, University of Tsukuba, Tsukuba, Ibaraki 305, Japan.
** Research Center for Advanced Science and Technology, University of Tokyo, 4-6-1 Komaba, Meguro-ku, Tokyo 156, Japan.
(Keywords) Location parameter family, Asymptotic efficiency, Asymptotic distribution, Asymptotically median unbiased estimator, Maximum likelihood estimator.

M. Akahira and K. Takeuchi

2. Assumptions and lemmas

Let $f_0(x)$ be a density function with respect to the Lebesgue measure and consider the location parameter family $f(x, \theta)$, $\theta \in R^1$, defined by $f(x, \theta) = f_0(x-\theta)$ for $x \in R^1$. We assume the following conditions on $f_0(x)$.

(A.1)
$$f_0(x) > 0 \qquad \text{for } |x| < 1,$$
$$f_0(x) = 0 \qquad \text{for } |x| \geq 1.$$

(A.2) $f_0(x)$ is twice continuously differentiable in the open interval $(-1, 1)$ and

$$\lim_{x \to -1+0} (1+x)^{1-\alpha} f_0(x) = A, \quad \lim_{x \to -1+0} (1+x)^{2-\alpha} f_0'(x) = A', \quad \lim_{x \to -1+0} (1+x)^{3-\alpha} f_0''(x) = A'',$$

$$\lim_{x \to 1-0} (1-x)^{1-\alpha} f_0(x) = B, \quad \lim_{x \to 1-0} (1-x)^{2-\alpha} f_0'(x) = B', \quad \lim_{x \to 1-0} (1-x)^{3-\alpha} f_0''(x) = B'',$$

where $1 < \alpha < 2$, $0 < A < \infty$, $0 < B < \infty$, $0 < |A'| < \infty$, $0 < |B'| < \infty$, $0 < |A''| < \infty$, $0 < |B''| < \infty$.

Suppose that $X_1, \ldots, X_n$ are independent and identically distributed real random variables according to the above density function $f_0(x-\theta)$ with a location parameter θ. Then we consider the estimation problem on the parameter. In the above formulation, it was shown by Akahira [1] that the order of consistency is equal to $n^{1/\alpha}$. Now, in order to get the bound for the asymptotic distribution of asymptotically median unbiased estimators $\hat{\theta}_n$ of θ, we shall obtain the moment generating function of

$$Z_h = \log f_0(X) - \log f_0(X-h) \qquad \text{for } -1 + h < X < 1,$$
$$= 0 \qquad \text{otherwise,} \tag{2.1}$$

where h is a sufficiently small positive number. Here, an estimator $\hat{\theta}_n$ of θ is called asymptotically median unbiased (AMU) if

$$\lim_{n \to \infty} n^{1/\alpha} \left| P_\theta \left\{ \hat{\theta}_n \leq \theta \right\} - (1/2) \right| = \lim_{n \to \infty} n^{1/\alpha} \left| P_\theta \left\{ \hat{\theta}_n \geq \theta \right\} - (1/2) \right| = 0$$

uniformly in some neighborhood of θ. From (2.1) we have the moment generating function

$$\varphi(t) = E_0 \left[\exp(t Z_h) \right] = \int_{-\infty}^{\infty} \left\{ \exp(t z_h) \right\} f_0(x) dx$$

$$= \int_{-1+h}^{1} \left\{ \exp(t z_h) \right\} f_0(x) \, dx + \int_{-1}^{-1+h} f_0(x) \, dx$$

$$= \int_{-1+h}^{1} \left[\exp\left\{ t \log(f_0(x)/f_0(x-h)) \right\} \right] f_0(x) \, dx + \int_{-1}^{-1+h} f_0(x) \, dx$$

$$= \int_{-1+h}^{1} \left\{ f_0(x)/f_0(x-h) \right\}^t f_0(x) \, dx + \int_{-1}^{-1+h} f_0(x) \, dx. \tag{2.2}$$

Asymptotic Efficiency for Densities with Bounded Support

Since

$$\{f_0(x) / f_0(x-h)\}^t = \exp\{thf'_0(x)/f_0(x)\} + o(h) = 1 + \{thf'_0(x)/f_0(x)\} + o(h)$$

we put

$$R_t(x) = \{f_0(x) / f_0(x-h)\}^t - 1 - \{thf'_0(x) / f_0(x)\}.$$

Then we have from (2.2)

$$\varphi(t) = \int_{-1+h}^{1} R_t(x) f_0(x)\, dx + \int_{-1+h}^{1} f_0(x)\, dx + th \int_{-1+h}^{1} f'_0(x)\, dx + \int_{-1}^{-1+h} f_0(x)\, dx . \quad (2.3)$$

For the first term of the right-hand side of (2.3) we have the following.

Lemma 2.1.   For $0<\delta<1$,

$$\int_{-1+h}^{1} R_t(x) f_0(x)\, dx = h^{\alpha}\{AG_{1h}(t) + BG_{2h}(t)\} + o(h^{\alpha\delta}),$$

where

$$G_{1h}(t) = \int_{0}^{h^{\delta-1}} \left[\{(u+1)/u\}^{(\alpha-1)t} - 1 - \{(A'/A)t/(u+1)\}\right] (u+1)^{\alpha-1}\, du,$$

$$G_{2h}(t) = \int_{1}^{1+h^{\delta-1}} \left[\{(u-1)/u\}^{(\alpha-1)t} - 1 + \{(B'/B)t/(u-1)\}\right] (u-1)^{\alpha-1}\, du.$$

Remark 2.1.   From Lemma 2.1 we see that

$$\lim_{h \to 0} G_{1h}(t) = \int_{0}^{\infty} \left[\{(u+1)/u\}^{(\alpha-1)t} - 1 - \{(A'/A)t/(u+1)\}\right] (u+1)^{\alpha-1}\, du$$

$$= \int_{0}^{\infty} \left[\{1 - (1/(u+1))\}^{(\alpha-1)t} - 1 - \{(A'/A)t/(u+1)\}\right] (u+1)^{\alpha-1}\, du = G_1(t) \text{ (say)},$$

$$\lim_{h \to 0} G_{2h}(t) = \int_{1}^{\infty} \left[\{(u-1)/u\}^{(\alpha-1)t} - 1 + \{(B'/B)t/(u-1)\}\right] (u-1)^{\alpha-1}\, du$$

$$= \int_{1}^{\infty} \left[\{1 + (1/(u+1))\}^{-(\alpha-1)t} - 1 + \{(B'/B)t/(u-1)\}\right] (u-1)^{\alpha-1}\, du = G_2(t) \text{ (say)}.$$

M. Akahira and K. Takeuchi

Proof of Lemma 2.1. First we have for $0<\delta<1$,

$$\int_{-1+h}^{1} R_t(x)\,f_0(x)\,dx = \left(\int_{-1+h}^{-1+h+h^{\delta}} + \int_{-1+h+h^{\delta}}^{1-h^{\delta}} + \int_{1-h^{\delta}}^{1}\right) R_t(x)\,f_0(x)\,dx = I_1 + I_2 + I_3 \ \ (\text{say}). \quad (2.4)$$

(i) I_1. From the conditions (A.1) and (A.2) we have for $0<\delta<1$,

$$I_1 = \int_{-1+h}^{-1+h+h^{\delta}} \left[\{(1+x)/(1+(x-h))\}^{(\alpha-1)\,t} - 1 + \{(A'/A)th/(1+x)\}\right] A\,(1+x)^{\alpha-1}\,dx.$$

Then it follows that

$$I_1 = A\int_{-1+h}^{-1+h+h^{\delta}} \left[\{(1+x)/(1+(x-h))\}^{(\alpha-1)\,t} - 1 - \{(A'/A)th/(1+x)\}\right] (1+x)^{\alpha-1}\,dx + o(h^{\alpha\delta})$$

$$= A\int_{0}^{h^{\delta}} \left[\{(y+h)/y\}^{(\alpha-1)\,t} - 1 - \{(A'/A)th/(y+h)\}\right] (y+h)^{\alpha-1}\,dy + o(h^{\alpha\delta})$$

$$= Ah^{\alpha}\int_{0}^{h^{\delta-1}} \left[\{(u+1)/u\}^{(\alpha-1)\,t} - 1 - \{(A'/A)th/(u+1)\}\right] (u+1)^{\alpha-1}\,du + o(h^{\alpha\delta})$$

$$= Ah^{\alpha}\,G_{1h}(t) + o(h^{\alpha\delta}) \ \ (\text{say}).$$

(ii) I_2. Putting $g(h) = \{f_0(x)/f_0(x-h)\}^{t}$, we have

$$g'(h)/g(h) = tf'_0(x-h)/f_0(x-h),$$

hence

$$\{g''(h)/g(h)\} - \{g'(h)/g(h)\}^2 = t\left[\{-f''_0(x-h)/f_0(x-h)\} + \{f'_0(x-h)/f_0(x-h)\}^2\right].$$

Since

$$g(h) = 1 + \{thf'_0(x)/f_0(x)\} + (h^2/2)g''(\rho h) \ \ \text{with} \ \ 0<\rho<1, \ \text{it follows that}$$

$$R_t(x) = (h^2/2)g''(\rho h)$$

and

$$g''(h) = t\left[\{-f''_0(x-h)/f_0(x-h)\} + (t+1)\{f'_0(x-h)/f_0(x-h)\}^2\right] g(h).$$

Since $f_0(x) \sim A\,(1+x)^{\alpha-1}$ as $x \to -1+0$, we have

$$g''(h) \sim t\left\{-(A''/A) + (t+1)(A'/A)^2\right\}(1+x-h)^{-2}\left\{1+(h/(1+x-h))\right\}^{(\alpha-1)t}.$$

as $x \to -1+h+0$. Letting $h\,(>0)$ with $h^{\delta} < \varepsilon$ for any fixed ε with $0 < \varepsilon < 1$, we obtain for $-1+h+h^{\delta} < x < -1+h+\varepsilon$,

$$0 < \{1+(h/(1+x-h))\}^{(\alpha-1)\,t} \le \max\{1, 2^{(\alpha-1)\,t}\}. \quad (2.5)$$

Asymptotic Efficiency for Densities with Bounded Support

We also have

$$\int_{-1+h+h^\delta}^{-1+h+\varepsilon} \left\{ (1+x)^{\alpha-1} / (1+x-\rho h)^2 \right\} dx$$

$$= \int_{h^\delta}^{\varepsilon} \left\{ (y+h)^{\alpha-1} / (y+(1-\rho)h)^2 \right\} dy \leq \int_{h^\delta}^{\varepsilon} (y+h)^{\alpha-1} / y^2 \, dy$$

$$\leq \int_{h^\delta}^{\varepsilon} (y+h^\delta)^{\alpha-1} / y^2 \, dy = \int_{1}^{\varepsilon h^{-\delta}} (h^\delta u + h^\delta)^{\alpha-1} / (h^\delta u^2) \, du$$

$$\leq \int_{1}^{\infty} h^{-\delta} h^{(\alpha-1)\delta} (u+1)^{\alpha-1} / u^2 \, du$$

$$= h^{(\alpha-2)\delta} \int_{1}^{\infty} (u+1)^{\alpha-1} / u^2 \, du$$

$$= O(h^{(\alpha-2)\delta}). \tag{2.6}$$

Then it follows from (2.5) and (2.6) that

$$\int_{-1+h+h^\delta}^{-1+h+\varepsilon} R_t(x) f_0(x) \, dx = O(h^{2+(\alpha-2)\delta}). \tag{2.7}$$

In a similar way to the above we have

$$\int_{1-h^\delta-\varepsilon}^{1-h^\delta} R_t(x) f_0(x) \, dx = O(h^{2+(\alpha-2)\delta}). \tag{2.8}$$

Since $|g''(h)| \leq M_\varepsilon$ with some constant M_ε for $-1+\varepsilon<x<1-\varepsilon$, we have

$$\int_{-1+h+\varepsilon}^{1-h-\varepsilon} R_t(x) f_0(x) \, dx = O(h^2). \tag{2.9}$$

From (2.7), (2.8) and (2.9) we obtain

$$I_2 = \int_{-1+h+h^\delta}^{1-h^\delta} R_t(x) f_0(x) \, dx = O(h^{2+(\alpha-2)\delta}).$$

(iii)　　I_3.　　In a similar way to the case (i), we have

$$I_3 = B \int_{1-h^\delta}^{1} \left[\{ (1-x)/(1-(x-h)) \}^{(\alpha-1)t} - 1 + \{ (B'/B)th/(1-x) \} \right] (1-x)^{\alpha-1} \, dx + o(h^{\alpha\delta})$$

$$= B \int_{h}^{h+h^\delta} \left[\{ (y-h)/y \}^{(\alpha-1)t} - 1 + \{ (B'/B)th/(y-h) \} \right] (y-h)^{\alpha-1} \, dy + o(h^{\alpha\delta})$$

$$= B h^\alpha \int_{1}^{1+h^{\delta-1}} \left[\{ (u-1)/u \}^{(\alpha-1)t} - 1 + \{ (B'/B)t/(u-1) \} \right] (u-1)^{\alpha-1} \, du + o(h^{\alpha\delta})$$

$$= B h^\alpha G_{2h}(t) + o(h^{\alpha\delta}) \text{ (say).}$$

M. Akahira and K. Takeuchi

From (2.4), (i), (ii) and (iii) we have for $0<\delta<1$,

$$\int_{-1+h}^{1} R_t(x) f_0(x)\, dx = h^\alpha \{ AG_{1h}(t) + BG_{2h}(t) \} + o(h^{\alpha\delta}).$$

Thus we complete the proof.

For the third term of the right-hand side of (2.3) we have the following.

Lemma 2.2. For $1<\alpha<2$

$$\int_{-1+h}^{1} f'_0(x)\, dx = -A' h^{\alpha-1}/(\alpha-1) + o(h^{\alpha-1}).$$

Proof. From the conditions (A.1) and (A.2), it follows that

$$\int_{-1+h}^{1} f'_0(x)\, dx = -\int_{-1}^{-1+h} A'(1+x)^{\alpha-2} dx + o(h^{\alpha-1}) = -A' h^{\alpha-1}/(\alpha-1) + o(h^{\alpha-1}),$$

which completes the proof.

From Lemmas 2.1 and 2.2 we have the following.

Lemma 2.3. The moment generating function $\varphi(t)$ of Z_h is given by

$$\varphi(t) = E_0\big[\exp(t Z_h)\big]$$

$$= 1 + h^\alpha \{ -(A'/(\alpha-1))t + AG_{1h}(t) + BG_{2h}(t) \} + o(h^\alpha).$$

Remark 2.2. Putting $H_h(t) = -(A'/(\alpha-1))t + AG_{1h}(t) + BG_{2h}(t)$, we see that $H_h(0) = 0$ since $G_{1h}(0) = G_{2h}(0) = 0$.

The proof of Lemma 2.3 is straighforward from Lemmas 2.1 and 2.2.

For each $i=1, \dots, n$, we define

$$\begin{aligned}
Z_{hi} &= \log f_0(X_i) - \log f_0(X_i - h) \quad &&\text{for } -1 + h < X_i < 1, \\
&= 0 &&\text{otherwise.}
\end{aligned}$$

Putting $W_n = \sum_{i=1}^{n} Z_{hi}$, from Lemma 2.3 and Remark 2.2 we have as the moment generating function $\varphi_n(t)$ of W_n

$$\varphi_n(t) = E_0\big[\exp(t W_n)\big] = \{ E_0[\exp(t Z_h)] \}^n = \{ 1 + h^\alpha H_h(t) + o(h^\alpha) \}^n.$$

We also put $h = an^{-1/\alpha}$, then

$$\varphi_n(t) = \big[1 + \{ a^\alpha H(t)/n \} + o(1/n) \big]^n = \exp\{ a^\alpha H(t) \} + O(1). \tag{2.10}$$

where $H(t) = -(A'/(\alpha-1))t + AG_{1n^{-1/\alpha}}(t) + BG_{2n^{-1/\alpha}}(t)$. A similar discussion to the above can be done for $h < 0$.

3. The bound for the asymptotic distribution of AMU estimators and its comparison with that of the MLE

In order to obtain the bound for asymptotic distributions of AMU estimators, we consider a problem of testing $H:\theta = an^{-1/\alpha} = h$ against $K:\theta = 0$, since the order of consistency is equal to $n^{1/\alpha}$ for this problem. Then the acceptance region is of the form

$$\prod_{i=1}^{n} Z_{hi} = 0 \quad \text{or} \quad \sum_{i=1}^{n} Z_{hi} \leq C,$$

where C is some constant. In a similar way to Akahira and Takeuchi [3] (e.g. pages 57, 84), it is seen that the upper bound for the asymptotic distribution $P_{\theta_0}\left\{n^{1/\alpha}\,(\hat{\theta}_n - \theta_0) \leq a\right\}$ of AMU estimators $\hat{\theta}_n$ for $a > 0$ is given by

$$P_0\left\{\prod_{i=1}^{n} Z_{hi} = 0\right\} + P_0\left\{\sum_{i=1}^{n} Z_{hi} \leq C\right\}. \tag{3.1}$$

Since, by Lemma 2.2, for $h > 0$,

$$P_0\left\{Z_{h1} = 0\right\} = \int_{-1}^{-1+h} f_0\,(x)dx = (A/\alpha)h^{\alpha} + o(h^{\alpha}),$$

it follows that, for $h > 0$,

$$P_0\left\{\prod_{i=1}^{n} Z_{hi} = 0\right\} = 1 - \left(1 - P_0\left\{Z_{h1} = 0\right\}\right)^n$$

$$= 1 - \left\{1 - (A/\alpha)h^{\alpha} + o(h^{\alpha})\right\}^n.$$

Letting $h = an^{-1/\alpha}$ with $a > 0$, we have

$$P_0\left\{\prod_{i=1}^{n} Z_{hi} = 0\right\} = 1 - \left\{1 - (Aa^{\alpha}/\alpha n) + o(1/n)\right\}^n = 1 - \exp\left\{-(Aa^{\alpha}/\alpha)\right\} + O(1). \tag{3.2}$$

On the other hand, it follows from (2.10) that the asymptotic density $p_n(x)$ of $W_n = \sum_{i=1}^{n} Z_{hi}$ is given by

$$p_n(x) = (1/2\pi) \int_{-\infty}^{\infty} e^{-itx}\,\varphi_n\,(it)dt.$$

where i denotes the imaginary unit, hence,

$$P_0\left\{W_n \leq C\right\} = \int_{-\infty}^{C} p_n\,(x)dx. \tag{3.3}$$

In consideration of the AMU condition, we determine the constant C so that, under the hypothesis $H:\theta = an^{-1/\alpha} = h$,

$$P_h\left\{W_n \leq C\right\} = 1/2 + o(1). \tag{3.4}$$

M. AKAHIRA and K. TAKEUCHI

Indeed, since the moment generating function of Z_h, under $H: \theta = h$, i.e.,

$$\psi(t) = E_h\left[\exp\left(t\,Z_h\right)\right] = \int_{-\infty}^{\infty} \left\{\exp\left(t\,z_h\right)\right\} f_0(x - h)dx,$$

it follows in a similar way to the one under $K: \theta = 0$ that it is calculated. Hence the above constant C can be obtained. From (3.2) to (3.4) we can calculate the upper bound, i.e. (3.1) for $a > 0$ and also similarly the lower bound for $a < 0$.

Next, we shall consider the maximum likelihood estimator (MLE). Denote by $\hat{\theta}_{ML}$ the MLE. We also define the likelihood function $L(\theta)$ by

$$L(\theta) = \prod_{i=1}^{n} f_0\,(X_i - \theta) > 0 \qquad \text{for } X_{(n)} - 1 < \theta < X_{(1)} + 1,$$
$$= 0 \qquad \text{otherwise,}$$

where $X_{(1)} = \min_{1 \le i \le n} X_i$, $X_{(n)} = \max_{1 \le i \le n} X_i$. It is known that the order of consistency is equal to $n^{1/\alpha}$.

When θ_0 is the true parameter, it can be shown that the event " $\hat{\theta}_{ML} < \theta_0 + an^{-1/\alpha}$ " is equivalent to the one " $(\partial/\partial\theta) \log L(\theta_0 + an^{-1/\alpha}) < 0$ ". We put $h = an^{-1/\alpha}$. Since

$$\log L(\theta) = \sum_{i=1}^{n} \log f_0\,(X_i - \theta)$$

it follows that

$$(\partial/\partial\theta)\log L(\theta_0 + h) = -\sum_{i=1}^{n} f_0'\,(X_i - \theta_0 - h)/f_0(X_i - \theta_0 - h) \text{ for } X_{(n)}-1 < \theta_0 + h < X_{(1)}+1.$$

Without loss of generality, we assume that $\theta_0 = 0$. We also put

$$U = -f_0'\,(X - h)/f_0\,(X - h) \qquad \text{for } |X - h| < 1,$$
$$= 0 \qquad\qquad\qquad \text{for } |X - h| \ge 1. \qquad (3.5)$$

Then it is shown that the density $\tilde{f}(u)$ of U is given by

$$\tilde{f}(u) = (A'/(\alpha - 1))/(1 + |u|^2)^{(\alpha+1)/2} + (A'/(\alpha - 1))\tilde{g}(u) \quad \text{for } -\infty < u < \infty \text{ and } 1 < \alpha < 2,$$

where $\tilde{g}(u) = O(u^{-(\alpha+2)})$. Since, as $|u| \to \infty$

$$\left\{1/(1+u^2)^{(\alpha+1)/2}\right\} - \left\{1/(1+|u|)^{\alpha+1}\right\} = O(|u|^{-(\alpha+2)}),$$

it follows that, as $|u| \to \infty$

$$\tilde{f}(u) = (A'/(\alpha - 1))/(1 + |u|)^{\alpha+1} + r\,(u), \qquad (3.6)$$

where $r\,(u) = O(|u|^{-(\alpha+2)})$. It is also seen that

$$\int_{-\infty}^{\infty} r\,(u)e^{i t u}\,du = O(|t|^2). \qquad (3.7)$$

In a similar way as in Takeuchi and Akahira [6], it follows that

$$\int_{-\infty}^{\infty} (\alpha/2)e^{itu}/(1 + |u|)^{\alpha+1}\,du = 1 + k_\alpha|t|^\alpha + o(|t|^\alpha), \qquad (3.8)$$

Asymptotic Efficiency for Densities with Bounded Support

where k_α is some constant. From (3.6), (3.7) and (3.8) it is seen that the characteristic function $\phi(t)$ of U is given by

$$\phi(t) = \int_{-\infty}^{\infty} e^{itu}\, \tilde{f}(u)du = 1 + k'_\alpha\, |t|^\alpha + o(\,|t|^\alpha),$$

where k'_α is some constant. We also see that the bias-adjusted MLE is not asymptotically efficient in the sense that its asymptotic distribution does not uniformly attain the bound, for Z_h of (2.1) and U of (3.5) are not equivalent random variables.

Remark 3.1. In this paper, the support of the density $f_0(x)$ is assumed to be the open interval (-1,1) in the conditions (A.1) and (A.2). However, the results of the paper still hold when the interval (-1,1) is replaced by a finite open interval (a,b) in the conditions.

REFERENCES

[1]   Akahira, M. (1975). Asymptotic theory for estimation of location in non-regular cases, I: Order of convergence of consistent estimators. *Rep. Stat. Appl. Res., JUSE, 22,* 8-26.

[2]   Akahira, M. (1975). Asymptotic theory for estimation of location in non-regular cases, II: Bounds of asymptotic distributions of consistent estimators. . *Rep. Stat. Appl. Res., JUSE, 22,* 99-115.

[3]   Akahira, M. and Takeuchi, K. (1981). *Asymptotic Efficiency of Statistical Estimators: Concepts and Higher Order Asymptotic Efficiency.* Lecture Notes in Statistics 7, Springer, New York.

[4]   Polfeldt, T. (1970). The order of minimum variance in a non-regular case. *Ann. Math. Statist., 41,* 667-672.

[5]   Smith, R. L. (1985). Maximum likelihood estimation in a class of non-regular cases. *Biometrika, 72,* 67-90.

[6]   Takeuchi, K. and Akahira, M. (1976). On Gram-Charlier-Edgeworth type expansion of the sums of random variables (II). *Rep. Univ. Electro-Comm., 27,*117-123.

[7]   Woodroofe, M. (1972). Maximum likelihood estimation of a translation parameter of a truncated distribution. *Ann. Math. Statist., 43,* 113-122.

Journal of Computing and Information

Vol. 2, No. 1, 1991, Pages 71-92

Institutum Gaussianum

A definition of information amount applicable to non-regular cases[*]

Masafumi Akahira

Institute of Mathematics
University of Tsukuba
Ibaraki 305, Japan

Kei Takeuchi

Research Center for Advanced Science and Technology
University of Tokyo
Komaba, Tokyo 156, Japan

---

[*]The results of this paper have been presented by the first author at the Second International Symposium on Probability and Information Theory at McMaster University of Canada, August 1985.

1 Introduction

Amount of information contained in a sample and in a statistic (or an estimator) plays an important role in the theory of statistical inference as was shown in papers and books by R.A. Fisher (1925,1934,1956), Kullback (1959) and others. There are, however, various ways of definitions of amount of information, some are more convenient (such as Fisher information) but more restricted in applications.

In this paper we shall discuss a definition of information amount between two distributions which is always well defined, symmetric and additive for independent samples and information contained in a statistic is always not greater than that in the whole sample and the equality holds if and only if the statistic is sufficient. Hence we can also discuss the asymptotic relative efficiency of a statistic (or an estimator) by the ratio of informations contained in the statistic and in the sample in a systematic and unified way both in regular and non-regular cases. We shall give several examples and discuss some details of the structure of information.

2 The definition of information amount

There are various definitions on the distance between two distributions or amounts of information for random variables. Here we consider the following quantity. Let X be a random variable defined over an abstract sample space χ and P and Q are absolutely continuous with respect to a σ-finite measure μ. We define an amount of information between P and Q as

$$I_X(P,Q) = -8\log \int \left(\frac{dP}{d\mu}\frac{dQ}{d\mu}\right)^{1/2} d\mu. \qquad (2.1)$$

Here the integral in the above is called affinity between P and Q (e.g. see Matusita, 1955). The above quantity is independent of a choice of the measure μ. Indeed, if ν is absolutely continuous w.r.t. μ, then

$$\int \left(\frac{dP}{d\nu}\frac{dQ}{d\nu}\right)^{1/2} d\nu = \int \left(\frac{dP}{d\mu}\frac{dQ}{d\mu}\right)^{1/2} d\mu.$$

Taking $P+Q$ as μ, we can see that $I_X(P,Q)$ is always well defined since P and Q are absolutely continuous w.r.t. μ. If P and Q are not equivalent, then $I_X(P,Q) > 0$. So far as P and Q are not disjoint, that is, for any measurable set A,

$$P(A)Q(A) + (1-P(A))(1-Q(A)) > 0,$$

it follows that $I_X(P,Q)$ is finite. When P and Q are disjoint, we define $I_X(P,Q) = \infty$.

If X and Y are independent random variables and have the distributions P_i and Q_i ($i = 1,2$), then it is easily seen that

$$I_{X,Y}(P_1 \times Q_1, P_2 \times Q_2) = I_X(P_1, P_2) + I_Y(Q_1, Q_2).$$

where $P_1 \times Q_1$ and $P_2 \times Q_2$ denote product measures of $P_i(i = 1,2)$ and Q_i ($i = 1,2$), respectively.

Let $T = t(X)$ be a statistic. We denote by $P_{X|T}$, $Q_{X|T}$ and P_T, Q_T the conditional distributions of X given T and the distributions of T, which are absolutely continuous w.r.t. σ-finite measures μ_1 and μ_2, respectively. Then we obtain the following:

Proposition. *It holds that*

$$I_T(P,Q) \le I_X(P,Q),$$

where the equality holds if and only if T is pairwise sufficient for P and Q.

Proof. We have

$$\int \left(\frac{dP_X}{d\mu} \cdot \frac{dQ_X}{d\mu} \right)^{1/2} d\mu$$

$$= \int \int \left(\frac{dP_{X|T}}{d\mu_1} \cdot \frac{dP_T}{d\mu_2} \right)^{1/2} \left(\frac{dQ_{X|T}}{d\mu_1} \cdot \frac{dQ_T}{d\mu_2} \right)^{1/2} d\mu_1 d\mu_2$$

$$= \int \left\{ \int \left(\frac{dP_{X|T}}{d\mu_1} \cdot \frac{dQ_{X|T}}{d\mu_1} \right)^{1/2} d\mu_1 \right\} \left(\frac{dP_T}{d\mu_2} \cdot \frac{dQ_T}{d\mu_2} \right)^{1/2} d\mu_2$$

$$\le \int \left(\frac{dP_T}{d\mu_2} \cdot \frac{dQ_T}{d\mu_2} \right)^{1/2} d\mu_2,$$

hence by (2.1)

$$I_T(P,Q) \le I_X(P,Q).$$

In the above the equality holds if and only if $P_{X|T} = Q_{X|T}$ a.a. T, that is, T is pairwise sufficient for P and Q. Thus we complete the proof.

Next we define a new distribution K_T of T as

$$\frac{dK_T}{d\mu_2} = c \left(\frac{dP_T}{d\mu_2} \cdot \frac{dQ_T}{d\mu_2} \right)^{1/2}, \quad \int dK_T = 1,$$

where c is some constant. Since

Journal of Computing and Information, 2, 71-92   75

$$\int \left(\frac{dP_X}{d\mu} \cdot \frac{dQ_X}{d\mu} \right)^{1/2} d\mu$$

$$= \left[\int \left\{ \exp\left(-\frac{1}{8} I_{X|T}(P,Q) \right) \right\} \frac{dK_T}{d\mu_2} d\mu_2 \right] \cdot \exp\left\{ -\frac{1}{8} I_T(P,Q) \right\},$$

it follows that

$$I_X(P,Q) = -8\log E_T^* \left[\exp\left\{ -\frac{1}{8} I_{X|T}(P,Q) \right\} \right] + I_T(P,Q), \qquad (2.2)$$

where $I_{X|T}(P,Q)$ is the amount of information between the conditional distributions of X given T, and E^* denotes the expectation w.r.t. the distribution K_T.

The constant 8 in the amount $I_X(P,Q)$ of information is given to have a connection with that of the Fisher information. Indeed, if the distributions depend on a real-valued parameter θ, we simply denote $I_X(P_{\theta_1}, P_{\theta_2})$ by $I(\theta_1, \theta_2)$. Suppose that for a neighborhood of some parameter θ_0, P_θ is absolutely continuous w.r.t. P_{θ_0} and $dP_\theta / dP_{\theta_0}$ is continuously differentiable w.r.t. θ. Letting $\mu = P_{\theta_0}$, we have for sufficiently small $\Delta\theta$

$$I(\theta_0, \theta_0 + \Delta\theta) = -8\log\left\{ \int \left(\frac{dP_{\theta_0 + \Delta\theta}}{dP_{\theta_0}} \right)^{1/2} dP_{\theta_0} \right\}$$

$$= -8\log\left[\int \left\{ 1 + \frac{d(P_{\theta_0 + \Delta\theta} - P_{\theta_0})}{dP_{\theta_0}} \right\}^{1/2} dP_{\theta_0} \right]$$

$$
= -8\log\left[1 - \frac{1}{8}\int\left\{\frac{d(P_{\theta_0+\Delta\theta} - P_{\theta_0})}{dP_{\theta_0}}\right\}^2 dP_{\theta_0}\right] + o((\Delta\theta)^2)
$$

$$
= \int\left\{\frac{d(P_{\theta_0+\Delta\theta} - P_{\theta_0})}{dP_{\theta_0}}\right\}^2 dP_{\theta_0} + o((\Delta\theta)^2)
$$

$$
= \left(\int\left\{\left[\frac{\partial}{\partial\theta}\left(\frac{dP_\theta}{dP_{\theta_0}}\right)\right]_{\theta=\theta_0}\right\}^2 dP_{\theta_0}\right)(\Delta\theta)^2 + o((\Delta\theta)^2)
$$

$$
= I(\theta_0)(\Delta\theta)^2 + o((\Delta\theta)^2),
$$

where $I(\theta_0)$ denotes the amount of Fisher information.

If P_θ is not absolutely continuous w.r.t. P_{θ_0}, we denote by $A(\theta)$ the support of the distribution P_θ. Putting

$$
\pi(\theta) = \int_{A(\theta_0)} dP_\theta
$$

we obtain

$$
\pi(\theta) \leq 1.
$$

If for sufficiently small $\Delta\theta$

$$
\pi(\theta_0 + \Delta\theta) = \pi(\theta_0) - \pi_0'|\Delta\theta| + o(|\Delta\theta|)
$$

then

$$
I(\theta_0, \theta_o + \Delta\theta) = -8\log\left(1 - \frac{1}{2}\pi_0'|\Delta\theta|\right) + o(|\Delta\theta|)
$$

$$
= 4\pi_0'|\Delta\theta| + o(|\Delta\theta|),
$$

where $\pi'_o = \pi'(\theta_0)$.

Suppose that $X_1,...,X_n$ are independent and identically distributed (i.i.d.) random variables with a density function $f(x,\theta)$, where θ is a real-valued parameter. Then, for θ and $\theta + \Delta\theta$, the amount of information for $(X_1,...,X_n)$ is given by $nI(\theta, \theta + \Delta\theta)$. Suppose furthermore that there exists an estimator $T_n = t(X_1,...,X_n)$ which is consistent with the order c_n. Then it follows that

$$\lim_{n\to\infty} I_{T_n}(\theta,\theta+c_n^{-1}\Delta) > 0$$

for some nonzero constant Δ, and

$$\lim_{n\to\infty} nI(\theta,\theta+c_n^{-1}\Delta) > 0 \tag{2.3}$$

which gives the bound for consistency (see Example 3.5).

3 Examples

In this section we have some examples on the amount of information.

Example 3.1. Suppose that for two distributions P_j ($j = 1,2$), $X_1,...,X_n$ are independently, identically and normally distributed (i.i.n.d.) with mean μ_j and variance σ_j^2. Then the amount of information between two distributions for each X_i is given by

$$I(1,2) = -8\log\left[\int \frac{1}{\sqrt{2\pi}\,\sigma_1\sigma_2} \exp\left\{-\frac{\sigma_2^2(x-\mu_1)^2 + \sigma_1^2(x-\mu_2)^2}{4\sigma_1^2\sigma_2^2}\right\}dx\right]$$

$$= -8\log\left[\sqrt{\frac{2\sigma_1\sigma_2}{\sigma_1^2+\sigma_2^2}}\,\exp\left\{-\frac{\mu_1-\mu_2}{4(\sigma_1^2+\sigma_2^2)}\right\}\right]$$

$$= 4\log\left(\frac{\sigma_1^2+\sigma_2^2}{2\sigma_1\sigma_2}\right) + \frac{2(\mu_1-\mu_2)^2}{\sigma_1^2+\sigma_2^2}.$$

Hence it is seen that the amount of information for $(X_1,...,X_n)$ is equal to n-times the above value. We put $T_1 = \bar{X} = \sum_{i=1}^{n} X_i/n$ and $T_2 = \sum_{i=1}^{n}(X_i - \bar{X})^2$. Since

$$I_{T_1}(1,2) = 4\log\left(\frac{\sigma_1^2+\sigma_2^2}{2\sigma_1\sigma_2}\right) + \frac{2n(\mu_1-\mu_2)^2}{\sigma_1^2+\sigma_2^2}$$

and

$$I_{T_2}(1,2) = 4(n-1)\log\left(\frac{\sigma_1^2+\sigma_2^2}{2\sigma_1\sigma_2}\right),$$

it follows that

$$I_{T_1}(1,2) + I_{T_2}(1,2) = nI(1,2).$$

This equality should hold since (T_1, T_2) is a sufficient statistic.

Example 3.2. Suppose that for two distributions P_{θ_j} $(j = 1,2)$, $X_1,...,X_n$ are i.i.d. random variables according to a t-distribution with 3 degrees of freedom i.e., with the density

$$f(x-\theta_j) = \frac{c}{(1+(x-\theta_j)^2)^2},$$

where c is some constant. Then the amount of information between two distributions for each X_i is given by

$$I(1,2) = -8\log\left(1+\frac{(\theta_1-\theta_2)^2}{4}\right).$$

Example 3.3. Suppose that for two distributions P_{θ_j} ($j = 1,2$), $X_1,...,X_n$ are i.i.d. random variables with an exponential density

$$f(x,\theta_j,\xi_j) = \begin{cases} \dfrac{1}{\theta_j}\exp\left(-\dfrac{x-\xi_j}{\theta_j}\right) & for\, x \geq \xi_j, \\ 0 & for\, x < \xi_j, \end{cases}$$

where $\theta_j > 0$ and $-\infty < \xi_j < \infty$ ($j = 1,2$). If $\xi_1 < \xi_2$, then the amount of information between two distributions for each X_i is given by

$$I(1,2) = -8\log\left[\int_{\xi_2}^{\infty}\frac{1}{(\theta_1\theta_2)^{1/2}}\exp\left\{-\frac{1}{2}\left(\frac{x-\xi_1}{\theta_1}+\frac{x-\xi_2}{\theta_2}\right)\right\}dx\right]$$

$$= -8\log\left\{\frac{2\sqrt{\theta_1\theta_2}}{\theta_1+\theta_2}\cdot\exp\left(-\frac{\xi_2-\xi_1}{2\theta_1}\right)\right\}$$

$$= 4\log\frac{(\theta_1+\theta_2)^2}{4\theta_1\theta_2}+\frac{4(\xi_2-\xi_1)}{\theta_1}.$$

We put $T_1 = \min_{1\leq i\leq n} X_i$ and $T_2 = \bar{X} - \min_{1\leq i\leq n} X_i$. Since T_1 is distributed according to an exponential distribution and it can be written that

$$nT_2 = \sum_{i=1}^{n} X_{(i)} - nX_{(1)} = \sum_{i=1}^{n-1} (n-i)(X_{(i+1)} - X_{(i)})$$

which is equal to the sum of $n-1$ i.i.d. exponential random variables, it follows that

$$I_{T_1}(1,2) = 4\log\frac{(\theta_1+\theta_2)^2}{4\theta_1\theta_2} + 4n\frac{\xi_2-\xi_1}{\theta_1}, \qquad (3.1)$$

$$I_{T_2}(1,2) = 4(n-1)\log\frac{(\theta_1+\theta_2)^2}{4\theta_1\theta_2}, \qquad (3.2)$$

where $X_{(1)} \leq X_{(2)} \leq \dots \leq X_{(n)}$. Then we have

$$I_{T_1,T_2}(1,2) = I_{T_1}(1,2) + I_{T_2}(1,2) = nI(1,2), \qquad (3.3)$$

which shows that the pair (T_1, T_2) is sufficient. And also (3.1) and (3.2) show that T_1 is sufficient when θ is known, i.e., $\theta = \theta_1 = \theta_2$, and T_1 has only $1/n$ of the total information when ξ is known, i.e., $\xi = \xi_1 = \xi_2$. This fact can be interpreted as showing that T_2 is asymptotically sufficient when ξ is known.

Example 3.4. Suppose that for two distributions P_j ($j=1,2$), $X_1,...,X_n$ are i.i.d. random variables according to an uniform distribution on the interval $[\theta_j - \tau_j/2, \theta_j + \tau_j/2]$. If $\tau_1 < \tau_2$, then the amount of information between two distributions for each X_i is given as

Journal of Computing and Information, 2, 71-92   81

$$
I(1,2) = \begin{cases}
\infty & \text{for } \theta_1 + \dfrac{\tau_1}{2} \le \theta_2 - \dfrac{\tau_2}{2} \text{ or } \theta_1 - \dfrac{\tau_1}{2} \ge \theta_2 + \dfrac{\tau_2}{2}, \\[2em]
-4\log\left\{ \dfrac{1}{\tau_1\tau_2}\left(\dfrac{\tau_1+\tau_2}{2} + \theta_2 - \theta_1 \right)^2 \right\} & \\[1em]
\qquad\qquad . & \text{for } \theta_1 - \dfrac{\tau_1}{2} < \theta_2 - \dfrac{\tau_2}{2} < \theta_1 + \dfrac{\tau_1}{2} < \theta_2 + \dfrac{\tau_2}{2}, \\[2em]
-4\log\left\{ \dfrac{1}{\tau_1\tau_2}\left(\dfrac{\tau_1+\tau_2}{2} + \theta_2 - \theta_1 \right)^2 \right\} & \\[1em]
\qquad\qquad . & \text{for } \theta_2 - \dfrac{\tau_2}{2} < \theta_1 - \dfrac{\tau_1}{2} < \theta_2 + \dfrac{\tau_2}{2} < \theta_1 + \dfrac{\tau_1}{2}, \\[2em]
-4\log\dfrac{\tau_1}{\tau_2} & \text{for } \theta_2 - \dfrac{\tau_2}{2} < \theta_1 - \dfrac{\tau_1}{2} < \theta_1 + \dfrac{\tau_1}{2} < \theta_2 + \dfrac{\tau_2}{2}.
\end{cases}
$$

We put $T_1 = \max_{1 \le i \le n} X_i - \min_{1 \le i \le n} X_i$ and $T_2 = (\max_{1 \le i \le n} X_i + \min_{1 \le i \le n} X_i)/2$. Then the joint density of $T_1, T_2)$ is given by

$$
f(t_1, t_2) = \begin{cases}
\dfrac{n(n-1)}{\tau^n} t_1^{n-2} & \text{for } t_2 - \dfrac{t_1}{2} > \theta - \dfrac{\tau}{2} \text{ or } t_2 + \dfrac{t_1}{2} < \theta + \dfrac{\tau}{2}, \\[1.5em]
0 & \text{otherwise}.
\end{cases}
$$

Hence the density of T_1 is obtained by

$$
f(t_1) = \begin{cases}
n(n-1)t_1^{n-2}(\tau - t_1)/\tau^n & \text{for } 0 < t_1 < \tau, \\[0.5em]
0 & \text{otherwise},
\end{cases}
$$

and the amount of information between two distributions is also given by

$$
I_{T_1}(1,2) = 4n\log\frac{\tau_2}{\tau_1} \text{ for } \tau_1 < \tau_2.
$$

If $\tau_1 = \tau_2 = \tau$, it is easily seen that $I_{T_1}(1,2) = 0$. However, it should be noted that T_2 alone can not be sufficient even when τ is known, i.e., $\tau = \tau_1 = \tau_2$ since T_1 and T_2 are not independent. Then the amount of information for $(X_1,...,X_n)$ is also given by

$$nI(1,2) = \begin{cases} -8n\log\left(1 - \dfrac{|\theta_1 - \theta_2|}{\tau}\right) & \textit{for } |\theta_1 - \theta_2| < \tau, \\ \infty & \textit{for } |\theta_1 - \theta_2| \geq \tau. \end{cases} \qquad (3.4)$$

Since the conditional distribution of T_2 given T_1 is an uniform distribution on the interval $[\theta - (\tau - T_1)/2, \theta + (\tau - T_1)/2]$, it follows that the conditional information amount is given by

$$I_{T_2|T_1}(1,2) = \begin{cases} -8\log\left(1 - \dfrac{|\theta_1 - \theta_2|}{\tau - T_1}\right) & \textit{for } |\theta_1 - \theta_2| < \tau - T_1, \\ \infty & \textit{for } |\theta_1 - \theta_2| \geq \tau - T_1. \end{cases}$$

Then we have

$$E\left[\exp\left\{-\frac{1}{8} I_{T_2|T_1}(1,2)\right\}\right] = E\left[\left(\frac{\tau - T_1 - |\theta_1 - \theta_2|}{\tau - T_1}\right)^{+}\right]$$

$$= \int_0^{\tau - |\theta_1 - \theta_2|} n(n-1)\tau - t_1 - \theta_1 - \theta_2) t_1^{n-2} / \tau^n dt_1$$

$$= \left(1 - \frac{|\theta_1 - \theta_2|}{\tau}\right)^n,$$

where $(\ )^{+}$ denotes a positive part of $(\ )$. Hence

$$-8\log E_{T_1}\left[\exp\left\{-\frac{1}{8} I_{T_2|T_1}(1,2)\right\}\right] = 8n\log\left(1 - \frac{|\theta_1 - \theta_2|}{\tau}\right).$$

for $|\theta_1 - \theta_2| < \tau$, which is equal to (3.4) as is expected from (2.2) since

$I_{T_1}(1,2) = 0$. We also obtain the density of T_2

$$f(t_2) = \int_0^{\tau - 2|t_2 - 0|} f(t_1, t_2)\, dt = \frac{n}{\tau^n}(\tau - 2\,|t_2 - \theta|)^{n-1}$$

for $|t_2 - \theta| < \tau/2$. Hence the amount of information for T_2 is given by

$$I_{T_2}(1,2) = -8\log\left[\int_{\theta_1 - \frac{\tau}{2}}^{\theta_2 + \frac{\tau}{2}} \frac{n}{\tau^n}(\tau - 2\,|t_2 - \theta_1|)^{\frac{n-1}{2}} (\tau - 2\,|t_2 - \theta_2|)^{\frac{n-1}{2}}\, dt_2\right]$$

for $0 < |\theta_2 - \theta_1| < \tau$. The explicit form of the above integral is complicated, so we consider the case when n is large and θ_1 and θ_2 are sufficiently close. Putting $\Delta/n = |\theta_2 - \theta_1|$, we have for sufficiently large n

$$\begin{aligned}
I_{T_2}(1,2) &\sim -8\log \int_{-\infty}^{\infty} \frac{1}{\tau} \exp\left\{-\frac{1}{\tau}(|t_2| + |t_2 - \Delta|)\right\} dt_2 \\
&= 8\left\{\frac{\Delta}{\tau} - \log\left(1 + \frac{\Delta}{\tau}\right)\right\}.
\end{aligned} \tag{3.5}$$

On the other hand we obtain from (3.4)

$$nI(1,2) \sim -8n\log\left(1 - \frac{\Delta}{n\tau}\right) \sim \frac{8\Delta}{\tau}$$

for sufficiently large n. Therefore the loss of information of T_2 is asymptotically equal to the value $8\log(1 + (\Delta/\tau))$.

Example 3.5. Suppose that $X_1,...,X_n$ are i.i.d. random variables with a triangular density function

$$f(x-\theta) = \begin{cases} 1 - |x-\theta| & \textit{for } |x-\theta| < 1, \\ 0 & \textit{for } |x-\theta| \geq 1. \end{cases}$$

For a small positive number $\Delta\theta$, we have

$$\begin{aligned}
\int_{-1+\Delta\theta}^{1} \{f(x)f(x-\Delta\theta)\}^{1/2}\,dx &= \int_{-1+\Delta\theta}^{0} \{(1+x)(1+x-\Delta\theta)\}^{1/2}\,dx \\
&\quad + \int_{0}^{\Delta\theta} \{(1-x)(1+x-\Delta\theta)\}^{1/2}\,dx \\
&\quad + \int_{\Delta\theta}^{1} \{(1-x)(1-x+\Delta\theta)\}^{1/2}\,dx \\
&= 2\int_{0}^{1-\Delta\theta} \sqrt{y^2 + y\Delta\theta}\,dy \\
&\quad + \int_{0}^{\Delta\theta} \{(1-x)(1+x-\Delta\theta)\}^{1/2}\,dx \\
&= 2\int_{\frac{\Delta\theta}{2}}^{1-\frac{\Delta\theta}{2}} \sqrt{u^2 - \frac{(\Delta\theta)^2}{4}}\,du \\
&\quad + \int_{-\frac{\Delta\theta}{2}}^{\frac{\Delta\theta}{2}} \sqrt{\left(1-\frac{\Delta\theta}{2}\right)^2 - u^2}\,du \\
&= \sqrt{\left(1-\frac{\Delta\theta}{2}\right)^2 - \frac{(\Delta\theta)^2}{4}} + \frac{(\Delta\theta)^2}{4}\log\frac{\Delta\theta}{4} \\
&\quad - \frac{(\Delta\theta)^2}{4}\log\frac{1}{2}\left|1 - \frac{\Delta\theta}{2}\sqrt{\left(1-\frac{\Delta\theta}{2}\right)^2 - \frac{(\Delta\theta)^2}{4}}\right| \\
&\quad + \left(1-\frac{\Delta\theta}{2}\right)^2 \sin^{-1}\frac{\frac{\Delta\theta}{2}}{1-\frac{\Delta\theta}{2}} \\
&= 1 + \frac{(\Delta\theta)^2}{4}\log\frac{\Delta\theta}{4} - \frac{3}{8}(\Delta\theta)^2 + o((\Delta\theta)^2).
\end{aligned}$$

In a similar way to the case $\Delta\theta > 0$, we obtain for $\Delta\theta < 0$,

$$\int_{-1}^{1+\Delta\theta} \{f(x)f(x-\Delta\theta)\}^{1/2}\,dx = 1 + \frac{(\Delta\theta)^2}{4}\log\frac{|\Delta\theta|}{4} - \frac{3}{8}(\Delta\theta)^2 + o((\Delta\theta)^2).$$

Hence we have for $\Delta\theta \neq 0$

$$\begin{aligned}
I(\theta,\theta+\Delta\theta) &= -8\log\int\{f(x)f(x-\Delta\theta)\}^{1/2}\,dx \\
&= -8\log\left(1 + \frac{(\Delta\theta)^2}{4}\log\frac{|\Delta\theta|}{4} - \frac{3}{8}(\Delta\theta)^2 + o((\Delta\theta)^2)\right) \\
&= -2(\Delta\theta)^2\log\frac{|\Delta\theta|}{4} + 3(\Delta\theta)^2 + o((\Delta\theta)^2).
\end{aligned}$$

In order that $nI(\theta+\Delta\theta) = O(1)$, i.e. $-n(\Delta\theta)^2\log|\Delta\theta| = O(1)$, we obtain $\Delta\theta = O((n\log n)^{-1/2})$. Letting $c_n = (n\log n)^{1/2}$, we see that (2.3) holds, i.e. $\lim_{n\to\infty} nI(\theta,\theta+c_n^{-1}\Delta) > 0$ for some nonzero constant. Hence it follows that the bound for consistency is equal to $(n\log n)^{1/2}$.

Example 3.6. Suppose that for a distribution P_θ, $X_1,...,X_n$ are i.i.d. random variables with a truncated normal density

$$f(x-\theta) = \begin{cases} ce^{-(x-\theta)^2/2} & for\ |x-\theta| < 1, \\ 0 & for\ |x-\theta| \geq 1, \end{cases}$$

where c is some constant. For $0 < \theta < 2$, the affinity $\rho(0,\theta)$ between P_0 and P_θ is given by

$$\begin{aligned}
\rho(0,\theta) &= \int_{-\infty}^{\infty}\left(\frac{dP_0}{dx}\cdot\frac{dP_\theta}{dx}\right)^{1/2}dx \\
&= \int_{\theta-1}^{1} c\exp\left[-\frac{1}{4}\{x^2+(x-\theta)^2\}\right]dx
\end{aligned}$$

$$= e^{-\theta^2/8} P_0 \left\{ |X_1| < 1 - \frac{\theta}{2} \right\}.$$

For small $\theta > 0$, the amount of information between P_0 and P_θ for X_1 is given by

$$
\begin{aligned}
I(0,\theta) &= \theta^2 - 8 \log P_0 \left\{ |X_1| < 1 - \frac{\theta}{2} \right\} \\
&= \theta^2 - 8 \log \left[1 - 2\sqrt{2\pi}\, c \left\{ \Phi(1) - \Phi\left(1 - \frac{\theta}{2} \right) \right\} \right] \\
&= \theta^2 - 8 \log \left(1 - k\theta - \frac{1}{4} k\theta^2 + o(\theta^2) \right) \\
&= 8k\theta + \{ 1 + 2k(2k+1) \} \theta^2 + o(\theta^2),
\end{aligned}
\qquad (3.6)
$$

where $\Phi(x)$ is the standard normal distribution function. Since $\bar{X} = \sum_{i=1}^{n} X_i / n$ is asymptotically and normally distributed with mean θ and variance $(1 - 2k)/n$, it follows that for large n the amount of information between two distributions for $\bar{X}$ is given by

$$
\begin{aligned}
I_{\bar{X}}(0,\theta) &\sim -8 \log \int_{-\infty}^{\infty} \frac{1}{\sqrt{2\pi}} \left(\sqrt{\frac{n}{1-2k}} \right)^{1/2} \exp\left[-\frac{n}{4(1-2k)} \{ x^2 + (x-\theta)^2 \} \right] dx \\
&= \frac{n\theta^2}{1-2k} + \frac{1}{4} \log(1-2k) - \frac{1}{4} \log n \\
&\sim n\theta^2 \{ 1 + 2k(2k+1) \} + \frac{1}{4} \log(1-2k) - \frac{1}{4} \log n,
\end{aligned}
$$

$$(3.7)$$

where $k = c e^{-1/2}$. We put $T_1 = n(X_{(1)} + 1 - \theta)$ and $T_2 = n(X_{(m)} - 1 - \theta)$, where $X_{(1)} = \min_{1 \le i \le n} X_i$ and $X_{(m)} = \max_{1 \le i \le n} X_i$. Since the asymptotic densities of T_1 and T_2 are given by

$$g_1(t) = \begin{cases} ke^{-kt} & \text{for } t>0; \\ 0 & \text{for } t \leq 0, \end{cases}$$

and

$$g_2(t) = \begin{cases} ke^{kt} & \text{for } t<0; \\ 0 & \text{for } t \geq 0, \end{cases}$$

respectively, it follows that for large n and $\theta > 0$, the amounts of information between two distributions for $X_{(1)}$ and $X_{(n)}$ are given by

$$I_{T_1}\left(0, \frac{\theta}{n}\right) \sim -8\log\left(\int_{\theta}^{\infty} ke^{-kt+(k\theta/2)}dt\right) = 4k\theta \qquad (3.8)$$

and

$$I_{T_2}\left(0, \frac{\theta}{n}\right) \sim -8\log\left(\int_{-\infty}^{0} ke^{kt-(k\theta/2)}dt\right) = 4k\theta \qquad (3.9)$$

It is noted that $\bar{X}$ is asymptotically independent of $X_{(1)}$ and $X_{(n)}$, and $(X_{(1)}, X_{(n)})$ is asymptotically sufficient. It follows from (3.6) to (3.9) that for large n and small $\theta > 0$ the amount of information between two distributions for $(X_1,...,X_n)$ has the following relationship:

$$\begin{aligned} I_{X_1,...,X_n}(0,\theta) &= nI_{X_1}(0,\theta) \\ &= 8kn\theta + \{1+2k(2k+1)\}n\theta^2 + o(n\theta^2) \\ &\sim I_{X_{(1)}}(0,\theta) + I_{X_{(n)}}(0,\theta) + I_{\bar{X}}(0,\theta), \end{aligned}$$

Suppose that for $n = 2m+1$, $X_1,...,X_n$ are independently and identically distributed random variables according to a density $f(x-\theta)$ w.r.t. the Lebesque measure. Let T be the median of $X_1,...,X_n$, symbolically

$T = \mathrm{med}\, X_i$. We shall obtain the amount of information between two distributions for T and the asymptotic relative efficiency of T w.r.t. $(X_1,...,X_n)$. First the density $g(t-\theta)$ of T is given by

$$g(t-\theta) \;=\; \frac{(2m+1)!}{m!\,m!}\,\{F(t-\theta)\}^m\,\{1-F(t-\theta)\}^m\,f(t-\theta),$$

where $F(x-\theta)$ denotes the distribution function of X_1. Then it follows that for $\theta_1 = \theta + \Delta$ and $\theta_2 = \theta - \Delta$, the affinity $\rho(\theta_1,\theta_2)$ between two distributions with densities $f(x-\theta_j)$ ($j = 1,2$) is given by

$$\rho(\theta_1,\theta_2) = \frac{(2m+1)!}{(m!\,m!)} \int_{-\infty}^{\infty} \{F(t-\Delta)\,F(t+\Delta)\,(1-F(t-\Delta))\,(1-F(t+\Delta))\}^{m/2}$$
$$\cdot\{f(t-\Delta)\,f(t+\Delta)\}^{1/2}\,dt.$$

Putting

$$H(t) \;=\; F(t-\Delta)\,F(t+\Delta)\,(1-F(t-\Delta))\,(1-F(t+\Delta)),$$

we have

$$\frac{d}{dt}\log H(t) \;=\; \frac{f(t-\Delta)}{F(t-\Delta)} + \frac{f(t+\Delta)}{F(t+\Delta)} - \frac{f(t-\Delta)}{1-F(t-\Delta)} - \frac{f(t+\Delta)}{1-F(t+\Delta)}.$$

If $f(-t) = f(t)$ for all real number t and $f(t)/1 - F(t)$ is a monotone increasing function of t, it follows that four terms of the right-hand side of the above are monotone decreasing functions of t, which implies that $(d/dt)\log H(t)$ is a monotone decreasing function of t. Since $f(-\Delta) = f(\Delta)$ and $F(\Delta) = 1 - F(-\Delta)$, it follows that

$$\frac{d}{dt}\log H(0) \;=\; \frac{f(-\Delta)}{F(-\Delta)} + \frac{f(\Delta)}{F(\Delta)} - \frac{f(-\Delta)}{1-F(-\Delta)} - \frac{f(\Delta)}{1-F(\Delta)} \;=\; 0,$$

implying that $\log H(t)$, i.e., $H(t)$ takes a maximum value at $t = 0$. We also have

$$\frac{d^2}{dt^2} \log H(0) = \frac{f'(-\Delta)}{F(-\Delta)} - \frac{f(\Delta)^2}{F(\Delta)^2} + \frac{f'(-\Delta)}{F(-\Delta)} - \frac{f(\Delta)^2}{F(\Delta)^2}$$

$$- \frac{f'(-\Delta)}{1-F(-\Delta)} - \frac{f(-\Delta)^2}{(1-F(-\Delta))^2} - \frac{f'(\Delta)}{1-F(\Delta)} - \frac{f(\Delta)^2}{1-F(\Delta))^2}$$

$$= \frac{2f'(\Delta)(1-2F(\Delta))}{F(\Delta)(1-F(\Delta))} - \frac{2f(\Delta)^2\{F(\Delta)^2+(1-F(\Delta))^2\}}{F(\Delta)^2(1-F(\Delta))^2}$$

$$= -\sigma^2 \ (say).$$

Then we obtain

$$\log H(t) = \log H(0) - \frac{\sigma^2}{2} t^2 + \dots.$$

Hence we have for a sufficient large m

$$\rho(\theta_1, \theta_2) \sim \frac{(2m+1)!}{m!\,m!} H(0)^{m/2} \int e^{-\frac{\sigma^2 m}{2} t^2} \{f(t-\Delta)f(t+\Delta)\}^{1/2} dt.$$

Since by the Stirling formula for large factorials

$$\frac{(2m)!}{m!\,m!}\left(\frac{1}{2}\right)^{2m} \sim \frac{1}{\sqrt{m\pi}}$$

it follows that

$$\rho(\theta_1,\theta_2) \sim \{4H(0)^{1/2}\}^m \frac{2m+1}{\sqrt{m\pi}} \int_{-\infty}^{\infty} e^{-\frac{\sigma^2 m}{2}t^2} \{f(t-\Delta)f(t+\Delta)\}^{1/2} dt$$

$$\sim \{4F(\Delta)(1-F(\Delta))\}^m \cdot 2\sqrt{2}f(\Delta)/\sigma$$

$$\sim \{2\sqrt{F(\Delta)(1-F(\Delta))}\}^n \cdot \frac{\sqrt{2}f(\Delta)}{\sigma\{F(\Delta)(1-F(\Delta))\}^{1/2}}$$

$$= \{2\sqrt{F(\Delta)(1-F(\Delta))}\}^n \{Q(\Delta)\}^{-1/2} \ (say).$$

Then we have

$$Q(\Delta) = \frac{F(\Delta)^2 + (1-F(\Delta))^2}{F(\Delta)(1-F(\Delta))} - \frac{f'(\Delta)}{f(\Delta)^2}(1-2F(\Delta)),$$

hence the amount of information between two distributions for $T = \mathrm{med}\, X_i$ is given by

$$I_T(1,2) = -4n\log(4F(\Delta)(1-F(\Delta))) + 4\log Q(\Delta) + o(1).$$

On the other hand the amount of information between two distributions for each X_i is given by

$$I(1,2) = -8\log \int f(x-\Delta)^{1/2} f(x+\Delta)^{1/2} dx$$

$$= I(\Delta) \ \ (say).$$

Then the asymptotic relative efficiency of $T = \mathrm{med}\, X_i$ w.r.t. $(X_1,...,X_n)$ is obtained by

$$\frac{I_T(1,2)}{nI(1,2)} = -\frac{4}{I(\Delta)}\log(4F(\Delta)(1-F(\Delta))). \qquad (3.10)$$

We also have for small $\Delta > 0$

$$I(\Delta) = -8\log \int f(x-\Delta)^{1/2} f(x+\Delta)^{1/2} dx$$

$$\sim -8\log \int f(x)\left\{1-\left(\frac{f'(x)}{f(x)}\right)^2 \Delta^2\right\}^{1/2} dx \qquad (3.11)$$

$$\sim 4\Delta^2 \int \frac{\{f'(x)\}^2}{f(x)} dx$$

$$= 4\Delta^2 I,$$

where I denotes the Fisher information of f. Further we obtain for small $\Delta > 0$

$$4F(\Delta)(1-F(\Delta)) = 1-4\left(F(\Delta)-\frac{1}{2}\right)^2 \qquad (3.12)$$

$$\sim 1-4f(0)^2\Delta^2,$$

Hence it follows from (3.10) to (3.12) that the asymptotic relative efficiency of $T = \mathrm{med} X_i$ w.r.t. $(X_1,...,X_n)$ is given by $4f(0)^2/I$, which is equal to the Pitman efficiency. In particular, letting $f(x) = e^{-|x|}/2$ we have for $\Delta > 0$

$$F(\Delta) = 1-\frac{1}{2}e^{-\Delta}$$

and

$$I(\Delta) = -8\log \int_{-\infty}^{\infty} \frac{1}{2} e^{-\{|x-\Delta|+|x+\Delta|\}/2} dx$$

$$= -8\log\left(\int_0^{\Delta} e^{-\Delta} dx + \int_{\Delta}^{\infty} e^{-x} dx\right)$$

$$= -8\log(e^{-\Delta}(1+\Delta))$$

$$= 8(\Delta - \log(1 + \Delta)).$$

Hence we see that the asymptotic relative efficiency of $T = \operatorname{med} X_i$ w.r.t. $(X_1,...,X_n)$ is given by

$$\frac{\Delta - \log(2 - e^{-\Delta})}{2\{\Delta - \log(1 + \Delta)\}},$$

which tends to 1 as $\Delta \to 0$.

References

[1]   Fisher, R.A. (1925). Theory of statistical estimation. *Proc. Camb. Phil. Soc.*, **22**, 700-725.

[2]   Fisher, R.A. (1934). Two new properties of mathematical likelihood. *Proc. Roy. Soc. London*, **A-144**, 285-307.

[3]   Fisher, R.A. (1956). Statistical Methods and Scientific Inference. Oliver & Boyd, Edinburgh.

[4]   Kullback, S. (1959). Information Theory and Statistics. John Wiley, New York.

[5]   Matusita, K. (1955). Decision rules based on the distance for problems of fit, two samples and estimation. *Ann. Math. Statist.*, **26**, 631-640.

Rep. Stat. Appl. Res., JUSE
Vol. 39, No. 4, 1992
pp. 1-13

A-Section

Unbiased Estimation in Sequential Binomial Sampling

Masafumi AKAHIRA* , Kei TAKEUCHI**, and Ken-ichi KOIKE*

Abstract

In the case of sequential Bernoulli trials, a sufficient condition for a parametric function to be unbiasedly estimable is given and the existence of a discontinuous unbiasedly estimable function is shown using non-randomized sample size procedures.

1. Introduction

It is obvious to observe that for usual "regular" case, any parametric function which has unbiased estimators must be continuous or differentiable in the original parameter (e.g. see Zacks (1971) and Lehmann (1983)). However, in a sequential case or a randomized sample size case when the sample size is not bounded, the continuity of estimable function does not necessarily follow.

The purpose of this paper is to show that for the simplest case of the Bernoulli sequence, we can construct an unbiased estimator of a discontinuous function of p using a sequential procedure with a stopping rule depending on the path and a non-randomized estimator. Singh (1964) claimed to have obtained the same results for a more general class of problems by randomizing the size of sample, but as will be discussed later, the authors found a flaw in his proof which seems to invalidate his main theorem. Our procedure is non-randomized, but it is not a procedure based on minimal sufficient statistics, hence in a sense equivalent to a randomized procedure based on a sufficient statistic. It is an open problem whether we can construct a non-randomized sequential unbiased estimator of a discontinuous function based on the minimal sufficient statistic alone. In view of our results the contention of Bhandari and Bose (1990) that the estimable parameter must be continuous in p in sequential estimation set up can not be valid since the general theoretical framework is essentially the same although the class of procedures in their paper is different from ours.

2. Definitions and Preliminaries

Now we have a sequence of independent and identically distributed Bernoulli random variables $X_1, X_2,..., X_n,...$ with $P\{X_1 = 1\} = p$ and $P\{X_1 = 0\} = 1 - p$, where $0 < p < 1$ and determine a stopping rule. The set of the paths are defined as follows. Let R_n be a set of the

Received September 3, 1992
*   Institute of Mathematics, University of Tsukuba, Tsukuba, Ibaraki 305, Japan
**   Research Center for Advanced Science and Technology, University of Tokyo, 4-6-1 Komaba, Meguro-ku, Tokyo 156, Japan
(Key words)   Sequential Bernoulli trial, Stopping rule, Discontinuous estimable function, Non-randomized procedure.

M. Akahira, K. Takeuchi and K. Koike

all possible Bernoulli sequences of the length n, that is, a set of 2^n sequences of 0 and 1, and denote $R = \cup_{n=1}^{\infty} R_n$. Now let S_n be a subset of R_n, and we stop sampling if the observed sequence of size n belongs to S_n. We denote $S = \cup_{n=1}^{\infty} S_n$. Suppose that $(i_1, ..., i_M) \in S_M$ for some positive integer M, then it is required that, for any $k < M$, $(i_1, ..., i_k) \notin S_k$, and this is equivalent to the condition that if $(i_1, ..., i_M) \in S_M$, then $(i_1, ..., i_M, i_{M+1}, ..., i_{M+m})$ $\notin S_{M+m}$ for any positive integer m and any $(i_{M+1}, ..., i_{M+m})$ with $i_{M+j} = 0$ or 1 ($j = 1, ..., m$). For each path $(X_1, ..., X_n) \in S_n$, we denote the terminal point by the pair (X, Y), where $X = \sum_{i=1}^{n} X_i$ and $Y = \sum_{i=1}^{n}(1 - X_i)$. We denote by T the terminal point, where sampling is stopped. Suppose that we have a stopping rule defined as above. Then we obtain the probability that the procedure stops at the point $T = (x, y)$ as

$$P\{T = (x, y)\} = N(x, y)p^x (1 - p)^y,$$

where N is the number of paths in S from the origin to the point and is independent of p. We assume that $N(x, y) > 0$ for all x and all y with $x + y \geq 1$ and the stopping rule is closed in the sense that $\sum_x \sum_y P\{T = (x, y)\} = 1$. Then we have

$$\sum_{x' \leq x, y' \leq y} N(x', y')\, _{x+y-x'-y'}C_{x-x'} \leq\, _{x+y}C_x .$$

First we have for any estimator $\pi(X, Y)$ depending on the stopping rule

$$E_p[\pi(X, Y)] = \sum_n \sum_{x+y=n} \pi(x, y)\, N(x, y)\, p^x (1 - p)^y .$$

Next we want a sufficient condition for $g(p)$ to be unbiasedly estimable, that is, there exists an unbiased estimator $\pi(X, Y)$ such that

$$E_p\big[\,|\pi(X, Y)|\,\big] < \infty .$$

Now suppose that there exists an asymptotically unbiased estimator $\hat{g}_n(X, Y)$ of $g(p)$, that is,

$$\lim_{n \to \infty} \sum_{x+y=n} \hat{g}_n(x, y)\, _nC_x\, p^x (1 - p)^y = g(p), \tag{2.1}$$

where $\hat{g}_n(x, y)$ is defined for $x + y = n$, $0 \leq x \leq n$. Define

$$g_n(p) = \sum_{x+y=n} \hat{g}_n(x, y)\, _nC_x\, p^x (1 - p)^y , \tag{2.2}$$

then $\lim_{n \to \infty} g_n(p) = g(p)$ for all p. Also define $h_n(p) = g_n(p) - g_{n-1}(p)$ for $n = 1, 2,.., h_0(p) = 0$ and $\hat{g}_0(0, 0) = \hat{g}_{n-1}(-1, n) = \hat{g}_{n-1}(n, -1) = 0$ for $n = 1, 2,...$ Then $\sum_{n=0}^{\infty} h_n(p) = g(p)$. We put

Unbiased Estimation in Sequential Binomial Sampling

$$G_n(x, y) = \hat{g}_n(x, y) - \frac{x}{n}\hat{g}_{n-1}(x-1, y) - \frac{y}{n}\hat{g}_{n-1}(x, y-1) \qquad (2.3)$$

for $n = 1, 2, \ldots$, and $G_0(0, 0) = 0$. Then we have the following.

Lemma 2.1.

$$h_n(p) = \sum_{x+y=n} G_n(x, y)\, _nC_x\, p^x(1-p)^y .$$

Proof. Since $h_n(p) = g_n(p) - g_{n-1}(p)$ for $n = 1, 2, \ldots$, it follows from (2.2) that

$$h_n(p) = \sum_{x+y=n} \hat{g}_{n-1}(x, y)\, _nC_x\, p^x(1-p)^y - \sum_{x+y=n-1} \hat{g}_{n-1}(x, y)\, _{n-1}C_x\, p^x(1-p)^y.$$

$$(2.4)$$

Since, for $x + y = n - 1$,

$$_{n-1}C_x\, p^x(1-p)^y = {}_{n-1}C_x\, p^x(1-p)^{n-1-x}\{p + (1-p)\}$$

$$= {}_{n-1}C_x\, p^{x+1}(1-p)^{n-1-x} + {}_{n-1}C_x\, p^x(1-p)^{n-x}$$

$$= {}_nC_{x+1}\, p^{x+1}(1-p)^{n-x-1}\, \frac{_{n-1}C_x}{_nC_{x+1}} + {}_nC_x\, p^x(1-p)^{n-x}\, \frac{_{n-1}C_x}{_nC_x}$$

$$= \frac{x+1}{n}\, _nC_{x+1}\, p^{x+1}(1-p)^{n-x-1} + \frac{n-x}{n}\, _nC_x\, p^x(1-p)^{n-x},$$

it follows that

$$\sum_{x+y=n-1} \hat{g}_{n-1}(x, y)\, _{n-1}C_x\, p^x(1-p)^y$$

$$= \sum_{x+y=n-1} \frac{x+1}{n}\, \hat{g}_{n-1}(x, y)\, _nC_{x+1}\, p^{x+1}(1-p)^{n-x-1}$$

$$+ \sum_{x+y=n-1} \frac{n-x}{n}\, \hat{g}_{n-1}(x, y)\, _nC_x\, p^x(1-p)^{n-x}$$

$$= \sum_{x=0}^{n-1} \frac{x+1}{n}\, \hat{g}_{n-1}(x, n-1-x)\, _nC_{x+1}\, p^{x+1}(1-p)^{n-x-1}$$

$$+ \sum_{x=0}^{n-1} \frac{n-x}{n}\, \hat{g}_{n-1}(x, n-1-x)\, _nC_x\, p^x(1-p)^{n-x}$$

$$= \sum_{x=1}^{n} \frac{x}{n}\, \hat{g}_{n-1}(x-1, n-x)\, _nC_x\, p^x(1-p)^{n-x}$$

M. Akahira, K. Takeuchi and K. Koike

$$+ \sum_{x=0}^{n} \frac{n-x}{n} \, \hat{g}_{n-1}(x, n-1-x) \,_n C_x \, p^x (1-p)^{n-x}$$

$$= \sum_{x+y=n} \frac{x}{n} \, \hat{g}_{n-1}(x-1, y) \,_n C_x \, p^x (1-p)^y$$
$$+ \sum_{x+y=n} \frac{y}{n} \, \hat{g}_{n-1}(x, y-1) \,_n C_x \, p^x (1-p)^y. \tag{2.5}$$

From (2.3), (2.4) and (2.5) we have

$$h_n(p) = \sum_{x+y=n} \left\{ \hat{g}_n(x, y) - \frac{x}{n} \hat{g}_{n-1}(x-1, y) - \frac{y}{n} \hat{g}_{n-1}(x, y-1) \right\} \,_n C_x \, p^x (1-p)^y$$

$$= \sum_{x+y=n} G_n(x, y) \,_n C_x \, p^x (1-p)^y.$$

This completes the proof.

Thus if we put

$$e(x, y) = G_n(x, y) \,_n C_x / N(x, y) \quad \text{for } x + y = n = 1, 2, \ldots ; \; e(0, 0) = 0$$

we have

$$\sum_{n} \sum_{x+y=n} e(x, y) N(x, y) \, p^x (1-p)^y = g(p).$$

Hence we established the following.

Lemma 2.2 *Suppose that there exists an asymptotically unbiased estimator of $g(p)$. Then $g(p)$ is unbiasedly estimable provided that*

$$\sum_{n} \sum_{x+y=n} |e(x, y)| N(x, y) \, p^x (1-p)^y < \infty. \tag{2.6}$$

A sufficient condition for (2.6) is that there exists a sequence $\{c_n\}$ of positive numbers with $\sum_n c_n < \infty$ such that

$$\sum_{x+y=n} |e(x, y)| N(x, y) \, p^x (1-p)^y = \sum_{x+y=n} |G_n(x, y)| \,_n C_x \, p^x (1-p)^y \le c_n \tag{2.7}$$

for all p.

3. The Existence of a Discontinuous Unbiasedly Estimable Function and its Related Results

In this section, under the setup of the previous one, we can show that a discontinuous unbiasedly estimable function exists, and also get its related results.

Unbiased Estimation in Sequential Binomial Sampling

Example 3.1   We consider the case when

$$
g\,(p) = \begin{cases} 1 & \text{for } p > 1/2, \\ 0 & \text{for } p = 1/2, \\ -1 & \text{for } p < 1/2. \end{cases}
$$

Then we may define

$$
\hat{g}_n\,(x,\,y) = \begin{cases} (1/a_n)\,(x-y) & \text{for } |x - y| \le a_n, \\ 1 & \text{for } x - y > a_n, \\ -1 & \text{for } x - y < -a_n, \end{cases} \tag{3.1}
$$

where $\{a_n\}$ is an increasing sequence of positive numbers in n such that $\lim_{n \to \infty} (a_n/n^{\gamma}) = c\,(> 0)$ for $1/2 < \gamma < 1$. We define

$$
g_n(p) = E_p\left[\hat{g}_n\,(X,\,Y)\right] = \sum_{x+y=n} \hat{g}_n(x,\,y)\,{}_nC_x\,p^x\,(1-p)^y.
$$

From (3.1) we obtain

$$
\lim_{n \to \infty} g_n\,(p) = \begin{cases} 1 & \text{for } p > 1/2, \\ 0 & \text{for } p = 1/2, \\ -1 & \text{for } p < 1/2. \end{cases}
$$

Indeed, we have, in view of the order of weak convergence of X/n to p,

$$
\hat{g}_n\,(X,\,Y) \xrightarrow{P} 1 \ \text{ as } n \to \infty \qquad \text{for } p > 1/2,
$$

$$
\hat{g}_n\,(X,\,Y) \xrightarrow{P} -1 \ \text{ as } n \to \infty \qquad \text{for } p < 1/2,
$$

where $\xrightarrow{P}$ means the convergence in probability. Since $\hat{g}_n\,(x,\,y)$ is bounded, we obtain

$$
g_n\,(p) = E_p\left[\hat{g}_n(X,\,Y)\right] \to 1 \qquad \text{as } n \to \infty \ \text{ for } p > 1/2,
$$

$$
g_n\,(p) = E_p\left[\hat{g}_n(X,\,Y)\right] \to -1 \qquad \text{as } n \to \infty \ \text{ for } p < 1/2.
$$

And when $p = 1/2$, we have

$$
g_n\,(p) = E_p\left[\hat{g}_n(X,\,Y)\right] \to 0 \qquad \text{as } n \to \infty,
$$

M. Akahira, K. Takeuchi and K. Koike

since $X - Y$ is symmetrically distributed around zero for $p = 1/2$ and g_n is an odd function

of $x - y$. Putting $h_n(p) = g_n(p) - g_{n-1}(p)$ for n = 1, 2, ..., we have $\sum_{n=1}^{\infty} h_n(p) = g(p)$. Setting $e(0,0) = 0$ and, for $x + y = n = 1, 2, ...,$

$$e(x, y) = \left\{ \hat{g}_n(x, y) - \frac{x}{n} \hat{g}_{n-1}(x-1, y) - \frac{y}{n} \hat{g}_{n-1}(x, y-1) \right\} / N(x, y),$$

where $\hat{g}_0(0,0) = \hat{g}_0(-1,1) = \hat{g}_0(1,-1) = 0$, we have by Lemma 2.1

$$h_n(p) = \sum_{x+y=n} e(x, y) N(x, y) p^x (1-p)^y$$

Hence it is enough to prove that

$$\sum_n \sum_{x+y=n} |e(x, y)| N(x, y) p^x (1-p)^y < \infty.$$

In order to do so we shall show that

$$\sum_{x+y=n} |e(x, y)| N(x, y) p^x (1-p)^y = O(n^{-\gamma-(1/2)})$$

from which the above follows immediately. First we have for sufficiently large n

$$\sum_{x+y=n} \left| \hat{g}_n(x, y) - \frac{x}{n} \hat{g}_{n-1}(x-1, y) - \frac{y}{n} \hat{g}_{n-1}(x, y-1) \right| {}_nC_x \, p^x (1-p)^y$$

$$\leq \begin{cases} \left(\sum_{\substack{|x-y| \leq a_{n-1}-1 \\ x+y=n}} + \sum_{\substack{a_{n-1}-1 < |x-y| \leq a_{n-1}+1 \\ x+y=n}} + \sum_{\substack{a_{n-1}+1 < |x-y| \leq a_n \\ x+y=n}} \right) \\ \qquad \times |\ldots|_n C_x \, p^x (1-p)^y \quad \text{if } a_{n-1}+1 < a_n, \\[2em] \left(\sum_{\substack{|x-y| \leq a_{n-1}-1 \\ x+y=n}} + \sum_{\substack{a_{n-1}-1 < |x-y| \leq a_n \\ x+y=n}} + \sum_{\substack{a_n < |x-y| \leq a_{n-1}+1 \\ x+y=n}} \right) \\ \qquad \times |\ldots|_n C_x \, p^x (1-p)^y \quad \text{if } a_n < a_{n-1}+1, \end{cases}$$

$$= \begin{cases} I + I_2 + I_3 \quad (\text{say}) \quad \text{if } a_{n-1}+1 < a_n, \\ I + I'_2 + I'_3 \quad (\text{say}) \quad \text{if } a_n < a_{n-1}+1, \end{cases}$$

(3.2)

Unbiased Estimation in Sequential Binomial Sampling

where $|\cdots|$ denotes the same one as that in the first term of (3.2). Second we have

$$I \leq \left(\left| \frac{1}{a_n} - \frac{1}{a_{n-1}} \right| + \frac{1}{na_{n-1}} \right) \sum_{\substack{|x-y| \leq a_{n-1}-1 \\ x+y=n}} |x-y| \, _nC_x \, p^x (1-p)^y$$

$$\leq \begin{cases} \left(\left| \frac{1}{a_n} - \frac{1}{a_{n-1}} \right| + \frac{1}{na_{n-1}} \right) E_p[|X-Y|] & \text{for } p=1/2, \\[2em] \left(\left| \frac{1}{a_n} - \frac{1}{a_{n-1}} \right| + \frac{1}{na_{n-1}} \right) (a_{n-1}-1) \sum_{\substack{|x-y| \leq a_{n-1}-1 \\ x+y=n}} {}_nC_x \, p^x (1-p)^y & \text{for } p \neq 1/2, \end{cases}$$

$$(3.3)$$

From the condition of a_n we have,

$$a_n^{-1} - a_{n-1}^{-1} \sim c \, (n^{-\gamma} - (n-1)^{-\gamma}) = O \, (n^{-\gamma-1}) \, ,$$

$$n^{-1} a_{n-1}^{-1} = O \, (n^{-\gamma-1}).$$

$$(3.4)$$

It is also seen that, in the terms I_i and I'_i $(i = 2,3)$,

$$\left| \hat{g}_n \, (x, y) - \frac{x}{n} \hat{g}_{n-1} \, (x-1, y) - \frac{y}{n} \hat{g}_{n-1} \, (x, y-1) \right| = O(n^{-\gamma}). \qquad (3.5)$$

We consider the case when $p = 1/2$. Since $E_p \, [|X - Y|] = O(n^{1/2})$ for $p = 1/2$, it follows from (3.3) and (3.4) that, for $p = 1/2$,

$$I = O(n^{-\gamma-(1/2)}).$$

Since, from (3.5),

$$\sum_{\substack{a_{n-1}-1 < |x-y| \leq a_{n-1}+1 \\ x+y=n}} |\cdots| \, _nC_x \, p^x (1-p)^y$$

$$\leq \sum_{\substack{a_{n-1}-1 < |x-y| \\ x+y=n}} |\cdots| \, _nC_x \, p^x (1-p)^y = O \left(n^{-\gamma} \exp \left\{ - c' \, n^{-1} (a_{n-1}-1)^2 \right\} \right)$$

for $a_n < a_{n-1} + 1$ and $p = 1/2$, it follows that

$$I'_2 + I'_3 = o(n^{-\gamma-(1/2)}),$$

where $|\cdots|$ means the left-hand side of (3.5) and c' is some positive constant. In a similar way to the above we have $I_2 + I_3 = o(n^{-\gamma-(1/2)})$ for $a_{n-1} + 1 < a_n$ and $p = 1/2$.

Next we consider the case when $p \neq 1/2$. Since $a_{n-1} = cn^{\gamma} + o(n^{\gamma})$ for $1/2 < \gamma < 1$, we have for $p < 1/2$

M. AKAHIRA, K. TAKEUCHI and K. KOIKE

$$\sum_{\substack{|x-y| \le a_{n-1}-1 \\ x+y=n}} {}_nC_x\, p^x\,(1-p)^y$$

$$= P_p\left\{ |2X-n| \le a_{n-1}-1 \right\}$$

$$= P_p\left\{ \frac{-a_{n-1}+1-n\,(2p-1)}{2\sqrt{np\,(1-p)}} \le \frac{X-n}{\sqrt{np\,(1-p)}} \le \frac{a_{n-1}-1-n\,(2p-1)}{2\sqrt{np\,(1-p)}} \right\}$$

$$= O\left(n^{\gamma-(1/2)}\exp\left(-\frac{(2p-1)^2 n}{8p\,(1-p)}\right) \right)$$

$$= o\left(n^{-\gamma+(1/2)} \right),$$

which implies that, for $p < 1/2$,

$$I \le \left(\left| \frac{1}{a_n} - \frac{1}{a_{n-1}} \right| + \frac{1}{na_{n-1}} \right)(a_{n-1}-1) \sum_{\substack{|x-y| \le a_{n-1}-1 \\ x+y=n}} {}_nC_x\, p^x\,(1-p)^y$$

$$= o\left(n^{-\gamma-(1/2)} \right).$$

Hence we have for $p < 1/2$

$$I = o\left(n^{-\gamma-(1/2)} \right).$$

In a similar way to the case $p < 1/2$, we also obtain for $p > 1/2$

$$I = o\left(n^{-\gamma-(1/2)} \right).$$

Since, from (3.5),

$$\sum_{\substack{a_{n-1}-1<|x-y| \le a_{n-1}+1 \\ x+y=n}} |\cdots|\, {}_nC_x\, p^x\,(1-p)^y$$

$$\le \sum_{\substack{|x-y| \le a_{n-1}+1 \\ x+y=n}} |\cdots|\, {}_nC_x\, p^x\,(1-p)^y = o\left(n^{-\gamma-(1/2)} \right)$$

for $a_n < a_{n-1}+1$ and $p \ne 1/2$, it follows that $I'_2 + I'_3 = o(n^{-\gamma-(1/2)})$, where $|\cdots|$ means the left-hand side of (3.5). In a similar way to the above, we have $I_2 + I_3 = o(n^{-\gamma-(1/2)})$ for $a_{n-1}+1 < a_n$ and $p \ne 1/2$. Hence we have

$$I + I_2 + I_3 = O(n^{-\gamma-(1/2)}), \quad I + I'_2 + I'_3 = O\left(n^{-\gamma-(1/2)} \right) \quad \text{for } p = 1/2,$$

$$I + I_2 + I_3 = o(n^{-\gamma-(1/2)}), \quad I + I'_2 + I'_3 = o(n^{-\gamma-(1/2)}) \qquad \text{for } p \ne 1/2.$$

Unbiased Estimation in Sequential Binomial Sampling

Therefore we obtain

$$\sum_{n} \sum_{x+y=n} \left| \hat{g}_n(x, y) - \frac{x}{n} \hat{g}_{n-1}(x-1, y) - \frac{y}{n} \hat{g}_{n-1}(x, y-1) \right| {}_nC_x \, p^x (1-p)^y < \infty,$$

which implies that $g(p)$ is discontinuous in p, but, by Lemma 2.2, it is unbiasedly estimable.

Example 3.2. Suppose that $g(p)$ is continuous in p on the closed interval $[0, 1]$ and that there is a constant $c > 0$ and $\delta > 0$ such that

$$\left| g(p) - \lambda \, g(p - (1-\lambda)\, \Delta) - (1-\lambda)\, g(p + \lambda\Delta) \right| \le K\lambda(1-\lambda)\, \Delta^{1+\delta}$$

for all $p \in [0, 1]$ and a sufficiently small $\Delta > 0$, where $0 < \lambda < 1$ and K is some positive constant. Then we may define

$$\hat{g}_n(x, y) = g\left(\frac{x}{x+y}\right) \qquad \text{for } x + y = n = 1, 2, \ldots,$$

and $\hat{g}_0(0, 0) = \hat{g}_{n-1}(-1, n) = \hat{g}_{n-1}(n, -1) = 0$ for $n = 1, 2, \ldots$ Obviously, (2.1) holds. From the condition and (2.3) we have for $x + y = n \ge 2$,

$$G_n(x, y) = g\left(\frac{x}{n}\right) - \frac{x}{n} g\left(\frac{x-1}{n-1}\right) - \frac{n-x}{n} g\left(\frac{x}{n-1}\right)$$

$$= g\left(\frac{x}{n}\right) - \frac{x}{n} g\left(\frac{x}{n} - \frac{1}{n-1}\left(\frac{n-x}{n}\right)\right) - \frac{n-x}{n} g\left(\frac{x}{n} + \frac{1}{n-1} \cdot \frac{x}{n}\right).$$

For a sufficiently large $n = x + y$, we have

$$|G_n(x, y)| \le K \, \frac{x\,(n-x)}{n^2} \left(\frac{1}{n-1}\right)^{1+\delta},$$

hence

$$\sum_{x+y=n} |G_n(x, y)| {}_nC_x \, p^x(1-p)^y \le K \left(\frac{1}{n-1}\right)^{1+\delta} \sum_{x=0}^{n} \frac{x\,(n-x)}{n^2} {}_nC_x \, p^x (1-p)^{n-x}$$

$$= \frac{Kp\,(1-p)}{n} \left(\frac{1}{n-1}\right)^{\delta}.$$

Since

$$\sum_{n=2}^{\infty} \frac{1}{n\,(n-1)^{\delta}} < \infty \qquad \text{for } \delta > 0,$$

it follows from (2.7) that $g(p)$ is unbiasedly estimable.

Now we have a sequence $\{g_\alpha(p)\}_{\alpha=1,2,\ldots}$ of functions of p on $[0, 1]$ for which an unbiased estimator $e_\alpha(x, y)$ of $g_\alpha(p)$ of the form

M. Akahira, K. Takeuchi and K. Koike

$$e_\alpha(x, y) = G_n^{\alpha}(x, y)\,{}_nC_x / N(x, y) \quad \text{for } x + y = n = 1, 2, \ldots; \quad e_\alpha(0, 0) = 0$$

is unbiased for each $\alpha = 1, 2, \ldots$, where, in a similar way to $G_n(x, y)$ of (2.3), for each α, $G_n^{\alpha}(x, y)$ are defined under the condition of the existence of an asymptotically unbiased estimator $\hat{g}_n^{\alpha}(X, Y)$ of $g\alpha(p)$. Suppose that $\lim_{\alpha \to \infty} g_\alpha(p) = g^*(p)$ and $\lim_{\alpha \to \infty} G_n^{\alpha}(x, y) = G_n^*(x, y)$.

Then we define

$$e^*(x, y) = G_n^*(x, y)\,{}_nC_x / N(x, y) \quad \text{for } x + y = n = 1, 2, \ldots; \quad e^*(0, 0) = 0.$$

If, for each n, there exists a function $H_n(x, y)$ for $x + y = n = 0, 1, 2, \ldots$ such that

$$\left| G_n^{\alpha}(x, y) \right| \le H_n(x, y) \quad \text{for } x + y = n = 0, 1, 2, \ldots. \text{ and } \alpha = 1, 2, \ldots,$$

$$\sum_{n=0}^{\infty} \sum_{x+y=n} H_n(x, y)\,{}_nC_x\, p^x (1 - p)^y < \infty,$$

then

$$\sum_{n=0}^{\infty} \sum_{x+y=n} e^*(x, y) N(x, y)\, p^x (1 - p)^y = \lim_{m \to \infty} \sum_{n=0}^{m} \sum_{x+y=n} G_n^*(x, y)\,{}_n C_x\, p^x (1 - p)^y$$

$$= \lim_{m \to \infty} \lim_{\alpha \to \infty} \sum_{n=0}^{m} \sum_{x+y=n} G_n^{\alpha}(x, y)\,{}_n C_x\, p^x (1 - p)^y$$

$$= \lim_{\alpha \to \infty} \sum_{n=0}^{\infty} \sum_{x+y=n} G_n^{\alpha}(x, y)\,{}_n C_x\, p^x (1 - p)^y$$

$$= \lim_{\alpha \to \infty} \sum_{n=0}^{\infty} \sum_{x+y=n} e_\alpha(x, y) N(x, y)\, p^x (1 - p)^y$$

$$= \lim_{\alpha \to \infty} g_\alpha(p)$$

$$= g(p),$$

hence $g(p)$ is also unbiasedly estimable.

Remark. From the above two examples and the above discussion, we can derive a fairly large class of unbiasedly estimable functions.

Still it seems to be difficult to have a clear-cut simple statement for necessary and sufficient conditions for a parameter to be unbiasedly estimable, since we need some condition or other to guarantee (2.6). Similar conditions must have been necessary for the main theorem of Singh (1964) giving a necessary and sufficient condition for unbiased-

Unbiased Estimation in Sequential Binomial Sampling

estimability with randomized sample size procedures. More precisely, let $g(\theta)$ be the limit of a sequence $\{p_n(\theta)\}$ of polynomials on Ω, and N be a stopping variable such that

$$P_\theta(N = n) = \frac{1}{2^n}, \qquad n = 1, 2, ..., \theta \in \Omega,$$

and then there is a sequence $Y_1, Y_2, ...$ such that

$$E_\theta(Y_n) = 2^n \{p_n(\theta) - p_{n-1}(\theta)\}, \quad n = 1, 2, ... , \theta \in \Omega,$$

where $p_0(\theta) \equiv 0$, and define $Y = Y_n$ if $N = n$. Then he obtains

$$E_\theta(Y) = \sum_{n=1}^{\infty} E(Y \mid N = n) P_\theta(N = n)$$

$$= \sum_{n=1}^{\infty} 2^n \{p_n(\theta) - p_{n-1}(\theta)\} \frac{1}{2^n}$$

$$= \lim_{n \to \infty} p_n(\theta) = g(\theta), \quad \theta \in \Omega.$$

But, in order that $E_\theta(Y)$ exists, we must have $E_\theta(|Y|) < \infty$. Since

$$E_\theta(|Y|) = \sum_{n=1}^{\infty} E(|Y| \mid N = n) P_\theta(N = n)$$

$$\geq \sum_{n=1}^{\infty} 2^n \left| p_n(\theta) - p_{n-1}(\theta) \right| \frac{1}{2^n}$$

$$= \sum_{n=1}^{\infty} \left| p_n(\theta) - p_{n-1}(\theta) \right|$$

a necessary (not sufficient) condition is that $\sum_{n=1}^{\infty} \left| p_n(\theta) - p_{n-1}(\theta) \right| < \infty$, which may not always be the case.

As an example, first define a function

$$f(x) = \begin{cases} -1 & \text{for} \ -1 < x < 0, \\ 0 & \text{for} \ x = 0, \pm 1, \\ 1 & \text{for} \ 0 < x < 1. \end{cases}$$

Then, it is easily shown that this function can be expanded into a Fourier series as

M. AKAHIRA, K. TAKEUCHI and K. KOIKE

$$f(x) = \frac{4}{\pi} \sum_{k=0}^{\infty} \frac{1}{2k+1} \sin (2k+1)\pi x, \tag{3.6}$$

where the convergence is pointwise or L^2-sense but not absolute nor uniform. Now, by transforming the variable x into θ by

$$\theta = \frac{1 + \sin \pi x}{2},$$

f is transformed into a discontinuous function $f^*(\theta)$ for $0 \le \theta \le 1$ by

$$f^*(\theta) = \begin{cases} -1 & \text{for} \quad 0 \le \theta < \frac{1}{2}, \\ 0 & \text{for} \quad \theta = \frac{1}{2}, \\ 1 & \text{for} \quad \frac{1}{2} < \theta \le 1. \end{cases}$$

Since $\sin(2k+1)\alpha$ can be expressed as a polynomial in terms of $\sin\alpha$ of degree $2k+1$, each term on the right-hand side of (3.6) can be expanded as

$$\frac{4}{\pi} \sum_{k=0}^{\infty} \frac{1}{2k+1} u_{2k+1}(\theta),$$

where $u_{2k+1}(\theta)$ is a polynomial of degree $2k+1$. Hence we may define

$$p_n(\theta) = \frac{4}{\pi} \sum_{j=1}^{n} \frac{1}{j} u_j(\theta),$$

where $u_j(\theta)$ is defined to be equal to 0 when j is even. Then, obviously,

$$\lim_{n \to \infty} p_n(\theta) = f^*(\theta) \qquad \text{for} \quad 0 < \theta < 1,$$

since

$$f^*(\theta) = \frac{4}{\pi} \sum_{k=0}^{\infty} \frac{1}{2k+1} u_{2k+1}(\theta),$$

but

$$\sum_{n=2}^{\infty} \left| p_n(\theta) - p_{n-1}(\theta) \right| = \frac{4}{\pi} \sum_{k=1}^{\infty} \frac{1}{2k+1} \left| u_{2k+1}(\theta) \right|$$

$$= \frac{4}{\pi} \sum_{k=1}^{\infty} \frac{1}{2k+1} \left| \sin (2k+1)\pi x \right|$$

which is not always finite (e.g. put $x = 1/2$).

Note that, for $0 \leq p \leq 1$, $f^*(p)$ is equal to $g(p)$ discussed in Example 3.1.

REFERENCES

[1] Bhandari, S. K. and A. Bose (1990). Existence of unbiased estimates in sequential binomial experiments. *Sankhyā* 52, Ser. A, 127-130.

[2] Lehmann, E. L. (1983). *Theory of Point Estimation*. Wiley, New York.

[3] Singh, R. (1964). Existence of unbiased estimates. *Sankhyā* 26, Ser. A, 93-96.

[4] Zacks, S. (1971). *The Theory of Statistical Inference*. Wiley, New York.

KEI TAKEUCHI (*) - MASAFUMI AKAHIRA (**)

Interval estimation with varying confidence levels

CONTENTS: 1. Introduction. — 2. Reconsideration of Neyman's procedure. — 3. Conditional procedures. — 4. Example of estimated confidence interval. — 5. Consideration of the lengths and levels of intervals. — 6. Solution to Wald's formulation. References. Summary. Riassunto. Key words and phrases.

1. INTRODUCTION

Interval estimation procedure is usually formalized, if we adopt non-Bayesian standpoint, as a procedure with a fixed confidence level covering the true value of the parameter with probability of preassigned value. Such a procedure is mathematically obtained from a class of test procedures. Thus the confidence level is predetermined and fixed. But when we use interval estimation procedures in practical cases, such a procedure may not be quite appropriate in a practical sense, especially when the width of the preassigned confidence interval depends on the unknown parameter and may become impractically wide, hence we may want to have a narrower interval even at the cost of lower confidence level. In such a case the confidence level may depend on the unknown parameter value, provided that we can estimate the actual confidence level. We would like to look into the matter from a more general and flexible viewpoint and discuss

(*) Research Center for Advanced Science and Technology, University of Tokyo, 4-6-1 Komaba, Meguro-ku, Tokyo 156, Japan.

(**) Institute of Mathematics, University of Tsukuba, Tsukuba, Ibaraki 305, Japan.

4

possible generalization of the procedure with varying confidence levels.

2. RECONSIDERATION OF NEYMAN'S PROCEDURE

Most commonly used interval estimation procedure is that of confidence intervals as formalized by Neyman (1937) and advocated by the so-called "frequentists" as opposed to Bayesian and Fisherians.

The confidence interval is a random interval $[\underline{\theta}(X), \overline{\theta}(X)]$ based on the sample $\mathbf{X}$ with the property that

$$P_\omega \{\underline{\theta}(\mathbf{X}) \le \theta(\omega) \le \overline{\theta}(\mathbf{X})\} \ge 1 - \alpha \qquad \text{for all } \omega \in \Omega.$$

Neyman (1937) emphasized that there is no meaning in the expression

$$P_\omega \{\underline{\theta}(\mathbf{X}) \le \theta(\omega) \le \overline{\theta}(\mathbf{X}) \mid \mathbf{X} = \mathbf{x}\},$$

that is, the confidence level (or coefficient) should be clearly distinguished from the probability of the inequality being true given the sample, a main point of disagreement with the Bayesians and Fisherians.

He argues that, once $\mathbf{X} = \mathbf{x}$ is observed, the statement

$$\underline{\theta}(\mathbf{x}) \le \theta(\omega) \le \overline{\theta}(\mathbf{x})$$

would be either true or false, hence the probability of the above statement could take the value only 0 or 1. Even without accepting Bayesian viewpoint that we could talk about the probability of the above type of statement given $\mathbf{X} = \mathbf{x}$ based on the subjective concept of probability, we can still point out apparent "contradictions" which are brought about by confidence interval procedures in some cases. For example, in case of the interval estimation of the ratio of the two means of normal populations, the interval obtained could be sometimes the whole line, then it is obvious that the obtained interval contains the true value with probability 1. In other cases calculated interval could be reduced to the null set, hence the probability must be equal to 0, and the confidence coefficient may become meaningless.

5

Now, more precisely analyzing the nature of the problem, we may proceed as follows. Let us define

$$\chi_\omega \left(\underline{\theta}\left(\mathbf{X}\right), \overline{\theta}\left(\mathbf{X}\right) \right) = \begin{cases} 1 & \text{if } \underline{\theta}\left(\mathbf{X}\right) \le \theta\left(\omega\right) \le \overline{\theta}\left(\mathbf{X}\right), \\ 0 & \text{otherwise,} \end{cases}$$

which is a random variable, and once $\mathbf{X} = \mathbf{x}$ is given, the value of χ_ω would be determined to be equal to either 1 or 0, just as Neyman (1937) insisted. But, since χ_ω dependes both on $\mathbf{X}$ and the unknown parameter θ, we can not know the value of χ_ω. Then we would like to "estimate" the value of χ_ω. If $p = E_\omega \left[\chi_\omega \left(\underline{\theta}\left(\mathbf{X}\right), \overline{\theta}\left(\mathbf{X}\right) \right) \right]$ is known, we can regard p as the estimator of the value of χ_ω.

Such a procedure can be justified by the following argument. Now suppose that the unobserved random value χ_ω is known to take only two values 0 and 1 with known probability

$$p = P_\omega \{ \chi_\omega = 1 \} = 1 - P_\omega \{ \chi_\omega = 0 \}$$

and we have no further information on $\mathbf{X}$. In this case let x with $0 \le x \le 1$ be the estimator of χ_ω (or predictor ?). Now let the loss due to such an estimator be

$$\phi_0\left(x\right) \text{ if } \chi_\omega = 0 \text{ and } \phi_1\left(x\right) \text{ if } \chi_\omega = 1 .$$

Naturally we can assume that $\phi_0\left(0\right) = \phi_1\left(1\right) = 0$ and $\phi_0\left(x\right)$ and $\phi_1\left(x\right)$ are monotone increasing and decreasing functions of x, respectively. Then the expected loss is equal to

$$E_x = p\,\phi_1\left(x\right) + \left(1 - p\right)\phi_0\left(x\right)$$

and the optimum value for x is given from the equation

$$\frac{\partial}{\partial x} E_x = p\,\phi_1'\left(x\right) + \left(1 - p\right)\phi_0'\left(x\right) = 0 \tag{2.1}$$

or

$$\phi_0'\left(x\right) / p = -\,\phi_1'\left(x\right) / \left(1 - p\right) .$$

6

In the case three conditions (i) $\phi_1(x) = \phi_0(1 - x)$, (ii) ϕ_0 is convex, and (iii) $\phi_0'(x)/x = \phi_0'(1 - x)/(1 - x)$ holds for all $0 < x < 1$, we have $x = p$ as the unique solution of (2.1). Two of such examples are $\phi_0(x) = x^2$ and $\phi_0(x) = -\log(1 - x)$, which are two of the most natural candidates for ϕ_0.

Now in the case of confidence interval, we may estimate χ_ω $(\underline{\theta}(\mathbf{X}), \overline{\theta}(\mathbf{X})$ by a function of $\mathbf{X}$, which we shall denote as $\hat{\chi}(\mathbf{X})$. If $p = P_\omega\{\underline{\theta}(\mathbf{X}) \leq \theta(\omega) \leq \overline{\theta}(\mathbf{X})\} = E_\omega(\chi_\omega)$ is known, we may put $\hat{\chi}(\mathbf{X}) = p$, but not necessarily. We may ask what will be the "best" estimator $\hat{\chi}(\mathbf{X})$ of $\chi_\omega(\underline{\theta}(\mathbf{X}), \overline{\theta}(\mathbf{X}))$. Before discussing the construction of such procedures we would like to point out that $\hat{\chi}$ can be regarded as the estimated posterior confidence coefficient based on the sample and would allow for such possibility as: (*a*) When an ancillary statistic T exists, $\hat{\chi}$ would be the estimator of the conditional probability $P_\omega\{\underline{\theta}(\mathbf{X}) \leq \theta(\omega) \leq \overline{\theta}(\mathbf{X})|T\}$. (*b*) If $P_\omega\{\underline{\theta}(\mathbf{X}) \leq \theta(\omega) \leq \overline{\theta}(\mathbf{X})\}$ is not constantly equal to $1 - \alpha$, we may estimate it based on the sample, such a procedure may be useful when we can not obtain a similar confidence interval.

3. Conditional procedures

Let us suppose that there exists a confidence procedure with the exact level $1 - \alpha$, and also a statistic T such that, given $T = t$, the conditional probability of the confidence interval including the true value $\theta(\omega)$ depends on T but independent of ω. Then let χ_ω be the function defined as above from the confidence procedure and let

$$\pi(T) = E_\omega[\chi_\omega|T] = P_\omega\{\underline{\theta}(\mathbf{X}) \leq \theta(\omega) \leq \overline{\theta}(\mathbf{X})|T\}.$$

Then for both of the choice $\phi_0(x) = x^2$ and $\phi_0(x) = -\log(1-x)$, it can be easily shown that $\pi(T)$ is a better estimator of χ_ω than the constant $\pi_0 \equiv 1 - \alpha$.

As a special case, suppose that there exists a transformation group G operating on the sample space and the parameter space with the property that

$$P_{g\omega}(\mathbf{X}) = P_\omega(g\mathbf{X}) \qquad \text{for all } \omega \text{ and all } g$$

and assume that $\theta(\omega) = \omega$. Then a confidence set $S(\mathbf{X})$ is said to be invariant with respect to G if $S(g\,\mathbf{X}) = g\,S(\mathbf{X})$. If we assume that G is transitive on Ω, i.e. for any ω_1, $\omega_2 \in \Omega$, there is a $g \in G$ such that $g\,\omega_1 = \omega_2$. Then $P_\omega\{S(\mathbf{X}) \ni \omega\}$ is independent of ω, since we can fix ω_0 and find $g \in G$ such that $g\,\omega_0 = \omega$, and we have

$$P_\omega\{S(\mathbf{X}) \ni \omega\} = P_{g\omega_0}\{S(\mathbf{X}) \ni g\,\omega_0\} = P_{\omega_0}\{S(g\,\mathbf{X}) \ni g\,\omega_0\}$$
$$= P_{\omega_0}\{g\,S(\mathbf{X}) \ni g\,\omega_0\} = P_{\omega_0}\{S(\mathbf{X}) \ni \omega_0\}\,.$$

Now let T be the maximal invariant statistic, that is, $T(g\,\mathbf{X}) = T(\mathbf{X})$ for all g. Then, under the above assumption, it follows that $\mathbf{X}$ can be one-to-one transformed into $(g(\mathbf{X}), T(\mathbf{X}))$ when $g(\mathbf{X})$ is an element of G, hence we may equate $\mathbf{X} = (\tilde{g}, T)$, where $\tilde{g} \in G$. Then, for an invariant confidence interval, we have

$$S(\mathbf{X}) = S(\tilde{g}, T) = \tilde{g}\,S(e, T)\,,$$

where e is the unit element of G. So, given T, we have

$$P_\omega\{S(\mathbf{X}) \ni \omega | T\} = P_{\omega_0}\{S(g\,\mathbf{X}) \ni g\,\omega_0 | T\} = P_{\omega_0}\{S(g\tilde{g}, T) \ni g\,\omega_0 | T\}$$
$$= P_{\omega_0}\{g\tilde{g}S(e, T) \ni g\,\omega_0 | T\} = P_{\omega_0}\{\tilde{g}S(e, T) \ni \omega_0 | T\}\,,$$

which is independent of g, hence of ω. Therefore we may use $P_{\omega_0}\{S(\mathbf{X}) \ni \omega_0 | T\}$ as the estimator of χ_ω derived from S.

In the simplest case of location parameter problem, let $\mathbf{X} = (X_1, \ldots, X_n)$ and a location invariant estimator $[\underline{\theta}, \overline{\theta}]$ of the parameter θ as $[\underline{\theta}(\mathbf{X}), \overline{\theta}(\mathbf{X})]$ which satisfies the condition that

$$\underline{\theta}(X_1 + a, \ldots, X_n + a) = \underline{\theta}(X_1, \ldots, X_n) + a\,,$$

$$\overline{\theta}(X_1 + a, \ldots, X_n + a) = \overline{\theta}(X_1, \ldots, X_n) + a\,.$$

Then the conditional level of $[\underline{\theta}, \overline{\theta}]$, given $T = (X_2 - X_1, \ldots, X_n - X_1)$, is computed in the following manner. Let

$$\underline{\theta}(X_1, \ldots, X_n) = X_1 + \underline{\theta}(0, X_2 - X_1, \ldots, X_n - X_1) = X_1 + \underline{\theta}_0(T)\,,$$

$$\overline{\theta}(X_1, \ldots, X_n) = X_1 + \overline{\theta}(0, X_2 - X_1, \ldots, X_n - X_1) = X_1 + \overline{\theta}_0(T)\,.$$

8

Then it can be shown that

$$P_\theta \{\underline{\theta}(\mathbf{X}) \le \theta \le \overline{\theta}(\mathbf{X})\} = P_0 \{X_1 + \underline{\theta}_0(T) \le 0 \le X_1 + \overline{\theta}_0(T) | T\}$$

$$= \int_{\underline{\theta}(\mathbf{X})}^{\overline{\theta}(\mathbf{X})} \prod_{i=1}^{n} f(X_i - u)\, du \bigg/ \int_{-\infty}^{\infty} \prod_{i=1}^{n} f(X_i - u)\, du \;.$$

Note that the above is equal to the fiducial probability of Fisher (1930) (see also Fraser (1961)).

4. Example of estimated confidence interval

Another case where "estimated confidence level" may be of value is that of discrete distribution. When the value $P\{\ \}$ can be well below the nominal level unless randomization is allowed.

Now suppose that the sample X can take only integer values and assume that

$$p_\theta(x) = P_\theta\{X = x\}\,, \qquad x = 0, 1, 2, \ldots,$$

where θ is a real parameter. We also assume that $p_\theta(x+1)/p_\theta(x)$ is monotone increasing in θ for all x. Then "good" confidence interval is given as

$$\underline{\theta}(x) \le \theta \le \overline{\theta}(x)$$

where $\overline{\theta}(x)$ and $\underline{\theta}(x)$ are monotone increasing functions with the condition that

$$P_\theta\{\underline{\theta}(X) \le \theta \le \overline{\theta}(X)\} \ge 1 - \alpha$$

or

$$P_\theta\{X > \underline{\theta}^{-1}(\theta)\} + P_\theta\{X > \overline{\theta}^{-1}(\theta)\} < \alpha$$

for all θ. For many practical cases the sum of the two terms in the above can well below α, thus giving unnecessarily conservative interval. In general, it is difficult to remedy the situation without voilating the condition for some value of θ, but we may adjust the conclusion by giving an estimator of the posterior confidence level given $X = x$.

Suppose that X is distributed according to the binomial distribution $B(n, \theta)$, $0 < \theta < 1$, and that an interval estimator of θ is required. Now suppose that a confidence interval estimation procedure $\underline{\theta}(X) \leq \theta \leq \bar{\theta}(X)$ is given. Then we may calculate

$$E_\theta[\underline{\theta}(X) \leq \theta \leq \bar{\theta}(X)] = \sum_{x=0}^{n} {}_nC_x \, \theta^x (1-\theta)^{n-x} \, \phi(x, \theta) = p(\theta) \quad \text{(say)}$$

for $0 < \theta < 1$. And we can construct an estimator $\hat{p} = \hat{p}(x)$ of $p(\theta)$ and may use it as an estimated confidence level. Since the function $p(\theta)$ is discontinuous (at the points where $\theta = \underline{\theta}(x)$ or $\theta(x)$ for some x) no unbiased estimator of $p(\theta)$ exists. We want to make

$$E_\theta[\{\hat{p} - p(\theta)\}^2] = \sum_{x=0}^{n} {}_nC_x \, (\hat{p}(x) - p(\theta))^2 \, \theta^x (1-\theta)^{n-x} = R(\theta) \quad \text{(say)}$$

as small as possible. Since $R(\theta)$ above depends on the unknown θ, $R(\theta)$ can not be uniformly minimized and we may resort to "Bayesian" approach and make

$$\int_0^1 R(\theta) \, \omega(\theta) \, d\theta = \text{minimum}$$

with some weight function $\omega(\theta)$ (or "prior" distribution if you like). Then the solution is given by

$$\hat{p}(x) = \int_0^1 \omega(\theta) \, p(\theta) \, \theta^x (1-\theta)^{n-x} \, d\theta \left/ \int_0^1 \omega(\theta) \, \theta^x (1-\theta)^{n-x} \, d\theta \right.$$

$$= \frac{1}{C_\omega(x)} \int_0^1 \omega(\theta) \, p(\theta) \, \theta^x (1-\theta)^{n-x} \, d\theta \quad \text{(say)}.$$

10

Recalling the definition of $p(\theta)$ we have

$$\hat{p}(x) = \frac{1}{C_\omega(x)} \sum_{y=0}^{n} {}_nC_y \int_0^1 \omega(\theta)\, \theta^{y+x}(1-\theta)^{2n-x-y}\, \phi(y,\theta)\, d\theta$$

$$= \frac{1}{C_\omega(x)} \sum_{y=0}^{n} {}_nC_y \int_{\underline{\theta}(y)}^{\overline{\theta}(y)} \omega(\theta)\, \theta^{y+x}(1-\theta)^{2n-x-y}\, d\theta\,.$$

Here $\omega(\theta)$ may be chosen arbitrarily but a natural choice will be the "beta function"

$$\omega(\theta) = \text{const. } \theta^{p-1}(1-\theta)^{q-1} \qquad \text{for } p > 0 \text{ and } q > 0\,.$$

Then $\hat{p}(x)$ is expressed in terms of incomplete beta function.

Usually (non-randomized) confidence interval is defined so that we have

$$p(\theta) \geq 1 - \alpha \qquad \text{for all } \theta\,,$$

but for small n the difference $p(\theta) - (1 - \alpha)$ is sometimes quite large. Then $\hat{p}(x)$ defined as above (which is the "posterior" mean of $p(\theta)$) becomes larger than $1 - \alpha$, and would give more relevant information of the accuracy of the confidence interval obtained. We may also define $\underline{\theta}(X)$ and $\overline{\theta}(X)$ so that the estimated confidence level $\hat{p}(x)$ is equal to $1 - \alpha$ for $x = 0, 1, ..., n$. Since there are $2(n+1)$ values of $\underline{\theta}(X)$ and $\overline{\theta}(X)$ to be determined while the condition $\hat{p}(x) = 1 - \alpha$ for $x = 0, 1, ..., n$, gives only $n+1$ equations to be satisfied, there will still be $n+1$ more conditions possibly to be imposed from other considerations.

5. CONSIDERATION OF THE LENGTHS AND LEVELS OF INTERVALS

We may consider a few types of confidence intervals with estimated confidence levels. One is the confidence interval with constant estimated levels. That is, we shall determine an interval

$[\underline{\theta}(X), \overline{\theta}(X)]$ with the property that the estimator $\hat{p}$ of $E_\omega[\chi_\omega(\underline{\theta}(X), \overline{\theta}(X))]$ is constant and equal to $1 - \alpha$, which means that $E_\omega[\chi_\omega(\underline{\theta}(X), \overline{\theta}(X))] = 1 - \alpha$ for all ω and that $\hat{p} = 1 - \alpha$ is a good estimator of the random variable χ_ω. In case of invariant estimation with transitive transformation group G, we may so choose $\underline{\theta}$ and $\overline{\theta}$ as to satisfy the condition that

$$P_{\omega_0}\{\underline{\theta}(X) \leq \theta(\omega_0) \leq \overline{\theta}(X)|T\} = 1 - \alpha,$$

in case of location parameter, $\underline{\theta}$ and $\overline{\theta}$ must satisfy the condition

$$\frac{\displaystyle\int_{\underline{\theta}}^{\overline{\theta}} \prod_{i=1}^{n} f(X_i - \theta)\, d\theta}{\displaystyle\int_{-\infty}^{\infty} \prod_{i=1}^{n} f(X_i - \theta)\, d\theta} = 1 - \alpha.$$

Another case is an interval with constant length, and estimated confidence levels.

As an example suppose that $X_1, ..., X_n$ are independently, identically and normally distributed with mean θ and unknown variance σ^2. Then it is intuitively clear (and may be rigorously proved) that the best invariant interval with fixed length l (irrespective of σ) is given by $[\overline{X} - l, \overline{X} + l]$, where $\overline{X} = \Sigma_{i=1}^{n} X_i / n$. Then

$$P\{\overline{X} - l \leq \theta \leq \overline{X} + l\} = 2\Phi(\sqrt{n}l / \sigma) - 1,$$

where

$$\Phi(u) = \int_{-\infty}^{u} (1 / \sqrt{2\pi})\, e^{-x^2/2}\, dx.$$

An unbiased estimator of $p = 2\Phi(\sqrt{n}l / \sigma) - 1$ is given by

$$\hat{p} = \begin{cases} 1 & \text{if } |X_1 - X_2| \leq \sqrt{2n}l, \\ 0 & \text{otherwise.} \end{cases}$$

12

By applying the Rao-Blackwell theorem we get the uniformly minimum variance unbiased estimator of p by

$$\hat{p}^* = E\left[\hat{p} \,\Big|\, \sum_{i=1}^{n} (X_i - \bar{X})^2\right] = c \int_{-a}^{a} (1 - r^2)^{(n-4)/2}\, dr \,,$$

where $a = \min\left(1, \sqrt{nl} \,/\, \sqrt{\Sigma_{i=1}^{n} (X_i - \bar{X})^2}\right)$.

As a second example, suppose that $X_1, \ldots, X_n$ are independently, identically and uniformly distributed on the interval $[\theta - (1/2), \theta + (1/2)]$. Then a sufficient statistic is given by the pair $(\min_{1 \leq i \leq n} X_i, \max_{1 \leq i \leq n} X_i)$ and a natural estimator is given by $\hat{\theta} = (\min_i X_i + \max_i X_i)/2$, and we may consider the interval of the type $[\hat{\theta} - l, \hat{\theta} + l]$. We shall denote $R = \max_i X_i - \min_i X_i$, then the distribution of R is independent of θ, hence it is an ancillary statistic. Given R, $\hat{\theta}$ is distributed according to the uniform distribution centered at θ and length equal to $1 - R$. Then

$$P\{\hat{\theta} - l \leq \theta \leq \hat{\theta} + l | R\} = \min\{1, 2l/(1-R)\}\,.$$

Therefore, for the fixed length interval, estimated confidence level is given by the above formula. And if we put $l = (1-\alpha)(1-R)/2$, we have a confidence interval with constant estimated level.

Now we shall consider the problem of determining the length of the interval simultaneously. In case of location parameter with known scale, we may first fix l and choose $\hat{\theta}_0$ and define an interval $[\hat{\theta}_0 - l, \hat{\theta}_0 + l]$ so that

$$\hat{p} = \int_{\hat{\theta}_0 - l}^{\hat{\theta}_0 + l} \prod_{i=1}^{n} f(X_i - \theta)\, d\theta \int_{-\infty}^{\infty} \prod_{i=1}^{n} f(X_i - \theta)\, d\theta$$

is maximized. More generally we may determine l so as to balance the probability of inclusion and the length of the interval. A simple approach is to define a loss function by

$$f(p) + g(l)\,,$$

where $f(p)$ is decreasing in p, the probability of inclusion of the true parameter and $g(l)$ is increasing in l of the interval. An intuitively appealing choice will be

$$f(p) = - \log p \qquad \text{and} \qquad g(l) = cl,$$

where c is a positive constant. Generally p is unknown, hence we may replace $f(p)$ by its unbiased estimator $\hat{f}(p)$ and we shall choose l so that

$$\hat{f}(p) + g(l)$$

is minimized. It is often difficult to obtain an unbiased estimator of $f(p)$, therefore we may replace $\hat{f}(p)$ by $f(\hat{p})$, where $\hat{p}$ is an unbiased estimator of p. For the above choice, the optimum value of l is given by

$$- \frac{dp}{dl} \frac{1}{p} + c = 0 \qquad \text{or} \qquad \frac{dp}{dl} = cp,$$

where c is some constant. In case of the normal distribution with known scale, we have

$$p = 2 \Phi (\sqrt{nl} / \sigma) - 1.$$

Hence $dp / dl = 2 (\sqrt{n} / \sigma) \phi (\sqrt{nl} / \sigma)$, and we obtain the optimum value of l from the equation

$$2 \frac{\sqrt{n}}{\sigma} \phi \left(\frac{\sqrt{nl}}{\sigma} \right) = 2\Phi \left(\frac{\sqrt{nl}}{\sigma} \right) - 1$$

or

$$\frac{\phi (u)}{2\Phi (u) - 1} = \frac{\sigma}{2 \sqrt{n}} \qquad\qquad (5.1)$$

14

and $l = \sigma u / \sqrt{n}$. Since the left-hand side of (5.1) is monotone decreasing in $u > 0$, and approaches to ∞ as $u \to 0$ and to 0 as $u \to \infty$, we always have a unique solution of (5.1) in $u > 0$. In case of location parameter with unknown scale, consideration of invariance leads the posterior estimator of the confidence level of the interval $[\hat{\theta}_0 - l, \hat{\theta}_0 + l]$ by

$$
\hat{p} = \frac{\displaystyle\int_0^{\infty} \int_{\hat{\theta}_0 - l}^{\hat{\theta}_0 + l} \frac{1}{\tau^{n+1}} \prod_{i=1}^{n} f\left(\frac{X_i - \theta}{\tau}\right) d\theta \, d\tau}{\displaystyle\int_0^{\infty} \int_{-\infty}^{\infty} \frac{1}{\tau^{n+1}} \prod_{i=1}^{n} f\left(\frac{X_i - \theta}{\tau}\right) d\theta \, d\tau} \ .
$$

In case of the normal distribution we have

$$
\hat{p} = \frac{\displaystyle\int_{\sqrt{n}\,(\hat{\theta}_0 - \bar{X} - l)/Z}^{\sqrt{n}\,(\hat{\theta}_0 - \bar{X} + l)/Z} g(w) \, dw}{\displaystyle\int_{-\infty}^{\infty} g(w) \, dw} \ ,
$$

where

$$
g(w) = C_n / (1 + w^2)^n \quad \text{for} \ -\infty < w < \infty \ ,
$$

with some constant C_n, and $Z = \sqrt{\Sigma_{i=1}^{n} (X_i - \bar{X})^2}$. We also have

$$
\frac{d\hat{p}}{dl} = \frac{\sqrt{n}}{Z} \left\{ g\left(\frac{\sqrt{n}\,(\hat{\theta}_0 - \bar{X} + l)}{Z}\right) + g\left(\frac{\sqrt{n}\,(\hat{\theta}_0 - \bar{X} - l)}{Z}\right) \right\}
$$

6. SOLUTION TO WALD'S FORMULATION

Another approach to the choice of levels and lengths is to start from the loss functions defined by Wald (1971) and others. Suppose that an interval $[\underline{\theta}, \overline{\theta}]$ is given. Then the loss from the statement that θ lies in between $\underline{\theta}$ and $\overline{\theta}$ may have the form

$$L(\underline{\theta}, \overline{\theta}, \theta) = \begin{cases} c(\underline{\theta}, \overline{\theta}) & \text{if } \underline{\theta} \le \theta \le \overline{\theta}, \\ \overline{d}(\underline{\theta}, \overline{\theta}) & \text{if } \theta > \overline{\theta}, \\ \underline{d}(\underline{\theta}, \overline{\theta}) & \text{if } \underline{\theta} < \theta. \end{cases}$$

A simple and plausible choice will be

$$c(\underline{\theta}, \overline{\theta}) = \overline{\theta} - \underline{\theta}, \quad \overline{d}(\underline{\theta}, \overline{\theta}) = c(\theta - \overline{\theta})^2 + \overline{\theta} - \underline{\theta},$$

and

$$\underline{d}(\underline{\theta}, \overline{\theta}) = c(\underline{\theta} - \theta)^2 + \overline{\theta} - \underline{\theta}$$

with a given constant c. Now our choice of $\underline{\theta}$ and $\overline{\theta}$ depends on the loss function $L(\underline{\theta}, \overline{\theta}, \theta)$ and we would like to choose $\underline{\theta}$ and $\overline{\theta}$ so that $E[L(\underline{\theta}, \overline{\theta}, \theta)]$ be minimized. We may use such $(\underline{\theta}, \overline{\theta})$ with the estimated risk $\hat{r}(\underline{\theta}, \overline{\theta})$ of $E[L(\underline{\theta}, \overline{\theta}, \theta)]$ and also with the estimated level $\hat{p}$.

Suppose that $X_1, ..., X_n$ are independently, identically and normally distributed with mean θ and known variance σ^2. Considering the interval of the type $[\overline{X} - l, \overline{X} + l]$ as $[\underline{\theta}, \overline{\theta}]$, we have

$$E[L(\underline{\theta}, \overline{\theta}, \theta)] = 2l + c \int_{\theta+l}^{\infty} (\overline{x} + l - \theta)^2 \frac{\sqrt{n}}{\sigma} \phi\left(\frac{\sqrt{n}(\overline{x} - \theta)}{\sigma}\right) d\overline{x}$$

$$+ c \int_{-\infty}^{\theta-l} (\theta - \overline{x} + l)^2 \frac{\sqrt{n}}{\sigma} \phi\left(\frac{\sqrt{n}(\overline{x} - \theta)}{\sigma}\right) d\overline{x}$$

$$= 2l + \frac{2c\sigma^2}{n} \int_{\sqrt{n}l/\sigma}^{\infty} \left(u - \frac{\sqrt{n}l}{\sigma}\right)^2 \phi(u) \, du$$

$$= 2l + \frac{2c\sigma^2}{n} \left(\frac{nl^2}{\sigma^2} + 1\right) \left\{1 - \Phi\left(\frac{\sqrt{n}l}{\sigma}\right)\right\} - \frac{2c\sigma l}{\sqrt{n}} \phi\left(\frac{\sqrt{n}l}{\sigma}\right).$$

16

The value of l minimizing $E[L(\underline{\theta}, \overline{\theta}, \theta)]$ is obtained from the equation

$$0 = 1 - \frac{2c\sigma}{\sqrt{n}} \int_{\sqrt{nl}/\sigma}^{\infty} \left(u - \frac{\sqrt{nl}}{\sigma}\right) \phi(u)\, du$$

$$= 1 - \frac{2c\sigma}{\sqrt{n}} \left\{ \phi\left(\frac{\sqrt{nl}}{\sigma}\right) - \frac{\sqrt{nl}}{\sigma}\left(1 - \Phi\left(\frac{\sqrt{nl}}{\sigma}\right)\right)\right\},$$

that is $l = \sigma u / \sqrt{u}$ and

$$\phi(u) - u\{1 - \Phi(u)\} = \sqrt{n}/(2c\sigma). \tag{6.1}$$

Since the left-hand side of (6.1) is not larger than $\phi(0) = 1/\sqrt{2\pi} = 0.3989$ and approaches to 0 as $u \to \infty$, we have a unique solution for l if $2c\sigma/\sqrt{n} > 1/0.3989 = 2.5069$, but if $2c\sigma/\sqrt{n} \leq 2.5069$, we should put $l = 0$.

In case of unknown scale σ, let us consider the standardized loss

$$L(\underline{\theta}, \overline{\theta}, \theta) = \begin{cases} \dfrac{\overline{\theta} - \underline{\theta}}{\sigma} & \text{if } \underline{\theta} \leq \theta \leq \overline{\theta}, \\[2ex] \dfrac{c(\theta - \overline{\theta})^2}{\sigma^2} + \dfrac{\overline{\theta} - \underline{\theta}}{\sigma} & \text{if } \theta > \overline{\theta}, \\[2ex] \dfrac{c(\theta - \underline{\theta})^2}{\sigma^2} + \dfrac{\overline{\theta} - \underline{\theta}}{\sigma} & \text{if } \theta < \underline{\theta}, \end{cases}$$

and the class of interval estimators of the type $[\underline{\theta}, \overline{\theta}]$ with $\underline{\theta} = \overline{X} - tS$ and $\overline{\theta} = \overline{X} + tS$, where c and t are constants, and $S = \sqrt{\Sigma_{i=1}^{n}(X_i - \overline{X})^2/n}$. Then we can show that

$$E[L(\underline{\theta}, \overline{\theta}, \theta)]$$

$$= \frac{2t}{\sigma} E(S) + \frac{2c}{n} E\left[\left(\frac{nt^2 S^2}{\sigma^2} + 1\right)\left\{1 - \Phi\left(\frac{\sqrt{n}tS}{\sigma}\right)\right\}\right.$$

$$\left. - \frac{2ctS}{\sqrt{n}\,\sigma} \phi\left(\frac{\sqrt{n}tS}{\sigma}\right)\right], \tag{6.2}$$

where the expectation is taken with respect to the distribution of S. Differentiating (6.2) with respect to t, we have

$$\frac{\partial}{\partial t} E[L(\underline{\theta}, \bar{\theta}, \theta)] = \frac{2}{\sigma} E(S) + \frac{2c}{n} E\left[\frac{2ntS^2}{\sigma^2}\left\{1 - \Phi\left(\frac{\sqrt{ntS}}{\sigma}\right)\right\} - \frac{2\sqrt{n}S}{\sigma} \phi\left(\frac{\sqrt{ntS}}{\sigma}\right)\right],$$

Hence, for optimum t, we obtain

$$E\left(\frac{S}{\sigma}\right) = \frac{2c}{n} E\left[\frac{\sqrt{n}S}{\sigma} \phi\left(\frac{\sqrt{ntS}}{\sigma}\right) - \frac{ntS^2}{\sigma^2}\left\{1 - \Phi\left(\frac{\sqrt{ntS}}{\sigma}\right)\right\}\right]. \quad (6.3)$$

Since S/σ is distributed independently of σ, both sides of (6.3) is independent of σ, hence t can be determined from the equation (6.3) dependent on c.

REFERENCES

FISHER, R.A. (1930). Inverse probability, *Proc. Camb. Phil. Soc.,* 26, 528-535.

FRASER, D.A.S. (1961). The fiducial method and invariance, *Biometrika* 48, 261-280.

NEYMAN, J. (1937). Outline of a theory of statistical estimation based on the classical theory of probability, *Phil. Trans. Roy. Soc.,* A236, 333-380.

WALD, A. (1971). *Statistical Decision Functions,* Chelsea Pub. Comp., Bronx, New York.

Interval estimation with varying confidence levels

SUMMARY

Usually confidence interval is defined as an interval with preassigned confidence level $1-\alpha$ for all the value of parameters. More generally, however we may consider interval estimation procedures with confidence coefficient varying according to the value of the unknown parameter, and associated procedure to estimate the actual level. Such a consideration leads to more general procedures including conditional procedures given the ancillary.

18

Stima per intervallo con livelli di confidenza variabili

RIASSUNTO

Un intervallo di confidenza è di solito definito come un intervallo con un prefissato livello di confidenza $1-\alpha$ per tutti i valori del parametro. Più in generale, però, si possono considerare metodi di stima per intervallo con un livello di confidenza variabile in funzione del valore del parametro incognito, a cui sono associate procedure per stimare l'effettivo livello di confidenza. Questa considerazione porta a procedimenti di carattere più generale, che includono il condizionamento rispetto a statistiche ancillari.

KEY WORDS AND PHRASES

Confidence interval; confidence level; ancillary statistic; invariance; loss function.

Stat. Sci. & Data Anal., pp. 375-382
K. Matsusita *et al.* (Eds)
© VSP 1993

Second Order Asymptotic Bound for the Variance of Estimators for the Double Exponential Distribution

MASAFUMI AKAHIRA and KEI TAKEUCHI

Institute of Mathematics, University of Tsukuba, Tsukuba, Ibaraki 305, Japan

Research Center for Advanced Science and Technology, University of Tokyo

4-6-1 Komaba, Meguro-ku, Tokyo 156, Japan

Abstract. The Bhattacharyya type bound for the variance of unbiased estimators of a location parameter of the double exponential distribution is obtained, and also the loss of information of the maximum likelihood estimator based on the distribution rounded off is discussed.

Key words: Bhattacharyya type bound, unbiased estimator, loss of information, maximum likelihood estimator.

1. INTRODUCTION

The double exponential distribution with unknown location parameter case is first order regular but second order non-regular, in the sense that the density admits the first order differentiability with respect to the parameter but not the second order. From that property it follows that the first order asymptotic theory of regular estimation can be applied but in the second order, the situation becomes non-regular. For example, R.A. Fisher, already in the paper [1], noted that the loss of information of the maximum likelihood estimator is of order $\sqrt{n}$ instead of constant order as in the regular case. In Akahira and Takeuchi [2], [3], Akahira [4], Sugiura and Naing [5] among others, the second order (or more precisely next to the first order) properties of the maximum likelihood estimator and other estimators were discussed. In this paper we shall discuss the problem from a different point of view, and shed lights on the situation. Indee !, we shall obtain the Bhattacharyya type bound for the variance of unbiased estimators of

M. Akahira and K. Takeuchi

a location parameter of the double exponential distribution and the loss of information of the maximum likelihood estimator based on the distribution rounded off.

2. THE BHATTACHARYYA TYPE BOUND FOR THE VARIANCE OF UNBIASED ESTIMATORS

Suppose that $X_1, \ldots, X_n$ are independent and identically distributed random variables with a double exponential (two-sided exponential) density

$$f(x - \theta) = (1/2) \exp(-|x - \theta|) \qquad (-\infty < x < \infty \; ; \; -\infty < \theta < \infty).$$

Then, for an unbiased estimator $\hat{\theta} = \hat{\theta}(X_1, \ldots, X_n)$ of the location parameter θ, we have the Cramér-Rao bound, i.e.

$$V_\theta(\hat{\theta}) \geq 1/nI(\theta) = 1/n,$$

where $V_\theta(\,\cdot\,)$ denotes the variance and

$$I(\theta) = E_\theta \left[\{(\partial/\partial\theta) \log f(X - \theta)\}^2 \right] = E_\theta \left[\{\mathrm{sgn}(X - \theta)\}^2 \right] = 1.$$

Since $f(x)$ is not twice differentiable, we can not further differentiate $\log f(x)$, hence the Bhattacharyya bound can not be obtained. However, we can obtain a similar bound as follows. Define

$$Z_1(\theta) = \frac{1}{\sqrt{n}} \sum_{i=1}^{n} \mathrm{sgn}(X_i - \theta).$$

Then, for an unbiased estimator $\hat{\theta}$ of θ, we have

$$\int \cdots \int Z_1(\theta)\hat{\theta}(\mathbf{x}) \prod_{i=1}^{n} f(x_i - \theta) \prod_{i=1}^{n} dx_i = \frac{1}{\sqrt{n}} \tag{2.1}$$

where $\mathbf{x} = (x_1, \ldots, x_n)$. Indeed, we obtain

$$\int \cdots \int \hat{\theta}(\mathbf{x}) \frac{\partial}{\partial\theta} \prod_{i=1}^{n} f(x_i - \theta) \prod_{i=1}^{n} dx_i = 1,$$

since the differentiation under the integral sign is allowed. Hence

$$1 = \int \cdots \int \hat{\theta}(\mathbf{x}) \left\{ \frac{\partial}{\partial\theta} \log \prod_{i=1}^{n} f(x_i - \theta) \right\} \prod_{i=1}^{n} f(x_i - \theta) \prod_{i=1}^{n} dx_i$$

$$= \int \cdots \int \hat{\theta}(\mathbf{x}) \left\{ \sum_{i=1}^{n} \frac{\partial}{\partial\theta} \log f(x_i - \theta) \right\} \prod_{i=1}^{n} f(x_i - \theta) \prod_{i=1}^{n} dx_i$$

$$= \int \cdots \int \sqrt{n} Z_1(\theta)\hat{\theta}(\mathbf{x}) \prod_{i=1}^{n} f(x_i - \theta) \prod_{i=1}^{n} dx_i,$$

which implies that (2.1) holds. From (2.1) it follows that

$$\int \cdots \int Z_1(\theta + \Delta\theta)\hat{\theta}(\mathbf{x}) \prod_{i=1}^{n} f(x_i - \theta - \Delta\theta) \prod_{i=1}^{n} dx_i = \frac{1}{\sqrt{n}}$$

from which we have

$$\int \cdots \int \left\{ \frac{Z_1(\theta + \Delta\theta) - Z_1(\theta)}{\Delta\theta} + \sqrt{n} Z_1^2(\theta) \right\} \hat{\theta}(\mathbf{x}) \prod_{i=1}^{n} f(x_i - \theta) \prod_{i=1}^{n} dx_i = o(1). \quad (2.2)$$

Indeed, from (2.1) and (2.2) we have

$$0 = \int \cdots \int \left\{ Z_1(\theta + \Delta\theta)\hat{\theta}(\mathbf{x}) \prod_{i=1}^{n} f(x_i - \theta - \Delta\theta) - Z_1(\theta)\hat{\theta}(\mathbf{x}) \prod_{i=1}^{n} f(x_i - \theta) \right\} \prod_{i=1}^{n} dx_i$$

$$= \int \cdots \int \left\{ Z_1(\theta + \Delta\theta) - Z_1(\theta) \right\} \hat{\theta}(\mathbf{x}) \prod_{i=1}^{n} f(x_i - \theta) \prod_{i=1}^{n} dx_i$$

$$+ \Delta\theta \int \cdots \int Z_1(\theta + \Delta\theta)\hat{\theta}(\mathbf{x}) \frac{\partial}{\partial\theta} \prod_{i=1}^{n} f(x_i - \theta) \prod_{i=1}^{n} dx_i + o(\Delta\theta)$$

$$= \int \cdots \int \left\{ Z_1(\theta + \Delta\theta) - Z_1(\theta) \right\} \hat{\theta}(\mathbf{x}) \prod_{i=1}^{n} f(x_i - \theta) \prod_{i=1}^{n} dx_i$$

$$+ \Delta\theta \int \cdots \int Z_1(\theta')\hat{\theta}(\mathbf{x}) \frac{\partial}{\partial\theta} \prod_{i=1}^{n} f(x_i - \theta' + \Delta\theta) \prod_{i=1}^{n} dx_i + o(\Delta\theta)$$

$$= \int \cdots \int \left\{ Z_1(\theta + \Delta\theta) - Z_1(\theta) \right\} \hat{\theta}(\mathbf{x}) \prod_{i=1}^{n} f(x_i - \theta) \prod_{i=1}^{n} dx_i$$

$$+ \Delta\theta \int \cdots \int Z_1(\theta')\hat{\theta}(\mathbf{x}) \left\{ \frac{\partial}{\partial\theta} \prod_{i=1}^{n} f(x_i - \theta') \right\} \prod_{i=1}^{n} dx_i + o(\Delta\theta)$$

$$= \int \cdots \int \left\{ Z_1(\theta + \Delta\theta) - Z_1(\theta) \right\} \hat{\theta}(\mathbf{x}) \prod_{i=1}^{n} f(x_i - \theta) \prod_{i=1}^{n} dx_i$$

$$+ \Delta\theta \int \cdots \int Z_1(\theta)\hat{\theta}(\mathbf{x}) \left\{ \frac{\partial}{\partial\theta} \log \prod_{i=1}^{n} f(x_i - \theta) \right\} \prod_{i=1}^{n} f(x_i - \theta) \prod_{i=1}^{n} dx_i + o(\Delta\theta)$$

$$= \int \cdots \int \left\{ Z_1(\theta + \Delta\theta) - Z_1(\theta) \right\} \hat{\theta}(\mathbf{x}) \prod_{i=1}^{n} f(x_i - \theta) \prod_{i=1}^{n} dx_i$$

$$+ \Delta\theta \int \cdots \int \sqrt{n}\hat{\theta}(\mathbf{x}) Z_1^2(\theta) \prod_{i=1}^{n} f(x_i - \theta) \prod_{i=1}^{n} dx_i + o(\Delta\theta),$$

which implies that (2.2) holds. Putting, for $\Delta\theta = O(1/\sqrt{n})$,

$$Z_2(\theta) = \frac{Z_1(\theta + \Delta\theta) - Z_1(\theta)}{\Delta\theta} + \sqrt{n} Z_1^2(\theta),$$

                    M. Akahira and K. Takeuchi

we have from (2.2)

$$\int \cdots \int Z_2(\theta)\hat{\theta}(\mathbf{x}) \prod_{i=1}^n f(x_i - \theta) \prod_{i=1}^n dx_i = o(1). \tag{2.3}$$

Note that $Z_2(\theta)$ depends on $\Delta\theta$, but for the sake of simplicity we omit $\Delta\theta$ from the expression. Then it follows from (2.1) and (2.3) that the Bhattacharyya type bound for the variance of unbiased estimators $\hat{\theta}$ of θ is given by

$$V_\theta(\hat{\theta}) \geq I^{11}(\theta)/n, \tag{2.4}$$

where $I^{11}(\theta)/n$ is the $(1,1)$-element of the matrix

$$\begin{pmatrix} nE_\theta[Z_1^2(\theta)] & \sqrt{n}E_\theta[Z_1(\theta)Z_2(\theta)] \\ \sqrt{n}E_\theta[Z_1(\theta)Z_2(\theta)] & E_\theta[Z_2^2(\theta)] \end{pmatrix}^{-1},$$

that is,

$$I^{11}(\theta) = \left[E_\theta[Z_1^2(\theta)] - \{E_\theta[Z_1(\theta)Z_2(\theta)]\}^2 / E_\theta[Z_2^2(\theta)] \right]^{-1} \tag{2.5}$$

(see, e.g., Zacks [6]). Since

$$Z_1(\theta) = \frac{1}{\sqrt{n}} \sum_{i=1}^n \operatorname{sgn}(X_i - \theta) = \frac{1}{\sqrt{n}} \sum_{i=1}^n X_i' \quad \text{(say)},$$

it follows that

$$E_\theta[Z_1^2(\theta)] = E_\theta\left[\left\{ \frac{1}{\sqrt{n}} \sum_{i=1}^n X_i' \right\}^2 \right] = \frac{1}{n} \sum_{i=1}^n E_\theta[X_i'^2] = 1. \tag{2.6}$$

Since

$$Z_2(\theta) = \frac{1}{\sqrt{n}} \sum_{i=1}^n \frac{1}{\Delta\theta} \{ \operatorname{sgn}(X_i - \theta - \Delta\theta) - \operatorname{sgn}(X_i - \theta) \} + \sqrt{n} Z_1^2(\theta)$$

$$= \frac{1}{\sqrt{n}} \sum_{i=1}^n Y_i + \sqrt{n} Z_1^2(\theta) \qquad \text{(say)}$$

and $E_\theta[Y_i] = -1 + (\Delta\theta/2) + o(\Delta\theta)$, $E_\theta[Y_i^2] = (2/\Delta\theta) - 1 + o(1)$, $E_\theta[X_i'Y_i] = -1 + (\Delta\theta/2) + o(\Delta\theta)$ $(i = 1, \ldots, n)$, it follows that

$$E_\theta[Z_1(\theta)Z_2(\theta)] = E_\theta\left[\frac{1}{\sqrt{n}} \sum_{i=1}^n X_i' \left(\frac{1}{\sqrt{n}} \sum_{i=1}^n Y_i + \sqrt{n} Z_1^2(\theta) \right) \right]$$

$$= \frac{1}{n} E_\theta\left[\sum_{i=1}^n X_i'Y_i + \sum_{i \neq j} X_i'Y_j \right] + \frac{1}{n} E_\theta\left[\sum_i \sum_j \sum_k X_i'X_j'X_k' \right]$$

$$= -1 + \frac{\Delta\theta}{2} + o(\Delta\theta). \tag{2.7}$$

Since

$$E_\theta\left[(\sum_{i=1}^n Y_i)^2\right] = nE_\theta[Y_i^2] + n(n-1)\{E_\theta(Y_i)\}^2$$

$$= n^2 + \frac{2n}{\Delta\theta} + o\left(\frac{n}{\Delta\theta}\right),$$

$$E_\theta\left[Z_1^2(\theta)\sum_{i=1}^n Y_i\right] = \frac{1}{n}E_\theta\left[\left(\sum_{i=1}^n X_i'^2 + \sum_{i\neq j}\sum X_i'X_j'\right)\sum_{k=1}^n Y_k\right]$$

$$= nE_\theta[Y_i] = -n + \frac{n\Delta\theta}{2} + o(n\Delta\theta),$$

$$E_\theta[Z_1^4(\theta)] = \frac{1}{n^2}E_\theta\left[\left(\sum_{i=1}^n X_i'\right)^4\right] = \frac{1}{n^2}E_\theta\left[\sum_{i=1}^n X_i'^4 + 3\sum_{i\neq k}\sum X_i'^2 X_k'^2\right]$$

$$= \frac{1}{n^2}\{n + 3n(n-1)\} = 3 - \frac{2}{n},$$

it follows that

$$E_\theta[Z_2^2(\theta)] = n + \frac{2}{\Delta\theta} + 2n\left(-1 + \frac{\Delta\theta}{2}\right) + n\left(3 - \frac{2}{n}\right) + o\left(\frac{1}{n}\right)$$

$$= 2n + o(n). \tag{2.8}$$

From (2.4) to (2.8) we have the following theorem.

Theorem. *The Bhattacharyya type bound for the variance of unbiased estimators* $\hat\theta$ *of* θ *is given by*

$$\mathrm{Var}_\theta(\hat\theta) \geq \frac{1}{n}\left(1 + \frac{1}{2n} + o\left(\frac{1}{n}\right)\right),$$

i.e.

$$\mathrm{Var}_\theta(\sqrt{n}\hat\theta) \geq \left(1 + \frac{1}{2n} + o\left(\frac{1}{n}\right)\right). \tag{2.9}$$

This gives the second order bound for the variance of unbiased estimators of θ. But the second term of the right-hand side of (2.9) is of order n^{-1}, while, for all estimators thus far discussed, the second order in the asymptotic variance is of order $n^{-1/2}$, and it is a situation still to be investigated and explained. The bound (2.9) may not be sharp. The related results can be found in Akahira and Takeuchi [3].

M. Akahira and K. Takeuchi

3. THE LOSS OF INFORMATION OF THE ESTIMATOR BASED ON THE DISTRIBUTION ROUNDED OFF

In a practical situation, all data are given only up to some decimal unit, hence all "real" distributions are not strictly continuous but are discrete distributions up to rounding off. And rounding off incurs loss of information, which tends to zero as the width of the rounding goes to zero, so the continuous distribution can be considered as the limiting case, hence a practical approximation for the actual situation where the width of round-off is small enough. In the regular case we do not need to go beyond the above consideration, but in a non-regular case, we must take another aspect into consideration. When the data is rounded, the class of distributions is no longer irregular, admitting differentiation with respect to the parameter as many times as we want. Hence the "asymptotic deficiency" of the maximum likelihood estimator (MLE) and other estimators becomes of order $O(1)$, but not of higher order as in the limiting non-regular case. Therefore, if we restrict our attention to the class of the MLE and other similar regular types of estimators, these must be optimum "compromise" for the width of rounding between the loss of information in the data due to rounding and the loss of information (asymptotic deficiency) due to estimators.

In this section we round off the double exponential distribution, consider the loss of information by the distribution and the loss of information of the MLE, and deal with their sum as the loss of information of the MLE based on the distribution.

In the case when the density is given by $f(x - \theta) = (1/2) \exp(-|x - \theta|)$ $(-\infty < x < \infty; -\infty < \theta < \infty)$, we consider to round off the density as follows. Without loss of generality we assume that $0 \leq \theta < h$. For any integer j we define $p_j(\theta)$ by

$$p_j(\theta) = \int_{jh}^{(j+1)h} f(x - \theta)dx.$$

Since

$$p_0(\theta) = \left(2 - e^{-\theta} - e^{\theta - h}\right)/2, \tag{3.1}$$

it follows that

$$p_0'(\theta) = \left(e^{-\theta} - e^{\theta - h}\right)/2 = (h - 2\theta)/2 + o(h), \tag{3.2}$$

$$p_0''(\theta) = -\left(e^{-\theta} - e^{\theta - h}\right)/2 = -1 + (h/2) + o(h). \tag{3.3}$$

Then it follows from (3.1) to (3.3) that the amount $I_h(\theta)$ of information, $J_h(\theta)$ and $M_h(\theta)$ on the distribution rounded off are defined and calculated by

$$
\begin{aligned}
I_h(\theta) &= \sum_j \left\{ \frac{\partial}{\partial \theta} \log p_j(\theta) \right\}^2 p_j(\theta) \\
&= \sum_{j \neq 0} \left\{ \frac{\partial}{\partial \theta} \log p_j(\theta) \right\}^2 p_j(\theta) + \left\{ \frac{\partial}{\partial \theta} \log p_0(\theta) \right\}^2 p_0(\theta) \\
&= \sum_{j \neq 0} p_j(\theta) + \{p_0'(\theta)/p_0(\theta)\}^2 p_0(\theta) = 1 - 2\theta \left(1 - \frac{\theta}{h} \right) + o(h),
\end{aligned}
\tag{3.4}
$$

$$
\begin{aligned}
J_h(\theta) &= \sum_j \left\{ \frac{\partial}{\partial \theta} \log p_j(\theta) \right\} \left\{ \frac{\partial^2}{\partial \theta^2} \log p_j(\theta) \right\} p_j(\theta) \\
&= \left\{ \frac{\partial}{\partial \theta} \log p_0(\theta) \right\} \left\{ \frac{\partial^2}{\partial \theta^2} \log p_0(\theta) \right\} p_0(\theta) \\
&= -1 + \frac{2\theta}{h} + o(1),
\end{aligned}
\tag{3.5}
$$

$$
\begin{aligned}
M_h(\theta) &= \sum_j \left\{ \frac{\partial^2}{\partial \theta^2} \log p_j(\theta) + I_h(\theta) \right\}^2 p_j(\theta) \\
&= \sum_{j \neq 0} I_h^2(\theta) p_j(\theta) + \left\{ \frac{\partial^2}{\partial \theta^2} \log p_0(\theta) + I_h(\theta) \right\}^2 p_0(\theta) \\
&= I_h^2(\theta)(1 - p_0(\theta)) + \{(p_0''(\theta)/p_0(\theta)) - (p_0'(\theta)/p_0(\theta))^2 + I_h(\theta)\}^2 p_0(\theta) \\
&= -1 + (2/h) + o(1).
\end{aligned}
\tag{3.6}
$$

Since the amount $I(\theta)$ of information on the double exponential density $f(x - \theta)$ is equal to 1, the loss of information by rounding off is defined by $n\{I(\theta) - I_h(\theta)\}$, and the loss $D_h(\theta)$ of information (asymptotic deficiency) of the MLE by rounding off is also given by $\{I_h(\theta)M_h(\theta) - J_h^2(\theta)\}/I_h^2(\theta)$ (see Akahira [7]). Hence the loss of information of the MLE based on the distribution rounded off is given by

$$
n\{I(\theta) - I_h(\theta)\} + D_h(\theta) = nI(\theta) - \{nI_h(\theta) - D_h(\theta)\}.
\tag{3.7}
$$

In order to get h minimizing it, we obtain h maximizing the average value of (3.7), i.e.

$$
(1/h) \int_0^h \{nI_h(\theta) - D_h(\theta)\}d\theta,
\tag{3.8}
$$

M. Akahira and K. Takeuchi

since θ is considered to be random in the interval $[0, h)$. From (3.4), (3.5) and (3.6) we have

$$D_h(\theta) = \{I_h(\theta)M_h(\theta) - J_h^2(\theta)\}/I_h^2(\theta) = \frac{2}{h} + \frac{4\theta}{h} - \frac{4\theta^2}{h^2} + o(1),$$

hence

$$nI_h(\theta) - D_h(\theta) = \left(n - \frac{2}{h}\right)\left\{1 - 2\theta\left(1 - \frac{\theta}{h}\right)\right\}.$$

Then we obtain

$$(1/h)\int_0^h \{nI_h(\theta) - D_h(\theta)\}d\theta = \left(n - \frac{2}{h}\right)\left(1 - \frac{h}{3}\right) + o(1).$$

Hence it is seen that the value of h maximizing (3.8) is given by $\sqrt{6/n}$, and also its value of (3.8) is done by $n - (2\sqrt{6n}/3) + (2/3)$. Now, in the second term, a quantity of order $\sqrt{n}$ appears, unlike the regular case when it should be of constant order, and the coefficient is not still satisfactorily explained in more general contexts.

REFERENCES

1. R. A. Fisher, *Proc. Cambridge Philos. Soc.*, **22**, 700-725 (1925).

2. M. Akahira and K. Takeuchi, *Asymptotic Efficiency of Statistical Estimators : Concepts and Higher Order Asymptotic Efficiency*, Lecture Notes in Statistics 7, Springer, New York (1981).

3. M. Akahira and K. Takeuchi, *Austral. J. Statist.*, **32**, 281-291 (1990).

4. M. Akahira, *Ann. Inst. Statist. Math.*, **40**, 311-328 (1988).

5. N. Sugiura and M. T. Naing, *Comm. Statist., A - Theory Methods* **18**, 541-554 (1989).

6. S. Zacks, *The Theory of Statistical Inference,* Wiley, New York (1971).

7. M. Akahira, *The Structure of Asymptotic Deficiency of Estimators,* Queen's Papers in Pure and Applied Mathematics **75**, Queen's University Press, Kingston, Ontario, Canada (1986).

Statistica Neerlandica (1993) Vol. 47, nr. 3, pp. 221–223

On the application of the Minkowski-Farkas theorem to sampling designs

M. Akahira

Institute of Mathematics
University of Tsukuba, Ibaraki 305, Japan

K. Takeuchi

Research Center for Advanced Science and Technology
University of Tokyo, 4-6-1 Komaba, Meguro-ku, Tokyo 156, Japan

In the paper, we consider the following problem: Let $\{\pi_k\}$ be a sequence satisfying $0 \leq \pi_k \leq 1$ $(k=1, ..., N)$ and $\sum_{k=1}^{N} \pi_k = n$. Then, is there an unordered sampling design such that, for each $k=1, ... N$, the inclusion probability of unit k is equal to π_k? It is shown that it can be solved by the straightforward application of the Minkowski-Farkas theorem.

Key words & Phrases: finite population, label, Minkowski-Farkas theorem, sampling design, unit, inclusion probability.

1 Introduction

In Chapter 1 of the book by CASSEL et al., (1977), discussing the finite population inference in survey sampling, the basic problem on determination of a sampling design given an inclusion probability is posed. A finite population is a collection of N units, where $N < \infty$. We use the label k to represent the physically existing unit u_k and talk about u_k as "the unit k" and thus denote the population as $\mathscr{U} = \{1, ..., k, ..., N\}$. A nonempty set s such that $s \subset \mathscr{U}$ is called an unordered sample. We assume that the number of elements of s is equal to n which is called the effective sample size. The set of all sets s of the effective sample size n will be denoted by $\mathscr{S}$. Then the number of components of $\mathscr{S}$ is equal to $_NC_n$ which is denoted by M, and also denote components $s_1, ..., s_M$. Let $C_k = \{s: k \in s\}$ $(k=1, ..., N)$. If a function $p(s)$ on $\mathscr{S}$ satisfies that $p(s) \geq 0$ for all $s \in \mathscr{S}$ and $\sum_{s \in \mathscr{S}} p(s) = 1$, then $p(s)$ is called an unordered sampling design.

Our problem is the following: Let $\{\pi_k\}$ be a sequence satisfying $0 \leq \pi_k \leq 1$ $(k=1, ..., N)$ and $\sum_{k=1}^{N} \pi_k = n$. Then is there an unordered sampling design $p(s)$ such that $\sum_{s \in C_k} p(s) = \pi_k$ $(k=1, ..., N)$? It is noted that $\sum_{s \in C_k} p(s)$ is called an inclusion probability of unit k. As a related result to this problem, a simple method to achieve the desired π_k $(k=1, ..., N)$ is given by MADOW (1949) (see also CASSEL et al., 1977). It is also shown in HANURAV (1966) that if a suitable drawing mechanism exists then the

222   *M. Akahira and K. Takeuchi*

existence of a design is established. Further, DUPACOVA (1979) answers the question of the existence of a design with given inclusion probabilities affirmatively. The related discussion is done by HAJEK (1981). However, the existence is proved simply by applying the Minkowski-Farkas theorem, which is given in the following section.

2   The existence theorem of sampling design

First, we state Minkowski-Farkas theorem to solve the above problem.

LEMMA (Minkowski-Farkas). *Let A βε α $N \times M$ matrix. In order that for a N-dimensional vector $\boldsymbol{n}$ there exists a nonnegative M-dimensional vector $\boldsymbol{p}$ such that $A\boldsymbol{p} = \boldsymbol{n}$ it is a necessary and sufficient condition that, for any N-dimensional vector $\boldsymbol{y}$ with $A'\boldsymbol{y} \geq \boldsymbol{0}$, $\boldsymbol{y}'\boldsymbol{n} \geq 0$, where "'" means the transposition and $\boldsymbol{0}$ denotes zero vector.*

The proof is omitted since it is a well-known theorem (e.g. see NIKAIDO, 1968). Using the above lemma we have the following theorem.

THEOREM. *Let $\{\pi_k\}$ be a sequence satisfying $0 \leq \pi_k \leq 1$ $(k = 1, ..., N)$ and $\sum_{k=1}^{N} \pi_k = n$. Then there exists an unordered sampling design $p(s)$ such that*

$$\sum_{s \in C_k} p(s) = \pi_k \qquad (k = 1, ..., N). \tag{1}$$

PROOF. Since $\sum_{i=1}^{M} p(s_i) = 1$, we denote a sampling plan with (1) as a nonnegative M-dimensional vector $\boldsymbol{p} = (p(s_1), ..., p(s_M))'$ satisfying $A\boldsymbol{p} = \boldsymbol{n}$, where A is a $(N + 1) \times M$ matrix and $\boldsymbol{n}$ is a $(N + 1)$-dimensional vector such that

$$A = \begin{bmatrix} a_{11} & a_{12} ... a_{1M} \\ \vdots & \cdots \quad \vdots \\ a_{N1} & a_{N2} ... a_{NM} \\ 1 & 1 \quad ... 1 \end{bmatrix}, \qquad \boldsymbol{n} = (\pi_1, ..., \pi_N, 1)',$$

where $a_{ki} = 1$ for $s_i \in C_k$ and $a_{ki} = 0$ for $s_i \notin C_k$. From the Lemma it follows that there exists a solution $p(s)$ with (1) if for any $(N + 1)$-dimensional vector $\boldsymbol{y}$ such that

$$A'\boldsymbol{y} \geq \boldsymbol{0} \tag{2}$$

it holds that $\boldsymbol{y}'\boldsymbol{n} \geq 0$. If we denote $\boldsymbol{y} = (y_1, ..., y_N, z)'$ with $y_1 \leq ... \leq y_N$, then (2) implies that $\sum_{k=1}^{N} a_{ki}y_k + z \geq 0$ for $i = 1, ..., M$. Since $z \geq - \min_{1 \leq i \leq M} \sum_{k=1}^{N} a_{ki}y_k = - \sum_{k=1}^{n} y_k$, we have

$$\mathbf{y}'\mathbf{n} = \sum_{k=1}^{N} \pi_k y_k + z \geq \sum_{k=1}^{N} \pi_k y_k - \sum_{k=1}^{n} y_k = \sum_{k=n+1}^{N} \pi_k y_k - \sum_{k=1}^{n} (1 - \pi_k) y_k \geq \sum_{k=n+1}^{N} \pi_k y_n -$$

$$\sum_{k=1}^{n} (1 - \pi_k) y_n = 0.$$

This completes the proof. $\qquad\qquad\square$

References

CASSEL, C. M., C. E. SARNDAL and J. H. WRETMAN (1977), *Foundations of inference in survey sampling,* Wiley, New York.

DUPACOVA, J. (1979), A note on rejective sampling, in: *Contributions to Statistics (Hajek Memorial Volume),* Academia Prague, 71–78.

HAJEK, J. (1981), *Sampling from a finite population,* Marcel Dekker, New York.

HANURAV, T. V. (1966), Some aspects of unified sampling theory, *Sankhya A 28,* 175–204.

MADOW, W. G. (1949), On the theory of systematic sampling II, *Annals of Mathematical Statistics 20,* 333–354.

NIKAIDO, H. (1968), *Convex structures and economic theory,* Academic Press, New York.

Received: March 1991, revised: February 1992.

COMMUN. STATIST.—SIMULA., 26(3), 1103–1128 (1997)

RANDOMIZED CONFIDENCE INTERVALS OF A PARAMETER FOR A FAMILY OF DISCRETE EXPONENTIAL TYPE DISTRIBUTIONS

Masafumi Akahira
Kunihiko Takahashi

Kei Takeuchi

Institute of Mathematics
University of Tsukuba
Ibaraki 305
Japan

Faculty of International Studies
Meiji-Gakuin University
Kamikuramachi 1598
Totsukaku, Yokohama 244
Japan

Key Words and Phrases: Randomized test; Exponential type distribution; Randomized confidence interval; Edgeworth expansion.

ABSTRACT

For a family of one-parameter discrete exponential type distributions, the higher order approximation of randomized confidence intervals derived from the optimum test is discussed. Indeed, it is shown that they can be asymptotically constructed by means of the Edgeworth expansion. The usefulness is seen from the numerical results in the case of Poisson and binomial distributions.

1. INTRODUCTION

For discrete distributions it is usually impossible to obtain a non-randomized test or confidence interval with given size, and an actual size is often quite different from the prescribed level. But randomized procedures, which is quite nice in theory, is not easily acceptable to practitioners. However, there is still something to promote randomized procedures. Although the practitioners and even theoreticians in actual applications may well feel it quite reluctant to apply randomized tests, the randomized confidence intervals may not seem so objectionable, if it is properly formulated. It should be noted that there are (infinitely) many ways to construct randomized confidence intervals from given randomized test, and it is necessary to choose an intuitively desirable method.

Suppose that our statistic T takes integer values and the "optimum" test for the real parameter $\theta = \theta_0$ (in some sense or other) is given by the test function ϕ of the form as

$$\phi(T) = \begin{cases} 1 & \text{for } T < c_1(\theta_0), \; T > c_2(\theta_0), \\ \gamma_1(\theta_0) & \text{for } T = c_1(\theta_0), \\ \gamma_2(\theta_0) & \text{for } T = c_2(\theta_0), \\ 0 & \text{for } c_1(\theta_0) < T < c_2(\theta_0), \end{cases}$$

where $c_1(\theta_0)$ and $c_2(\theta_0)$ are integers and $0 < \gamma_i(\theta_0) < 1$, (i=1, 2) and all are increasing in θ_0 (see, e.g. Lehmann (1986)). Then the test is equivalent to the non-randomized test ϕ^* based on $Y := T + U$, where U is an uniformly distributed random variable distributed uniformly over $(0, 1)$ and independent of T, and ϕ^* is given by

$$\phi^*(Y) = \begin{cases} 1 & \text{for } Y < c_1(\theta_0) + \gamma_1(\theta_0) \text{ or } Y > c_2(\theta_0) - \gamma_2(\theta_0) + 1, \\ 0 & \text{otherwise.} \end{cases}$$

We can write $\underline{y}(\theta_0) := c_1(\theta_0) + \gamma_1(\theta_0)$ and $\overline{y}(\theta_0) := c_2(\theta_0) - \gamma_2(\theta_0) + 1$. Then $\underline{y}(\theta)$ and $\overline{y}(\theta)$ are both monotone increasing in θ and we may assume that they are strictly monotone and also continuous. This yields that a randomized confidence interval $\left[\underline{\theta}(Y), \overline{\theta}(Y)\right]$ can be directly obtained from the equations

$$\underline{y}(\overline{\theta}) = Y \text{ and } \overline{y}(\underline{\theta}) = Y.$$

Such a confidence interval has the property that it depends on the random variable U and moves to left or right depending on it. And such confidence intervals do not look so obnoxious, we should think. Such a situation exists surely for the case of discrete exponential type distributions (including the Poisson, binomial, negative binomial, etc.) (see, e.g. Kendall, Stuart and Ord (1994), Johnson, Kotz and Kemp (1992), Molenaar (1973), Takeuchi and Fujino (1981)). The randomized confidence interval is approximated up to the higher order by means of the Edgeworth expansion for a family of one-parameter exponential type distributions. It is also seen from numerical results that the approximation is very useful in the case of Poisson and binomial distributions.

2. CONFIDENCE INTERVALS OF A PARAMETER FOR DISCRETE EXPONENTIAL TYPE DISTRIBUTIONS

Suppose that $X_1, \ldots, X_n$ are independent and identically distributed non-negative integer-valued random variables according to a one-parameter exponential type distribution with a discrete density function

$$f(x, \theta) = h(x) \exp\{\eta(\theta)x - C(\theta)\}$$

for $x = 0, 1, 2, \ldots$, and $\theta \in \mathbf{R}^1$, where $h(x), \eta(\theta)$ and $C(\theta)$ are real-valued functions. Then, for $r = 1, 2, \ldots$, we denote by $\kappa_r(\theta)$ the r-th cumulant for the distribution. In particular, we denote $\kappa_1(\theta)$ and $\kappa_2(\theta)$ by $\mu(\theta)$ and $\sigma^2(\theta)$, respectively. Letting $T = \sum_{i=1}^n X_i$, we standardize as

$$Z_n := \frac{T - n\mu(\theta)}{\sqrt{n}\sigma(\theta)}.$$

Then the r-th cumulant of Z_n is given by $n^{-(r/2)+1}\beta_r(\theta)$, where

$$\beta_r(\theta) = \kappa_r(\theta)/\{\sigma(\theta)\}^r$$

for $r = 3, 4, \ldots$. Hence the Edgeworth expansion of the distribution of T is given by

$$P_\theta\{T \leq t\} \doteqdot P_\theta\{Z_n \leq z\}$$
$$= \Phi(z) - \phi(z)\left\{\frac{\beta_3(\theta)}{6\sqrt{n}}(z^2 - 1) + \frac{\beta_4(\theta)}{24n}(z^3 - 3z)\right.$$

$$+ \frac{\beta_3{}^2(\theta)}{72n}(z^5 - 10z^3 + 15z) - \frac{1}{24n\sigma^2(\theta)}z\right\} + o\left(\frac{1}{n}\right),$$

where

$$z = \{t + \frac{1}{2} - \mu(\theta)\}/\{\sqrt{n}\sigma(\theta)\},$$

and $\Phi(z) = \int_{-\infty}^{z} \phi(t)dt$ with $\phi(t) = (1/\sqrt{2\pi})e^{-t^2/2}$. Then we consider a problem of testing the hypothesis $H : \theta = \theta_0$ against the alternative hypothesis $K : \theta \neq \theta_0$. For the problem we obtain the uniformly most powerful unbiased randomized test function ϕ given by

$$\phi(T) = \begin{cases} 1 & \text{for } T < t_1, \ T > t_2, \\ u_i & \text{for } T = t_i \ (i = 1, 2), \\ 0 & \text{for } t_1 < T < t_2, \end{cases} \tag{2.1}$$

where t_1, t_2, u_1 and u_2 are determined from the equations

$$\begin{aligned} E_{\theta_0}[\phi(T)] &= \alpha, \\ E_{\theta_0}[T\phi(T)] &= n\alpha\mu(\theta_0). \end{aligned} \tag{2.2}$$

Suppose that U is uniformly distributed on the interval $[0, 1]$ and is independent of T. Letting $Y := T + U$, we can regard (2.1) as a function of Y like

$$\phi^*(Y) = \begin{cases} 1 & \text{for } Y < t_1 + u_1, \ Y > t_2 - u_2 + 1, \\ 0 & \text{for } t_1 + u_1 \leq Y \leq t_2 - u_2 + 1. \end{cases}$$

Taking the adjustment into consideration since U is uniformly distributed on the interval $[0, 1]$, we put $\underline{y}(\theta_0) := t_1 + u_1 - (1/2)$ and $\overline{y}(\theta_0) := t_2 - u_2 + (1/2)$. Then we get a randomized confidence interval $\left[\underline{\theta}(Y), \overline{\theta}(Y)\right]$ of θ at level $1 - \alpha$ by solving the equations $\underline{y}(\overline{\theta}) = Y$ and $\overline{y}(\underline{\theta}) = Y$.

Theorem *The functions $\underline{y}(\theta)$ and $\overline{y}(\theta)$ are approximated up to the third order, i.e. the order $o(1/n)$ as*

$$\begin{aligned} \underline{y}(\theta) &= n\mu + \sqrt{n}\sigma(-u_{\alpha/2} + \Delta_1), \\ \overline{y}(\theta) &= n\mu + \sqrt{n}\sigma(u_{\alpha/2} + \Delta_2), \end{aligned} \tag{2.3}$$

where

$$\Delta_1 = \frac{\beta_3}{6\sqrt{n}}u_{\alpha/2}^2 + \frac{\beta_3^2}{72n}(4u_{\alpha/2}^3 - 15u_{\alpha/2})$$
$$- \frac{\beta_4}{24n}(u_{\alpha/2}^3 - 3u_{\alpha/2}) - \frac{1}{24n\sigma^2}\left\{12u_1(1-u_1)-1\right\}u_{\alpha/2}$$
$$- \frac{1}{4n\sigma^2 u_{\alpha/2}}\left\{u_2(1-u_2) - u_1(1-u_1)\right\} + o\left(\frac{1}{n}\right), \tag{2.4}$$

$$\Delta_2 = \frac{\beta_3}{6\sqrt{n}}u_{\alpha/2}^2 - \frac{\beta_3^2}{72n}(4u_{\alpha/2}^3 - 15u_{\alpha/2})$$
$$+ \frac{\beta_4}{24n}(u_{\alpha/2}^3 - 3u_{\alpha/2}) + \frac{1}{24n\sigma^2}\left\{12u_2(1-u_2)-1\right\}u_{\alpha/2}$$
$$- \frac{1}{4n\sigma^2 u_{\alpha/2}}\left\{u_2(1-u_2) - u_1(1-u_1)\right\} + o\left(\frac{1}{n}\right) \tag{2.5}$$

with the upper 100α *percentile* u_α.

The proof is given in Section 4.

Corollary 1 *A randomized confidence interval* $[\underline{\theta}(Y), \overline{\theta}(Y)]$ *of* θ *at level* $1-\alpha$ *is approximated up to the third order as solutions* $\underline{\theta}(Y)$ *and* $\overline{\theta}(Y)$ *of the equations*

$$n\mu + \sqrt{n}\sigma(-u_{\alpha/2} + \Delta_1) = Y$$

and

$$n\mu + \sqrt{n}\sigma(u_{\alpha/2} + \Delta_2) = Y$$

respectively, where Δ_1 *and* Δ_2 *are given by (2.4) and (2.5).*

The proof is straightforward from the Theorem.

Corollary 2 *The first and second order approximations of the functions* $\underline{y}(\theta)$ *and* $\overline{y}(\theta)$ *are given by*

$$\begin{aligned}
\underline{y}(\theta) &= n\mu - \sqrt{n}\sigma u_{\alpha/2} + o(\sqrt{n}), \\
\overline{y}(\theta) &= n\mu + \sqrt{n}\sigma u_{\alpha/2} + o(\sqrt{n}),
\end{aligned} \tag{2.6}$$

and

$$\begin{aligned}
\underline{y}(\theta) &= n\mu - \sqrt{n}\sigma u_{\alpha/2} + \frac{1}{6}\sigma\beta_3 u_{\alpha/2}^2 + o(1), \\
\overline{y}(\theta) &= n\mu + \sqrt{n}\sigma u_{\alpha/2} + \frac{1}{6}\sigma\beta_3 u_{\alpha/2}^2 + o(1),
\end{aligned} \tag{2.7}$$

respectively.

The proof is straightfoward from (2.3), (2.4) and (2.5).

Remark *The third order approximations (2.3) of $\underline{y}(\theta)$ and $\overline{y}(\theta)$ are computed in the following way. Let θ_0 be any fixed in R^1. We can determine positive integers t_1 and t_2 so that $E_{\theta_0}[\phi^*(Y)] = \alpha$, since $0 < u_i < 1$ for $i = 1, 2$. Substituting them for (2.3), we obtain solutions u_1 and u_2 from the equations (2.3) with $\underline{y}(\theta_0) = t_1 + u_1 - (1/2)$ and $\overline{y}(\theta_0) = t_2 - u_2 + (1/2)$, and get the third order approximations (2.3) of $\underline{y}(\theta_0)$ and $\overline{y}(\theta_0)$.*

In order to compare the approximations of $\underline{y}(\theta)$ and $\overline{y}(\theta)$, we also consider two versions (I) and (II) of the third order approximation (2.3) as follows.

$$(I) \qquad \begin{aligned} \underline{y}(\theta) &= n\mu + \sqrt{n}\sigma(-u_{\alpha/2} + \Delta_1^0), \\ \overline{y}(\theta) &= n\mu + \sqrt{n}\sigma(u_{\alpha/2} + \Delta_2^0), \end{aligned} \qquad (2.8)$$

where

$$\Delta_1^0 = \frac{\beta_3}{6\sqrt{n}}u_{\alpha/2}^2 + \frac{\beta_3^2}{72n}(4u_{\alpha/2}^3 - 15u_{\alpha/2}) - \frac{\beta_4}{24n}(u_{\alpha/2}^3 - 3u_{\alpha/2}),$$

$$\Delta_2^0 = \frac{\beta_3}{6\sqrt{n}}u_{\alpha/2}^2 - \frac{\beta_3^2}{72n}(4u_{\alpha/2}^3 - 15u_{\alpha/2}) + \frac{\beta_4}{24n}(u_{\alpha/2}^3 - 3u_{\alpha/2}).$$

$$(II) \qquad \begin{aligned} \underline{y}(\theta) &= n\mu + \sqrt{n}\sigma(-u_{\alpha/2} + \hat{\Delta}_1), \\ \overline{y}(\theta) &= n\mu + \sqrt{n}\sigma(u_{\alpha/2} + \hat{\Delta}_2), \end{aligned} \qquad (2.9)$$

$$\hat{\Delta}_1 = \Delta_1^0 - \frac{1}{24n\sigma^2}\left\{12\hat{u}_1(1 - \hat{u}_1) - 1\right\}u_{\alpha/2}$$
$$- \frac{1}{4n\sigma^2 u_{\alpha/2}}\left\{\hat{u}_2(1 - \hat{u}_2) - \hat{u}_1(1 - \hat{u}_1)\right\} + o\left(\frac{1}{n}\right),$$

$$\hat{\Delta}_2 = \Delta_2^0 + \frac{1}{24n\sigma^2}\left\{12\hat{u}_2(1 - \hat{u}_2) - 1\right\}u_{\alpha/2}$$
$$- \frac{1}{4n\sigma^2 u_{\alpha/2}}\left\{\hat{u}_2(1 - \hat{u}_2) - \hat{u}_1(1 - \hat{u}_1)\right\} + o\left(\frac{1}{n}\right).$$

with $\hat{u}_1$ and $\hat{u}_2$ obtained as the solutions of the equations (2.7), with $\underline{y}(\theta) = t_1 + u_1 - (1/2)$ and $\overline{y}(\theta) = t_2 - u_2 + (1/2)$, of u_1 and u_2.

The numerical comparison of the approximations (2.3), (2.6), (2.7), (2.8) and (2.9) are obtained in Section 3 where the underlying distribution is Poisson and binomial. Using the approximations (2.3) and (2.8) we can also obtain randomized confidence intervals which are derived from $\underline{y}(\bar{\theta}) = Y$ and $\bar{y}(\underline{\theta}) = Y$. Indeed, we get numerical results in the case of Poisson and binomial distributions in Section 3.

3. RANDOMIZED CONFIDENCE INTERVALS IN THE CASE OF POISSON AND BINOMIAL DISTRIBUTIONS

In this section we consider the cases when the Poisson distribution with a density function

$$f_1(x, \lambda) = \frac{e^{-\lambda}\lambda^x}{x!} \qquad (x = 0, 1, 2, \cdots ; \lambda > 0)$$

and the binomial distribution with a density function

$$f_2(x, p) = {}_nC_x p^x (1 - p)^{n-x} \qquad (x = 0, 1, \cdots, n; 0 < p < 1).$$

In the Poisson case with $\lambda = n$, the errors of the approximations (2.3), (2.6), (2.7), (2.8) and (2.9) are given for $n = 1, 2, \cdots, 40$ and $\alpha = 0.05$ in Tables 3.1 and 3.2, where the true values of $t_1 + u_1 - (1/2)$ and $t_2 - u_2 + (1/2)$ are derived from (2.2). From the Tables it is seen that the third order approximations (2.3) and (2.9) are nearer to the true values than the others. In particular the approximation (2.9) may be recommended since the computation of (2.9) is easier than that of (2.3).

From Corollary 1 we can asymptotically obtain a randomized confidence interval $[\underline{\lambda}, \bar{\lambda}]$ of λ at level $1 - \alpha$ by solutions $\underline{\lambda}$ and $\bar{\lambda}$ of the equations

$$Y = \lambda + \sqrt{\lambda}(u_{\alpha/2} + \Delta_2^*), \tag{3.1}$$

$$Y = \lambda + \sqrt{\lambda}(-u_{\alpha/2} + \Delta_1^*), \tag{3.2}$$

respectively, where

$$\begin{aligned}
\Delta_1^* &= \frac{1}{6\sqrt{\lambda}}u_{\alpha/2}^2 + \frac{1}{72\lambda}(4u_{\alpha/2}^3 - 15u_{\alpha/2}) - \frac{1}{24\lambda}(u_{\alpha/2}^3 - 3u_{\alpha/2}) + \frac{1}{24\lambda}u_{\alpha/2} \\
&= \frac{1}{6\sqrt{\lambda}}u_{\alpha/2}^2 + \frac{1}{72\lambda}(u_{\alpha/2}^3 - 3u_{\alpha/2}),
\end{aligned} \tag{3.3}$$

$$\Delta_2^* = \frac{1}{6\sqrt{\lambda}}u_{\alpha/2}^2 - \frac{1}{72\lambda}(4u_{\alpha/2}^3 - 15u_{\alpha/2}) + \frac{1}{24\lambda}(u_{\alpha/2}^3 - 3u_{\alpha/2}) - \frac{1}{24\lambda}u_{\alpha/2}$$

$$= \frac{1}{6\sqrt{\lambda}}u_{\alpha/2}^2 - \frac{1}{72\lambda}(u_{\alpha/2}^3 - 3u_{\alpha/2}), \tag{3.4}$$

We also have a randomized confidence interval $[\underline{\lambda}_0, \overline{\lambda}_0]$ of λ at level $1 - \alpha$ by the solutions $\underline{\lambda}_0$ and $\overline{\lambda}_0$ of the equations (3.1) and (3.2) with Δ_1^0 and Δ_2^0 instead of Δ_1^* and Δ_2^* in them, respectively, i.e. what one disregards the final terms in Δ_1^* and Δ_2^*. The graphs of $[\underline{\lambda}, \overline{\lambda}]$ and $[\underline{\lambda}_0, \overline{\lambda}_0]$ are given in Figures 3.1 and 3.2, respectively, and they are very close to the acceptance region $[t_1 + u_1 - (1/2), t_2 - u_2 + (1/2)]$. Hence they are useful in practice. In the binomial case, the errors of the approximations (2.3), (2.6), (2.7), (2.8) and (2.9) are given $n = 1 \sim 50(5)$, $50 \sim 100(10)$, $p = 0.1 \sim 0.5(0.1)$ and $\alpha = 0.05$ in Tables 3.3 to 3.12, where the values of $t_1 + u_1 - (1/2)$ and $t_2 - u_2 + (1/2)$ are derived from (2.2). From the Tables it is seen that the third order approximations (2.3) and (2.9) are nearer to the true values than the others. In particular the approximation (2.9) may be recommended since computation of (2.9) is easier than that of (2.3).

From Corollary 1 we can asymptotically obtain a randomized confidence interval $[\underline{p}, \overline{p}]$ of p at level $1 - \alpha$ by solutions of the equations

$$Y = np + \sqrt{np(1 - p)}\left(-u_{\alpha/2} + \Delta_1^0 + \frac{u_{\alpha/2}}{24np(1 - p)}\right), \tag{3.5}$$

$$Y = np + \sqrt{np(1 - p)}\left(u_{\alpha/2} + \Delta_2^0 - \frac{u_{\alpha/2}}{24np(1 - p)}\right), \tag{3.6}$$

where Δ_1^0 and Δ_2^0 are given in the Remark and $\beta_3 = (q - p)\big/\sqrt{pq}$, $\beta_4 = (1 - 6pq)\big/(pq)^{3/2}$ with $q = 1 - p$. We also have a randomized confidence interval $[\underline{p}_0, \overline{p}_0]$ of p at level $1 - \alpha$ by the solutions $\underline{p}_0$ and $\overline{p}_0$ of the equations (3.5) and (3.6) with Δ_1^0 and Δ_2^0 instead of $\Delta_1^0 + u_{\alpha/2}\big/(24np(1-p))$ and $\Delta_2^0 - u_{\alpha/2}\big/(24np(1-p))$ in them, respectively. The graphs of $[\underline{p}, \overline{p}]$ and $[\underline{p}_0, \overline{p}_0]$ are given in Figures 3.3, 3.4, 3.5 and 3.6 respectively, when $n = 20, 30$, and they are very close to the acceptance region $[t_1 + u_1 - (1/2), t_2 - u_2 + (1/2)]$. Hence they are useful in practice.

TABLE 3.1 The errors of the approximations of $\underline{y}(\lambda)$ for $\alpha = 0.05$ in the Poisson case.

λ	t_1	t_2	true value	1st approx. (2.6)	2nd approx. (2.7)	3rd approx. (2.3)	3rd approx. (2.8)	3rd approx. (2.9)
1	0	4	——	——	——	——	——	——
2	0	6	——	——	——	——	——	——
3	0	7	0.1232	——	0.1223	0.0037	0.0884	0.0238
4	1	9	0.6662	−0.5861	0.0541	0.0008	0.0247	−0.0112
5	1	10	1.1755	−0.5581	0.0821	−0.0054	0.0559	0.0081
6	2	11	1.7708	−0.5717	0.0685	0.0062	0.0444	−0.0030
7	2	13	2.4522	−0.6378	0.0025	−0.0211	−0.0197	−0.0134
8	3	14	3.0191	−0.5627	0.0775	0.0018	0.0568	0.0034
9	4	16	3.7151	−0.5950	0.0453	0.0033	0.0257	−0.0027
10	4	17	4.4340	−0.6319	0.0083	−0.0140	−0.0103	−0.0081
11	5	18	5.0763	−0.5767	0.0635	−0.0005	0.0458	0.0023
12	6	19	5.8001	−0.5896	0.0506	0.0033	0.0337	−0.0012
13	7	21	6.5644	−0.6311	0.0091	−0.0044	−0.0072	−0.0064
14	7	22	7.2689	−0.6024	0.0378	−0.0057	0.0221	−0.0001
15	8	23	7.9934	−0.5843	0.0560	0.0011	0.0408	0.0009
16	9	24	8.7618	−0.6016	0.0386	0.0024	0.0239	−0.0011
17	10	26	9.5530	−0.6341	0.0061	−0.0039	−0.0081	−0.0052
18	10	27	10.2943	−0.6097	0.0305	−0.0050	0.0167	−0.0006
19	11	28	11.0480	−0.5913	0.0490	−0.0001	0.0355	0.0008
20	12	29	11.8336	−0.5988	0.0414	0.0020	0.0283	−0.0006
21	13	31	12.6389	−0.6206	0.0197	−0.0009	0.0069	−0.0031
22	13	32	13.4408	−0.6338	0.0064	−0.0062	−0.0061	−0.0038
23	14	33	14.2042	−0.6038	0.0364	−0.0025	0.0242	0.0004
24	15	34	14.9945	−0.5963	0.0439	0.0006	0.0319	0.0004
25	16	35	15.8059	−0.6057	0.0345	0.0016	0.0228	−0.0006
26	17	37	16.6298	−0.6237	0.0166	−0.0010	0.0050	−0.0027
27	17	38	17.4515	−0.6358	0.0045	−0.0051	−0.0068	−0.0034
28	18	39	18.2387	−0.6098	0.0304	−0.0025	0.0193	0.0001
29	19	40	19.0495	−0.6006	0.0396	0.0000	0.0287	0.0004
30	20	41	19.8695	−0.6046	0.0356	0.0012	0.0249	−0.0003
31	21	43	20.7050	−0.6176	0.0226	0.0004	0.0121	−0.0013
32	22	44	21.5485	−0.6357	0.0045	−0.0023	−0.0059	−0.0029
33	22	45	22.3655	−0.6246	0.0156	−0.0035	0.0054	−0.0013
34	23	46	23.1801	−0.6085	0.0317	−0.0015	0.0216	0.0003
35	24	47	24.0087	−0.6040	0.0363	0.0002	0.0263	0.0002
36	25	48	24.8492	−0.6090	0.0313	0.0011	0.0215	−0.0003
37	26	50	25.6981	−0.6201	0.0202	0.0003	0.0105	−0.0006
38	27	51	26.5533	−0.6353	0.0049	−0.0019	−0.0046	−0.0025
39	27	52	27.3882	−0.6281	0.0121	−0.0031	0.0027	−0.0013
40	28	53	28.2172	−0.6131	0.0272	−0.0016	0.0179	0.0001

TABLE 3.2. The errors of the approximations of $\bar{y}(\lambda)$ for $\alpha = 0.05$ in the Poisson case.

λ	t_1	t_2	true value	1st approx. (2.6)	2nd approx. (2.7)	3rd approx. (2.3)	3rd approx. (2.8)	3rd approx. (2.9)
1	0	4	3.9942	−1.0342	−0.3940	−0.2464	−0.3352	———
*2	0	6	5.5735	−0.8017	−0.1615	−0.1287	−0.1199	———
3	0	7	7.1729	−0.7781	−0.1379	−0.0221	−0.1042	−0.0147
4	1	9	8.6403	−0.7204	−0.0801	−0.0465	−0.0508	−0.0565
5	1	10	10.1244	−0.7418	−0.1016	−0.0090	−0.0753	−0.0062
6	2	11	11.4788	−0.6779	−0.0377	−0.0222	−0.0137	−0.0016
7	2	13	12.8924	−0.7068	−0.0666	0.0050	−0.0444	−0.0024
8	3	14	14.2559	−0.7123	−0.0721	−0.0115	−0.0513	−0.0041
9	4	16	15.5476	−0.6677	−0.0275	−0.0203	−0.0079	−0.0213
10	4	17	16.8953	−0.6974	−0.0571	0.0032	−0.0385	−0.0018
11	5	18	18.2041	−0.7036	−0.0634	−0.0067	−0.0457	−0.0023
12	6	19	19.4546	−0.6651	−0.0249	−0.0107	−0.0079	−0.0070
13	7	21	20.7479	−0.6812	−0.0409	−0.0014	−0.0246	−0.0061
14	7	22	22.0312	−0.6977	−0.0575	−0.0007	−0.0418	−0.0015
15	8	23	23.2786	−0.6877	−0.0475	−0.0064	−0.0323	−0.0020
16	9	24	24.4918	−0.6520	−0.0117	−0.0079	0.0030	−0.0066
17	10	26	25.7571	−0.6760	−0.0357	−0.0004	−0.0215	−0.0039
18	10	27	27.0055	−0.6901	−0.0499	0.0000	−0.0360	−0.0011
19	11	28	28.2288	−0.6855	−0.0453	−0.0043	−0.0318	−0.0012
20	12	29	29.4269	−0.6617	−0.0215	−0.0063	−0.0083	−0.0034
21	13	31	30.6463	−0.6646	−0.0244	−0.0033	−0.0116	−0.0058
22	13	32	31.8709	−0.6779	−0.0376	0.0016	−0.0251	−0.0008
23	14	33	33.0851	−0.6945	−0.0452	−0.0012	−0.0330	−0.0007
24	15	34	34.2783	−0.6765	−0.0363	−0.0040	−0.0243	−0.0012
25	16	35	35.4542	−0.6544	−0.0142	−0.0051	−0.0024	−0.0034
26	17	37	36.6565	−0.6626	−0.0224	−0.0023	−0.0109	−0.0043
27	17	38	37.8577	−0.6734	−0.0332	0.0014	−0.0219	−0.0006
28	18	39	39.0522	−0.6811	−0.0408	−0.0006	−0.0297	−0.0006
29	19	40	40.2307	−0.6760	−0.0358	−0.0028	−0.0248	−0.0007
30	20	41	41.3957	−0.6606	−0.0203	−0.0040	−0.0096	−0.0019
31	21	43	42.5651	−0.6525	−0.0123	−0.0038	−0.0017	−0.0047
32	22	44	43.7528	−0.6656	−0.0254	0.0001	−0.0150	−0.0018
33	22	45	44.9388	−0.6747	−0.0344	0.0007	−0.0242	−0.0005
34	23	46	46.1055	−0.6771	−0.0368	−0.0010	−0.0267	−0.0004
35	24	47	47.2657	−0.6704	−0.0302	−0.0026	−0.0202	−0.0007
36	25	48	48.4158	−0.6560	−0.0158	−0.0034	−0.0060	−0.0018
37	26	50	49.5740	−0.6520	−0.0118	−0.0030	−0.0021	−0.0076
38	27	51	50.7451	−0.6631	−0.0229	0.0000	−0.0133	−0.0015
39	27	52	51.9110	−0.6711	−0.0308	0.0007	−0.0214	−0.0005
40	28	53	53.0702	−0.6743	−0.0341	−0.0006	−0.0248	−0.0003

* Note that $t_2 = 5$ is taken in (2.3) for $\lambda = 2$.

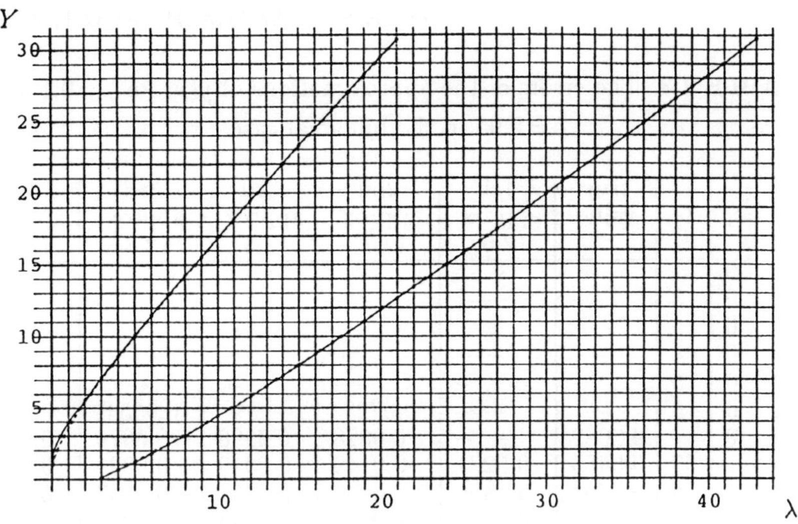

FIGURE 3.1. The graphs of the acceptance region $[t_1+u_1-(1/2), t_2-u_2+(1/2)]$ and randomized confidence interval $[\underline{\lambda}, \overline{\lambda}]$ for Poisson distribution.

——— the acceptance region

- - - - - the randomized confidence interval

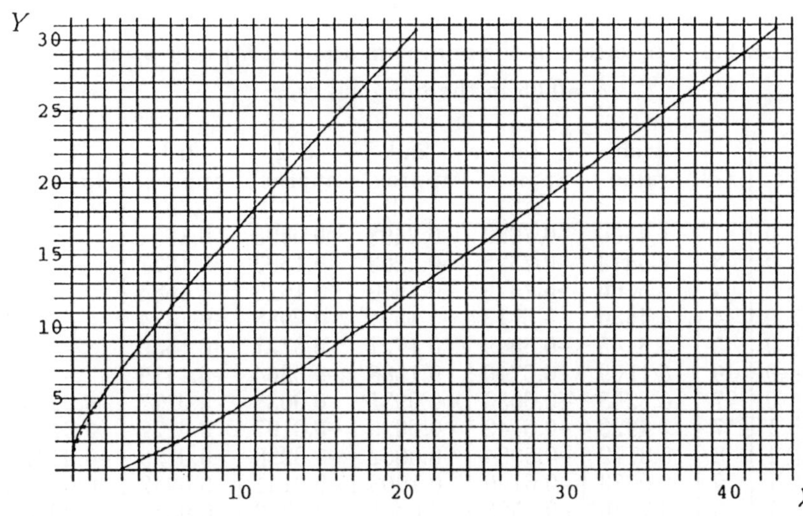

FIGURE 3.2. The graphs of the acceptance region $[t_1+u_1-(1/2), t_2-u_2+(1/2)]$ and randomized confidence interval $[\underline{\lambda}_0, \overline{\lambda}_0]$ for Poisson distribution.

——— the acceptance region

- - - - - the randomized confidence interval

TABLE 3.3. The errors of the approximations of $\underline{y}(p)$ for $p=0.1$ and $\alpha=0.05$ in the binomial case.

n	t_1	t_2	true value	1st approx. (2.6)	2nd approx. (2.7)	3rd approx. (2.3)	3rd approx. (2.8)	3rd approx. (2.9)
1	0	1	——	——	——	——	——	——
5	0	3	——	——	——	——	——	——
10	0	4	——	——	——	——	——	——
15	0	4	——	——	——	——	——	——
20	0	5	——	——	——	——	——	——
25	0	6		——	(0.0723)	——	(0.0554)	——
30	0	7	0.2089	——	0.0828	−0.0090	0.0674	0.0177
35	1	7	0.5485	−0.5271	−0.0149	−0.0049	−0.0291	0.0007
40	1	8	0.7268	−0.4456	0.0666	0.0051	0.0533	−0.0100
45	1	9	0.9703	−0.4146	0.0976	0.0043	0.0850	0.0059
50	1	10	1.3243	−0.4820	0.0302	−0.0158	0.0183	−0.0016
60	2	11	1.8782	−0.4327	0.0795	0.0044	0.0686	−0.0008
70	3	12	2.5866	−0.5061	0.0061	−0.0011	−0.0039	−0.0032
80	3	14	3.2030	−0.4621	0.0501	−0.0066	0.0407	0.0026
90	4	15	3.8709	−0.4490	0.0632	0.0027	0.0543	−0.0014
100	5	16	4.6176	−0.4975	0.0147	0.0002	0.0063	−0.0032

TABLE 3.4. The errors of the approximations of $\overline{y}(p)$ for $p=0.1$ and $\alpha=0.05$ in the binomial case.

n	t_1	t_2	true value	1st approx. (2.6)	2nd approx. (2.7)	3rd approx. (2.3)	3rd approx. (2.8)	3rd approx. (2.9)
1	0	1	1.4500	−0.7620	−0.2498	−0.0696	−0.1657	——
*5	0	3	2.5473	−0.7325	−0.2203	−0.1583	−0.1827	——
*10	0	4	3.5666	−0.7072	−0.1950	−0.1629	−0.1684	——
15	0	4	4.4486	−0.6713	−0.1591	−0.0988	−0.1374	——
20	0	5	5.3145	−0.6849	−0.1727	−0.0737	−0.1539	——
25	0	6	6.1109	−0.6710	−0.1588	−0.0364	−0.1419	——
30	0	7	6.8258	−0.6053	−0.0931	0.0035	−0.0777	−0.0220
35	1	7	7.4928	−0.5142	−0.0020	−0.0187	0.0122	−0.0271
40	1	8	8.3172	−0.5984	−0.0862	−0.0290	−0.0729	−0.0137
45	1	9	9.0633	−0.6190	−0.1068	−0.0141	−0.0942	−0.0142
50	1	10	9.7170	−0.5593	−0.0471	0.0050	−0.0352	−0.0096
60	2	11	11.1542	−0.5997	−0.0875	−0.0143	−0.0766	−0.0081
70	3	12	12.4455	−0.5260	−0.0138	−0.0114	−0.0038	−0.0104
80	3	14	13.8292	−0.5701	−0.0579	0.0014	−0.0485	−0.0073
90	4	15	15.1582	−0.5801	−0.0679	−0.0090	−0.0590	−0.0042
100	5	16	16.4114	−0.5315	−0.0193	−0.0087	−0.0109	−0.0059

* Note that $t_2=2$ and $t_2=3$ are taken for $n=5$ and $n=10$, respectively, in (2.3).

RANDOMIZED CONFIDENCE INTERVALS 1115

TABLE 3.5. The errors of the approximations of $\underline{y}(p)$ for $p=0.2$ and $\alpha=0.05$ in the binomial case.

n	t_1	t_2	true value	1st approx. (2.6)	2nd approx. (2.7)	3rd approx. (2.3)	3rd approx. (2.8)	3rd approx. (2.9)
1	0	1	——	——	——	——	——	——
5	0	3	——	——	——	——	——	——
10	0	5	——	——	——	——	——	——
15	0	6	0.3205	——	0.0273	−0.0100	0.0278	0.0035
20	1	8	0.7999	−0.3060	0.0782	0.0054	0.0786	−0.0047
*25	2	9	1.5040	−0.4239	−0.0398	−0.0105	−0.0495	0.0072
30	2	11	2.0194	−0.3135	0.0707	0.0012	0.0711	0.0057
35	3	12	2.7066	−0.3447	0.0394	0.0018	0.0398	−0.0060
40	3	13	3.4391	−0.3974	−0.0133	−0.0088	−0.0130	−0.0105
45	4	15	4.0719	−0.3310	0.0531	−0.0006	0.0534	0.0036
50	5	16	4.7940	−0.3376	0.0465	0.0017	0.0468	−0.0030
60	6	18	6.2934	−0.3661	0.0180	−0.0028	0.0183	0.0009
70	8	21	7.7867	−0.3460	0.0382	0.0011	0.0384	−0.0025
80	9	23	9.3670	−0.3792	0.0050	−0.0032	0.0052	−0.0019
90	11	26	10.9061	−0.3436	0.0405	0.0003	0.0404	−0.0006
100	13	28	12.5501	−0.3899	−0.0058	−0.0011	−0.0056	−0.0006

* Note that $t_1=1$ is taken in (2.3) for $n=25$.

TABLE 3.6. The errors of the approximations of $\overline{y}(p)$ for $p=0.2$ and $\alpha=0.05$ in the binomial case.

n	t_1	t_2	true value	1st approx. (2.6)	2nd approx. (2.7)	3rd approx. (2.3)	3rd approx. (2.8)	3rd approx. (2.9)
1	0	1	1.4500	−0.4660	−0.0819	−0.0503	−0.0840	——
5	0	3	3.3516	−0.5986	−0.2144	−0.1104	−0.2153	——
10	0	5	5.0395	−0.5603	−0.1762	−0.0517	−0.1768	——
15	0	6	6.4285	−0.3921	−0.0080	−0.0082	−0.0085	−0.0104
20	1	8	7.9946	−0.4885	−0.1044	−0.0165	−0.1048	−0.0203
25	2	9	9.3304	−0.4105	−0.0263	−0.0056	−0.0267	−0.0028
30	2	11	10.7246	−0.4305	−0.0464	−0.0022	−0.0468	−0.0130
35	3	12	12.0932	−0.4551	−0.0709	−0.0094	−0.0713	−0.0059
40	3	13	13.3582	−0.3998	−0.0157	−0.0033	−0.0160	−0.0001
45	4	15	14.6699	−0.4108	−0.0266	−0.0012	−0.0269	−0.0072
50	5	16	15.9878	−0.4442	−0.0601	−0.0046	−0.0603	−0.0053
60	6	18	18.4532	−0.3805	0.0037	−0.0031	0.0034	−0.0053
70	8	21	20.9933	−0.4340	−0.0499	−0.0031	−0.0501	−0.0033
80	9	23	23.4012	−0.3890	−0.0049	−0.0021	−0.0051	−0.0018
90	11	26	25.8608	−0.4233	−0.0391	−0.0010	−0.0394	−0.0035
100	13	28	28.2475	−0.4077	−0.0235	−0.0022	−0.0237	−0.0002

TABLE 3.7. The errors of the approximations of $\underline{y}(p)$ for $p=0.3$ and $\alpha=0.05$ in the binomial case.

n	t_1	t_2	true value	1st approx. (2.6)	2nd approx. (2.7)	3rd approx. (2.3)	3rd approx. (2.8)	3rd approx. (2.9)
1	0	1	——	——	——	——	——	——
5	0	4	——	——	——	——	——	——
10	0	6	0.4578	−0.2981	−0.0420	−0.0324	−0.0285	−0.0387
15	1	8	1.2334	−0.2120	0.0441	−0.0068	0.0551	0.0040
20	2	10	2.1928	−0.2095	0.0466	−0.0029	0.0561	0.0048
25	3	12	3.2299	−0.2207	0.0354	−0.0042	0.0439	0.0022
30	4	14	4.3222	−0.2417	0.0144	−0.0068	0.0222	−0.0028
35	5	16	5.4632	−0.2768	−0.0207	−0.0081	−0.0135	−0.0120
40	7	18	6.5798	−0.2603	−0.0042	−0.0015	0.0025	−0.0007
45	8	20	7.7084	−0.2335	0.0226	0.0007	0.0289	−0.0021
50	9	22	8.8690	−0.2200	0.0361	0.0020	0.0421	0.0002
60	11	25	11.2817	−0.2389	0.0172	−0.0026	0.0227	−0.0002
70	14	29	13.7220	−0.2366	0.0195	0.0005	0.0246	−0.0014
80	16	32	16.1995	−0.2329	0.0232	−0.0009	0.0279	0.0010
90	19	36	18.7186	−0.2394	0.0167	0.0003	0.0212	−0.0012
100	21	39	21.2584	−0.2401	0.0160	−0.0013	0.0203	0.0002

TABLE 3.8. The errors of the approximations of $\overline{y}(p)$ for $p=0.3$ and $\alpha=0.05$ in the binomial case.

n	t_1	t_2	true value	1st approx. (2.6)	2nd approx. (2.7)	3rd approx. (2.3)	3rd approx. (2.8)	3rd approx. (2.9)
1	0	1	1.4500	−0.2518	0.0043	−0.0365	−0.0382	——
5	0	4	3.9458	−0.4374	−0.1813	−0.0647	−0.2003	——
10	1	7	6.1639	−0.3236	−0.0675	0.0000	−0.0810	0.0110
15	2	10	8.2807	−0.3021	−0.0460	−0.0051	−0.0570	0.0031
20	4	12	10.3077	−0.2910	−0.0349	−0.0050	−0.0444	0.0008
25	5	15	12.2816	−0.2908	−0.0347	−0.0035	−0.0432	0.0015
30	7	17	14.2139	−0.2944	−0.0383	−0.0014	−0.0461	0.0027
35	8	20	16.1087	−0.2951	−0.0390	0.0006	−0.0462	0.0033
40	10	22	17.9790	−0.2985	−0.0424	−0.0007	−0.0491	−0.0019
45	12	25	19.8158	−0.2907	−0.0346	−0.0008	−0.0409	−0.0041
50	13	27	21.6133	−0.2623	−0.0062	−0.0023	−0.0122	−0.0028
60	17	32	25.2386	−0.2814	−0.0253	−0.0011	−0.0308	0.0011
70	20	36	28.7968	−0.2822	−0.0261	−0.0004	−0.0312	−0.0025
80	24	41	32.3057	−0.2723	−0.0162	−0.0015	−0.0209	0.0000
90	27	45	35.7999	−0.2791	−0.0230	−0.0001	−0.0275	−0.0018
100	31	50	39.2561	−0.2744	−0.0183	−0.0009	−0.0226	0.0004

TABLE 3.9. The errors of the approximations of $\underline{y}(p)$ for $p=0.4$ and $\alpha=0.05$ in the binomial case.

n	t_1	t_2	true value	1st approx. (2.6)	2nd approx. (2.7)	3rd approx. (2.3)	3rd approx. (2.8)	3rd approx. (2.9)
1	0	1	——	——	——	——	——	——
5	0	4	——	——	——	——	(0.0090)	——
10	1	7	1.0034	−0.0398	0.0883	0.0068	0.1081	0.0095
15	2	10	2.4375	−0.1563	−0.0282	−0.0187	−0.0121	−0.0213
20	4	12	3.7917	−0.0858	0.0423	0.0074	0.0563	0.0028
25	5	15	5.3145	−0.1154	0.0127	−0.0090	0.0252	−0.0040
30	7	17	6.8302	−0.0893	0.0387	0.0050	0.0501	0.0019
35	8	20	8.4735	−0.1540	−0.0260	−0.0064	−0.0154	−0.0112
40	10	22	10.0134	−0.0861	0.0419	0.0014	0.0518	0.0016
45	12	25	11.6788	−0.1199	−0.0918	0.0019	0.0175	0.0004
50	13	27	13.3310	−0.1205	0.0075	−0.0084	0.0164	−0.0023
60	17	32	16.6831	−0.1206	0.0074	0.0015	0.0155	0.0003
70	20	36	20.0640	−0.0974	0.0306	0.0032	0.0381	0.0009
80	24	41	23.5517	−0.1398	−0.0118	−0.0007	−0.0048	0.0015
90	27	45	26.9914	−0.1005	0.0276	0.0008	0.0342	0.0006
100	31	50	30.5388	−0.1406	−0.0126	−0.0007	−0.0063	0.0013

TABLE 3.10. The errors of the approximations of $\overline{y}(p)$ for $p=0.4$ and $\alpha=0.05$ in the binomial case.

n	t_1	t_2	true value	1st approx. (2.6)	2nd approx. (2.7)	3rd approx. (2.3)	3rd approx. (2.8)	3rd approx. (2.9)
1	0	1	0.9495	0.4107	−0.5387	0.4674	0.4762	——
5	0	4	4.3411	−0.1941	−0.0660	−0.0245	−0.0940	——
10	1	7	7.2323	−0.1959	−0.0679	−0.0096	−0.0877	0.0022
15	2	10	9.8911	−0.1723	−0.0443	0.0094	−0.0604	0.0038
20	4	12	12.4210	−0.1269	0.0011	−0.0073	−0.0129	−0.0093
25	5	15	14.9742	−0.1733	−0.0452	0.0039	−0.0578	0.0016
30	7	17	17.3929	−0.1338	−0.0057	−0.0053	−0.0171	−0.0047
35	8	20	19.8326	−0.1521	−0.0240	0.0043	−0.0346	0.0021
40	10	22	22.2320	−0.1593	−0.0312	−0.0029	−0.0411	0.0004
45	12	25	24.5560	−0.1149	0.0131	−0.0021	0.0038	0.0015
50	13	27	26.9491	−0.1596	−0.0316	0.0022	−0.0404	0.0009
60	17	32	31.5537	−0.1162	0.0119	−0.0016	0.0038	0.0012
70	20	36	36.1878	−0.1544	−0.0263	−0.0013	−0.0338	0.0004
80	24	41	40.7248	−0.1367	−0.0086	0.0012	−0.0156	−0.0002
90	27	45	45.2554	−0.1463	−0.0183	−0.0016	−0.0249	0.0000
100	31	50	49.7385	−0.1367	−0.0086	0.0011	−0.0149	−0.0001

1118   AKAHIRA, TAKAHASHI, AND TAKEUCHI

TABLE 3.11. The errors of the approximations of $\underline{y}(p)$ for $p=0.5$ and $\alpha=0.05$ in the binomial case.

n	t_1	t_2	true value	1st approx. (2.6)	2nd approx. (2.7)	3rd approx. (2.3)	3rd approx. (2.8)	3rd approx. (2.9)
1	0	1	———	———	———	———	———	———
5	0	5	0.3000	0.0087	0.0087	−0.0533	0.0394	−0.0231
10	2	8	1.8244	0.0766	0.0766	0.0117	0.0984	0.0011
15	4	11	3.6782	0.0264	0.0264	0.0092	0.0441	0.0039
20	6	14	5.6165	0.0009	0.0009	0.0058	0.0163	0.0074
25	8	17	7.6042	−0.0041	−0.0041	0.0044	0.0096	0.0070
30	10	20	9.6291	0.0033	0.0033	0.0043	0.0159	0.0046
35	12	23	11.6861	0.0163	0.0163	0.0044	0.0279	0.0020
40	14	26	13.7730	0.0291	0.0291	0.0037	0.0399	0.0004
45	16	29	15.8891	0.0370	0.0370	0.0020	0.0472	0.0001
50	18	32	18.0351	0.0354	0.0354	−0.0008	0.0451	0.0003
60	22	38	22.4229	−0.0138	−0.0138	−0.0023	−0.0049	−0.0047
70	27	43	26.7791	0.0218	0.0218	0.0022	0.0300	0.0003
80	31	49	31.2185	0.0163	0.0163	−0.0023	0.0240	−0.0005
90	36	54	35.6933	0.0098	0.0098	0.0018	0.0170	0.0008
100	40	60	40.1824	0.0178	0.0178	−0.0016	0.0247	−0.0001

TABLE 3.12. The errors of the approximations of $\overline{y}(p)$ for $p=0.5$ and $\alpha=0.05$ in the binomial case.

n	t_1	t_2	true value	1st approx. (2.6)	2nd approx. (2.7)	3rd approx. (2.3)	3rd approx. (2.8)	3rd approx. (2.9)
1	0	1	1.4500	0.0300	0.0300	−0.0406	−0.0387	———
5	0	5	4.7000	−0.0087	−0.0087	0.0533	−0.0394	0.0231
10	2	8	8.1756	−0.0766	−0.0766	−0.0117	−0.0984	−0.0011
15	4	11	11.3218	−0.0264	−0.0264	−0.0092	−0.0441	−0.0039
20	6	14	14.3835	−0.0009	−0.0009	−0.0058	−0.0163	−0.0074
25	8	17	17.3958	0.0041	0.0041	−0.0044	−0.0096	−0.0070
30	10	20	20.2709	−0.0033	−0.0033	−0.0043	−0.0159	−0.0046
35	12	23	23.3139	−0.0163	−0.0163	−0.0044	−0.0279	−0.0020
40	14	26	26.2270	−0.0291	−0.0291	−0.0037	−0.0399	−0.0004
45	16	29	29.1109	−0.0370	−0.0370	−0.0020	−0.0472	−0.0001
50	18	32	31.9649	−0.0354	−0.0354	0.0008	−0.0451	−0.0003
60	22	38	37.5771	0.0138	0.0138	0.0023	0.0049	0.0047
70	27	43	43.2209	−0.0218	−0.0218	−0.0022	−0.0300	−0.0003
80	31	49	48.7815	−0.0161	−0.0161	0.0023	−0.0240	0.0005
90	36	54	54.3067	−0.0098	−0.0098	−0.0018	−0.0170	−0.0008
100	40	60	59.8176	−0.0178	−0.0178	0.0016	−0.0247	0.0001

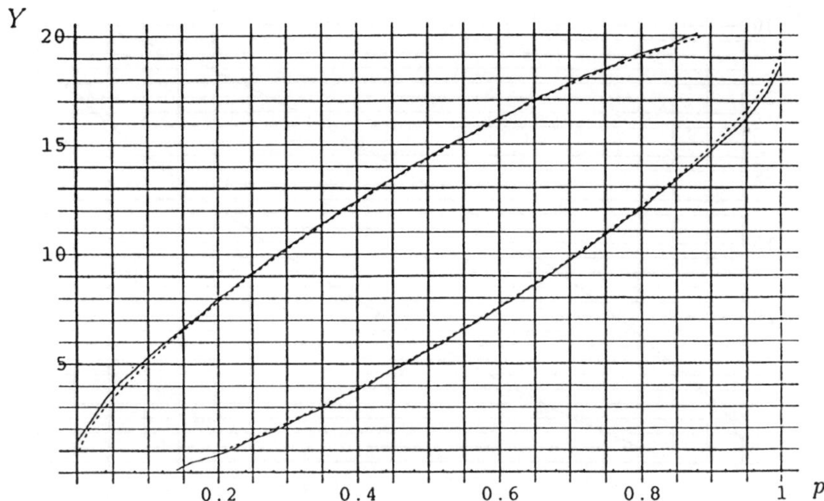

FIGURE 3.3. The graphs of the acceptance region $[t_1+u_1-(1/2), t_2-u_2+(1/2)]$ and randomized confidence interval $[\underline{p}, \overline{p}]$ for binomial distribution when $n=20$
———— the acceptance region
- - - - - the randomized confidence interval

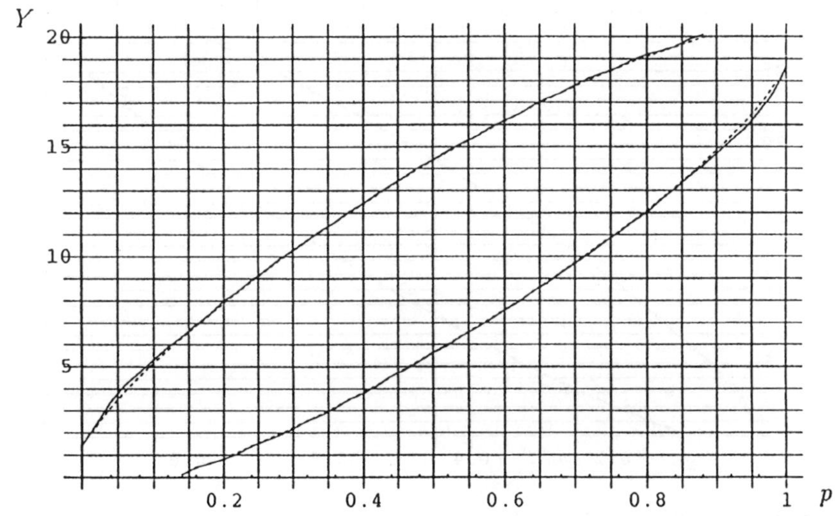

FIGURE 3.4. The graphs of the acceptance region $[t_1+u_1-(1/2), t_2-u_2+(1/2)]$ and randomized confidence interval $[\underline{p}_0, \overline{p}_0]$ for binomial distribution when $n=20$
———— the acceptance region
- - - - - the randomized confidence interval

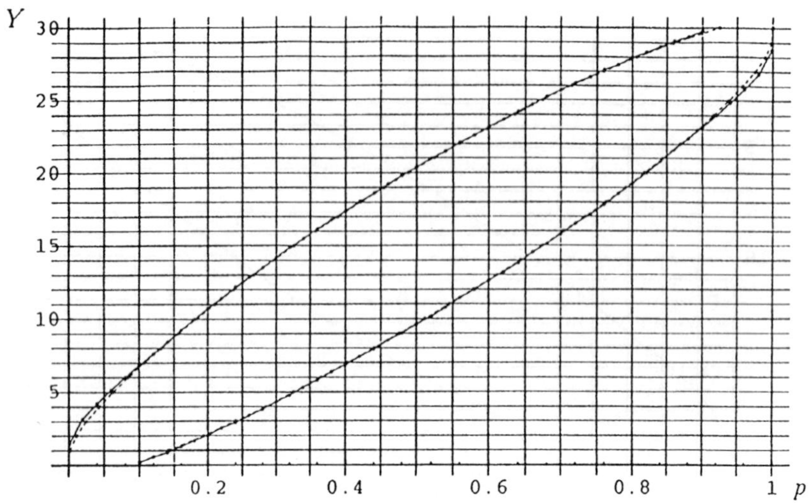

FIGURE 3.5. The graphs of the acceptance region $[t_1+u_1-(1/2), t_2-u_2+(1/2)]$ and randomized confidence interval $[\underline{p}, \overline{p}]$ for binomial distribution when $n=30$

———— the acceptance region

- - - - - the randomized confidence interval

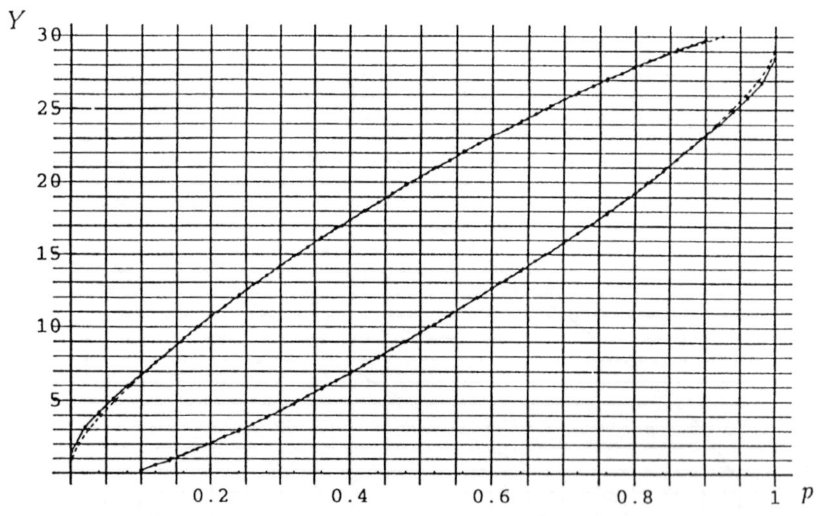

FIGURE 3.6. The graphs of the acceptance region $[t_1+u_1-(1/2), t_2-u_2+(1/2)]$ and randomized confidence interval $[\underline{p}_0, \overline{p}_0]$ for binomial distribution when $n=30$

———— the acceptance region

- - - - - the randomized confidence interval

4. PROOF

In this section we give a proof of the Theorem. Now, letting t be a nonnegative integer and $0 \leq u < 1$, we define two functions $F_n(y)$ and $M_n(y)$ of $y := t + u$ by

$$
\begin{aligned}
F_n(y) &:= P_\theta\{Y \leq t + u\} = P_\theta\{T_n \leq t - 1\} + uP_\theta\{T_n = t\} \\
&= (1 - u)P_\theta\{T_n \leq t - 1\} + uP_\theta\{T_n \leq t\}, \\
M_n(y) &:= \sum_{x=0}^{t-1} xP_\theta\{T_n = x\} + utP_\theta\{T_n = t\},
\end{aligned}
$$

respectively. Then the condition (2.2) turns out

$$
\begin{aligned}
F_n(t_2 - u_2 + 1) - F_n(t_1 + u_1) &= 1 - \alpha, \\
M_n(t_2 - u_2 + 1) - M_n(t_1 + u_1) &= n\mu(\theta_0)(1 - \alpha).
\end{aligned}
$$

For simplicity we denote $\mu(\theta_0)$, $\sigma(\theta_0)$ and $\beta_r(\theta_0)$ by μ, σ and β_r, respectively. If n is large and $|t - n\mu|/\sqrt{n}\sigma$ is small, then we have

$$
\begin{aligned}
F_n(y) &\doteq (1 - u)P\{Z_n \leq z'\} + uP\{Z_n \leq z\} \\
&= (1 - u)\left[\Phi(z') - \phi(z')\left\{\frac{\beta_3}{6\sqrt{n}}(z'^2 - 1) + \frac{\beta_4}{24n}(z'^3 - 3z')\right.\right. \\
&\qquad \left.\left. + \frac{\beta_3^2}{72n}(z'^5 - 10z'^3 + 15z') - \frac{1}{24n\sigma^2}z'\right\}\right] \\
&\quad + u\left[\Phi(z) - \phi(z)\left\{\frac{\beta_3}{6\sqrt{n}}(z^2 - 1) + \frac{\beta_4}{24n}(z^3 - 3z)\right.\right. \\
&\qquad \left.\left. + \frac{\beta_3^2}{72n}(z^5 - 10z^3 + 15z) - \frac{1}{24n\sigma^2}z\right\}\right] + o\left(\frac{1}{n}\right), \quad (4.1)
\end{aligned}
$$

where

$$
z' = (t - \frac{1}{2} - n\mu)/(\sqrt{n}\sigma), \quad z = (t + \frac{1}{2} - n\mu)/(\sqrt{n}\sigma).
$$

On the other hand we have

$$
\begin{aligned}
&M_n(y) - n\mu F_n(y) \\
&= (1 - u)\sum_{x=0}^{t-1}(x - n\mu)P_\theta\{T_n = x\} + u\sum_{x=0}^{t}(x - n\mu)P_\theta\{T_n = x\}. \quad (4.2)
\end{aligned}
$$

Then we obtain for a large n

$$\sum_{x=0}^{t}(x - n\mu)P_\theta\{T_n = x\}$$

$$= \sum_{x=0}^{t}\frac{x - n\mu}{\sqrt{n}\sigma}\left[\phi(z_x) + \phi(z_x)\left\{\frac{\beta_3}{6\sqrt{n}}(z_x^3 - 3z_x) + \frac{\beta_4}{24n}(z_x^4 - 6z_x^2 + 3)\right.\right.$$

$$\left.\left. + \frac{\beta_3^2}{72n}(z_x^6 - 15z_x^4 + 45z_x^2 - 15)\right\}\right] + o\left(\frac{1}{n}\right), \quad (4.3)$$

where $z_x = (x - n\mu)/(\sqrt{n}\sigma)$. For the first term of the right-hand side of (4.3), we have as its integral approximation

$$\frac{1}{\sqrt{n}\sigma}\sum_{x=0}^{t} z_x\phi(z_x)$$

$$\doteqdot \int_{-\infty}^{z} x\phi(x)dx - \frac{1}{24n\sigma^4}\int_{-\infty}^{z}(x\phi(x))''dx + o\left(\frac{1}{n}\right)$$

$$= -\phi(z) + \frac{1}{24n\sigma^4}(z^2 - 1)\phi(z) + o\left(\frac{1}{n}\right),$$

where $z = \{t + (1/2) - n\mu\}/(\sqrt{n}\sigma)$. Since the other terms of the RHS of (4.3) can be also approximated as integrals from sums, we have from (4.3)

$$\sum_{x=0}^{t}(x - n\mu)P_\theta\{T_n = x\}$$

$$= \sqrt{n}\sigma\left[\int_{-\infty}^{z} x\phi(x)\left\{1 + \frac{\beta_3}{6\sqrt{n}}(x^3 - 3x) + \frac{\beta_4}{24n}(x^4 - 6x^2 + 3)\right.\right.$$

$$\left. + \frac{\beta_3^2}{72n}(x^6 - 15x^4 + 45x^2 - 15)\right\} dx$$

$$\left. + \frac{1}{24n\sigma^4}(x^2 - 1)\phi(x) + o\left(\frac{1}{n}\right)\right]. \quad (4.4)$$

Letting $H_j(x)$'s be Hermite polynomials, i.e. $H_j(x) = \{(-d/dx)^j\phi(x)\}/\phi(x)$ ($j = 0, 1, 2, \ldots$), we obtain

$$\int x\phi(x)H_j(x)dx = -x\phi(x)H_{j-1}(x) + \int \phi(x)H_{j-1}(x)dx$$

$$= -x\phi(x)H_{j-1}(x) - \phi(z)H_{j-2}(x)$$

for $j = 2, 3, \ldots$. Then we have from (4.4)

RANDOMIZED CONFIDENCE INTERVALS 1123

$$\sum_{x=0}^{t} (x - n\mu) P_\theta\{T_n = x\}$$

$$= \sqrt{n}\sigma \left[-\phi(z) \left\{ 1 + \frac{\beta_3}{6\sqrt{n}} z^3 + \frac{\beta_4}{24n}(z^4 - 2z^2 - 1) + \frac{\beta_3{}^2}{72n}(z^6 - 9z^4 + 9z^2 + 3) \right. \right.$$

$$\left. \left. - \frac{1}{24n\sigma^4}(z^2 - 1) \right\} \right] + o\left(\frac{1}{\sqrt{n}}\right). \tag{4.5}$$

Since, for given smooth function G and a small h

$$(1 - u)G(t) + uG(t + h)$$
$$= G(t + uh) + \frac{1}{2}u(1 - u)h^2 G''(t + uh) + o(h^2),$$

it follows from (4.1), (4.2) and (4.5) that

$$F_n(t + u)$$
$$= \Phi(z_0) - \phi(z_0) \left\{ \frac{\beta_3}{6\sqrt{n}}(z_0{}^2 - 1) + \frac{\beta_4}{24n}(z_0{}^3 - 3z_0) \right.$$
$$\left. + \frac{\beta_3{}^2}{72n}(z_0{}^5 - 10z_0{}^3 + 15z_0) + \frac{1}{24n\sigma^2}(12u(1 - u) - 1)z_0 \right\}$$
$$+ o\left(\frac{1}{n}\right), \tag{4.6}$$

$$M_n(t + u) - n\mu F_n(t + u)$$
$$= -\sqrt{n}\sigma\phi(z_0) \left\{ 1 + \frac{\beta_3}{6\sqrt{n}} z_0{}^3 + \frac{\beta_4}{24n}(z_0{}^4 - 2z_0{}^2 - 1) \right.$$
$$\left. + \frac{\beta_3{}^2}{72n}(z_0{}^6 - 9z_0{}^4 + 9z_0{}^2 + 3) + \frac{1}{24n\sigma^4}(12u(1 - u) - 1)(z_0{}^2 - 1) \right\}$$
$$+ o\left(\frac{1}{\sqrt{n}}\right), \tag{4.7}$$

where $z_0 = \{t + u - (1/2) - n\mu\}/(\sqrt{n}\sigma)$. Then the condition (2.2) is represented by

$$F_n(t_2 + (1 - u_2)) - F_n(t_1 + u_1) = 1 - \alpha, \tag{4.8}$$
$$M_n(t_2 + (1 - u_2)) - n\mu F_n(t_2 + (1 - u_2))$$
$$- \{M_n(t_1 + u_1) - n\mu F_n(t_1 + u_1)\} = 0. \tag{4.9}$$

                    AKAHIRA, TAKAHASHI, AND TAKEUCHI

Let

$$z_1^* := (t_1 + u_1 - \frac{1}{2} - n\mu)/(\sqrt{n}\sigma),$$

$$z_2^* := (t_2 - u_2 + \frac{1}{2} - n\mu)/(\sqrt{n}\sigma) = \{t_2 + (1 - u_2) - \frac{1}{2} - n\mu\}/(\sqrt{n}\sigma).$$

Putting $\Delta_1 := z_1^* + u_{\alpha/2}$ and $\Delta_2 := z_2^* - u_{\alpha/2}$, we see that $\Delta_1 = O(1/\sqrt{n})$ and $\Delta_2 = O(1/\sqrt{n})$, where u_α is the upper 100α percentile. Then we have

$$\Phi(z_1^*) = \Phi(-u_{\alpha/2}) + \phi(-u_{\alpha/2})\Delta_1 + \frac{1}{2}\phi'(-u_{\alpha/2})\Delta_1{}^2 + o\left(\frac{1}{n}\right)$$

$$= \frac{\alpha}{2} + \phi(u_{\alpha/2})\Delta_1 + \frac{1}{2}u_{\alpha/2}\phi(u_{\alpha/2})\Delta_1{}^2 + o\left(\frac{1}{n}\right),$$

$$\Phi(z_2^*) = 1 - \frac{\alpha}{2} + \phi(u_{\alpha/2})\Delta_2 - \frac{1}{2}u_{\alpha/2}\phi(u_{\alpha/2})\Delta_2{}^2 + o\left(\frac{1}{n}\right),$$

$$(z_1^{*2} - 1)\phi(z_1^*) = (u_{\alpha/2}^2 - 1)\phi(u_{\alpha/2}) + (u_{\alpha/2}^3 - 3u_{\alpha/2})\phi(u_{\alpha/2})\Delta_1 + o\left(\frac{1}{\sqrt{n}}\right),$$

$$(z_2^{*2} - 1)\phi(z_2^*) = (u_{\alpha/2}^2 - 1)\phi(u_{\alpha/2}) - (u_{\alpha/2}^3 - 3u_{\alpha/2})\phi(u_{\alpha/2})\Delta_2 + o\left(\frac{1}{\sqrt{n}}\right).$$

From (4.6) we have

$$F_n(t_2 + (1 - u_2)) - F_n(t_1 + u_1)$$

$$= \Phi(z_2^*) - \Phi(z_1^*) - \frac{\beta_3}{6\sqrt{n}}\left\{(z_2^{*2} - 1)\phi(z_2^*) - (z_1^{*2} - 1)\phi(z_1^*)\right\}$$

$$- \phi(u_{\alpha/2})\left\{\frac{\beta_4}{24n}(u_{\alpha/2}^3 - 3u_{\alpha/2}) + \frac{\beta_3{}^2}{72n}(u_{\alpha/2}^5 - 10u_{\alpha/2}^3 + 15u_{\alpha/2})\right.$$

$$\left. + \frac{1}{24n\sigma^2}(12u_2(1 - u_2) - 1)u_{\alpha/2}\right\}$$

$$+ \phi(u_{\alpha/2})\left\{\frac{\beta_4}{24n}(-u_{\alpha/2}^3 + 3u_{\alpha/2}) + \frac{\beta_3{}^2}{72n}(-u_{\alpha/2}^5 + 10u_{\alpha/2}^3 - 15u_{\alpha/2})\right.$$

$$\left. - \frac{1}{24n\sigma^2}(12u_1(1 - u_1) - 1)u_{\alpha/2}\right\} + o\left(\frac{1}{n}\right)$$

$$= 1 - \alpha + (\Delta_2 - \Delta_1)\phi(u_{\alpha/2}) - \frac{1}{2}(\Delta_1{}^2 + \Delta_2{}^2)u_{\alpha/2}\phi(u_{\alpha/2})$$

$$+ \frac{\beta_3}{6\sqrt{n}}(\Delta_1 + \Delta_2)(u_{\alpha/2}^3 - 3u_{\alpha/2})\phi(u_{\alpha/2})$$

$$- \phi(u_{\alpha/2})\left\{\frac{\beta_4}{12n}(u_{\alpha/2}^3 - 3u_{\alpha/2}) + \frac{\beta_3{}^2}{36n}(u_{\alpha/2}^5 - 10u_{\alpha/2}^3 + 15u_{\alpha/2})\right.$$

RANDOMIZED CONFIDENCE INTERVALS   1125

$$+ \frac{1}{12n\sigma^2}(6u_1(1-u_1) + 6u_2(1-u_2) - 1)u_{\alpha/2}\Big\} + o\left(\frac{1}{n}\right).$$

From (4.8) we obtain

$$0 = \Delta_2 - \Delta_1 - \frac{1}{2}u_{\alpha/2}(\Delta_1{}^2 + \Delta_2{}^2) + \frac{\beta_3}{6\sqrt{n}}(u_{\alpha/2}^3 - 3u_{\alpha/2})(\Delta_1 + \Delta_2)$$

$$- \frac{\beta_4}{12n}(u_{\alpha/2}^3 - 3u_{\alpha/2}) - \frac{\beta_3{}^2}{36n}(u_{\alpha/2}^5 - 10u_{\alpha/2}^3 + 15u_{\alpha/2})$$

$$- \frac{u_{\alpha/2}}{12n\sigma^2}\{6u_1(1-u_1) + 6u_2(1-u_2) - 1\} + o\left(\frac{1}{n}\right), \qquad (4.10)$$

hence

$$\Delta_2 - \Delta_1 = \frac{1}{2}u_{\alpha/2}(\Delta_1{}^2 + \Delta_2{}^2)$$

$$- \frac{\beta_3}{6\sqrt{n}}(u_{\alpha/2}^3 - 3u_{\alpha/2})(\Delta_1 + \Delta_2) + \frac{\beta_4}{12n}(u_{\alpha/2}^3 - 3u_{\alpha/2})$$

$$+ \frac{\beta_3{}^2}{36n}(u_{\alpha/2}^5 - 10u_{\alpha/2}^3 + 15u_{\alpha/2})$$

$$+ \frac{1}{12n\sigma^2}\{6u_1(1-u_1) + 6u_2(1-u_2) - 1\}u_{\alpha/2}$$

$$+ o\left(\frac{1}{n}\right). \qquad (4.11)$$

From (4.7) and (4.9) we have

$$0 = M_n(t_2 + (1 - u_2)) - n\mu F_n(t_2 + (1 - u_2))$$

$$- \{M_n(t_1 + u_1) - n\mu F_n(t_1 + u_1)\}$$

$$= -\sqrt{n}\sigma\left[\phi(z_2^*) - \phi(z_1^*) + \frac{\beta_3}{6\sqrt{n}}\left\{z_2^{*3}\phi(z_2^*) - z_1^{*3}\phi(z_1^*)\right\}\right.$$

$$\left. + \frac{1}{2n\sigma^2}\phi(u_{\alpha/2})(u_{\alpha/2}^2 - 1)\{u_2(1 - u_2) - u_1(1 - u_1)\}\right] + o\left(\frac{1}{\sqrt{n}}\right).$$

$$(4.12)$$

Since

$$z_1^{*3}\phi(z_1^*) = -u_{\alpha/2}^3\phi(u_{\alpha/2}) - (u_{\alpha/2}^4 - 3u_{\alpha/2}^2)\phi(u_{\alpha/2})\Delta_1 + o\left(\frac{1}{\sqrt{n}}\right),$$

$$z_2^{*3}\phi(z_2^*) = u_{\alpha/2}^3\phi(u_{\alpha/2}) - (u_{\alpha/2}^4 - 3u_{\alpha/2}^2)\phi(u_{\alpha/2})\Delta_2 + o\left(\frac{1}{\sqrt{n}}\right),$$

it follows that

                    AKAHIRA, TAKAHASHI, AND TAKEUCHI

$$z_2^{*3}\phi(z_2^*) - z_1^{*3}\phi(z_1^*)$$
$$= 2u_{\alpha/2}^3\phi(u_{\alpha/2}) - (u_{\alpha/2}^4 - 3u_{\alpha/2}^2)\phi(u_{\alpha/2})(\Delta_2 - \Delta_1) + o\left(\frac{1}{\sqrt{n}}\right).$$

$$(4.13)$$

Since

$$\phi(z_2^*) - \phi(z_1^*) = -(\Delta_1 + \Delta_2)u_{\alpha/2}\phi(u_{\alpha/2})$$
$$+ \frac{1}{2}(\Delta_2{}^2 - \Delta_1{}^2)(u_{\alpha/2}^2 - 1)\phi(u_{\alpha/2}) + o\left(\frac{1}{n}\right),$$

it folloes from (4.12) and (4.13) that

$$0 = -u_{\alpha/2}(\Delta_1 + \Delta_2) + \frac{1}{2}(u_{\alpha/2}^2 - 1)(\Delta_2{}^2 - \Delta_1{}^2)$$
$$+ \frac{\beta_3}{3\sqrt{n}}u_{\alpha/2}^3 - \frac{\beta_3}{6\sqrt{n}}(u_{\alpha/2}^4 - 3u_{\alpha/2}^2)(\Delta_2 - \Delta_1)$$
$$+ \frac{1}{2n\sigma^2}(u_{\alpha/2}^2 - 1)\left\{u_2(1 - u_2) - u_1(1 - u_1)\right\} + o\left(\frac{1}{n}\right),$$

hence

$$\Delta_1 + \Delta_2$$
$$= \frac{1}{2}\left(u_{\alpha/2} - \frac{1}{u_{\alpha/2}}\right)(\Delta_2{}^2 - \Delta_1{}^2)$$
$$+ \frac{\beta_3}{3\sqrt{n}}u_{\alpha/2}^2 - \frac{\beta_3}{6\sqrt{n}}(u_{\alpha/2}^3 - 3u_{\alpha/2})(\Delta_2 - \Delta_1)$$
$$+ \frac{1}{2n\sigma^2}\left(u_{\alpha/2} - \frac{1}{u_{\alpha/2}}\right)\left\{u_2(1 - u_2) - u_1(1 - u_1)\right\} + o\left(\frac{1}{n}\right). \quad (4.14)$$

From (4.11) and (4.14) we have

$$\Delta_1 = \frac{\beta_3}{6\sqrt{n}}u_{\alpha/2}^2 - \frac{u_{\alpha/2}}{2}\Delta_1{}^2 - \frac{1}{4u_{\alpha/2}}(\Delta_2{}^2 - \Delta_1{}^2)$$
$$+ \frac{\beta_3}{6\sqrt{n}}(u_{\alpha/2}^3 - 3u_{\alpha/2})\Delta_1 - \frac{\beta_4}{24n}(u_{\alpha/2}^3 - 3u_{\alpha/2})$$
$$- \frac{\beta_3{}^2}{72n}(u_{\alpha/2}^5 - 10u_{\alpha/2}^3 + 15u_{\alpha/2}) - \frac{1}{24n\sigma^2}\left\{12u_1(1 - u_1) - 1\right\}u_{\alpha/2}$$
$$- \frac{1}{4n\sigma^2 u_{\alpha/2}}\left\{u_2(1 - u_2) - u_1(1 - u_1)\right\} + o\left(\frac{1}{n}\right),$$
$$\Delta_2 = \frac{\beta_3}{6\sqrt{n}}u_{\alpha/2}^2 + \frac{u_{\alpha/2}}{2}\Delta_2{}^2 - \frac{1}{4u_{\alpha/2}}(\Delta_2{}^2 - \Delta_1{}^2)$$

$$- \frac{\beta_3}{6\sqrt{n}}(u_{\alpha/2}^3 - 3u_{\alpha/2})\Delta_2 + \frac{\beta_4}{24n}(u_{\alpha/2}^3 - 3u_{\alpha/2})$$

$$+ \frac{\beta_3{}^2}{72n}(u_{\alpha/2}^5 - 10u_{\alpha/2}^3 + 15u_{\alpha/2}) + \frac{1}{24n\sigma^2}\left\{12u_2(1 - u_2) - 1\right\}u_{\alpha/2}$$

$$- \frac{1}{4n\sigma^2 u_{\alpha/2}}\left\{u_2(1 - u_2) - u_1(1 - u_1)\right\} + o\left(\frac{1}{n}\right),$$

hence

$$\Delta_1 = \frac{\beta_3}{6\sqrt{n}}u_{\alpha/2}^2 + \frac{\beta_3{}^2}{72n}(4u_{\alpha/2}^3 - 15u_{\alpha/2})$$

$$- \frac{\beta_4}{24n}(u_{\alpha/2}^3 - 3u_{\alpha/2}) - \frac{1}{24n\sigma^2}\left\{12u_1(1 - u_1) - 1\right\}u_{\alpha/2}$$

$$- \frac{1}{4n\sigma^2 u_{\alpha/2}}\left\{u_2(1 - u_2) - u_1(1 - u_1)\right\} + o\left(\frac{1}{n}\right),$$

$$\Delta_2 = \frac{\beta_3}{6\sqrt{n}}u_{\alpha/2}^2 - \frac{\beta_3{}^2}{72n}(4u_{\alpha/2}^3 - 15u_{\alpha/2})$$

$$+ \frac{\beta_4}{24n}(u_{\alpha/2}^3 - 3u_{\alpha/2}) + \frac{1}{24n\sigma^2}\left\{12u_2(1 - u_2) - 1\right\}u_{\alpha/2}$$

$$- \frac{1}{4n\sigma^2 u_{\alpha/2}}\left\{u_2(1 - u_2) - u_1(1 - u_1)\right\} + o\left(\frac{1}{n}\right).$$

Thus we complete the proof.

ACKNOWLEDGEMENTS

This research was supported in part by Venture Business Laboratory, University of Tsukuba.

REFERENCES

Johnson, N. L., Kotz. S. and Kemp, A. W. (1992). *Univariate Discrete Distributions* (2nd ed.). Wiley & Sons, New York.

Kendall, Sir M., Stuart, A. and Ord, J. K. (1994). *Kendall's Advanced Theory of Statistics Volume 1: Distribution Theory* (6th ed.). Edward Arnold, London.

Lehmann, E. L. (1986). *Testing Statistical Hypotheses* (2nd ed.). Wiley & Sons, New York.

Molenaar, W. (1973). *Approximations to the Poisson, Binomial and Hyper-Geometric Distribution Functions*. Mathematical Centre, Amsterdam.

Takeuchi, K. and Fujino, Y. (1981). *Binomial Distribution and Poisson Distribution*. (In Japanese), Tokyo Univ. Press.

Received October, 1996.

MASAFUMI AKAHIRA (*) - KEI TAKEUCHI (**)

The existence of a test with the largest order of consistency in the case of a two-sided Gamma type distribution

CONTENTS: 1. Introduction. — 2. Order of consistency in a general case. — 3. The largest order of consistency in the case of a two-sided Gamma type distribution. — 4. The existence of a test with the largest order n^2 of consistency. References. Summary. Riassunto. Key words and phrases.

1. INTRODUCTION

The order of consistency, i.e. the order of convergence of consistent estimators based on a random sample of size n, is discussed in Akahira (1975a, 1975b), Akahira and Takeuchi (1981, 1995) and others. In the regular case, the order is $\sqrt{n}$, but, in non-regular cases, the order is not always so. For example, let $X_1, ..., X_n$ be independent and identically distributed (i.i.d.) real random variables with a density $f_0(x - \theta)$ (with a location parameter θ) satisfying $f_0(x) > 0$ for $a < x < b$, $f_0(x) = 0$ otherwise, $f_0(x) \sim A'(x - a)^{\alpha - 1}$ as $x \to a + 0$, and $f_0(x) \sim B'(b - x)^{\beta - 1}$ as $x \to b - 0$, where $\alpha \leq \beta$, $0 < A' < \infty$ and $0 < B' < \infty$. Then order of consistency in $n^{1/\alpha}$ for $0 < \alpha < 2$, $(n \log n)^{1/2}$ for $\alpha = 2$ and $n^{1/2}$ for $\alpha > 2$ (see, e.g. Akahira (1975a), Akahira and Takeuchi (1981, 1991, 1995) and Woodroofe (1972)). Hence they are not always $\sqrt{n}$ in such cases. In each case, the above is also shown to be the largest order of consistency in the sense that there does not exist a consistent estimator with order greater than the above (see Akahira (1975a)). As one of other non-regular cases, we con-

---

(*) Institute of Mathematics, University of Tsukuba, Ibaraki 305, Japan.

(**) Faculty of International Studies, Meiji-Gakuin University, Kamikuramachi 1598, Totsuka-ku, Yokohama 244, Japan.

94

sider the case when the density $f_0(x)$ satisfies $f_0(x) \sim C \mid x \mid^{-\alpha}$ as $\mid x \mid \to 0$, where $0 < \alpha < 1$, where C is some positive constant. In the estimation problem of a location parameter θ, it is shown in Akahira (1975b) that the largest order of consistency in $n^{1/(1-\alpha)}$ for $0 < \alpha < 1/2$, the median of $X_1, \ldots, X_n$ is a $\{n^{1/(1-\alpha)}\}$ - consistent estimator for $0 < \alpha < 1$, and its asymptotic distribution is also given. Related results are found in Polfeldt (1970a, 1970b), Ibragimov and Has'minskii (1981), Smith (1985, 1989) and Woodroofe (1974).

In this paper we deal with the case of a two-sided Gamma type distribution with a density $f_0(x)$ satisfying $f_0(x) = \mid x \mid^{-1/2} g(x)$, where $g(x)$ is a bounded continuous positive-valued function on $\mathbf{R}^1 - \{0\}$, and $g'(x)/g(x)$ is bounded on $\mathbf{R}^1$. For a random sample $X_1, \ldots, X_n$ of size n from the distribution, it is shown that the largest order of consistency is n^2. In a problem of testing the hypothesis H: $\theta = \theta_0$ against K: $\theta = \theta_0 + tn^{-2}$, a test with a rejection region $\{\min_{1 \leq i \leq n} \mid X_i \mid > kn^{-2}\}$ is given as one with the order, where $t \neq 0$ and k is a positive constant. The power function of the test is asymptotically given.

2. ORDER OF CONSISTENCY IN A GENERAL CASE

Let χ ben an abstract sample space whose generic point is denoted by x, β a σ-field of subsets of χ and $\{P_\theta : \theta \in \Theta\}$ a set of probability measures on β, where Θ is called a parameter space. We assume that Θ is an open subset of Euclidean 1-space R^1. We denote by $(\chi^{(n)}, \beta^{(n)})$ the n-fold direct products of (χ, β). For each $n = 1, 2, \ldots$, the points of $\chi^{(n)}$ will be denoted by $x = (x_1, \ldots, x_n)$. Consider n-fold product measures P_θ^n of P_θ. An estimator of θ is defined to be a sequence $\{\hat{\theta}_n\}$ of $\beta^{(n)}$-measurable functions $\hat{\theta}_n$ on $\chi^{(n)}$ into Θ. For simplicity we denote $\{\hat{\theta}_n\}$ by $\hat{\theta}_n$.

For an increasing sequence of positive numbers $\{c_n\}$ (c_n tending to infinity) an estimator $\hat{\theta}_n$ is called $\{c_n\}$-consistent if for any $\eta \in \Theta$ there exists a sufficiently small positive number δ such that

$$\lim_{L \to \infty} \overline{\lim_{n \to \infty}} \sup_{\theta : |\theta - \eta| < \delta} P_\theta^n \left\{ c_n |\hat{\theta}_n - \theta| \geq L \right\} = 0.$$

For any two points θ_1 and θ_2 in Θ there exists a σ-finite measure μ_n such that $P_{\theta_1}^n$ and $P_{\theta_2}^n$ are absolutely continuous with respect to μ_n. Then for any points θ_1 and θ_2 in Θ we define

95

$$I_n(\theta_1, \theta_2) = \int_{x^{(n)}} \frac{dP^n_{\theta_1}}{d\mu_n} \log \frac{dP^n_{\theta_1}}{dP^n_{\theta_2}} \, d\mu_n$$

which is called the amount of the Kullback-Leibler information in $(X_1,\ldots, X_n)$ for discriminating between $P^n_{\theta_1}$ and $P^n_{\theta_2}$ when $P^n_{\theta_1}$ represents the true distribution. It is easily seen that, for each n, I_n is independent of μ_n. The following theorem shows that a necessary condition for the existence of a consistent estimator is that the limit of the amount of the Kullback-Leibler information is infinite.

THEOREM 2.1. Suppose that, for each $n = 1, 2,\ldots$, the support $\{x:(dP^n_\theta/d\mu_n)(x)>0\}$ does not depend on θ. If there exists a $\{c_n\}$-consistent estimator, then for any $\theta \in \Theta$ and any nonzero a

$$\lim_{L\to\infty} \varliminf_{n\to\infty} I_n(\theta, \theta+aLc_n^{-1}) = \infty.$$

The proof is omitted since it is quite similar to that of Theorem 2.2.2 in Akahira and Takeuchi (1981). The above theorem is useful to obtain the largest order of consistency.

3. THE LARGEST ORDER OF CONSISTENCY IN THE CASE OF A TWO-SIDED GAMMA TYPE DISTRIBUTION

Let $X = \Theta = R^1$, and we suppose that, for each $\theta \in \Theta$, P_θ is absolutely continuous with respect to the Lebesgue measure and constitutes a location parameter family. Then we denote the density dP_θ/dx by $f_\theta(x)$ and $f_\theta(x) = f_0(x-\theta)$. Let $X_1, X_2,\ldots, X_n,\ldots$ be a sequence of i.i.d. real random variables with the density $f_0(x-\theta)$ with respect to the Lebesgue measure. We assume that $f_0(x) = |x|^{-1/2}g(x)$, where $g(x)$ is a bounded continuous positive-valued function on R^1, symmetric around the origin, twice differentiable on $R^1-\{0\}$, and $g'(x)/g(x)$ is bounded on R^1. Here, for $x = 0$, $g'(0\pm0)$ is taken instead of $g'(0)$. A distribution with the above density $f_0(x)$ is called a two-sided Gamma type distribution. For example, when $g(x) = \{1/(2\sqrt{\pi})\} e^{-|x|}$, $f_0(x)$ is called a density of the two-sided Gamma distribution.

For θ_1 and θ_2 in Θ, the amount of the Kullback-Leibler (K-L) information in X_1 is

96

$$I_1(\theta_1, \theta_2) = \int_{-\infty}^{\infty} f_0(x - \theta_1) \log \frac{f_0(x - \theta_1)}{f_0(x - \theta_2)} \, dx,$$

and that of the K-L information I_n $(\theta_1\ \theta_2)$ in $(X_1, \ldots, X_n)$ becomes nI_1 (θ_1, θ_2), since $X_1, \ldots, X_n$ are i.i.d. random variables. Then we have the following.

Lemma 3.1. The value of the K-L information is given by

$$I_1(0, \Delta) = K \sqrt{|\Delta|} + O(|\Delta|^{3/2})$$

for small $|\Delta|$, where

$$K = \frac{1}{2} g(0) \left[\int_0^{\infty} \left\{ \log\left(1 + \frac{1}{u}\right) \right\} \left\{ (1 + u)^{-1/2} - u^{-1/2} \right\} du \right.$$

$$\left. - \int_0^1 \left(\log \frac{1 - u}{u} \right) u^{-1/2} du \right].$$

Proof. Let Δ be a small positive number. Since f_0 (x) is symmetric about $x = 0$, it follows that

$$I_1(0, \Delta) = -\int_0^{\infty} \left\{ \log\ f_0(x + \Delta) - \log\ f_0(x) \right\} \left\{ f_0(x + \Delta) - f_0(x) \right\} dx$$

$$+ \int_0^{\Delta} \left\{ \log\ f_0(\Delta - x) - \log\ f_0(x) \right\} f_0(x) dx = J_1 + J_2 \text{ (say).} \qquad (3.1)$$

Since, by the boundedness of g' $(x)/g(x)$,

$$\log f_0(x + \Delta) - \log f_0(x)$$

$$= -\frac{1}{2} \log\left(1 + \frac{\Delta}{x}\right) + \log g(x + \Delta) - \log g(x)$$

97

$$= -\frac{1}{2}\log\left(1+\frac{\Delta}{x}\right) + \Delta\frac{g'(x+\xi\Delta)}{g(x+\xi\Delta)}$$

$$= -\frac{1}{2}\log\left(1+\frac{\Delta}{x}\right) + O(\Delta),$$

we have

$$J_1 = \int_0^\infty \left\{\frac{1}{2}\log\left(1+\frac{\Delta}{x}\right) + O(\Delta)\right\}$$

$$\cdot\left\{(x+\Delta)^{-1/2}g(x+\Delta) - x^{-1/2}g(x)\right\}dx$$

$$= \frac{1}{2}\int_0^\infty \left\{\log\left(1+\frac{\Delta}{x}\right)\right\}\left\{(x+\Delta)^{-1/2}g(x+\Delta) - x^{-1/2}g(x)\right\}dx$$

$$+ O(\Delta)\int_0^\infty \left\{(x+\Delta)^{-1/2}g(x+\Delta) - x^{-1/2}g(x)\right\}dx, \qquad (3.2)$$

where $0 < \xi < 1$. First we obtain

$$\int_0^\infty \left\{\log\left(1+\frac{\Delta}{x}\right)\right\}\left\{(x+\Delta)^{-1/2}g(x+\Delta) - x^{-1/2}g(x)\right\}dx$$

$$= \int_0^\infty \left\{\log\left(1+\frac{\Delta}{x}\right)\right\}\left\{(x+\Delta)^{-1/2} - x^{-1/2}\right\}g(x)dx$$

$$+ \int_0^\infty \left\{\log\left(1+\frac{\Delta}{x}\right)\right\}(x+\Delta)^{-1/2}\left\{g(x+\Delta) - g(x)\right\}dx$$

$$= \sqrt{\Delta}\int_0^\infty \left\{\log\left(1+\frac{1}{u}\right)\right\}\left\{(1+u)^{-1/2} - u^{-1/2}\right\}g(\Delta u)du$$

$$+ \sqrt{\Delta}\int_0^\infty \left\{\log\left(1+\frac{1}{u}\right)\right\}(1+u)^{-1/2}\left\{g(\Delta u+\Delta) - g(\Delta u)\right\}du. \qquad (3.3)$$

98

Since

$$-\infty < \int_0^\infty \left\{ \log\left(1+\frac{1}{u}\right)\right\} \left\{(1+u)^{-1/2} - u^{-1/2}\right\} u\, du < 0,$$

it follows from the boundedness of $g'(u)$ that

$$\int_0^\infty \left\{ \log\left(1+\frac{1}{u}\right)\right\} \left\{(1+u)^{-1/2} - u^{-1/2}\right\} g(\Delta u)\, du$$

$$= g(0)\int_0^\infty \left\{ \log\left(1+\frac{1}{u}\right)\right\} \left\{(1+u)^{-1/2} - u^{-1/2}\right\} du$$

$$+\Delta \int_0^\infty \left\{ \log\left(1+\frac{1}{u}\right)\right\} \left\{(1+u)^{-1/2} - u^{-1/2}\right\} u g'(\xi \Delta u)\, du$$

$$= g(0)\int_0^\infty \left\{ \log\left(1+\frac{1}{u}\right)\right\} \left\{(1+u)^{-1/2} - u^{-1/2}\right\} du + O(\Delta), \qquad (3.4)$$

where $0 < \xi < 1$. Since

$$0 < \int_0^\infty \left\{ \log\left(1+\frac{1}{u}\right)\right\} (1+u)^{-1/2}\, du < \infty,$$

it follows from the boundedness of $g'(u)$ that

$$\int_0^\infty \left\{ \log\left(1+\frac{1}{u}\right)\right\} (1+u)^{-1/2} \left\{ g(\Delta u + \Delta) - g(\Delta u) \right\} du = O(\Delta). \qquad (3.5)$$

From (3.3), (3.4) and (3.5) we have

$$\int_0^\infty \left\{ \log\left(1+\frac{\Delta}{x}\right)\right\} \left\{(x+\Delta)^{-1/2} g(x+\Delta) - x^{-1/2} g(x) \right\} dx$$

$$= g(0)\sqrt{\Delta} \int_0^\infty \left\{ \log\left(1+\frac{1}{u}\right)\right\} \left\{(1+u)^{-1/2} - u^{-1/2}\right\} du + O\left(\Delta^{3/2}\right). \qquad (3.6)$$

99

Since $g'(x)/g(x)$ is bounded, we obtain

$$\int_0^\infty \left\{ (x+\Delta)^{-1/2} g(x+\Delta) - x^{-1/2} g(x) \right\} dx$$

$$= \int_\Delta^\infty t^{-1/2} g(t) dt - \int_0^\infty x^{-1/2} g(x) dx$$

$$= -\int_0^\Delta t^{-1/2} g(t) dt = O\left(\Delta^{1/2}\right),$$

hence

$$O(\Delta) \int_0^\infty \left\{ (x+\Delta)^{-1/2} g(x+\Delta) - x^{-1/2} g(x) \right\} dx = O\left(\Delta^{3/2}\right). \qquad (3.7)$$

From (3.2), (3.6) and (3.7) we have

$$J_1 = \frac{1}{2} g(0) \sqrt{\Delta} \int_0^\infty \left\{ \log\left(1 + \frac{1}{u}\right) \right\} \left\{ (1+u)^{-1/2} - u^{-1/2} \right\} du + O\left(\Delta^{3/2}\right). \quad (3.8)$$

Note that

$$-\infty < \int_0^\infty \left\{ \log\left(1 + \frac{1}{u}\right) \right\} \left\{ (1+u)^{-1/2} - u^{-1/2} \right\} du < 0.$$

Next we have

$$J_2 = \int_0^\Delta \left\{ \log f_0(\Delta - x) - \log f_0(x) \right\} f_0(x) dx$$

$$= -\frac{1}{2} \int_0^\Delta \left\{ \log\left(\frac{\Delta}{x} - 1\right) \right\} x^{-1/2} g(x) dx$$

$$+ \int_0^\Delta \left\{ \log g(\Delta - x) - \log g(x) \right\} x^{-1/2} g(x) dx. \qquad (3.9)$$

Since

$$\int_0^\Delta \left\{ \log\left(\frac{\Delta}{x} - 1\right) \right\} x^{-1/2} g(x) dx$$

100

$$= \Delta^{1/2}\left[\int_0^1\{\log(1-u)\}u^{-1/2}g(\Delta u)du - \int_0^1(\log u)u^{-1/2}g(\Delta u)du\right]$$

and $g(x)$ is bounded, we have

$$\int_0^\Delta\left\{\log\left(\frac{\Delta}{x}-1\right)\right\}x^{-1/2}g(x)dx = O\left(\Delta^{1/2}\right). \tag{3.10}$$

Since $g(x)$ and $g'(x)/g(x)$ are bounded, it follows that

$$\int_0^\Delta\{\log g(\Delta-x)-\log g(x)\}x^{-1/2}g(x)dx = O\left(\Delta^{3/2}\right). \tag{3.11}$$

Since $g'(x)$ is bounded and

$$\left|\int_0^1\left(\log\frac{1-u}{u}\right)u^{1/2}du\right| < \infty,$$

it follows from (3.9), (3.10) and (3.11) that

$$J_2 = -\frac{\sqrt{\Delta}}{2}\int_0^1\left(\log\frac{1-u}{u}\right)u^{-1/2}g(\Delta u)du + O\left(\Delta^{3/2}\right)$$

$$= -\frac{\sqrt{\Delta}}{2}\left\{g(0)\int_0^1\left(\log\frac{1-u}{u}\right)u^{-1/2}du + \Delta\int_0^1\left(\log\frac{1-u}{u}\right)u^{1/2}g'(\xi\Delta u)du\right\}$$

$$+O\left(\Delta^{3/2}\right)$$

$$= -\frac{1}{2}g(0)\sqrt{\Delta}\int_0^1\left(\log\frac{1-u}{u}\right)u^{-1/2}du + O\left(\Delta^{3/2}\right). \tag{3.12}$$

Note that

$$\left|\int_0^1\left(\log\frac{1-u}{u}\right)u^{-1/2}du\right| < \infty.$$

From (3.1), (3.8) and (3.12) we have

$$I_1(0,\Delta) = \frac{1}{2}g(0)\sqrt{\Delta}\left[\int_0^\infty\left\{\log\left(1+\frac{1}{u}\right)\right\}\left\{(1+u)^{-1/2}-u^{-1/2}\right\}du\right.$$

101

$$-\int_0^1 \left(\log \frac{1-u}{u} \right) u^{-1/2} du \right] + O\left(\Delta^{3/2} \right)$$

$$= K\sqrt{\Delta} + O\left(\Delta^{3/2} \right) \text{ (say)}.$$

In a similar way to the case $\Delta > 0$, we obtain for $\Delta < 0$

$$I_1(0,\Delta) = K\sqrt{-\Delta} + O\left(|\Delta|^{3/2} \right).$$

Thus we complete the proof.

THEOREM 3.1. Suppose that $X_1, X_2, \ldots, X_n, \ldots$ is a sequence of i.i.d. random variables with the density $f_0(x-\theta)$. Then the largest order of consistency is n^2 in the sense that there does not exist a consistent estimator with order greater than order n^2.

Proof. From Lemma 3.1 we have for sufficiently large n and any nonzero a

$$I_n\left(\theta, \theta + aLc_n^{-1} \right) = nI_1\left(\theta, \theta + aLc_n^{-1} \right) = nK\sqrt{|a|\,Lc_n^{-1}} + O\left(nL^{3/2} c_n^{-3/2} \right).$$

If order c_n is greater than order n^2, then

$$\lim_{n\to\infty} I_n\left(\theta, \theta + aLc_n^{-1} \right) = \lim_{n\to\infty} K\sqrt{|a|L} \sqrt{\frac{n^2}{c_n}} = 0.$$

Hence it follows from Theorem 2.1 that there does not exist a consistent estimator of θ with the order greater than order n^2. Thus we complete the proof.

4. THE EXISTENCE OF A TEST WITH THE LARGEST ORDER n^2 OF CONSISTENCY

In the previous section it is shown that the largest order of consistency is n^2. In this section we show that there exists a test with order n^2 of consistency.

We consider a problem of testing the hypothesis H: $\theta=\theta_0$ against the alternative K: $\theta = \theta_0 + tn^{-2}$, where $t \neq 0$. Without loss of generality we

102

assume that $\theta_0 = 0$. In the problem we take a test with a rejection region

$$R_n := \left\{ (X_1, \ldots, X_n) \Big| \min_{1 \leq i \leq n} |X_i| > C_n \right\}, \quad \text{where } C_n \text{ is a small positive constant}$$

depending on n. Since for sufficiently large n

$$P_0 \left\{ |X_1| \leq C_n \right\} = \int_{-C_n}^{C_n} f_0(x)dx = 2\int_0^{C_n} x^{-1/2} g(x)dx \sim 4g(0)C_n^{1/2},$$

it follows that

$$P_0^n \left\{ \min_{1 \leq i \leq n} |X_i| > C_n \right\} = \left[1 - P_0 \left\{ |X_1| \leq C_n \right\} \right]^n$$

$$\sim \left\{ 1 - 4g(0)C_n^{1/2} \right\}^n. \tag{4.1}$$

Letting $C_n = kn^{-2}$ with a positive constant k, from (4.1) we have for sufficiently large n

$$P_0^n \left\{ \min_{1 \leq i \leq n} |X_i| > C_n \right\} \sim \left(1 - 4g(0)\sqrt{k}\, n^{-1} \right)^n \sim e^{-4g(0)\sqrt{k}}.$$

Taking $0 < \alpha < 1$ as a level of the test, we obtain

$$e^{-4g(0)\sqrt{k}} = \alpha,$$

hence

$$k = \frac{(\log \alpha)^2}{16g^2(0)} = k_0 \quad \text{(say)}.$$

Putting $C_n^* = k_0 n^{-2}$, we have the following.

THEOREM 4.1. In testing the hypothesis $H: \theta = 0$ against the alternative $K: \theta = tn^{-2}$, the level α test with the rejection region

$$R_n^* := \left\{ (X_1, \ldots, X_n) \Big| \min_{1 \leq i \leq n} |X_i| > C_n^* \right\} \text{ is a test with order } n^2 \text{ of consistency and}$$

has a power function

$$\beta_n(t) := P^n_{tn^{-2}}\left\{\min_{1\le i\le n}|X_i| > C_n^*\right\}$$

$$\sim \begin{cases} \exp\left[-2g(0)\left\{(k_0+t)^{1/2} + (k_0-t)^{1/2}\right\}\right] & \text{for } |t| \le k_0, \\[2ex] \exp\left[-2g(0)\left\{(|t|+k_0)^{1/2} - (|t|-k_0)^{1/2}\right\}\right] & \text{for } |t| > k_0 \end{cases}$$

for sufficiently large n, where $t \ne 0$.

Proof. In the testing problem, we consider a test

$$\varphi_n^*(x_1,\ldots,x_n) = \begin{cases} 1 & \text{for } \min_{1\le i\le n}|x_i| > C_n^*, \\[1ex] 0 & \text{for } \min_{1\le i\le n}|x_i| < C_n^*. \end{cases}$$

Since $C_n^* = k_0 n^{-2}$, it follows that φ_n^* is a test of level α with order n^2 of consistency. Then the power function $\beta_n(t)$ of the level α test φ_n^* is given by

$$\beta_n(t) = E_{tn^{-2}}\left[\varphi_n^*(X_1,\ldots,X_n)\right]$$

$$= P^n_{tn^{-2}}\left\{\min_{1\le i\le n}|X_i| > C_n^*\right\}$$

$$= \left[1 - P_{tn^{-2}}\left\{|X_1| \le k_0 n^{-2}\right\}\right]^n$$

$$= \left[1 - P_{tn^{-2}}\left\{-(k_0+t)n^{-2} \le X_1 - tn^{-2} \le (k_0-t)n^{-2}\right\}\right]^n$$

$$= \left[1 - P_0\left\{-(k_0+t)n^{-2} \le X_1 \le (k_0-t)n^{-2}\right\}\right]^n. \tag{4.2}$$

Putting $a_n := -(k_0+t)n^{-2}$ and $b_n := (k_0-t)n^{-2}$, we have for $|t| < k_0$ and sufficiently large n

$$P_0\left\{a_n \le X_1 \le b_n\right\} = \int_{a_n}^{b_n} f_0(x)\,dx = \int_{a_n}^{b_n} |x|^{-1/2} g(x)\,dx$$

$$= \left(\int_{a_n}^0 + \int_0^{b_n}\right)|x|^{-1/2} g(x)\,dx$$

104

$$= \int_0^{-a_n} x^{-1/2} g(x)\,dx + \int_0^{b_n} x^{-1/2} g(x)\,dx$$

$$\sim 2g(0)\left\{(-a_n)^{1/2} + b_n^{1/2}\right\}$$

$$= 2g(0)\left\{(k_0+t)^{1/2} n^{-1} + (k_0-t)^{1/2} n^{-1}\right\}. \qquad (4.3)$$

From (4.2) and (4.3) we obtain for $|t| < k_0$ and sufficiently large n

$$\beta_n(t) \sim \left[1 - 2g(0)\left\{(k_0+t)^{1/2} + (k_0-t)^{1/2}\right\}n^{-1}\right]^n$$

$$\sim \exp\left[-2g(0)\left\{(k_0+t)^{1/2} + (k_0-t)^{1/2}\right\}\right]$$

$$\geq e^{-4g(0)\sqrt{k_0}} = \alpha. \qquad (4.4)$$

Since for $t > k_0$ and sufficiently large n

$$P_0\left\{a_n \leq X_1 \leq b_n\right\} = \int_{a_n}^{b_n} f_0(x)\,dx = \int_{-b_n}^{-a_n} x^{-1/2} g(x)\,dx$$

$$= \left(\int_0^{-a_n} - \int_0^{-b_n}\right) x^{-1/2} g(x)\,dx$$

$$\sim 2g(0)\left\{(-a_n)^{1/2} - (-b_n)^{1/2}\right\}$$

$$= 2g(0)\left\{(k_0+t)^{1/2} n^{-1} - (t-k_0)^{1/2} n^{-1}\right\},$$

we have from (2.2)

$$\beta_n(t) \sim \left[1 - 2g(0)\left\{(k_0+t)^{1/2} - (t-k_0)^{1/2}\right\}n^{-1}\right]^n$$

$$\sim \exp\left[-2g(0)\left\{(k_0+t)^{1/2} - (t-k_0)^{1/2}\right\}\right]$$

$$\geq e^{-2\sqrt{2k_0}\,g(0)} > e^{-4g(0)\sqrt{k_0}} = \alpha. \qquad (4.5)$$

In a similar way to the above we obtain for $t \leq -k_0$ and sufficiently large n

105

$$\beta_n(t) \sim \left[1 - 2g(0)\left\{(k_0 - t)^{1/2} - (-k_0 - t)^{1/2}\right\}n^{-1}\right]^n$$

$$\sim \exp\left[-2g(0)\left\{(k_0 - t)^{1/2} - (-k_0 - t)^{1/2}\right\}\right]$$

$$\geq e^{-2\sqrt{2k_0}\,g(0)} > e^{-4g(0)\sqrt{k_0}} = \alpha. \tag{4.6}$$

The desired result follows form (4.4), (4.5) and (4.6).

Remark 4.1. The power function $\beta_n(t)$ of the level α test with the rejection region $\{\min_{1 \leq i \leq n} |X_i| > C_n^*\}$ is asymptotically given as Figure 4.1.

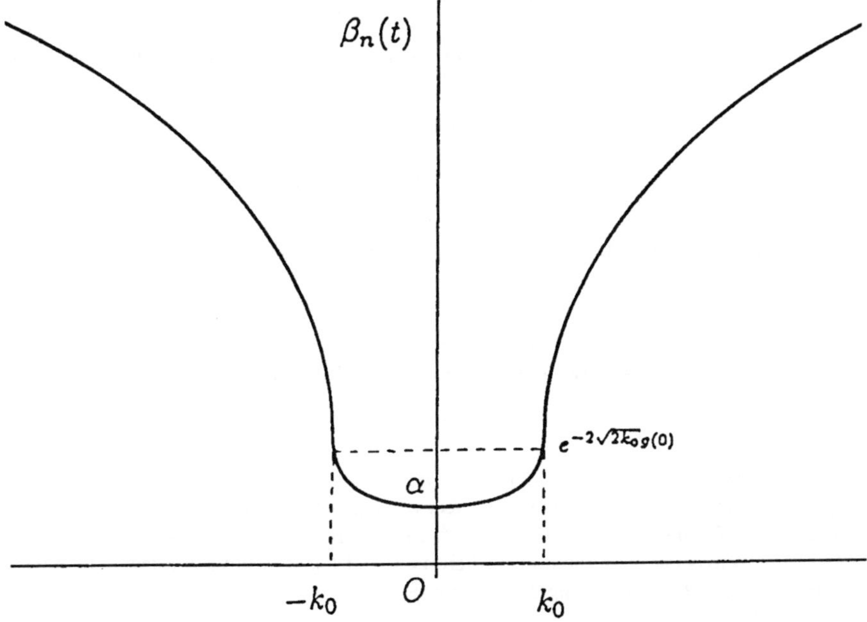

Fig. 4.1 – Asymptotic power function of the level α test.

REFERENCES

AKAHIRA, M. (1975a) Asymptotic theory for estimation of location in non-regular cases, I: Order of convergence of consistent estimators. *Rep. Stat. Appl. Res., JUSE,* 22, 8-26.

106

AKAHIRA, M. (1975b) Asymptotic theory for estimation of location in non-regular cases, II: Bounds of asymptotic distributions of consistent estimators. *Rep. Stat. Appl. Res., JUSE*, 22, 99-115.

AKAHIRA, M. and TAKEUCHI, K. (1981) *Asymptotic Efficiency of Estimators: Concepts and Higher Order Asymptotic Efficiency*. Lecture Notes in Statistics 7, Springer, New York.

AKAHIRA, M. and TAKEUCHI, K. (1991) Asymptotic efficiency of estimators for a location parameter family of densities with the bounded support. *Rep. Stat. Appl. Res., JUSE*, 38, 1-9.

AKAHIRA, M. and TAKEUCHI, K. (1995) *Non-Regular Statistical Estimation*. Lecture Notes in Statistics 107, Springer, New York.

IBRAGIMOV, I.A. and HAS'MINSKII, R.Z. (1981). *Statistical Estimation: Asymptotic Theory*. Springer, New York.

POLFELDT, T. (1970a) Asymptotic results in non-regular estimation. *Skand. Akt. Tidskr. Suppl.*, 1-2.

POLFELDT, T. (1970b) The order of the minimum variance in a non-regular case. *Ann. Math. Statist.*, 41, 667-672.

SMITH, R.L. (1985) Maximum likelihood estimation in a class of non-regular cases. *Biometrica* 72, 67-90.

SMITH, R.L. (1989) A survey of non-regular problems. *Bull. Int. Statist. Inst.*, 53, 353-372.

WOODROOFE, M. (1972) Maximum likelihood estimation of a translation parameter of a truncated distribution. *Ann. Math. Statist.*, 43, 113-122.

WOODROOFE, M. (1974) Maximum likelihood estimation of a translation parameter of a truncated distribution II. *Ann. Statist.*, 2, 474-488.

The existence of a test with the largest order of consistency in the case of a two-sided Gamma type distribution

SUMMARY

For a random sample $(X_1,\ldots,X_n)$ of size n from a two-sided Gamma type distribution, it is shown that the largest order of consistency is equal to n^2. In a problem of testing the hypothesis $H : \theta = \theta_0$ against $K : \theta = \theta_0 + tn^{-2}$, a test with the rejection region $\{\min_{1 \le i \le n} |X_i| > kn^{-2}\}$ is given as one with the order, where $t \ne 0$ and k is a positive constant. The power function of the test is also asymptotically given.

Esistenza di un test con il più ampio ordine di consistenza nel caso di distribuzione del tipo Gamma a due lati

RIASSUNTO

Per un campione casuale $(X_1, \ldots, X_n)$ di ampiezza n estratto da una distribuzione del tipo Gamma bilaterale, si dimostra che l'ordine di consistenza più grande è uguale a n^2. In un problema di verifica dell'ipotesi $H : \theta = \theta_0$ contro $K : \theta = \theta_0 + tn^{-2}$, viene fornito un test con la regione di rifiuto $\{\min_{1 \le i \le n} |X_i| > kn^{-2}\}$ dove $t \ne 0$ e K è una costante positiva. Viene anche data, asintoticamente, la funzione di potenza del test.

KEY WORDS AND PHRASES

Order of consistency, two-sided Gamma distribution, Kullback-Leibler information.

[Manuscript received October, 1997; final version received December, 1997]

COMMUN. STATIST.—THEORY METH., 28(3&4), 705–726 (1999)

THE HIGHER ORDER LARGE-DEVIATION APPROXIMATION FOR THE DISTRIBUTION OF THE SUM OF INDEPENDENT DISCRETE RANDOM VARIABLES

Masafumi Akahira
Kunihiko Takahashi

Kei Takeuchi

Institute of Mathematics
University of Tsukuba
Ibaraki 305-8571
Japan

Faculty of International Studies
Meiji-Gakuin University
Kamikurata-cho 1598
Totsuka-ku, Yokohama 244-0816
Japan

Key Words and Phrases: Large-deviation approximation; Normal approximation; Edgeworth expansion; Binomal case; Negative-binomial case.

ABSTRACT

For a sum of not identically but independently distributed discrete random variables, its higher order large-deviation approximations are given. They are compared with the normal and Edgeworth type approximations in various cases. Consequently, the large-deviation approximations give sufficiently accurate results.

1. INTRODUCTION

The Edgeworth and allied expansions are discussed by Barndorff-Nielsen and Cox (1989). In particular the tilted Edgeworth expansion is closely connected with the large deviation (see Daniels (1954, 1987)). Related results are found in Barndorff-Nielsen and Cox (1979), Booth and Wood (1995), Gatto and Ronchetti (1996), Harvill and Newton (1995) and Lieberman (1994).

The distributions of a sum of not identically but independently distributed random variables are difficult to calculate exactly. Obviously the normal and Edgeworth type approximations can be applied when the number of independent random variables is not too small, but it is not always sufficiently accurate especially for the tail part.

In such cases, large-deviation approximations could give better approximations especially for tails. In this paper, we discuss large-deviation approximations for the distribution of the sum of discrete random variables and show in various cases they give sufficiently accurate results.

2. LARGE-DEVIATION APPROXIMATIONS

Suppose that $X_1, \ldots, X_n, \ldots$ is a sequence of independent integer-valued random variables and, for each $j = 1, \ldots, n, \ldots$, X_j is distributed according to a probability function

$$p_j(x) = P\{X_j = x\} \quad \text{for } x = 0, \pm 1, \pm 2, \ldots.$$

Putting $S_n := \sum_{j=1}^{n} X_j$, we denote a probability function of S_n by

$$p_n^*(y) := P\{S_n = y\} \tag{2.1}$$

for $y = 0, \pm 1, \pm 2, \ldots$. Denote the moment generating function (m.g.f.) of X_j by

$$M_j(\theta) := E[\, e^{\theta X_j}\,] \tag{2.2}$$

for each $j = 1, \ldots, n, \ldots$, assuming that $M_j(\theta)$'s exist for values of θ in an open interval Θ which includes 0. Now, for each j, we consider a discrete exponential family $\mathcal{P}_j := \{p_{j,\theta}(x) : \theta \in \Theta\}$ of probability functions

$$p_{j,\theta}(x) := P_\theta\{X_j = x\} = p_j(x)e^{\theta x}M_j(\theta)^{-1} \tag{2.3}$$

for $x = 0, \pm 1, \pm 2, \ldots$, where $p_{j,0}(x) = p_j(x)$. Denote a probability function of S_n by

$$p_{n,\theta}^*(y) := P_\theta\{S_n = y\} \tag{2.4}$$

for $y = 0, \pm 1, \pm 2, \ldots$, where $p_{n,0}^*(y) = p_n^*(y)$.

From (2.1) to (2.4) we have

$$p_{n,\theta}^*(y) = p_n^*(y) e^{\theta y} \prod_{j=1}^{n} M_j(\theta)^{-1}. \tag{2.5}$$

On the other hand, it follows from (2.2) that, for each j, the characteristic function of X_j, under the family $\mathcal{P}_j$, is given by

$$E_\theta \left[e^{itX_j} \right] = \sum_x e^{itx} p_{j,\theta}(x) = M_j(\theta)^{-1} M_j(\theta + it). \tag{2.6}$$

Since, by (2.6), the characteristic function of S_n, under the families $\{\mathcal{P}_j\}$, is

$$E_\theta \left[e^{itS_n} \right] = \prod_{j=1}^{n} M_j(\theta + it) \prod_{j=1}^{n} M_j(\theta)^{-1}.$$

By the Fourier inverse transform, we have

$$p_{n,\theta}^*(y) = \frac{1}{2\pi} \int_{-\pi}^{\pi} \prod_{j=1}^{n} M_j(\theta + it) \prod_{j=1}^{n} M_j(\theta)^{-1} e^{-ity} dt. \tag{2.7}$$

From (2.5) and (2.7) we obtain

$$p_n^*(y) = e^{-\theta y} \prod_{j=1}^{n} M_j(\theta) p_{n,\theta}^*(y)$$

$$= e^{-\theta y} \prod_{j=1}^{n} M_j(\theta) \cdot \frac{1}{2\pi} \int_{-\pi}^{\pi} \prod_{j=1}^{n} M_j(\theta + it) \prod_{j=1}^{n} M_j(\theta)^{-1} e^{-ity} dt. \tag{2.8}$$

Letting $K_n(\theta) := \sum_{j=1}^{n} \log M_j(\theta)$, we have

$$\prod_{j=1}^{n} M_j(\theta) = e^{K_n(\theta)}. \tag{2.9}$$

From (2.8) and (2.9) we obtain

$$p_n^*(y) = e^{K_n(\theta) - \theta y} \cdot \frac{1}{2\pi} \int_{-\pi}^{\pi} e^{K_n(\theta + it) - K_n(\theta) - ity} dt. \tag{2.10}$$

For small $|t|$, we have by the Taylor expansion

$$K_n(\theta + it) - K_n(\theta) = K_n^{(1)}(\theta) it + \frac{1}{2} K_n^{(2)}(\theta)(it)^2 + \frac{1}{6} K_n^{(3)}(\theta)(it)^3 + \cdots,$$

where $K_n^{(\alpha)}(\theta) = (d^\alpha/d\theta^\alpha)\,K_n(\theta)$ for $\alpha = 1, 2, \ldots$. Then we consider an estimator $\hat{\theta} := \hat{\theta}(S_n)$ for θ such that

$$K_n^{(1)}(\hat{\theta}) = y \tag{2.11}$$

when $S_n = y$ for $y = 0, \pm 1, \pm 2, \ldots$. Using the estimator $\hat{\theta}$, we have the following expansion of $p_n^*(y)$.

Theorem 1. *If $K_n^{(j)}(\hat{\theta}) = O(n)$ $(j = 2, 3, \ldots)$, then the probability function $p_n^*(y)$ of the sum S_n is asymptotically given by*

$$p_n^*(y) = \frac{1}{\sqrt{2\pi K_n^{(2)}(\hat{\theta})}} e^{K_n(\hat{\theta}) - \hat{\theta}y} \left[1 + \frac{K_n^{(4)}(\hat{\theta})}{8\{K_n^{(2)}(\hat{\theta})\}^2} - \frac{5\{K_n^{(3)}(\hat{\theta})\}^2}{24\{K_n^{(2)}(\hat{\theta})\}^3} + O\left(\frac{1}{n^2}\right)\right].$$

The proof is omitted, since it is essentially similar to Chapter 2 of Jensen (1995).

Next we define the amount of the Kullback–Leibler information between probability functions $p_{j,\theta}(\cdot)$ and $p_j(\cdot)$ as

$$I_j(\theta, 0) := \sum_x p_{j,\theta}(x) \log \frac{p_{j,\theta}(x)}{p_j(x)}$$

for $j = 1, \ldots, n, \ldots$. From (2.3) we have for each j

$$I_j(\theta, 0) = \sum_x p_{j,\theta}(x) \log \left(e^{\theta x} M_j(\theta)^{-1}\right)$$

$$= \theta \frac{\partial}{\partial \theta} \log M_j(\theta) - \log M_j(\theta),$$

hence, in a similar way to the above, it follows from (2.5) that the amount of the Kullback–Leibler information between $p_{n,\theta}^*(\cdot)$ and $p_n^*(\cdot)$ is given by

$$I_n^*(\theta, 0) = \sum_y p_{n,\theta}^*(y) \log \frac{p_{n,\theta}^*(y)}{p_n^*(y)}$$

$$= \sum_{j=1}^n I_j(\theta, 0) = \theta K_n^{(1)}(\theta) - K_n(\theta). \tag{2.12}$$

From (2.10) and (2.11) we have

$$I_n^*(\hat{\theta}, 0) = \hat{\theta}y - K_n(\hat{\theta}). \tag{2.13}$$

Then we obtain the following.

Corollary. *If* $K_n^{(j)}(\hat{\theta}) = O(n)$ $(j = 2, 3, \ldots)$, *then the probability function* $p_n^*(y)$ *of the sum* S_n *has the asymptotic form*

$$p_n^*(y) = \frac{1}{\sqrt{2\pi K_n^{(2)}(\hat{\theta})}} e^{-I_n^*(\hat{\theta},0)} \left[1 + \frac{K_n^{(4)}(\hat{\theta})}{8\{K_n^{(2)}(\hat{\theta})\}^2} - \frac{5\{K_n^{(3)}(\hat{\theta})\}^2}{24\{K_n^{(2)}(\hat{\theta})\}^3} + O\left(\frac{1}{n^2}\right) \right].$$

The proof is straighforward from Theorem 1 and (2.13).

We also have the tail probability of S_n as follows.

Theorem 2. *If* $K_n^{(j)}(\hat{\theta}) = O(n)$ $(j = 2, 3, \ldots)$, *then*

$$P\{S_n \geq y\} = \frac{1}{\sqrt{2\pi K_n^{(2)}(\hat{\theta})}} e^{K_n(\hat{\theta}) - \hat{\theta}y} \sum_{z=0}^{\infty} e^{-\hat{\theta}z} \left[1 - \frac{z^2}{2K_n^{(2)}(\hat{\theta})} - \frac{K_n^{(3)}(\hat{\theta})z}{2\{K_n^{(2)}(\hat{\theta})\}^2} \right.$$

$$\left. + \frac{K_n^{(4)}(\hat{\theta})}{8\{K_n^{(2)}(\hat{\theta})\}^2} - \frac{5\{K_n^{(3)}(\hat{\theta})\}^2}{24\{K_n^{(2)}(\hat{\theta})\}^3} + O\left(\frac{1}{n^2}\right) \right]$$

$$(2.14)$$

for all $y > E(S_n)$.

The proof is given in Section 5.

Note that the formula (2.14) does not include any term in which $1 - \Phi(x)$ appears, as are the formulas discussed by Jensen (1995) and others, where $\Phi(x)$ denotes the standard normal distribution function. Both can be related to each other through

$$1 - \Phi(x) = \phi(x) \left\{ \frac{1}{x} - \frac{1}{x^3} + \frac{3}{x^5} - \frac{3 \cdot 5}{x^7} \right.$$

$$\left. + \cdots + (-1)^n \frac{1 \cdot 3 \cdot 5 \cdots (2n - 1)}{x^{2n+1}} + O\left(\frac{1}{x^{2n+3}}\right) \right\}$$

as $x \to \infty$, where $\phi(x)$ is the standard normal density function.

It is to be remarked that in (2.14) the right-hand side is meaningful only when $\hat{\theta}$ is not too small, which implies that $y - E(S_n)$ is of order n. This is in contrast to the usual large-deviation approximation (such as discussed in Lugannani and Rice (1980) and Robinson (1982)) which is based on the normal density, where $\hat{\theta}$ is assumed to be small, that is $y - E(S_n)$ is of $o(n)$. The formula (2.14) can be regarded as *truely large-deviation* approximation.

We could further improve the approximation (2.14) by setting the limit of summation by actual possible largest value.

3. BINOMIAL CASES

Suppose that $X_1, \ldots, X_n, \ldots$ is a sequence of independent random variables and, for each $j = 1, \ldots, n, \ldots$, X_j is distributed according to the binomial distribution $B(1, p_j)$ with a probability function

$$f_{X_j}(x) = P\{X_j = x\} = p_j^x q_j^{1-x}$$

for $x = 0, 1$, where $0 < p_j < 1$ and $q_j = 1 - p_j$. Let $S_n := \sum_{j=1}^{n} X_j$. Then we consider the exact distribution, large-deviation approximations, normal approximation and Edgeworth expansion for S_n and compare them numerically. (i) *Exact distribution.* We obtain the exact distribution of S_n recusively as follows.

$$P\{S_k = 0\} = P\left\{\sum_{j=1}^{k} X_j = 0\right\} = P\{X_1 = 0, \ldots, X_k = 0\} = q_1 \cdots q_k$$

for $k = 1, \ldots, n$.

$$P\{S_k = y\} = P\{X_k = 1\}P\{S_k = y \,|\, X_k = 1\} + P\{X_k = 0\}P\{S_k = y \,|\, X_k = 0\}$$
$$= p_k P\{S_{k-1} = y - 1\} + q_k P\{S_{k-1} = y\}$$

for $y = 1, \ldots, k$ and $k = 1, \ldots .n$, where $P\{S_k = y\} = 0$ for $y \geq k + 1$. (ii) *Large-deviation approximations.* Since the m.g.f. of X_j is given by $M_j(\theta) = p_j e^\theta + q_j$ for each $j = 1, \ldots, n$, it follows that

$$p_{j,\theta}(x) = P_\theta\{X_j = x\} = (p_j e^\theta)^x q_j^{1-x} \big/ (p_j e^\theta + q_j)$$

for $j = 1, \ldots, n$. Since

$$K_n(\theta) = \sum_{j=1}^{n} \log M_j(\theta) = \sum_{j=1}^{n} \log(p_j e^\theta + q_j),$$

we have

$$K_n^{(1)}(\theta) = \sum_{j=1}^n \frac{p_j e^\theta}{p_j e^\theta + q_j} = \sum_{j=1}^n p_{j,\theta}(1).$$

Then we take $\hat\theta$ such that

$$\sum_{j=1}^n p_{j,\hat\theta}(1) = y. \tag{3.1}$$

Putting $\hat\tau := e^{\hat\theta}$, we have from (3.1)

$$y = \sum_{j=1}^n p_{j,\hat\theta}(1) = \sum_{j=1}^n \frac{p_j e^{\hat\theta}}{p_j e^{\hat\theta} + q_j} = \sum_{j=1}^n \frac{p_j \hat\tau}{p_j \hat\tau + q_j}, \tag{3.2}$$

and also

$$n - y = \sum_{j=1}^n \frac{q_j}{p_j \hat\tau + q_j}. \tag{3.3}$$

From (3.2) or (3.3) we can obtain the value of $\hat\tau$. We put $\hat p_j := p_{j,\hat\theta}(1)$ and $\hat q_j := 1 - \hat p_j$. Then we have

$$\begin{aligned}
I_n^\star(\hat\theta, 0) &= \sum_{j=1}^n \left\{ p_{j,\hat\theta}(1) \log \frac{p_{j,\hat\theta}(1)}{p_j(1)} + p_{j,\hat\theta}(0) \log \frac{p_{j,\hat\theta}(0)}{p_j(0)} \right\} \\
&= \sum_{j=1}^n \left(\hat p_j \log \frac{\hat p_j}{p_j} + \hat q_j \log \frac{\hat q_j}{q_j} \right).
\end{aligned} \tag{3.4}$$

Since

$$\begin{aligned}
K_n^{(2)}(\theta) &= \sum_{j=1}^n \left\{ \frac{p_j e^\theta}{p_j e^\theta + q_j} - \frac{p_j^2 e^{2\theta}}{(p_j e^\theta + q_j)^2} \right\} \\
&= \sum_{j=1}^n p_{j,\theta}(1) \left\{ 1 - p_{j,\theta}(1) \right\},
\end{aligned}$$

it follows that

$$K_n^{(2)}(\hat\theta) = \sum_{j=1}^n \hat p_j \hat q_j. \tag{3.5}$$

712   AKAHIRA, TAKAHASHI, AND TAKEUCHI

In a similar way to the above, we obtain

$$K_n^{(3)}(\hat\theta) = \sum_{j=1}^{n} \hat p_j \hat q_j (\hat q_j - \hat p_j), \tag{3.6}$$

$$K_n^{(4)}(\hat\theta) = \sum_{j=1}^{n} \hat p_j \hat q_j (1 - 6\hat p_j \hat q_j). \tag{3.7}$$

Since

$$e^{K(\hat\theta)} = \prod_{j=1}^{n} M_j(\hat\theta) = \prod_{j=1}^{n}(p_j e^{\hat\theta} + q_j) = \prod_{j=1}^{n} \frac{q_j}{\hat q_j}$$

it follows from Theorem 1 and (3.5) that the first order large-deviation approximation LD_1 of $p_n^*(y)$ is given by

$$p_n^*(y) = \frac{1}{\sqrt{2\pi K_n^{(2)}(\hat\theta)}} e^{K_n(\hat\theta) - \hat\theta y}\,(1 + o(1))$$

$$= \frac{1}{\sqrt{2\pi \sum_{j=1}^{n} \hat p_j \hat q_j}} \left(\prod_{j=1}^{n} \frac{q_j}{\hat q_j}\right) \hat\tau^{-y}\,(1 + o(1)) \tag{3.8}$$

for $y = 0, 1, \ldots, n$. From the Corollary we have the second order large-deviation approximation LD_2 of $p_n^*(y)$ is given by

$$p_n^*(y) = \frac{1}{\sqrt{2\pi K_n^{(2)}(\hat\theta)}} e^{-I_n^*(\hat\theta,0)} \left[1 + \frac{K_n^{(4)}(\hat\theta)}{8\{K_n^{(2)}(\hat\theta)\}^2} - \frac{5\{K_n^{(3)}(\hat\theta)\}^2}{24\{K_n^{(2)}(\hat\theta)\}^3} + O\left(\frac{1}{n^2}\right) \right]$$

$$= \frac{1}{\sqrt{2\pi \sum_{j=1}^{n} \hat p_j \hat q_j}} \left(\prod_{j=1}^{n} \frac{q_j}{\hat q_j}\right) \hat\tau^{-y} \left[1 + \frac{K_n^{(4)}(\hat\theta)}{8\left(\sum_{j=1}^{n} \hat p_j \hat q_j\right)^2} \right.$$

$$\left. - \frac{5\{K_n^{(3)}(\hat\theta)\}^2}{24\left(\sum_{j=1}^{n} \hat p_j \hat q_j\right)^3} + O\left(\frac{1}{n^2}\right) \right] \tag{3.9}$$

for $y = 0, 1, \ldots, n$, where $K_n^{(3)}(\hat\theta)$ and $K_n^{(4)}(\hat\theta)$ are given by (3.6) and (3.7), respectively. In the special cases $y = 0$ and $y = n$, it is easily seen that $p_n^*(0) = q_1 \cdots q_n$ and $p_n^*(n) = p_1 \cdots p_n$.

Since the cumulant of S_n are given by

$$\mu_n := E(S_n) = \sum_{j=1}^{n} p_j, \quad v_n := V(S_n) = \sum_{j=1}^{n} p_j q_j,$$

$$\kappa_{3,n} := \kappa_3(S_n) = \sum_{j=1}^{n} p_j q_j (q_j - p_j),$$

$$\kappa_{4,n} := \kappa_4(S_n) = \sum_{j=1}^{n} p_j q_j (1 - 6 p_j q_j),$$

it follows that the Edgeworth expansion of the distribution of S_n is given by

$$P\{S_n = t\} = P\left\{\frac{S_n - \mu_n}{\sqrt{v_n}} = \frac{t - \mu_n}{\sqrt{v_n}}\right\}$$

$$= \frac{1}{\sqrt{v_n}} \phi(y) \left\{ 1 + \frac{\kappa_{3,n}}{6 v_n^{3/2}} (y^3 - 3y) + \frac{\kappa_{4,n}}{24 v_n^2} (y^4 - 6y^2 + 3) \right.$$

$$\left. + \frac{\kappa_{3,n}^2}{72 v_n^3} (y^6 - 15y^4 + 45y^2 - 15) \right\} + o\left(\frac{1}{n\sqrt{n}}\right), \qquad (3.10)$$

where $y := (t - \mu_n)/\sqrt{v_n}$. The first term of the final expression of (3.10) is called the normal approximation of the distribution of S_n. We consider the cases when (i) $\{p_j\}_{j=1}^{19} = 0.05(0.05)0.95$, (ii) $\{p_j\}_{j=1}^{20} = 0.03(0.03)0.60$, (iii) $\{p_j\}_{j=1}^{20}$ is uniformly distributed on the interval $(0,1)$. In each case, we numerically compare the exact distribution of S_n, the normal approximation, the Edgeworth one (3.10), the first order large-deviation one LD_1 (3.8) and the second order large-deviation one LD_2 (3.9) (see Tables 3.1 to 3.3). It is seen that the large-deviation approximation LD_1 and LD_2 are much better than the others and are also sufficiently accurate especially for the tail part. Note that the normal approximation and the Edgeworth one correspond to LD_1 and LD_2, respectively, in view of the order.

Next we consider the one-sided probability of $S_n = \sum_{j=1}^{n} X_j$ the binomial case. In a similar way to the proof of Theorem 2 we have

$$P\{S_n \geq y\} = \sum_{z=0}^{n-y} p_n^*(y + z)$$

$$= \frac{1}{\sqrt{2\pi K_n^{(2)}(\hat{\theta})}} e^{K_n(\hat{\theta}) - \hat{\theta}y} \sum_{z=0}^{n-y} e^{-\hat{\theta}z} \left[1 - \frac{z^2}{2 K_n^{(2)}(\hat{\theta})} - \frac{K_n^{(3)}(\hat{\theta}) z}{2\{K_n^{(2)}(\hat{\theta})\}^2} \right.$$

$$\left. + \frac{K_n^{(4)}(\hat{\theta})}{8\{K_n^{(2)}(\hat{\theta})\}^2} - \frac{5\{K_n^{(3)}(\hat{\theta})\}^2}{24\{K_n^{(2)}(\hat{\theta})\}^3} + O\left(\frac{1}{n^2}\right) \right] \qquad (3.11)$$

for $y = 0, 1, \ldots, n$. Putting $m := n - y$, we obtain

$$\sum_{z=0}^{m} e^{-\hat{\theta}z} = \frac{1 - e^{-(m+1)\hat{\theta}}}{1 - e^{-\hat{\theta}}},$$

$$\sum_{z=0}^{m} z e^{-\hat{\theta}z} = -\frac{(m+1)e^{-(m+1)\hat{\theta}}}{1 - e^{-\hat{\theta}}} + \frac{(1 - e^{-(m+1)\hat{\theta}})e^{-\hat{\theta}}}{(1 - e^{-\hat{\theta}})^2},$$

$$\sum_{z=0}^{m} z^2 e^{-\hat{\theta}z} = -\frac{(m+1)^2 e^{-(m+1)\hat{\theta}}}{1 - e^{-\hat{\theta}}} + \frac{1}{(1 - e^{-\hat{\theta}})^3} \left\{ -(2m+3)e^{-(m+2)\hat{\theta}} \right.$$
$$\left. + (2m+1)e^{-(m+3)\hat{\theta}} + e^{-\hat{\theta}}(1 + e^{-\theta}) \right\}.$$

From (3.11) we have

$$P\{S_n \geq y\}$$
$$= \frac{1}{\sqrt{2\pi K_n^{(2)}(\hat{\theta})}} e^{K_n(\hat{\theta}) - \hat{\theta}y} \left[\frac{1 - e^{-(m+1)\hat{\theta}}}{1 - e^{-\hat{\theta}}} \left\{ 1 + \frac{K_n^{(4)}(\hat{\theta})}{8(K_n^{(2)}(\hat{\theta}))^2} - \frac{5(K_n^{(3)}(\hat{\theta}))^2}{24(K_n^{(2)}(\hat{\theta}))^3} \right\} \right.$$
$$- \frac{1}{2K_n^{(2)}(\hat{\theta})} \left\{ -\frac{(m+1)^2 e^{-(m+1)\hat{\theta}}}{1 - e^{-\hat{\theta}}} \right.$$
$$\left. + \frac{1}{(1 - e^{-\hat{\theta}})^3} \left(-(2m+3)e^{-(m+2)\hat{\theta}} + (2m+1)e^{-(m+3)\hat{\theta}} + e^{-\hat{\theta}}(1 + e^{-\theta}) \right) \right\}$$
$$\left. - \frac{K_n^{(3)}(\hat{\theta})}{2\{K_n^{(2)}(\hat{\theta})\}^2} \left\{ -\frac{(m+1)e^{-(m+1)\hat{\theta}}}{1 - e^{-\hat{\theta}}} + \frac{(1 - e^{-(m+1)\hat{\theta}})e^{-\hat{\theta}}}{(1 - e^{-\hat{\theta}})^2} \right\} + O\left(\frac{1}{n^2}\right) \right].$$

Letting $\hat{\tau} = e^{\hat{\theta}}$, we have

$$P\{S_n \geq y\}$$
$$= \frac{1}{\sqrt{2\pi \sum_{j=1}^{n} \hat{p}_j \hat{q}_j}} \left(\prod_{j=1}^{n} \frac{q_j}{\hat{q}_j} \right) \hat{\tau}^{-y}$$
$$\cdot \left[\frac{1 - \hat{\tau}^{-(m+1)}}{1 - \hat{\tau}^{-1}} \left\{ 1 + \frac{K_n^{(4)}(\hat{\theta})}{8(\sum_{j=1}^{n} \hat{p}_j \hat{q}_j)^2} - \frac{5\{K_n^{(3)}(\hat{\theta})\}^2}{24(\sum_{j=1}^{n} \hat{p}_j \hat{q}_j)^3} \right\} \right.$$
$$- \frac{1}{2\sum_{j=1}^{n} \hat{p}_j \hat{q}_j} \left\{ -\frac{(m+1)^2 \hat{\tau}^{-(m+1)}}{1 - \hat{\tau}^{-1}} \right.$$
$$\left. + \frac{1}{(1 - \hat{\tau}^{-1})^3} \left(-(2m+3)\hat{\tau}^{-(m+2)} + (2m+1)\hat{\tau}^{-(m+3)} + \hat{\tau}^{-1}(1 + \hat{\tau}^{-1}) \right) \right\}$$

HIGHER ORDER LARGE-DEVIATION APPROXIMATION 715

TABLE 3.1. The values of the exact distribution $P\{S_n = y\}$ of S_n, and the relative errors of the normal approximation, the Edgeworth one (3.10), the first order large-deviation one LD_1 (3.8), the second order large-deviation one LD_2 (3.9) when $\{p_j\}_{j=1}^{19} = 0.05(0.05)0.95$.

y	Exact(%)	Normal	Edgeworth	LD_1	LD_2
0	0.0000	11.0363	——	0.0000	0.0000
1	0.0001	2.4706	−0.5734	0.0742	−0.0080
2	0.0025	0.8305	−0.0435	0.0321	−0.0024
3	0.0293	0.2991	0.0076	0.0196	−0.0010
4	0.2122	0.0905	0.0040	0.0141	−0.0005
5	1.0319	0.0091	−0.0000	0.0112	−0.0002
6	3.5182	−0.0144	−0.0006	0.0095	−0.0002
7	8.6510	−0.0120	−0.0000	0.0085	−0.0001
8	15.6199	−0.0140	0.0001	0.0079	−0.0001
9	20.9349	0.0065	−0.0000	0.0076	−0.0001
10	20.9349	0.0065	−0.0000	0.0076	−0.0001
11	15.6199	−0.0140	0.0001	0.0079	−0.0001
12	8.6510	−0.0120	−0.0000	0.0085	−0.0001
13	3.5182	−0.0144	−0.0006	0.0095	−0.0002
14	1.0319	0.0091	−0.0000	0.0112	−0.0002
15	0.2122	0.0905	0.0040	0.0141	−0.0005
16	0.0293	0.2991	0.0076	0.0196	−0.0010
17	0.0025	0.8305	−0.0435	0.0321	−0.0024
18	0.0001	2.4706	−0.5734	0.0742	−0.0080
19	0.0000	11.0363	——	0.0000	0.0000

TABLE 3.2. The values of the exact distribution $P\{S_n = y\}$ of S_n, and the relative errors of the normal approximation, the Edgeworth one (3.10), the first order large-deviation one LD_1 (3.8), the second order large-deviation one LD_2 (3.9) when $\{p_j\}_{j=1}^{20} = 0.03(0.03)0.60$.

y	Exact(%)	Normal	Edgeworth	LD_1	LD_2
0	0.0263	2.7822	0.0933	0.0000	0.0000
1	0.2970	0.5926	0.0512	0.0840	−0.0059
2	1.5463	0.1126	0.0032	0.0416	−0.0013
3	4.9248	−0.0290	−0.0058	0.0273	−0.0006
4	10.7461	−0.0548	−0.0014	0.0201	−0.0003
5	17.0540	−0.0334	0.0018	0.0159	−0.0002
6	20.3938	0.0024	0.0005	0.0132	−0.0002
7	18.7873	0.0312	−0.0014	0.0114	−0.0002
8	13.5171	0.0378	−0.0002	0.0102	−0.0001
9	7.6552	0.0139	0.0025	0.0094	−0.0001
10	3.4236	−0.0416	0.0011	0.0089	−0.0001
11	1.2081	−0.1226	−0.0071	0.0087	−0.0001
12	0.3348	−0.2185	−0.0131	0.0089	−0.0001
13	0.0722	−0.3169	0.0087	0.0094	−0.0001
14	0.0120	−0.4057	0.1017	0.0105	−0.0002
15	0.0015	−0.4748	0.3331	0.0122	−0.0003
16	0.0001	−0.5146	0.8263	0.0152	−0.0005
17	0.0000	−0.5092	1.8805	0.0210	−0.0010
18	0.0000	−0.4175	4.4459	0.0339	−0.0022
19	0.0000	−0.0774	12.7190	0.0764	−0.0075
20	0.0000	1.6449	60.6571	0.0000	0.0000

TABLE 3.3. The values of the exact distribution $P\{S_n = y\}$ of S_n, and the relative errors of the normal approximation, the Edgeworth one (3.10), the first order large-deviation one LD_1 (3.8), the second order large-deviation one LD_2 (3.9) when $\{p_j\}_{j=1}^{20}$ is uniformly distributed on the interval $(0,1)$, that is, $p_1, \ldots, p_{20}$ are given by 0.305146, 0.715095, 0.612101, 0.672283, 0.447648, 0.268358, 0.434328, 0.552620, 0.608603, 0.130255, 0.941095, 0.141198, 0.164085, 0.693920, 0.565611, 0.977985, 0.0513902, 0.877854, 0.451323, 0.0628465, respectively.

y	Exact(%)	Normal	Edgeworth	LD_1	LD_2
0	0.0000	36.8118	——	0.0000	0.0000
1	0.0001	5.5998	——	0.0564	−0.0122
2	0.0015	1.5567	−0.0944	0.0186	−0.0036
3	0.0210	0.5223	0.0311	0.0123	−0.0012
4	0.1670	0.1642	0.0138	0.0119	−0.0000
5	0.8584	0.0291	0.0001	0.0120	0.0003
6	3.0390	−0.0142	−0.0025	0.0117	0.0003
7	7.7151	−0.0172	−0.0007	0.0110	0.0003
8	14.4053	−0.0057	0.0007	0.0102	0.0002
9	20.0831	0.0055	0.0006	0.0093	0.0001
10	21.0635	0.0087	−0.0002	0.0085	0.0000
11	16.6429	0.0025	−0.0006	0.0077	−0.0000
12	9.8669	−0.0091	−0.0003	0.0072	−0.0002
13	4.3488	−0.0167	0.0006	0.0070	−0.0003
14	1.4041	−0.0059	0.0019	0.0074	−0.0004
15	0.3251	0.0457	0.0024	0.0091	−0.0005
16	0.0524	0.1788	−0.0065	0.0125	−0.0007
17	0.0056	0.4856	−0.0668	0.0192	−0.0011
18	0.0004	1.2369	−0.3813	0.0338	−0.0022
19	0.0000	3.5355	——	0.0789	−0.0064
20	0.0000	15.4703	——	0.0000	0.0000

TABLE 3.4. The values of the exact tail probability $P\{S_n \geq y\}$ of S_n, and the relative errors of its Edgeworth approximation (3.13) and its second order large-deviation approximation LD_2 (3.12) when $\{p_j\}_{j=1}^{19} = 0.05(0.05)0.95$.

y	Exact(%)	Edgeworth	LD_2
10	50.0000	0.0000	——
11	29.0651	−0.0000	−0.8356
12	13.4452	0.0000	−0.2010
13	4.7942	0.0004	−0.0627
14	1.2761	0.0014	−0.0232
15	0.2442	−0.0006	−0.0097
16	0.0320	−0.0293	−0.0048
17	0.0027	0.2154	−0.0038
18	0.0001	——	−0.0083
19	0.0000	——	——

TABLE 3.5. The values of the exact tail probability $P\{S_n \geq y\}$ of S_n, and the relative errors of its Edgeworth approximation (3.13) and its second order large-deviation approximation LD_2 (3.12) when $\{p_j\}_{j=1}^{20} = 0.03(0.03)0.60$.

y	Exact(%)	Edgeworth	LD_2
7	45.0120	0.0002	—
8	26.2247	0.0003	−0.9933
9	12.7076	−0.0003	−0.2467
10	5.0524	−0.0032	−0.0840
11	1.6288	−0.0067	−0.0346
12	0.4207	0.0004	−0.0161
13	0.0859	0.0451	−0.0082
14	0.0136	0.1758	−0.0044
15	0.0016	0.4772	−0.0026
16	0.0001	1.1191	−0.0018
17	0.0000	2.5290	−0.0017
18	0.0000	6.0939	−0.0025
19	0.0000	18.1006	−0.0076
20	0.0000	91.0675	—

$$-\frac{K_n^{(3)}(\hat{\theta})}{2(\sum_{j=1}^{n} \hat{p}_j\hat{q}_j)^2}\left\{-\frac{(m+1)\hat{\tau}^{-(m+1)}}{1-\hat{\tau}^{-1}} + \frac{(1-\hat{\tau}^{-(m+1)})\hat{\tau}^{-1}}{(1-\hat{\tau}^{-1})^2}\right\} + O\left(\frac{1}{n^2}\right)\Bigg],$$

$$(3.12)$$

where

$$K_n^{(3)}(\hat{\theta}) = \sum_{j=1}^{n} \hat{p}_j\hat{q}_j(\hat{q}_j - \hat{p}_j), \quad K_n^{(4)}(\hat{\theta}) = \sum_{j=1}^{n} \hat{p}_j\hat{q}_j(1 - 6\hat{p}_j\hat{q}_j).$$

On the other hand, by the Edgeworth expansion we have

$$P\{S_n \geq y\} = 1 - \Phi(z) + \phi(z)\left\{\frac{\kappa_{3,n}}{6v_n^{3/2}}(z^2 - 1) + \frac{\kappa_{4,n}}{24v_n^2}(z^3 - 3z)\right.$$
$$\left. + \frac{\kappa_{3,n}^2}{72v_n^3}(z^5 - 10z^3 + 15z) - \frac{1}{24v_n}z + o\left(\frac{1}{n}\right)\right\},$$

$$(3.13)$$

where $z = (y - 0.5 - \mu_n)/\sqrt{v_n}$, In the cases (i) to (iii), we compare the Edgeworth expansion (3.13) of the tail probability of S_n and the second order large-deviation approximation (3.12) (see Tables 3.4 to 3.6).

TABLE 3.6. The values of the exact tail probability $P\{S_n \geq y\}$ of S_n, and the relative errors of its Edgeworth approximation (3.13) and its second order large-deviation approximation LD_2 (3.12) when $\{p_j\}_{j=1}^{20}$ is uniformly distributed on the interval $(0,1)$, that is, $p_1, \ldots, p_{20}$ are given by 0.305146, 0.715095, 0.612101, 0.672283, 0.447648, 0.268358, 0.434328, 0.552620, 0.608603, 0.130255, 0.941095, 0.141198, 0.164085, 0.693920, 0.565611, 0.977985, 0.0513902, 0.877854, 0.451323, 0.0628465, respectively.

y	Exact(%)	Edgeworth	LD_2
10	53.7097	−0.0002	——
11	32.6462	−0.0003	——
12	16.0033	0.0001	−0.2793
13	6.1365	0.0012	−0.0836
14	1.7877	0.0023	−0.0306
15	0.3836	−0.0014	−0.0129
16	0.0585	−0.0307	−0.0061
17	0.0060	0.1676	−0.0036
18	0.0004	−0.7757	−0.0032
19	0.0000	——	−0.0067
20	0.0000	——	——

TABLE 3.7. The values of the exact tail probability $P\{S_n \geq y\}$ of S_n, and the relative errors of its normal approximation, the saddlepoint approximation with the first term only, the saddlepoint (sp.) expansion, the Lugannani-Rice formula and the large-deviation approximation LD_2 when $p_j = p = 0.15$ $(j = 1, \ldots, n)$ for $n = 10, 20$. The values except the exact and the large-deviation approximation LD_2 are calculated from Table 2.4.4 of Jensen (1995, page 44).

$n = 10, p = 0.15$

y	2	4	5	8	9
Exact(%)	45.57	5.00	0.987	8.67×10^{-4}	3.33×10^{-5}
Normal	0.097	−0.234	−0.601	−0.994	−1.000
Saddlepoint	−0.137	−0.050	−0.035	−0.007	0.021
Sp.-expansion	0.004	0.004	0.005	0.010	0.024
Lugannani-Rice	0.004	0.008	0.013	0.038	0.081
LD_2	—	−0.070	−0.021	−0.003	−0.009

$n = 20, p = 0.15$

y	6	8	18	19
Exact(%)	6.73	0.592	2.07×10^{-11}	3.80×10^{-13}
Normal	−0.128	−0.591	−1.000	−1.000
Saddlepoint	−0.043	−0.027	0.010	0.039
Sp.-expansion	0.001	0.002	0.005	0.021
Lugannani-Rice	0.003	0.005	0.039	0.084
LD_2	−0.126	−0.022	−0.005	−0.008

In the independently, identically and binomially distributed case, Jensen (1995, page 44) gives the table to compare the exact tail probability with the normal approximation, the saddlepoint approximation with the first term only, the saddlepoint expansion and the Lugannani-Rice formula. So, we add our large-deviation approximation LD_2 to the table when $p_j = p = 0.15$ ($j = 1, \ldots, n$) for $n = 10, 20$ (see Table 3.7). The LD_2 seems to be better than the others for the tail part in the case.

4. NEGATIVE-BINOMIAL CASE

Suppose that $X_1, \ldots, X_n, \ldots$ is a sequence of independent random variables and, for each $j = 1, \ldots, n, \ldots$, X_j is distributed according to the negative-binomial distribution $NB(r_j, p_j)$ with a probability function

$$f_{X_j}(x) = P\{X_j = x\} = \binom{x + r_j - 1}{x} p_j^{r_j} q_j^x$$

for $x = 0, 1, 2, \ldots$, where $0 < p_j < 1$ and $q_j = 1 - p_j$. Then we consider the exact distribution, large-deviation approximations, normal approximation and Edgeworth expansion for $S_n = \sum_{j=1}^{n} X_j$ and compare them numerically.

(i) *Exact distribution.* We obtain the exact distribution of S_n recursively follows : Let $n = 20$ and $r_j = r$ for $j = 1, \ldots, 20$. Let $q_1 = \cdots = q_5 = a$, $q_6 = \cdots = q_{10} = b$, $q_{11} = \cdots = q_{15} = c$, $q_{16} = \cdots = q_{20} = d$. Then we have

$$P\{X_j = x\} = \frac{r(r+1)\cdots(r+x-1)}{x!} a^r (1-a)^x$$
$$(x = 0, 1, 2, \ldots ; j = 1, \ldots, 5),$$

$$P\{X_j = x\} = \frac{r(r+1)\cdots(r+x-1)}{x!} b^r (1-b)^x$$
$$(x = 0, 1, 2, \ldots ; j = 6, \ldots, 10),$$

$$P\{X_j = x\} = \frac{r(r+1)\cdots(r+x-1)}{x!} c^r (1-c)^x$$
$$(x = 0, 1, 2, \ldots ; j = 11, \ldots, 15),$$

$$P\{X_j = x\} = \frac{r(r+1)\cdots(r+x-1)}{x!} d^r (1-d)^x$$
$$(x = 0, 1, 2, \ldots ; j = 16, \ldots, 20).$$

Let $S_1' := S_5 = \sum_{j=1}^{5} X_j$, $S_2' := \sum_{j=6}^{10} X_j$, $S_3' := \sum_{j=11}^{15} X_j$ and $S_4' := \sum_{j=16}^{20} X_j$. Then it is seen that S_1', S_2', S_3' and S_4' are distributed according to the negative-binomial distributions $NB(5r, a)$, $NB(5r, b)$, $NB(5r, c)$ and $NB(5r, d)$, respectively. Since S_1' and S_2' are independent, the probability function of $S_1' + S_2'$ is given by

$$P\{S_1' + S_2' = y\} = \sum_{k=0}^{y} P\{S_1' = k\} P\{S_2' = y - k\}$$

for $y = 0, 1, 2, \ldots$. Letting $S_{12}' := S_1' + S_2'$, we have

$$S_1' + S_2' + S_3' = S_{12}' + S_3'.$$

Since S_{12}' and S_3' are independent, in a similar way the probability function of $S_{12}' + S_3' = S_1' + S_2' + S_3'$ can be given. Repeating the above process, we also get the probability function of $S_1' + S_2' + S_3' + S_4'$.

(ii) *Large-deviation approximations.* Since the m.g.f. of X_j is given by

$$M_j(\theta) = E(e^{\theta X_j}) = p_j^{r_j}(1 - q_j e^{\theta})^{-r_j}$$

for small $|\theta|$ and $j = 1, \ldots, n$. Since

$$K_n(\theta) = \sum_{j=1}^{n} \log M_j(\theta) = \sum_{j=1}^{n} \left\{ r_j \log p_j - r_j \log(1 - q_j e^{\theta}) \right\},$$

it follows that

$$K_n^{(1)}(\theta) = \sum_{j=1}^{n} \frac{r_j q_j e^{\theta}}{1 - q_j e^{\theta}}.$$

Then we take $\hat{\theta}$ such that

$$\sum_{j=1}^{n} \frac{r_j q_j e^{\hat{\theta}}}{1 - q_j e^{\hat{\theta}}} = y. \tag{4.1}$$

Putting $\hat{\tau} = e^{\hat{\theta}}$, we have from (4.1)

$$y = \sum_{j=1}^{n} \frac{r_j q_j \hat{\tau}}{1 - q_j \hat{\tau}}.$$

We also obtain

$$K_n^{(2)}(\hat{\theta}) = \sum_{j=1}^{n} \frac{r_j q_j e^{\hat{\theta}}}{(1 - q_j e^{\hat{\theta}})^2} = \sum_{j=1}^{n} \frac{r_j q_j \hat{\tau}}{(1 - q_j \hat{\tau})^2}. \tag{4.2}$$

From Theorem 1, it follows that the first order large-deviation approximation LD_1 of $p_n^*(y)$ is given by

$$
\begin{aligned}
p_n^*(y) &= \frac{1}{\sqrt{2\pi K_n^{(2)}(\hat\theta)}} e^{K_n(\hat\theta) - \hat\theta y}(1 + o(1)) \\
&= \frac{1}{\sqrt{2\pi K_n^{(2)}(\hat\theta)}} \left\{ \prod_{j=1}^n M_j(\hat\theta) \right\} e^{-\hat\theta y}(1 + o(1)) \\
&= \frac{1}{\sqrt{2\pi \sum_{j=1}^n \frac{r_j q_j \hat\tau}{(1 - q_j \hat\tau)^2}}} \left\{ \prod_{j=1}^n \left(\frac{p_j}{1 - q_j \hat\tau} \right)^{r_j} \right\} \hat\tau^{-y}(1 + o(1))
\end{aligned}
\tag{4.3}
$$

for $y = 0, 1, 2, \ldots$. Since

$$
K^{(3)}(\hat\theta) = \sum_{j=1}^n \frac{r_j q_j \hat\tau (1 + q_j \hat\tau)}{(1 - q_j \hat\tau)^3},
\tag{4.4}
$$

$$
K^{(4)}(\hat\theta) = \sum_{j=1}^n \frac{r_j q_j (1 + 4 q_j \hat\tau + q_j^2 \hat\tau^2)}{(1 - q_j \hat\tau)^4},
\tag{4.5}
$$

it follows from Theorem 1 that the second order large-deviation approximation LD_2 of p_n^* is given by

$$
\begin{aligned}
p_n^*(y) &= \frac{1}{\sqrt{2\pi K_n^{(2)}(\hat\theta)}} \left\{ \prod_{j=1}^n \left(\frac{p_j}{1 - q_j \hat\tau} \right)^{r_j} \right\} \hat\tau^{-y} \\
&\quad \cdot \left[1 + \frac{K_n^{(4)}(\hat\theta)}{8\{K_n^{(2)}(\hat\theta)\}^2} - \frac{5\{K_n^{(3)}(\hat\theta)\}^2}{24\{K_n^{(2)}(\hat\theta)\}^3} + O\left(\frac{1}{n^2} \right) \right],
\end{aligned}
\tag{4.6}
$$

where $K_n^{(2)}(\hat\theta)$, $K_n^{(3)}(\hat\theta)$ and $K_n^{(4)}(\hat\theta)$ are given by (4.2), (4.4) and (4.5), respectively.

Since the cumulants of S_n are given by

$$
\mu_n := E(S_n) = \sum_{j=1}^n \frac{q_j r_j}{p_j}, \qquad v_n := V(S_n) = \sum_{j=1}^n \frac{q_j r_j}{p_j^2},
$$

$$
\kappa_{3,n} := \kappa_3(S_n) = \sum_{j=1}^n \frac{q_j(1 + q_j) r_j}{p_j^3},
$$

$$
\kappa_{4,n} := \kappa_4(S_n) = \sum_{j=1}^n \frac{q_j r_j}{p_j^4}(1 + 4q_j + q_j^2),
$$

it follows that the Edgeworth expansion of the distribution of S_n is given by

722   AKAHIRA, TAKAHASHI, AND TAKEUCHI

TABLE 4.1. The values of the exact distribution $P\{S_n = y\}$ of S_n, and the relative errors of the normal approximation, the Edgeworth one (4.7), the first order large-deviation one LD_1 (4.1) and second order large-deviation one LD_2 (4.6), when $n = 20$, $r_j = 1(j = 1, \ldots, 20)$, $q_1 = \cdots = q_5 = 0.1$, $q_6 = \cdots = q_{10} = 0.2$, $q_{11} = \cdots = q_{15} = 0.3$, $q_{16} = \cdots = q_{20} = 0.4$.

y	Exact(%)	Normal	Edgeworth	LD_1	LD_2
0	0.2529	3.1201	0.3185	0.0000	0.0000
1	1.2644	0.5443	0.1379	0.0845	-0.0059
2	3.3506	-0.0044	0.0296	0.0422	-0.0012
3	6.2587	-0.1701	-0.0061	0.0281	-0.0005
4	9.2491	-0.2028	-0.0090	0.0210	-0.0002
5	11.5098	-0.1711	-0.0024	0.0168	-0.0001
6	12.5389	-0.1026	0.0017	0.0139	-0.0001
7	12.2775	-0.0147	0.0007	0.0119	-0.0000
8	11.0112	0.0766	-0.0015	0.0103	-0.0000
9	9.1744	0.1543	0.0002	0.0091	-0.0000
10	7.1794	0.2011	0.0059	0.0081	-0.0000
11	5.3226	0.2025	0.0080	0.0073	-0.0000
12	3.7646	0.1503	-0.0028	0.0066	-0.0000
13	2.5548	0.0453	-0.0267	0.0060	-0.0000
14	1.6716	-0.1018	-0.0475	0.0055	-0.0000
15	1.0586	-0.2733	-0.0387	0.0050	-0.0000
16	0.6512	-0.4481	0.0164	0.0046	-0.0000
17	0.3902	-0.6079	0.1057	0.0043	-0.0000

TABLE 4.2. The values of the exact distribution $P\{S_n = y\}$ of S_n, and the relative errors of the normal approximation, the Edgeworth one (4.7), the first order large-deviation one LD_1 (4.1) and second order large-deviation one LD_2 (4.6), when $n = 20$, $r_j = 2(j = 1, \ldots, 20)$, $q_1 = \cdots = q_5 = 0.1$, $q_6 = \cdots = q_{10} = 0.2$, $q_{11} = \cdots = q_{15} = 0.3$, $q_{16} = \cdots = q_{20} = 0.4$.

y	Exact(%)	Normal	Edgeworth	LD_1	LD_2
0	0.0006	97.8639	—	0.0000	0.0000
1	0.0064	17.9615	—	0.0844	-0.0060
2	0.0329	5.7426	-0.6928	0.0422	-0.0012
3	0.1164	2.3357	-0.0674	0.0281	-0.0005
4	0.3173	1.0422	0.0456	0.0210	-0.0003
5	0.7115	0.4515	0.0460	0.0168	-0.0002
6	1.3660	0.1504	0.0266	0.0140	-0.0001
7	2.3082	-0.0110	0.0101	0.0120	-0.0001
8	3.5026	-0.0961	0.0006	0.0105	-0.0001
9	4.8463	-0.1349	-0.0031	0.0093	-0.0000
10	6.1873	-0.1433	-0.0030	0.0084	-0.0000
11	7.3591	-0.1306	-0.0015	0.0076	-0.0000
12	8.2185	-0.1028	0.0000	0.0069	-0.0000
13	8.6742	-0.0647	0.0006	0.0064	-0.0000
14	8.7003	-0.0205	0.0004	0.0059	-0.0000
15	8.3318	0.0259	-0.0003	0.0055	-0.0000
16	7.6491	0.0700	-0.0006	0.0051	-0.0000
17	6.7559	0.1076	0.0001	0.0048	-0.0000
18	5.7583	0.1342	0.0015	0.0045	-0.0000
19	4.7493	0.1460	0.0028	0.0043	-0.0000
20	3.7996	0.1397	0.0027	0.0040	-0.0000

TABLE 4.3. The values of the exact distribution $P\{S_n = y\}$ of S_n, and the relative errors of the normal approximation, the Edgeworth one (4.7), the first order large-deviation one LD_1 (4.1) and second order large-deviation one LD_2 (4.6), when $n = 20$, $r_j = 1(j = 1, \ldots, 20)$, $q_1 = \cdots = q_5 = 0.2$, $q_6 = \cdots = q_{10} = 0.4$, $q_{11} = \cdots = q_{15} = 0.6$, $q_{16} = \cdots = q_{20} = 0.8$.

y	Exact(%)	Normal	Edgeworth	LD_1	LD_2
0	0.0000	7135.4238	—	0.0000	0.0000
1	0.0001	916.1823	—	0.0845	−0.0059
2	0.0004	219.6504	—	0.0422	−0.0012
3	0.0017	73.7118	—	0.0281	−0.0005
4	0.0049	30.7225	—	0.0210	−0.0002
5	0.0122	14.8687	—	0.0168	−0.0001
6	0.0265	7.9959	—	0.0139	−0.0001
7	0.0519	4.6291	−0.2211	0.0119	−0.0001
8	0.0931	2.8151	0.1071	0.0103	−0.0000
9	0.1551	1.7612	0.2037	0.0091	−0.0000
10	0.2427	1.1110	0.2100	0.0081	−0.0000
11	0.3599	0.6900	0.1831	0.0073	−0.0000
12	0.5091	0.4070	0.1469	0.0066	−0.0000
13	0.6910	0.2111	0.1113	0.0060	−0.0000
14	0.9042	0.0728	0.0801	0.0055	−0.0000
15	1.1454	−0.0261	0.0543	0.0050	−0.0000
16	1.4091	−0.0969	0.0341	0.0046	−0.0000
17	1.6887	−0.1471	0.0188	0.0043	−0.0000
18	1.9764	−0.1817	0.0078	0.0040	−0.0001
19	2.2640	−0.2043	0.0004	0.0037	−0.0001
20	2.5432	−0.2171	−0.0042	0.0035	−0.0001
21	2.8062	−0.2221	−0.0067	0.0033	−0.0001
22	3.0462	−0.2205	−0.0075	0.0032	−0.0001
23	3.2573	−0.2134	−0.0072	0.0030	−0.0002
24	3.4351	−0.2015	−0.0062	0.0029	−0.0002
25	3.5766	−0.1854	−0.0049	0.0028	−0.0002
26	3.6801	−0.1658	−0.0036	0.0028	−0.0002
27	3.7452	−0.1432	−0.0024	0.0027	−0.0003
28	3.7729	−0.1179	−0.0014	0.0027	−0.0003
29	3.7648	−0.0905	−0.0008	0.0027	−0.0003
30	3.7237	−0.0614	−0.0005	0.0027	−0.0004
31	3.6526	−0.0310	−0.0004	0.0027	−0.0004
32	3.5553	0.0001	−0.0004	0.0027	−0.0004
33	3.4357	0.0316	−0.0004	0.0028	−0.0004
34	3.2975	0.0628	−0.0003	0.0029	−0.0005
35	3.1447	0.0932	0.0000	0.0030	−0.0005
36	2.9810	0.1223	0.0007	0.0031	−0.0005
37	2.8098	0.1497	0.0018	0.0031	−0.0005
38	2.6344	0.1746	0.0032	0.0033	−0.0005
39	2.4575	0.1966	0.0048	0.0034	−0.0005
40	2.2815	0.2151	0.0064	0.0035	−0.0005
41	2.1086	0.2297	0.0078	0.0036	−0.0006
42	1.9404	0.2399	0.0087	0.0038	−0.0006
43	1.7785	0.2454	0.0087	0.0039	−0.0006
44	1.6237	0.2458	0.0075	0.0040	−0.0006
45	1.4770	0.2409	0.0048	0.0042	−0.0006

$$P\{S_n = t\} = P\left\{\frac{S_n - \mu_n}{\sqrt{v_n}} = \frac{t - \mu_n}{\sqrt{v_n}}\right\}$$

$$= \frac{1}{\sqrt{v_n}}\phi(y)\left\{1 + \frac{\kappa_{3,n}}{6v_n^{3/2}}(y^3 - 3y) + \frac{\kappa_{4,n}}{24v_n^2}(y^4 - 6y^2 + 3)\right.$$

$$\left. + \frac{\kappa_{3,n}^2}{72v_n^3}(y^6 - 15y^4 + 45y^2 - 15)\right\} + o\left(\frac{1}{n\sqrt{n}}\right), \qquad (4.7)$$

where $y := (t - \mu_n)/\sqrt{v_n}$. The first term of the final expansion of (4.7) is called the normal approximation of the distribution of S_n.

We consider the cases when (i) $n = 20$, $r_j = 1$ $(j = 1, \dots, 20)$, $q_1 = \cdots = q_5 = 0.1$, $q_6 = \cdots = q_{10} = 0.2$, $q_{11} = \cdots = q_{15} = 0.3$, $q_{16} = \cdots = q_{20} = 0.4$, (ii) $n = 20$, $r_j = 2$ $(j = 1, \dots, 20)$, $q_1 = \cdots = q_5 = 0.1$, $q_6 = \cdots = q_{10} = 0.2$, $q_{11} = \cdots = q_{15} = 0.3$, $q_{16} = \cdots = q_{20} = 0.4$, (iii) $n = 20$, $r_j = 1$ $(j = 1, \dots, 20)$, $q_1 = \cdots = q_5 = 0.2$, $q_6 = \cdots = q_{10} = 0.4$, $q_{11} = \cdots = q_{15} = 0.6$, $q_{16} = \cdots = q_{20} = 0.8$. In each case, we compare the exact distribution of S_n, the normal approximation, the Edgeworth one (4.7), the first order large-deviation one LD_1 (4.3) and the second order large-deviation one LD_2 (4.6) (see Tables 4.1 to 4.3). It is seen that the large-deviation approximation LD_1 and LD_2 are much better than the others. Note that the normal approximation and the Edgeworth one correspond to LD_1 and LD_2, respectively, in view of the order.

5. PROOF

In this section we give the proof of Theorems 2 in Section 2.

Proof of Theorem 2. From (2.10) we have for any $z \geq 0$

$$p_n^*(y + z) = e^{K_n(\hat{\theta}) - \hat{\theta}(y+z)} \cdot \frac{1}{2\pi}\int_{-\pi}^{\pi} e^{K_n(\hat{\theta}+it) - K_n(\hat{\theta}) - it(y+z)}dt.$$

We also obtain

$$\frac{1}{2\pi}\int_{-\pi}^{\pi} e^{K_n(\hat{\theta}+it) - K_n(\hat{\theta}) - it(y+z)}dt$$

$$= \frac{1}{2\pi\sqrt{K_n^{(2)}(\hat{\theta})}}\int_{-\pi\sqrt{K_n^{(2)}(\hat{\theta})}}^{\pi\sqrt{K_n^{(2)}(\hat{\theta})}} \exp\left\{K_n\left(\hat{\theta} + \frac{iu}{\sqrt{K_n^{(2)}(\hat{\theta})}}\right)\right.$$

$$\left. - K_n(\hat{\theta}) - \frac{iu(y+z)}{\sqrt{K_n^{(2)}(\hat{\theta})}}\right\}du$$

$$
= \frac{1}{2\pi\sqrt{K_n^{(2)}(\hat{\theta})}} \int_{-\pi\sqrt{K_n^{(2)}(\hat{\theta})}}^{\pi\sqrt{K_n^{(2)}(\hat{\theta})}} \exp\left\{ -\frac{iuz}{\sqrt{K_n^{(2)}(\hat{\theta})}} + \frac{1}{2}(iu)^2 \right.
$$

$$
\left. + \frac{1}{6}K_n^{(3)}(\hat{\theta})\left(\frac{iu}{\sqrt{K_n^{(2)}(\hat{\theta})}}\right)^3 + \frac{1}{24}K_n^{(4)}(\hat{\theta})\left(\frac{iu}{\sqrt{K_n^{(2)}(\hat{\theta})}}\right)^4 \right\} du + O\left(\frac{1}{n^2}\right)
$$

$$
= \frac{1}{\sqrt{2\pi K_n^{(2)}(\hat{\theta})}} \left[\int_{-\infty}^{\infty} \phi(u)\left\{ 1 - \frac{izu}{\sqrt{K_n^{(2)}(\hat{\theta})}} + \frac{1}{6}K_n^{(3)}(\hat{\theta})\left(\frac{iu}{\sqrt{K_n^{(2)}(\hat{\theta})}}\right)^3 \right.\right.
$$

$$
+ \frac{1}{24}\left(\frac{iu}{\sqrt{K_n^{(2)}(\hat{\theta})}}\right)^4 - \frac{z^2u^2}{2K_n^{(2)}(\hat{\theta})} + \frac{1}{72}\{K_n^{(3)}(\hat{\theta})\}^2\left(\frac{iu}{\sqrt{K_n^{(2)}(\hat{\theta})}}\right)^6
$$

$$
\left.\left. - \frac{K_n^{(3)}(\hat{\theta})zu^4}{6\{K_n^{(2)}(\hat{\theta})\}^2} \right\} du + O\left(\frac{1}{n^2}\right) \right]
$$

$$
= \frac{1}{\sqrt{2\pi K_n^{(2)}(\hat{\theta})}} \left\{ 1 - \frac{z^2}{2K_n^{(2)}(\hat{\theta})} - \frac{K_n^{(3)}(\hat{\theta})z}{2\{K_n^{(2)}(\hat{\theta})\}^2} + \frac{K_n^{(4)}(\hat{\theta})}{8\{K_n^{(2)}(\hat{\theta})\}^2} \right.
$$

$$
\left. - \frac{5\{K_n^{(3)}(\hat{\theta})\}^2}{24\{K_n^{(2)}(\hat{\theta})\}^3} + O\left(\frac{1}{n^2}\right) \right\},
$$

where $\phi(u) = (1/\sqrt{2\pi})e^{-u^2/2}$. Hence we have

$$
P\{S_n \geq y\} = \sum_{z=0}^{\infty} p_n^*(y+z)
$$

$$
= \frac{1}{\sqrt{2\pi K_n^{(2)}(\hat{\theta})}} e^{K_n(\hat{\theta}) - \hat{\theta}y} \sum_{z=0}^{\infty} e^{-\hat{\theta}z}\left\{ 1 - \frac{z^2}{2K_n^{(2)}(\hat{\theta})} - \frac{K_n^{(3)}(\hat{\theta})z}{2\{K_n^{(2)}(\hat{\theta})\}^2} \right.
$$

$$
\left. + \frac{K_n^{(4)}(\hat{\theta})}{8\{K_n^{(2)}(\hat{\theta})\}^2} - \frac{5\{K_n^{(3)}(\hat{\theta})\}^2}{24\{K_n^{(2)}(\hat{\theta})\}^3} + O\left(\frac{1}{n^2}\right) \right\} \tag{5.1}
$$

for all integer $y > E(S_n)$. Thus we complete the proof.

ACKNOWLEDGEMENTS

The authors wish to thank the referee for his comment on the independently, identically and binomially distributed case in Jensen (1995).

REFERENCES

Barndorff-Nielsen, O. E. and Cox, D. R. (1979). Edgeworth and saddlepoint approximations with statistical applications (with discussion). *J. Roy. Statist. Soc. Ser. B*, **41**, 279–312.

Barndorff-Nielsen, O. E. and Cox, D. R. (1989). *Asymptotic Techniques for Use in Statistics*. Chapman and Hall, London.

Booth, J. G. and Wood, A. T. A. (1995). An example in which th Lugannani-Rice saddlepoint formula fails. *Statist. Probab. Lett.*, **23**, 53–61.

Daniels, H. E. (1954). Saddlepoint approximations in statistics. *Ann. Math. Statist.*, **25**, 631–650.

Daniels, H. E. (1987). Tail probability approximations. *Int. Statist. Review*, **55**, 37–48.

Gatto, R. and Ronchetti, E. (1996). General saddlepoint approximations of marginal densities and tail probabilities. *J. Amer. Statist. Assoc.*, **91**, 666-673.

Harvill, J. L. and Newton, H. J. (1995). Saddlepoint approximations for the difference of order statistics. *Biometrika* **82**, 226–231.

Jensen, J. L. (1995). *Saddlepoint Approximations*. Clarendon Press, Oxford.

Lieberman, O. (1994). On the approximation of saddlepoint expansions in statistics. *Econometric Theory* **10**, 900–916.

Lugannani, R. and Rice, S. (1980). Saddlepoint approximation for the distribution of the sum of independent random variables. *Adv. Appl. Prob.*, **12**, 475–490.

Robinson, J. (1982). Saddlepoint approximations for permutation tests and confidence intervals. *J. Roy. Statist. Soc. Ser. B*, **44**, 91–101.

Received February, 1998; Revised October, 1998.

Ann. Inst. Statist. Math.
Vol. 53, No. 3, 427–435 (2001)
©2001 The Institute of Statistical Mathematics

INFORMATION INEQUALITIES IN A FAMILY OF UNIFORM DISTRIBUTIONS

MASAFUMI AKAHIRA[1] AND KEI TAKEUCHI[2]

[1]*Institute of Mathematics, University of Tsukuba, Ibaraki 305–8571, Japan*
[2]*Faculty of International Studies, Meiji-Gakuin University, Kamikurata-cho 1598,
Totsuka-ku, Yokohama 244–0816, Japan*

(Received June 2, 1999; revised January 21, 2000)

Abstract. For a family of uniform distributions, it is shown that for any small $\varepsilon > 0$ the average mean squared error (MSE) of any estimator in the interval of θ values of length ε and centered at θ_0 can not be smaller than that of the midrange up to the order $o(n^{-2})$ as the size n of sample tends to infinity. The asymptotic lower bound for the average MSE is also shown to be sharp.

Key words and phrases: Best location equivariant estimator, average mean squared error, sufficient statistic.

1. Introduction

Estimation of the mean θ of the uniform distribution with known range, which may be assumed to be equal to 1, is simple but a typical case of non-regular estimation. It is known that the variance of the locally best unbiased estimator at any $\theta = \theta_0$ is equal to zero even when the sample size is equal to one (see, e.g. Akahira and Takeuchi (1995)), while the best location equivariant estimator

$$\hat{\theta}^* = \frac{1}{2} \left(\min_{1 \leq i \leq n} X_i + \max_{1 \leq i \leq n} X_i \right)$$

has the variance equal to $1/\{2(n+1)(n+2)\}$ for all θ. There are several ways to construct unbiased estimators with zero variance at a specified value $\theta = \theta_0$, but they all have variance larger than that of $\hat{\theta}^*$ for other values of θ, and when the size of sample is large, even for values arbitrarily close to θ. The purpose of this paper is to show that for any small $\varepsilon > 0$ the average mean squared error of any estimator $\hat{\theta}$ in the interval of θ values of length ε and centered at θ_0 can not be smaller than that of $\hat{\theta}^*$. More precisely we shall prove that for any estimator $\hat{\theta}$ based on the sample of size n

$$\varliminf_{\varepsilon \to 0} \lim_{n \to \infty} \frac{1}{\varepsilon} \int_{\theta_0 - \varepsilon/2}^{\theta_0 + \varepsilon/2} n^2 E_\theta[(\hat{\theta} - \theta)^2] d\theta \geq \frac{1}{2}.$$

This means that in a sense asymptotically $\hat{\theta}^*$ can be regarded as uniformly best.

The result can be generalized to the case of estimation of the unknown location parameter θ with the density $f(x - \theta)$, where f has the following conditions:

(i) $f(x) > 0$ for $a < x < b$, $f(x) = 0$ otherwise.
(ii) $\lim_{x \to a+0} f(x) = \lim_{x \to b-0} f(x) = A > 0$.

428   MASAFUMI AKAHIRA AND KEI TAKEUCHI

(iii) f is continuously differentiable in the interval (a, b).

Indeed, it is derived from the fact that the estimator which minimizes

$$\int_{\theta_0 - \varepsilon/2}^{\theta_0 + \varepsilon/2} E_\theta[(\hat\theta - \theta)^2] d\theta$$

is the Bayes estimator with respect to the uniform prior over the interval $[\theta_0 - (\varepsilon/2), \theta_0 + (\varepsilon/2)]$, and that is asymptotically equivalent to the one with respect to the prior over the entire interval and asymptotically equal to the estimator $\hat\theta^* = (\min_{1 \le i \le n} X_i + \max_{1 \le i \le n} X_i)/2$. The logic of the proof is nearly the same as is shown, in this paper, from the rectangular distribution.

The related results to the above are found in Vincze (1979), Khatri (1980) and Móri (1983). In particular, the Móri type inequality is shown to be derived from the information inequality in this paper.

2. Information inequalities

Suppose that X be distributed uniformly over the interval $[\theta - (1/2), \theta + (1/2)]$. Let $\hat\theta(X)$ be an unbiased estimator of θ, i.e.

$$\int_{\theta - 1/2}^{\theta + 1/2} \hat\theta(x) dx = \theta \qquad \text{for all} \quad \theta \in \mathbb{R}.$$

Denote the variance of $\hat\theta$ by

$$(2.1) \qquad v(\theta) := V_\theta\big(\hat\theta(X)\big) = \int_{\theta - 1/2}^{\theta + 1/2} \{\hat\theta(x) - \theta\}^2 dx.$$

Now we have the following.

THEOREM 2.1.   *For any $\theta \in \mathbb{R}$*

$$\int_{\theta - 1/2}^{\theta + 1/2} v(t) dt = \int_{-1/2}^{1/2} v(t) dt = \int_{-1/2}^{1/2} \{\hat\theta(x) - x\}^2 dx + \int_{-1/2}^{1/2} x^2 dx \ge V_\theta(X) = \frac{1}{12}.$$

PROOF.   Denote

$$\psi(x) := \hat\theta(x) - x.$$

Then $\psi(x)$ is a periodic function with periodicity 1, i.e. $\psi(x + 1) = \psi(x)$ for almost all x. Indeed, since

$$\int_{\theta - 1/2}^{\theta + 1/2} \psi(x) dx = 0 \qquad \text{for all} \quad \theta \in \mathbb{R},$$

by differentiation we have

$$\psi\left(\theta + \frac{1}{2}\right) - \psi\left(\theta - \frac{1}{2}\right) = 0 \qquad \text{a.e.}$$

This shows the periodicity of $\psi(x)$.

Now we have

$$(2.2) \quad v(\theta) = \int_{\theta-1/2}^{\theta+1/2} \{\psi(x) + x - \theta\}^2 dx$$

$$= \int_{\theta-1/2}^{\theta+1/2} \psi^2(x)dx + 2\int_{\theta-1/2}^{\theta+1/2} (x-\theta)\psi(x)dx + \int_{\theta-1/2}^{\theta+1/2} (x-\theta)^2 dx.$$

Since ψ is a periodic function, if we express θ by $n+p$ with an integer n and $0 \le p < 1$, we have

$$(2.3) \quad \int_{\theta-1/2}^{\theta+1/2} \psi^2(x)dx = \int_{p-1/2}^{p+1/2} \psi^2(x)dx = \int_{p-1/2}^{1/2} \psi^2(x)dx + \int_{1/2}^{p+1/2} \psi^2(x)dx$$

$$= \int_{p-1/2}^{1/2} \psi^2(x)dx + \int_{-1/2}^{p-1/2} \psi^2(x)dx = \int_{-1/2}^{1/2} \psi^2(x)dx$$

and

$$(2.4) \quad \int_{\theta-1/2}^{\theta+1/2} (x-\theta)\psi(x)dx = \int_{p-1/2}^{p+1/2} (x-p)\psi(x)dx = \int_{p-1/2}^{p+1/2} x\psi(x)dx$$

$$= \int_{p-1/2}^{1/2} x\psi(x)dx + \int_{1/2}^{p+1/2} x\psi(x)dx$$

$$= \int_{p-1/2}^{1/2} x\psi(x)dx + \int_{-1/2}^{p-1/2} (x+1)\psi(x)dx$$

$$= \int_{-1/2}^{1/2} x\psi(x)dx + \int_{-1/2}^{p-1/2} \psi(x)dx.$$

From (2.2), (2.3) and (2.4) we have

$$v(\theta) = v(p) = \int_{-1/2}^{1/2} \psi^2(x)dx + \int_{-1/2}^{1/2} x^2 dx + 2\left\{\int_{-1/2}^{1/2} x\psi(x)dx + \int_{-1/2}^{p-1/2} \psi(x)dx\right\}.$$

Therefore, if we can prove that

$$(2.5) \quad \int_0^1 \left\{\int_{-1/2}^{1/2} x\psi(x)dx + \int_{-1/2}^{p-1/2} \psi(x)dx\right\} dp = 0,$$

then the theorem is established. We denote

$$(2.6) \quad \Psi(p) := \int_{-1/2}^{p-1/2} \psi(x)dx.$$

Then we have $\Psi(0) = \Psi(1) = 0$, and it is shown that

$$(2.7) \quad \int_0^1 \left\{\int_{-1/2}^{1/2} x\psi(x)dx\right\} dp = \int_{-1/2}^{1/2} x\psi(x)dx$$

$$= \int_0^1 \left(p - \frac{1}{2}\right) \psi\left(p - \frac{1}{2}\right) dp = \int_0^1 (p-1)\Psi'(p)dp$$

$$= -\int_0^1 \Psi(p)dp.$$

From (2.6) and (2.7) we get (2.5). $\square$

For estimators not necessarily unbiased, we have the following.

THEOREM 2.2.   *Let $M(\theta)$ be the mean squared error (MSE) of an estimator $\hat{\theta}(X)$ of θ, i.e.*

$$M(\theta) := \int_{\theta-1/2}^{\theta+1/2} \{\hat{\theta}(x) - \theta\}^2 dx.$$

Then for any $\theta_0 \in \mathbb{R}$

$$(2.8) \qquad \int_{\theta_0-\varepsilon/2}^{\theta_0+\varepsilon/2} M(\theta)d\theta \geq \begin{cases} \dfrac{1}{12}\left(\varepsilon - \dfrac{1}{2}\right) & for \quad \varepsilon > 1, \\[3mm] \dfrac{\varepsilon^3}{12}\left(1 - \dfrac{\varepsilon}{2}\right) & for \quad \varepsilon \leq 1. \end{cases}$$

PROOF.   We assume $\theta_0 = 0$ without loss of generality. Then we have for $\varepsilon > 1$

$$(2.9) \qquad \int_{-\varepsilon/2}^{\varepsilon/2} M(\theta)d\theta$$

$$= \int_{-\varepsilon/2}^{\varepsilon/2} d\theta \int_{\theta-1/2}^{\theta+1/2} \{\hat{\theta}(x) - \theta\}^2 dx$$

$$= \int_{-\varepsilon/2}^{\varepsilon/2} d\theta \int_{\theta-1/2}^{\theta+1/2} \{\hat{\theta}^2(x) - 2\theta\hat{\theta}(x) + \theta^2\} dx$$

$$= \iint_{\{(x,y)||x-y|<1/2,\, 0<x<(\varepsilon/2)+(1/2)\}} \{\hat{\theta}^2(x) - 2y\hat{\theta}(x)\} dxdy$$

$$+ \iint_{\{(x,y)||x-y|<1/2,\, -(\varepsilon/2)-(1/2)<x<0\}} \{\hat{\theta}^2(x) - 2y\hat{\theta}(x)\} dxdy + \frac{\varepsilon^3}{12}$$

$$=: I_1 + I_2 + \frac{\varepsilon^3}{12},$$

where

$$I_1 = \iint_{\{(x,y)||x-y|<1/2,\, 0<x<(\varepsilon/2)+(1/2)\}} \{\hat{\theta}^2(x) - 2y\hat{\theta}(x)\} dydx,$$

$$I_2 = \iint_{\{(x,y)||x-y|<1/2,\, -(\varepsilon/2)-(1/2)<x<0\}} \{\hat{\theta}^2(x) - 2y\hat{\theta}(x)\} dydx.$$

Then we have

$$(2.10) \quad I_1 = \left(\int_0^{\varepsilon/2-1/2} \int_{x-1/2}^{x+1/2} + \int_{\varepsilon/2-1/2}^{\varepsilon/2+1/2} \int_{x-1/2}^{\varepsilon/2}\right) \{\hat{\theta}^2(x) - 2y\hat{\theta}(x)\} dydx$$

$$= \int_0^{\varepsilon/2-1/2} \{\hat{\theta}^2(x) - 2x\hat{\theta}(x)\} dx$$

$$+ \int_{\varepsilon/2-1/2}^{\varepsilon/2+1/2} \left\{\left(\frac{\varepsilon}{2} - x + \frac{1}{2}\right)\hat{\theta}^2(x) - \left(\frac{\varepsilon}{2} - x + \frac{1}{2}\right)\left(\frac{\varepsilon}{2} + x - \frac{1}{2}\right)\hat{\theta}(x)\right\} dx$$

$$\geq -\int_0^{\varepsilon/2-1/2} x^2 dx - \int_{\varepsilon/2-1/2}^{\varepsilon/2+1/2} \frac{1}{4}\left(\frac{\varepsilon}{2} - x + \frac{1}{2}\right)\left(\frac{\varepsilon}{2} + x - \frac{1}{2}\right)^2 dx$$

$$= -\frac{1}{24}(\varepsilon - 1)^3 - \frac{1}{48}(6\varepsilon^2 - 8\varepsilon + 3)$$

$$= \frac{1}{48}(-2\varepsilon^3 + 2\varepsilon - 1),$$

where the equality holds for

$$\hat{\theta}(x) = \begin{cases} x & \text{for} \quad 0 < x \leq \dfrac{\varepsilon}{2} - \dfrac{1}{2}, \\ \dfrac{1}{2}\left(\dfrac{\varepsilon}{2} + x - \dfrac{1}{2}\right) & \text{for} \quad \dfrac{\varepsilon}{2} - \dfrac{1}{2} < x \leq \dfrac{\varepsilon}{2} + \dfrac{1}{2}. \end{cases}$$

Similarly we obtain

$$(2.11) \qquad I_2 \geq \frac{1}{48}(-2\varepsilon^3 + 2\varepsilon - 1).$$

From (2.9), (2.10) and (2.11) we have the inequality (2.8) for $\varepsilon > 1$. For $\varepsilon \leq 1$, we have

$$(2.12) \qquad \int_{-\varepsilon/2}^{\varepsilon/2} M(\theta)d\theta = \iint_{\{(x,\theta)||x-\theta|<1/2,|\theta|<\varepsilon/2\}} \{\hat{\theta}^2(x) - 2\theta\hat{\theta}(x)\}dxd\theta$$

$$= I_1' + I_2' + \frac{\varepsilon^3}{12},$$

where I_1' and I_2' denote I_1 and I_2, respectively. Then we obtain

$$(2.13) \qquad I_1' = \int_0^{\varepsilon/2} \int_{\theta-1/2}^{\theta+1/2} \{\hat{\theta}^2(x) - 2\theta\hat{\theta}(x)\}dxd\theta$$

$$= \left(\int_0^{-\varepsilon/2+1/2} \int_{-\varepsilon/2}^{\varepsilon/2} + \int_{-\varepsilon/2+1/2}^{\varepsilon/2+1/2} \int_{x-1/2}^{\varepsilon/2}\right) \{\hat{\theta}^2(x) - 2y\hat{\theta}(x)\}dydx$$

$$= \int_0^{-\varepsilon/2+1/2} \varepsilon\hat{\theta}^2(x)dx - \int_{-\varepsilon/2+1/2}^{\varepsilon/2+1/2} \left(\frac{\varepsilon}{2} - x + \frac{1}{2}\right)$$

$$\cdot \left\{\hat{\theta}^2(x) - \left(\frac{\varepsilon}{2} + x - \frac{1}{2}\right)\hat{\theta}(x)\right\} dx$$

$$\geq -\frac{1}{4}\int_{-\varepsilon/2+1/2}^{\varepsilon/2+1/2} \left(\frac{\varepsilon}{2} - x + \frac{1}{2}\right)\left(\frac{\varepsilon}{2} + x - \frac{1}{2}\right)^2 dx$$

$$= -\frac{\varepsilon^3}{48}.$$

Similarly we have

$$(2.14) \qquad I_2' \geq -\frac{\varepsilon^3}{48}.$$

From (2.12), (2.13) and (2.14) we get the inequality (2.8) for $\varepsilon \leq 1$. $\square$

COROLLARY 2.1. *For the case when the range is equal to ℓ instead of 1, let $M_\ell(\theta)$ be the MSE of an estimator $\hat{\theta}(X)$ of θ, i.e.*

$$M_\ell(\theta) := \int_{\theta-\ell/2}^{\theta+\ell/2} \{\hat{\theta}(x) - \theta\}^2 dx.$$

          MASAFUMI AKAHIRA AND KEI TAKEUCHI

Then for any $\theta_0 \in \mathbb{R}$

$$
\int_{\theta_0 - \varepsilon/2}^{\theta_0 + \varepsilon/2} M_\ell(\theta)\,d\theta \geq
\begin{cases}
\dfrac{1}{12}\left(\ell^2 \varepsilon - \dfrac{\ell^3}{2}\right) & \text{for } \varepsilon > \ell, \\[2ex]
\dfrac{\varepsilon^3}{12}\left(1 - \dfrac{\varepsilon}{2\ell}\right) & \text{for } \varepsilon \leq \ell.
\end{cases}
$$

OUTLINE OF THE PROOF. Let $Y := X/\ell$ and $\theta' := \theta/\ell$. Then it follows that Y is uniformly distributed on the interval $[\theta' - (1/2), \theta' + (1/2)]$. Letting $\theta_0' := \theta_0/\ell$ and $\varepsilon' := \varepsilon/\ell$, from Theorem 2.2 we have

$$
\int_{\theta_0 - \varepsilon/2}^{\theta_0 + \varepsilon/2} M_\ell(\theta)\,d\theta \geq
\begin{cases}
\dfrac{\ell^3}{12}\left(\dfrac{\varepsilon}{\ell} - \dfrac{1}{2}\right) & \text{for } \varepsilon > \ell, \\[2ex]
\dfrac{\ell^3}{12}\left(\dfrac{\varepsilon}{\ell}\right)^3 \left(1 - \dfrac{\varepsilon}{2\ell}\right) & \text{for } \varepsilon \leq \ell
\end{cases}
$$

$$
=
\begin{cases}
\dfrac{1}{12}\left(\ell^2 \varepsilon - \dfrac{\ell^3}{2}\right) & \text{for } \varepsilon > \ell, \\[2ex]
\dfrac{\varepsilon^3}{12}\left(1 - \dfrac{\varepsilon}{2\ell}\right) & \text{for } \varepsilon \leq \ell.
\end{cases}
\qquad \square
$$

3. Asymptotic lower bound for the average mean squared error

Now suppose that $X_1, \ldots, X_n$ are independently, identically and uniformly distributed over the interval $[\theta - (1/2), \theta + (1/2)]$. Let $Y := (X_{(1)} + X_{(n)})/2$ and $R = X_{(n)} - X_{(1)}$, where $X_{(1)} := \min_{1 \leq i \leq n} X_i$ and $X_{(n)} := \max_{1 \leq i \leq n} X_i$. Then it is shown that the pair $(X_{(1)}, X_{(n)})$ is a sufficient statistic, and given R, Y is uniformly distributed over the interval $[\theta - \{(1-R)/2\}, \theta + \{(1-R)/2\}]$. Let $\hat{\theta} := \hat{\theta}(X_{(1)}, X_{(n)})$ be an estimator of θ. Define

$$
J_\varepsilon := \int_{-\varepsilon/2}^{\varepsilon/2} E_\theta\left[(\hat{\theta} - \theta)^2\right] d\theta
$$

$$
= E^R\left[\int_{-\varepsilon/2}^{\varepsilon/2} E_\theta\left[(\hat{\theta} - \theta)^2 \mid R\right] d\theta\right].
$$

From Corollary 2.1, we have

$$
(3.1) \quad \int_{-\varepsilon/2}^{\varepsilon/2} E_\theta\left[(\hat{\theta} - \theta)^2 \mid R\right] d\theta \geq
\begin{cases}
\dfrac{1}{12}\left\{(1-R)^2 \varepsilon - \dfrac{1}{2}(1-R)^3\right\} & \text{for } \varepsilon > 1 - R, \\[2ex]
\dfrac{\varepsilon^3}{12}\left\{1 - \dfrac{\varepsilon}{2(1-R)}\right\} & \text{for } \varepsilon \leq 1 - R.
\end{cases}
$$

The density of R is given by

$$
(3.2) \quad f(R) =
\begin{cases}
n(n-1)R^{n-2}(1-R) & \text{for } 0 < R < 1, \\[1ex]
0 & \text{otherwise,}
\end{cases}
$$

hence it follows that for any $\varepsilon \leq 1$

$$(3.3) \quad J_\varepsilon \geq n(n-1)\int_{1-\varepsilon}^{1} \frac{1}{12}\left\{\varepsilon(1-R)^2 - \frac{1}{2}(1-R)^3\right\} R^{n-2}(1-R)dR$$

$$+n(n-1)\int_0^{1-\varepsilon} \frac{\varepsilon^3}{12}\left\{1 - \frac{\varepsilon}{2(1-R)}\right\} R^{n-2}(1-R)dR$$

$$= \frac{\varepsilon}{2(n+1)(n+2)} - \frac{n}{12}\varepsilon(1-\varepsilon)^{n-1} + \frac{1}{4}(n-1)\varepsilon(1-\varepsilon)^n$$

$$-\frac{n(n-1)}{4(n+1)}\varepsilon(1-\varepsilon)^{n+1} + \frac{n(n-1)}{12(n+2)}\varepsilon(1-\varepsilon)^{n+2}$$

$$-\frac{1}{(n+1)(n+2)(n+3)} + \frac{n}{24}(1-\varepsilon)^{n-1} - \frac{1}{6}(n-1)(1-\varepsilon)^n$$

$$+\frac{n(n-1)}{4(n+1)}(1-\varepsilon)^{n+1} - \frac{n(n-1)}{6(n+2)}(1-\varepsilon)^{n+2} + \frac{n(n-1)}{24(n+3)}(1-\varepsilon)^{n+3}$$

$$+\frac{n}{12}\varepsilon^3(1-\varepsilon)^{n-1} - \frac{1}{12}(n-1)\varepsilon^3(1-\varepsilon)^n - \frac{n}{24}\varepsilon^4(1-\varepsilon)^{n-1}.$$

Then we have the following.

THEOREM 3.1. *For any estimator* $\hat{\theta} = \hat{\theta}(X_{(1)}, X_{(n)})$

$$(3.4) \qquad \lim_{\varepsilon \to 0}\lim_{n \to \infty} n^2 \frac{J_\varepsilon}{\varepsilon} = \lim_{\varepsilon \to 0}\lim_{n \to \infty} \frac{n^2}{\varepsilon}\int_{-\varepsilon/2}^{\varepsilon/2} E_\theta\left[(\hat{\theta} - \theta)^2\right]d\theta \geq \frac{1}{2}.$$

The proof is straightforwardly derived from (3.3). We also have somewhat weaker result.

COROLLARY 3.1. *For any estimator* $\hat{\theta} = \hat{\theta}(X_{(1)}, X_{(n)})$

$$(3.5) \qquad \lim_{\varepsilon \to 0}\lim_{n \to \infty}\sup_{|\theta|<\varepsilon/2} \frac{n^2}{\varepsilon} E_\theta\left[(\hat{\theta} - \theta)^2\right] \geq \frac{1}{2}.$$

The proof is omitted since (3.5) is easily derived from Theorem 3.1. Note that the equalities in (3.4) and (3.5) are attained by

$$\hat{\theta}^* = \frac{1}{2}(X_{(1)} + X_{(n)}).$$

Indeed, since the probability density function of $\hat{\theta}^*$ is given by

$$f_{\hat{\theta}^*}(y) = \begin{cases} n(1 - 2|y - \theta|)^{n-1} & \text{for} \quad \theta - \frac{1}{2} \leq y \leq \theta + \frac{1}{2}, \\ 0 & \text{otherwise}, \end{cases}$$

it follows that

$$(3.6) \qquad E_\theta\left[(\hat{\theta}^* - \theta)^2\right] = \frac{1}{2(n+1)(n+2)}.$$

434   MASAFUMI AKAHIRA   AND   KEI TAKEUCHI

Since $J_\varepsilon = \varepsilon/\{2(n+1)(n+2)\}$ by (3.6), it is easily seen that

$$\lim_{\varepsilon \to 0} \lim_{n \to \infty} n^2 \frac{J_\varepsilon}{\varepsilon} = \lim_{\varepsilon \to 0} \lim_{n \to \infty} \frac{n^2}{2(n+1)(n+2)} = \frac{1}{2},$$

which implies that the equality in (3.4) is attained by $\hat{\theta}^*$. Since by (3.6)

$$\lim_{n \to \infty} n^2 \sup_{|\theta| < \varepsilon/2} E_\theta\big[(\hat{\theta}^* - \theta)^2\big] = \lim_{n \to \infty} \frac{n^2}{2(n+1)(n+2)} = \frac{1}{2},$$

the equality in (3.5) is seen to be attained by $\hat{\theta}^*$.

Finally we shall show that the Móri type inequality is easily derived from the inequality (3.1). Letting $c = \varepsilon/2$, we have from (3.1)

$$\int_{-c}^{c} E_\theta\big[(\hat{\theta} - \theta)^2 \mid R\big] d\theta \geq \frac{1}{12}\left\{ 2c(1-R)^2 - \frac{1}{2}(1-R)^3 \right\} \quad \text{for} \quad c > (1-R)/2,$$

hence

$$(3.7) \qquad \int_{-c}^{c} E_\theta\big[(\hat{\theta} - \theta)^2\big] d\theta = E^R\left[\int_{-c}^{c} E_\theta\big[(\hat{\theta} - \theta)^2 \mid R\big] d\theta \right]$$

$$\geq \frac{1}{12} E^R\left[2c(1-R)^2 - \frac{1}{2}(1-R)^3 \right]$$

for large c. Since by (3.2)

$$E^R[(1-R)^2] = \frac{6}{(n+1)(n+2)},$$

$$E^R[(1-R)^3] = \frac{24}{(n+1)(n+2)(n+3)},$$

it follows from (3.7) that

$$\frac{1}{2c} \int_{-c}^{c} E_\theta[(\hat{\theta} - \theta)^2] d\theta \geq \frac{1}{2(n+1)(n+2)} - \frac{1}{2c(n+1)(n+2)(n+3)}$$

for large c, which implies

$$\lim_{c \to \infty} \frac{1}{2c} \int_{-c}^{c} E_\theta\big[(\hat{\theta} - \theta)^2\big] d\theta \geq \frac{1}{2(n+1)(n+2)}.$$

This type inequality is given by Móri (1983).

Acknowledgement

      The authors thank the referee for the kind comments.

INFORMATION INEQUALITIES IN UNIFORM CASES 435

REFERENCES

Akahira, M. and Takeuchi, K. (1995). *Non-Regular Statistical Estimation*, Lecture Notes in Statistics, No. 107, Springer, New York.

Khatri, C. G. (1980). Unified treatment of Cramér-Rao bound for the nonregular density functions, *J. Statist. Plann. Inference*, **4**, 75–79.

Móri, T. F. (1983). Note on the Cramér-Rao inequality in the nonregular case: The family of uniform distributions, *J. Statist. Plann. Inference*, **7**, 353–358.

Vincze, I. (1979). On the Cramér-Fréchet-Rao inequality in the non-regular case, *Contributions to Statistics, The J. Hájek Memorial Volume*, 253–262, Academia, Prague.

Permissions

The editors and World Scientific Publishing Co. Pte. Ltd. would like to thank the original publishers of the joint papers of Akahira and Takeuchi for granting permissions to reprint specific papers in this volume. The following list contains the credit lines for those articles.

[1] Reprint from *Annals of Statistics* **3** ©1975 by the Institute of Mathematical Statistics.

[2] Reprint from *Rep. Univ. Electro-Commun.*, **26** ©1976 by The University of Electro-Communications.

[3] Reprint from *Rep. Univ. Electro-Commun.*, **27** ©1976 by The University of Electro-Communications.

[4] Reprint from *Rep. Univ. Electro-Commun.*, **27** ©1976 by The University of Electro-Communications.

[5] Reprint from *Lecture Notes in Mathematics* **550** ©1976 by Springer-Verlag.

[6] Reprint from *Ann. Inst. Statist. Math.*, **29** ©1977 by The Institute of Statistical Mathematics.

[7] Reprint from *Rep. Univ. Electro-Commun.*, **28** ©1978 by The University of Electro-Communications.

[8] Reprint from *Rep. Univ. Electro-Commun.*, **28** ©1978 by The University of Electro-Communications.

[9] Reprint from *Rep. Univ. Electro-Commun.*, **29** ©1978 by The University of Electro-Communications.

[10] Reprint from *Rep. Stat. Appl. Res., JUSE*, **26** ©1979 by Union of Japanese Scientists and Engineers.

[11] Reprint from *Rep. Stat. Appl. Res., JUSE*, **26** ©1979 by Union of Japanese Scientists and Engineers.

[12] Reprint from *Ann. Inst. Statist. Math.*, **31** ©1979 by The Institute of Statistical Mathematics.

[13] Reprint from *Ann. Inst. Statist. Math.*, **31** ©1979 by The Institute of Statistical Mathematics.

[14] Reprint from *Rep. Univ. Electro-Commun.*, **30** ©1979 by The University of Electro-Communications.

[15] Reprint from *Austral. J. Statist.*, **22** ©1980 by Blackwell Publishing Asia.

[16] Reprint from *Rep. Univ. Electro-Commun.*, **31** ©1980 by The University of Electro-Communications.

[17] Reprint from *Statistics & Decisions* **1** ©1982 by Oldenbourg Wissenschaftsverlag GmbH.

[18] Reprint from *Ann. Inst. Statist. Math.*, **37** ©1985 by The Institute of Statistical Mathematics.

[19] Reprint from *Rep. Stat. Appl. Res., JUSE*, **32** ©1985 by Union of Japanese Scientists and Engineers.

[20] Reprint from *Ann. Inst. Statist. Math.*, **38** ©1986 by The Institute of Statistical Mathematics.

[21] Reprint from *Metrika* **33** ©1986 by Physica-Verlag.

[22] Reprint from *Metrika* **33** ©1986 by Physica-Verlag.

[23] Reprint from *Publ. Inst. Stat. Univ. Paris* **31** ©1986 by LSTA, Université Pierre et Marie Curie.

[24] Reprint from *Foundations of Statistical Inference, Advances in the Statistical Sciences* Vol. 2, (ed. McNeill) ©1987 by Kluwer Academic Publishers.

[25] Reprint from *Metrika* **34** ©1987 by Physica-Verlag.

[26] Reprint from *Ann. Inst. Statist. Math.*, **39** ©1987 by The Institute of Statistical Mathematics.

[28] Reprint from *Lecture Notes in Mathematics* **1299** ©1988 by Springer-Verlag.

[29] Reprint from *Ann. Inst. Statist. Math.*, **41** ©1989 by The Institute of Statistical Mathematics.

[31] Reprint from *Publ. Inst. Stat. Univ. Paris* **35** ©1990 by LSTA, Université Pierre et Marie Curie.

[32] Reprint from *Austral. J. Statist.*, **32** ©1990 by Blackwell Publishing Asia.

[33] Reprint from *Ann. Inst. Statist. Math.*, **43** ©1991 by The Institute of Statistical Mathematics.

[35] Reprint from *Rep. Stat. Appl. Res., JUSE*, **38** ©1991 by Union of Japanese Scientists and Engineers.

[37] Reprint from *Rep. Stat. Appl. Res., JUSE*, **39** ©1992 by Union of Japanese Scientists and Engineers.

[38] Reprint from *Metron* **50** ©1992 by Dipartimento di Statistica, Probabilità e Statistiche Applicate, Università degli Studi di Roma.

[39] Reprint from *Statistical Sciences and Data Analysis* (ed. Matusita et al.) ©1993 by VSP International Science Publishers.

[40] Reprint from *Statistica Neerlandica* **47** ©1993 by Blackwell Science Ltd.

[42] Reprint from *Metron* **55** ©1997 by Dipartimento di Statistica, Probabilità e Statistiche Applicate, Università degli Studi di Roma.

[44] Reprint from *Ann. Inst. Statist. Math.*, **53** ©2001 by The Institute of Statistical Mathematics.